Lecture Notes in Computer Science 13435

More information about this series at https://link.springer.com/bookseries/558

Linwei Wang · Qi Dou · P. Thomas Fletcher ·
Stefanie Speidel · Shuo Li (Eds.)

Medical Image Computing and Computer Assisted Intervention – MICCAI 2022

25th International Conference
Singapore, September 18–22, 2022
Proceedings, Part V

Springer

Editors
Linwei Wang
Rochester Institute of Technology
Rochester, NY, USA

P. Thomas Fletcher ⓘ
University of Virginia
Charlottesville, VA, USA

Shuo Li ⓘ
Case Western Reserve University
Cleveland, OH, USA

Qi Dou ⓘ
Chinese University of Hong Kong
Hong Kong, Hong Kong

Stefanie Speidel ⓘ
National Center for Tumor Diseases
(NCT/UCC)
Dresden, Germany

ISSN 0302-9743   ISSN 1611-3349 (electronic)
Lecture Notes in Computer Science
ISBN 978-3-031-16442-2   ISBN 978-3-031-16443-9 (eBook)
https://doi.org/10.1007/978-3-031-16443-9

This Springer imprint is published by the registered company Springer Nature Switzerland AG
The registered company address is: Gewerbestrasse 11, 6330 Cham, Switzerland

Preface

We are pleased to present the proceedings of the 25th International Conference on Medical Image Computing and Computer-Assisted Intervention (MICCAI) which – after two difficult years of virtual conferences – was held in a hybrid fashion at the Resort World Convention Centre in Singapore, September 18–22, 2022. The conference also featured 36 workshops, 11 tutorials, and 38 challenges held on September 18 and September 22. The conference was also co-located with the 2nd Conference on Clinical Translation on Medical Image Computing and Computer-Assisted Intervention (CLINICCAI) on September 20.

MICCAI 2022 had an approximately 14% increase in submissions and accepted papers compared with MICCAI 2021. These papers, which comprise eight volumes of Lecture Notes in Computer Science (LNCS) proceedings, were selected after a thorough double-blind peer-review process. Following the example set by the previous program chairs of past MICCAI conferences, we employed Microsoft's Conference Managing Toolkit (CMT) for paper submissions and double-blind peer-reviews, and the Toronto Paper Matching System (TPMS) to assist with automatic paper assignment to area chairs and reviewers.

From 2811 original intentions to submit, 1865 full submissions were received and 1831 submissions reviewed. Of these, 67% were considered as pure Medical Image Computing (MIC), 7% as pure Computer-Assisted Interventions (CAI), and 26% as both MIC and CAI. The MICCAI 2022 Program Committee (PC) comprised 107 area chairs, with 52 from the Americas, 33 from Europe, and 22 from the Asia-Pacific or Middle East regions. We maintained gender balance with 37% women scientists on the PC.

Each area chair was assigned 16–18 manuscripts, for each of which they were asked to suggest up to 15 suggested potential reviewers. Subsequently, over 1320 invited reviewers were asked to bid for the papers for which they had been suggested. Final reviewer allocations via CMT took account of PC suggestions, reviewer bidding, and TPMS scores, finally allocating 4–6 papers per reviewer. Based on the double-blinded reviews, area chairs' recommendations, and program chairs' global adjustments, 249 papers (14%) were provisionally accepted, 901 papers (49%) were provisionally rejected, and 675 papers (37%) proceeded into the rebuttal stage.

During the rebuttal phase, two additional area chairs were assigned to each rebuttal paper using CMT and TPMS scores. After the authors' rebuttals were submitted, all reviewers of the rebuttal papers were invited to assess the rebuttal, participate in a double-blinded discussion with fellow reviewers and area chairs, and finalize their rating (with the opportunity to revise their rating as appropriate). The three area chairs then independently provided their recommendations to accept or reject the paper, considering the manuscript, the reviews, and the rebuttal. The final decision of acceptance was based on majority voting of the area chair recommendations. The program chairs reviewed all decisions and provided their inputs in extreme cases where a large divergence existed between the area chairs and reviewers in their recommendations. This process resulted in the acceptance of a total of 574 papers, reaching an overall acceptance rate of 31% for MICCAI 2022.

In our additional effort to ensure review quality, two Reviewer Tutorials and two Area Chair Orientations were held in early March, virtually in different time zones, to introduce the reviewers and area chairs to the MICCAI 2022 review process and the best practice for high-quality reviews. Two additional Area Chair meetings were held virtually in July to inform the area chairs of the outcome of the review process and to collect feedback for future conferences.

For the MICCAI 2022 proceedings, 574 accepted papers were organized in eight volumes as follows:

- Part I, LNCS Volume 13431: Brain Development and Atlases, DWI and Tractography, Functional Brain Networks, Neuroimaging, Heart and Lung Imaging, and Dermatology
- Part II, LNCS Volume 13432: Computational (Integrative) Pathology, Computational Anatomy and Physiology, Ophthalmology, and Fetal Imaging
- Part III, LNCS Volume 13433: Breast Imaging, Colonoscopy, and Computer Aided Diagnosis
- Part IV, LNCS Volume 13434: Microscopic Image Analysis, Positron Emission Tomography, Ultrasound Imaging, Video Data Analysis, and Image Segmentation I
- Part V, LNCS Volume 13435: Image Segmentation II and Integration of Imaging with Non-imaging Biomarkers
- Part VI, LNCS Volume 13436: Image Registration and Image Reconstruction
- Part VII, LNCS Volume 13437: Image-Guided Interventions and Surgery, Outcome and Disease Prediction, Surgical Data Science, Surgical Planning and Simulation, and Machine Learning – Domain Adaptation and Generalization
- Part VIII, LNCS Volume 13438: Machine Learning – Weakly-supervised Learning, Machine Learning – Model Interpretation, Machine Learning – Uncertainty, and Machine Learning Theory and Methodologies

We would like to thank everyone who contributed to the success of MICCAI 2022 and the quality of its proceedings. These include the MICCAI Society for support and feedback, and our sponsors for their financial support and presence onsite. We especially express our gratitude to the MICCAI Submission System Manager Kitty Wong for her thorough support throughout the paper submission, review, program planning, and proceeding preparation process – the Program Committee simply would not have be able to function without her. We are also grateful for the dedication and support of all of the organizers of the workshops, tutorials, and challenges, Jianming Liang, Wufeng Xue, Jun Cheng, Qian Tao, Xi Chen, Islem Rekik, Sophia Bano, Andrea Lara, Yunliang Cai, Pingkun Yan, Pallavi Tiwari, Ingerid Reinertsen, Gongning Luo, without whom the exciting peripheral events would have not been feasible. Behind the scenes, the MICCAI secretariat personnel, Janette Wallace and Johanne Langford, kept a close eye on logistics and budgets, while Mehmet Eldegez and his team from Dekon Congress & Tourism, MICCAI 2022's Professional Conference Organization, managed the website and local organization. We are especially grateful to all members of the Program Committee for their diligent work in the reviewer assignments and final paper selection, as well as the reviewers for their support during the entire process. Finally, and most importantly, we thank all authors, co-authors, students/postdocs, and supervisors, for submitting and presenting their high-quality work which made MICCAI 2022 a successful event.

We look forward to seeing you in Vancouver, Canada at MICCAI 2023!

September 2022

Linwei Wang
Qi Dou
P. Thomas Fletcher
Stefanie Speidel
Shuo Li

Organization

General Chair

Shuo Li   Case Western Reserve University, USA

Program Committee Chairs

Linwei Wang   Rochester Institute of Technology, USA
Qi Dou   The Chinese University of Hong Kong, China
P. Thomas Fletcher   University of Virginia, USA
Stefanie Speidel   National Center for Tumor Diseases Dresden, Germany

Workshop Team

Wufeng Xue   Shenzhen University, China
Jun Cheng   Agency for Science, Technology and Research, Singapore
Qian Tao   Delft University of Technology, the Netherlands
Xi Chen   Stern School of Business, NYU, USA

Challenges Team

Pingkun Yan   Rensselaer Polytechnic Institute, USA
Pallavi Tiwari   Case Western Reserve University, USA
Ingerid Reinertsen   SINTEF Digital and NTNU, Trondheim, Norway
Gongning Luo   Harbin Institute of Technology, China

Tutorial Team

Islem Rekik   Istanbul Technical University, Turkey
Sophia Bano   University College London, UK
Andrea Lara   Universidad Industrial de Santander, Colombia
Yunliang Cai   Humana, USA

Clinical Day Chairs

Jason Chan   The Chinese University of Hong Kong, China
Heike I. Grabsch   University of Leeds, UK and Maastricht
                            University, the Netherlands
Nicolas Padoy   University of Strasbourg & Institute of
                            Image-Guided Surgery, IHU Strasbourg,
                            France

Young Investigators and Early Career Development Program Chairs

Marius Linguraru   Children's National Institute, USA
Antonio Porras   University of Colorado Anschutz Medical
                           Campus, USA
Nicole Rieke   NVIDIA, Deutschland
Daniel Racoceanu   Sorbonne University, France

Social Media Chairs

Chenchu Xu   Anhui University, China
Dong Zhang   University of British Columbia, Canada

Student Board Liaison

Camila Bustillo   Technische Universität Darmstadt, Germany
Vanessa Gonzalez Duque   Ecole centrale de Nantes, France

Submission Platform Manager

Kitty Wong   The MICCAI Society, Canada

Virtual Platform Manager

John Baxter   INSERM, Université de Rennes 1, France

Program Committee

Ehsan Adeli   Stanford University, USA
Pablo Arbelaez   Universidad de los Andes, Colombia
John Ashburner   University College London, UK
Ulas Bagci   Northwestern University, USA
Sophia Bano   University College London, UK
Adrien Bartoli   Université Clermont Auvergne, France
Kayhan Batmanghelich   University of Pittsburgh, USA

Xiaoxiao Li	University of British Columbia, Canada
Jianming Liang	Arizona State University, USA
Herve Lombaert	ETS Montreal, Canada
Marco Lorenzi	Inria Sophia Antipolis, France
Le Lu	Alibaba USA Inc., USA
Klaus Maier-Hein	German Cancer Research Center (DKFZ), Germany
Anne Martel	Sunnybrook Research Institute, Canada
Diana Mateus	Centrale Nantes, France
Mehdi Moradi	IBM Research, USA
Hien Nguyen	University of Houston, USA
Mads Nielsen	University of Copenhagen, Denmark
Ilkay Oksuz	Istanbul Technical University, Turkey
Tingying Peng	Helmholtz Zentrum Muenchen, Germany
Caroline Petitjean	Université de Rouen, France
Gemma Piella	Universitat Pompeu Fabra, Spain
Chen Qin	University of Edinburgh, UK
Hedyeh Rafii-Tari	Auris Health Inc., USA
Tammy Riklin Raviv	Ben-Gurion University of the Negev, Israel
Hassan Rivaz	Concordia University, Canada
Michal Rosen-Zvi	IBM Research, Israel
Su Ruan	University of Rouen, France
Thomas Schultz	University of Bonn, Germany
Sharmishtaa Seshamani	Allen Institute, USA
Feng Shi	United Imaging Intelligence, China
Yonggang Shi	University of Southern California, USA
Yang Song	University of New South Wales, Australia
Rachel Sparks	King's College London, UK
Carole Sudre	University College London, UK
Tanveer Syeda-Mahmood	IBM Research, USA
Qian Tao	Delft University of Technology, the Netherlands
Tolga Tasdizen	University of Utah, USA
Pallavi Tiwari	Case Western Reserve University, USA
Mathias Unberath	Johns Hopkins University, USA
Martin Urschler	University of Auckland, New Zealand
Maria Vakalopoulou	University of Paris Saclay, France
Harini Veeraraghavan	Memorial Sloan Kettering Cancer Center, USA
Satish Viswanath	Case Western Reserve University, USA
Christian Wachinger	Technical University of Munich, Germany
Hua Wang	Colorado School of Mines, USA
Hongzhi Wang	IBM Research, USA
Ken C. L. Wong	IBM Almaden Research Center, USA

Reviewers

Yingying Zhu
Yuemin Zhu
Alaa Eldin Abdelaal
Amir Abdi
Mazdak Abulnaga
Burak Acar
Iman Aganj
Priya Aggarwal
Ola Ahmad
Seyed-Ahmad Ahmadi
Euijoon Ahn
Faranak Akbarifar
Cem Akbaş
Saad Ullah Akram
Tajwar Aleef
Daniel Alexander
Hazrat Ali
Sharib Ali
Max Allan
Pablo Alvarez
Vincent Andrearczyk
Elsa Angelini
Sameer Antani
Michela Antonelli
Ignacio Arganda-Carreras
Mohammad Ali Armin
Josep Arnal
Md Ashikuzzaman
Mehdi Astaraki
Marc Aubreville
Chloé Audigier
Angelica Aviles-Rivero
Ruqayya Awan
Suyash Awate
Qinle Ba
Morteza Babaie
Meritxell Bach Cuadra
Hyeon-Min Bae
Junjie Bai
Wenjia Bai
Ujjwal Baid
Pradeep Bajracharya
Yaël Balbastre
Abhirup Banerjee
Sreya Banerjee

Shunxing Bao
Adrian Barbu
Sumana Basu
Deepti Bathula
Christian Baumgartner
John Baxter
Sharareh Bayat
Bahareh Behboodi
Hamid Behnam
Sutanu Bera
Christos Bergeles
Jose Bernal
Gabriel Bernardino
Alaa Bessadok
Riddhish Bhalodia
Indrani Bhattacharya
Chitresh Bhushan
Lei Bi
Qi Bi
Gui-Bin Bian
Alexander Bigalke
Ricardo Bigolin Lanfredi
Benjamin Billot
Ryoma Bise
Sangeeta Biswas
Stefano B. Blumberg
Sebastian Bodenstedt
Bhushan Borotikar
Ilaria Boscolo Galazzo
Behzad Bozorgtabar
Nadia Brancati
Katharina Breininger
Rupert Brooks
Tom Brosch
Mikael Brudfors
Qirong Bu
Ninon Burgos
Nikolay Burlutskiy
Michał Byra
Ryan Cabeen
Mariano Cabezas
Hongmin Cai
Jinzheng Cai
Weidong Cai
Sema Candemir

Qing Cao
Weiguo Cao
Yankun Cao
Aaron Carass
Ruben Cardenes
M. Jorge Cardoso
Owen Carmichael
Alessandro Casella
Matthieu Chabanas
Ahmad Chaddad
Jayasree Chakraborty
Sylvie Chambon
Yi Hao Chan
Ming-Ching Chang
Peng Chang
Violeta Chang
Sudhanya Chatterjee
Christos Chatzichristos
Antong Chen
Chao Chen
Chen Chen
Cheng Chen
Dongdong Chen
Fang Chen
Geng Chen
Hanbo Chen
Jianan Chen
Jianxu Chen
Jie Chen
Junxiang Chen
Junying Chen
Junyu Chen
Lei Chen
Li Chen
Liangjun Chen
Liyun Chen
Min Chen
Pingjun Chen
Qiang Chen
Runnan Chen
Shuai Chen
Xi Chen
Xiaoran Chen
Xin Chen
Xinjian Chen

Xuejin Chen
Yuanyuan Chen
Zhaolin Chen
Zhen Chen
Zhineng Chen
Zhixiang Chen
Erkang Cheng
Jianhong Cheng
Jun Cheng
Philip Chikontwe
Min-Kook Choi
Gary Christensen
Argyrios Christodoulidis
Stergios Christodoulidis
Albert Chung
Özgün Çiçek
Matthew Clarkson
Dana Cobzas
Jaume Coll-Font
Toby Collins
Olivier Commowick
Runmin Cong
Yulai Cong
Pierre-Henri Conze
Timothy Cootes
Teresa Correia
Pierrick Coupé
Hadrien Courtecuisse
Jeffrey Craley
Alessandro Crimi
Can Cui
Hejie Cui
Hui Cui
Zhiming Cui
Kathleen Curran
Claire Cury
Tobias Czempiel
Vedrana Dahl
Tareen Dawood
Laura Daza
Charles Delahunt
Herve Delingette
Ugur Demir
Liang-Jian Deng
Ruining Deng

Yang Deng
Cem Deniz
Felix Denzinger
Adrien Depeursinge
Hrishikesh Deshpande
Christian Desrosiers
Neel Dey
Anuja Dharmaratne
Li Ding
Xinghao Ding
Zhipeng Ding
Ines Domingues
Juan Pedro Dominguez-Morales
Mengjin Dong
Nanqing Dong
Sven Dorkenwald
Haoran Dou
Simon Drouin
Karen Drukker
Niharika D'Souza
Guodong Du
Lei Du
Dingna Duan
Hongyi Duanmu
Nicolas Duchateau
James Duncan
Nicha Dvornek
Dmitry V. Dylov
Oleh Dzyubachyk
Jan Egger
Alma Eguizabal
Gudmundur Einarsson
Ahmet Ekin
Ahmed Elazab
Ahmed Elnakib
Amr Elsawy
Mohamed Elsharkawy
Ertunc Erdil
Marius Erdt
Floris Ernst
Boris Escalante-Ramírez
Hooman Esfandiari
Nazila Esmaeili
Marco Esposito
Théo Estienne

Christian Ewert
Deng-Ping Fan
Xin Fan
Yonghui Fan
Yubo Fan
Chaowei Fang
Huihui Fang
Xi Fang
Yingying Fang
Zhenghan Fang
Mohsen Farzi
Hamid Fehri
Lina Felsner
Jianjiang Feng
Jun Feng
Ruibin Feng
Yuan Feng
Zishun Feng
Aaron Fenster
Henrique Fernandes
Ricardo Ferrari
Lukas Fischer
Antonio Foncubierta-Rodríguez
Nils Daniel Forkert
Wolfgang Freysinger
Bianca Freytag
Xueyang Fu
Yunguan Fu
Gareth Funka-Lea
Pedro Furtado
Ryo Furukawa
Laurent Gajny
Francesca Galassi
Adrian Galdran
Jiangzhang Gan
Yu Gan
Melanie Ganz
Dongxu Gao
Linlin Gao
Riqiang Gao
Siyuan Gao
Yunhe Gao
Zeyu Gao
Gautam Gare
Bao Ge

Rongjun Ge
Sairam Geethanath
Shiv Gehlot
Yasmeen George
Nils Gessert
Olivier Gevaert
Ramtin Gharleghi
Sandesh Ghimire
Andrea Giovannini
Gabriel Girard
Rémi Giraud
Ben Glocker
Ehsan Golkar
Arnold Gomez
Ricardo Gonzales
Camila Gonzalez
Cristina González
German Gonzalez
Sharath Gopal
Karthik Gopinath
Pietro Gori
Michael Götz
Shuiping Gou
Maged Goubran
Sobhan Goudarzi
Alejandro Granados
Mara Graziani
Yun Gu
Zaiwang Gu
Hao Guan
Dazhou Guo
Hengtao Guo
Jixiang Guo
Jun Guo
Pengfei Guo
Xiaoqing Guo
Yi Guo
Yuyu Guo
Vikash Gupta
Prashnna Gyawali
Stathis Hadjidemetriou
Fatemeh Haghighi
Justin Haldar
Mohammad Hamghalam
Kamal Hammouda

Bing Han
Liang Han
Seungjae Han
Xiaoguang Han
Zhongyi Han
Jonny Hancox
Lasse Hansen
Huaying Hao
Jinkui Hao
Xiaoke Hao
Mohammad Minhazul Haq
Nandinee Haq
Rabia Haq
Michael Hardisty
Nobuhiko Hata
Ali Hatamizadeh
Andreas Hauptmann
Huiguang He
Nanjun He
Shenghua He
Yuting He
Tobias Heimann
Stefan Heldmann
Sobhan Hemati
Alessa Hering
Monica Hernandez
Estefania Hernandez-Martin
Carlos Hernandez-Matas
Javier Herrera-Vega
Kilian Hett
David Ho
Yi Hong
Yoonmi Hong
Mohammad Reza Hosseinzadeh Taher
Benjamin Hou
Wentai Hou
William Hsu
Dan Hu
Rongyao Hu
Xiaoling Hu
Xintao Hu
Yan Hu
Ling Huang
Sharon Xiaolei Huang
Xiaoyang Huang

Yangsibo Huang

Yi-Jie Huang

Yijin Huang

Yixing Huang

Yue Huang

Zhi Huang

Ziyi Huang

Arnaud Huaulmé

Jiayu Huo

Raabid Hussain

Sarfaraz Hussein

Khoi Huynh

Seong Jae Hwang

Ilknur Icke

Kay Igwe

Abdullah Al Zubaer Imran

Ismail Irmakci

Benjamin Irving

Mohammad Shafkat Islam

Koichi Ito

Hayato Itoh

Yuji Iwahori

Mohammad Jafari

Andras Jakab

Amir Jamaludin

Mirek Janatka

Vincent Jaouen

Uditha Jarayathne

Ronnachai Jaroensri

Golara Javadi

Rohit Jena

Rachid Jennane

Todd Jensen

Debesh Jha

Ge-Peng Ji

Yuanfeng Ji

Zhanghexuan Ji

Haozhe Jia

Meirui Jiang

Tingting Jiang

Xiajun Jiang

Xiang Jiang

Zekun Jiang

Jianbo Jiao

Jieqing Jiao

Zhicheng Jiao

Chen Jin

Dakai Jin

Qiangguo Jin

Taisong Jin

Yueming Jin

Baoyu Jing

Bin Jing

Yaqub Jonmohamadi

Lie Ju

Yohan Jun

Alain Jungo

Manjunath K N

Abdolrahim Kadkhodamohammadi

Ali Kafaei Zad Tehrani

Dagmar Kainmueller

Siva Teja Kakileti

John Kalafut

Konstantinos Kamnitsas

Michael C. Kampffmeyer

Qingbo Kang

Neerav Karani

Turkay Kart

Satyananda Kashyap

Alexander Katzmann

Anees Kazi

Hengjin Ke

Hamza Kebiri

Erwan Kerrien

Hoel Kervadec

Farzad Khalvati

Bishesh Khanal

Pulkit Khandelwal

Maksim Kholiavchenko

Ron Kikinis

Daeseung Kim

Jae-Hun Kim

Jaeil Kim

Jinman Kim

Won Hwa Kim

Andrew King

Atilla Kiraly

Yoshiro Kitamura

Stefan Klein

Tobias Klinder

Chuanbin Liu
Daochang Liu
Dong Liu
Dongnan Liu
Fenglin Liu
Han Liu
Hao Liu
Haozhe Liu
Hong Liu
Huafeng Liu
Huiye Liu
Jianfei Liu
Jiang Liu
Jingya Liu
Kefei Liu
Lihao Liu
Mengting Liu
Peirong Liu
Peng Liu
Qin Liu
Qun Liu
Shenghua Liu
Shuangjun Liu
Sidong Liu
Tianrui Liu
Xiao Liu
Xingtong Liu
Xinwen Liu
Xinyang Liu
Xinyu Liu
Yan Liu
Yanbei Liu
Yi Liu
Yikang Liu
Yong Liu
Yue Liu
Yuhang Liu
Zewen Liu
Zhe Liu
Andrea Loddo
Nicolas Loménie
Yonghao Long
Zhongjie Long
Daniel Lopes
Bin Lou

Nicolas Loy Rodas
Charles Lu
Huanxiang Lu
Xing Lu
Yao Lu
Yuhang Lu
Gongning Luo
Jie Luo
Jiebo Luo
Luyang Luo
Ma Luo
Xiangde Luo
Cuong Ly
Ilwoo Lyu
Yanjun Lyu
Yuanyuan Lyu
Sharath M S
Chunwei Ma
Hehuan Ma
Junbo Ma
Wenao Ma
Yuhui Ma
Anderson Maciel
S. Sara Mahdavi
Mohammed Mahmoud
Andreas Maier
Michail Mamalakis
Ilja Manakov
Brett Marinelli
Yassine Marrakchi
Fabio Martinez
Martin Maška
Tejas Sudharshan Mathai
Dimitrios Mavroeidis
Pau Medrano-Gracia
Raghav Mehta
Felix Meissen
Qingjie Meng
Yanda Meng
Martin Menten
Alexandre Merasli
Stijn Michielse
Leo Milecki
Fausto Milletari
Zhe Min

Tadashi Miyamoto
Sara Moccia
Omid Mohareri
Tony C. W. Mok
Rodrigo Moreno
Kensaku Mori
Lia Morra
Aliasghar Mortazi
Hamed Mozaffari
Pritam Mukherjee
Anirban Mukhopadhyay
Henning Müller
Balamurali Murugesan
Tinashe Mutsvangwa
Andriy Myronenko
Saad Nadeem
Ahmed Naglah
Usman Naseem
Vishwesh Nath
Rodrigo Nava
Nassir Navab
Peter Neher
Amin Nejatbakhsh
Dominik Neumann
Duy Nguyen Ho Minh
Dong Ni
Haomiao Ni
Hannes Nickisch
Jingxin Nie
Aditya Nigam
Lipeng Ning
Xia Ning
Sijie Niu
Jack Noble
Jorge Novo
Chinedu Nwoye
Mohammad Obeid
Masahiro Oda
Steffen Oeltze-Jafra
Ayşe Oktay
Hugo Oliveira
Sara Oliveira
Arnau Oliver
Emanuele Olivetti
Jimena Olveres

Doruk Oner
John Onofrey
Felipe Orihuela-Espina
Marcos Ortega
Yoshito Otake
Sebastian Otálora
Cheng Ouyang
Jiahong Ouyang
Xi Ouyang
Utku Ozbulak
Michal Ozery-Flato
Danielle Pace
José Blas Pagador Carrasco
Daniel Pak
Jin Pan
Siyuan Pan
Yongsheng Pan
Pankaj Pandey
Prashant Pandey
Egor Panfilov
Joao Papa
Bartlomiej Papiez
Nripesh Parajuli
Hyunjin Park
Sanghyun Park
Akash Parvatikar
Magdalini Paschalı
Diego Patiño Cortés
Mayank Patwari
Angshuman Paul
Yuchen Pei
Yuru Pei
Chengtao Peng
Jialin Peng
Wei Peng
Yifan Peng
Matteo Pennisi
Antonio Pepe
Oscar Perdomo
Sérgio Pereira
Jose-Antonio Pérez-Carrasco
Fernando Pérez-García
Jorge Perez-Gonzalez
Matthias Perkonigg
Mehran Pesteie

Jorg Peters
Terry Peters
Eike Petersen
Jens Petersen
Micha Pfeiffer
Dzung Pham
Hieu Pham
Ashish Phophalia
Tomasz Pieciak
Antonio Pinheiro
Kilian Pohl
Sebastian Pölsterl
Iulia A. Popescu
Alison Pouch
Prateek Prasanna
Raphael Prevost
Juan Prieto
Federica Proietto Salanitri
Sergi Pujades
Kumaradevan Punithakumar
Haikun Qi
Huan Qi
Buyue Qian
Yan Qiang
Yuchuan Qiao
Zhi Qiao
Fangbo Qin
Wenjian Qin
Yanguo Qin
Yulei Qin
Hui Qu
Kha Gia Quach
Tran Minh Quan
Sandro Queirós
Prashanth R.
Mehdi Rahim
Jagath Rajapakse
Kashif Rajpoot
Dhanesh Ramachandram
Xuming Ran
Hatem Rashwan
Daniele Ravì
Keerthi Sravan Ravi
Surreerat Reaungamornrat
Samuel Remedios

Yudan Ren
Mauricio Reyes
Constantino Reyes-Aldasoro
Hadrien Reynaud
David Richmond
Anne-Marie Rickmann
Laurent Risser
Leticia Rittner
Dominik Rivoir
Emma Robinson
Jessica Rodgers
Rafael Rodrigues
Robert Rohling
Lukasz Roszkowiak
Holger Roth
Karsten Roth
José Rouco
Daniel Rueckert
Danny Ruijters
Mirabela Rusu
Ario Sadafi
Shaheer Ullah Saeed
Monjoy Saha
Pranjal Sahu
Olivier Salvado
Ricardo Sanchez-Matilla
Robin Sandkuehler
Gianmarco Santini
Anil Kumar Sao
Duygu Sarikaya
Olivier Saut
Fabio Scarpa
Nico Scherf
Markus Schirmer
Alexander Schlaefer
Jerome Schmid
Julia Schnabel
Andreas Schuh
Christina Schwarz-Gsaxner
Martin Schweiger
Michaël Sdika
Suman Sedai
Matthias Seibold
Raghavendra Selvan
Sourya Sengupta

Carmen Serrano
Ahmed Shaffie
Keyur Shah
Rutwik Shah
Ahmed Shahin
Mohammad Abuzar Shaikh
S. Shailja
Shayan Shams
Hongming Shan
Xinxin Shan
Mostafa Sharifzadeh
Anuja Sharma
Harshita Sharma
Gregory Sharp
Li Shen
Liyue Shen
Mali Shen
Mingren Shen
Yiqing Shen
Ziyi Shen
Luyao Shi
Xiaoshuang Shi
Yiyu Shi
Hoo-Chang Shin
Boris Shirokikh
Suprosanna Shit
Suzanne Shontz
Yucheng Shu
Alberto Signoroni
Carlos Silva
Wilson Silva
Margarida Silveira
Vivek Singh
Sumedha Singla
Ayushi Sinha
Elena Sizikova
Rajath Soans
Hessam Sokooti
Hong Song
Weinan Song
Youyi Song
Aristeidis Sotiras
Bella Specktor
William Speier
Ziga Spiclin

Jon Sporring
Anuroop Sriram
Vinkle Srivastav
Lawrence Staib
Johannes Stegmaier
Joshua Stough
Danail Stoyanov
Justin Strait
Iain Styles
Ruisheng Su
Vaishnavi Subramanian
Gérard Subsol
Yao Sui
Heung-Il Suk
Shipra Suman
Jian Sun
Li Sun
Liyan Sun
Wenqing Sun
Yue Sun
Vaanathi Sundaresan
Kyung Sung
Yannick Suter
Raphael Sznitman
Eleonora Tagliabue
Roger Tam
Chaowei Tan
Hao Tang
Sheng Tang
Thomas Tang
Youbao Tang
Yucheng Tang
Zihao Tang
Rong Tao
Elias Tappeiner
Mickael Tardy
Giacomo Tarroni
Paul Thienphrapa
Stephen Thompson
Yu Tian
Aleksei Tiulpin
Tal Tlusty
Maryam Toloubidokhti
Jocelyne Troccaz
Roger Trullo

Chialing Tsai
Sudhakar Tummala
Régis Vaillant
Jeya Maria Jose Valanarasu
Juan Miguel Valverde
Thomas Varsavsky
Francisco Vasconcelos
Serge Vasylechko
S. Swaroop Vedula
Roberto Vega
Gonzalo Vegas Sanchez-Ferrero
Gopalkrishna Veni
Archana Venkataraman
Athanasios Vlontzos
Ingmar Voigt
Eugene Vorontsov
Xiaohua Wan
Bo Wang
Changmiao Wang
Chunliang Wang
Clinton Wang
Dadong Wang
Fan Wang
Guotai Wang
Haifeng Wang
Hong Wang
Hongkai Wang
Hongyu Wang
Hu Wang
Juan Wang
Junyan Wang
Ke Wang
Li Wang
Liansheng Wang
Manning Wang
Nizhuan Wang
Qiuli Wang
Renzhen Wang
Rongguang Wang
Ruixuan Wang
Runze Wang
Shujun Wang
Shuo Wang
Shuqiang Wang
Tianchen Wang

Tongxin Wang
Wenzhe Wang
Xi Wang
Xiangdong Wang
Xiaosong Wang
Yalin Wang
Yan Wang
Yi Wang
Yixin Wang
Zeyi Wang
Zuhui Wang
Jonathan Weber
Donglai Wei
Dongming Wei
Lifang Wei
Wolfgang Wein
Michael Wels
Cédric Wemmert
Matthias Wilms
Adam Wittek
Marek Wodzinski
Julia Wolleb
Jonghye Woo
Chongruo Wu
Chunpeng Wu
Ji Wu
Jianfeng Wu
Jie Ying Wu
Jiong Wu
Junde Wu
Pengxiang Wu
Xia Wu
Xiyin Wu
Yawen Wu
Ye Wu
Yicheng Wu
Zhengwang Wu
Tobias Wuerfl
James Xia
Siyu Xia
Yingda Xia
Lei Xiang
Tiange Xiang
Deqiang Xiao
Yiming Xiao

Hongtao Xie
Jianyang Xie
Lingxi Xie
Long Xie
Weidi Xie
Yiting Xie
Yutong Xie
Fangxu Xing
Jiarui Xing
Xiaohan Xing
Chenchu Xu
Hai Xu
Hongming Xu
Jiaqi Xu
Junshen Xu
Kele Xu
Min Xu
Minfeng Xu
Moucheng Xu
Qinwei Xu
Rui Xu
Xiaowei Xu
Xinxing Xu
Xuanang Xu
Yanwu Xu
Yanyu Xu
Yongchao Xu
Zhe Xu
Zhenghua Xu
Zhoubing Xu
Kai Xuan
Cheng Xue
Jie Xue
Wufeng Xue
Yuan Xue
Faridah Yahya
Chaochao Yan
Jiangpeng Yan
Ke Yan
Ming Yan
Qingsen Yan
Yuguang Yan
Zengqiang Yan
Baoyao Yang
Changchun Yang

Chao-Han Huck Yang
Dong Yang
Fan Yang
Feng Yang
Fengting Yang
Ge Yang
Guanyu Yang
Hao-Hsiang Yang
Heran Yang
Hongxu Yang
Huijuan Yang
Jiawei Yang
Jinyu Yang
Lin Yang
Peng Yang
Pengshuai Yang
Xiaohui Yang
Xin Yang
Yan Yang
Yifan Yang
Yujiu Yang
Zhicheng Yang
Jiangchao Yao
Jiawen Yao
Li Yao
Linlin Yao
Qingsong Yao
Chuyang Ye
Dong Hye Ye
Huihui Ye
Menglong Ye
Youngjin Yoo
Chenyu You
Haichao Yu
Hanchao Yu
Jinhua Yu
Ke Yu
Qi Yu
Renping Yu
Thomas Yu
Xiaowei Yu
Zhen Yu
Pengyu Yuan
Paul Yushkevich
Ghada Zamzmi

Ramy Zeineldin
Dong Zeng
Rui Zeng
Zhiwei Zhai
Kun Zhan
Bokai Zhang
Chaoyi Zhang
Daoqiang Zhang
Fa Zhang
Fan Zhang
Hao Zhang
Jianpeng Zhang
Jiawei Zhang
Jingqing Zhang
Jingyang Zhang
Jiong Zhang
Jun Zhang
Ke Zhang
Lefei Zhang
Lei Zhang
Lichi Zhang
Lu Zhang
Ning Zhang
Pengfei Zhang
Qiang Zhang
Rongzhao Zhang
Ruipeng Zhang
Ruisi Zhang
Shengping Zhang
Shihao Zhang
Tianyang Zhang
Tong Zhang
Tuo Zhang
Wen Zhang
Xiaoran Zhang
Xin Zhang
Yanfu Zhang
Yao Zhang
Yi Zhang
Yongqin Zhang
You Zhang
Youshan Zhang
Yu Zhang
Yubo Zhang
Yue Zhang

Yulun Zhang
Yundong Zhang
Yunyan Zhang
Yuxin Zhang
Zheng Zhang
Zhicheng Zhang
Can Zhao
Changchen Zhao
Fenqiang Zhao
He Zhao
Jianfeng Zhao
Jun Zhao
Li Zhao
Liang Zhao
Lin Zhao
Qingyu Zhao
Shen Zhao
Shijie Zhao
Tianyi Zhao
Wei Zhao
Xiaole Zhao
Xuandong Zhao
Yang Zhao
Yue Zhao
Zixu Zhao
Ziyuan Zhao
Xingjian Zhen
Haiyong Zheng
Hao Zheng
Kang Zheng
Qinghe Zheng
Shenhai Zheng
Yalin Zheng
Yinqiang Zheng
Yushan Zheng
Tao Zhong
Zichun Zhong
Bo Zhou
Haoyin Zhou
Hong-Yu Zhou
Huiyu Zhou
Kang Zhou
Qin Zhou
S. Kevin Zhou
Sihang Zhou

Tao Zhou
Tianfei Zhou
Wei Zhou
Xiao-Hu Zhou
Xiao-Yun Zhou
Yanning Zhou
Yaxuan Zhou
Youjia Zhou
Yukun Zhou
Zhiguo Zhou
Zongwei Zhou
Dongxiao Zhu
Haidong Zhu
Hancan Zhu

Lei Zhu
Qikui Zhu
Xiaofeng Zhu
Xinliang Zhu
Zhonghang Zhu
Zhuotun Zhu
Veronika Zimmer
David Zimmerer
Weiwei Zong
Yukai Zou
Lianrui Zuo
Gerald Zwettler
Reyer Zwiggelaar

Outstanding Area Chairs

Ester Bonmati   University College London, UK
Tolga Tasdizen   University of Utah, USA
Yanwu Xu   Baidu Inc., China

Outstanding Reviewers

Seyed-Ahmad Ahmadi   NVIDIA, Germany
Katharina Breininger   Friedrich-Alexander-Universität
                           Erlangen-Nürnberg, Germany
Mariano Cabezas   University of Sydney, Australia
Nicha Dvornek   Yale University, USA
Adrian Galdran   Universitat Pompeu Fabra, Spain
Alexander Katzmann   Siemens Healthineers, Germany
Tony C. W. Mok   Hong Kong University of Science and
                           Technology, China
Sérgio Pereira   Lunit Inc., Korea
David Richmond   Genentech, USA
Dominik Rivoir   National Center for Tumor Diseases (NCT)
                           Dresden, Germany
Fons van der Sommen   Eindhoven University of Technology,
                           the Netherlands
Yushan Zheng   Beihang University, China

Honorable Mentions (Reviewers)

Chloé Audigier   Siemens Healthineers, Switzerland
Qinle Ba   Roche, USA

Meritxell Bach Cuadra	University of Lausanne, Switzerland
Gabriel Bernardino	CREATIS, Université Lyon 1, France
Benjamin Billot	University College London, UK
Tom Brosch	Philips Research Hamburg, Germany
Ruben Cardenes	Ultivue, Germany
Owen Carmichael	Pennington Biomedical Research Center, USA
Li Chen	University of Washington, USA
Xinjian Chen	Soochow University, Taiwan
Philip Chikontwe	Daegu Gyeongbuk Institute of Science and Technology, Korea
Argyrios Christodoulidis	Centre for Research and Technology Hellas/Information Technologies Institute, Greece
Albert Chung	Hong Kong University of Science and Technology, China
Pierre-Henri Conze	IMT Atlantique, France
Jeffrey Craley	Johns Hopkins University, USA
Felix Denzinger	Friedrich-Alexander University Erlangen-Nürnberg, Germany
Adrien Depeursinge	HES-SO Valais-Wallis, Switzerland
Neel Dey	New York University, USA
Guodong Du	Xiamen University, China
Nicolas Duchateau	CREATIS, Université Lyon 1, France
Dmitry V. Dylov	Skolkovo Institute of Science and Technology, Russia
Hooman Esfandiari	University of Zurich, Switzerland
Deng-Ping Fan	ETH Zurich, Switzerland
Chaowei Fang	Xidian University, China
Nils Daniel Forkert	Department of Radiology & Hotchkiss Brain Institute, University of Calgary, Canada
Nils Gessert	Hamburg University of Technology, Germany
Karthik Gopinath	ETS Montreal, Canada
Mara Graziani	IBM Research, Switzerland
Liang Han	Stony Brook University, USA
Nandinee Haq	Hitachi, Canada
Ali Hatamizadeh	NVIDIA Corporation, USA
Samra Irshad	Swinburne University of Technology, Australia
Hayato Itoh	Nagoya University, Japan
Meirui Jiang	The Chinese University of Hong Kong, China
Baoyu Jing	University of Illinois at Urbana-Champaign, USA
Manjunath K N	Manipal Institute of Technology, India
Ali Kafaei Zad Tehrani	Concordia University, Canada
Konstantinos Kamnitsas	Imperial College London, UK

Pulkit Khandelwal	University of Pennsylvania, USA
Andrew King	King's College London, UK
Stefan Klein	Erasmus MC, the Netherlands
Ender Konukoglu	ETH Zurich, Switzerland
Ivica Kopriva	Rudjer Boskovich Institute, Croatia
David Kügler	German Center for Neurodegenerative Diseases, Germany
Manuela Kunz	National Research Council Canada, Canada
Gilbert Lim	National University of Singapore, Singapore
Tiancheng Lin	Shanghai Jiao Tong University, China
Bin Lou	Siemens Healthineers, USA
Hehuan Ma	University of Texas at Arlington, USA
Ilja Manakov	ImFusion, Germany
Felix Meissen	Technische Universität München, Germany
Martin Menten	Imperial College London, UK
Leo Milecki	CentraleSupelec, France
Lia Morra	Politecnico di Torino, Italy
Dominik Neumann	Siemens Healthineers, Germany
Chinedu Nwoye	University of Strasbourg, France
Masahiro Oda	Nagoya University, Japan
Sebastian Otálora	Bern University Hospital, Switzerland
Michal Ozery-Flato	IBM Research, Israel
Egor Panfilov	University of Oulu, Finland
Bartlomiej Papiez	University of Oxford, UK
Nripesh Parajuli	Caption Health, USA
Sanghyun Park	DGIST, Korea
Terry Peters	Robarts Research Institute, Canada
Theodoros Pissas	University College London, UK
Raphael Prevost	ImFusion, Germany
Yulei Qin	Tencent, China
Emma Robinson	King's College London, UK
Robert Rohling	University of British Columbia, Canada
José Rouco	University of A Coruña, Spain
Jerome Schmid	HES-SO University of Applied Sciences and Arts Western Switzerland, Switzerland
Christina Schwarz-Gsaxner	Graz University of Technology, Austria
Liyue Shen	Stanford University, USA
Luyao Shi	IBM Research, USA
Vivek Singh	Siemens Healthineers, USA
Weinan Song	UCLA, USA
Aristeidis Sotiras	Washington University in St. Louis, USA
Danail Stoyanov	University College London, UK

Ruisheng Su	Erasmus MC, the Netherlands
Liyan Sun	Xiamen University, China
Raphael Sznitman	University of Bern, Switzerland
Elias Tappeiner	UMIT - Private University for Health Sciences, Medical Informatics and Technology, Austria
Mickael Tardy	Hera-MI, France
Juan Miguel Valverde	University of Eastern Finland, Finland
Eugene Vorontsov	Polytechnique Montreal, Canada
Bo Wang	CtrsVision, USA
Tongxin Wang	Meta Platforms, Inc., USA
Yan Wang	Sichuan University, China
Yixin Wang	University of Chinese Academy of Sciences, China
Jie Ying Wu	Johns Hopkins University, USA
Lei Xiang	Subtle Medical Inc, USA
Jiaqi Xu	The Chinese University of Hong Kong, China
Zhoubing Xu	Siemens Healthineers, USA
Ke Yan	Alibaba DAMO Academy, China
Baoyao Yang	School of Computers, Guangdong University of Technology, China
Changchun Yang	Delft University of Technology, the Netherlands
Yujiu Yang	Tsinghua University, China
Youngjin Yoo	Siemens Healthineers, USA
Ning Zhang	Bloomberg, USA
Jianfeng Zhao	Western University, Canada
Tao Zhou	Nanjing University of Science and Technology, China
Veronika Zimmer	Technical University Munich, Germany

Mentorship Program (Mentors)

Ulas Bagci	Northwestern University, USA
Kayhan Batmanghelich	University of Pittsburgh, USA
Hrvoje Bogunovic	Medical University of Vienna, Austria
Ninon Burgos	CNRS - Paris Brain Institute, France
Hao Chen	Hong Kong University of Science and Technology, China
Jun Cheng	Institute for Infocomm Research, Singapore
Li Cheng	University of Alberta, Canada
Aasa Feragen	Technical University of Denmark, Denmark
Zhifan Gao	Sun Yat-sen University, China
Stamatia Giannarou	Imperial College London, UK
Sharon Huang	Pennsylvania State University, USA

Contents – Part V

Integration of Imaging with Non-imaging Biomarkers

Image Segmentation II

Automatic Segmentation of Hip Osteophytes in DXA Scans Using U-Nets

Raja Ebsim[1]([✉]), Benjamin G. Faber[2,3], Fiona Saunders[4], Monika Frysz[2,3],
Jenny Gregory[4], Nicholas C. Harvey[5,6], Jonathan H. Tobias[2,3],
Claudia Lindner[1], and Timothy F. Cootes[1]

[1] Division of Informatics, Imaging and Data Sciences, University of Manchester,
Manchester, UK
raja.ebsim@manchester.ac.uk
[2] Musculoskeletal Research Unit, University of Bristol, Bristol, UK
[3] Medical Research Council Integrative Epidemiology Unit, University of Bristol,
Bristol, UK
[4] Centre for Arthritis and Musculoskeletal Health, University of Aberdeen,
Aberdeen, UK
[5] Medical Research Council Lifecourse Epidemiology Centre,
University of Southampton, Southampton, UK
[6] NIHR Southampton Biomedical Research Centre, University of Southampton
and University Hospitals Southampton NHS Foundation Trust, Southampton, UK

Abstract. Osteophytes are distinctive radiographic features of osteoarthritis (OA) in the form of small bone spurs protruding from joints
that contribute significantly to symptoms. Identifying the genetic determi-
nants of osteophytes would improve the understanding of their biological
pathways and contributions to OA. To date, this has not been possible
due to the costs and challenges associated with manually outlining osteophytes in sufficiently large datasets. Automatic systems that can segment
osteophytes would pave the way for this research and also have potential
clinical applications. We propose, to the best of our knowledge, the first
work on automating pixel-wise segmentation of osteophytes in hip dual-
energy x-ray absorptiometry scans (DXAs). Based on U-Nets, we developed an automatic system to detect and segment osteophytes at the superior and the inferior femoral head, and the lateral acetabulum. The system
achieved sensitivity, specificity, and average Dice scores (±std) of (0.98,
0.92, 0.71 ± 0.19) for the superior femoral head [793 DXAs], (0.96, 0.85,
0.66 ± 0.24) for the inferior femoral head [409 DXAs], and (0.94, 0.73,
0.64 ± 0.24) for the lateral acetabulum [760 DXAs]. This work enables
large-scale genetic analyses of the role of osteophytes in OA, and opens
doors to using low-radiation DXAs for screening for radiographic hip OA.

Keywords: Computational anatomy · U-Nets · Osteophytes
segmentation · Osteophytes detection · Automated osteoarthritis risk
assessment

Supplementary Information The online version contains supplementary material
available at https://doi.org/10.1007/978-3-031-16443-9_1.

L. Wang et al. (Eds.): MICCAI 2022, LNCS 13435, pp. 3–12, 2022.
https://doi.org/10.1007/978-3-031-16443-9_1

1 Introduction

Osteoarthritis (OA) is a degenerative disease in which bones and surrounding soft tissue of the affected joint deteriorate, leading to pain and loss of function. Hip OA incidence is rising [5,6] due to increasing aging and obesity in populations [3,4]. Total hip replacement (THR) is the preferred treatment for end-stage disease which is associated with substantial care costs. For example, in England and Wales there are 150 THRs per 100,000 of population per year [6]. Semi-quantitative grading of hip OA can be done using Kellgren-Lawrence (KL) [7] or Croft [8] scoring which examine the presence and severity of radiographic characteristic features, including joint-space narrowing (JSN), osteophytes, subchondral scleroisis, and cysts. These criteria are inherently subjective [9], making clinical application difficult [10–12]. Although severity grading is traditionally performed by manual inspection of standard radiographs, Yoshida et al. [16] found that this could be performed as accurately using DXA scans. A recent semi-automated scoring system [13] on DXA-derived radiographic hip OA (rHOA) showed strong relationships with symptoms and was predictive of THR. The classifier considered automated joint space measurements and semi-automated osteophyte area calculations. Other semi-automated approaches [14,15], incorporating osteophytes within a statistical shape model (based on manually checked and corrected point positions) also showed an association with future THR and with change in hip shape over 6–12 months. Semi-automated scoring is time-consuming and costly. Fully automated methods would overcome this and remove, as far as possible, the element of subjectivity.

This work is based on the U-Net architecture [25] which has been widely-used with great success in the domain of image segmentation. A U-Net is a deep convolutional encoder-decoder architecture with various layers at different levels forming a U shape. In addition to the upsampling operators used to increase the resolution on their outputs, U-Nets differ from conventional fully convolutional neural network for having skip connections between the encoding and decoding paths, which improves the use of context and localisation during learning [25]. Recent work on nnU-Net [29] showed that a U-Net with carefully tuned parameters outperforms other architectures on several datasets.

Contributions: To our knowledge, this is the first work on automating pixel-based hip osteophyte segmentation. Our system automatically detects osteophytes with high accuracy and segments osteophytes at the superior and the inferior femoral head, and the lateral acetabulum. It has the following potential:

- It is an essential step towards identifying osteophytes genetic variants which have not been explored before. Manual osteophyte detection/segmentation for genetic studies is not feasible. That is because genetic studies need very large-scale datasets given the low prevalence of osteophytes.
- Due to the low radiation dose and strong relationships recently-associated [13] between DXA-derived rHOA features (JSN and osteophyte sizes) with OA symptoms and prediction of THR, DXAs could potentially be used to screen low risk clinical populations and assess their risk of subsequent THR. This

could help stratify patients who should be targeted with weight loss and physiotherapy interventions. Previous research automated JSN calculations in hip DXAs, but osteophyte segmentation has not been automated before.

2   Related Work

There are few published methods in the area of automated rHOA classification. Some research focused on detecting rHOA (i.e. binary classification of healthy vs. diseased) as shown in radiographs [17–19] and in CTs [19] by fine-tuning pre-trained deep learning models. Other work [20] also used transfer learning to evaluate multiple individual features of rHOA, including femoral osteophytes (FOs), and acetabular osteophytes (AOs). Each of these two features were graded as absent, mild, moderate, or severe. Lateral/medial and superior/inferior sides were combined by picking the most severe grade for each feature. The reported performance on two test sets was as follows: (1) sensitivity [74.6–83% for FOs, 69.9–76% for AOs], (2) specificity [91–91.1% for FOs, 76–76.4% for AOs], and (3) F1-score(%) [82.6–87.4% for FOs, 65.1–72.1% for AOs].

Few studies considered segmenting osteophytes in particular. The KOACAD algorithm [21] considered only one of the regions in which knee osteophytes develop. The determined osteophyte area was on the medial tibial margin and found as the medial prominence over the smoothly extended outline of the tibia. The area under ROC curve (AUC) for detecting OA by the osteophyte area was 0.65. Thomson et al. [22,23] detected knee OA by first segmenting knee bones including osteophytes implicitly [22] and explicitly [23] using a Random Forest Regression Voting Constrained Local Model [24] and then trained random forest classifiers on derived shape and texture information to detect OA. Their system achieved an AUC of 0.85 and 0.93 for detecting knee osteophytes and OA, respectively. A recent work [30], which used a U-Net based approach to segment knee osteophytes from MRI images, achieved a reasonable estimate of osteophyte volumes with average Dice scores < 0.5.

3   Method

An example of a left hip DXA with no osteophytes is given in Fig. 1(a, b) with its outline annotated with 85 points. The areas where osteophytes develop are around point 15 (inferior femoral head osteophytes), point 30 (superior femoral head osteophytes), and point 78 (lateral acetabulum osteophytes). A diseased hip having osteophytes on all three sites is shown in Fig. 1(c, d).

U-Nets were trained to predict the segmentation mask for a given patch from a radiograph. Patches were cropped around the three sites. One U-Net was trained from scratch for each site. Our implementation has only small changes from the original U-Net architecture [25]. We used an input image size of 128×128 pixels, outputting segmentation maps of the same size as the input image with an osteophytes class probability at each pixel.

To eliminate the effect of different hip sizes/scales in cropping the areas of interest, points {12,39} in Fig. 1(b) were used to define a reference length. The width/height of each pixel in the reference frame is a proportion (dp) of the reference length. By visually inspecting the resulting patches of different dp values and ensuring a good coverage of the area of interest, we set dp to 0.005 for the inferior femoral head, 0.006 for the superior femoral head, and 0.01 for the acetabulum. All cropped patches were normalised. Examples of input patches and their corresponding ground truth masks are shown in Fig. 2.

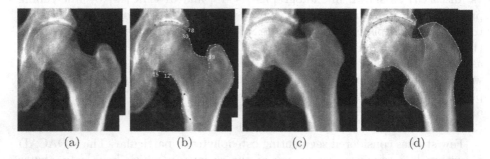

(a)   (b)   (c)   (d)

Fig. 1. (a, b): Example DXA of healthy hip showing the joint outline annotated with 85 points. Red points are key anatomical landmarks. Points {12,39} define a reference length. Each pixel in the normalised cropped patches has a width/height equal to a proportion (dp) of the reference length. (c, d): Hip DXA with features of radiographic hip OA (blue: inferior femoral head osteophyte; red: superior femoral head osteophyte; green: acetabular osteophyte). (Color figure online)

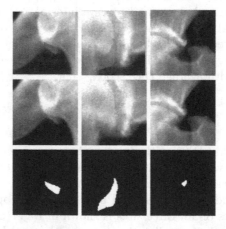

Fig. 2. Top: Osteophyte sites of the diseased hip in Fig. 1(d). Middle: Input patches to U-Nets. Bottom: Their corresponding segmentation masks (ground truth).

4 Experiments

4.1 Data

The UK Biobank (UKB) is a prospective cohort study with phenotypic and genetic data collected on approximately 500,000 individuals from the UK, aged between 40 and 69 years at the time of recruitment (between 2006–2010)[1]. High resolution iDXA (iDXA GE-Lunar, Madison, WI, USA) scans are being collected as part of the Imaging Enhancement study [2]. As of April 2021, DXA images were available for 42,441 participants, of which 41,160 had left hip images. We excluded 820 left hip scans/participants due to either poor image quality, image error or withdrawal of consent, resulting in a dataset of 40,340 participants [mean age 63.7 years (range 44–82 years)]. Osteophytes were present in 4,013 (10 %) participants/images. Manual osteophyte markups/segmentations were agreed by two experienced annotators (BGF & FS). The markup repeatability was tested after > two months from the first review on a set of 500 DXAs selected to include 20% with osteophytes. The intra-rater kappa values were between 0.80–0.91 for the presence of osteophytes, and the concordance correlation coefficients were between 0.87–0.92 for osteophyte size, depending on osteophyte site.

Data partitions are summarised in Table 1. Per osteophyte site, we randomly selected negative examples to match the number of images with osteophytes to balance the dataset. We split the data randomly into training, validation and test sets. Manual point placements were available for all data. Automatic point placements were available for a subset of all images; obtained fully automatically by applying a Random Forest-based landmark detection system as described in [13] (available from www.bone-finder.com). The training and validation sets used the manual point placements and were balanced (i.e. positive examples (with osteophytes) = negative examples). The focus was on developing a fully automatic system and we thus only included images with automatic point placements in the test sets, leading to slightly imbalanced test sets.

Table 1. Data partitions used to train, validate, and test a U-Net for each osteophyte site. Values are expressed as number of left hip scans/participants (OPs = osteophytes). Some images contained multiple osteophytes and were used for more than one site.

Osteophyte Site	Training set (with OPs)	Validation set (with OPs)	Test set (with OPs)
Inferomedial femoral head	1400 (700)	230 (115)	409 (200)
Superolateral femoral head	1700 (850)	286 (143)	793 (351)
Lateral acetabulum	3760 (1880)	400 (200)	760 (335)

4.2 Implementation Details

U-Nets were trained for 15 epoches on NVIDIA Tesla V100 SXM2 16GB, with Keras/Tensorflow implementation (Python 3.7 for GPUs). The best performing models on the validation sets during training were saved. The sparse categorical cross entropy loss function was optimized with Adam [27] (default parameter values used), and weights were initialised with the normalizer from He et al. [26]. We used dropout with probability 0.3 to reduce overfitting, and zero padding to control the shrinkage of dimension due to convolution. The training and validation datasets were augmented with seven additional samples per image including random rotation (within $\pm\, 0.1\,$rad) and scaling (in the range of $\pm\, 10\%$). No augmentation was performed for the test sets.

4.3 Results

The performance of the trained U-Nets were evaluated for (1) osteophyte detection (i.e. osteophyte presence vs absence) in terms of sensitivity and specificity; and (2) osteophyte segmentation by average Dice scores ($\pm$std). The detection threshold was set to > 0 pixels. The results are summarised in Table 2.

Table 2. Performance of osteophyte segmentation by U-Nets (1–4), each trained/tested for one osteophyte site. All U-Nets (1–4) had the same architecture, and were trained/-validated on their corresponding datasets with the same augmentation procedure. U-Net(4) only differs in introducing random pixel displacements when augmenting the training set.

	Model	Sensitivity	Specificity	F1-score	Dice score ($\pm$std)
1	Inferior femoral head	0.96	0.85	0.91	0.66 ± 0.24
2	Superior femoral head	0.98	0.92	0.94	0.71 ± 0.19
3	Lateral acetabulum	0.67	0.84	0.71	0.46 ± 0.38
4	Lateral acetabulum	0.94	0.73	0.82	0.64 ± 0.24

Our initial results for the lateral acetabulum osteophytes (U-Net(3) in Table 2) were poorer than the performance we achieved at the other sites. Therefore, we investigated the reason by studying two possibilities: *Firstly, could it be overfitting due to insufficient training examples?* To test this, we conducted experiments with an increased number of samples in the training set (by decreasing the size of the test set). However, the low performance pattern was persistent even with a significantly larger training set. *Secondly, could it be related to the automatic point placements?* The U-Net was trained on patches centered on the manual placement of point 78, but tested on new images in which all points were placed automatically. To accommodate for the differences between manual and automatic point placements, we conducted experiments including additional data augmentation. We augmented the training data with 19 additional samples per image including the above described rotation and scaling augmentations as well as random displacements of the centres (point 78) of the training patches. A random displacement was set to be within 0.05 of the patch width/height (about $\pm$6 pixels) in x- and y-axes. The latter model (U-Net(4) in Table 2) yielded an increase in performance for the lateral acetabulum osteophytes (sensitivity $= 0.94$, specificity $= 0.73$, Dice $= 0.64 \pm 0.24$). Figure 3 shows the binary confusion matrices for detecting osteophytes at the three different sites. Examples of segmentation outputs are given in Fig. 4 for various Dice scores.

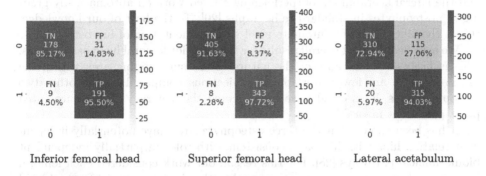

Inferior femoral head   Superior femoral head   Lateral acetabulum

Fig. 3. Performance of three U-Nets on detecting osteophytes at three specific sites. For the lateral acetabulum, random pixel displacements were introduced when augmenting the training set.

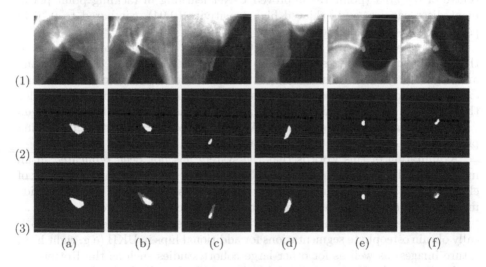

Fig. 4. Examples of osteophyte segmentation results. To visualise the overlap between true and predicted masks, the summations of true masks with two times of predicted masks were plotted. Row (1): Input test patches to U-Nets (based on automatic point placements) with manual masks (ground truth) in colours. Row (2): Automatically predicted masks. Row (3): The overlap of true and predicted mask are the brightest pixels. (a, b) are examples of inferior femoral head osteophytes with Dice scores (0.95, 0.64). (c, d) are examples of superior femoral head osteophytes with Dice scores (0.42, 0.83). (e, f) are examples of lateral acetabulum osteophytes with Dice scores (0.93, 0.59).

5   Discussion and Conclusion

We have developed an automatic system to detect and segment osteophytes at three sites in hip DXAs. Our results show that the system achieves high performance in detecting osteophytes at the superior and the inferior femoral head, and the lateral acetabulum, outperforming related work on automatically grading hip osteophytes in standard radiographs [20]. To the best of our knowledge, this is the first system to automate pixel-wise segmentation of hip osteophytes.

We achieved the best overall results for the superior femoral head osteophytes. This is likely to be because of the nature of the surrounding area within the image as there are fewer overlapping projections compared to the other two osteophyte sites. This could have also affected the accuracy of the automated point placement favorably.

It has been shown that the three osteophyte sites have potentially independent relationships with hip pain, consistent with roles in partially-independent biomechanical pathways [28]. For this reason our work considered each femoral head osteophyte site separately as opposed to the single combined femoral head osteophyte feature in [20]. For the acetabular osteophytes, we showed that enriching the training set with augmented examples of random displacements from the centre of the area (point 78) improved U-Net learning in tackling point placement discrepancies, and achieved high sensitivity (0.94), compared to 0.76 in [20], as well as an improved Dice score (0.64).

In future work, we will explore whether such an augmentation as applied to the acetabular osteophytes would also improve the performance for detecting and segmenting femoral head osteophytes. In addition, we will conduct further experiments to identify whether there is room for performance improvements for the three sites by performing hyperparameter optimisation to find patch sizes, dp values, and augmentation procedures. We would also like to explore fine-tuning pre-trained U-Net models to our data and compare results.

A limitation of this work is that the system has not been replicated in an independent dataset. Further, it would be of interest to analyse the agreement of clinical experts with the automatically segmented osteophytes. This will provide insight into what Dice score is needed to yield clinical relevant results.

With respect to epidemiological studies, the system can be used to automatically obtain osteophyte segmentations for additional hips in UKB (e.g. right hips, future images) as well as for other large cohort studies such as the Rotterdam Study (RS) or the Osteoarthritis Initiative (OAI). It will also be interesting to explore how the automatically segmented osteophyte areas are related to manual osteophyte grading as well as clinical outcomes.

Acknowledgements. RE, FS and MF are funded by a Wellcome Trust collaborative award (reference number 209233). BGF is supported by a Medical Research Council (MRC) clinical research training fellowship (MR/S021280/1). CL was funded by the MRC, UK (MR/S00405X/1) as well as a Sir Henry Dale Fellowship jointly funded by the Wellcome Trust and the Royal Society (223267/Z/21/Z). NCH is supported by the UK Medical Research Council [MC_PC_21003; MC_PC_21001].

References

1. Bycroft, C., et al.: The UK Biobank resource with deep phenotyping and genomic data. Nature **562**, 203–209 (2018)
2. Littlejohns, T.J., et al.: The UK Biobank imaging enhancement of 100,000 participants: rationale, data collection, management and future directions. Nat. Commun. **11**, 1–12 (2020)
3. Prieto-Alhambra, D., Judge, A., Javaid, M.K., Cooper, C., Diez-Perez, A., Arden, N.K.: Incidence and risk factors for clinically diagnosed knee, hip and hand osteoarthritis: influences of age, gender and osteoar thritis affecting other joints. Ann. Rheum. Dis. **73**(9), 1659–1664 (2014)
4. Yu, D., Peat, G., Bedson, J., Jordan, K.P.: Annual consultation incidence of osteoarthritis estimated from population-based health care data in England. Rheumatol. (Oxf. Engl.) **54**(11), 2051–2060 (2015)
5. Kim, C., Linsenmeyer, K.D., Vlad, S.C., et al.: Prevalence of radiographic and symptomatic hip osteoarthritis in an urban United States community: the Framingham osteoarthritis study. Arthritis. Rheumatol. **66**(11), 3013–3017 (2014)
6. National Joint Registry 17th Annual Report 2020. https://www.njrcentre.org.uk/njrcentre/Reports-Publications-and-Minutes/Annual-reports. Accessed Feb 2022
7. Kellgren, J.H., Lawrence, J.: Radiological assessment of osteoarthrosis. Ann. Rheum. Dis. **16**(4), 494 (1957)
8. Croft, P., Cooper, C., Wickham, C., Coggon, D.: Defining osteoarthritis of the hip for epidemiologic studies. Am. J. Epidemiol. **132**(3), 514–522 (1990)
9. Reijman, M., Hazes, J.M., Pols, H.A., Bernsen, R.M., Koes, B.W., Bierma-Zeinstra, S.M.: Validity and reliability of three definitions of hip osteoarthritis: cross sectional and longitudinal approach. Ann. Rheum. Dis. **63**(11), 1427–1433 (2004)
10. Birrell, F., Lunt, M., Macfarlane, G., Silman, A.: Association between pain in the hip region and radiographic changes of osteoarthritis: results from a population-based study. Rheumatology **44**(3), 337–341 (2005)
11. Iidaka, T., et al.: Prevalence of radiographic hip osteoarthritis and its association with hip pain in Japanese men and women: the ROAD study. Osteoarthr. Cartil. **24**(1), 117–123 (2016)
12. Arden, N.K., et al.: Defining incident radiographic hip osteoarthritis for epidemiologic studies in women. Arthritis Rheum. **60**(4), 1052–1059 (2009)
13. Faber, B.G., et al.: A novel semi-automated classifier of hip osteoarthritis on DXA images shows expected relationships with clinical outcomes in UK Biobank. Rheumatology **61**(9), 3586–3595 (2022). https://doi.org/10.1093/rheumatology/keab927
14. Barr, R.J., et al.: Predicting OA progression to total hip replacement: can we do better than risk factors alone using active shape modelling as an imaging biomarker? Rheumatology **51**(3), 562–570 (2012)
15. Barr, R.J., Gregory, J.S., Yoshida, K., Alesci, S., Aspden, R.M., Reid, D.M.: Significant morphological change in osteoarthritic hips identified over 6–12 months using statistical shape modelling. Osteoarthr. Cartil. **26**(6), 783–789 (2018)
16. Yoshida, K., Barr, R.J., Galea-Soler, S., Aspden, R.M., Reid, D.M., Gregory, J.S.: Reproducibility and diagnostic accuracy of Kellgren-Lawrence grading for osteoarthritis using radiographs and dual-energy X-ray absorptiometry images. J. Clin. Densitom. **18**(2), 239–244 (2015)
17. Xue, Y., Zhang, R., Deng, Y., Chen, K., Jiang, T.: A preliminary examination of the diagnostic value of deep learning in hip osteoarthritis. PloS One **12**(6), e0178992 (2017)

18. Üreten, K., Arslan, T., Gültekin, K.E., Demir, A.N.D., Özer, H.F., Bilgili, Y.:
    Detection of hip osteoarthritis by using plain pelvic radiographs with deep learning
    methods. Skelet. Radiol. **49**(9), 1369–1374 (2020). https://doi.org/10.1007/s00256-
    020-03433-9
19. Gebre, R.K., et al.: Detecting hip osteoarthritis on clinical CT: a deep learning
    application based on 2-D summation images derived from CT. Osteoporos. Int.
    33, 355–365 (2022). https://doi.org/10.1007/s00198-021-06130-y
20. von Schacky, C.E., et al.: Development and validation of a multitask deep learning
    model for severity grading of hip osteoarthritis features on radiographs. Radiology
    295(1), 136–145 (2020)
21. Oka, H., Muraki, S., Akune, T., Nakamura, K., Kawaguchi, H., Yoshimura, N.:
    Normal and threshold values of radiographic parameters for knee osteoarthritis
    using a computer-assisted measuring system (KOACAD): the ROAD study. J.
    Orthop. Sci. **15**(6), 781–789 (2010)
22. Thomson, J., O'Neill, T., Felson, D., Cootes, T.: Automated shape and texture
    analysis for detection of osteoarthritis from radiographs of the knee. In: Navab, N.,
    Hornegger, J., Wells, W.M., Frangi, A.F. (eds.) MICCAI 2015. LNCS, vol. 9350, pp.
    127–134. Springer, Cham (2015). https://doi.org/10.1007/978-3-319-24571-3_16
23. Thomson, J., O'Neill, T., Felson, D., Cootes, T.: Detecting osteophytes in radio-
    graphs of the knee to diagnose osteoarthritis. In: Wang, L., Adeli, E., Wang, Q.,
    Shi, Y., Suk, H.-I. (eds.) MLMI 2016. LNCS, vol. 10019, pp. 45–52. Springer, Cham
    (2016). https://doi.org/10.1007/978-3-319-47157-0_6
24. Cootes, T.F., Ionita, M.C., Lindner, C., Sauer, P.: Robust and accurate shape
    model fitting using random forest regression voting. In: Fitzgibbon, A., Lazebnik,
    S., Perona, P., Sato, Y., Schmid, C. (eds.) ECCV 2012. LNCS, vol. 7578, pp. 278–
    291. Springer, Heidelberg (2012). https://doi.org/10.1007/978-3-642-33786-4_21
25. Ronneberger, O., Fischer, P., Brox, T.: U-Net: convolutional networks for biomed-
    ical image segmentation. In: Navab, N., Hornegger, J., Wells, W.M., Frangi, A.F.
    (eds.) MICCAI 2015. LNCS, vol. 9351, pp. 234–241. Springer, Cham (2015).
    https://doi.org/10.1007/978-3-319-24574-4_28
26. He, K., Zhang, X., Ren, S., Sun, J.: Delving deep into rectifiers: surpassing human-
    level performance on ImageNet classification. In: Proceedings of the IEEE Inter-
    national Conference on Computer Vision, pp. 1026–1034 (2015)
27. Kingma, D.P., Ba, J.: Adam: a method for stochastic optimization. arXiv preprint
    arXiv:1412.6980 (2014)
28. Faber, B.G., et al.: Osteophyte size and location on hip DXA scans are associated
    with hip pain: findings from a cross sectional study in UK Biobank. Bone **153**,
    116146 (2021)
29. Isensee, F., Jaeger, P.F., Kohl, S., Petersen, J., Maier-Hein, K.H.: nnU-Net: a self-
    configuring method for deep learning-based biomedical image segmentation. Nat.
    Methods **18**(2), 203–211 (2021). https://doi.org/10.1038/s41592-020-01008-z
30. Schachinger, J.K., et al.: A deep-learning-based technique for the quantitative anal-
    ysis of femorotibial osteophyte and bone volumes-data from the osteoarthritis ini-
    tiative. Osteoarthr. Cartil. **29**, S328–S329 (2021)

CIRDataset: A Large-Scale Dataset for Clinically-Interpretable Lung Nodule Radiomics and Malignancy Prediction

Wookjin Choi[1](✉), Navdeep Dahiya[2], and Saad Nadeem[3](✉)

[1] Department of Radiation Oncology, Thomas Jefferson University Hospital,
Philadelphia, USA
wookjin.choi@jefferson.edu
[2] School of Electrical and Computer Engineering, Georgia Institute of Technology,
Atlanta, USA
[3] Department of Medical Physics, Memorial Sloan Kettering Cancer Center,
New York, USA
nadeems@mskcc.org

Abstract. Spiculations/lobulations, sharp/curved spikes on the surface of lung nodules, are good predictors of lung cancer malignancy and hence, are routinely assessed and reported by radiologists as part of the standardized Lung-RADS clinical scoring criteria. Given the 3D geometry of the nodule and 2D slice-by-slice assessment by radiologists, manual spiculation/lobulation annotation is a tedious task and thus no public datasets exist to date for probing the importance of these clinically-reported features in the SOTA malignancy prediction algorithms. As part of this paper, we release a large-scale Clinically-Interpretable Radiomics Dataset, CIRDataset, containing 956 radiologist QA/QC'ed spiculation/lobulation annotations on segmented lung nodules from two public datasets, LIDC-IDRI (N = 883) and LUNGx (N = 73). We also present an end-to-end deep learning model based on multi-class Voxel2Mesh extension to segment nodules (while preserving spikes), classify spikes (sharp/spiculation and curved/lobulation), and perform malignancy prediction. Previous methods have performed malignancy prediction for LIDC and LUNGx datasets but without robust attribution to any clinically reported/actionable features (due to known hyperparameter sensitivity issues with general attribution schemes). With the release of this comprehensively-annotated CIRDataset and end-to-end deep learning baseline, we hope that malignancy prediction methods can validate their explanations, benchmark against our baseline, and provide clinically-actionable insights. Dataset, code, pretrained models, and docker containers are available at https://github.com/nadeemlab/CIR.

Keywords: Lung nodule · Spiculation · Malignancy prediction

1 Introduction

In the United States, lung cancer is the leading cause of cancer death [14]. Recently, radiomics and deep learning studies have been proposed for a variety

© The Author(s), under exclusive license to Springer Nature Switzerland AG 2022
L. Wang et al. (Eds.): MICCAI 2022, LNCS 13435, pp. 13–22, 2022.
https://doi.org/10.1007/978-3-031-16443-9_2

of clinical applications, including lung cancer screening nodule malignancy prediction [5,8,9,11]. The likelihood of malignancy is influenced by the radiographic edge characteristics of a pulmonary nodule, particularly spiculation. Benign nodule borders are usually well-defined and smooth, whereas malignant nodule borders are frequently blurry and irregular. The American College of Radiology (ACR) created the Lung Imaging Reporting and Data System (Lung-RADS) to standardize lung cancer screening on CT images based on size, appearance type, and calcification [7]. Spiculation has been proposed as an additional image finding that raises the suspicion of malignancy and allows for more precise prediction. Spiculation is caused by interlobular septal thickness, fibrosis caused by pulmonary artery obstruction, or lymphatic channels packed with tumor cells (also known as sunburst or corona radiata sign). It has a good positive predictive value for malignancy with a positive predictive value of up to 90%. Another feature significantly linked to malignancy is lobulation, which is associated with varied or uneven development rates [15].

Spiculation/lobulation quantification has previously been studied [8,10,13] but not in an end-to-end deep learning malignancy prediction context. Similarly, previous methods [17] have performed malignancy prediction alone but without robust attribution to clinically-reported/actionable features (due to known hyperparameter sensitivity issues and variability in general attribution/explanation schemes [3,4]). To probe the importance of spiculation/lobulation in the context of malignancy prediction and bypass reliance on sensitive/variable saliency maps, first we release a large-scale Clinically-Interpretable Radiomics Dataset, CIRDataset, containing 956 QA/QC'ed spiculation/lobulation annotations on segmented lung nodules for two public datasets, LIDC-IDRI (with visual radiologist malignancy RM scores for the entire cohort and pathology-proven malignancy PM labels for a subset) and LUNGx (with pathology-proven size-matched benign/malignant nodules to remove the effect of size on malignancy prediction). Second, we present a multi-class Voxel2Mesh [16] extension to provide a good baseline for end-to-end deep learning lung nodule segmentation (while preserving spikes), spikes classification (lobulation/spiculation), and malignancy prediction; Voxel2Mesh [16] is the only published method to our knowledge that preserves spikes during segmentation and hence its use as our base model. With the release of this comprehensively-annotated dataset and end-to-end deep learning baseline, we hope that malignancy prediction methods can validate their explanations, benchmark against our baseline, and provide clinically-actionable insights. Dataset, code, pretrained models, and docker containers are available at https://github.com/nadeemlab/CIR.

2   CIRDataset

Rather than relying on traditional radiomics features that are difficult to reproduce and standardize across same/different patient cohorts [12], this study focuses on standardized/reproducible Lung-RADS clinically-reported and interpretable features (spiculation/lobulation, sharp/curved spikes on the surface of the nodule). Given the 3D geometry of the nodule and 2D slice-by-slice assessment by radiologists, manual spiculation/lobulation annotation is a tedious task and thus no public datasets exist to date for probing the importance of these clinically-reported features in the SOTA malignancy prediction algorithms.

We release a large-scale dataset with high-quality lung nodule segmentation masks and spiculation/lobulation annotations for LIDC (N = 883) and LUNGx (N = 73) datasets. The spiculation/lobulation annotations were computed automatically and QA/QC'ed by an expert on meshes generated from nodule segmentation masks using negative area distortion metric from spherical parameterization [8]. Specifically, (1) the nodule segmentation masks were rescaled to isotropic voxel size with the CT image's finest spacing to preserve the details, (2) isosurface was extracted from the rescaled segmentation masks to construct a 3D mesh model, (3) spherical parameterization was then applied to extract the area distortion map of the nodule (computed from the log ratio of the input and the spherical mapped triangular mesh faces), and (4) spikes are detected on the mesh's surface (negative area distortion) and classified into spiculation, lobulation, and other. Because the area distortion map and spikes classification map were generated on a mesh model, these must be voxelized before deep learning model training. The voxelized area distortion map is divided into two masks: the nodule base ($\varepsilon > 0$) and the spikes ($\varepsilon \leq 0$). Then, in the spikes mask, the spiculation and lobulation classes were voxelized from the vertices classification map, while the other classifications were ignored and treated as nodule bases.

Following [8], we applied semi-auto segmentation for the largest nodules in each LIDC-IDRI patient scan [1,2] for more reproducible spiculation quantification, as well as calculated consensus segmentation using STAPLE to combine multiple contours by the radiologists. LUNGx only provides the nodule's location but no the segmentation mask. We applied the same semi-automated segmentation method on nodules as LIDC to obtain the segmentation masks. All these segmentation masks are released in CIRDataset. Complete pipeline for generating annotations from scratch on LIDC/LUNGx or private datasets can also be found on our CIR GitHub along with preprocessed data for different stages. Samples of the dataset, including area distortion maps (computed from our spherical parameterization method), are shown in Fig. 1.

3 Method

Several deep learning voxel/pixel segmentation algorithms have been proposed in the past, but most of these algorithms tend to smooth out the high-frequency spikes that constitute spiculation and lobulation features (Voxel2Mesh [16] is the only exception to date that preserves these spikes). The Jaccard index for nodule segmentation on a random LIDC training/validation split via UNet, FPN, and Voxel2Mesh was 0.775/0.537, 0.685/0.592, and 0.778/0.609, and for peaks segmentation it was 0.450/0.203, 0.332/0.236, and 0.493/0.476.

Using the Voxel2Mesh as our based model, we present a multi-class Voxel2Mesh extension that takes as input 3D CT volume and returns segmented

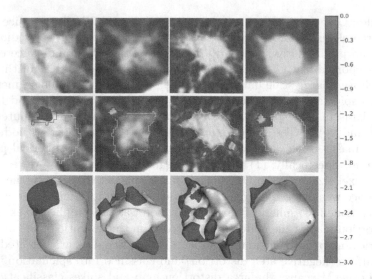

Fig. 1. Nodule spiculation quantification dataset samples; the first row - input CT image; the second row - superimposed area distortion map [8] and contours of each classifications on the input CT image; the third row - 3D mesh model with vertices classifications; red: spiculations, blue: lobulations, white: nodule (Color figure online)

3D nodule surface mesh (preserving spikes), vertex-level spiculation/lobulation classification, and binary benign/malignancy prediction. Implementation details, code, and trained models can be found on our CIR GitHub.

3.1   Multi-class Mesh Decoder

Voxel2Mesh [16] can generate a mesh model and voxel segmentation of the target item simultaneously. However, Voxel2Mesh only allows for multi-class segmentation in different disjoint objects. This paper presents a multi-class mesh decoder that enables multi-class segmentation in a single object. The multi-class decoder segments a baseline model first, then deforms it to include spiculation and lobulation spikes. Traditional voxel segmentation and mesh decoders were unable to capture nodule surface spikes because they attempted to provide a smooth and tight surface of the target object. To capture spikes and classify these into lobulations and spiculations, we added extra deformation modules to the mesh decoder. The mesh decoder deforms the input sphere mesh to segment the nodule with the deformation being controlled by chamfer distance between the vertices on the mesh and the ground truth nodule vertices with regularization (laplacian, edge, and normal consistency). Following the generation of a nodule surface by the mesh decoder at each level, the model deforms the nodule surface to capture lobulations and spiculations. The chamfer distance loss (chamfer_loss) between the ground truth lobulation and spiculation vertices and the deformed mesh is used to assess the extra deformations. To capture spikes, we reduced the regular-

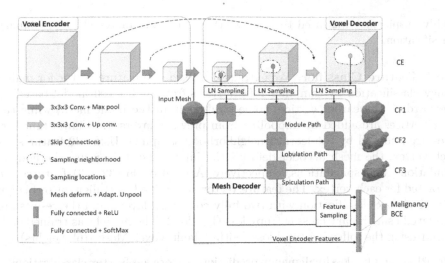

Fig. 2. Depiction of end-to-end deep learning architecture based on multi-class Voxel2Mesh extension. The standard UNet based voxel encoder/decoder (top) extracts features from the input CT volumes while the mesh decoder deforms an initial spherical mesh into increasing finer resolution meshes matching the target shape. The mesh deformation utilizes feature vectors sampled from the voxel decoder through the Learned Neighborhood (LN) Sampling technique and also performs adaptive unpooling with increased vertex counts in high curvature areas. We extend the architecture by introducing extra mesh decoder layers for spiculation and lobulation classification. We also sample vertices (shape features) from the final mesh unpooling layer as input to Fully Connected malignancy prediction network. We optionally add deep voxel-features from the last voxel encoder layer to the malignancy prediction network.

ization for the extra deformation to allow free deformation. For lobulation and spiculation, we classified each vertex based on the distance between the same vertex on the nodule surface and the deformed surface. The mean cross entropy loss (ce_loss) between the final mesh vertices and the ground truth vertices is used to evaluate their vertex classification. Deep shape features were extracted during the multi-class mesh decoding for spiculation quantification and subsequently used to predict malignancy (Fig. 2).

3.2 Malignancy Prediction

LIDC provides pathological malignancy (strong label) for a small subset of the data (LIDC-PM, $N = 72$), whereas LUNGx provides it for the entire dataset ($N = 73$). Unlike PM, LIDC provides weakly labeled radiological malignancy scores (RM) for the entire dataset ($N = 883$). Due to the limited number of strong labeled datasets, these can not be used to train a deep learning model. In contrast, RM cases are enough to train a data-intensive deep learning model. We used LIDC-RM to train and validate the model. In addition, because the RM score is graded on a five-point scale, $RM > 3$ (moderately suspicious to highly suspicious) was used to binarize the scores and matched to PM binary classification.

Mesh Feature Classifier. We extracted a fixed-size feature vector for malignancy classification by sampling 1000 vertices from each mesh model based on their order. The earlier vertices come straight from the input mesh, while the later vertices are added by unpooling from previous layers. Less important vertices are removed by the learned neighborhood sampling. Using 32 features for each vertex, the mesh decoder deforms the input mesh to capture the nodule, lobulations, and spiculations, respectively. A total of 96 (32 × 3) features are extracted for each vertex. The feature vector is classified as malignant or benign using Softmax classification with two fully connected layers, and the results are evaluated using binary cross entropy loss (bce_loss). The model was trained end-to-end using the following total loss (with default Voxel2Mesh [16] weights):

$$
\begin{aligned}
\text{total loss} =\ & 1 \times \text{bce\_loss } [\textbf{malignancy prediction}] + 1 \times \text{ce\_loss } [\textbf{vertex classification}] \\
& + 1 \times \text{chamfer\_loss } [\textbf{nodule mesh}] + 1 \times \text{chamfer\_loss } [\textbf{spiculation mesh}] \\
& + 1 \times \text{chamfer\_loss } [\textbf{lobulation mesh}] + (0.1 \times \text{laplacian\_loss} + 1 \times \text{edge\_loss} \\
& + 0.1 \times \text{normal\_consistency\_loss}) \ [\textbf{regularization}]
\end{aligned}
$$

Hybrid (voxel+mesh) Feature Classifier. The features from the last UNet encoder layer (256 × 4 × 4 × 4 = 16384) were flattened and then concatenated with the mesh features and fed into the last three fully connected layers to predict malignancy, as shown in Table 1. This leads to a total of 112384 (16384 + 96000) input features to the classifier which remains otherwise same as before. The motivation behind this hybrid feature classifier was to test using low level voxel-based deep features from the encoder in addition to the higher level shape features extracted from the mesh decoder for the task of malignancy prediction.

Table 1. Malignancy prediction model using mesh features only and using mesh and encoder features

Network	Layer	Input	Output	Activation
Mesh Only	FC Layer1	96000	512	RELU
Mesh Only	FC Layer2	512	128	RELU
Mesh Only	FC Layer3	128	2	Softmax
Mesh+Encoder	FC Layer1	112384	512	RELU
Mesh+Encoder	FC Layer2	512	128	RELU
Mesh+Encoder	FC Layer3	128	2	Softmax

4   Results and Discussion

All implementations were created using Pytorch. After separating the 72 strongly labeled datasets (LIDC-PM) for testing, we divided the remaining LIDC dataset

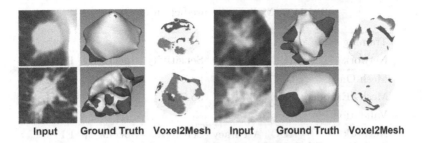

Input   Ground Truth   Voxel2Mesh   Input   Ground Truth   Voxel2Mesh

Fig. 3. Results of nodule segmentation and vertex classification; the first column - input CT image; the second column - 3D mesh model with vertices classifications (ground truth); the third column - 3D mesh model with vertices classifications (predictions); red: spiculations, blue: lobulations, white: nodule (Color figure online)

Table 2. Nodule (Class0), spiculation (Class1), and lobulation (Class2) peak classification metrics

Training						
Network	Chamfer Weighted Symmetric↓			Jaccard Index↑		
	Class0	Class1	Class2	Class0	Class1	Class2
Mesh Only	0.009	0.010	0.013	0.507	0.493	0.430
Mesh+Encoder	0.008	0.009	0.011	0.488	0.456	0.410
Validation						
Network	Chamfer Weighted Symmetric↓			Jaccard Index↑		
	Class0	Class1	Class2	Class0	Class1	Class2
Mesh Only	0.010	0.011	0.014	0.526	0.502	0.451
Mesh+Encoder	0.014	0.015	0.018	0.488	0.472	0.433
Testing LIDC-PM N = 72						
Network	Chamfer Weighted Symmetric↓			Jaccard Index↑		
	Class0	Class1	Class2	Class0	Class1	Class2
Mesh Only	0.011	0.011	0.014	0.561	0.553	0.510
Mesh+Encoder	0.009	0.010	0.012	0.558	0.541	0.507
Testing LUNGx N = 73						
Network	Chamfer Weighted Symmetric↓			Jaccard Index↑		
	Class0	Class1	Class2	Class0	Class1	Class2
Mesh Only	0.029	0.028	0.030	0.502	0.537	0.545
Mesh+Encoder	0.017	0.017	0.019	0.506	0.523	0.525

into train and validation subsets and trained on NVIDIA HPC clusters ($4 \times$ RTX A6000 (48 GB), $2 \times$ AMD EPYC 7763 CPUs (256 threads), and 768 GB RAM) for a maximum of 200 epochs. We saved the best model during training based on the Jaccard Index on the validation set. Once fully trained, we tested both the trained networks on LIDC-PM (N = 72) and LUNGx (N = 73) hold out test sets. For estimating the mesh classification (nodule, spiculation and lobulation)

Table 3. Malignancy prediction metrics.

Training					
Network	AUC	Accuracy	Sensitivity	Specificity	F1
Mesh Only	0.885	80.25	54.84	93.04	65.03
Mesh+Encoder	0.899	80.71	55.76	93.27	65.94
Validation					
Network	AUC	Accuracy	Sensitivity	Specificity	F1
Mesh Only	0.881	80.37	53.06	92.11	61.90
Mesh+Encoder	0.808	75.46	42.86	89.47	51.22
Testing LIDC-PM N = 72					
Network	AUC	Accuracy	Sensitivity	Specificity	F1
Mesh Only	0.790	70.83	56.10	90.32	68.66
Mesh+Encoder	0.813	79.17	70.73	90.32	79.45
Testing LUNGx N = 73					
Network	AUC	Accuracy	Sensitivity	Specificity	F1
Mesh Only	0.733	68.49	80.56	56.76	71.60
Mesh+Encoder	0.743	65.75	86.11	45.95	71.26

performance, we computed Jaccard Index and Chamfer Weighted Symmetric index, and for measuring the malignancy classification performance we computed standard metrics including Area Under ROC Curve (AUC), Accuracy, Sensitivity, Specificity, and F1 score.

Table 2 reports the mesh classification results for the two models. On the LIDC-PM test set, the mesh-only model produces better Jaccard Index for nodule (0.561 vs 0.558), spiculation (0.553 vs 0.541), and lobulation classification (0.510 vs 0.507). Opposite trend is observed in the Chamfer distance metric. On the external LUNGx testing dataset (N = 73), the hybrid voxel classifier model does better in terms of Chamfer distance metric for all three classes. The results for malignancy prediction are reported in Table 3. On the LIDC-PM test dataset, the hybrid features network produces an excellent AUC of 0.813 with an accuracy of 79.17%. The mesh-only features model, on the other hand, does slightly worse in terms of AUC (=0.790) and produces worse accuracy (70.83%). On the external LUNGx test dataset, the hybrid features network does better in terms of AUC and sensitivity metrics. This is likely due to the fact that the hybrid features' model uses voxel-level deep features in classification and no datasets from the LUNGx are used in training. There may also be differences in CT scanning protocols and/or different scanner properties which are never seen during training.

Figure 3 shows segmentation results using the proposed method, as well as ground truth mesh models with spiculation and lobulation classifications for comparison. The proposed method segmented nodule accurately while also detecting spiculations and lobulations at the vertex level. The vertex-level classification however detects only pockets of spiculations and lobulations rather than a contiguous whole. In the future, we will use mesh segmentation to solve this problem by exploiting the features of classified vertices and the relationship between neighboring vertices in the mesh model.

Previous works have performed malignancy prediction on LIDC and LUNGx datasets but again without any robust attribution to clinically-reported features. For reference, NoduleX [6] reported results only on the LIDC RM cohort, not the PM subset. When we ran the NoduleX pre-trained model (http://bioinformatics. astate.edu/NoduleX) on the LIDC PM subset, the AUC, accuracy, sensitivity, and specificity were 0.68, 0.68, 0.78, and 0.55 respectively versus ours 0.73, 0.68. 0.81 and 0.57. On LUNGx, AUC for NoduleX was 0.67 vs ours 0.73. MV-KBC [17] (implementation not available) reported the best malignancy prediction numbers with 0.77 AUC on LUNGx and 0.88 on LIDC RM (NOT PM).

In this work, we have focused on lung nodule spiculation/lobulation quantification via the Lung-RADS scoring criteria. In the future, we will extend our framework to breast nodule spiculation/lobulation quantification and malignancy prediction via BI-RADS scoring criteria (which has similar features). We will also extend our framework for advanced lung/breast cancer recurrence and outcomes prediction via spiculation/lobulation quantification.

Acknowledgements. This project was supported by MSK Cancer Center Support Grant/Core Grant (P30 CA008748) and by the Sidney Kimmel Cancer Center Support Grant (P30 CA056036).

References

1. Armato, S.G., McLennan, G., Bidaut, L., McNitt-Gray, M.F., Meyer, C.R., et al.: The lung image database consortium (LIDC) and image database resource initiative (IDRI): a completed reference database of lung nodules on CT scans. Med. Phys. **38**(2), 915–931 (2011). https://doi.org/10.1118/1.3528204
2. Armato, S.G., McLennan, G., Bidaut, L., McNitt-Gray, M.F., Meyer, C.R., et al.: Data from LIDC-IDRI. Cancer Imaging Arch. (2015). https://doi.org/10.7937/K9/TCIA.2015.LO9QL9SX
3. Arun, N., et al.: Assessing the trustworthiness of saliency maps for localizing abnormalities in medical imaging. Radiol. Artif. Intell. **3**(6), e200267 (2021)
4. Bansal, N., Agarwal, C., Nguyen, A.: SAM: he sensitivity of attribution methods to hyperparameters. In: Proceedings of the IEEE/CVF Conference on Computer Vision and Pattern Recognition, pp. 8673–8683 (2020)
5. Buty, M., Xu, Z., Gao, M., Bagci, U., Wu, A., Mollura, D.J.: Characterization of lung nodule malignancy using hybrid shape and appearance features. In: Ourselin, S., Joskowicz, L., Sabuncu, M.R., Unal, G., Wells, W. (eds.) MICCAI 2016. LNCS, vol. 9900, pp. 662–670. Springer, Cham (2016). https://doi.org/10.1007/978-3-319-46720-7_77
6. Causey, J.L., et al.: Highly accurate model for prediction of lung nodule malignancy with CT scans. Sci. Rep. **8**(1), 1–12 (2018)

7. Chelala, L., Hossain, R., Kazerooni, E.A., Christensen, J.D., Dyer, D.S., White, C.S.: Lung-RADS version 1.1: challenges and a look ahead, from the AJR special series on radiology reporting and data systems. Am. J. Roentgenol. **216**(6), 1411–1422 (2021). https://doi.org/10.2214/AJR.20.24807. pMID: 33470834 10:2214/AJR.20.24807, https://doi.org/10.2214/AJR.20.24807, pMID: 33470834

8. Choi, W., Nadeem, S., Alam, S.R., Deasy, J.O., Tannenbaum, A., Lu, W.: Reproducible and interpretable spiculation quantification for lung cancer screening. Comput. Methods Programs Biomed. **200**, 105839 (2021). https://doi.org/10.1016/j.cmpb.2020.105839. https://www.sciencedirect.com/science/article/pii/S0169260720316722

9. Choi, W., et al.: Radiomics analysis of pulmonary nodules in low-dose CT for early detection of lung cancer. Med. Phys. (2018). https://doi.org/10.1002/mp.12820

10. Dhara, A.K., Mukhopadhyay, S., Saha, P., Garg, M., Khandelwal, N.: Differential geometry-based techniques for characterization of boundary roughness of pulmonary nodules in CT images. Int. J. Comput. Assist. Radiolo. Surg. **11**(3), 337–349 (2016)

11. Hawkins, S., et al.: Predicting malignant nodules from screening CT scans. J. Thorac. Oncol. **11**(12), 2120–2128 (2016)

12. Meyer, M., et al.: Reproducibility of CT radiomic features within the same patient: influence of radiation dose and CT reconstruction settings. Radiology **293**(3), 583–591 (2019)

13. Niehaus, R., Raicu, D.S., Furst, J., Armato, S.: Toward understanding the size dependence of shape features for predicting spiculation in lung nodules for computer-aided diagnosis. J. Digit. Imaging **28**(6), 704–717 (2015)

14. Siegel, R.L., Miller, K.D., Jemal, A.: Cancer statistics, 2019. CA Cancer J. Clin. **69**(1), 7–34 (2019). https://doi.org/10.3322/caac.21551. https://acsjournals.onlinelibrary.wiley.com/doi/abs/10.3322/caac.21551

15. Snoeckx, A., et al.: Evaluation of the solitary pulmonary nodule: size matters, but do not ignore the power of morphology. Insights Imaging **9**, 73–86 (2017)

16. Wickramasinghe, U., Remelli, E., Knott, G., Fua, P.: Voxel2Mesh: 3D mesh model generation from volumetric data. In: Martel, A.L., Abolmaesumi, P., Stoyanov, D., Mateus, D., Zuluaga, M.A., Zhou, S.K., Racoceanu, D., Joskowicz, L. (eds.) MICCAI 2020. LNCS, vol. 12264, pp. 299–308. Springer, Cham (2020). https://doi.org/10.1007/978-3-030-59719-1_30

17. Xie, Y., et al.: Knowledge-based collaborative deep learning for benign-malignant lung nodule classification on chest CT. IEEE Trans. Med. Imaging **38**(4), 991–1004 (2018)

UNeXt: MLP-Based Rapid Medical Image Segmentation Network

Jeya Maria Jose Valanarasu[✉] and Vishal M. Patel

Johns Hopkins University, Baltimore, USA
jeyamariajose7@gmail.com

Abstract. UNet and its latest extensions like TransUNet have been the leading medical image segmentation methods in recent years. However, these networks cannot be effectively adopted for rapid image segmentation in point-of-care applications as they are parameter-heavy, computationally complex and slow to use. To this end, we propose UNeXt which is a Convolutional multilayer perceptron (MLP) based network for image segmentation. We design UNeXt in an effective way with an early convolutional stage and a MLP stage in the latent stage. We propose a tokenized MLP block where we efficiently tokenize and project the convolutional features and use MLPs to model the representation. To further boost the performance, we propose shifting the channels of the inputs while feeding in to MLPs so as to focus on learning local dependencies. Using tokenized MLPs in latent space reduces the number of parameters and computational complexity while being able to result in a better representation to help segmentation. The network also consists of skip connections between various levels of encoder and decoder. We test UNeXt on multiple medical image segmentation datasets and show that we reduce the number of parameters by **72x**, decrease the computational complexity by **68x**, and improve the inference speed by **10x** while also obtaining better segmentation performance over the state-of-the-art medical image segmentation architectures. Code is available at https://github.com/jeya-maria-jose/UNeXt-pytorch.

Keywords: Medical image segmentation · MLP · Point-of-care

1 Introduction

Medical imaging solutions have played a pivotal role for diagnosis and treatment in the healthcare sector. One major task in medical imaging applications is segmentation as it is essential for computer-aided diagnosis and image-guided surgery systems. Over the past decade, many works in the literature have focused on developing efficient and robust segmentation methods. UNet [18] is a landmark work which showed how efficient an encoder-decoder convolutional network with skip connections can be for medical image segmentation. UNet has became the backbone of almost all the leading methods for medical image segmentation in recent years. Following UNet, a number of key extensions like UNet++

© The Author(s), under exclusive license to Springer Nature Switzerland AG 2022
L. Wang et al. (Eds.): MICCAI 2022, LNCS 13435, pp. 23–33, 2022.
https://doi.org/10.1007/978-3-031-16443-9_3

[30], UNet3+ [13], 3D UNet [7], V-Net [16], Y-Net [15] and KiUNet [23] have been proposed. Recently, many transformer-based networks have been proposed for medical image segmentation as they learn a global understanding of images which can be helpful in segmentation. TransUNet [6] modifies the ViT architecture [10] into an UNet for 2D medical image segmentation. Other transformer-based networks like MedT [22], TransBTS [26], and UNETR [11] have also been proposed for medical image segmentation. Note that almost all the above works have focused on improving the performance of the network but do not focus much on the computational complexity, inference time or the number of parameters, which are essential in many real-world applications. As most of these are used for analysis in laboratory settings, they are tested using machines with high compute power (like GPUs). This helps accelerate the speed of inference and also help accommodate a large number of parameters.

In recent times, there has been a translation of medical imaging solutions from laboratory to bed-side settings. This is termed as point-of-care imaging as the testing and analysis is done by the side of the patient. Point-of-care imaging [24] helps clinicians with expanded service options and improved patient care. It helps in reducing the time and procedures involved in patients having to go visit radiology centers. Technology improvements around point-of-care imaging are leading to greater patient satisfaction. The use of point-of-care devices has been increasing in recent years. For example, point-of-care ultrasound (POCUS) devices [1] have shown to be useful to quickly check pleural irregularities in lungs, cardiac hemodynamic flow and automatic bladder volume calculation. Phone-camera based images are also being used to detect and diagnose skin conditions [2]. Magnetic resonance imaging (MRI) machines have also been developed for bed-side operation and fast analysis [3]. These recent diagnostic developments have helped in clear and fast acquisition of medical images at point-of-care as seen in Fig. 1. Tasks like segmentation, classification and registration are also being integrated along with these appliances to help patients and clinicians accelerate the diagnosis process. The major deep-learning based solutions for these tasks (like UNet and TransUNet) come with an inherent computation overhead and a large number of parameters making them difficult to use in point-of-care

Fig. 1. Motivation for UNeXt: As medical imaging solutions become more applicable at point-of-care, it is important to focus on making the deep networks light-weight and fast while also being efficient. (a) Point-of-Care medical intervention workflow. (b) Recent medical imaging developments: POCUS device [1] and (c) Phone-based skin lesion detection and identification application [2].

applications. In this work, we focus on solving this problem and design an efficient network that has less computational overhead, low number of parameters, a faster inference time while also maintaining a good performance. Designing such a network is essential to suit the shifting trends of medical imaging from laboratory to bed-side. To this end, we propose UNeXt which is designed using convolutional networks and (multilayer perceptron) MLPs.

Recently, MLP-based networks [14,20,21,28] have also been found to be competent in computer vision tasks. Especially MLP-Mixer [20], an all-MLP based network which gives comparable performance with respect to transformers with less computations. Inspired by these works, we propose UNeXt which is a convolutional and MLP-based network. We still follow a 5-layer deep encoder-decoder architecture of UNet with skip connections but change the design of each block. We have two stages in UNeXt- a convolutional stage followed by an MLP stage. We use convolutional blocks with less number of filters in the initial and final blocks of the network. In the bottleneck, we use a novel Tokenized MLP (Tok-MLP) block which is effective at maintaining less computation while also being able to model a good representation. Tokenized MLP projects the convolutional features into an abstract token and then uses MLPs to learn meaningful information for segmentation. We also introduce shifting operation in the MLPs to extract local information corresponding to different axial shifts. As the tokenized features are of the less dimensions and MLPs are less complicated than convolution or self-attention and transformers; we are able to reduce the number of parameters and computational complexity significantly while also maintaining a good performance. We evaluate UNeXt on ISIC skin lesion dataset [8] and Breast UltraSound Images (BUSI) dataset [4] and show that it obtains better performance than recent generic segmentation architectures. More importantly, we reduce the number of parameters by **72x**, decrease the computational complexity by **68x** and increase the inference speed by **10x** when compared to TransUNet making it suitable for point-of-care medical imaging applications.

In summary, this paper makes the following contributions: 1) We propose UNeXt, the first convolutional MLP-based network for image segmentation. 2) We propose a novel tokenized MLP block with axial shifts to efficiently learn a good representation at the latent space. 3) We successfully improve the performance on medical image segmentation tasks while having less parameters, high inference speed, and low computational complexity.

2   UNeXt

Network Design: UNeXt is an encoder-decoder architecture with two stages: 1) Convolutional stage, and a 2) Tokenized MLP stage. The input image is passed through the encoder where the first 3 blocks are convolutional and the next 2 are Tokenized MLP blocks. The decoder has 2 Tokenized MLP blocks followed by 3 convolutional blocks. Each encoder block reduces the feature resolution by 2 and each decoder block increases the feature resolution by 2. Skip connections are also included between the encoder and decoder. The number of channels across each block is a hyperparameter denoted as $C1$ to $C5$. For the experiments using UNeXt architecture, we follow $C1 = 32$, $C2 = 64$, $C3 = 128$, $C4 = 160$, and $C5 = 256$ unless stated otherwise. Note that these numbers are actually less than the number of filters of UNet and its variants contributing to the first change to reduce parameters and computation (Fig. 2).

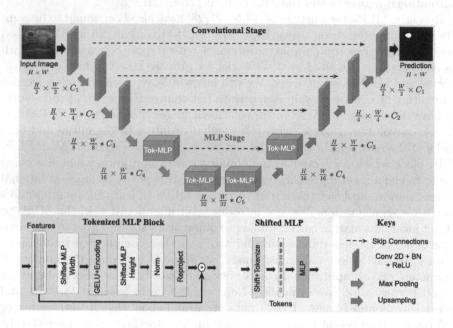

Fig. 2. Overview of the proposed UNeXt architecture.

Convolutional Stage: Each convolutional block is equipped with a convolution layer, a batch normalization layer and ReLU activation. We use a kernel size of 3×3, stride of 1 and padding of 1. The conv blocks in the encoder use a max-pooling layer with pool window 2×2 while the ones in the decoder consist of a bilinear interpolation layer to upsample the feature maps. We use bilinear interpolation instead of transpose convolution as transpose convolution is basically learnable upsampling and contributes to more learnable parameters.

Shifted MLP: In shifted MLP, we first shift the axis of the channels of conv features before tokenizing. This helps the MLP to focus on only certain locations of the conv features thus inducing locality to the block. The intuition here is similar to Swin transformer [5] where window-based attention is introduced to add more locality to an otherwise completely global model. As the Tokenized MLP block has 2 MLPs, we shift the features across width in one and across height in another like in axial-attention [25]. We split the features to h different partitions and shift them by $j = 5$ locations according to the specified axis. This helps us create random windows introducing locality along an axis (Fig. 3).

Tokenized MLP Stage: In the tokenized MLP block, we first shift the features and project them into tokens. To tokenize, we first use a kernel size of 3 and change the number of channels to E, where E is the embedding dimension (number of tokens) which is a hyperparameter. We then pass these tokens to a shifted MLP (across width) where hidden dimensions of the MLP is a hyperparameter H. Next, the features are passed through a depth wise convolutional layer (DWConv). We use DWConv in this block for two reasons: 1) It helps to encode a positional information of the MLP features. It is shown in [27] that Conv layer in an MLP block is enough to encode the positional information and it actually performs better than the standard positional encoding techniques. Positional encoding techniques like the ones in ViT need to be interpolated when the test and training resolutions are not the same often leading to reduced performance. 2) DWConv uses less number of parameters and hence increases efficiency. We then use a GELU [12] activation layer. We use GELU instead of RELU as it is a more smoother alternative and is found to perform better. In addition, most recent architectures like ViT [10] and BERT [9] have successfully used GELU to obtain improved results. We then pass the features through another shifted MLP (across height) that converts the dimensions from H to O. We use a residual connection here and add the original tokens as residuals. We then apply a layer normalization (LN) and pass the output features to the next block. LN is preferred over BN as it makes more sense to normalize along the tokens instead of normalizing across the batch in the Tokenized MLP block.

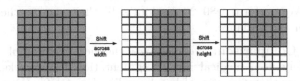

Fig. 3. Shifting operation. The features are shifted sequentially across height and width before tokenizing to induce window locality in the network.

The computation in the Tokenized MLP block can be summarized as:

$$X_{shift} = Shift_W(X); T_W = Tokenize(X_{shift}), \tag{1}$$

$$Y = f(DWConv((MLP(T_W)))), \tag{2}$$

$$Y_{shift} = Shift_H(Y); T_H = Tokenize(Y_{shift}), \tag{3}$$

$$Y = f(LN(T_W + MLP(GELU(T_H)))), \tag{4}$$

where T denotes the tokens, H denotes height, W denotes width, $DWConv$ denotes depth-wise convolution and LN denotes layer normalization. Note that all of these computations are performed across the embedding dimension H which is significantly less than the dimensionality of the feature maps $\frac{H}{N} \times \frac{H}{N}$ where N is a factor of 2 depending on the block. In our experiments, we set H to 768 unless stated otherwise. This way of designing the Tokenized MLP block helps in encoding meaningful feature information and not contribute much in terms of computation or parameters.

3   Experiments and Results

Datasets: To make our experiments as close to point-of-care imaging as possible, we pick International Skin Imaging Collaboration (ISIC 2018) [8] and Breast UltraSound Images (BUSI) [4] datasets to benchmark our results. The ISIC dataset contains camera-acquired dermatologic images and corresponding segmentation maps of skin lesion regions. The ISIC 2018 dataset consists of 2594 images. We resize all the images to a resolution of 512×512. BUSI consists of ultrasound images of normal, benign and malignant cases of breast cancer along with the corresponding segmentation maps. We use only benign and mailgnant images which results in a total of 647 images resized to a resolution of 256×256.

Implementation Details: We develop UNeXt using Pytorch framework. We use a combination of binary cross entropy (BCE) and dice loss to train UNeXt. The loss $\mathcal{L}$ between the prediction $\hat{y}$ and the target y is formulated as:

$$\mathcal{L} = 0.5BCE(\hat{y}, y) + Dice(\hat{y}, y) \tag{5}$$

We use an Adam optimizer with a learning rate of 0.0001 and momentum of 0.9. We also use a cosine annealing learning rate scheduler with a minimum learning rate upto 0.00001. The batch size is set equal to 8. We train UNeXt for a total of 400 epochs. We perform a 80–20 random split thrice across the dataset and report the mean and variance.

Performance Comparison: We compare the performance of UNeXt with recent and widely used medical image segmentation frameworks. In particular, we compare with convolutional baselines like UNet [18], UNet++ [30] and ResUNet [29]. We also compare with very recent transformer baselines like TransUNet [6] and MedT [22]. Note that we have focused on comparing against the baselines in terms of segmentation performance (F1 score and IoU) as well as number of parameters, computational complexity (in GFLOPs) and inference time (in ms).

We tabulate the results in Table 1. It can be observed that UNeXt obtains better segmentation performance than all the baselines with close second being TransUNet. The improvements are statistically significant with $p < 10^{-5}$. However, the most compelling point to note here is that UNeXt has very less number of computation compared to TransUNet as UNeXt does not have any attention blocks. The computation is calculated in terms of the number of floating point operators (FLOPs). We note that UNeXt has the least GFLOPs of 0.57 compared to TransUNet's 38.52 and UNet's 55.84. It is also the most light-weight network compared to all baselines. In particular, we note that UNeXt has only 1.58 M parameters compared to 105.32 M parameters of TransUNet. We also present the average inference time while operating on a CPU. Note that we have specifically bench-marked the inference time in CPU instead of GPU as point-of-care devices mostly operate on low-compute power and often do not have the computing advantage of GPU. We perform feed forward for 10 images of resolution 256 × 256 and report the average inference time. The CPU used for bench-marking was an Intel Xeon Gold 6140 CPU operating at 2.30 GHz. It can be noted that we experimented with Swin-UNet [5] but found have problems with convergence on small datasets resulting in poor performance. However, Swin-UNet is heavy with 41.35 M parameters and also computationally complex with 11.46 GFLOPs.

In Fig. 4, we plot the comparison charts of F1 score vs. GLOPs, F1 score vs. Inference time and F1 Score vs. Number of Parameters. The F1 score used here corresponds to the ISIC dataset. It can be clearly seen from the charts that UNeXt and TransUNet are the best performing methods in terms of the segmentation performance. However, UNeXt clearly outperforms all the other networks in terms of computational complexity, inference time and number of parameters which are all important characteristics to consider for point-of-care imaging applications. In Fig. 5, we present sample qualitative results of UNeXt along with other baselines. It can be observed that UNeXt produces competitive segmentation predictions compared to the other methods.

Table 1. Performance comparison with convolutional and transformer baselines.

Networks	Params (in M)	Inference speed (in ms)	GFLOPs	ISIC [8]		BUSI [4]	
				F1	IoU	F1	IoU
UNet [18]	31.13	223	55.84	84.03 ± 0.87	74.55 ± 0.96	76.35 ± 0.89	63.85 ± 1.12
UNet++ [30]	9.16	173	34.65	84.96 ± 0.71	75.12 ± 0.65	77.54 ± 0.74	64.33 ± 0.75
FastSCNN [17]	1.14	60	2.17	− ± −	− ± −	70.14 ± 0.64	54.98 ± 1.21
MobileNetv2 [19]	6.63	28	7.64	− ± −	− ± −	80.65 ± 0.34	68.95 ± 0.46
ResUNet [29]	62.74	333	94.56	85.60 ± 0.68	75.62 ± 1.11	78.25 ± 0.74	64.89 ± 0.83
MedT [22]	1.60	751	21.24	87.35 ± 0.18	79.54 + 0.26	76.93 ± 0.11	63.89 ± 0.55
TransUNet [6]	105.32	246	38.52	88.91 ± 0.63	80.51 ± 0.72	79.30 ± 0.37	66.92 ± 0.75
UNeXt	**1.47**	**25**	**0.57**	**89.70 ± 0.96**	**81.70 ± 1.53**	**79.37 ± 0.57**	**66.95 ± 1.22**

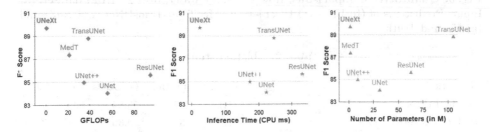

Fig. 4. Comparison charts. Y-axis corresponds to F1 score (higher the better). X-axis corresponds to GFLOPs, inference time and number of parameters (lower the better). It can be seen that UNeXt is the most efficient network compared to the others.

4   Discussion

Ablation Study: We conduct an ablation study (shown in Table 2) to understand the individual contribution of each module in UNeXt. We first start with the original UNet and then just reduce the number of filters to reduce the number of parameters and complexity. We see a reduction of performance with not much reduction in parameters. Next, we reduce the depth and use only a 3-level deep architecture which is basically the Conv stage of UNeXt. This reduces the number of parameters and complexity significantly but also reduces the performance by 4%. Now, we introduce the tokenized MLP block which improves the performance significantly while increasing the complexity and parameters by a minimal value. Next, we add the positional embedding method using DWConv as in [27] and see some more improvement. Next, we add the shifting operation in the MLPs and show that shifting the features before tokenizing improves the performance without any addition to parameters or complexity. As the shift operation does not contribute to any addition or multiplication it does not add on to any FLOPs. We note that shifting the features across both axes results in the best performance which is the exact configuration of UNeXt with min-

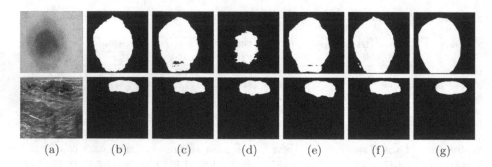

(a)   (b)   (c)   (d)   (e)   (f)   (g)

Fig. 5. Qualitative comparisons. Row 1 - ISIC dataset, Row 2 - BUSI dataset. (a) Input. Predictions of (b) UNet (c) UNet++ (d) MedT (e) TransUNet (f) UNeXt and (g) Ground Truth.

Table 2. Ablation study.

Network	Params	Inf. time	GFLOPs	F1	IoU
Original UNet	31.19	223	55.84	84.03	74.55
Reduced UNet	7.65	38	9.36	83.65	72.54
Conv Stage	0.88	9	0.36	80.12	67.75
Conv Stage + Tok-MLP w/o PE	1.46	22	0.57	88.78	79.32
Conv Stage + Tok-MLP + PE	1.47	23	0.57	89.25	80.76
Conv Stage + Shifted Tok-MLP (W) + PE	1.47	24	0.57	89.38	82.01
Conv Stage + Shifted Tok-MLP (H) + PE	1.47	24	0.57	89.25	81.94
Conv Stage + Shifted Tok-MLP (H+W) + PE	1.47	25	0.57	90.41	82.78

imal parameters and complexity. Note that all of the above experiments were conducted using a single fold of the ISIC dataset.

Analysis on Number of Channels: The number of channels is a main hyperparamter of UNeXt which affects the number of parameters, complexity and the performance of the network. In Table 3, we conduct experiments on single fold of ISIC to show two more different configurations of UNeXt. It can be observed that increasing the channels (UNeXt-L) further improves the performance while adding on to computational overhead. Although decreasing it (UNeXt-S) reduces the performance (the reduction is not drastic) but we get a very lightweight model.

Table 3. Analysis on the number of channels.

Network	C1	C2	C3	C4	C5	Params	Inf speed	GFLOPs	F1	IoU
UNeXt-S	8	16	32	64	128	0.32	22	0.10	89.62	81.40
UNeXt-L	32	64	128	256	512	3.99	82	1.42	90.65	83.10
UNeXt	16	32	128	160	256	1.47	25	0.57	90.41	82.78

Difference from MLP-Mixer: MLP-Mixer uses an all-MLP architecture for image recognition. UNeXt is a convolutional and MLP-based network for image segmentation. MLP-Mixer focuses on channel mixing and token mixing to learn a good representation. In contrast, we extract convolutional features and then tokenize the channels and use a novel tokenized MLPs using shifted MLPs to model the representation. It is worthy to note that we experimented with MLP-Mixer as encoder and a normal convolutional decoder. The performance was not optimal for segmentation and it was still heavy with around 11 M parameters.

5   Conclusion

In this work, we have proposed a new deep network architecture UNeXt for medical image segmentation focussed for point-of-care applications. UNeXt is a convolutional and MLP-based architecture where there is an initial conv stage followed by MLPs in the latent space. Specifically, we propose a tokenized MLP block with shifted MLPs to efficiently model the representation with minimal complexity and parameters. We validated UNeXt on multiple datasets where we achieve faster inference, reduced complexity and less number of parameters while also achieving the state-of-the-art performance.

References

1. https://www.butterflynetwork.com/iq
2. https://blog.google/technology/health/ai-dermatology-preview-io-2021/
3. https://hyperfine.io/

4. Al-Dhabyani, W., Gomaa, M., Khaled, H., Fahmy, A.: Dataset of breast ultrasound images. Data Brief **28**, 104863 (2020)
5. Cao, H., et al.: Swin-UNet: UNet-like pure transformer for medical image segmentation. arXiv preprint arXiv:2105.05537 (2021)
6. Chen, J., et al.: TransUNet: transformers make strong encoders for medical image segmentation. arXiv preprint arXiv:2102.04306 (2021)
7. Çiçek, Ö., Abdulkadir, A., Lienkamp, S.S., Brox, T., Ronneberger, O.: 3D U-Net: learning dense volumetric segmentation from sparse annotation. In: Ourselin, S., Joskowicz, L., Sabuncu, M.R., Unal, G., Wells, W. (eds.) MICCAI 2016. LNCS, vol. 9901, pp. 424–432. Springer, Cham (2016). https://doi.org/10.1007/978-3-319-46723-8_49
8. Codella, N.C., et al.: Skin lesion analysis toward melanoma detection: a challenge at the 2017 international symposium on biomedical imaging (ISBI), hosted by the international skin imaging collaboration (ISIC). In: 2018 IEEE 15th International Symposium on Biomedical Imaging (ISBI 2018), pp. 168–172. IEEE (2018)
9. Devlin, J., Chang, M.W., Lee, K., Toutanova, K.: BERT: pre-training of deep bidirectional transformers for language understanding. arXiv preprint arXiv:1810.04805 (2018)
10. Dosovitskiy, A., et al.: An image is worth 16×16 words: transformers for image recognition at scale. arXiv preprint arXiv:2010.11929 (2020)
11. Hatamizadeh, A., et al.: UNETR: transformers for 3D medical image segmentation. In: Proceedings of the IEEE/CVF Winter Conference on Applications of Computer Vision, pp. 574–584 (2022)
12. Hendrycks, D., Gimpel, K.: Gaussian error linear units (GELUS). arXiv preprint arXiv:1606.08415 (2016)
13. Huang, H., et al.: UNet 3+: a full-scale connected UNet for medical image segmentation. In: ICASSP 2020–2020 IEEE International Conference on Acoustics, Speech and Signal Processing (ICASSP), pp. 1055–1059. IEEE (2020)
14. Lian, D., Yu, Z., Sun, X., Gao, S.: As-MLP: an axial shifted MLP architecture for vision. arXiv preprint arXiv:2107.08391 (2021)
15. Mehta, S., Mercan, E., Bartlett, J., Weaver, D., Elmore, J.G., Shapiro, L.: Y-Net: joint segmentation and classification for diagnosis of breast biopsy images. In: Frangi, A.F., Schnabel, J.A., Davatzikos, C., Alberola-López, C., Fichtinger, G. (eds.) MICCAI 2018. LNCS, vol. 11071, pp. 893–901. Springer, Cham (2018). https://doi.org/10.1007/978-3-030-00934-2_99
16. Milletari, F., Navab, N., Ahmadi, S.A.: V-Net: fully convolutional neural networks for volumetric medical image segmentation. In: 2016 Fourth International Conference on 3D Vision (3DV), pp. 565–571. IEEE (2016)
17. Poudel, R.P., Liwicki, S., Cipolla, R.: Fast-SCNN: fast semantic segmentation network. arXiv preprint arXiv:1902.04502 (2019)
18. Ronneberger, O., Fischer, P., Brox, T.: U-Net: convolutional networks for biomedical image segmentation. In: Navab, N., Hornegger, J., Wells, W.M., Frangi, A.F. (eds.) MICCAI 2015. LNCS, vol. 9351, pp. 234–241. Springer, Cham (2015). https://doi.org/10.1007/978-3-319-24574-4_28
19. Sandler, M., Howard, A., Zhu, M., Zhmoginov, A., Chen, L.C.: MobileNetV2: inverted residuals and linear bottlenecks. In: Proceedings of the IEEE Conference on Computer Vision and Pattern Recognition, pp. 4510–4520 (2018)
20. Tolstikhin, I.O., et al.: MLP-mixer: an all-MLP architecture for vision. In: Advances in Neural Information Processing Systems 34 (2021)
21. Touvron, H., et al.: ResMLP: feedforward networks for image classification with data-efficient training. arXiv preprint arXiv:2105.03404 (2021)

22. Valanarasu, J.M.J., Oza, P., Hacihaliloglu, I., Patel, V.M.: Medical transformer: gated axial-attention for medical image segmentation. In: de Bruijne, M., et al. (eds.) MICCAI 2021. LNCS, vol. 12901, pp. 36–46. Springer, Cham (2021). https:// doi.org/10.1007/978-3-030-87193-2_4

23. Valanarasu, J.M.J., Sindagi, V.A., Hacihaliloglu, I., Patel, V.M.: KiU-Net: towards accurate segmentation of biomedical images using over-complete representations. In: Martel, A.L., et al. (eds.) MICCAI 2020. LNCS, vol. 12264, pp. 363–373. Springer, Cham (2020). https://doi.org/10.1007/978-3-030-59719-1_36

24. Vashist, S.K.: Point-of-care diagnostics: recent advances and trends. Biosensors 7(4), 62 (2017)

25. Wang, H., Zhu, Y., Green, B., Adam, H., Yuille, A., Chen, L.-C.: Axial-DeepLab: stand-alone axial-attention for panoptic segmentation. In: Vedaldi, A., Bischof, H., Brox, T., Frahm, J.-M. (eds.) ECCV 2020. LNCS, vol. 12349, pp. 108–126. Springer, Cham (2020). https://doi.org/10.1007/978-3-030-58548-8_7

26. Wang, W., Chen, C., Ding, M., Yu, H., Zha, S., Li, J.: TransBTS: multimodal brain tumor segmentation using transformer. In: de Bruijne, M., et al. (eds.) MICCAI 2021. LNCS, vol. 12901, pp. 109–119. Springer, Cham (2021). https://doi.org/10. 1007/978-3-030-87193-2_11

27. Xie, E., Wang, W., Yu, Z., Anandkumar, A., Alvarez, J.M., Luo, P.: SegFormer: simple and efficient design for semantic segmentation with transformers. arXiv preprint arXiv:2105.15203 (2021)

28. Yu, T., Li, X., Cai, Y., Sun, M., Li, P.: S2-MLP: spatial-shift MLP architecture for vision. In: Proceedings of the IEEE/CVF Winter Conference on Applications of Computer Vision, pp. 297–306 (2022)

29. Zhang, Z., Liu, Q., Wang, Y.: Road extraction by deep residual U-Net. IEEE Geosci. Remote Sens. Lett. 15(5), 749–753 (2018)

30. Zhou, Z., Rahman Siddiquee, M.M., Tajbakhsh, N., Liang, J.: UNet++: a nested U-Net architecture for medical image segmentation. In: Stoyanov, D., et al. (eds.) DLMIA/ML-CDS -2018. LNCS, vol. 11045, pp. 3–11. Springer, Cham (2018). https://doi.org/10.1007/978-3-030-00889-5_1

Exploring Smoothness and Class-Separation for Semi-supervised Medical Image Segmentation

Yicheng Wu[1(✉)], Zhonghua Wu[2], Qianyi Wu[1], Zongyuan Ge[3,4], and Jianfei Cai[1]

[1] Department of Data Science & AI, Faculty of Information Technology, Monash University, Melbourne, VIC 3800, Australia
yicheng.wu@monash.edu
[2] School of Computer Science and Engineering, Nanyang Technological University, Singapore 639798, Singapore
[3] Monash-Airdoc Research, Monash University, Melbourne, VIC 3800, Australia
[4] Monash Medical AI, Monash eResearch Centre, Melbourne, VIC 3800, Australia

Abstract. Semi-supervised segmentation remains challenging in medical imaging since the amount of annotated medical data is often scarce and there are many blurred pixels near the adhesive edges or in the low-contrast regions. To address the issues, we advocate to firstly constrain the consistency of pixels with and without strong perturbations to apply a sufficient smoothness constraint and further encourage the class-level separation to exploit the low-entropy regularization for the model training. Particularly, in this paper, we propose the SS-Net for semi-supervised medical image segmentation tasks, via exploring the pixel-level **S**moothness and inter-class **S**eparation at the same time. The pixel-level smoothness forces the model to generate invariant results under adversarial perturbations. Meanwhile, the inter-class separation encourages individual class features should approach their corresponding high-quality prototypes, in order to make each class distribution compact and separate different classes. We evaluated our SS-Net against five recent methods on the public LA and ACDC datasets. Extensive experimental results under two semi-supervised settings demonstrate the superiority of our proposed SS-Net model, achieving new state-of-the-art (SOTA) performance on both datasets. The code is available at https://github.com/ycwu1997/SS-Net.

Keywords: Semi-supervised segmentation · Pixel-level smoothness · Inter-class separation

1 Introduction

Most of deep learning-based segmentation models rely on large-scale dense annotations to converge and generalize well. However, it is extremely expensive and

Supplementary Information The online version contains supplementary material available at https://doi.org/10.1007/978-3-031-16443-9_4.

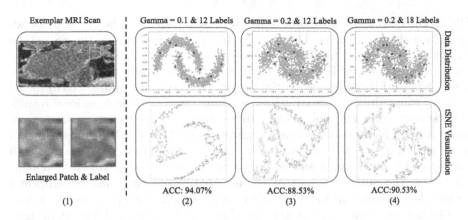

Fig. 1. Exemplar MRI scan (Top Left) and illustrations of three two-moon datasets (Top Right), equipped with the enlarged patch/label (Bottom Left) and their tSNE [12] visualized features (Bottom Right). The gamma is used to control the class dispersion.

labor-consuming to obtain adequate per-pixel labels for the model training. Such a heavy annotation cost has motivated the community to study the semi-supervised segmentation methods [9,15,22], aiming to train a model with few labeled data and abundant unlabeled data while still achieving a satisfactory segmentation performance.

Existing semi-supervised segmentation methods are usually based on the two assumptions: smoothness and low entropy. The smoothness assumption encourages the model to generate invariant results under small perturbations at the data level [3,8,14,19], the feature level [4,16,23,25] and the model level [7,10,11,24,28]. Its success implies that the local distributional smoothness (LDS) is crucial to leverage abundant unlabeled data for the model training. The low-entropy assumption further constrains that the decision boundary should lie in the low-density regions [6,17,20,21]. In other words, the semi-supervised models are expected to output highly confident predictions even without the supervision of corresponding labels, aiming at the inter-class separation.

Despite the progress, the semi-supervised segmentation in the medical imaging area remains challenging due to two factors: fewer labels and blurred targets. As illustrated in Fig. 1, there are many ambiguous pixels near the adhesive edges or low-contrast regions and the amount of labeled data is usually limited. Such concomitant challenges may lead to poor performance of the existing models [7,10,11,21,28]. We further use the synthetic two-moon dataset as a toy example to illustrate this scenario. Specifically, we trained a four-layer MLP in a supervised manner and then used the tSNE tool [12] to visualize the deep features. We can see that, with fewer labels and more blurred data (larger gamma), the performance drops significantly and the model cannot distinguish different classes well since the feature manifolds of different classes are inter-connected, see the 3rd column of Fig. 1, where the low-entropy assumption is violated.

To alleviate the problems, we advocate: (1) the *strong perturbations* are needed to sufficiently regularize the large amounts of unlabeled medical data

[3]; (2) the *class-level separation* is also suggested to pursue the decision boundary in low-density regions. Therefore, in this paper, we propose the SS-Net, to explore the pixel-level **S**moothness and inter-class **S**eparation at the same time for semi-supervised medical image segmentation. Specifically, our SS-Net has two novel designs. First, inspired by the virtually adversarial training (VAT) model [14], we adopt the adversarial noises as strong perturbations to enforce a sufficient smoothness constraint. Second, to encourage the inter-class separation, we select a set of high-quality features from the labeled data as the prototypes and force other features to approach their corresponding prototypes, making each class distribution compact and pushing away different classes. We evaluated our SS-Net on the public LA and ACDC datasets [2,26]. Extensive experimental results demonstrate our model is able to achieve significant performance gains.

Overall, our contributions are three-fold: (1) we point out two challenges, *i.e.*, fewer labels and blurred targets, for the semi-supervised medical image segmentation and show our key insight that it is crucial to employ the strong perturbations to sufficiently constrain the pixel-level smoothness while at the same time encouraging the inter-class separation, enabling the model to produce low-entropy predictions; (2) in our SS-Net, we introduce the adversarial noises as strong perturbations. To our knowledge, it is one of the first to apply this technique to perturb medical data for semi-supervised tasks. Then, the inter-class separation is encouraged via shrinking each class distribution, which leads to better performance and complements the pixel-level smoothness; (3) via utilizing both techniques, our SS-Net outperforms five recent semi-supervised methods and sets the new state of the art on the LA and ACDC datasets.

2 Method

Figure 2 shows the overall pipeline of our SS-Net. We propose two designs to encourage the pixel-level smoothness and the inter-class separation, respectively. First, the pixel-level smoothness is enforced via applying a consistency constraint between an original image $x \in X$ and its perturbed sample with the per-pixel adversarial noises. Second, we compute a set of feature prototypes Z from the labeled data X_L, and then encourage high dimensional features F to be close to the prototypes Z so as to separate different classes in the feature space. We now delve into the details.

2.1 Pixel-level Smoothness

It is nowadays widely recognized that the LDS is critical for semi-supervised learning [15]. This kind of regularization can be formulated as:

$$\mathcal{LDS}(x;\theta) = D\left[\hat{y}, p(x+r)\right], \|r\| \leq \epsilon \tag{1}$$

where D is used to compute the discrepancy between the prediction of a perturbed sample and its true label $\hat{y} \in Y$, and ϵ controls the magnitude of the

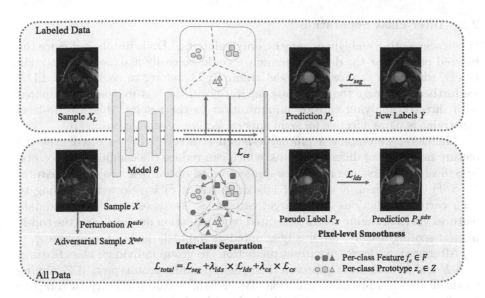

Fig. 2. Pipeline of our proposed SS-Net. The pixel-level smoothness is constrained at the pixel level via applying adversarial noises as strong perturbations while the inter-class separation is performed at the feature level by a prototype-based strategy.

enforced perturbation r. Since adequate true labels are not available in the semi-supervised scenario, $\hat{y}$ is usually set as the pseudo label $p(\hat{y}|x)$. Essentially, LDS regularizes the model to be robust or consistent with small perturbations of data.

Meanwhile, to apply strong perturbations, following the VAT model [14], we use the gradient g as the direction of r^{adv} to perturb the original sample x, which is at the pixel level and can be estimated as:

$$g = \nabla_{r^{ini}} D\left[p(\hat{y}|x), p(y|x + r^{ini})\right]$$
$$r^{adv} = \epsilon \times g/\|g\|_2, \tag{2}$$

where r^{ini} can be set as a random noise vector, g denotes the fastest-changing direction at the measurement of D, and $r^{adv} \in R^{adv}$ is the corresponding adversarial noise vector. Note that, in the original VAT model [14], D is adopted as the K-L Divergence. However, through experiments, we found that the K-L Divergence might not be a suitable one for segmentation tasks. Therefore, we utilize the Dice loss as D to generate adversarial noises and the LDS loss becomes

$$\mathcal{L}_{lds} = \frac{1}{C}\sum_{c=1}^{C}[1 - \frac{2\|p(\hat{y_c}|x) \cap p(y_c|x + r^{adv})\|}{\|p(\hat{y_c}|x)\| + \|p(y_c|x + r^{adv})\|}] \tag{3}$$

where C is the number of classes. $p(\hat{y_c}|x)$ and $p(y_c|x+r^{adv})$ denote the predictions of x with and without strong perturbations in the c-th class, respectively.

In this way, g can be efficiently computed via the back-propagation scheme. Compared to random noises, such adversarial noises can provide a stronger smoothness constraint to facilitate the model training [14].

2.2 Inter-class Separation

When segmenting ambiguous targets, only enforcing LDS is insufficient since the blurred pixels near the decision boundary could be easily assigned to uncertain labels, which can confuse the model training. Therefore, to complement LDS, we further encourage the inter-class separation in the feature space. Compared with directly applying entropy minimization to the results, this feature-level constraint is more effective for semi-supervised image segmentation [16].

Therefore, we employ a prototype-based strategy [1,27] to disconnect the feature manifolds of different classes, which can reduce the computational costs. Specifically, we first use non-linear projectors to obtain the projected features $F = F_l \cup F_u$. Then, a subset of labeled features F_l is selected according to their correct predictions in the target categories. Next, we sort these candidate features via the ranking scores generated by C attention modules, and the top-K highest-scoring features are finally adopted as the high-quality prototypes Z.

Afterward, we leverage current predictions to group individual class features $f_c \in F$ and force them to approach their corresponding prototype $z_c \in Z$, aiming to shrink the intra-class distribution. We use the cosine similarity to compute the distance between z_c and f_c with the loss $\mathcal{L}_{cs}$ defined as

$$\mathcal{L}_{cs} = \frac{1}{C}\frac{1}{N}\frac{1}{M}\sum_{c=1}^{C}\sum_{i=1}^{N}\sum_{j=1}^{M} w_{ij}\left(1 - \frac{\langle z_c^i, f_c^j \rangle}{\|z_c^i\|_2 \cdot \|f_c^j\|_2}\right) \tag{4}$$

where w_{ij} is the weight for normalization as [1], and N or M respectively denotes the number of prototypes or projected features in the c-th class. Here, $\mathcal{L}_{cs}$ can align the labeled/unlabeled features and make each class distribution compact, resulting in a good separation of different classes in the feature space.

Finally, the overall loss is a weighted sum of the segmentation loss $\mathcal{L}_{seg}$ and the other two losses:

$$\mathcal{L}_{total} = \mathcal{L}_{seg} + \lambda_{lds} \times \mathcal{L}_{lds} + \lambda_{cs} \times \mathcal{L}_{cs} \tag{5}$$

where λ_{lds} and λ_{cs} are the corresponding weights to balance the losses. Note that, $\mathcal{L}_{seg}$ is a Dice loss, which is applied for the few labeled data. The other two losses are applied for all data to regularize the model training.

3 Experiment and Results

3.1 Dataset

We evaluate the proposed SS-Net on the LA[1] dataset [26] and the ACDC[2] dataset [2]. The LA dataset consists of 100 gadolinium-enhanced MRI scans, with a fixed split[3] of 80 samples for training and 20 samples for validation. We report the performance on the validation set for fair comparisons as [7,10,21,28]. On the ACDC dataset, the data split[4] is also fixed with 70, 10, and 20 patients'

---

[1] http://atriaseg2018.cardiacatlas.org.
[2] https://www.creatis.insa-lyon.fr/Challenge/acdc/databases.html.
[3] https://github.com/yulequan/UA-MT/tree/master/data.
[4] https://github.com/HiLab-git/SSL4MIS/tree/master/data/ACDC.

scans for training, validation, and testing, respectively. All experiments follow
the identical setting for fair comparisons and we consider the challenging semi-
supervised settings to verify our model, where 5% and 10% labels are used and
the rest in the training set are regarded as unlabeled data.

3.2 Implementation Details

Following the public methods [7,10,11,21,28] on both datasets, all inputs were
normalized as zero mean and unit variance. We used the rotation and flip operations to augment data and trained our model via a SGD optimizer with a learn-
ing rate 0.01. The loss weights λ_{lds} and λ_{cs} were set as an iteration-dependent
warming-up function [5] and w_{ij} was obtained by normalizing the learnable
attention weights as [1]. We updated each class-level feature prototype of size
32×32 in an online fashion. On LA, ϵ was set as 10 and we chose the V-Net [13]
as the backbone. For training, we randomly cropped $112 \times 112 \times 80$ patches and
the batch size was 4, containing two labeled patches and two unlabeled patches.
We trained the model via 15K iterations. For testing, we employed a fixed stride
$(18 \times 18 \times 4)$ to extract patches and the entire predictions were recomposed from
patch-based outputs. On ACDC, we set ϵ as 6 and chose the U-Net model [18]
as the backbone to process 2D patches of size 256×256. The batch size was 24
and the total training iterations were 30K. All experiments in this paper were
conducted on the same environments with fixed random seeds (Hardware: Single
NVIDIA Tesla V100 GPU; Software: PyTorch 1.8.0+cu111, and Python 3.8.10).

3.3 Results

Performance on the LA Dataset. In Table 1, we use the metrics of Dice,
Jaccard, 95% Hausdorff Distance (95HD), and Average Surface Distance (ASD)
to evaluate the results. It reveals that: (1) compared to the lower bounds, *i.e.*,
only with 5%/10% labeled data training, our proposed SS-Net can effectively
leverage the unlabeled data and produce impressive performance gains of all
metrics; (2) when trained with fewer labels, *i.e.*, 5%, our SS-Net significantly
outperforms other models on the LA dataset. It indicates enforcing the adversarial perturbations is useful to sufficiently regularize the unlabeled data when
the amount of labeled data is limited. Furthermore, we post-processed all the
results via selecting the largest connected component as [7] for fair comparisons.
Note that we did not enforce any boundary constraint to train our model and our
model naturally achieves satisfied shape-related performance on the LA dataset.
The visualized results in Fig. 3 indicates that our SS-Net can detect most organ
details, especially for the blurred edges and thin branches (highlighted by yellow
circles), which are critical attributes for clinical applications.

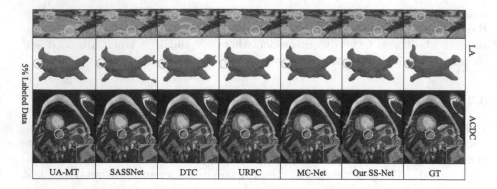

Fig. 3. Exemplar results of several semi-supervised segmentation methods with 5% labeled data training and corresponding ground truth (GT) on the LA dataset (Top two rows) and ACDC dataset (Bottom row).

Table 1. Comparisons with five recent methods on the LA dataset. All results were reproduced as [7, 10, 11, 21, 28] in the identical experimental setting for fair comparisons.

Method	# Scans used		Metrics				Complexity	
	Labeled	Unlabeled	Dice (%)↑	Jaccard (%)↑	95HD (voxel)↓	ASD (voxel)↓	Para. (M)	MACs (G)
V-Net	4 (5%)	0	52.55	39.60	47.05	9.87	9.44	47.02
V-Net	8 (10%)	0	82.74	71.72	13.35	3.26	9.44	47.02
V-Net	80 (All)	0	91.47	84.36	5.48	1.51	9.44	47.02
UA-MT [28] (MICCAI'19)	4 (5%)	76 (95%)	82.26	70.98	13.71	3.82	9.44	47.02
SASSNet [7] (MICCAI'20)			81.60	69.63	16.16	3.58	9.44	47.05
DTC [10] (AAAI'21)			81.25	69.33	14.90	3.99	9.44	47.05
URPC [11] (MICCAI'21)			82.48	71.35	14.65	3.65	5.88	69.43
MC-Net [21] (MICCAI'21)			83.59	72.36	14.07	2.70	12.35	95.15
SS-Net (Ours)			**86.33**	**76.15**	**9.97**	**2.31**	9.46	47.17
UA-MT [28] (MICCAI'19)	8 (10%)	72 (90%)	87.79	78.39	8.68	2.12	9.44	47.02
SASSNet [7] (MICCAI'20)			87.54	78.05	9.84	2.59	9.44	47.05
DTC [10] (AAAI'21)			87.51	78.17	8.23	2.36	9.44	47.05
URPC [11] (MICCAI'21)			86.92	77.03	11.13	2.28	5.88	69.43
MC-Net [21] (MICCAI'21)			87.62	78.25	10.03	**1.82**	12.35	95.15
SS-Net (Ours)			**88.55**	**79.62**	**7.49**	1.90	9.46	47.17

Performance on the ACDC Dataset. Table 2 gives the averaged performance of three-class segmentation including the myocardium, left and right ventricles on the ACDC dataset. We can see that, with 5% labeled data training, the performance of UA-MT model [28] decreases significantly and is even worse than the lower bound, i.e., 46.04% vs. 47.83% in Dice. Since [28] filters highly uncertain regions during training, such a performance drop suggests that it is needed to make full use of the ambiguous pixels, especially in the regime with extremely scarce labels. On the contrary, as Table 2 shows, our SS-Net surpasses all other methods and achieves the best segmentation performance. The bottom row of Fig. 3 also shows that our model can output good segmentation results and effectively eliminate most of the false-positive predictions on ACDC.

Table 2. Comparisons with five recent methods on the ACDC dataset. All results were re-produced as [7,10,11,21,28] in the identical experimental setting for fair comparisons.

Method	# Scans used		Metrics				Complexity	
	Labeled	Unlabeled	Dice (%)↑	Jaccard (%)↑	95HD (voxel)↓	ASD (voxel)↓	Para. (M)	MACs (G)
U-Net	3 (5%)	0	47.83	37.01	31.16	12.62	1.81	2.99
U-Net	7 (10%)	0	79.41	68.11	9.35	2.70	1.81	2.99
U-Net	70 (All)	0	91.44	84.59	4.30	0.99	1.81	2.99
UA-MT [28] (MICCAI'19)	3 (5%)	67 (95%)	46.04	35.97	20.08	7.75	1.81	2.99
SASSNet [7] (MICCAI'20)			57.77	46.14	20.05	6.06	1.81	3.02
DTC [10] (AAAI'21)			56.90	45.67	23.36	7.39	1.81	3.02
URPC [11] (MICCAI'21)			55.87	44.64	13.60	3.74	1.83	3.02
MC-Net [21] (MICCAI'21)			62.85	52.29	7.62	2.33	2.58	5.39
SS-Net (Ours)			**65.82**	**55.38**	**6.67**	**2.28**	1.83	2.99
UA-MT [28] (MICCAI'19)	7 (10%)	63 (90%)	81.65	70.64	6.88	2.02	1.81	2.99
SASSNet [7] (MICCAI'20)			84.50	74.34	5.42	1.86	1.81	3.02
DTC [10] (AAAI'21)			84.29	73.92	12.81	4.01	1.81	3.02
URPC [11] (MICCAI'21)			83.10	72.41	**4.84**	1.53	1.83	3.02
MC-Net [21] (MICCAI'21)			86.44	77.04	5.50	1.84	2.58	5.39
SS-Net (Ours)			**86.78**	**77.67**	6.07	**1.40**	1.83	2.99

Table 3. Ablation studies on the LA dataset.

# Scans used		Loss			Metrics			
Labeled	Unlabeled	$\mathcal{L}_{seg}$	$\mathcal{L}_{lds}$	$\mathcal{L}_{cs}$	Dice (%)↑	Jaccard (%)↑	95HD (voxel)↓	ASD (voxel)↓
4 (5%)	0	✓			52.55	39.60	47.05	9.87
	76 (95%)	✓	✓*		82.27	70.46	13.82	3.48
	76 (95%)	✓	✓		84.31	73.50	12.91	3.42
	76 (95%)	✓		✓	84.10	73.36	11.85	3.36
	76 (95%)	✓	✓	✓	**86.33**	**76.15**	**9.97**	**2.31**
8 (10%)	0	✓			82.74	71.72	13.35	3.26
	72 (90%)	✓	✓*		87.48	77.98	8.95	2.11
	72 (90%)	✓	✓		87.50	77.98	9.44	2.08
	72 (90%)	✓		✓	87.68	78.23	8.84	2.12
	72 (90%)	✓	✓	✓	**88.55**	**79.62**	**7.49**	**1.90**

* We adopt the K-L Divergence as D, following the traditional VAT model [14].

Ablation Study. We conducted ablation studies on LA to show the effects of individual components. Table 3 indicates that either using $\mathcal{L}_{lds}$ to pursue the pixel-level smoothness or applying $\mathcal{L}_{cs}$ to encourage the inter-class separation is effective to improve the semi-supervised segmentation performance. Table 3 also gives the results of using K-L Divergence to estimate the adversarial noises as [14], whose results suggest that adopting the Dice loss as D can achieve better performance (*e.g.*, 84.31% vs. 82.27% in Dice with 5% labeled training data).

4 Conclusion

In this paper, we have presented the SS-Net for semi-supervised medical image segmentation. Given that fewer labels and blurred targets in the medical domain, our key idea is that it is important to simultaneously apply the adversarial noises for a sufficient smoothness constraint and shrink each class distribution to separate different classes, which can effectively exploit the unlabeled training data. The experimental results on the LA and ACDC datasets have demonstrated our proposed SS-Net outperforms other methods and achieves a superior performance for semi-supervised medical image segmentation. Future work will include the adaptive selections of the perturbation magnitude and the prototype size.

Acknowledgements. This work was supported by Monash FIT Start-up Grant. We also appreciate the efforts to collect and share the LA and ACDC datasets [2, 26] and several public repositories [7, 10, 11, 21, 28].

References

1. Alonso, I., Sabater, A., Ferstl, D., Montesano, L., Murillo, A.C.: Semi-supervised semantic segmentation with pixel-level contrastive learning from a class-wise memory bank. In: ICCV 2021, pp. 8219–8228 (2021)
2. Bernard, O., et al.: Deep learning techniques for automatic MRI cardiac multi-structures segmentation and diagnosis: is the problem solved? IEEE Trans. Med. Imaging **37**(11), 2514–2525 (2018)
3. French, G., Laine, S., Aila, T., Mackiewicz, M., Finlayson, G.: Semi-supervised semantic segmentation needs strong, varied perturbations. arXiv preprint arXiv:1906.01916 (2019)
4. Lai, X., et al..: Semi-supervised semantic segmentation with directional context-aware consistency. In: CVPR 2021, pp. 1205–1214 (2021)
5. Laine, S., Aila, T.: Temporal ensembling for semi-supervised learning. arXiv preprint arXiv:1610.02242 (2016)
6. Lee, D.H., et al.: Pseudo-label: the simple and efficient semi-supervised learning method for deep neural networks. In: ICML 2013, vol. 3, no. 2 (2013)
7. Li, S., Zhang, C., He, X.: Shape-aware semi-supervised 3D semantic segmentation for medical images. In: Martel, A.L., et al. (eds.) MICCAI 2020. LNCS, vol. 12261, pp. 552–561. Springer, Cham (2020). https://doi.org/10.1007/978-3-030-59710-8_54
8. Li, X., Yu, L., Chen, H., Fu, C.W., Xing, L., Heng, P.A.: Transformation-consistent self-ensembling model for semisupervised medical image segmentation. IEEE Trans. Neural Netw. Learn. Syst. **32**(2), 523–534 (2020)
9. Liu, W., Wu, Z., Ding, H., Liu, F., Lin, J., Lin, G.: Few-shot segmentation with global and local contrastive learning. arXiv preprint arXiv:2108.05293 (2021)
10. Luo, X., Chen, J., Song, T., Wang, G.: Semi-supervised medical image segmentation through dual-task consistency. In: AAAI 2021, vol. 35, no. 10, pp. 8801–8809 (2021)
11. Luo, X., et al.: Efficient semi-supervised gross target volume of nasopharyngeal carcinoma segmentation via uncertainty rectified pyramid consistency. In: de Bruijne, M., et al. (eds.) MICCAI 2021. LNCS, vol. 12902, pp. 318–329. Springer, Cham (2021). https://doi.org/10.1007/978-3-030-87196-3_30
12. Van der Maaten, L., Hinton, G.: Visualizing data using t-SNE. J. Mach. Learn. Res. **9**(11), 2579–2605 (2008)
13. Milletari, F., Navab, N., Ahmadi, S.A.: V-Net: fully convolutional neural networks for volumetric medical image segmentation. In: 3DV 2016, pp. 565–571 (2016)

14. Miyato, T., Maeda, S.i., Koyama, M., Ishii, S.: Virtual adversarial training: a regularization method for supervised and semi-supervised learning. IEEE Trans. Pattern Anal. Mach. Intell. **41**(8), 1979–1993 (2018)
15. Ouali, Y., Hudelot, C., Tami, M.: An overview of deep semi-supervised learning. arXiv preprint arXiv:2006.05278 (2020)
16. Ouali, Y., Hudelot, C., Tami, M.: Semi-supervised semantic segmentation with cross-consistency training. In: CVPR 2020, pp. 12674–12684 (2020)
17. Pham, H., Dai, Z., Xie, Q., Le, Q.V.: Meta pseudo labels. In: CVPR 2021, pp. 11557–11568 (2021)
18. Ronneberger, O.: Invited talk: U-net convolutional networks for biomedical image segmentation. In: Maier-Hein, geb. Fritzsche, K., Deserno, geb. Lehmann, T., Handels, H., Tolxdorff, T. (eds.) Bildverarbeitung für die Medizin 2017. I, p. 3. Springer, Heidelberg (2017). https://doi.org/10.1007/978-3-662-54345-0_3
19. Sohn, K., Berthelot, D., et al.: FixMatch: simplifying semi-supervised learning with consistency and confidence. In: NeurIPS 2020, vol. 33, pp. 596–608 (2020)
20. Wu, Y., et al.: Mutual consistency learning for semi-supervised medical image segmentation. Med. Image Anal. **81**, 102530 (2022)
21. Wu, Y., Xu, M., Ge, Z., Cai, J., Zhang, L.: Semi-supervised left atrium segmentation with mutual consistency training. In: de Bruijne, M., et al. (eds.) MICCAI 2021. LNCS, vol. 12902, pp. 297–306. Springer, Cham (2021). https://doi.org/10. 1007/978-3-030-87196-3_28
22. Wu, Z., Lin, G., Cai, J.: Keypoint based weakly supervised human parsing. Image Vis. Comput. **91**, 103801 (2019)
23. Wu, Z., Shi, X., Lin, G., Cai, J.: Learning meta-class memory for few-shot semantic segmentation. In: ICCV 2021, pp. 517–526 (2021)
24. Xia, Y., et al.: 3D semi-supervised learning with uncertainty-aware multi-view co-training. In: WACV 2020, pp. 3646–3655 (2020)
25. Xie, Y., Zhang, J., Liao, Z., Verjans, J., Shen, C., Xia, Y.: Intra-and inter-pair consistency for semi-supervised gland segmentation. IEEE Trans. Image Process. **31**, 894–905 (2021)
26. Xiong, Z., et al.: A global benchmark of algorithms for segmenting the left atrium from late gadolinium-enhanced cardiac magnetic resonance imaging. Med. Image Anal. **67**, 101832 (2021)
27. Xu, Z., et al.: All-around real label supervision: cyclic prototype consistency learning for semi-supervised medical image segmentation. IEEE J. Biomed. Health Inform. **26**(7), 3174–3184 (2022)
28. Yu, L., Wang, S., Li, X., Fu, C.-W., Heng, P.-A.: Uncertainty-aware self-ensembling model for semi-supervised 3D left atrium segmentation. In: Shen, D., et al. (eds.) MICCAI 2019. LNCS, vol. 11765, pp. 605–613. Springer, Cham (2019). https:// doi.org/10.1007/978-3-030-32245-8_67

Uncertainty-Guided Lung Nodule Segmentation with Feature-Aware Attention

Han Yang[2], Lu Shen[2], Mengke Zhang[2], and Qiuli Wang[1(✉)]

[1] Center for Medical Imaging, Robotics, Analytic Computing and Learning (MIRACLE), School of Biomedical Engineering and Suzhou Institute for Advanced Research, University of Science and Technology of China, Suzhou, China
wangqiuli@ustc.edu.cn

[2] School of Big Data and Software Engineering, Chongqing University, Chongqing, China
{yang_han,shenlu,mkzhang}@cqu.edu.cn

Abstract. Since radiologists have different training and clinical experiences, they may provide various segmentation annotations for a lung nodule. Conventional studies choose a single annotation as the learning target by default, but they waste valuable information of consensus or disagreements ingrained in the multiple annotations. This paper proposes an Uncertainty-Guided Segmentation Network (UGS-Net), which learns the rich visual features from the regions that may cause segmentation uncertainty and contributes to a better segmentation result. With an Uncertainty-Aware Module, this network can provide a Multi-Confidence Mask (MCM), pointing out regions with different segmentation uncertainty levels. Moreover, this paper introduces a Feature-Aware Attention Module to enhance the learning of the nodule boundary and density differences. Experimental results show that our method can predict the nodule regions with different uncertainty levels and achieve superior performance in the LIDC-IDRI dataset.

Keywords: Lung nodule · Segmentation · Uncertainty · Attention mechanism · Computed tomography

1 Introduction

Lung nodule segmentation plays a crucial role in Computer-Aided Diagnosis (CAD) systems for lung nodules [1]. Traditional methods like [2–6] usually use a single annotation as the learning target, which is consistent with conventional deep learning strategy. However, in the clinical situation, a sample lung nodule may be evaluated by several radiologists, and datasets like the LIDC-IDRI [7] also provide multiple annotations for a lung nodule. As a result, these methods cannot make the most of the information in multiple annotations.

H. Yang and L. Shen—Equal contribution.

Supplementary Information The online version contains supplementary material available at https://doi.org/10.1007/978-3-031-16443-9_5.

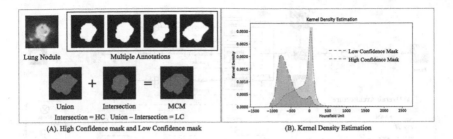

(A). High Confidence mask and Low Confidence mask   (B). Kernel Density Estimation

Fig. 1. (A): Overview of High Certainty (HC) mask, Low Certainty (LC) mask, and Multi-Confidence Mask (MCM). **(B)**: Kernel density estimation of HC and LC masks in the LIDC-IDRI. The HU values in LC are mainly distributed around *HU -750*, and the HU values in HC are mainly distributed around *HU 0*.

The challenge of learning multiple annotations is that there might be uncertain regions, which are annotated as 'nodule tissues' by some doctors and annotated as 'not nodule tissues' by the other. To this end, studies like [8–10] propose to model these uncertainties as probabilistic distributions and produce segmentation masks with the random variables in latent space. Simon A. A. Kohl [9] proposed a model based on probabilistic U-Net [11] with a conditional variational auto-encoder that used a hierarchical latent space decomposition. X. Long [10] proposed to learn the probabilistic distributions of multiple 3D annotations for a lung nodule with a probabilistic V-Net. But these methods cannot have stable segmentation results relying on random variables.

This paper proposes that regions causing segmentation uncertainty are not *random* and introduces a Multi-Confidence Mask (MCM). As shown in Fig. 1(A), the MCM defines two categories for the mask pixels: High Confidence (HC) mask and Low Confidence (LC) mask. The HC mask indicates the regions annotated by all radiologists (intersection of all annotations), and the LC mask shows the regions that the radiologists have disagreements (the difference of union and intersection). *The question is, what's the difference between the HC and LC masks?*

To demonstrate the difference between HC and LC masks, we further calculate the HU (Hounsfield Unit) Kernel Estimations in HC and LC of LIDC-IDRI [7]. As shown in Fig. 1(B), HC and LC masks have quite different HU distributions: HU values in HC masks are mainly distributed around *HU 0*, while HU values in LC masks are mainly distributed around *HU -750*. HU reflects tissues' density [12]. Additionally, we can find that LC is basically around the edge of nodules from Fig. 1(A). These phenomenons indicate that the radiologists may have more diverse opinions about low-dense nodule tissues and boundaries.

Therefore, our study focuses on two challenges: (1) Making good use of all annotations during model training. (2) Utilizing radiologists' consensus and disagreements to improve segmentation performance. To tackle these challenges, we propose to learn the consensus and disagreements in all annotations directly without changing the end-to-end training strategy. Our contributions can be summarized as: (1) To enable the network to learn the information of multiple annotations, this paper introduces a Multi-Confidence Mask (MCM) composed of all annotations' union and intersection,

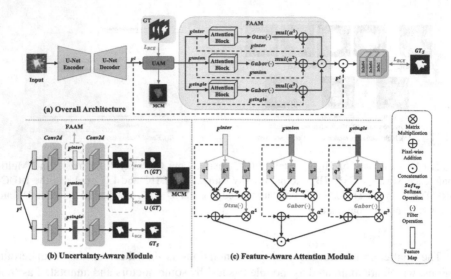

Fig. 2. (a): Overview of uncertainty-guided segmentation network. **(b):** Uncertainty-aware module. It takes the feature maps F^I as the input, and then fuses them into three branches with three learning targets: $\cup(GT)$, $\cap(GT)$, and GT_S. **(c):** Feature-aware attention module.

reflecting lung nodule regions with different uncertainty levels, which can be used as references for clinical use. (2) This paper proposes an Uncertainty-Guided Segmentation Network (UGS-Net), which contains an Uncertainty-Aware Module (UAM) and a Feature-Aware Attention Module (FAAM). The UAM introduces the consensus and disagreements in the annotation sets into the learning process. At the same time, the FAAM guides the network to learn more specific features about ambiguous regions, like boundaries and low-dense tissues, further improving the segmentation performance.

2   Methods

Figure 2 shows the architecture of the proposed Uncertainty-Guided Segmentation Network (UGS-Net). UGS-Net's input is the lung nodule CT image X, and its final learning target keeps consistent with the current mainstream methods, which is a single annotation GT_S. Moreover, we introduce two additional auxiliary learning targets, multiple annotation's union $\cup(GT)$ and intersection $\cap(GT)$, to assist the proposed network in learning richer information from disagreements among multiple annotations. The input images and their masks have the same size 64×64.

The UGS-Net contains a three-stage processing pipeline: **(1)** UGS-Net uses an Attention U-Net [13] to capture deep feature maps F^I with 64 channels. This Attention U-Net has five down-sampling and up-sampling layers. Each up-sampling layer is composed of two convolutional layers and an attention block. **(2)** the Uncertainty-Aware Module (UAM) analyzes the F^I and generates $\cup(GT)'$, $\cap(GT)'$, and GT_S' under the guidance of $\cup(GT)$, $\cap(GT)$, and GT_S. Additionally, three visual feature variables

F^{union}, F^{inter}, and F^{single} are obtained in UAM and will be fed into the next module. UAM can generate MCM simultaneously, pointing out lung nodule regions with different uncertainty levels. **(3)** the Feature-Aware Attention Module (FAAM) further analyzes the F^{union}, F^{inter}, and F^{single} with different learning preferences by introducing the visual filters to the self-attention mechanism. The FAAM output is concatenated with F^I. Then they will be fed into the output CNN module, which contains three convolutional layers. The output of this module is the final segmentation result S.

2.1 Uncertainty-Aware Module

The consensus and disagreements among multiple experts reflect the uncertainty levels of different nodule tissues in CT images. How to better take advantage of this valuable uncertainty information to further improve the overall segmentation performance is an important research problem. In this paper, we propose an Uncertainty-Aware Module (UAM) and introduce two auxiliary learning targets to take full utilization of all the annotation information.

Specifically, we define a *Union Mask* $\cup(GT)$ and an *Intersection Mask* $\cap(GT)$ as two auxiliary learning targets besides the final learning target *Single Mask* GT_S, where GT is a collection of all the annotations. As shown in Fig. 2 (b), the backbone of UAM is a three-branch CNN network and each branch has two convolutional layers. This module takes the deep feature maps F^I with 64 channels as the input and its loss is computed as $L_{UAM} = L_{BCE}(\cup(GT)', \cup(GT)) + L_{BCE}(\cap(GT)', \cap(GT)) + L_{BCE}(GT_S', GT_S)$, where the $\cup(GT)'$, $\cap(GT)'$, and GT_S' represent the results of the three branches. L_{BCE} denotes the binary cross entropy loss.

The outputs of UMA contain two parts. The first part is Multi-Confidence Mask defined as $normalization(\cup(GT)' + \cap(GT)')$, which can show different uncertainty levels in different nodule regions. Moreover, under the guidance of different learning targets, three feature maps F^{union}, F^{inter}, and F^{single} with different latent visual features will be obtained after the first convolutional layer. These three feature maps are the second part of outputs, and they will be fed into the next module for further analysis.

2.2 Feature-Aware Attention Module

To further distinguish three feature maps (F^{union}, F^{inter}, and F^{single}) and extract more features with high uncertainty, a Feature-Aware Attention Module (FAAM) is introduced to capture more density differences and boundaries of nodules.

As shown in Fig. 2 (c), the Feature-Aware Attention Module (FAAM) has three attention blocks, and each block contains a self-attention block [14] and a feature-aware filter. These attention blocks process F^{union}, F^{inter}, and F^{single} with different feature-aware filters, which enable the network to formulate different learning preferences for different learning objectives, to obtain more image features that are helpful for segmentation. More specifically, assuming the input is F^z, for each attention block, we have $o_z = \Gamma(A(F^z))$, where $z \in \{union, inter, single\}$, A indicates the self-attention architecture, and Γ is a feature-aware filter. The process of A please refer to the study

[14]. The output of FAAM is $\odot\{o_{union}, o_{inter}, o_{single}\}$, where $\odot$ is the concatenation operation. Each o_z has 32 channels, so $\odot\{o_{union}, o_{inter}, o_{single}\}$ has 96 channels in total.

As mentioned above, each learning objective has its own filter setting. We use Otsu [15] as its feature-aware filter when learning the F^{inter}, because Otsu is more sensitive to high dense nodule tissues [16]. We choose Gabor as their feature-aware filter when learning F^{union} and F^{single} for two reasons. Firstly, Gabor filter is sensitive to image edges and can provide good directional selection and scale selection features, so that it can better capture the nodule boundary features in F^{union}. Secondly, the Gabor filter can capture the local structural features of multiple directions in the image's local area, which is very suitable for extracting the overall general features in F^{single}. To improve the calculation speed, we use GConv in Gabor CNN [17], which can also enhance the adaptability of deep learning features to changes in orientation and scale.

Finally, the output CNN module that mentioned above processes the concatenation of $\odot\{o_{union}, o_{inter}, o_{single}\}$ and F^I, and the final segmentation result S is generated at the end of this module. The total training loss L for the proposed UGS-Net can be represented as $L = L_{BCE}(S, GT_S) + L_{UAM}$.

3   Experiments and Results

3.1   Dataset and Experimental Setup

We use the LIDC-IDRI [7] dataset, which has 2635 annotated lung nodules with malignancy labels. We extract 1859 nodules with multiple annotations (more than 1) and their annotations' union and intersection. The LIDC-IDRI didn't provide the information about which annotation was better. So the target annotation for the final output is the first annotation that appeared in the annotation list. To eliminate the effect of data split, we use five-cross validation in this study and ensure each fold has the same malignant/benign nodule distributions. The source code is available on GitHub.

We use the Stochastic Gradient Descent with Warm Restarts (SGDR) [18] to optimize parameters and build the network with PyTorch. All experiments are run on a GPU of NVIDIA Tesla V100. The UGS-Net needs 23 GB GPU memory, and is trained with 150 epochs. The batch size is set as 32. More details can be found in supplementary materials. We will release the source code for both data pre-processing and the proposed network.

This paper uses Dice, IoU, and NSD as the evaluation metrics. For the convenience of explanation, we temporarily define the ground truth as G and segmentation result as S. Dice is used to calculate the similarity of G and S. IoU is the ratio of G and S [19, 20]. Normalized surface Dice (NSD) [21] is used to evaluate how close G and S are to each other at a specified tolerance τ. The NSD is defined as:

$$NSD(G, S) = \frac{|\partial G \cap B_{\partial S}^{(\tau)}| + |\partial S \cap B_{\partial G}^{(\tau)}|}{|\partial G| + |\partial S|} \tag{1}$$

where $|\partial G|$ and $|\partial S|$ are the number of pixels of G and S. $B_{\partial G}^{(\tau)}, B_{\partial S}^{(\tau)} \subset R^3$ and $B_{\partial G}^{(\tau)} = \{x \in R^3 | \exists \tilde{x} \in \partial G, ||x - \tilde{x}|| \leq \tau\}$, $B_{\partial S}^{(\tau)} = \{x \in R^3 | \exists \tilde{x} \in \partial S, ||x - \tilde{x}|| \leq \tau\}$.

3.2 Performance of General Segmentation

Lung Nodule Segmentation. We compare our network with the state-of-the-art (SOTA) methods for medical image segmentation which are trained with single annotation in Table 1. We run experiments on normal U-Net, nnU-Net, R2U-Net, Attention U-Net, Attention U-Net+UAM (V1), V1+self-attention block (V2), and UGS-Net. V1, V2, and UGS-Net are trained with our settings, other methods (no-multiple-annotation methods) are trained with traditional training strategy (each nodule had a single annotation).

Table 1. Comparison between our framework and existing methods on the LIDC-IDRI dataset.

Method	Dice(%)	IoU(%)	NSD(%)
FCN V-Net [1]	79.59	–	–
Dual-Branch ResNet [22]	82.74	–	–
nnU-Net [23]	85.82	–	–
R2U-Net [24]	82.10	71.18	93.07
Attention U-Net [13]	85.37	75.45	94.80
Multi-Orientation U-Net [25]	83.00	76.00	–
3D DCNN [26]	83.10	71.85	–
Channel U-Net [27]	84.02	73.53	–
Nested U-Net [28]	83.44	72.72	–
Baseline (U-Net)	85.05	75.27	94.43
Attention U-Net+UAM (V1)	85.94	76.20	95.06
V1+self-attention block (V2)	85.97	76.18	95.22
UGS-Net	**86.12**	**76.44**	**95.36**

According to Table 1, the average Dice score of our network is 86.12%, which achieves superior performance in all compared methods. At the same time, our average IoU and NSD are 76.44% and 95.36%, which are also better than other methods. The first seven columns of Fig. 3 show the segmentation results of UGS-Net and other networks. It is obvious that the segmentation results generated by the proposed network are better than other methods, especially for the ambiguous regions among different experts, such as low-dense tissues, edge areas, and spiculations. We choose Attention U-Net as our first component even though nnU-Net achieves better performance. That's because the nnU-Net needs over 150 GB of system memory and takes too much system resources, but the performance is slightly better than the Attention U-Net.

Ablation Studies. This section will evaluate the performance of each component in the UGS-Net. Firstly, by integrating the UAM into Attention U-Net, the Dice score increases by 0.89%. It indicates that the calibration prediction ability of the model can be improved by introducing the consensus and disagreements of doctors' professional knowledge to guide model learning. Secondly, to better use the uncertain clues provided by doctors, the FAAM is designed to conduct targeted learning of nodules' different features under the guidance of these clues, which improves Dice, IoU, and NSD scores, respectively. Compared with integrating the self-attention block (Table 1 (V2)) into the model directly, the FAAM (Table 1 (UGS-Net)) achieves better performance, which indicates filtering can enhance the differentiation of attention in different branches and enable the model to extract features more comprehensively.

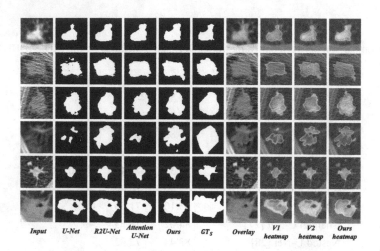

| Input | U-Net | R2U-Net | Attention U-Net | Ours | GT$_S$ | Overlay | V1 heatmap | V2 heatmap | Ours heatmap |

Fig. 3. Segmentation results of different methods.

To explore the influence of UAM and FAAM, we visualize the network's last convolution layer in different configurations with Grad CAM [29], and show the results in Fig. 3 (column 8–10). According to this figure: (1) our network can identify cavities more clearly (Fig. 3 (row6)); (2) our network can notice more low-density tissues (Fig. 3 (row 1, 2, 4)); (3) our network is more sensitive to spiculation (Fig. 3 (row 3, 5)); (4) for nodules with higher density, model's attention will gradually turn to the boundary (Fig. 3 (row 3)). More results are provided in the supplementary materials.

3.3 Performance of Multi-Confidence Mask

Prediction of Intersection and Union Mask. The quality of the predicted union and intersection directly affects the final estimation of Multi-Confidence Mask (MCM). To demonstrate the UGS-Net's ability to predict union and intersection, we design a comparative experiment. The experimental results are shown in Table 2.

According to Table 2, the U-Net baseline achieves an average Dice score of 85.05% with *Single Mask*. Meanwhile, it achieves average Dice scores of 84.08% and 82.66% with *Intersection Mask* and *Union Mask*. The proposed UGS-Net shows the optimal performance in most indicators. Compared with baseline, UGS-Net improves all indicators of *Single mask* and *Intersection mask*, in which Dice scores are improved by

1.07% and 0.63%, IoU scores are improved by 1.17% and 0.55%, and NSD scores are improved 0.83% and 0.35%, respectively. This indicates that the proposed network enhances the recognition of most nodule tissues, and has a more accurate calibration of nodule boundaries. Additionally, UGS-Net makes greater progress in *Union mask*, whose Dice score improved by 1.7% compared to the baseline, IoU improved by 2.03%, and NSD is improved by 1.1%. Experimental results show that with the help of UAM and FAAM, our network can effectively utilize the disagreements of doctors to learn more high uncertainty nodule regions. However, traditional training strategy could not let U-Net have the ability to focus on uncertain regions.

Table 2. Performances of UGS-Net and other methods on the $\cup(GT)$ and $\cap(GT)$ segmentation.

Method	Single mask			Intersection mask			Union mask		
	Dice	IoU	NSD	Dice	IoU	NSD	Dice	IoU	NSD
U-Net [30]	85.05	75.27	94.43	84.08	73.92	95.20	82.66	71.75	92.62
R2U-Net [24]	82.10	71.18	93.07	77.90	65.70	92.06	83.12	72.28	92.71
Attention U-Net [13]	85.37	75.45	94.80	84.07	73.71	95.10	83.48	**73.93**	93.30
UGS-Net	**86.12**	**76.44**	**95.26**	**84.71**	**74.47**	**95.55**	**84.36**	73.78	**93.72**

Prediction of Multi-Confidence Mask. Figure 4(A) shows the generated MCM and the prediction result of multiple annotations' intersection, union. Colors are used to show the differences, and the original outputs are grayscale. As can be seen, our MCM has several advantages: (1) MCM can better show the significant cavity features of lung nodules (Fig. 4(A).(a)(b)). On the prediction of the intersection, the cavity is obviously larger, and we believe it is because our network can better capture the density differences about nodule tissues, and keep more features about cavity. (2) MCM can better show the spiculation, which is an important feature for the diagnosis of the malignant nodule (Fig. 4(A).(c)(d)). The spiculation is a stellate distortion caused by the intrusion of nodules into surrounding tissue, which is low-dense and distributed around the nodule edges. The network pays more attention to the low-dense tissues and boundaries when learning the union of multiple annotations. As a result, it naturally has better performance on spiculation segmentation. (3) MCM can segment the part-solid nodule (Fig. 4(A).(e)) better, which means the network can take advantage of the multiple annotations since the experts tend to have different opinions in the low-dense regions.

Most evaluation metrics can not directly measure the quality of MCM prediction due to the small LC regions. So we calculate the HU distributions in predicted HC and LC masks, and compare them with the real ones. If our network can predict the regions with different uncertainty levels well, the HU distributions of predicted HC and LC should be similar to the real one. As shown in this Fig. 4(B), the predicted and real curves are almost the same, which means our predicted uncertainty levels of regions are convincing from the perspective of statistics.

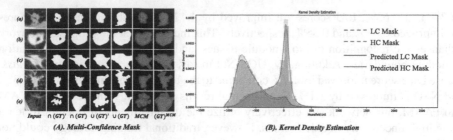

(A). *Multi-Confidence Mask*   (B). *Kernel Density Estimation*

Fig. 4. (A): Prediction of the Multi-Confidence Mask. For each nodule, we provide the predicted intersection ($\cap(GT)'$), the standard intersection ($\cap(GT)$), the predicted union ($\cup(GT)'$), the standard union ($\cup(GT)$), and the predicted MCM, the standard MCM ($(GT)^{MCM}$). **(B)**: Kernel density estimation for the predicted HC and LC.

4   Conclusions

This paper presents a novel Uncertainty-Guided framework for lung nodule segmentation, which focuses on utilizing richer uncertainty information from experts' diverse opinions and reflects better segmentation performance. We introduce an Uncertainty-Aware Module and a Feature-Aware Attention Module to fully learn the corresponding features under the guidance of three different learning targets and improve lung nodule segmentation performance. Additionally, the network can generate a Multi-Confidence Mask to show nodule regions with different uncertainty levels. Experimental results show that our method can predict lung nodule regions with high uncertainty levels and outperform many state-of-the-art networks.

References

1. Liu, H., Cao, H., Song, E., et al.: A cascaded dual-pathway residual network for lung nodule segmentation in CT images. Physica Medica **63**, 112–121 (2019)
2. Zhu, W., Liu, C., Fan, W., et al.: DeepLung: deep 3D dual path nets for automated pulmonary nodule detection and classification. In: Proceedings of IEEE WACV, pp. 673–681. IEEE (2018)
3. Xie, H., Yang, D., Sun, N., et al.: Automated pulmonary nodule detection in CT images using deep convolutional neural networks. Pattern Recogn. **85**, 109–119 (2019)
4. Gonçalves, L., Novo, J., Campilho, A.: Hessian based approaches for 3D lung nodule segmentation. Expert Syst. Appl. **61**, 1–15 (2016)
5. Wu, D., Lu, L., Bi, J., et al.: Stratified learning of local anatomical context for lung nodules in CT images. In: Proceedings of IEEE CVPR, pp. 2791–2798 (2010)
6. Pezzano, G., Ripoll, V.R., Radeva, P.: CoLe-CNN: context-learning convolutional neural network with adaptive loss function for lung nodule segmentation. Comput. Meth. Prog. Biomed. **198**, 105792 (2021)
7. Armato, S., III., McLennan, G., Bidaut, L., et al.: The lung image database consortium (LIDC) and image database resource initiative (IDRI): a completed reference database of lung nodules on CT scans. Med. Phys. **38**(2), 915–931 (2011)

8. Hu, S., Worrall, D., Knegt, S., Veeling, B., Huisman, H., Welling, M.: Supervised uncertainty quantification for segmentation with multiple annotations. In: Shen, D., et al. (eds.) MICCAI 2019. LNCS, vol. 11765, pp. 137–145. Springer, Cham (2019). https://doi.org/10.1007/978-3-030-32245-8_16

9. Kohl, S.A., Romera-Paredes, B., Maier-Hein, K.H., et al.: A hierarchical probabilistic U-Net for modeling multi-scale ambiguities. arXiv preprint arXiv:1905.13077 (2019)

10. Long, X., Chen, W., Wang, Q., et al.: A probabilistic model for segmentation of ambiguous 3D lung nodule. In: Proceedings of IEEE ICASSP, pp. 1130–1134. IEEE (2021). https://doi.org/10.1109/ICASSP39728.2021.9415006

11. Kohl, S., Romera-Paredes, B., Meyer, C., et al.: A probabilistic U-Net for segmentation of ambiguous images. In: Proceedings of NIPS, vol. 31 (2018)

12. Gao, M., Bagci, U., Lu, L., et al.: Holistic classification of CT attenuation patterns for interstitial lung diseases via deep convolutional neural networks. Comput. Methods Biomech. Biomed. Eng. Imaging. Vis. **6**(1), 1–6 (2015)

13. Oktay, O., Schlemper, J., Folgoc, L.L., et al.: Attention U-Net: learning where to look for the pancreas. arXiv preprint arXiv:1804.03999 (2018)

14. Vaswani, A., Shazeer, N., Parmar, N., et al.: Attention is all you need. In: Proceedings of NIPS, vol. 30 (2017)

15. Otsu, N.: A threshold selection method from gray-level histograms. IEEE Trans. Syst. Man. Cybern. **9**(1), 62–66 (2007)

16. Wang, Q., Zhang, X., Zhang, W., et al.: Realistic lung nodule synthesis with multi-target co-guided adversarial mechanism. IEEE Trans. Med. Imaging **40**(9), 2343–2353 (2021)

17. Luan, S., Chen, C., Zhang, B., Han, J., et al.: Gabor convolutional networks. IEEE Trans. Image Process. **27**(9), 4357–4366 (2018)

18. Loshchilov, I., Hutter, F.: SGDR: stochastic gradient descent with warm restarts (2016)

19. Chen, W., Wang, K., Yang, D., et al.: MTGAN: mask and texture-driven generative adversarial network for lung nodule segmentation. In: Proceedings of IEEE ICPR, pp. 1029–1035. IEEE (2020). https://doi.org/10.1109/ICPR48806.2021.9413064

20. Kamrul Hasan, S.M., Linte, C.A.: A modified U-nNt convolutional network featuring a nearest-neighbor re-sampling-based elastic-transformation for brain tissue characterization and segmentation. In: Proceedings of IEEE WNYISPW, pp. 1–5 (2018). https://doi.org/10.1109/WNYIPW.2018.8576421

21. Ma, J., Zhang, Y., Gu, S., et al.: AbdomenCT-1K: is abdominal organ segmentation a solved problem. IEEE Trans. Pattern Anal. (2021)

22. Cao, H., et al.: Dual-branch residual network for lung nodule segmentation. Appl. Soft Comput. **86**, 105934 (2020)

23. Isensee, F., Jaeger, P.F., Kohl, S.A., et al.: NNU-Net: a self-configuring method for deep learning-based biomedical image segmentation. Nat. Methods **18**(2), 203–211 (2021)

24. Alom, M.Z., Hasan, M., Yakopcic, C., Taha, T.M., Asari, V.K.: Recurrent residual convolutional neural network based on U-Net (R2U-Net) for medical image segmentation. arXiv preprint arXiv:1802.06955 (2018)

25. Amorim, P., Moraes, T.F., Silva, J., et al.: Lung nodule segmentation based on convolutional neural networks using multi-orientation and patchwise mechanisms. In: Proceedings of VipIMAGE, pp. 286–295 (2019)

26. Tang, H., Zhang, C., Xie, X.: NoduleNet: decoupled false positive reduction for pulmonary nodule detection and segmentation. In: Shen, D., et al. (eds.) MICCAI 2019. LNCS, vol. 11769, pp. 266–274. Springer, Cham (2019). https://doi.org/10.1007/978-3-030-32226-7_30

27. Tolooshams, B., Giri, R., Song, A., et al.: Channel-attention dense U-Net for multichannel speech enhancement. In: Proceedings of ICASSP, pp. 836–840 (2020)

28. Zhou, Z., Siddiquee, R., Mahfuzur, M., et al.: Unet++: a nested U-Net architecture for medical image segmentation. In: Proceedings of DLMIA Workshop, pp. 3–11 (2018)

29. Selvaraju, R.R., Cogswell, M., Das, A., et al.: Grad-CAM: visual explanations from deep networks via gradient-based localization. Int. J. Comput. Vision **128**(2), 336–359 (2020)
30. Ronneberger, O., Fischer, P., Brox, T.: U-Net: convolutional networks for biomedical image segmentation. In: Navab, N., Hornegger, J., Wells, W.M., Frangi, A.F. (eds.) MICCAI 2015. LNCS, vol. 9351, pp. 234–241. Springer, Cham (2015). https://doi.org/10.1007/978-3-319-24574-4_28

Thoracic Lymph Node Segmentation in CT Imaging via Lymph Node Station Stratification and Size Encoding

Dazhou Guo[1]([✉]), Jia Ge[2], Ke Yan[3], Puyang Wang[3], Zhuotun Zhu[4], Dandan Zheng[2], Xian-Sheng Hua[3], Le Lu[1], Tsung-Ying Ho[5], Xianghua Ye[2], and Dakai Jin[1]

[1] Alibaba DAMO Academy USA, New York, NY, USA
guo2004131@gmail.com
[2] The First Affiliated Hospital Zhejiang University, Hangzhou, China
[3] Alibaba DAMO Academy, Hangzhou, China
[4] Johns Hopkins University, Baltimore, USA
[5] Chang Gung Memorial Hospital, Linkou, Taiwan
hye1982@zju.edu.cn

Abstract. Visible lymph node (i.e., LN, short axis $\geq$ 5 mm) assessment and delineation in thoracic computed tomography (CT) images is an indispensable step in radiology and oncology workflows. The high demanding of clinical expertise and prohibitive laboring cost motivate the automated approaches. Previous works focus on extracting effective LN imaging features and/or exploiting the anatomical priors to help LN segmentation. However, the performance in general is struggled with low recall/precision due to LN's low contrast in CT and tumor-induced shape and size variations. Given that LNs reside inside the lymph node station (LN-station), it is intuitive to directly utilize the LN-station maps to guide LN segmentation. We propose a stratified LN-station and LN size encoded segmentation framework by casting thoracic LN-stations into three super lymph node stations and subsequently learning the station-specific LN size variations. Four-fold cross-validation experiments on the public NIH 89-patient dataset are conducted. Compared to previous leading works, our framework produces significant performance improvements, with an average 74.2% (9.9% increases) in Dice score and 72.0% (15.6% increases) in detection recall at 4.0 (1.9 reduces) false positives per patient. When directly tested on an external dataset of 57 esophageal cancer patients, the proposed framework demonstrates good generalizability and achieves 70.4% in Dice score and 70.2% in detection Recall at 4.4 false positives per patient.

1 Introduction

Lymph node (LN) involvement assessment is an essential predictive or prognostic bio-marker in radiology and oncology. Precision quantitative LN analysis is an

---

Supplementary Information The online version contains supplementary material available at https://doi.org/10.1007/978-3-031-16443-9_6.

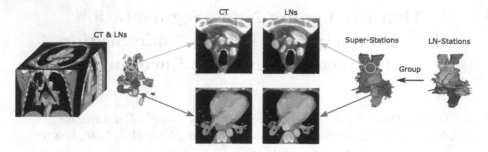

Fig. 1. An illustration of LN, LN-Station, and Super-Station. The top and bottom row show the LN contexts in LN-Station 2 (clear boundaries) and 8 (hard to discern).

indispensable step in staging, treatment planning, and disease progression monitoring of cancers in thoracic region [10,23]. Visual identification, measurement or delineation of thoracic LNs from CT scans are performed in the current clinical practice, which is tedious, time-consuming, expensive and expertise-demanding. Due to LN's low contrast to the surrounding anatomies and tumor-induced shape and size variations, it can be easily visually confused with vessels or muscles. Although enlarged LNs with short axis larger than 10 mm are often considered as pathologically targeted lesions to be assessed [20], studies show that enlarged size alone is not the most reliable predictive factor for LN malignancy with only 60%–80% recall in lung cancer patients [6,22]. Smaller LNs potentially involving metastasis should also be included for improved diagnostic accuracy [5,21]. Thus an automated segmentation framework for both enlarged and smaller (short axis ≥ 5 mm) throacic LNs is of high clinical importance.

Thoracic LN detection and segmentation has been exploited for more than a decade mainly focusing on extracting effective LN features, incorporating organ priors or utilizing advanced learning models [1–4,7–9,13–17,25]. Early work often adopted the model-based or statistical learning-based methods. Feuerstein et al. built a statistical atlas for lymph node detection in chest CT [7]. Liu et al. detected and segmented LNs by first locating the approximate region of interest (mediastinum) and then integrating the spatial priors, intensity and shape features into a random forest classification model [15]. Recent deep learning based approaches have also been explored. Nogues et al. combined the holistically-nested neural networks and structured optimization to segment the LNs [16]. Bouget et al. segmented LNs in each cropped slabs and further ensembled with a full 3D UNet model that incorporated the organ priors [3]. Although being extensively studied, the general performance has been struggling with low recall/precision and is unable to deploy to clinical practice: e.g., for enlarged LNs (short axis ≥ 10 mm), 70.4% recall at 4 false positives (FPs) per patient in [15]; for both enlarged and smaller LNs, 52.4% recall at 6 FPs per patient [3].

To tackle this tremendously challenging task, important anatomic knowledge and clinical reasoning insights from physicians should be leveraged. 1) almost all LNs reside in the lymph node station (LN-station) defined according to key anatomic organs or landmarks, i.e., thoracic LN-station recommended by International Association for the Study of Lung Cancer (IASLC) [19] and head & neck LN-station outlined by American Academy of OtolaryngologyHead and Neck Surger (AAO-HN) [18]. Confining LN identification within the LN-station is beneficial, since the constrained searching and learning space can reduce FPs occurring at vessels, muscles, and soft-tissues in other body regions with similar local appearance. 2) LNs in different LN-stations often have distinct contexts and exhibit different levels of identification uncertainties. For example, thoracic LNs in stations 2 and 4 usually have clearer boundaries, while those in stations 7 and 8 can be more easily confused with adjacent vessels or esophagus (Fig. 1). Moreover, enlarged LNs often yield different texture patterns (e.g., calcified/necrosis) and shapes (e.g., tree/star-shape) to the smaller LNs. Further stratifying the LN segmentation task via different LN-stations and LN sizes could be beneficial.

Given the above key observations, we utilize the LN-station priors and propose a novel LN-station- and size-aware LN segmentation framework by explicitly incorporating the LN-station prior and learning the LN size variance. To achieve this, we first segment thoracic LN-stations 1 to 14 using a robust Deep-Stationing model in [11]. According to the LN-station context and physician's clinical experiences, we stratify and group the 14 LN-stations into 3 super lymph node stations (Super-Stations), i.e., LN-stations 1–4 (upper level), LN-stations 5–9 (lower level) and LN-stations 10–14 (lung regions). Then, a new deep segmentation network with multi-encoding paths are designed with each to focus on learning the LN features in a specific Super-Station (Fig. 2). Next, for explicitly learning LN's size variance, two decoding branches are adopted concentrating on the small and large LNs, respectively (Fig. 2). Results from the two decoding branches are then merged using a post-fusion module. For experimental evaluations, 1) we conduct 4-fold cross validations on a public LN dataset with 89 lung cancer patients and more than 2000 recently annotated LNs ($\geq$5 mm) [3]. Our framework achieves an average 74.2% in Dice score and 72.0% in the detection recall at 4.0 false positives per patient (FP-PW), which significantly improves the segmentation and detection accuracy with at least 10% absolute Dice score improvement and 13% recall increase at 4.0 FP-PW, compared to previous leading LN segmentation/detection methods [3,12,16,24]. 2) We further apply our pre-trained model to an external testing dataset of 57 esophageal caner patients with 360 thoracic LNs (>5 mm labeled). We demonstrate good generalizability by obtaining 70.4% Dice score (Dice) (13.8% increases) and 70.2% Recall (17.0% increases) at 4.4 FP-PW as compared to the strong nnUNet baseline [12].

2   Method

Figure 2 depicts the overview of our proposed LN-station-specific and size-aware LN segmentation framework. It contains three independent encoding paths based on the stratified Super-Stations and two decoding branches to learn size-specific LN's features. Together with the original CT image, post-fusion blocks leverage the predicted big and small LNs to generate the final prediction.

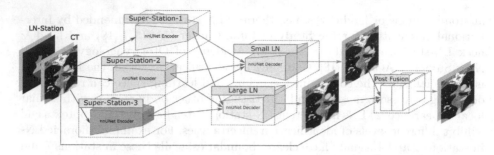

Fig. 2. Overall workflow of our proposed framework, which consists of Super-Station-based stratified encoders, size-aware decoder branches, and a post-fusion module.

2.1 LN-Station Segmentation and Stratification

To utilize the LN-station priors, we first segment a set of 14 thoracic LN-stations. Motivated by [11], we adopt the key referencing organ guided LN-station segmentation model. As mentioned in [11], 6 key referencing organs are used: *esophagus, aortic arch, ascending aorta, heart, spine,* and *sternum.* Assuming N data instances, the training data is denoted as $\mathbf{D} = \{X_n, Y_n^K, Y_n^S, Y_n^L\}_{n=1}^N$, where X_n, Y_n^K, Y_n^S, Y_n^L denote the input CT image and ground-truth masks for the key referencing organs, LN-stations, and LNs, respectively. Let C_K and C_S denote the number of key organs and LN-stations, respectively. Dropping n for clarity, the key organ and LN-station segmentation models predict every voxel location j with a class c:

$$\hat{Y}_{c_k}^K(j) = f^K\left(Y^K(j) = c_k \mid X; \mathbf{W}^K\right), \quad \hat{\mathbf{Y}}^K = \left[\hat{Y}_1^K \dots \hat{Y}_{C_K}^K\right], \qquad (1)$$

$$\hat{Y}_{c_s}^S(j) = f^S\left(Y^S(j) = c_s \mid X, \hat{\mathbf{Y}}^K; \mathbf{W}^S\right), \quad \hat{\mathbf{Y}}^S = \left[\hat{Y}_1^S \dots \hat{Y}_{C_S}^S\right], \qquad (2)$$

where $f^{(*)}(.)$ denotes the network functions, $\mathbf{W}^{(*)}$ represents the corresponding network parameters, and $\hat{Y}_{c_*}^{(*)}$ for the predicted segmentation maps.

According to LN-station context and physician's clinical experience, we combine the predicted LN-stations into three Super-Stations, i.e., LN-stations 1–4, 5–9 and 10–14. To avoid potential LN-station under-segmentation, we dilate each Super-Station with a diameter of 15 mm. We then use the dilated Super-Station binary maps to 'mask' Super-Station covered CT images (abbreviated as mCT), while setting other voxel values to a constant of -1024. We thereafter obtain three Super-Station-masked mCT images. Please note that the inputs of three encoders are independent. For instance, for Super-Station-1 encoder, its input is the mCT masked using dilated Super-Station-1 and the LN-stations 1–4 maps. Let $\hat{Y}_1^{S+} = [\hat{Y}_c^S]_{c=1}^4$, $\hat{Y}_2^{S+} = [\hat{Y}_c^S]_{c=5}^9$, and $\hat{Y}_3^{S+} = [\hat{Y}_c^S]_{c=10}^{14}$ denote the grouped LN-stations maps. Each encoder targets Super-Station-specific LN features.

$$\hat{\mathbf{H}}_i^{S+} = f_i^{S+}\left(X_i^m, \hat{Y}_i^{S+}; \mathbf{W}_i^{S+}\right), \quad i \in \{1, 2, 3\}, \qquad (3)$$

where X_i^m denotes the mCT and $\hat{\mathbf{H}}_{(*)}^{S+}$ denotes the output feature maps of the stratified encoder.

2.2   LN Size Stratification and Post-fusion

Considering that enlarged LNs often yield different texture patterns (e.g., calcified/necrosis) or shapes (e.g., tree/star-shape) as compared to smaller ones, we further introduce two decoding branches to learn size-specific LN features. For the annotated LNs, we manually separate the enlarged LNs ($L+$) whose short-axes are greater than 10 mm, and vice versa ($L-$). Each decoder is supervised using the respect enlarged/small LN labels. The input of each decoder branch is the channel-wise concatenation of the feature maps from three Super-Stations encoding paths: $\hat{\mathbf{H}}^{S+} = \left[\hat{\mathbf{H}}_1^{S+}, \hat{\mathbf{H}}_2^{S+}, \hat{\mathbf{H}}_3^{S+} \right]$.

$$\hat{Y}^{L+} = f^{L+}\left(\hat{\mathbf{H}}^{S+}; \mathbf{W}^{L+} \right), \quad \hat{Y}^{L-} = f^{L-}\left(\hat{\mathbf{H}}^{S+}; \mathbf{W}^{L-} \right), \tag{4}$$

where $\hat{Y}^{L+}$ and $\hat{Y}^{L-}$ denote the output prediction maps of the large- and small-LN decoders, respectively. The output feature maps of each decoder are additionally combined with the original CT image and input to a simple post fusion module. The post fusion module is created using the first two nnUNet convolutional blocks without the pooling function:

$$\hat{Y}^L = f^P\left(X, \hat{Y}^{L+}, \hat{Y}^{L-}; \mathbf{W}^{L-} \right). \tag{5}$$

The proposed segmentation framework explicitly encodes both the LN-station- and size-specific information in the model training. The final prediction is $\hat{Y}^L$.

3   Experimental Results

Dataset. *For the LN model development and validation,* we used 89 patients with contrast-enhanced venous-phase CT scans from the pubic NIH dataset[1], where more than 2000 thoracic LN instance labels were recently annotated by [3]. The average CT image size is $512 \times 512 \times 616$ voxels with the average voxel resolution of $0.8 \times 0.8 \times 1.2$ mm^3. The average LN size is $7.7 \times 6.6 \times 9.4$ mm^3. *For the external testing,* we collected additional 57 contrast-enhanced venous-phase CT scans of esophageal cancer patients underwent surgery and/or radiotherapy treatment. The average CT image size is $512 \times 512 \times 80$ voxels with the average voxel resolution of $0.7 \times 0.7 \times 5.0$ mm^3. A board-certified radiation oncologist with more than 15 years of experience labeled each patient with visible LNs (≥ 5 mm). The average LN size of external testing set is $9.5 \times 9.0 \times 9.1$ mm. Moreover, to develop the LN-station segmentation model, 3D masks of thoracic LN-stations 1–14, and 6 key referencing organs were annotated in this esophageal dataset according to the IASLC guideline [19].

---

[1] https://wiki.cancerimagingarchive.net/display/Public/CT+Lymph+Nodes.

Table 1. LN segmentation performance on the NIH dataset using 4-fold cross-validation. We abbreviate Super-Station-based stratified encoder as S.S.E. and size-aware decoder branches as S.A.D. We evaluate the segmentation performance for all annotated LNs (upper) and only enlarged LNs (lower). The best performance scores are highlighted in **bold**. Our *full version* is: 'mCT + S.S.E. + S.A.D. + post fusion module'. The 1st row 'CT Only' is the default nnUNet [12] performance.

	Dice	Recall	Recall-PW	FP-PW
All LNs – Short Axis ≥ 5 mm				
CT Only (Isensee et al. [12])	58.2±17.1	55.3	56.9±21.4	6.2±4.4
CT + LNS (Ours)	66.3±13.5	60.1	61.5±19.7	4.9±3.8
mCT + LNS (Ours)	70.4±12.6	66.2	67.4±20.1	**3.8±2.8**
mCT + S.S.E. (Ours)	72.1±13.8	69.4	70.8±19.1	4.1±2.6
mCT + S.S.E. + S.A.D. (Ours)	74.1±14.8	71.2	71.9±20.4	4.9±3.6
Nogues et al. 2016 [16]	57.7±19.4	57.2	62.2±20.7	4.9±4.1
Yan et al. 2020 [24]	60.6±15.1	59.3	64.8±18.3	6.1±2.5
Bouget et al. 2021 [3]	64.3±14.5	56.4	55.2±18.0	5.9±3.7
Full version	**74.2±14.7**	**72.0**	**72.4±19.0**	4.0±2.9
Enlarged LNs – Short Axis ≥ 10 mm				
Nogues et al. 2016 [16]	58.6±14.2	79.2	81.4±18.5	3.1±3.5
Yan et al. 2020 [24]	65.6±12.1	83.3	89.8±15.3	4.0±3.5
Isensee et al. 2021 [12]	61.8±12.4	85.3	89.5±15.4	3.1±3.9
Bouget et al. 2021 [3]	-	82.7	88.6±15.6	-
Full version	**79.1±9.8**	**89.8**	**92.4±12.6**	**2.4±2.3**

Implementation Details. We adopt '3d-fullres' version of nnU-Net [12] with Dice+CE losses as our backbone modules. Each encoder is the same as the default nnUNet encoder, and each block contains two "Conv+InstanceNorm+IReLU" layers. With additional two skip-connections, each decoder block receives 2x its original input channels. The default nnUNet's deep-supervision is not used in our experiment. Instead, we apply two side supervisions for the two decoding branches using the enlarged and small LN labels, respectively. We use the default nnUNet data augmentation settings for our model training, and set the patch size to $96 \times 128 \times 32$. We implemented our framework using PyTorch and trained on an NVIDIA Tesla V100. The total training epochs is 1000. The average training time is 5.5 GPU days.

Evaluation. For the NIH dataset, extensive four-fold cross-validation (CV), separated at the patient level, was conducted. The esophageal dataset was held out as an external testing dataset. We follow the LN evaluation metrics in [3] and calculate Dice, instance detection Recall, patient-wise detection recall (Recall-PW) and the patient-wise false positive numbers (FP-PW).

Results of LN-Station Segmentation. We first evaluate the performance of the LN-station segmentation model. The average LN-station segmentation performance is: Dice $81.2 \pm 5.8\%$, Hausdorff distance (HD) 9.6 ± 4.2 mm, and average surface distance (ASD) 0.9 ± 0.6 mm. In our experiment, we select a diameter of 15 mm to dilate the predicted and grouped Super-Stations to cover the thoracic LNs, which might be missed in the original LN-station prediction due to the under-segmentation. Meanwhile, the dilated Super-Stations should not include too many similar tissues such as vessels. The quantitative instance/volume LN coverage using predicted LN-stations is reported in the supplementary materials.

Table 2. LN segmentation performance on the collected in-house 57 esophageal cancer patients testing set. The best performance scores are highlighted in **bold**.

	Dice	Recall	Recall-PW	FP-PW
All LNs – Short Axis ≥ 5 mm				
Nogues et al. 2016 [16]	55.9±20.6	53.4	54.5±25.7	6.3±4.5
Yan et al. 2020 [24]	54.2±17.2	60.3	62.9±21.3	9.5±6.4
Isensee et al. 2021 [12]	56.6±18.4	53.2	55.3±22.4	5.8±3.9
Full version	**70.4±14.7**	**70.2**	**70.8±20.2**	**4.4±3.3**
Enlarged LNs – Short Axis ≥ 10 mm				
Nogues et al. 2016 [16]	56.5±17.6	77.4	80.5±17.9	4.5±3.6
Yan et al. 2020 [24]	64.2±16.2	80.1	86.9±15.3	7.6±4.4
Isensee et al. 2021 [12]	59.2±15.2	79.6	83.2±16.7	3.7±4.1
Full version	**74.9±12.4**	**85.8**	**91.8±13.5**	**2.4±2.7**

Quantitative Evaluation in NIH Dataset. Table 1 outlines the quantitative comparisons of different input and model setups when evaluated in the NIH dataset: 1) only CT images, 2) early fusion of CT and LN-station (CT + LNS), 3) CT masked using the dilation of the whole LN-station region (mCT + LNS), 4) Super-Station-stratified encoders with the default single UNet decoder (mCT + S.S.E.), 5) Super-Station-stratified encoders + size-aware decoders *without* post fusion (mCT + S.S.E. + S.A.D.). Several observations can be drawn. First, LN segmentation exhibits the lowest performance with an average 58.2% Dice and 55.3% Recall at 6.2 FP-PW when using only CT. When using LN-station as an additional input channel, all metrics show remarked improvements: 8.1% and 4.8% increase in Dice and Recall, and 1.3 reduction in FP-PW. This demonstrates the importance and effectiveness of using LN-stations for LNs segmentation. Second, when adopting the 'mCT + LN-station' setup, the performance is markedly improved with another 6.1% boosted Recall and 1.1 reduced FP-PW. This indicates that constraining the learning space within the mCT region (hence, eliminating the confusing anatomy tissues in the irrelevant regions) make the LN identification task much easier. Third, the LN-station- and size-aware LN segmentation schemes are effective, since both S.S.E. and S.A.D. modules yield marked improvements boosting the Dice and Recall to 74.1% and 71.2, respectively. Finally, equipped with a simple post-fusion module, the final proposed model can further reduce the FP-PW from 4.9 to 4.0 while preserving the high Dice and Recall as compared to the mCT + S.S.E. + S.A.D. model.

Table 1 also shows the performance comparisons in NIH dataset between our proposed framework and four leading methods [3,12,16,24]. Among the comparison methods, the best Recall of 59.3% at 6.1 FP-WP is achieved by [24], which is the leading approach for 3D universal lesion detection. It can be seen the proposed framework significantly outperforms [24] by 14.4% Dice, 12.7% Recall, and 2.1 in FP-PW. When analyzing the performance of enlarged LNs (short axis > 10 mm) commonly studied in previous works, our framework achieves a high Recall of 89.8% at 2.4 FP-PW. In comparison, previous leading methods exhibit inferior performance with the best 85.3% Recall at 3.1 FP-PW [12].

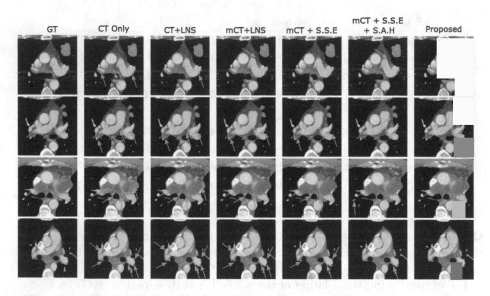

Fig. 3. Examples of LN segmentation results using different setups. *green* and *blue* arrows are used to depict under-segmentations and FPs. It can be observed that severe under-segmentation exists in the CT only method leading to low Dice and Recall (2nd column). In contrast, when LN-station information is explicitly incorporated into the model, more LNs can be correctly identified (from 3rd to 5th columns). Size-stratification (S.A.D) can further improve the LNs segmentation Recall, while the proposed final model suppresses some FPs while maintaining the high Recall. (Color figure online)

External Testing on Esophageal Dataset. The independent external testing results on the esophageal dataset are illustrated in Table 2. The proposed framework demonstrates good generalizability by achieving 70.4% Dice and 70.2% Recall at 4.4 FP-PW, which are comparable to those in the NIH dataset. For the failure cases, under-segmentation along the z-direction for LNs in the inferior mediastinal region is observed. The assumed reasons might be: 1) unclear boundaries of the inferior mediastinal LNs, and 2) most LNs are relatively short in z-direction and the model might bias toward the majority average. For the enlarged LNs, our framework also shows robust performance of 74.9% Dice and 85.8 Recall at 2.4 FP-PW. The assumed reasons of achieving good generalizability might be that segmenting LNs in a much confined Super-Station region is comparably easy and robust. In contrast, the previous second best performing detection method [24] yields low generalizability as its FP-PW significantly increased from 6.1 (NIH) to 9.5 (external) for all LNs and from 4.0 (NIH) to 7.6 (external) for enlarged LNs.

4   Conclusion

In this paper, we propose a novel LN-station-specific and size-aware LN segmentation framework by explicitly utilizing the LN-station priors and learning the LN size variance. We first segment thoracic LN-stations and then group the LN-stations into 3 super lymph node stations, based on which a multi-encoder deep network is designed to learn LN-station-specific LN features. For learning LN's size variance, we further stratify decoding path into two decoding branches to concentrate on learning the small and large LNs, respectively. Validated on the public NIH dataset and further tested on the external esophageal dataset, the proposed framework demonstrates high LN segmentation performance while preserving good generalizability. Our work represents an important step towards the reliable and automated LN segmentation.

References

1. Barbu, A., Suehling, M., Xu, X., Liu, D., Zhou, S.K., Comaniciu, D.: Automatic detection and segmentation of lymph nodes from CT data. IEEE Trans. Med. Imaging **31**(2), 240–250 (2011)
2. Bouget, D., Jørgensen, A., Kiss, G., Leira, H.O., Langø, T.: Semantic segmentation and detection of mediastinal lymph nodes and anatomical structures in CT data for lung cancer staging. Int. J. Comput. Assist. Radiol. Surg. **14**(6), 977–986 (2019)
3. Bouget, D., Pedersen, A., Vanel, J., Leira, H.O., Langø, T.: Mediastinal lymph nodes segmentation using 3d convolutional neural network ensembles and anatomical priors guiding. arXiv preprint arXiv:2102.06515 (2021)
4. Chao, C.-H., et al.: Lymph node gross tumor volume detection in oncology imaging via relationship learning using graph neural network. In: Martel, A.L., et al. (eds.) MICCAI 2020. LNCS, vol. 12267, pp. 772–782. Springer, Cham (2020). https://doi.org/10.1007/978-3-030-59728-3_75
5. Choi, S.B., Han, H.J., Park, P., Kim, W.B., Song, T.J., Choi, S.Y.: Systematic review of the clinical significance of lymph node micrometastases of pancreatic adenocarcinoma following surgical resection. Pancreatology **17**(3), 342–349 (2017)
6. De Langen, A.J., Raijmakers, P., Riphagen, I., Paul, M.A., Hoekstra, O.S.: The size of mediastinal lymph nodes and its relation with metastatic involvement: a meta-analysis. Eur. J. Cardiothorac. Surg. **29**(1), 26–29 (2006)

7. Feuerstein, M., Glocker, B., Kitasaka, T., Nakamura, Y., Iwano, S., Mori, K.: Mediastinal atlas creation from 3-d chest computed tomography images: application to automated detection and station mapping of lymph nodes. Med. Image Anal. **16**(1), 63–74 (2012)
8. Feulner, J., Zhou, S.K., Hammon, M., Hornegger, J., Comaniciu, D.: Lymph node detection and segmentation in chest CT data using discriminative learning and a spatial prior. Med. Image Anal. **17**(2), 254–270 (2013)
9. Feulner, J., Zhou, S.K., Huber, M., Hornegger, J., Comaniciu, D., Cavallaro, A.: Lymph node detection in 3-d chest CT using a spatial prior probability. In: 2010 IEEE Computer Society Conference on Computer Vision and Pattern Recognition, pp. 2926–2932. IEEE (2010)
10. Goldstraw, P., et al.: The IASLC lung cancer staging project: proposals for the revision of the TNM stage groupings in the forthcoming (seventh) edition of the TNM classification of malignant tumours. J. Thorac. Oncol. **2**(8), 706–714 (2007)
11. Guo, D., et al.: DeepStationing: thoracic lymph node station parsing in CT scans using anatomical context encoding and key organ auto-search. In: de Bruijne, M., et al. (eds.) MICCAI 2021. LNCS, vol. 12905, pp. 3–12. Springer, Cham (2021). https://doi.org/10.1007/978-3-030-87240-3_1
12. Isensee, F., Jaeger, P.F., Kohl, S.A., Petersen, J., Maier-Hein, K.H.: nnU-Net: a self-configuring method for deep learning-based biomedical image segmentation. Nat. Meth. **18**, 203–211 (2021)
13. Iuga, A.I., et al.: Automated detection and segmentation of thoracic lymph nodes from CT using 3d foveal fully convolutional neural networks. BMC Med. Imaging **21**(1), 1–12 (2021)
14. Li, Z., Xia, Y.: Deep reinforcement learning for weakly-supervised lymph node segmentation in CT images. IEEE J. Biomed. Health Inform. **25**(3), 774–783 (2020)
15. Liu, J., et al.: Mediastinal lymph node detection and station mapping on chest CT using spatial priors and random forest. Med. Phys. **43**(7), 4362–4374 (2016)
16. Nogues, I., et al.: Automatic lymph node cluster segmentation using holistically-nested neural networks and structured optimization in CT images. In: Ourselin, S., Joskowicz, L., Sabuncu, M.R., Unal, G., Wells, W. (eds.) MICCAI 2016. LNCS, vol. 9901, pp. 388–397. Springer, Cham (2016). https://doi.org/10.1007/978-3-319-46723-8_45
17. Roth, H.R., et al.: Improving computer-aided detection using convolutional neural networks and random view aggregation. IEEE Trans. Med. Imaging **35**(5), 1170–1181 (2016)
18. Roubbins, K.T., et al.: Neck dissection classification update: revisions proposed by the American head and neck society and the American academy of otolaryngology-head and neck surgery. Arch. Otolaryngol. Head Neck Surg. **128**(7), 751–758 (2002)
19. Rusch, V.W., Asamura, H., Watanabe, H., Giroux, D.J., Rami-Porta, R., Goldstraw, P.: The IASLC lung cancer staging project: a proposal for a new international lymph node map in the forthcoming seventh edition of the TNM classification for lung cancer. J. Thorac. Oncol. **4**(5), 568–577 (2009)
20. Schwartz, L., et al.: Evaluation of lymph nodes with RECIST 1.1. Eur. J. Cancer **45**(2), 261–267 (2009)
21. Stanley Leong, P., Tseng, W.W.: Micrometastatic cancer cells in lymph nodes, bone marrow, and blood: clinical significance and biologic implications. Cancer J. Clin. (CA) **64**(3), 195–206 (2014)
22. McLoud, T.C., et al.: Bronchogenic carcinoma: analysis of staging in the mediastinum with CT by correlative lymph node mapping and sampling. Radiology **182**(2), 319–323 (1992)

23. Terán, M.D., Brock, M.V.: Staging lymph node metastases from lung cancer in the mediastinum. J. Thorac. Dis. **6**(3), 230 (2014)

24. Yan, K., et al.: Learning from multiple datasets with heterogeneous and partial labels for universal lesion detection in CT. IEEE Trans. Med. Imaging **40**(10), 2759–2770 (2020)

25. Zhu, Z., et al.: Lymph node gross tumor volume detection and segmentation via distance-based gating using 3d CT/PET imaging in radiotherapy. In: Martel, A.L., et al. (eds.) MICCAI 2020. LNCS, vol. 12267, pp. 753–762. Springer, Cham (2020). https://doi.org/10.1007/978-3-030-59728-3_73

ACT: Semi-supervised Domain-Adaptive Medical Image Segmentation with Asymmetric Co-training

Xiaofeng Liu[1(✉)], Fangxu Xing[1], Nadya Shusharina[2], Ruth Lim[1],
C.-C. Jay Kuo[3], Georges El Fakhri[1], and Jonghye Woo[1]

[1] Gordon Center for Medical Imaging, Department of Radiology, Massachusetts General Hospital and Harvard Medical School, Boston, MA 02114, USA
xliu61@mgh.harvard.edu
[2] Division of Radiation Biophysics, Department of radiation Oncology, Massachusetts General Hospital and Harvard Medical School, Boston, MA 02114, USA
[3] Department of Electrical and Computer Engineering, University of Southern California, Los Angeles, CA 90007, USA

Abstract. Unsupervised domain adaptation (UDA) has been vastly explored to alleviate domain shifts between source and target domains, by applying a well-performed model in an unlabeled target domain via supervision of a labeled source domain. Recent literature, however, has indicated that the performance is still far from satisfactory in the presence of significant domain shifts. Nonetheless, delineating a few target samples is usually manageable and particularly worthwhile, due to the substantial performance gain. Inspired by this, we aim to develop semi-supervised domain adaptation (SSDA) for medical image segmentation, which is largely underexplored. We, thus, propose to exploit both labeled source and target domain data, in addition to unlabeled target data in a unified manner. Specifically, we present a novel asymmetric co-training (ACT) framework to integrate these subsets and avoid the domination of the source domain data. Following a divide-and-conquer strategy, we explicitly decouple the label supervisions in SSDA into two asymmetric sub-tasks, including semi-supervised learning (SSL) and UDA, and leverage different knowledge from two segmentors to take into account the distinction between the source and target label supervisions. The knowledge learned in the two modules is then adaptively integrated with ACT, by iteratively teaching each other, based on the confidence-aware pseudo-label. In addition, pseudo label noise is well-controlled with an exponential MixUp decay scheme for smooth propagation. Experiments on cross-modality brain tumor MRI segmentation tasks using the BraTS18 database showed, even with limited labeled target samples, ACT yielded marked improvements over UDA and state-of-the-art SSDA methods and approached an "upper bound" of supervised joint training.

1 Introduction

Accurate delineation of lesions or anatomical structures is a vital step for clinical diagnosis, intervention, and treatment planning [24]. While recently flourished

L. Wang et al. (Eds.): MICCAI 2022, LNCS 13435, pp. 66–76, 2022.
https://doi.org/10.1007/978-3-031-16443-9_7

deep learning methods excel at segmenting those structures, deep learning-based segmentors cannot generalize well in a heterogeneous domain, e.g., different clinical centers, scanner vendors, or imaging modalities [4,14,16,20]. To alleviate this issue, unsupervised domain adaptation (UDA) has been actively developed, by applying a well-performed model in an unlabeled target domain via supervision of a labeled source domain [5,15,18,19]. Due to diverse target domains, however, the performance of UDA is far from satisfactory [9,17,31]. Instead, labeling a small set of target domain data is usually more feasible [25]. As such, semi-supervised domain adaptation (SSDA) has shown great potential as a solution to domain shifts, as it can utilize both labeled source and target data, in addition to unlabeled target data. To date, while several SSDA classification methods have been proposed [8,13,23,29], based on discriminative class boundaries, they cannot be directly applied to segmentation, since segmentation involves complex and dense pixel-wise predictions.

Recently, while a few works [6,10,26] have been proposed to extend SSDA for segmentation on natural images, to our knowledge, no SSDA for medical image segmentation has yet been explored. For example, a depth estimation for natural images is used as an auxiliary task as in [10], but that approach cannot be applied to medical imaging data, e.g., MRI, as they do not have perspective depth maps. Wang et al. [26] simply added supervision from labeled target samples to conventional adversarial UDA. Chen et al. [6] averaged labeled source and target domain images at both region and sample levels to mitigate the domain gap. However, source domain supervision can easily dominate the training, when we directly combine the labeled source data with the target data [23]. In other words, the extra small amount of labeled target data has not been effectively utilized, because the volume of labeled source data is much larger than labeled target data, and there is significant divergence across domains [23].

To mitigate the aforementioned limitations, we propose a practical asymmetric co-training (ACT) framework to take each subset of data in SSDA in a unified and balanced manner. In order to prevent a segmentor, jointly trained by both domains, from being dominated by the source data only, we adopt a divide-and-conquer strategy to decouple the label supervisions for the two asymmetric segmentors, which share the same objective of carrying out a decent segmentation performance for the unlabeled data. By "asymmetric," we mean that the two segmentors are assigned different roles to utilize the labeled data in either source or target domain, thereby providing a complementary view for the unlabeled data. That is, the first segmentor learns on the labeled source domain data and unlabeled target domain data as a conventional UDA task, while the other segmentor learns on the labeled and unlabeled target domain data as a semi-supervised learning (SSL) task. To integrate these two asymmetric branches, we extend the idea of co-training [1,3,22], which is one of the most established multi-view learning methods. Instead of modeling two views on the same set of data with different feature extractors or adversarial sample generation in conventional co-training [1,3,22], our two cross-domain views are explicitly provided by the segmentors with the correlated and complementary UDA and SSL tasks.

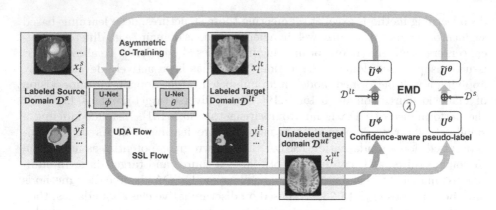

Fig. 1. Illustration of our proposed ACT framework for SSDA cross-modality (e.g., T2-weighted to T1-weighted MRI) image segmentation. Note that only target domain specific segmentor θ will be used in testing.

Specifically, we construct the pseudo label of the unlabeled target sample based on the pixel-wise confident predictions of the other segmentor. Then, the segmentors are trained on the pseudo labeled data iteratively with an exponential MixUp decay (EMD) scheme for smooth propagation. Finally, the target segmentor carries out the target domain segmentation.

The contributions of this work can be summarized as follows:

- We present a novel SSDA segmentation framework to exploit the different supervisions with the correlated and complementary asymmetric UDA and SSL sub-tasks, following a divide-and-conquer strategy. The knowledge is then integrated with confidence-aware pseudo-label based co-training.
- An EMD scheme is further proposed to mitigate the noisy pseudo label in early epochs of training for smooth propagation.
- To our knowledge, this is the first attempt at investigating SSDA for medical image segmentation. Comprehensive evaluations on cross-modality brain tumor (i.e., T2-weighted MRI to T1-weighted/T1ce/FLAIR MRI) segmentation tasks using the BraTS18 database demonstrate superiority performance over conventional source-relaxed/source-based UDA methods.

2   Methodology

In our SSDA setting for segmentation, we are given a labeled source set $\mathcal{D}^s = \{(x_i^s, y_i^s)\}_{i=1}^{N^s}$, a labeled target set $\mathcal{D}^{lt} = \{(x_i^{lt}, y_i^{lt})\}_{i=1}^{N^{lt}}$, and an unlabeled target set $\mathcal{D}^{ut} = \{(x_i^{ut})\}_{i=1}^{N^{ut}}$, where N^s, N^{lt}, and N^{ut} are the number of samples for each set, respectively. Note that the slice x_i^s, x_i^{lt}, and x_i^{ut}, and the segmentation mask labels y_i^s, and y_i^{lt} have the same spatial size of $H \times W$. In addition, for each pixel $y_{i:n}^s$ or $y_{i:n}^{lt}$ indexed by $n \in \mathbb{R}^{H \times W}$, the label has C classes, i.e.,

$y_{i:n}^s, y_{i:n}^{lt} \in \{1, \cdots, C\}$. There is a distribution divergence between source domain samples, $\mathcal{D}^s$, and target domain samples, $\mathcal{D}^{lt}$ and $\mathcal{D}^{ut}$. Usually, N^{lt} is much smaller than N^s. The learning objective is to perform well in the target domain.

2.1 Asymmetric Co-training for SSDA Segmentation

To decouple SSDA via a divide-and-conquer strategy, we integrate $\mathcal{D}^{ut}$ with either $\mathcal{D}^s$ or $\mathcal{D}^{lt}$ to form the correlated and complementary sub-tasks of UDA and SSL. We configure a cross-domain UDA segmentor ϕ and a target domain SSL segmentor θ, which share the same objective of achieving a decent segmentation performance in $\mathcal{D}^{ut}$. The knowledge learned from the two segmentors is then integrated with ACT. The overall framework of this work is shown in Fig. 1.

Conventional co-training has focused on two independent views of the source and target data or generated artificial multi-views with adversarial examples, which learns two classifiers for each of the views and teaches each other on the unlabeled data [3,22]. By contrast, in SSDA, without multiple views of the data, we propose to leverage the distinct yet correlated supervision, based on the inherent discrepancy of the labeled source and target data. We note that the sub-tasks and datasets adopted are different for the UDA and SSL branches. Therefore, all of the data subsets can be exploited, following well-established UDA and SSL solutions without interfering with each other.

To achieve co-training, we adopt a simple deep pseudo labeling method [27], which assigns the pixel-wise pseudo label $\hat{y}_{i:n}$ for $x_{i:n}^{ut}$. Though UDA and SSL can be achieved by different advanced algorithms, deep pseudo labeling can be applied to either UDA [32] or SSL [27]. Therefore, we can apply the same algorithm to the two sub-tasks, thereby greatly simplifying our overall framework. We note that while a few methods [28] can be applied to either SSL or UDA like pseudo labeling, they have not been jointly adopted in the context of SSDA.

Specifically, we assign the pseudo label for each pixel $x_{i:n}^{ut}$ in $\mathcal{D}^{ut}$ with the prediction of either ϕ or θ, therefore constructing the pseudo labeled sets U^ϕ and U^θ for the training of another segmentor θ and ϕ, respectively:

$$U^\phi = \{(x_{i:n}^{ut}, \hat{y}_{i:n}^\phi = \arg\max_c p(c|x_{i:n}^{ut}; \phi)); \text{ if } \max_c p(c|x_{i:n}^{ut}; \phi) > \epsilon\}, \quad (1)$$

$$U^\theta = \{(x_{i:n}^{ut}, \hat{y}_{i:n}^\theta = \arg\max_c p(c|x_{i:n}^{ut}; \theta)); \text{ if } \max_c p(c|x_{i:n}^{ut}; \theta) > \epsilon\}, \quad (2)$$

where $p(c|x_{i:n}^{ut}; \theta)$ and $p(c|x_{i:n}^{ut}; \phi)$ are the predicted probability of class $c \in \{1, \cdots, C\}$ w.r.t. $x_{i:n}^{ut}$ using θ and ϕ, respectively. ϵ is a confidence threshold. Note that the low softmax prediction probability indicates the low confidence for training [18,32]. Then, the pixels in the selected pseudo label sets are merged with the labeled data to construct $\{\mathcal{D}^s, U^\theta\}$ and $\{\mathcal{D}^{lt}, U^\phi\}$ for the training of ϕ and θ with a conventional supervised segmentation loss, respectively. Therefore, the two segmentors with asymmetrical tasks act as teacher and student of each other to distillate the knowledge with highly confident predictions.

Algorithm 1: An iteration of the ACT algorithm.

Input: batch size N, λ, η, ϵ, $\mathcal{D}^s$, $\mathcal{D}^{lt}$, $\mathcal{D}^{ut}$, current network parameters ω_ϕ, ω_θ;
Sample $\{(x_i^s, y_i^s)\}_{i=1}^N$, $\{(x_i^{lt}, y_i^{lt})\}_{i=1}^N$, and $\{(x_i^{ut})\}_{i=1}^N$ from $\mathcal{D}^s$, $\mathcal{D}^{lt}$, and $\mathcal{D}^{ut}$, respectively;
Initialize $U^\phi = \emptyset$, $U^\theta = \emptyset$;
for $i \leftarrow 1$ **to** N **do**
    $\hat{y}_{i:n}^\phi = \arg\max_c p(c|x_{i:n}^{ut}; \phi)$; and $\hat{y}_{i:n}^\theta = \arg\max_c p(c|x_{i:n}^{ut}; \theta)$
    if $\max_c p(c|x_{i:n}^{ut}; \phi) > \epsilon$: **update** $U^\phi \leftarrow U^\phi \cup \{(x_{i:n}^{ut}, \hat{y}_{i:n}^\theta)$ with Eq. (1);
    if $\max_c p(c|x_{i:n}^{ut}; \theta) > \epsilon$: **update** $U^\theta \leftarrow U^\theta \cup \{(x_{i:n}^{ut}, \hat{y}_{i:n}^\theta)$ with Eq. (2);
end
Obtain $\tilde{U}^\phi = \{\text{EMD}(U_i^\phi, \{(x_i^{lt}, y_i^{lt})\}_{i=1}^N; \lambda)\}_{i=1}^{|U^\phi| \times N}$ with Eq. (3);
Obtain $\tilde{U}^\theta = \{\text{EMD}(U_i^\theta, \{(x_i^s, y_i^s)\}_{i=1}^N; \lambda)\}_{i=1}^{|U^\theta| \times N}$ with Eq. (4);
Update $\omega_\phi \leftarrow \omega_\phi - \eta\nabla(\mathcal{L}(\omega_\phi, \mathcal{D}^s) + \mathcal{L}(\omega_\phi, \tilde{U}^\theta))$; $\omega_\theta \leftarrow \omega_\theta - \eta\nabla(\mathcal{L}(\omega_\theta, \mathcal{D}^{lt}) + \mathcal{L}(\omega_\theta, \tilde{U}^\phi))$;
Output: Updated network parameters ω_ϕ and ω_θ.

2.2 Pseudo-label with Exponential MixUp Decay

Initially generated pseudo labels with the two segmentors are typically noisy, which is significantly acute in the initial epochs, thus leading to a deviated solution with propagated errors. Numerous conventional co-training methods relied on simple assumptions that there is no domain shift, and the predictions of the teacher model can be reliable and be simply used as ground truth. Due to the domain shift, however, the prediction of ϕ in the target domain could be noisy and lead to an aleatoric uncertainty [7,11,12]. In addition, insufficient labeled target domain data can lead to an epistemic uncertainty related to the model parameters [7,11,12].

To smoothly exploit the pseudo labels, we propose to adjust the contribution of the supervision signals from both labels and pseudo labels as the training progresses. Previously, vanilla MixUp [30] was developed for efficient data augmentation, by combining both samples and their labels to generate new data for training. We note that the MixUp used in SSL [2,6] adopted a constant sampling, and did not take the decay scheme for gradual co-training. Thus, we propose to gradually exploit the pseudo label by mixing up $\mathcal{D}^s$ or $\mathcal{D}^{lt}$ with pseudo labeled $\mathcal{D}^{ut}$, and adjust their ratio with the EMD scheme. For the selected U^ϕ and U^θ with the number of slices $|U^\phi|$ and $|U^\theta|$, we mix up each pseudo labeled image with all images from $\mathcal{D}^s$ or $\mathcal{D}^{lt}$ to form the mixed pseudo labeled sets $\tilde{U}^\theta$ and $\tilde{U}^\phi$. Specifically, our EMD can be formulated as:

$$\tilde{U}^\phi = \{(\tilde{x}_{i:n}^{lt} = \lambda x_{i:n}^{lt} + (1-\lambda)x_{i:n}^{ut}, \lambda\tilde{y}_{i:n}^{lt} = \lambda y_{i:n}^{lt} + (1-\lambda)\hat{y}_{i:n}^\theta)\}_i^{|U^\theta| \times N}, \quad (3)$$

$$\tilde{U}^\theta = \{(\tilde{x}_{i:n}^s = \lambda x_{i:n}^s + (1-\lambda)x_{i:n}^{ut}, \lambda\tilde{y}_{i:n}^s = \lambda y_{i:n}^s + (1-\lambda)\hat{y}_{i:n}^\phi)\}_i^{|U^\phi| \times N}, \quad (4)$$

where $\lambda = \lambda^0\exp(-I)$ is the MixUp parameter with the exponential decay w.r.t. iteration I. λ^0 is the initial weight of ground truth samples and labels, which is empirically set to 1. Therefore, along with the increase over iteration I, we have smaller λ, which adjusts the contribution of the ground truth label to be large at the start of the training, while utilizing the pseudo labels at the later training epochs. Therefore, $\tilde{U}^\phi$ and $\tilde{U}^\theta$ gradually represent the pseudo label sets

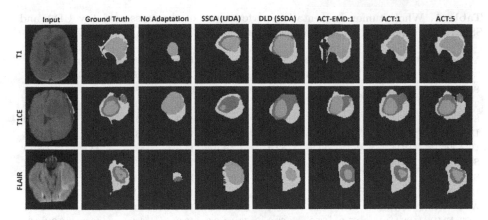

Fig. 2. Comparisons with other UDA/SSDA methods and ablation studies for the cross-modality tumor segmentation. We show target test slices of T1, T1ce, and FLAIR MRI from three subjects.

of U^ϕ and U^θ. We note that the mixup operates on the image level, which is indicated by i. The number of generated mixed samples depends on the scale of U^ϕ and U^θ in each iteration and batch size N. With the labeled $\mathcal{D}^s$, $\mathcal{D}^{lt}$, as well as the pseudo labeled sets with EMD $\tilde{U}^\phi$ and $\tilde{U}^\phi$, we update the parameters of the segmentors ϕ and θ, i.e., ω_ϕ and ω_θ with SGD as:

$$\omega_\phi \leftarrow \omega_\phi - \eta \nabla (\mathcal{L}(\omega_\phi, \mathcal{D}^s) + \mathcal{L}(\omega_\phi, \tilde{U}^\theta)), \tag{5}$$

$$\omega_\theta \leftarrow \omega_\theta - \eta \nabla (\mathcal{L}(\omega_\theta, \mathcal{D}^{lt}) + \mathcal{L}(\omega_\theta, \tilde{U}^\phi)), \tag{6}$$

where η indicates the learning rate, and $\mathcal{L}(\omega_\phi, \mathcal{D}^s)$ denotes the learning loss on $\mathcal{D}^s$ with the current segmentor ϕ parameterized by ω_ψ. The training procedure is detailed in Algorithm 1. After training, only the target domain specific SSL segmentor θ is used for testing.

3   Experiments and Results

To demonstrate the effectiveness of our proposed SSDA method, we evaluated our method on T2-weighted MRI to T1-weighted/T1ce/FLAIR MRI brain tumor segmentation using the BraTS2018 database [21]. We denote our proposed method as ACT, and used ACT-EMD for an ablation study of an EMD-based pseudo label exploration.

Of note, the BraTS2018 database contains a total of 285 patients [21] with the MRI scannings, including T1-weighted (T1), T1-contrast enhanced (T1ce), T2-weighted (T2), and T2 Fluid Attenuated Inversion Recovery (FLAIR) MRI. For the segmentation labels, each pixel belongs to one of four classes, i.e., enhancing tumor (EnhT), peritumoral edema (ED), necrotic and non-enhancing tumor core (CoreT), and background. In addition, the whole tumor covers CoreT, EnhT,

Table 1. Whole tumor segmentation performance of the cross-modality UDA and SSDA. The supervised joint training can be regarded as an "upper bound".

Method		DICE Score (DSC) [%] ↑				Hausdorff Distance (HD) [mm] ↓			
	Task	T1	FLAIR	T1CE	Ave	T1	FLAIR	T1CE	Ave
Source only	No DA	4.2	65.2	6.3	27.7 ± 1.2	55.7	28.0	49.8	39.6 ± 0.5
Target only	SSL:5	43.8	54.6	47.5	48.6 ± 1.7	31.9	29.6	35.4	32.3 ± 0.8
SIFA [5]	UDA	51.7	68.0	58.2	59.3 ± 0.6	19.6	16.9	15.0	17.1 ± 0.4
DSFN [31]	UDA	57.3	78.9	62.2	66.1 ± 0.8	17.5	13.8	15.5	15.6 ± 0.3
DSA [9]	UDA	57.7	81.8	62.0	67.2 ± 0.7	14.2	8.6	13.7	12.2 ± 0.4
SSCA [17]	UDA	59.3	82.9	63.5	68.6 ± 0.6	12.5	7.9	11.2	11.5 ± 0.3
SLA [26]	SSAD:1	64.7	82.3	66.1	71.0 ± 0.5	12.2	7.1	10.5	9.9 ± 0.3
DLD [6]	SSAD:1	65.8	81.5	66.5	71.3 ± 0.6	12.0	7.1	10.3	9.8 ± 0.2
ACT	SSAD:1	**69.7**	**84.5**	**69.7**	**74.6 ± 0.3**	**10.5**	**5.8**	**10.0**	**8.8 ± 0.1**
ACT-EMD	SSAD:1	67.4	83.9	69.0	73.4 ± 0.6	10.9	6.4	10.3	9.2 ± 0.2
ACT	SSAD:5	**71.3**	**85.0**	**70.8**	**75.7 ± 0.5**	**10.0**	**5.2**	**9.8**	**8.3 ± 0.1**
ACT-EMD	SSAD:5	70.3	84.4	69.8	74.8 ± 0.4	10.4	5.7	10.2	8.8 ± 0.2
Joint training	Supervised	73.2	85.6	72.6	77.1 ± 0.5	9.5	4.6	9.2	7.7 ± 0.2

and ED. We follow the conventional cross-modality UDA (i.e., T2-weighted to T1-weighted/T1ce/FLAIR) evaluation protocols [9, 17, 31] for 8/2 splitting for training/testing, and extend it to our SSDA task, by accessing the labels of 1–5 target domain subjects at the adaptation training stage. All of the data were used in a subject-independent and unpaired manner. We used SSDA:1 or SSDA:5 to denote that one or five target domain subjects are labeled in training.

For a fair comparison, we used the same segmentor backbone as in DSA [9] and SSCA [17], which is based on Deeplab-ResNet50. Without loss of generality, we simply adopted the cross-entropy loss as $\mathcal{L}$, and set the learning rate $\eta = 1e-3$ and confidence threshold $\epsilon = 0.5$. Both ϕ and θ have the same network structure. For the evaluation metrics, we adopted the widely used DSC (the higher, the better) and Hausdorff distance (HD: the lower, the better) as in [9, 17]. The standard deviation was reported over five runs.

The quantitative evaluation results of the whole tumor segmentation are provided in Table 1. We can see that SSDA largely improved the performance over the compared UDA methods [9, 17]. For the T2-weighted to T1-weighted MRI transfer task, we were able to achieve more than 10% improvements over [9, 17] with only one labeled target sample. Recent SSDA methods for natural image segmentation [6, 26] did not take the balance between the two labeled supervisions into consideration, easily resulting in a source domain-biased solution in case of limited labeled target domain data, and thus did not perform well on target domain data [23]. In addition, the depth estimation in [10] cannot be applied to the MRI data. Thus, we reimplemented the aforementioned methods [6, 26] with the same backbone for comparisons, which is also the first attempt at the medical image segmentation. Our ACT outperformed [6, 26] by a DSC of 3.3%

Table 2. Detailed comparison of Core/EnhT/ED segmentation. Results are averaged over three tasks including T2-weighted to T1-weighted, T1CE, and FLAIR MRI with the backbone as in [9,17].

Method		DICE	Score (DSC)	[%] ↑	Hausdorff	Distance (HD)	[mm] ↓
	Task	**CoreT**	**EnhT**	**ED**	**CoreT**	**EnhT**	**ED**
Source only	No DA	20.6±1.0	39.5±0.8	41.3±0.9	54.7±0.4	55.2±0.6	42.5±0.4
Target only	SSL:5	27.3±1.1	38.0±1.0	40.2.3±1.3	51.8±0.7	52.3±0.9	46.4±0.6
DSA [9]	UDA	57.8±0.6	44.0±0.6	56.8±0.5	25.8±0.4	34.2±0.3	25.6±0.5
SSCA [17]	UDA	58.2±0.4	44.5±0.5	60.7±0.4	26.4±0.2	32.8±0.2	23.4±0.3
SLA [26]	SSDA:1	58.9±0.6	48.1±0.5	65.4±0.4	24.5±0.1	27.6±0.3	20.3±0.2
DLD [6]	SSDA:1	60.3±0.6	48.2±0.5	66.0±0.3	24.2±0.2	27.8±0.1	19.7±0.2
ACT	SSDA:1	**64.5±0.3**	**52.7±0.4**	**69.8±0.6**	**20.0±0.2**	**24.6±0.1**	**16.2±0.2**
ACT	SSDA:5	**66.9±0.3**	**54.0±0.3**	**71.2±0.5**	**18.4±0.4**	**23.7±0.2**	**15.1±0.2**
Joint training	Supervised	70.4±0.3	62.5±0.2	75.1±0.4	15.8±0.2	22.7±0.1	13.0±0.2

w.r.t. the averaged whole tumor segmentation in SSDA:1 task. The better performance of ACT over ACT-EMD demonstrated the effectiveness of our EMD scheme for smooth adaptation with pseudo-label. We note that we did not manage to outperform the supervised joint training, which accesses all of the target domain labels, which can be considered an "upper bound" of UDA and SSDA. Therefore, it is encouraging that our ACT can approach joint training with five labeled target subjects. In addition, the performance was stable for the setting of λ from 1 to 10.

In Table 2, we provide the detailed comparisons for more fine-grained segmentation w.r.t. CoreT, EnhT, and ED. The improvements were consistent with the whole tumor segmentation. The qualitative results of three target modalities in Fig. 2 show the superior performance of our framework, compared with the comparison methods.

In Fig. 3(a), we analyzed the testing pixel proportion change along with the training that has both, only one, and none of two segmentor pseudo-labels, i.e., the maximum confidence is larger than ϵ as in Eq. (1). We can see that the consensus of the two segmentors keeps increasing, by teaching each other in the co-training scheme for knowledge integration. "Both" low rates, in the beginning, indicate ϕ and θ may provide a different view based on their asymmetric tasks, which can be complementary to each other. The sensitivity studies of using a different number of labeled target domain subjects are shown in Fig. 3(b). Our ACT was able to effectively use $\mathcal{D}^{lt}$. In Fig. 3(c), we show that using more EMD pairs improves the performance consistently.

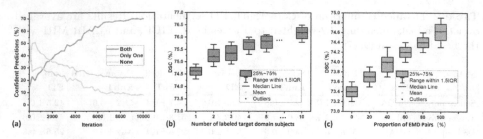

Fig. 3. Analysis of our ACT-based SSDA on the whole tumor segmentation task. (a) The proportion of testing pixels that both, only one, or none of the segmentors have high confidence on (b) the performance improvements with a different number of labeled target domain training subjects, and (c) a sensitivity study of changing different proportion of EMD pairs of $|\tilde{U}^\phi| \times N$ and $|\tilde{U}^\theta| \times N$.

4 Conclusion

This work proposed a novel and practical SSDA framework for the segmentation task, which has the great potential to improve a target domain generalization with a manageable labeling effort in clinical practice. To achieve our goal, we resorted to a divide-and-conquer strategy with two asymmetric sub-tasks to balance between the supervisions from source and target domain labeled samples. An EMD scheme is further developed to exploit the pseudo-label smoothly in SSDA. Our experimental results on the cross-modality SSDA task using the BraTS18 database demonstrated that the proposed method surpassed the state-of-the-art UDA and SSDA methods.

Acknowledgement. This work is supported by NIH R01DC018511, R01DE027989, and P41EB022544.

References

1. Balcan, M.F., Blum, A., Yang, K.: Co-training and expansion: towards bridging theory and practice. Adv. Neural Inf. Process. Syst. **17**, 89–96 (2005)
2. Berthelot, D., Carlini, N., Goodfellow, I., Papernot, N., Oliver, A., Raffel, C.: Mixmatch: a holistic approach to semi-supervised learning. arXiv preprint arXiv:1905.02249 (2019)
3. Blum, A., Mitchell, T.: Combining labeled and unlabeled data with co-training. In: Proceedings of the Eleventh Annual Conference on Computational Learning Theory, pp. 92–100 (1998)
4. Che, T., et al.: Deep verifier networks: verification of deep discriminative models with deep generative models. ArXiv (2019)
5. Chen, C., Dou, Q., Chen, H., Qin, J., Heng, P.A.: Synergistic image and feature adaptation: towards cross-modality domain adaptation for medical image segmentation. In: Proceedings of the AAAI Conference on Artificial Intelligence, vol. 33, pp. 865–872 (2019)

6. Chen, S., Jia, X., He, J., Shi, Y., Liu, J.: Semi-supervised domain adaptation based on dual-level domain mixing for semantic segmentation. In: CVPR, pp. 11018–11027 (2021)
7. Der Kiureghian, A., Ditlevsen, O.: Aleatory or epistemic? Does it matter? Struct. Saf. **31**(2), 105–112 (2009)
8. Donahue, J., Hoffman, J., Rodner, E., Saenko, K., Darrell, T.: Semi-supervised domain adaptation with instance constraints. In: Proceedings of the IEEE Conference on Computer Vision and Pattern Recognition, pp. 668–675 (2013)
9. Han, X., et al.: Deep symmetric adaptation network for cross-modality medical image segmentation. IEEE Trans. Med. Imaging (2022)
10. Hoyer, L., Dai, D., Wang, Q., Chen, Y., Van Gool, L.: Improving semi-supervised and domain-adaptive semantic segmentation with self-supervised depth estimation. arXiv preprint arXiv:2108.12545 (2021)
11. Hu, S., Worrall, D., Knegt, S., Veeling, B., Huisman, H., Welling, M.: Supervised uncertainty quantification for segmentation with multiple annotations. In: Shen, D., et al. (eds.) MICCAI 2019. LNCS, vol. 11765, pp. 137–145. Springer, Cham (2019). https://doi.org/10.1007/978-3-030-32245-8_16
12. Kendall, A., Gal, Y.: What uncertainties do we need in Bayesian deep learning for computer vision? arXiv preprint arXiv:1703.04977 (2017)
13. Kim, T., Kim, C.: Attract, perturb, and explore: learning a feature alignment network for semi-supervised domain adaptation. In: Vedaldi, A., Bischof, H., Brox, T., Frahm, J.-M. (eds.) ECCV 2020. LNCS, vol. 12359, pp. 591–607. Springer, Cham (2020). https://doi.org/10.1007/978-3-030-58568-6_35
14. Kong, L., Hu, B., Liu, X., Lu, J., You, J., Liu, X.: Constraining pseudo-label in self-training unsupervised domain adaptation with energy-based model. Int. J. Intell. Syst. (2022)
15. Liu, X., et al.: Domain generalization under conditional and label shifts via variational Bayesian inference. IJCAI (2021)
16. Liu, X., Li, S., Ge, Y., Ye, P., You, J., Lu, J.: Recursively conditional gaussian for ordinal unsupervised domain adaptation. In: International Conference on Computer Vision (ICCV), October 2021
17. Liu, X., Xing, F., Fakhri, G.E., Woo, J.: Self-semantic contour adaptation for cross modality brain tumor segmentation. In: IEEE International Symposium on Biomedical Imaging (ISBI) (2022)
18. Liu, X., et al.: Generative self-training for cross-domain unsupervised tagged-to-cine MRI synthesis. In: de Bruijne, M., et al. (eds.) MICCAI 2021. LNCS, vol. 12903, pp. 138–148. Springer, Cham (2021). https://doi.org/10.1007/978-3-030-87199-4_13
19. Liu, X., et al.: Unsupervised black-box model domain adaptation for brain tumor segmentation. Front. Neurosci., 341 (2022)
20. Liu, X., et al.: Deep unsupervised domain adaptation: a review of recent advances and perspectives. APSIPA Trans. Signal Inf. Process. (2022)
21. Menze, B.H., et al.: The multimodal brain tumor image segmentation benchmark (BRATS). IEEE Trans. Medical Imaging **34**(10), 1993–2024 (2014)
22. Qiao, S., Shen, W., Zhang, Z., Wang, B., Yuille, A.: Deep co-training for semi-supervised image recognition. In: Ferrari, V., Hebert, M., Sminchisescu, C., Weiss, Y. (eds.) ECCV 2018. LNCS, vol. 11219, pp. 142–159. Springer, Cham (2018). https://doi.org/10.1007/978-3-030-01267-0_9
23. Saito, K., Kim, D., Sclaroff, S., Darrell, T., Saenko, K.: Semi-supervised domain adaptation via minimax entropy. In: Proceedings of the IEEE/CVF International Conference on Computer Vision, pp. 8050–8058 (2019)

24. Tajbakhsh, N., Jeyaseelan, L., Li, Q., Chiang, J.N., Wu, Z., Ding, X.: Embracing imperfect datasets: a review of deep learning solutions for medical image segmentation. Med. Image Anal. **63**, 101693 (2020)

25. van Engelen, J.E., Hoos, H.H.: A survey on semi-supervised learning. Mach. Learn. **109**(2), 373–440 (2019). https://doi.org/10.1007/s10994-019-05855-6

26. Wang, Z., et al.: Alleviating semantic-level shift: a semi-supervised domain adaptation method for semantic segmentation. In: Proceedings of the IEEE/CVF Conference on Computer Vision and Pattern Recognition Workshops, pp. 936–937 (2020)

27. Wei, C., Shen, K., Chen, Y., Ma, T.: Theoretical analysis of self-training with deep networks on unlabeled data. arXiv preprint arXiv:2010.03622 (2020)

28. Xia, Y., et al.: Uncertainty-aware multi-view co-training for semi-supervised medical image segmentation and domain adaptation. Med. Image Anal. **65**, 101766 (2021)

29. Yao, T., Pan, Y., Ngo, C.W., Li, H., Mei, T.: Semi-supervised domain adaptation with subspace learning for visual recognition. In: Proceedings of the IEEE conference on Computer Vision and Pattern Recognition, pp. 2142–2150 (2015)

30. Zhang, H., Cisse, M., Dauphin, Y.N., Lopez-Paz, D.: mixup: Beyond empirical risk minimization. arXiv preprint arXiv:1710.09412 (2017)

31. Zou, D., Zhu, Q., Yan, P.: Unsupervised domain adaptation with dualscheme fusion network for medical image segmentation. In: Proceedings of the Twenty-Ninth International Joint Conference on Artificial Intelligence, IJCAI-20, International Joint Conferences on Artificial Intelligence Organization, pp. 3291–3298 (2020)

32. Zou, Y., Yu, Z., Liu, X., Kumar, B., Wang, J.: Confidence regularized self-training. In: Proceedings of the IEEE/CVF International Conference on Computer Vision, pp. 5982–5991 (2019)

A Sense of Direction in Biomedical Neural Networks

Zewen Liu[✉] and Timothy F. Cootes

Division of Informatics, Imaging and Data Science, Stopford Building,
The University of Manchester, Manchester M13 9PT, UK
zewen.liu@postgrad.manchester.ac.uk

Abstract. We describe an approach to making a model be aware of not only intensity but also properties such as feature direction and scale. Such properties can be important when analysing images containing curvilinear structures such as vessels or fibres. We propose the General Multi-Angle Scale Convolution (G-MASC), whose kernels are arbitrarily rotatable and also fully differentiable. The model manages its directional detectors in sets, and supervises a set's rotation symmetricity with a novel rotation penalty called PoRE. The algorithm works on pyramid representations to enable scale search. Direction and scale can be extracted from the output maps, encoded and analysed separately. Tests were conducted on three public datasets, MoNuSeg, DRIVE, and CHASE-DB1. Good performance is observed while the model requires 1% or fewer parameters compared to other approaches such as U-Net.

Keywords: Steerable filter · Intepretability · Image Segmentation

1 Introduction

Examination of sets of filters trained in a CNN often shows sets of kernels have arisen which are approximately rotated versions of each other, suggesting a lot of redundancy[1]. It has been shown [8] that explicitly constructing banks of filters to deal with rotation using inter-filter constraints during training can lead to good results with only a fraction of the number of overall parameters.

We build on that work, demonstrating that additional constraints can be used to encourage filters to adopt well spaced orientations. We also show that the resulting filters at different orientations have outputs which are directly comparable, an important property when seeking to select the one which best represents the orientation (or scale) of local structure at a point.

We describe a new approach, General Multi-Angle Scale Convolution (G-MASC), which integrates the direction and scale selection with convolution. As shown in Fig. 1 the resulting kernels cover a wide range of orientations equally, picking out key structures in the training set. We describe the approach in detail and show experiments demonstrating its performance on three datasets.

---

[1] Our codes are available at https://github.com/Zewen-Liu/MASC-Unit.

© The Author(s), under exclusive license to Springer Nature Switzerland AG 2022
L. Wang et al. (Eds.): MICCAI 2022, LNCS 13435, pp. 77–86, 2022.
https://doi.org/10.1007/978-3-031-16443-9_8

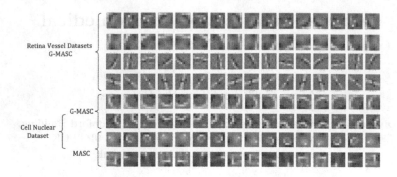

Retina Vessel Datasets
G-MASC

G-MASC

Cell Nuclear
Dataset

MASC

Fig. 1. Examples of the directional kernels trained on different tasks.

2   Related Works

Models which deal with scale variations can be constructed using pyramid pooling approaches. Achieving invariance to rotations is more challenging and has not yet been completely solved in neural networks.

Augmenting a training set with rotated versions [15] is perhaps the simplest solution but is also very expensive. This greedy method can cover all orientated patterns but inevitably leads to detector redundancy.

Another approach is steerable functions [11] in which complex rotation detectors are created by combining some bases with weights. These bases are analytically determined as a function of orientation [4], e.g. harmonic atoms [18]. The atoms are usually symmetric [19] which can be a limitation. A steerable kernel may require a certain size to express variation. A scalable atom function such as [5] can be made at the cost of significantly increased parameters. In addition, the bases must be predefined.

Rotational equivalent kernel groups [12] is a different idea. A seed work of this genre is [3], where rotatable kernels are created by matrix actions. The method suffers from an approximation issue in continuous domain and thus was only practiced on $\frac{\pi}{2}i$ directions. A detailed discussion is given in [2].

Hybrid kernels that combine a steerable function with convolutional kernels are studied in [9,10]. Initialisation is a decisive factor of them but no sophisticated solution was proposed.

Multi-angle and Scale Convolution (MASC) [8] is fully differentiable and found to be effective at segmenting curvilinear structures. A set of rotated versions of a kernel are learnt, constrained by a set of n equations relating how pairs of kernels interact. If n is less than the kernel size, the model may not be well defined. MASC also assumes patterns to have 180° rotational symmetry about the origin. One motivation of G-MASC is to expand this idea on common tasks by exploring new ways of imposing constraints to encourage rotational symmetries in the kernels.

All the above methods focus on finding invariant measures rather than trying to estimate the actual orientation.

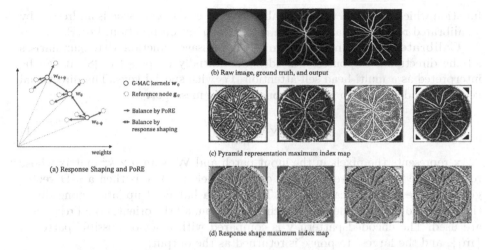

(b) Raw image, ground truth, and output

O G-MAC kernels $\mathbf{w}_\theta$
O Reference node $\mathbf{g}_\theta$

→ Balance by PoRE
↔ Balance by response shaping

weights

(c) Pyramid representation maximum index map

(a) Response Shaping and PoRE

(d) Response shape maximum index map

Fig. 2. (a)The relation and intension of response shaping and PoRE (in kernel space). (b-d) Orientation and scale index maps.

3   Methods

Inspired by MASC [8], G-MASC is developed with orientation and scale awareness. The model uses a supervision function, called Phase-offset Rotational Error (PoRE), to manage rotation consistency. An additional module called index map featurisation (IMF) is introduced for analysing the resulting property labels.

3.1   Rotation Mechanism

Directional Kernels are the basic function unit in a G-MASC layer (Fig. 3(f)). A number of c direction kernel sets can be arranged in parallel. Each G-MASC layer has c input channels and c output channels. Each set contains n directional kernels (including reflections as special directions). c and n are hyperparameters, n has to be a multiple of 8. For a n-kernel set, there are $\frac{1}{8}n$ base kernels which are actually maintained, and the other $\frac{7}{8}n$ are the virtual kernels created by rotations and reflections [3]. We call a basic kernel and its variants a kernel group, and a set contains $\frac{1}{8}n$ groups.

Suppose that vector $\mathbf{w}$ represents the pixels in a $k \times k$ kernel defining a structure at an angle $\theta \in [0, \pi/2)$. Rotating it by $90°, 180°, 270°$ leads to four variants at angles $\theta, \pi/2 + \theta, \pi + \theta$ and $3\pi/2 + \theta$. If we reflect the kernels about the diagonal, we get a set of four more reflected kernels.

Thus for $n = 16$ we have two base kernels at $\theta = \{\frac{\pi}{8}, \frac{3}{8}\pi\}$ leading to 8 overall directional kernels plus 8 reflections. Each base kernel is a Hadamard product of a trainable Gabor function (g_θ) and a free kernel, as $\mathbf{g}_\theta \odot \mathbf{c}_\theta$, which is then normalised so that $|\mathbf{w}| = 1$.

During training gradients are shared by the group. The challenge is in managing the intra-group balance. The original MASC used the response shaping function which we found to be unstable in some cases. The issue is addressed by a calibrated response shaping and a new regularisation function, PoRE.

Calibrated Response Shaping is a message function that summarises all the directional responses. The idea was initially proposed in [8]. It can be interpreted as a multi-head self-attention [17] with special keys. The directional kernels encode inputs to get a response's query message $\mathbf{v}$.

$$\mathbf{v} = \frac{\mathbf{W}\mathbf{x}}{||\mathbf{x}||} \tag{1}$$

$\mathbf{x}$ represents the pixels in the input patch and $\mathbf{W}$ is an $n \times k^2$ matrix whose rows are the vector representations of the kernels $(\mathbf{w}_i)$. $\mathbf{v}$ is then a n-D vector giving the response to each kernel. To ensure a balanced update among directions, instead of using one best-matched direction, all the orientational responses are used. The encoded pattern $\mathbf{v}$ is compared with a set of possible patterns, $\{\mathbf{m}_i\}$, and the largest response is returned as the output.

$$R(\mathbf{x}) = \max_i(\mathbf{m}_i^T\mathbf{v}) \tag{2}$$

The vectors defining the patterns, $\mathbf{m}_i$, evolve as training progresses. The index which maximises the above indicates the main direction of the structure at that point (i indexes the rotated/reflected kernel set).

Let $\mathbf{M}$ be an $n \times n$ matrix with rows $\mathbf{m}_i$. Let $\mathbf{W}_0$ be the matrix containing the kernels on initialisation as its rows. The initial value of $\mathbf{M}$ is set to $\mathbf{M}_0 = \mathbf{W}_0\mathbf{W}_0^T$.

To encourage filters to be equally spaced in orientation, $\mathbf{M}$ should be a circulant matrix, such that each row is a shifted version of the next, $M(i+1, j+1) = M(i, j)$, with wrap-around on the indices.

Such a structure is enforced by averaging along diagonals in the following steps. At training epoch t, let

$$\hat{\mathbf{M}} = \mathbf{W}_t\mathbf{W}_t^T$$

$$\Omega(\hat{M}(i, j)) = \frac{1}{n}\sum_{k=1}^{n} \hat{M}((i + k)|_n, (j + k)|_n) \tag{3}$$

$$\mathbf{M}_t = a\mathbf{M}_{t-1} + (1 - a)\Omega(\hat{\mathbf{M}})$$

$\Omega()$ averages $\hat{\mathbf{M}}$ along its diagonals. $a = 0.99$ controls the momentum.

This response shaping makes directional kernels rotate adaptively. However, there may be a problem of too few constraints when $n < k \times k$.

Phase-Offset Rotational Error (PoRE) is proposed to provided extra regularisation and supervision of rotation. The idea of PoRE is like local linear embedding [14]; the global kernel distribution is preserved by keeping local relationships. Suppose we have a steerable function which can generate a kernel $\mathbf{g}(\theta)$ at angle θ. Suppose the n kernels $\mathbf{w}_i$ are ordered so that $\mathbf{w}_0$ corresponds to angle zero, and $\mathbf{w}_i$ corresponds to the same feature at angle $4i\pi/n$. The first half of the set are the rotated kernels, the second half their reflections. To ensure consistency we would like the distance between $\mathbf{w}_0$ and $\mathbf{g}(\psi)$ to be similar to that between $\mathbf{w}_i$ and $\mathbf{g}(4i\pi/n + \psi)$, and ψ is a small angle. We can measure this consistency by calculating the variance of that distance over all kernels;

$$d_i = \frac{1}{k^2}\|\mathbf{w}_i - \mathbf{g}(4i\pi/n + \psi)\| \tag{4}$$

$$\sigma^2(\mathbf{W}, \mathbf{G}) = \frac{1}{n}\sum_{i=0}^{n/2-1}(d_i - \bar{d})^2 + \frac{1}{n}\sum_{i=n/2}^{n-1}(d_i - \bar{d}_r)^2 \tag{5}$$

where $\bar{d}$ is the mean value of d_i over the first $n/2$ (unreflected) kernels, $\bar{d}_r$ is the mean over the reflected kernels. In practice we use a set of Gabor reference functions, $\mathbf{g}_m(\theta)$, $m = 1..n_{\text{ref}}$. The PoRE is only applied on the base kernels rather than all kernels, which is equivalent to Eq. 5 up to a scaling factor. PoRE and the ultimate loss function is then,

$$\text{PoRE}(\mathbf{W}) = \frac{1}{n_{\text{ref}}}\sum_{m=1}^{n_{\text{ref}}}\sigma^2(\mathbf{W}, \mathbf{G_m}) \tag{6}$$

$$L = BCE + \lambda_{\text{PoRE}}\text{PoRE} \tag{7}$$

Here, BCE is binary cross entropy. The characteristics of response shaping and PoRE are visualised in Fig. 2 (a), where response shaping locates a kernel by other directional kernels, while PoRE uses reference nodes.

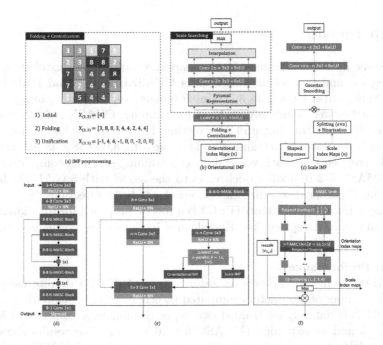

Fig. 3. The Structure of a G-MASC 5-2-8 model and the module details.

3.2   Orientation and Scale Embedding

For a human, both pattern orientation and strength are critical for decision making. Just as vision transformers [1] have positional encoding, in G-MASC, the orientation and scale labels are retrieved. Examples of maps of orientation and scale indexes are displayed in Fig. 2(c) and (d).

The i in Eq. (2) indicates the pattern orientation. The process of encoding them is called Orientational Index Map Featurisation (Orientational IMF). It has two steps, 1) preprocessing (extracting a patch around each point then subtracting the mean), and 2) encoding. The IMF module is illustrated in Fig. 3(b). In the preprocessing step, the indexes are converted to a relative value depending on the central orientation (Fig. 3(a)). Then the relative orientation is down-sampled to its pyramid representations with average pooling. The featurised index maps are processed by two convolutional layers and then projected back to the original scale with interpolation. Only the largest intensity among the scale channels will be output.

Similarly, the scale labels generated by the scale searching (pyramid pooling in Fig. 3(f)) are obtained as shown in Fig. 3(c). The scale index maps are split into $s \times n$ binary maps (s is the number of scales considered). Each channel highlights the area that has the strongest $R(\mathbf{x})$ at the corresponding scale. The binary maps are blurred by a Gaussian kernel and then multiplied with $R(\mathbf{x})$.

The structure of a full G-MASC module is shown in Fig. 3(d).

4   Experiments

Three public datasets were used in this paper, a multi-organ histology dataset MoNuSeg [6], and two retina vessel tasks CHASEDB1 [13] and DRIVE [16]. The MoNuSeg dataset contains 30 training images, and 14 testing images. 8,000 72×72 patches were randomly cropped for training. CHASE-DB1 (999×960, 8-bit) has 20 images training, and 8 for testing. DRIVE has 40 images (565×584 8-bit) evenly separated into training and testing sets. 8,000 48×48 patches were randomly cropped from each vessel dataset's training set. The comparisons a

A G-MASC 5-2-16 5×5 model refers to the model with 5 G-MASC blocks, each having 2 parallel 5×5 directional kernel sets in 16 directions.

Phase-Offset Correlation (PoC) is a metric proposed for rotational symmetry evaluation. It replaces the Euclidean d in Eq. (5) by normalised correlation.

General Performance
The G-MASC achieved competitive results on CHASE-DB1 despite using only 1% of the number of parameters compared to other of the benchmark methods. On the DRIVE dataset, we trained two types of models, a compact G-MASC 5-2-16 5×5 and an extended G-MASC 5-6-16 11×11. The results show that if both models use a compact setup (fewer parameters), a G-MASC model can outperform its MASC counterpart by 1.0% sensitivity and 0.6% AUC(area under curve).

On the non-curvilinear detection task, G-MASC outperformed the MASC with 76.3%(73.6%) F1 score and 94.5%(93.6%) AUC. These show the advantage of using G-MASC comparing to the original MASC in different tasks. The margin can be expanded using fewer directional kernel sets and smaller kernels.

These results show that the new method performs at least as well as the old method, which is what one would expect. Putting the retina vessel experiments together with the cell nucleus experiment can illustrate when the additional rotational symmetry protection is required and the significance of its positive impact.

Table 1. Comparison with other methods on CHASE_DB1 (*results obtained from [7])

Datasets	CHASE- DB1	F1	Se	Sp	Acc	AUC	Para
CHASE-DB1	2nd Observer	76.9	74.3	97.9	94.7	–	–
	U-Net	79.2	79.1	97.4	95.4	95.3	57.4K
	VesselNet*	79.1	78.2	**98.1**	**96.2**	**97.6**	585.8K
	MASC 5-2-16 5×5	**80.2**	**80.7**	97.8	95.6	97.4	3.8K
	G-MASC 5-2-16 5×5	79.4	76.4	98.0	95.6	97.4	5.4K
DRIVE	2nd Observer	78.8	77.6	97.3	94.7	–	–
	U-Net*	**81.4**	75.4	98.2	95.3	97.6	30.6M
	VesselNet*	81.3	76.3	**98.4**	**95.6**	**97.7**	585.8K
	MASC 5-2-16 5×5	79.1	77.1	97.4	94.7	96.4	3.8K
	G-MASC 5-2-16 5×5	79.6	**78.1**	97.7	95.1	97.0	5.4K
	G-MASC 5-6-16 11×11	81.0	78.0	97.9	95.5	97.5	33.6K
MoNuSeg	U-Net(Double Conv)	71.4	72.5	90.1	86.7	91.4	57.4K
	MASC5-2-16 7×7	73.6	**86.3**	85.8	85.9	93.6	4.3K
	G-MASC5-2-16 7×7	**76.3**	81.4	**90.5**	**88.5**	**94.5**	5.9K

Another characteristic of G-MASC is that it helps maintain a stable orientation relationship amongst its kernels. To demonstrate the benefits of this, we did another experiment on non-hybrid G-MASC (conv-only). The Gabor part in the hybrid MASC kernel is for enhancing rotation. In a cross-validation test on DRIVE, removing it decreases the AUC from 96.9% to 91.6%. The best conv-only model has an AUC of 94.4%. Compared to the standard algorithm, the conv-only model is more adaptive to gradients, but its flexibility makes it vulnerable to overtraining. The objective of PoRE is to encourage every base kernel to move towards the closest niche position (the rotation-aligned position in the kernel space).

Ablation Study

An ablation test was done on the DRIVE data. A smaller set of training patches (4,000 samples) was used in the ablation experiments to show differences more clearly. See Table 2. G score is the geometric mean of Se and Sp.

The merit of G-MASC can be shown when comparing it to the case with neither IMF nor PoRE. The full G-MASC has a clear advantage, as the F1 score shows a 4.6% improvement, and 9.3% gain in averaged sensitivity. The fully ablated model can still produce a good result in its best form but is not as stable as the one with PoRE whose error range is much smaller. On the other hand, the complete version has a better class balance of G equal to 86.3 when the number for the original MASC is 81.0. We realised the value of the label maps as a side product and their benefits. With IMF modules, the new model's sensitivity increased 8.7% compared to the baseline. It was found that, with limited training examples, AUC saturated at around 96.7%. The model without either PoRE or IMF reached this level. However, there were differences in sensitivities (Se) and specificities (Sp), on which PoRE contributes more than the scale and orientation IMF modules do. The scale IMF helps least in the vessel segmentation, where most vessels are of the same scale (Fig. 2)(c).

Table 2. Ablation Study on DRIVE (4,000 samples for 5-fold cross validation)

Methods	F1	Se	Sp	Acc	AUC	G
G-MASC-5-2-16 5×5	79.4 ± 0.1	76.2 ± 0.7	97.7 ± 0.1	95.0 ± 0.1	96.9 ± 0.1	86.3 ± 0.3
w/o IMF, $\lambda_{PoRE} = 0$	74.8 ± 1.2	66.9 ± 2.8	98.3 ± 0.4	94.3 ± 0.3	95.9 ± 0.3	81.0 ± 1.5
$\lambda_{PoRE} = 0$	79.1 ± 0.2	75.6 ± 0.7	97.7 ± 0.2	94.9 ± 0.1	96.7 ± 0.1	86.0 ± 0.4
w/o IMF	79.1 ± 0.5	76.1 ± 0.6	97.6 ± 0.3	94.9 ± 0.2	96.7 ± 0.1	86.2 ± 0.3
w/o Scale-IMF	79.3 ± 0.2	76.1 ± 0.5	97.7 ± 0.1	95.0 ± 0.0	96.8 ± 0.3	86.2 ± 0.3
w/o Orientation-IMF	79.4 ± 0.2	75.6 ± 0.6	97.8 ± 0.1	95.0 ± 0.1	96.8 ± 0.1	86.0 ± 0.3

The Effect of the Hyperparameters

In [8], a larger kernel size was shown to improve performance, but made it harder to maintain consistent rotation kernels. In G-MASC, PoRE as a solution needs to be given a set of n_{ref} reference functions (such as Gabor). The difference between the performance with small and large kernels was not as large for our approach. A smaller kernel is an economic choice. The general POC level is stable, this indicates that the rotation for larger kernels is well controlled. λ_{PoRE} is the penalty weights. The λ_{PoRE} experiment shows that both too small or large values of PoRE are detrimental to the model's ability to detect true inputs (see the Se columns).

Compared to CHASE-DB1, DRIVE has a biased vessel centre and many tiny branches. We trained a model without PoRE on CHASE-DB1, and found the result is very close to the model with PoRE, and the PoC is steady as well. Again, this indicates that for a type of datasets on which a model can achieve good set of rotated kernels with the simpler loss function, the benefit from PoRE may not be significant.

Table 3. Hyperparameter Study on DRIVE (4,000 samples for 5-fold cross validation)

Kernel sizes	F1	Se	Sp	Acc	AUC	G	PoC
$k = 5 \times 5$, $n_{\text{ref}} = 25$	79.5 ± 0.1	75.1 ± 0.4	98.0 ± 0.1	95.1 ± 0.1	96.9 ± 0.1	85.8 ± 0.2	4.6 ± 0.1
$k = 5 \times 5$, $n_{\text{ref}} = 33$	79.4 ± 0.1	76.2 ± 0.7	97.7 ± 0.1	95.0 ± 0.1	96.9 ± 0.1	86.3 ± 0.3	$\mathbf{3.8 \pm 0.1}$
$k = 7 \times 7$, $n_{\text{ref}} = 33$	79.6 ± 0.3	75.6 ± 0.2	97.9 ± 0.1	95.1 ± 0.1	97.0 ± 0.1	86.0 ± 0.1	5.4 ± 0.3
$k = 7 \times 7$, $n_{\text{ref}} = 49$	78.8 ± 0.2	75.0 ± 0.7	97.8 ± 0.1	94.9 ± 0.1	96.6 ± 0.1	85.6 ± 0.4	5.3 ± 0.5
$k = 9 \times 9$, $n_{\text{ref}} = 49$	$\mathbf{80.0 \pm 0.3}$	$\mathbf{76.9 \pm 0.3}$	97.8 ± 0.1	95.1 ± 0.1	97.0 ± 0.1	$\mathbf{86.7 \pm 0.1}$	4.5 ± 0.1
$k = 9 \times 9$, $n_{\text{ref}} = 81$	79.6 ± 0.2	74.9 ± 1.0	$\mathbf{98.1 \pm 0.2}$	$\mathbf{95.2 \pm 0.0}$	$\mathbf{97.1 \pm 0.1}$	85.7 ± 0.5	4.0 ± 0.2
Weights	F1	Se	Sp	Acc	AUC	G	PoC
$\lambda_{\text{PoRE}} = 1$	79.0 ± 0.3	74.8 ± 0.6	97.9 ± 0.1	95.0 ± 0.1	96.8 ± 0.1	85.6 ± 0.3	$\mathbf{2.6 \pm 0.2}$
$\lambda_{\text{PoRE}} = 10^{-1}$	$\mathbf{79.4 \pm 0.1}$	76.2 ± 0.7	97.7 ± 0.1	$\mathbf{95.0 \pm 0.1}$	$\mathbf{96.9 \pm 0.1}$	86.3 ± 0.3	3.8 ± 0.1
$\lambda_{\text{PoRE}} = 10^{-2}$	79.3 ± 0.3	$\mathbf{76.3 \pm 0.6}$	97.7 ± 0.1	95.0 ± 0.1	96.7 ± 0.1	$\mathbf{86.3 \pm 0.3}$	7.5 ± 0.3
$\lambda_{\text{PoRE}} = 10^{-3}$	78.9 ± 0.3	74.7 ± 0.4	$\mathbf{97.9 \pm 0.1}$	94.9 ± 0.1	96.6 ± 0.1	85.5 ± 0.2	11.1 ± 0.6

5 Conclusion

We introduce a model called G-MASC which explicitly creates a set of rotated and reflected kernels which can be used to estimate the orientation and scale of local structures. The approach requires the estimation of far fewer parameters than other methods (such as U-Net), but can achieve comparable performance. The model resolved problems from the original MASC, and achieved good performance on public datasets.

References

1. Chu, X., et al.: Conditional positional encodings for vision transformers. arXiv preprint arXiv:2102.10882 (2021)
2. Cohen, T., Welling, M.: Group equivariant convolutional networks. In: International Conference on Machine Learning, pp. 2990–2999. PMLR (2016)
3. Cohen, T.S., Welling, M.: Steerable CNNs. arXiv preprint arXiv:1612.08498 (2016)
4. Freeman, W.T., et al.: The design and use of steerable filters. IEEE Trans. Pattern Anal. Mach. Intell. **13**(9), 891–906 (1991)
5. Ghosh, R., Gupta, A.K.: Scale steerable filters for locally scale-invariant convolutional neural networks. arXiv preprint arXiv:1906.03861 (2019)
6. Kumar, N., et al.: A multi-organ nucleus segmentation challenge. IEEE Trans. Med. Imaging **39**(5), 1380–1391 (2019)
7. Liu, B., Gu, L., Lu, F.: Unsupervised ensemble strategy for retinal vessel segmentation. In: Shen, D., et al. (eds.) MICCAI 2019. LNCS, vol. 11764, pp. 111–119. Springer, Cham (2019). https://doi.org/10.1007/978-3-030-32239-7_13
8. Liu, Z., Cootes, T.: MASC-units: training oriented filters for segmenting curvilinear structures. In: de Bruijne, M., et al. (eds.) MICCAI 2021. LNCS, vol. 12906, pp. 590–599. Springer, Cham (2021). https://doi.org/10.1007/978-3-030-87231-1_57
9. Liu, Z., Cootes, T., Ballestrem, C.: An end to end system for measuring axon growth. In: Liu, M., et al. (eds.) MLMI 2020. LNCS, vol. 12436, pp. 455–464. Springer, Cham (2020). https://doi.org/10.1007/978-3-030-59861-7_46

10. Luan, S., Chen, C., Zhang, B., Han, J., Liu, J.: Gabor convolutional networks. IEEE Trans. Image Process. **27**(9), 4357–4366 (2018)
11. Mehrotra, R., Namuduri, K.R., Ranganathan, N.: Gabor filter-based edge detection. Pattern Recogn. **25**(12), 1479–1494 (1992)
12. Mroueh, Y., Voinea, S., Poggio, T.: Learning with group invariant features: a kernel perspective. arXiv preprint arXiv:1506.02544 (2015)
13. Owen, C.G., et al.: Measuring retinal vessel tortuosity in 10-year-old children: validation of the computer-assisted image analysis of the retina (CAIAR) program. Investig. Ophthalmol. Vis. Sci. **50**(5), 2004–2010 (2009)
14. Roweis, S.T., Saul, L.K.: Nonlinear dimensionality reduction by locally linear embedding. Science 290(5500), 2323–2326 (2000)
15. Shorten, C., Khoshgoftaar, T.M.: A survey on image data augmentation for deep learning. J. Big Data **6**(1), 1–48 (2019)
16. Staal, J., Abràmoff, M.D., Niemeijer, M., Viergever, M.A., Van Ginneken, B.: Ridge-based vessel segmentation in color images of the retina. IEEE Trans. Med. Imaging **23**(4), 501–509 (2004)
17. Vaswani, A., et al.: Attention is all you need. In: Advances in Neural Information Processing Systems, pp. 5998–6008 (2017)
18. Weiler, M., Hamprecht, F.A., Storath, M.: Learning steerable filters for rotation equivariant CNNs. In: Proceedings of the IEEE Conference on Computer Vision and Pattern Recognition, pp. 849–858 (2018)
19. Worrall, D.E., Garbin, S.J., Turmukhambetov, D., Brostow, G.J.: Harmonic networks: deep translation and rotation equivariance. In: Proceedings of the IEEE Conference on Computer Vision and Pattern Recognition, pp. 5028–5037 (2017)

Stroke Lesion Segmentation from Low-Quality and Few-Shot MRIs via Similarity-Weighted Self-ensembling Framework

Dong Zhang[1,2(✉)], Raymond Confidence[3], and Udunna Anazodo[1,3,4]

[1] Lawson Health Research Institute, London, ON, Canada
zhangdong9612@gmail.com
[2] Department of Electrical and Computer Engineering, University of British Columbia, Vancouver, BC, Canada
[3] Department of Medical Biophysics, Western University, London, ON, Canada
[4] Department of Neurology and Neurosurgery, McGill University, Montreal, QC, Canada

Abstract. Ischemic stroke lesion is one of the prevailing diseases with the highest mortality in low- and middle-income countries. Although deep learning-based segmentation methods have the great potential to improve the medical resource imbalance and reduce stroke risk in these countries, existing segmentation studies are difficult to be deployed in these low-resource settings because they have such high requirements for the data amount (plenty-shot) and quality (high-field and high resolution) that are usually unavailable in these countries. In this paper, we propose a **SimIlarity-weiGhed self-eNsembling** framework (SIGN) to segment stroke lesions from low-quality and few-shot MRI data by leveraging publicly available glioma data. To overcome the low-quality challenge, a novel Identify-to-Discern Network employs attention mechanisms to identify lesions from a global perspective and progressively refine the coarse prediction via focusing on the ambiguous regions. To overcome the few-shot challenge, a new Soft Distribution-aware Updating strategy trains the Identify-to-Discern Network in the direction beneficial to tumor segmentation via respective optimizing schemes and adaptive similarity evaluation on glioma and stroke data. The experiment indicates our method outperforms existing few-shot methods and achieves the Dice of 76.84% after training with 14-case low-quality stroke lesion data, illustrating the effectiveness of our method and the potential to be deployed in low resource settings. Code is available in: https://github.com/MINDLAB1/SIGN.

Keywords: Few-shot · Stroke lesion segmentation · Low-quality · MRI

1 Introduction

Ischemic stroke is the most common cerebrovascular disease, and one of the most common causes of death and disability worldwide [1], especially in Low- and Middle-Income Countries (LMICs). In ischemic stroke, an obstruction of the cerebral blood supply causes tissue hypoxia (hyper-perfusion) and advancing tissue death over the next hours [2]. Over 13.7 million each year worldwide suffer from ischemic stroke and 5.8

© The Author(s), under exclusive license to Springer Nature Switzerland AG 2022
L. Wang et al. (Eds.): MICCAI 2022, LNCS 13435, pp. 87–96, 2022.
https://doi.org/10.1007/978-3-031-16443-9_9

million a year die as a consequence [3]. The majority (85%) of these stroke deaths occur in LMICs, largely due to several overlapping factors, including lack of awareness which leads to patients presenting days after the first signs of stroke, poor diagnostic capacity, including imaging, and lack of therapies to treat stroke [4]. Disability-adjusted life years lost in LMICs is seven times those lost in high-income countries [4]. Especially in the Sub-Saharan Africa area where poor access to high-value diagnostic imaging (i.e., high image quality and trained neuroradiologists) contributes to the region bearing the highest indicators of stroke worldwide, with an estimated age-standardized stroke incidence rate of 316 per 100,000, the prevalence rate of 1460 per 100,000 population, 1-month mortality of up to 40% and a 3-year mortality rate of 84% since 2013 [5–8].

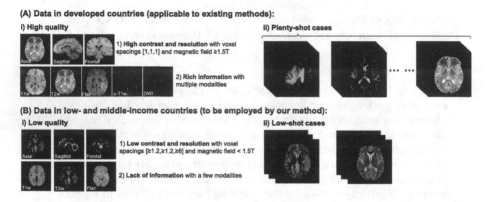

Fig. 1. The data comparison between high-resourced (A) and low-resourced (B) settings. (A) 3T MRI with T1-weighted (T1w), T2-weighted (T2w), and enhanced T1-weighted (e-T1w) images of individuals from SISS and BraTS databases collected largely from Europe and North America. (B) T1-weighted (T1w), T2-weighted (T2w), and T2-FLAIR (Flair) images of a representative scan acquired from a 0.30T MRI in Nigeria.

Deep learning-based automatic stroke lesion segmentation has the great potential to improve the medical resource imbalance, thereby significantly reducing the ischemic stroke mortality rate in LMICs. Automated stroke lesion segmentation will allow clinics in LMICs to make rapid diagnoses on magnetic resonance imaging (MRI). More specifically, the diagnosis of stroke lesions needs to promptly identify the lesion, quantify the lesion volume, and determine the features (i.e., core, penumbra) to effectively characterize the evolution of the stroke and speedily decide the course of treatment to prevent further brain tissue injury/damage. While rapid diagnosis can be made in well-resourced settings, in LMICs, the lack of trained neuroradiologists, combined with the late presentation of patients to the clinic, shortens the effective time window to make a life-saving diagnosis. Compared with manual delineation taking several minutes per case, if deep-learning-based automatic segmentation can identify the stroke lesion's contour to quantify the volume and predict the evolution in a few seconds, treatment can be planned rapidly to save the brain.

However, no current automated stroke lesion segmentation method is designed to include or specifically handle MRI data challenges in low-resourced settings characteristic of LMICs. The large gap between data produced in LMICs and developed countries makes existing segmentation methods exclusively designed using MRI data from developed countries inapplicable to LMICs [9]. As shown in Fig. 1, developed countries produce high-quality data with high resolution (voxel spacings [1 mm, 1 mm, 1 mm]) and high image contrast (magnetic field $\geq 1.5T$) for clinical diagnosis, while LMICs facing low-resources produce quite low-quality data with low resolution (voxel spacings [≥ 1.2 mm, ≥ 1.2 mm, ≥ 6.0 mm]) and low image contrast (with magnetic field $<1.5T$). Developed countries are equipped with high-technical imaging devices that present stroke lesions clearly and allow existing segmentation methods to identify stroke lesions easily. As a comparison, in LMICs, imaging devices are usually in short supply or antiquated, which leads to the noisier, low-contrast, and low-resolution images. Hence, it can be challenging for existing methods to accurately distinguish stroke lesions that are blurred with surrounding normal tissues. In addition, well-resourced developed countries have the capability to produce plenty-shot, well-annotated data [10]. At the same time, LMICs are burdened with a prohibitive shortage of experienced radiologists, which significantly contributes to the lack of annotated imaging data for deep learning applications from LMICs [11], especially from resource-starved areas such as Sub-Saharan Africa. Without enough annotated data, existing stroke lesion segmentation methods tend to overfit, thus resulting in poor performance.

In this paper, we propose a SimIlarity-weiGhed self-eNsembling framework (SIGN) to segment stroke lesions from low-quality and few-shot MRIs by leveraging publicly available glioma data. SIGN mainly builds a transfer learning from the plenty-shot glioma data to the few-shot stroke data to address the stroke lesion segmentation. To classify every lesion voxel correctly and overcome the low-quality challenge, an Identify-to-Discern Network (IDN) is designed to mimic radiologists' process that identifies lesions first and then focuses on ambiguous areas to refine the coarse prediction progressively. IDN first identifies lesions roughly from a global perspective by encoding the long-range anatomical dependencies with channel and spatial attention. Then, IDN refines the identification and removes false-predicted areas by performing multi-scale anatomy exploration based on foreground- and background-attention in every decoding layer. To achieve an effective transferring and overcome the few-shot challenge, a Soft Distribution-aware Updating strategy (SDU) is proposed to train the above IDN. The SDU determines the direction beneficial to tumor segmentation via respective schemes on glioma and stroke data. Particularly, the SDU firstly employs stroke data to give a rough optimizing direction and then employs multiple glioma data to specify the detailed direction. To avoid a negative transfer and its resulting optimizing divergence, SDU evaluates the distribution similarity of every glioma case to stroke lesions and thus adaptively determines their optimizing weights. In the end, the SDU updates the IDN with the overall direction through exponential moving average (EMA) to keep the training stability. Our SIGN was extensively evaluated on the dataset simulated with ISLES Challenge 2015 [12] and BraTS Challenge 2020 [10]. The results demonstrate that our SIGN segments stroke lesions accurately from the low-quality

MRIs with few-shot training cases and outperforms other state-of-the-art few-shot segmentation methods.

2   Method

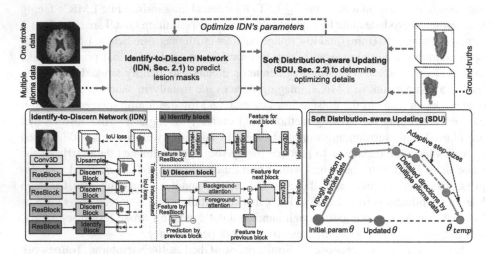

Fig. 2. SIGN employs the Soft Distribution-aware Updating strategy (SDU) to train the Identify-to-Discern Network (IDN), where IDN predicts segmentation for the inputted stroke data and glioma data, SDU determines the optimizing direction maximally beneficial to tumor segmentation based on the segmentation and ground-truths.

SIGN deploys the Soft Distribution-aware Updating strategy (SDU) to train the Identify-to-Discern Network (IDN) by transferring learning from the glioma to the stroke data, thereby accurately segmenting stroke lesions from low-quality MRIs with few-shot training data. As shown in Fig. 2, in each iteration during training, IDN predicts lesion segmentation for the low-quality inputs, i.e., one stroke data and multiple glioma data. SDU determines the optimizing direction maximally beneficial to tumor segmentation based on the segmentation and ground truths. SIGN updates the IDN's parameters based on the above optimizing direction at the end of this iteration. During testing, the well-trained IDN is able to segment stroke lesions independently.

2.1   Identify-to-Discern Network (IDN)

Identify-to-Discern Network (IDN) encodes the input brain images to identify lesions from a global perspective and then progressively refines the identified lesions from-coarse-to-fine during decoding, thereby predicting lesions accurately from the low-quality input. As shown in Fig. 2, IDN mainly employs the proposed identify block and discern block to determine and refine the lesion segmentation after deploying Conv3D and ResBlocks to encode brain anatomy.

Identify block collects high-level anatomical features from the deepest ResBlock, generates an initial lesion prediction, and conveys the features and prediction to the discern block. As Fig. 2 illustrates, it consists of a channel-wise self-attention and spatial-wise self-attention. These two different attentions facilitate the identify block to capture lesions' anatomical dependencies in channel consistency and spatial context [13], thereby enabling the identify block to determine the lesion's location and predict an initial mask from the global level.

Discern block takes the current ResBlock features and last-discern/identify block features and prediction as input and outputs the refined features and a finer prediction. As Fig. 2 demonstrates, the discern block firstly employs the prediction and the current ResBlock features to result in foreground-attentive features and background-attentive features. The discern block then builds two Context Exploration Blocks [14] to discover the false prediction in the foreground-attentive and background-attentive features. After this, the discern block uses subtraction and addition operations to suppress the ambiguous background (i.e., false-positive normal tissues) and augment the missing foregrounds (i.e., false-negative lesions). Finally, the discern block outputs a more accurate lesion prediction by 3D convolution on the refined features.

Multi-layered supervision is employed to constrain the identify block and the discern block to output a more accurate lesion prediction and a purer feature representation. Particularly, the loss function in IDN is formulated as:

$$\mathcal{L} = \sum_{i=1}^{5} 2^{i-1} * I(\hat{y}^i, T(y, \hat{y}^i)) \tag{1}$$

where $\hat{y}^i$ is the lesion prediction by the *i-th* decode layer, and $I(\hat{y}, y)$ is the IoU loss. $T(y, \hat{y}^i)$ is a tri-linear interpolation function that re-scales the ground-truth y to the size of $\hat{y}^i$.

2.2 Soft Distribution-aware Updating (SDU)

Soft Distribution-aware Updating (SDU) utilizes the plenty-shot glioma data to anxiliarily determine optimizing directions for the few-shot stroke data, thereby facilitating the few-shot stroke lesion segmentation. Mainly, in each iteration, the SDU employs one stroke data to give a rough optimization direction via a large step size for the IDN to ensure that the IDN will be optimized to the direction beneficial stroke lesion segmentation. Then, based on the above rough direction, SDU employs multiple glioma data to specify the detailed optimization directions, fully utilizing the plenty-shot glioma to assist the optimizing exploration. Ultimately, the SDU updates the IDN with the above overall optimizing direction through exponential moving average (EMA) to keep the training stability. Therefore, the entire process can be formulated as follows:

$$\begin{cases} \theta' = \theta - \alpha * \nabla[\mathcal{L}((x^s, \theta), y^s] \\ \theta' = \mathbb{F}_n(...\mathbb{F}_2(\mathbb{F}_1(\theta', x_1^g, y_1^g), x_2^g, y_2^g), x_n^g, y_n^g) \\ \theta = \epsilon * \theta + (1 - \epsilon) * \theta' \end{cases} \tag{2}$$

where θ and θ' are the IDN's parameters and temporal parameters. $\{(x_i^s, y_i^s)\}_{i=1}^{N}$ and $\{(x_i^g, y_i^g)\}_{i=1}^{M}$ are stroke lesion data and glioma data with N and M cases, where M is much larger than N. α indicates the step-size to update IDN's parameters on stroke data. n means the number of glioma data employed in each iteration, which is set as $n = \frac{M}{N}$ in this paper. ϵ is a hyper-parameter to control the updating speed of θ, which is set as a time-dependent Gaussian warming up function in this paper (same as [15]). $\mathbb{F}$ is a weight-adaptive gradient-decreasing function that updates temporal parameters on glioma data, namely:

$$\mathbb{F} = \theta' - \beta * \nabla[\mathcal{D}(x, \theta, \theta') * \mathcal{L}((x, \theta'), y)] \tag{3}$$

where $\mathcal{D}$ is a weight determined by the similar distance between glioma data and stroke data, β is the step-size that much small than α to update temporal parameters on glioma data. As Eq. (4) illustrates, $\mathcal{D}$ is 1 if the distance d is less than a threshold T, otherwise is a time-dependent function ϵ' (same as ϵ). d is to measure the distance between the predictions by θ and θ' for the same glioma data. Namely, $d = \sum \mathbb{I}(|\hat{y}_x^\theta - \hat{y}_x^{\theta'}| > \mu)$, where $\mathbb{I}(\cdot)$ is an indicator function and μ is a tolerance. And T is a distance threshold determined by the mean distance in last epoch.

$$\mathcal{D}(x, \theta, \theta') = \begin{cases} 1, & \text{if } d < T \\ \epsilon' & \text{otherwise} \end{cases} \tag{4}$$

3   Experiment

Dataset Acquirement. We decreased the quality of the SISS dataset in Ischemic Stroke Lesion Segmentation (ISLES) Challenge 2015, where the segmentation was conducted completely manual by an experienced MD [12], to simulate the general data quality collected from low-resourced settings in LMICs. In detail, we employed three steps to make the SISS data have a similar quality as the LMIC data: 1) exclusion of SISS DWI and only reserved three staple modalities (T1-w, T2-w, and T2-Flair) to reduce imaging information considering some LMICs cannot perform high-quality DWI; 2) resliced every image from $1 \times 1 \times 1$ mm into $1 \times 1 \times 4$ mm [16] to reduce spatial consistency and simulate larger spatial resolution and slice thickness based on our data collected from LMICs; 3) smoothed every image with Gaussian filters and added Gaussian noises to reduce local contrast [17] and simulate the low-field of LMIC data. We leveraged the Brain Tumor Segmentation (BraTS) Challenge 2020 [10] dataset as our glioma dataset and decreased its quality as described above. We also merged the three foreground labels into one on the BraTS to keep consistency with the SISS, which facilitates our method to focus on the difference between normal and abnormal (lesion or tumor) brain tissues instead of the interior structures of abnormal tissues.

Implementation Details. All data was divided into 50%/25%/25% for training, validation, and testing. All the MRI was cropped into $48 \times 192 \times 192$ by cutting and padding the surrounding air area. Batch normalization and ReLU were employed after convolution layers. Hyper-parameters were tuned to obtain the best results. Adam optimizer (tuned from SGD and Adam) was deployed. The learning rates α and β started with

1e−4 (tuned from [1e−3, 1e−4, 1e−5]) and 5e−4 (tuned from [1e−4, 5e−4, 5e−5]), decayed by 20 every 10 epochs. The batch size was set as 2. SIGN was implemented with PyTorch 1.7.1[1] on 1 NVIDIA Tesla V100 and 24 Intel Xeon CPUs under Ubuntu 16.04.

Evaluation Metrics. Three matrices were deployed to evaluate the performance: 1) Dice, measuring the overlap rate between the predicted lesion voxels and the ground truth. 2) Accuracy measures the rate between the correctly-detected lesion voxels out of the predicted voxels. 3) Hausdorff Distance (HD), evaluating the respective distance between the predicted lesion surface and the ground-truth surface.

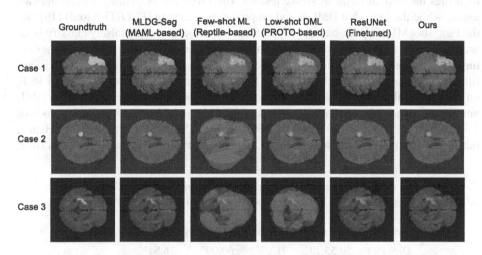

Fig. 3. Our method obtains the best performance based on the visualization of stroke lesion segmentation by different methods, where red and blue colors show the ground truth and predictions. (Color figure online)

Accurate Stroke Lesion Segmentation. As illustrated in Fig. 3 and Table 1, our method segments stroke lesions accurately from the few-shot low-quality data and achieves an overall Dice of 76.84%. Particularly, as Fig. 3 displayed, the visualized segmentation results in three cases from different transverse slices demonstrate our SIGN identifies the stroke lesions successfully and segments them accurately despite the stroke lesions located in various areas of the brain with various shapes and presents a similar contrast in some cases with surrounding normal tissues. As shown in Table 1, quantitatively, our method achieves the Dice of 76.84%, the Accuracy of 77.30%, and the HD of 0.98 voxels, which indicates that our method is capable of low-quality and few-shot stroke lesion segmentation.

Outperformance Compared with Other Few-shot Methods. As shown in Fig. 3 and Table 1, our method achieves the best performance in this low-quality stroke lesion segmentation task compared with three state-of-art few-shot medical image segmentation

---

[1] https://pytorch.org/.

Table 1. Comparison between our method and various methods.

Methods	Dice [%]	Accuracy [%]	HD [voxels]
MLDG-Seg (MAML-based)	66.36 ± 6.29	65.23 ± 6.90	1.49 ± 0.11
Few-shot DML (PROTO-based)	29.42 ± 4.07	19.32 ± 1.91	2.66 ± 0.70
Few-shot ML (Reptile-based)	21.63 ± 1.10	12.33 ± 0.38	5.12 ± 0.53
SIGN (ours)	$\mathbf{76.84 \pm 1.31}$	$\mathbf{77.30 \pm 2.63}$	$\mathbf{0.98 \pm 0.19}$

methods. Particularly, in Fig. 3, the MLDG-Seg (MAML-based) [18] false-positively identifies the normal tissue as stroke lesions when they are presenting similar appearances, while the Few-shot DML based on prototype-learning (PROTO-based) [19] and the Few-shot ML (Reptile-based) [20] perform worse because they falsely classify large area of normal tissues as stroke lesion in these three cases. In comparison, our method has no false-positive prediction among the three cases. Such performance is also quantitatively evaluated in Table 1, where the MLDG-Seg obtains a Dice value of 66.36%, an Accuracy value of 65.23%, and a HD value of 1.49 voxels. While the few-shot DML and the Few-shot ML obtain Dice values of 29.42% and 21.63%, Accuracy values of 19.32% and 12.33%, and HD values of 2.66 voxels and 5.12 voxels, respectively. Hence, our method outperforms these state-of-art few-shot methods by at least 10.00% in Dice.

Table 2. Control experiments for each components. The $*$ denotes $p < 0.05$ and $**$ denotes $p < 0.01$ for the paired t-test of the Dice values between the baseline approach and other components.

Methods	Baseline	IDN only	SDU only	IDN+SDU (SIGN)
Dice [%]	55.20	71.69^*	60.09^*	$\mathbf{76.84}^{**}$
Accuracy [%]	52.44	48.22	49.87	**77.30**
HD [voxels]	2.00	**0.93**	2.08	0.98

Analysis of Control Experiments. As shown in Table 2, our proposed IDN and SDU are effective in increasing the segmentation performance to varying degrees. Particularly, the control experiments take the ResUNet [21] as the baseline method, where the ResUNet is pretrained with glioma data and then finetuned with stroke lesion data. As shown in Fig. 3, this finetuned ResUNet identifies stroke lesions successfully but falsely classifies some normal tissues as the lesion. Quantitatively, as shown in Fig. 2, the finetuned ResUNet achieves Dice, Accuracy and HD values of 55.20%, 52.44%, and 2.00 voxels, respectively. By pretraining and fine-tuning the IDN in the same way (i.e., IDN only), the Dice and HD values are improved/decreased to 71.69% and 0.93 voxels. By optimizing the ResUNet with Our SDU (i.e., SDU only), the Dice value is improved to 60.09%. By integrating the IDN and the SDU, our SIGN achieves the best performance which is 76.84% Dice and 77.30%. The statistical tests also indicate the improvement by each component is significant. Overall, the control experiments demonstrate that our proposed components are adequate for this low-quality few-shot stroke lesion segmentation task.

4 Conclusion

This paper presents a novel similarity-weighed self-ensembling framework (SIGN) for few-shot and low-quality stroke lesion segmentation. Our SIGN consists of two newly-proposed components. The IDN mimics the radiologists first identifying lesions and then focusing on the ambiguous areas to overcome the low-quality challenge. The SDU selects the glioma data with similar distributions as stroke lesion data to avoid a negative transfer. The experimental result illustrates the outperformance of the SIGN to existing few-shot medical object segmentation methods in this low-quality and few-shot stroke lesion segmentation task, indicating our SIGN has great potential as an effective clinical tool to assist rapid stroke diagnosis in low-resourced settings, predominating low- and middle-income countries. Our further work will improve the lesion segmentation accuracy, thereby further increasing clinical impact.

Acknowledgement. The authors are grateful for the single-subject data contribution from the Stroke Investigative Research & Educational Network (SIREN) to illustrate the lower quality MRI from low-resourced settings (Fig. 1. (B)). The authors would particularly like to thank Prof. Mayowa Owolabi, SIREN co-Principal Investigator at the College of the Medicine University of Ibadan, Prof. Bruce Ovbiagle, SIREN co-Principal Investigator at the University of California San Francisco, and Dr. Godwin Ogbole, College of Medicine, University of Ibadan, Nigeria.

References

1. Organization, W.H., et al.: Cause-specific mortality, estimates for 2000–2012 (2012). https://www.who.int/healthinfo/global_burden_disease/estimates/en/index1.html
2. Randolph, S.A.: Ischemic stroke. Workplace Heal. Saf. **64**(9), 444–444 (2016)
3. Phipps, M.S., Cronin, C.A.: Management of acute ischemic stroke. BMJ. **368**, 1693 (2020)
4. Sarfo, F.S., et al.: Stroke among young west Africans: evidence from the siren (stroke investigative research and educational network) large multisite case-control study. Stroke **49**(5), 1116–1122 (2018)
5. Owolabi, M.O., et al.: Dominant modifiable risk factors for stroke in Ghana and Nigeria (siren): a case-control study. Lancet Glob. Health **6**(4), e436–e446 (2018)
6. Sarfo, F.S., et al.: Unraveling the risk factors for spontaneous intracerebral hemorrhage among west Africans. Neurology **94**(10), e998–e1012 (2020)
7. Kayima, J., Wanyenze, R.K., Katamba, A., Leontsini, E., Nuwaha, F.: Hypertension awareness, treatment and control in Africa: a systematic review. BMC Cardiovasc. Disorders **13**(1), 1–11 (2013)
8. Ataklte, F., et al.: Burden of undiagnosed hypertension in sub-Saharan Africa: a systematic review and meta-analysis. Hypertension **65**(2), 291–298 (2015)
9. Anazodo, U.C., Adewole, M., Dako, F.: AI for population and global health in radiology (2022)
10. Menze, B.H., et al.: The multimodal brain tumor image segmentation benchmark (brats). IEEE Trans. Med. Imaging **34**(10), 1993–2024 (2014)
11. Li, J., et al.: Association between white matter hyperintensities and stroke in a west African patient population: Evidence from the stroke investigative research and educational network study. NeuroImage **215**, 116789 (2020)
12. Maier, O., et al.: Isles 2015-a public evaluation benchmark for ischemic stroke lesion segmentation from multispectral MRI. Med. Image Anal. **35**, 250–269 (2017)

13. Xu, C., Gao, Z., Zhang, H., Li, S., de Albuquerque, V.H.C.: Video salient object detection using dual-stream spatiotemporal attention. Appl. Soft Comput. **108**, 107433 (2021)
14. Mei, H., Ji, G.P., Wei, Z., Yang, X., Wei, X., Fan, D.P.: Camouflaged object segmentation with distraction mining. In: Proceedings of the IEEE/CVF Conference on Computer Vision and Pattern Recognition, pp. 8772–8781 (2021)
15. Yu, L., Wang, S., Li, X., Fu, C.-W., Heng, P.-A.: Uncertainty-aware self-ensembling model for semi-supervised 3d left atrium segmentation. In: Shen, D., et al. (eds.) MICCAI 2019. LNCS, vol. 11765, pp. 605–613. Springer, Cham (2019). https://doi.org/10.1007/978-3-030-32245-8_67
16. Yan, C.G., Wang, X.D., Zuo, X.N., Zang, Y.F.: DPABI: data processing & analysis for (resting-state) brain imaging. Neuroinformatics **14**(3), 339–351 (2016)
17. Zeng, Y., et al.: Magnetic resonance image denoising algorithm based on cartoon, texture, and residual parts. In: Computational and Mathematical Methods in Medicine 2020 (2020)
18. Khandelwal, P., Yushkevich, P.: Domain generalizer: a few-shot meta learning framework for domain generalization in medical imaging. In: Albarqouni, S., et al. (eds.) DART/DCL -2020. LNCS, vol. 12444, pp. 73–84. Springer, Cham (2020). https://doi.org/10.1007/978-3-030-60548-3_8
19. Cui, H., Wei, D., Ma, K., Gu, S., Zheng, Y.: A unified framework for generalized low-shot medical image segmentation with scarce data. IEEE Trans. Med. Imaging **40**(10), 2656–2671 (2020)
20. Dawoud, Y., Hornauer, J., Carneiro, G., Belagiannis, V.: Few-shot microscopy image cell segmentation. In: Dong, Y., Ifrim, G., Mladenić, D., Saunders, C., Van Hoecke, S. (eds.) ECML PKDD 2020. LNCS (LNAI), vol. 12461, pp. 139–154. Springer, Cham (2021). https://doi.org/10.1007/978-3-030-67670-4_9
21. Diakogiannis, F.I., Waldner, F., Caccetta, P., Wu, C.: ResuNet-a: a deep learning framework for semantic segmentation of remotely sensed data. ISPRS J. Photogram. Rem. Sens. **162**, 94–114 (2020)

Edge-Oriented Point-Cloud Transformer for 3D Intracranial Aneurysm Segmentation

Yifan Liu, Jie Liu, and Yixuan Yuan[✉]

Department of Electrical Engineering, City University of Hong Kong, Kowloon, Hong Kong, China
yifliu3-c@my.cityu.edu.hk, yxyuan.ee@cityu.edu.hk

Abstract. Point-based 3D intracranial aneurysm segmentation is fundamental for automatic aneurysm diagnosis. Though impressive performances, existing point-based 3D segmentation frameworks still perform poorly around the edge between vessels and aneurysms, which is extremely harmful for the clipping surgery process. To address the issue, we propose an Edge-oriented Point-cloud Transformer Network (EPT-Net) to produce precise segmentation predictions. The framework consists of three paradigms, i.e., dual stream transformer (DST), outer-edge context dissimilation (OCD) and inner-edge hard-sample excavation (IHE). In DST, a dual stream transformer is proposed to jointly optimize the semantics stream and the edge stream, where the latter imposes more supervision around the edge and help the semantics stream produce sharper boundaries. In OCD, aiming to refine features outside the edge, an edge-separation graph is constructed where connections across the edge are prohibited, thereby dissimilating contexts of points belonging to different categories. Upon that, graph convolution is performed to refine the confusing features via information exchange with dissimilated contexts. In IHE, to further refine features inside the edge, triplets (i.e. anchor, positive and negative) are built up around the edge, and contrastive learning is employed. Differently from previous contrastive methods of point clouds, we only select points nearby the edge as hard-negatives, providing informative clues for discriminative feature learning. Extensive experiments on the 3D intracranial aneurysm dataset IntrA demonstrate the superiority of our EPT-Net compared with state-of-the-art methods. Code is available at https://github.com/CityU-AIM-Group/EPT.

Keywords: Intracranial Aneurysm segmentation · 3D point cloud segmentation · Graph convolution

---

Supplementary Information The online version contains supplementary material available at https://doi.org/10.1007/978-3-031-16443-9_10.

1   Introduction

Rupturing intracranial aneurysm is the most frequent cause of subarachnoid hemorrhage, associated with high fatality and morbidity rates [3,9]. In order to avoid the sinister outcome of aneurysmal subarachnoid hemorrhage, many unruptured intracranial aneurysms are treated prophylactically before they rupture, e.g., aneurysm clipping. This surgery procedure heavily relies on the accurate segmentation of aneurysms [1,17], where the edges, i.e. the clipping lines between aneurysms and vessels, can be extracted.

A variety of methods [6,8,14] based on medical images are proposed to address the aneurysm segmentation and achieve satisfactory performance. However, these methods are confined to 2D/3D convolutional networks based on 2D CT/MRI images, which require high computational cost and lack complex topology information. In contrast, 3D segmentation techniques based on point clouds can benefit the surgery significantly with less computational cost, and point clouds sampled from 3D surface models can precisely describe the complicated biological structures of vessels and aneurysms.

Existing point-based 3D segmentation frameworks are pioneered by Point-Net [12], which achieves permutation invariance by utilizing cascaded MLPs and symmetric pooling functions. To further capture local patterns, PointNet++ [13] introduces a hierarchical structure by applying PointNet recursively on a nested partitioning of the input set. After that, PointConv [15] is proposed to apply convolutions over irregular points, where convolution kernels are treated as nonlinear functions of local coordinates. RandLA-Net [4] advocates to replace farthest point sampling with random sampling, and based on that, a local feature aggregation module is elaborated to enlarge the receptive field. Recently, inspired by the powerful modeling capacity of transformer in 2D vision [2], Point Transformer [19] is customized for point-based 3D segmentation. Though achieving impressive performances, these methods are devised for urban and indoor scenarios, omitting the biological properties of the aneurysm segmentation task, e.g., complex typology, thus directly applying these methods to the aneurysm segmentation may lead to biased performance. More recently, 3DMedPT [18] is proposed for aneurysm segmentation to model complex biological structures via tailored attention and graph reasoning, achieving remarkable performance compared with general methods. However, 3DMedPT can still not perform well around the edge between vessels and aneurysms due to the less supervision and ambiguous features, where is extremely harmful for the clipping surgery process.

To address this challenge, we propose a novel point-based 3D segmentation framework, termed Edge-oriented Point-cloud Transformer Network (EPT-Net), for 3D intracranial aneurysm segmentation. Aiming to obtain more precise segmentation results around the edge, we propose three novel paradigms, i.e., *Dual Stream Transformer* (DST), *Outer-edge Context Dissimilation* (OCD) and *Inner-edge Hard-sample Excavation* (IHE). The main contributions are summarized as: (1) We propose DST to jointly optimize the semantics and the edge streams, which are composed of stacked transformer blocks. Complementary to the semantic segmentation, the edge detection can impose more supervision

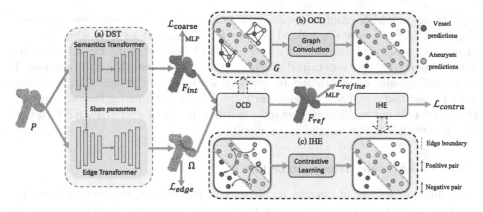

Fig. 1. Overview of EPT-Net framework. (a) Dual Stream Transformer. (b) Outer-edge Context Dissimilation. (c) Inner-edge Hard-sample Excavation. During training period, three paradigms work together to generate refined segmentation results. In the testing stage, IHE is removed and output from DST and OCD is used for evaluation.

around the edge, and help the semantic stream produce sharper boundaries. (2) To refine ambiguous features caused by similar contexts around the edge, we propose OCD to construct an edge-separation graph with nodes being non-edge points nearby the edge and connections across the edge being prohibited, and perform graph convolution to propagate the information. By doing so, contexts of points belonging to different categories can be dissimilated, and ambiguous features outside the edge can be refined with dissimilated contexts. (3) To further rectify confusing features inside the edge, IHE is proposed to build up triplets (i.e., anchor, positive and negative), where negatives are specifically sampled from nearby the edge, and contrastive learning is used to regularize the feature space. Compared with previous contrastive methods of point clouds [5,16] that equally treat unpaired points as negatives, we specifically sample points around the edge as hard-negatives, which are closer to the decision boundary and thus can provide more informative clues for discriminative feature learning. (4) Extensive experiments on the 3D intracranial aneurysm dataset IntrA [17] demonstrate the superiority of our EPT-Net compared with state-of-the-art methods.

2 Approach

As illustrated in Fig. 1, we present EPT-Net for the point-based 3D aneurysm segmentation task. Formally, we denote a point cloud as $P \in \mathcal{R}^{N \times C}$ of N points with C-dimensional features composed of 3-dimensional coordinates and other attributes like normals. First in DST, the input point cloud P is passed through the dual stream transformer to generate intermediate semantic feature $F_{int} \in \mathcal{R}^{N \times D}$ and edge $\Omega \in \mathcal{R}^{N_e \times 3}$, where N_e refers to the number of edge points. Then in OCD, an edge-separation graph G is constructed outside edge Ω, upon which graph convolution is performed to produce refined feature $F_{ref} \in \mathcal{R}^{N \times D}$. Finally the refined feature F_{ref} is passed to IHE where points inside and nearby Ω are treated as anchors and positive/hard negative candidates respectively. With obtained triplets, a contrastive loss $\mathcal{L}_{contra}$ is computed to encourage anchors attracted by positive samples and repelled by negative ones in the feature space. In summary, by jointly optimizing two streams in DST and refining ambiguous semantic features in OCD and IHE, EPT-Net is able to produce better segmentation results.

2.1   Dual Stream Transformer (DST)

Segmentation results of current point-based segmentation frameworks tend to be poor around the edge, since edge points only constitute a small portion of the whole segment and possess less supervision during training. To remedy the defect, we propose to simultaneously conduct the edge detection as an auxiliary task, which can put more focus on edge points and help the semantic network produce sharper boundaries. Inspired by that, we design a dual stream transformer (DST) as shown in Fig. 1(a). DST consists of semantics and edge streams responsible for segmentation and edge detection respectively, and each stream utilizes the cascaded transformer blocks.

Semantics Stream. Input points X is passed to the semantics stream and produce the intermediate feature F_{int}. The semantics stream adopts the U-net style encoder-decoder architecture, and the encoder and decoder both consist of five layers with each layer containing one downsampling/upsampling layer and one transformer blocks proposed by [19]. F_{int} is also passed to MLP layers to generate coarse predictions $Y_{int} \in \mathcal{R}^{N \times K}$, supervised by cross entropy loss $\mathcal{L}_{coarse} = -\frac{1}{N} \sum_{i=1}^{N} \sum_{k=1}^{K} Y_{gt}^i(k) log Y_{int}^i(k)$ with semantics labels $Y_{gt} \in \mathcal{R}^{N \times K}$, where K is the class number.

Edge Stream. Same as the semantics stream, the edge stream also adopts the cascaded transformer blocks design. The encoder of the edge stream is shared with the semantics stream to jointly shape the feature space, and the decoder of the edge stream outputs the edge distribution $D \in \mathcal{R}^{N \times 2}$, supervised by the cross entropy loss $\mathcal{L}_{edge} = -\frac{1}{N} \sum_{i=1}^{N} \sum_{k=1}^{2} D_{gt}^i(k) log D_i(k)$ with edge labels $D^{gt} \in \mathcal{R}^{N \times 2}$. Note that D^{gt} is extracted from the semantic labels Y^{gt} by an intuitive principle: for a specific point, if its 4 neighbors have different semantic categories, we label this point as the edge point. With predicted edge distribution D, the edge set Ω containing predicted edge points can be extracted, formulated as $\Omega = \{P_i | argmax_k D_i(k) = 2, D_i \in D\}$, where i and k refer to the i-th point and k-th category respectively.

2.2   Outer-edge Context Dissimilation (OCD)

Due to the ambiguous features generated from similar contexts, points around the edge are easily misclassified, which is harmful for the surgery process. To rectify these confusing features, OCD is proposed to construct an edge-separation

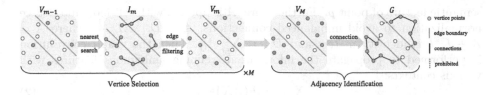

Fig. 2. Illustration of ES-graph construction procedure.

graph (ES-Graph). Nodes of ES-Graph are points outside the edge, selected in an iterative manner to contain more candidates, offering ambiguous features with more supportive neighbors. Connections of ES-Graph across the edge are prohibited, thereby providing points of diverse categories with diverse contexts. Then graph convolution is performed to propagate information, refining the ambiguous features with dissimilated contexts, as shown in Fig. 1(b).

ES-Graph Vertice Selection. The vertice set V of ES-Graph $G = (V, E)$ is supposed to contain points around the edge Ω, except edge points in Ω, since points inside Ω are indistinguishable without annotations. To achieve this, we iteratively use k-nearest search based on the geodesic metric, which represents the length of shortest path between points on a manifold surface.

We first initialize the vertice set V_0 as Ω. Then in step $m = 1, 2, ..., M$, as shown in Fig. 2, an intermediate neighbor set I_m is obtained by nearest search: $I_m = \{p_i | p_i \in \mathcal{N}_K(p_j), p_j \in V_{m-1}\}$, where $\mathcal{N}_K(p_j)$ denotes the set of K nearest points of point p_j including the point p_j itself. V_m is obtained from I_m by the edge filtering process: $V_m = I_m - \Omega$. By doing so iteratively, the updated V_m contains more points outside the edge prepared for feature rectification, and the final vertice set V is obtained via $V = V_M$.

ES-Graph Adjacency Identification. We further identify the connections of elements in V by the adjacency matrix A, aiming to provide points belonging to different categories with dissimilar contexts. Considering vessel and aneurysm points are separated by the edge, a reasonable strategy is to prohibit connections across the edge, i.e., connect points only to points belonging to the same category. Based on this rule, A_{ij} is 1 if point p_i and p_j satisfy the two conditions below, otherwise is 0:

$$p_j \in \mathcal{N}_K(p_i), \ d(p_i, p_j) < \min_{p_m \in \Omega} d(p_i, p_m), \tag{1}$$

where $d(p_i, p_j)$ denotes the geodesic distance between point p_i and p_j. The first term $p_j \in \mathcal{N}_K(p_i)$ provides p_i connections with adjacent points. The second term $d(p_i, p_j) < \min d(p_i, p_m)$ constrains that the geodesic distance of p_i with adjacent points can not exceed the minimum distance of p_i with edge points in Ω. The two terms together guarantee connections across the edge are prohibited, and points of different categories possess diverse contexts.

ES-Graph Feature Rectification. Corresponding features of points in the vertice set V are obtained by $X = \{f_{p_i} \in F_{int} | p_i \in V\}$, where f_{p_i} is the associate semantic feature of point p_i. Furthermore, A is applied with self-loop and degree normalization to avoid imbalanced connections, $\hat{A} = \tilde{D}^{-\frac{1}{2}} \tilde{A} \tilde{D}^{-\frac{1}{2}}$, where $\tilde{A} = A + I$, I is identity matrix and $\tilde{D}_{ii} = \sum_j \tilde{A}_{ij}$. Then the graph convolution is performed to propagate the information, refining ambiguous features with various contexts:

$$X_{ref} = \mathcal{F}(\hat{A}XW), \qquad (2)$$

where $W \in \mathcal{R}^{D \times D}$ is the trainable parameters, $\mathcal{F}$ is the ReLU function, and X_{ref} is the refined vertice feature. With refined X_{ref}, feature F_{int} can be updated to refined feature F_{ref} by replacing features with X_{ref} at the corresponding location. And F_{ref} is further passed to MLP layers to obtain precise predictions Y, which is supervised by cross entropy loss $\mathcal{L}_{refine} = -\frac{1}{N} \sum_{i=1}^{N} \sum_{k=1}^{K} Y_{gt}^i(k) log Y_i(k)$. After graph convolution, ambiguous feature outside the edge are refined with dissimilated contexts, leading to a better segmenation result around the edge.

2.3   Inner-Edge Hard-Sample Excavation (IHE)

To further refine indistinguishable features inside the edge, contrastive learning techniques is utilized to optimize the feature space based on triplets (i.e., anchor, positive and negative). Unlike previous contrastive methods of point clouds [5,16] where negative samples are equally chosen, we specifically select hard-negative samples nearby the edge, as shown in Fig. 1(c). These hard-negatives are closer to the decision boundary in the feature space and can provide more informative clues for discriminative feature learning compared to easy ones.

Edge-Guided Triplets Sampling. Points inside the edge set Ω are viewed as anchors, with corresponding ambiguous features to be rectified. Then, to obtain hard-negatives which are nearby the decision boundary in the feature space, we select points nearby Ω in a similar way as the vertice selection process in Sect. 2.2, composing the candidate set C with iteration times T. Each candidate in C is treated as positive or hard-negative based on the agreement of candidate and anchor's labels. For each anchor $p_i \in \Omega$ with label Y_{gt}^i, the set of hard-negative samples can be extracted by $C_i^- = \{p_j \in C | Y_{gt}^j \neq Y_{gt}^i\}$, and the rest elements in C compose the positive set $C_i^- = C - C_i^+$.

Point-Based Contrastive Loss. With obtained triplets, the InfoNCE [10] style contrastive loss l_{contra} is computed in the feature space:

$$\mathcal{L}_{contra} = - \sum_{p_i \in \Omega} \sum_{p_j \in C_i^+} log \frac{exp(f_{p_i}^\top f_{p_j}/\tau)}{exp(f_{p_i}^\top f_{p_j}/\tau) + \sum_{p_k \in C_i^-} exp(f_{p_i}^\top f_{p_k}/\tau)}, \qquad (3)$$

where τ is the temperature that controls the concentration level, and $f_p \in F_{ref}$ is the normalized feature at the corresponding location p. $\mathcal{L}_{contra}$ encourages anchor features being away from hard-negative features and close to same-category ones, contributing to a more compact intra-class and more distinguishable inter-class feature space with clearer decision boundary.

Table 1. Performance comparison with different methods under 512 sampling scheme. Results under 1024 and 2048 schemes are shown in supplementary materials.

Methods	Vessel		Aneurysm		mIoU(%)	mDSC(%)
	IoU(%)	DSC(%)	IoU(%)	DSC(%)		
PointNet++ (17') [13]	93.42	96.48	76.22	83.92	84.82	90.20
SO-Net (18') [7]	94.22	96.95	80.14	87.90	87.18	92.43
PointConv (19') [15]	94.16	96.89	79.09	86.01	86.63	91.45
RandLA-Net (20') [4]	92.02	95.84	69.33	81.89	80.68	88.87
3DMedPT (21') [18]	94.82	97.29	81.80	89.25	88.31	93.27
PT (21') [19]	95.10	97.49	82.11	90.18	88.61	93.84
EPT-Net (Ours)	**95.46**	**97.68**	**87.58**	**93.38**	**91.52**	**95.53**
EPT-Net w/o DST	94.90	97.38	86.48	92.75	90.69	95.07
EPT-Net w/o ODC	94.87	97.37	86.79	92.93	90.83	95.15
EPT-Net w/o IHE	95.03	97.45	87.13	93.12	91.08	95.29

2.4 Training and Testing

During training, the weights of dual stream transformer and graph convolution layer are optimized by the loss function defined as:

$$\mathcal{L} = \mathcal{L}_{coarse} + \lambda_1 \mathcal{L}_{edge} + \lambda_2 \mathcal{L}_{refine} + \lambda_3 \mathcal{L}_{contra}, \tag{4}$$

where $\mathcal{L}_{coarse}$ and $\mathcal{L}_{edge}$ are cross entropy loss supervised by semantic labels and edge labels respectively, guiding DST to generate intermediate features and edges. $\mathcal{L}_{refine}$ is the cross entropy loss supervised by semantic labels, imposing supervisions on the refined segmentation results. Moreover, $\mathcal{L}_{contra}$ is the contrastive loss imposed on the built triplets, regularizing the feature space in a contrastive manner. At the testing stage, we discard IHE since it requires ground truth annotations, and use trained DST and OCD to generate outputs for evaluation.

3 Experiment

Dataset and Implementation Details. We train and evaluate the proposed EPT-Net on the publicly available dataset IntrA [17] with a single RTX2080Ti GPU in PyTorch [11]. The dataset contains 1909 blood vessel point clouds generated from reconstructed 3D models of real patients, including 1694 healthy vessel segments and 215 aneurysm segments. 116 in 215 aneurysm segments are further divided and annotated by medical experts to evaluate segmentation methods, and geodesic matrices of each segment is also provided. Following [17], we adopt 512, 1024 and 2048 sampling schemes to train the framework, and 5 cross-validation strategy is also used. The framework is trained for 200 epochs with batch size 8, using the SGD optimizer with momentum 0.9 and initial learning rate 1e–3, dropped by 10x at epochs 120 and 160. The balance weights λ_1, λ_2

PointNet++   SO-Net   PointConv   RandLA-Net   PT   Ours   GroundTruth

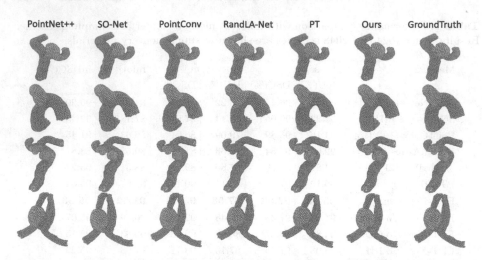

Fig. 3. Illustration examples compared with other methods.

and λ_3 are set to $1, 0.5$ and 0.1 respectively. The iteration times M in OCD is 2, T in IHE is 1, and the temperature τ in IHE is 1. Following [17], the evaluation of each testing fold is repeated 3 times to avoid randomness, and the final results are averaged over the repeated times and splits of folds. For evaluation metrics, we use the Intersection over Union (IOU) and Dice Coefficient (DSC).

Comparison with State-of-the-Art Methods. We assessed the performance of the proposed EPT-Net by comparing it with state-of-the-art point-based methods [4,7,13,15,18,19] and the results under 512 sampling scheme are shown in Table 1 (1st-7th rows). The proposed EPT-Net outperforms these methods [4,7,13,15,18,19] with a large margin of 6.70%, 4.34%, 4.89%, 10.84%, 3.21%, 2.91% in mIoU, and 5.33%, 3.10%, 4.08%, 6.66%, 2.26%, 1.69% in mDSC. The proposed EPT-Net delivers huge improvement in aneurysm category, which is more correlated with the clipping surgery process. Moreover, since the accuracy of aneurysm segmentation highly depends on the quality of predictions around the edge, the results also demonstrate the effectiveness of edge-oriented design of EPT-Network. Visualization examples are illustrated in Fig. 3, where the edge produced by EPT-Net is more accurate compared with other methods, intuitively testifying the effective design of EPT-Net. Note that experiment results under 1024 and 2048 scheme are recorded in supplementary materials, and they have similar conclusions as the 512 sampling scheme.

Ablation Study. Furthermore, we conduct comprehensive ablation studies to examine the effectiveness of each component in EPT-Net as shown in Table 1 (8th-10th rows). Compared with the 'w/o DST' case, where the edge stream is removed and the edge used in ODC and IHE is replaced with edge labels, EPT-Net obtains 0.83% mIoU and 0.46% mDSC gain, revealing that the edge stream can impose more supervisions around the edge and help semantics stream generate better segmentation results. Then, under the 'w/o ODC' and 'w/o IHE' case, the performance degrades 0.67%, 0.44% in mIoU, and 0.38%, 0.24% in mDSC, which quantitatively demonstrate that the ambiguous features around the edge can be refined to promote the segmentation accuracy. Overall, the proposed EPT-Net is able to generate more precise predictions around the edge with three designed paradigms DST, ODC and IHE, leading to a better segmentation performance.

4   Conclusions

In this work, we present a novel Edge-oriented Point-cloud Transformer (EPT-Net) framework for point-based 3D aneurysm segmentation. Against the challenges of poor predictions around the edge, we propose three novel paradigms, i.e., DST, OCD and IHE, to impose more supervisions and refine ambiguous features around the edge. Experiments on IntrA dataset testify the effectiveness of the proposed DST, OCD and IHE, and results indicate that the proposed EPT-Net can outperform the state-of-the-art methods with a large margin.

Acknowledgements. This work is supported by the Hong Kong Innovation and Technology Commission (InnoHK Project CIMDA).

References

1. Alaraj, A., et al.: Virtual reality cerebral aneurysm clipping simulation with real-time haptic feedback. Oper. Neurosur. **11**(1), 52–58 (2015)
2. Dosovitskiy, A., et al.: An image is worth 16x16 words: transformers for image recognition at scale. arXiv preprint arXiv:2010.11929 (2020)
3. Frösen, J., et al.: Saccular intracranial aneurysm: pathology and mechanisms. Acta Neuropathol. **123**(6), 773–786 (2012)
4. Hu, Q., et al.: Randla-net: Efficient semantic segmentation of large-scale point clouds. In: Proceedings of the IEEE/CVF Conference on Computer Vision and Pattern Recognition. pp. 11108–11117 (2020)
5. Jiang, L., et al.: Guided point contrastive learning for semi-supervised point cloud semantic segmentation. In: Proceedings of the IEEE/CVF International Conference on Computer Vision, pp. 6423–6432 (2021)
6. Lareyre, F., Adam, C., Carrier, M., Dommerc, C., Mialhe, C., Raffort, J.: A fully automated pipeline for mining abdominal aortic aneurysm using image segmentation. Sci. Rep. **9**(1), 1–14 (2019)
7. Li, J., Chen, B.M., Lee, G.H.: So-net: Self-organizing network for point cloud analysis. In: Proceedings of the IEEE Conference on Computer Vision and Pattern Recognition. pp. 9397–9406 (2018)
8. López-Linares, K., García, I., García-Familiar, A., Macía, I., Ballester, M.A.G.: 3D convolutional neural network for abdominal aortic aneurysm segmentation. arXiv preprint arXiv:1903.00879 (2019)
9. Nieuwkamp, D.J., Setz, L.E., Algra, A., Linn, F.H., de Rooij, N.K., Rinkel, G.J.: Changes in case fatality of aneurysmal subarachnoid haemorrhage over time, according to age, sex, and region: a meta-analysis. Lancet Neurol. **8**(7), 635–642 (2009)

10. Oord, A.v.d., Li, Y., Vinyals, O.: Representation learning with contrastive predictive coding. arXiv preprint arXiv:1807.03748 (2018)
11. Paszke, A., et al.: Pytorch: an imperative style, high-performance deep learning library. Adv. Neural Inf. Process. Syst. **32**, 8026–8037 (2019)
12. Qi, C.R., Su, H., Mo, K., Guibas, L.J.: Pointnet: Deep learning on point sets for 3d classification and segmentation. In: Proceedings of the IEEE conference on Computer Vision and Pattern Recognition, pp. 652–660 (2017)
13. Qi, C.R., Yi, L., Su, H., Guibas, L.J.: Pointnet++: Deep hierarchical feature learning on point sets in a metric space. Adv. Neural Inf. Process. Syst. **30** (2017)
14. Sichtermann, T., Faron, A., Sijben, R., Teichert, N., Freiherr, J., Wiesmann, M.: Deep learning-based detection of intracranial aneurysms in 3D TOF-MRA. Am. J. Neuroradiol. **40**(1), 25–32 (2019)
15. Wu, W., Qi, Z., Fuxin, L.: Pointconv: Deep convolutional networks on 3d point clouds. In: Proceedings of the IEEE/CVF Conference on Computer Vision and Pattern Recognition. pp. 9621–9630 (2019)
16. Xie, S., Gu, J., Guo, D., Qi, C.R., Guibas, L., Litany, O.: PointContrast: unsupervised pre-training for 3d point cloud understanding. In: Vedaldi, A., Bischof, H., Brox, T., Frahm, J.-M. (eds.) ECCV 2020. LNCS, vol. 12348, pp. 574–591. Springer, Cham (2020). https://doi.org/10.1007/978-3-030-58580-8_34
17. Yang, X., Xia, D., Kin, T., Igarashi, T.: Intra: 3D intracranial aneurysm dataset for deep learning. In: Proceedings of the IEEE/CVF Conference on Computer Vision and Pattern Recognition, pp. 2656–2666 (2020)
18. Yu, J., et al.: 3D medical point transformer: introducing convolution to attention networks for medical point cloud analysis. arXiv preprint arXiv:2112.04863 (2021)
19. Zhao, H., Jiang, L., Jia, J., Torr, P.H., Koltun, V.: Point transformer. In: Proceedings of the IEEE/CVF International Conference on Computer Vision, pp. 16259–16268 (2021)

mmFormer: Multimodal Medical Transformer for Incomplete Multimodal Learning of Brain Tumor Segmentation

Yao Zhang[1,2], Nanjun He[3], Jiawei Yang[4], Yuexiang Li[3], Dong Wei[3],
Yawen Huang[3(✉)], Yang Zhang[5], Zhiqiang He[5(✉)], and Yefeng Zheng[3]

[1] Institute of Computing Technology, Chinese Academy of Sciences, Beijing, China
[2] University of Chinese Academy of Sciences, Beijing, China
zhangyao215@mails.ucas.ac.cn
[3] Jarvis Lab, Tencent, Shenzhen, China
yawenhuang@tencent.com
[4] Electrical and Computer Engineering, University of California, Los Angeles, USA
[5] Lenovo Research, Beijing, China

Abstract. Accurate brain tumor segmentation from Magnetic Resonance Imaging (MRI) is desirable to joint learning of multimodal images. However, in clinical practice, it is not always possible to acquire a complete set of MRIs, and the problem of missing modalities causes severe performance degradation in existing multimodal segmentation methods. In this work, we present *the first attempt* to exploit the Transformer for multimodal brain tumor segmentation that is robust to any combinatorial subset of available modalities. Concretely, we propose a novel multimodal Medical Transformer (**mmFormer**) for *incomplete multimodal learning* with three main components: the hybrid modality-specific encoders that bridge a convolutional encoder and an intra-modal Transformer for both local and global context modeling within each modality; an inter-modal Transformer to build and align the long-range correlations across modalities for modality-invariant features with global semantics corresponding to tumor region; a decoder that performs a progressive up-sampling and fusion with the modality-invariant features to generate robust segmentation. Besides, auxiliary regularizers are introduced in both encoder and decoder to further enhance the model's robustness to incomplete modalities. We conduct extensive experiments on the public BraTS 2018 dataset for brain tumor segmentation. The results demonstrate that the proposed mmFormer outperforms the state-of-the-art methods for incomplete multimodal brain tumor segmentation on almost all subsets of incomplete modalities, especially by an average 19.07% improvement of Dice on tumor segmentation with only one available modality. The code is available at https://github.com/YaoZhang93/mmFormer.

Keywords: Incomplete multimodal learning · Brain tumor segmentation · Transformer

Y. Zhang and N. He—Equal contribution. This work is done when Yao Zhang is an intern at Jarvis Lab, Tencent.

L. Wang et al. (Eds.): MICCAI 2022, LNCS 13435, pp. 107–117, 2022.
https://doi.org/10.1007/978-3-031-16443-9_11

1   Introduction

Automated and accurate segmentation of brain tumors plays an essential role in clinical assessment and diagnosis. Magnetic Resonance Imaging (MRI) is a common neuroimaging technique for the quantitative evaluation of brain tumors in clinical practice, where multiple imaging modalities, i.e., T1-weighted (T1), contrast-enhanced T1-weighted (T1c), T2-weighted (T2), and Fluid Attenuated Inversion Recovery (FLAIR) images, are provided. Each imaging modality provides a distinctive contrast of the brain structure and pathology. The joint learning of multimodal images for brain tumor segmentation is essential and can significantly boost the segmentation performance. Plenty of methods have been widely explored to effectively fuse multimodal MRIs for brain tumor segmentation by, for example, concatenating multimodal images in channel dimension as the input or fusing features in the latent space [17,23]. However, in clinical practice, it is not always possible to acquire a complete set of MRIs due to data corruption, various scanning protocols, and unsuitable conditions of patients. In this situation, most existing multimodal methods may fail to deal with incomplete imaging modalities and face a severe degradation in segmentation performance. Consequently, a robust multimodal method is highly desired for a flexible and practical clinical application with one or more missing modalities.

Incomplete multimodal learning, also known as hetero-modal learning [8], aims at designing methods that are robust with any subset of available modalities at inference. A straightforward strategy for incomplete multimodal learning of brain tumor segmentation is synthesizing the missing modalities by generative models [18]. Another stream of methods explores knowledge distillation from complete modalities to incomplete ones [2,10,21]. Although promising results are obtained, such methods have to train and deploy a specific model for each subset of missing modalities, which is complicated and burdensome in clinical application. Zhang et al. [22] proposed an ensemble learning of single-modal models with adaptive fusion to achieve multimodal segmentation. However, it only works when one or all modalities are available. Meanwhile, all these methods require complete modalities during the training process.

Recent methods focused on learning a unified model, instead of a bunch of distilled networks, for incomplete multimodal segmentation [8,16]. For example, HeMIS [8] learns an embedding of multimodal information by computing mean and variance across features from any number of available modalities. U-HVED [4] further introduces multimodal variational auto-encoder to benefit incomplete multimodal segmentation with generation of missing modalities. More recent methods also proposed to exploit feature disentanglement [1] and attention mechanism [3] for robust multimodal brain tumor segmentation. Fully Convolutional Network (FCN) [11,15] has achieved great success in medical image segmentation and is widely used for feature extraction in the methods mentioned above. Despite its excellent performance, the inductive bias of convolution, i.e., the locality, makes FCN difficult to build long-range dependencies explicitly. In incomplete multimodal learning of brain tumor segmentation, the features extracted with limited receptive fields tend to be biased when dealing with varying modalities.

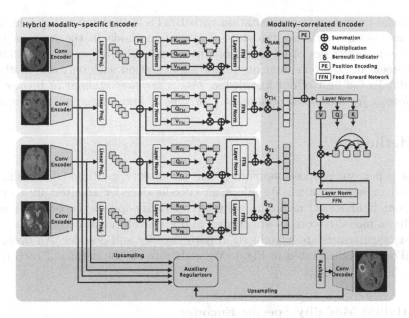

Fig. 1. Overview of the proposed mmFormer, which is composed of four hybrid modality-specific encoders, a modality-correlated encoder, and a convolutional decoder. Meanwhile, auxiliary regularizers are introduced in both encoder and decoder. The skip connections between the convolutional encoder and decoder are hidden for clear display.

In contrast, a modality-invariant embedding with global semantic information of tumor region across different modalities may contribute to more robust segmentation, especially when one or more modalities are missing.

Transformer was originally proposed to model long-range dependencies for sequence-to-sequence tasks [19], and also shows state-of-the-art performance on various computer vision tasks [5]. Concurrent works [7,14,20] exploited Transformer for brain tumor segmentation from the view of backbone network. However, the dedicated Transformer for multimodal modeling of brain tumor segmentation has not been carefully tapped yet, letting alone the incomplete multimodal segmentation.

This paper aims to exploit Transformer to build a unified model for incomplete multimodal learning of brain tumor segmentation. We propose Multimodal Medical Transformer (**mmFormer**) that leverages hybrid modality-specific encoders and a modality-correlated encoder to build the long-range dependencies both within and across different modalities. With the modality-invariant representations extracted by explicitly building and aligning global correlations between different modalities, the proposed mmFormer demonstrates superior robustness to incomplete multimodal learning of brain tumor segmentation. Meanwhile, auxiliary regularizers are introduced into mmFormer to encourage both encoder and decoder to learn discriminative features even when a certain number of modalities are missing. We validate mmFormer on the task of multimodal brain tumor segmentation with BraTS 2018 dataset [12]. The proposed method outperforms the state-of-the-art methods in the average Dice metric over all settings of missing modalities, especially by an average 19.07% improvement in Dice on enhancing tumor segmentation with only one available modality. To the best of our knowledge, *this is the first attempt to involve the Transformer for incomplete multimodal learning of brain tumor segmentation.*

2   Method

In this paper, we propose mmFormer for incomplete multimodal learning of brain tumor segmentation. We adopt an encoder-decoder architecture to construct our mmFormer, including a hybrid modality-specific encoder for each modality, a modality-correlated encoder, and a convolutional decoder. Besides, auxiliary regularizers are introduced in both encoder and decoder. An overview of mmFormer is illustrated in Fig. 1. We elaborate on the details of each component in the followings.

2.1   Hybrid Modality-Specific Encoder

The hybrid modality-specific encoder aims to extract both local and global context information within a specific modality by bridging a convolutional encoder and an intra-modal Transformer. We denote the complete set of modalities by $M = \{FLAIR, T1c, T1, T2\}$. Given an input of $\mathbf{X}_m \in \mathbb{R}^{1 \times D \times H \times W}$ with a size of $D \times H \times W$, $m \in M$, we first utilize the convolutional encoder to generate compact feature maps with the local context and then leverage the intra-modal Transformer to model the long-range dependency in a global space.

Convolutional Encoder. The convolutional encoder is constructed by stacking convolutional blocks, similar to the encoder part of U-Net [15]. The feature maps with the local context within each modality produced by the convolutional encoder $\mathcal{F}_m^{conv}$ can be formulated as

$$\mathbf{F}_m^{local} = \mathcal{F}_m^{conv}(\mathbf{X}_m; \theta_m^{conv}) \tag{1}$$

where $\mathbf{F}_m^{local} \in \mathbb{R}^{C \times \frac{D}{2^{l-1}} \times \frac{H}{2^{l-1}} \times \frac{W}{2^{l-1}}}$, C is the channel dimension, and l is the number of the stages in the encoder. Concretely, we build a five-stage encoder, and each stage consists of two convolutional blocks. Each block contains cascaded group normalization, ReLU, and convolutional layers with kernel size of 3, while the first convolutional block in the first stage only contains a convolutional layer. Between two consecutive blocks, a convolutional layer with stride of 2 is employed to downsample the feature maps. The number of filters at each level of the encoder is 16, 32, 64, 128, and 256, respectively.

Intra-modal Transformer. Limited by the intrinsic locality of the convolutional network, the convolutional encoder fails to effectively build the long-range dependency within each modality. Therefore, we exploit the Intra-modal Transformer for explicitly long-range contextual modeling. The Intra-modal Transformer contains a tokenizer, a Multi-head Self Attention (MSA), and a Feed-Forward Network (FFN).

As Transformer processes the embeddings in a sequence-to-sequence manner, the local feature maps $\mathbf{F}_m^{local}$ produced by the convolutional encoder is first flattened into a 1D sequence and transformed into token space by a linear projection. However, the flattening operation inevitably collapses the spatial information, which is critical to image segmentation. To address this issue, we introduce a learnable position embedding $\mathbf{P_m}$ to supplement the flattened features via element-wise summation, which is formulated as

$$\mathbf{F}_m^{token} = \mathbf{F}_m^{local}\mathbf{W}_m + \mathbf{P}_m, \tag{2}$$

where $\mathbf{F}_m^{token} \in \mathbb{R}^{C' \times \frac{DHW}{2^{3(l-1)}}}$ denotes the token and $\mathbf{W}_m$ denotes the weights of linear projection. The MSA builds the relationship within each modality by looking over all possible locations in the feature map, which is formulated as

$$head_m^i = Attention(\mathbf{Q}_m^i, \mathbf{K}_m^i, \mathbf{V}_m^i) = softmax(\frac{\mathbf{Q}_m^i \mathbf{K}_m^{iT}}{\sqrt{d_k}})\mathbf{V}_m^i, \tag{3}$$

$$MSA_m = [head_m^1, ..., head_m^N]\mathbf{W}_m^o, \tag{4}$$

where $\mathbf{Q}_m^i = LN(\mathbf{F}_m^{token})\mathbf{W}_m^{Qi}$, $\mathbf{K}_m^i = LN(\mathbf{F}_m^{token})\mathbf{W}_m^{Ki}$, $\mathbf{V}_m^i = LN(\mathbf{F}_m^{token})\mathbf{W}_m^{Vi}$, $LN(\cdot)$ is layer normalization, d_k is the dimension of $\mathbf{K}_m$, $N - 8$ is the number of attention heads, and $[\cdot, \cdot]$ is a concatenation operation. The FFN is a two-layer perceptron with GELU [9] activation. The feature maps with global context within each modality produced by the intra-modal Transformer is defined as

$$\mathbf{F}_m^{global} = FFN_m(LN(z)) + z, z = MSA_m(LN(\mathbf{F}_m^{token})) + \mathbf{F}_m^{token}, \tag{5}$$

where $\mathbf{F}_m^{global} \in \mathbb{R}^{C' \times \frac{DHW}{2^{3(l-1)}}}$.

2.2 Modality-Correlated Encoder

The modality-correlated encoder is designed to build the long-range correlations across modalities for modality-invariant features with global semantics corresponding to the tumor region. It is implemented as an inter-modal Transformer.

Inter-modal Transformer. In contrast to the intra-modal Transformer, the inter-modal Transformer combines the embeddings from all modality-specific encoders by concatenation as the input multimodal token, which is defined as

$$\mathbf{F}^{token} = [\delta_{FLAIR}\mathbf{F}_{FLAIR}^{global}, \delta_{T1c}\mathbf{F}_{T1c}^{global}, \delta_{T1}\mathbf{F}_{T1}^{global}, \delta_{T2}\mathbf{F}_{T2}^{global}]\mathbf{W} + \mathbf{P}, \tag{6}$$

where $\delta_m \in \{0,1\}$ is a Bernoulli indicator that aims to grant robustness when building long-range dependencies between different modalities even when some modalities are missing. This kind of modality-level dropout is randomly conducted during training by setting δ_m to 0. In case of missing modalities, the multimodal token for the missing modalities will be held by a zero vector. Subsequently, it is processed by MSD and FFN for modality-invariant features across modalities, which is formulated as

$$\mathbf{F}^{global} = FFN(LN(z)) + z, z = MSA(LN(\mathbf{F}^{token})) + \mathbf{F}^{token}, \tag{7}$$

where $\mathbf{F}^{global} \in \mathbb{R}^{C' \times \frac{DHW}{2^{(l-1)}}}$.

2.3   Convolutional Decoder

The convolutional decoder is designed to progressively restore the spatial resolution from high-level latent space to original mask space. The output sequence $\mathbf{F}^{global}$ of the modality-correlated Transformer is reshaped into feature maps corresponding to the size before flattening. The convolutional decoder has a symmetric architecture of convolutional encoder, similar to U-Net [15]. Besides, the skip connections between encoder and decoder are also added to keep more low-level details for better segmentation. The features from convolutional encoders of different modalities at a specific level are concatenated and forwarded as skip features to the convolutional decoder.

2.4   Auxiliary Regularizer

Conventional multimodal learning models tend to recognize brain tumors relying on the discriminative modalities [1,3]. Such models are likely to face severe degradation when the discriminative modalities are missing. Therefore, it is critical to encourage each convolutional encoder to segment brain tumors even without the assistance of other modalities. To this end, the outputs of convolutional encoders are upsampled by a shared-weight decoder to segment tumors from each modality separately. The shared-weight decoder has the same architecture with the convolutional decoder. Besides, we also introduce auxiliary regularizers in the convolutional decoder to force the decoder to generate accurate segmentation even when certain modalities are missing. It is achieved by interpolating the feature maps in each stage of the convolutional decoder to segment tumors via deep supervision [6]. Dice loss [13] is employed as the regularizer. Combining the training loss of the network's output with the auxiliary regularizers, the overall loss function is defined as

$$\mathcal{L} = 1 - Dice = 1 - \frac{2\sum_{c=1}^{C}\sum_{i=1}^{N_c} g_i^c p_i^c}{\sum_{c=1}^{C}\sum_{i=1}^{N_c} g_i^{c2} + \sum_{c=1}^{C}\sum_{i=1}^{N_c} p_i^{c2}}, \tag{8}$$

$$\mathcal{L}_{total} = \sum_{i \in M} \mathcal{L}_i^{encoder} + \sum_{i=1}^{l-1} \mathcal{L}_i^{decoder} + \mathcal{L}^{output}, \tag{9}$$

Table 1. Results of the proposed method and state-of-the-art unified models, i.e., HeMIS [8] and U-HVED [4], on BraTS 2018 dataset [12]. Dice similarity coefficient (DSC) [%] is employed for evaluation with every combination settings of modalities. • and ○ denote available and missing modalities, respectively.

Modalities				Enhancing Tumor			Tumor Core			Whole Tumor		
F	T1c	T1	T2	U-HeMIS	U-HVED	Ours	U-HeMIS	U-HVED	Ours	U-HeMIS	U-HVED	Ours
•	○	○	○	11.78	23.80	**39.33**	26.06	57.90	**61.21**	52.48	84.39	**86.10**
○	•	○	○	62.02	57.64	**72.60**	65.29	59.59	**75.41**	61.53	53.62	**72.22**
○	○	•	○	10.16	8.60	**32.53**	37.39	33.90	**56.55**	57.62	49.51	**67.52**
○	○	○	•	25.63	22.82	**43.05**	57.20	54.67	**64.20**	80.96	79.83	**81.15**
•	•	○	○	66.10	68.36	**75.07**	71.49	75.07	**77.88**	68.99	85.93	**87.30**
•	○	•	○	10.71	27.96	**42.96**	41.12	61.14	**65.91**	64.62	85.71	**87.06**
•	○	○	•	30.22	32.31	**47.52**	57.68	62.70	**69.75**	82.95	87.58	**87.59**
○	•	•	○	66.22	61.11	**74.04**	72.46	67.55	**78.59**	68.47	64.22	**74.42**
○	•	○	•	67.83	67.83	**74.51**	76.64	73.92	**78.61**	82.48	81.32	**82.99**
○	○	•	•	32.39	24.29	**44.99**	60.92	56.26	**69.42**	**82.41**	81.56	82.20
•	•	•	○	68.54	68.60	**75.47**	76.01	77.05	**79.80**	72.31	86.72	**87.33**
•	•	○	•	68.72	68.93	**75.67**	77.53	76.75	**79.55**	83.85	88.09	**88.14**
•	○	•	•	31.07	32.34	**47.70**	60.32	63.14	**71.52**	83.43	**88.07**	87.75
○	•	•	•	69.92	67.75	**74.75**	78.96	75.28	**80.39**	83.94	82.32	82.71
•	•	•	•	70.24	69.03	**77.61**	79.48	77.71	**85.78**	84.74	88.46	**89.64**
Average				46.10	46.76	**59.85**	62.57	64.84	**72.97**	74.05	79.16	**82.94**

where C is the number of segmentation classes, and N_c is the number of voxels of class c, g_i^c is a binary indicator if class label c is the correct classification for pixel i, p_i^c is the corresponding predicted probability, $M = \{FLAIR, T1c, T1, T2\}$, and l is the number of stages in the convolutional decoder.

3 Experiments and Results

Dataset and Implementation. The experiments are conducted on BraTS 2018 dataset[1] [12], which consists of 285 multi-contrast MRI scans with four modalities: T1, T1c, T2, and FLAIR. Different subregions of brain tumors are combined into three nested subregions: whole tumor, tumor core, and enhancing tumor. All the volumes have been co-registered to the same anatomical template and interpolated to the same resolution by the organizers. Dice Similarity Coefficient (DSC) as defined in Eq. (8) is employed for evaluation. The framework is implemented with PyTorch 1.7 on four NVIDIA Tesla V100 GPUs. The input size is $128 \times 128 \times 128$ voxels and batch size is 1. Random flip, crop, and intensity shifts are employed for data augmentation. The mmFormer has 106M parameters and 748G FLOPs. The network is trained with the Adam optimizer with an initial learning rate of 0.0002 for 1000 epochs. The model is trained for about 25 h with 17G memory on each GPU.

---

[1] https://www.med.upenn.edu/sbia/brats2018/data.html.

Performance of Incomplete Multimodal Segmentation. We evaluate the robustness of our method to incomplete multimodal segmentation. The absence of modality is implemented by setting $\delta_i, i \in \{FLAIR, T1c, T1, T2\}$ to be zero for dropping the specific modalities at inference. We compare our method with two representative models using shared latent space, i.e., HeMIS [8] and U-HVED [4]. For a fair comparison, we use the same data split in [21] and directly reference the results. In Table 1, our method significantly outperforms HeMIS and U-HVED on the segmentation of enhancing tumor and tumor core on all the 15 possible combinantions of available modalities and the segmentation of the whole tumor on 12 out of 15. In Table 2, we show that with the increased number of missing modalities, the average improvement obtained by mmFormer is more considerable. Meanwhile, it is observed that mmFormer gains more improvement when the target is more difficult to segment. These results demonstrate the effectiveness of mmFormer for incomplete multimodal learning of brain tumor segmentation. Figure 2 shows that even with one modality available, mmFormer can achieve proper segmentation for brain tumor.

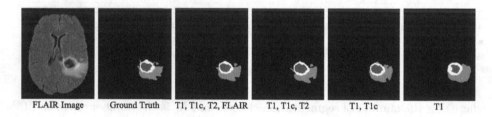

FLAIR Image   Ground Truth   T1, T1c, T2, FLAIR   T1, T1c, T2   T1, T1c   T1

Fig. 2. Segmentation results of mmFormer with various available modalities.

We also compare mmFormer with ACN [21]. ACN relies on knowledge distillation for incomplete multimodal brain tumor segmentation. In the case of N modalities in total, ACN has to train $2^4 - 2$ times to distill $2^N - 2$ student models for all conditions of missing modalities, while our mmFormer only learns once by a unified model. Specifically, ACN is trained for 672 h with 144M parameters for 1 teacher and 14 student models, while mmFormer requires only 25 h with 106 M parameters. Nevertheless, the average DSC for enhancing tumor, tumor core, and whole tumor of mmFormer (59.85, 72.97 and 82.94, respectively) is still close to it of ACN (61.21, 77.62, and 85.92, respectively).

Performance of Complete Multimodal Segmentation. We compare our method with a recent Transformer-based method, i.e., TransBTS [20], for multimodal brain tumor segmentation with full modalities. We reproduce the results with the official repository. TransBTS obtains DSC of 72.66%, 72.69%, and 79.99% on enhancing tumor, tumor core, and the whole tumor, respectively. Our mmFormer outperforms TransBTS on all subregions of brain tumor with DSC of 77.61%, 85.78%, and 89.64%, demonstrating the effectiveness of mmFormer even for complete multimodal brain tumor segmentation.

Ablation Study. We investigate the effectiveness of intra-modal Transformer, inter-modal Transformer, and auxiliary regularizer as three critical components in our method. We analyze the effectiveness of each component by excluding one of them from mmFormer. In Table 3, we compare the performance of the three variants to mmFormer with DSC, averaging over the 15 possible combinations of input modalities. It shows that intra-modal Transformer, inter-modal Transformer, and auxiliary regularizer bring performance improvement across all the tumor subregions.

Table 2. Average improvements of mmFormer upon HeMIS [8] and U-HVED [4] with different numbers of missing modalities evaluated by DSC [%].

Regions	# of missing modalities			
	0	1	2	3
Enhancing	+7.98	+8.91	+13.57	+19.07
Core	+7.19	+4.68	+8.62	+15.34
Whole	+3.04	+2.89	+5.57	+11.75

Table 3. Ablation study of critical components of mmFormer.

Methods	Average DSC [%]		
	Enhancing	Core	Whole
mmFormer	59.85	72.97	82.94
w/o IntraTrans	56.98	71.83	81.32
w/o InterTrans	56.05	70.28	81.12
w/o Aux. Reg	55.78	69.33	81.65

4 Conclusion

We proposed a Transformer-based method for incomplete multimodal learning of brain tumor segmentation. The proposed mmFormer bridges Transformer and CNN to build the long-range dependencies both within and across different modalities of MRI images for a modality-invariant representation. We validated our method on brain tumor segmentation under various combinations of missing modalities, and it outperformed state-of-the-art methods on the BraTS benchmark. Our method gains more improvements when more modalities are missing and/or the target ones are more difficult to segment.

References

1. Chen, C., Dou, Q., Jin, Y., Chen, H., Qin, J., Heng, P.-A.: Robust multimodal brain tumor segmentation via feature disentanglement and gated fusion. In: Shen, D., et al. (eds.) MICCAI 2019. LNCS, vol. 11766, pp. 447–456. Springer, Cham (2019). https://doi.org/10.1007/978-3-030-32248-9_50
2. Chen, C., Dou, Q., Jin, Y., Liu, Q., Heng, P.A.: Learning with privileged multimodal knowledge for unimodal segmentation. IEEE Trans. Med. Imaging **41**, 621–632 (2021)
3. Ding, Y., Yu, X., Yang, Y.: RFNet: Region-aware fusion network for incomplete multi-modal brain tumor segmentation. In: Proceedings of the IEEE/CVF International Conference on Computer Vision, pp. 3975–3984 (2021)

4. Dorent, R., Joutard, S., Modat, M., Ourselin, S., Vercauteren, T.: Hetero-modal variational encoder-decoder for joint modality completion and segmentation. In: Shen, D., et al. (eds.) MICCAI 2019. LNCS, vol. 11765, pp. 74–82. Springer, Cham (2019). https://doi.org/10.1007/978-3-030-32245-8_9

5. Dosovitskiy, A., et al.: An image is worth 16x16 words: Transformers for image recognition at scale. arXiv preprint arXiv:2010.11929 (2020)

6. Dou, Q., et al.: 3D deeply supervised network for automated segmentation of volumetric medical images. Med. Image Anal. **41**, 40–54 (2017)

7. Hatamizadeh, A., et al.: UNETR: transformers for 3D medical image segmentation. In: Proceedings of the IEEE/CVF Winter Conference on Applications of Computer Vision, pp. 574–584 (2022)

8. Havaei, M., Guizard, N., Chapados, N., Bengio, Y.: HeMIS: hetero-modal image segmentation. In: Ourselin, S., Joskowicz, L., Sabuncu, M.R., Unal, G., Wells, W. (eds.) MICCAI 2016. LNCS, vol. 9901, pp. 469–477. Springer, Cham (2016). https://doi.org/10.1007/978-3-319-46723-8_54

9. Hendrycks, D., Gimpel, K.: Gaussian error linear units (GELUs). arXiv preprint arXiv:1606.08415 (2016)

10. Hu, M., et al.: Knowledge distillation from multi-modal to mono-modal segmentation networks. In: Martel, A.L., et al. (eds.) MICCAI 2020. LNCS, vol. 12261, pp. 772–781. Springer, Cham (2020). https://doi.org/10.1007/978-3-030-59710-8_75

11. Long, J., Shelhamer, E., Darrell, T.: Fully convolutional networks for semantic segmentation. In: Proceedings of the IEEE Conference on Computer Vision and Pattern Recognition, pp. 3431–3440 (2015)

12. Menze, B.H., et al.: The multimodal brain tumor image segmentation benchmark (BRATS). IEEE Trans. Med. Imaging **34**(10), 1993–2024 (2014)

13. Milletari, F., Navab, N., Ahmadi, S.A.: V-Net: fully convolutional neural networks for volumetric medical image segmentation. In: Fourth International Conference on 3D Vision, pp. 565–571. IEEE (2016)

14. Peiris, H., Hayat, M., Chen, Z., Egan, G., Harandi, M.: A volumetric transformer for accurate 3D tumor segmentation. arXiv preprint arXiv:2111.13300 (2021)

15. Ronneberger, O., Fischer, P., Brox, T.: U-Net: convolutional networks for biomedical image segmentation. In: Navab, N., Hornegger, J., Wells, W.M., Frangi, A.F. (eds.) MICCAI 2015. LNCS, vol. 9351, pp. 234–241. Springer, Cham (2015). https://doi.org/10.1007/978-3-319-24574-4_28

16. Shen, Y., Gao, M.: Brain tumor segmentation on MRI with missing modalities. In: Chung, A.C.S., Gee, J.C., Yushkevich, P.A., Bao, S. (eds.) IPMI 2019. LNCS, vol. 11492, pp. 417–428. Springer, Cham (2019). https://doi.org/10.1007/978-3-030-20351-1_32

17. Tseng, K.L., Lin, Y.L., Hsu, W., Huang, C.Y.: Joint sequence learning and cross-modality convolution for 3D biomedical segmentation. In: Proceedings of the IEEE conference on Computer Vision and Pattern Recognition, pp. 6393–6400 (2017)

18. van Tulder, G., de Bruijne, M.: Why does synthesized data improve multi-sequence classification? In: Navab, N., Hornegger, J., Wells, W.M., Frangi, A.F. (eds.) MICCAI 2015. LNCS, vol. 9349, pp. 531–538. Springer, Cham (2015). https://doi.org/10.1007/978-3-319-24553-9_65

19. Vaswani, A., et al.: Attention is all you need. In: Advances in Neural Information Processing Systems 30 (2017)

20. Wang, W., Chen, C., Ding, M., Yu, H., Zha, S., Li, J.: TransBTS: multimodal brain tumor segmentation using transformer. In: de Bruijne, M., et al. (eds.) MICCAI 2021. LNCS, vol. 12901, pp. 109–119. Springer, Cham (2021). https://doi.org/10.1007/978-3-030-87193-2_11

21. Wang, Y., et al.: ACN: adversarial co-training network for brain tumor segmentation with missing modalities. In: de Bruijne, M., et al. (eds.) MICCAI 2021. LNCS, vol. 12907, pp. 410–420. Springer, Cham (2021). https://doi.org/10.1007/978-3-030-87234-2_39
22. Zhang, Y., et al.: Modality-aware mutual learning for multi-modal medical image segmentation. In: de Bruijne, M., et al. (eds.) MICCAI 2021. LNCS, vol. 12901, pp. 589–599. Springer, Cham (2021). https://doi.org/10.1007/978-3-030-87193-2_56
23. Zhou, C., Ding, C., Lu, Z., Wang, X., Tao, D.: One-pass multi-task convolutional neural networks for efficient brain tumor segmentation. In: Frangi, A.F., Schnabel, J.A., Davatzikos, C., Alberola-López, C., Fichtinger, G. (eds.) MICCAI 2018. LNCS, vol. 11072, pp. 637–645. Springer, Cham (2018). https://doi.org/10.1007/978-3-030-00931-1_73

Multimodal Brain Tumor Segmentation Using Contrastive Learning Based Feature Comparison with Monomodal Normal Brain Images

Huabing Liu[1], Dong Nie[2], Dinggang Shen[3,4], Jinda Wang[5], and Zhenyu Tang[1(✉)]

[1] School of Computer Science and Engineering, Beihang University, Beijing 100191, China
tangzhenyu@buaa.edu.cn
[2] Alibaba Inc., Hangzhou, China
[3] School of Biomedical Engineering, ShanghaiTech University, Shanghai 201210, China
[4] Shanghai United Imaging Intelligence Co., Ltd., Shanghai 200230, China
[5] Sixth Medical Center, Chinese PLA General Hospital, Beijing 100853, China

Abstract. Many deep learning (DL) based methods for brain tumor segmentation have been proposed. Most of them put emphasis on elaborating deep network's internal structure to enhance the capacity of learning tumor-related features, while other valuable related information, such as normal brain appearance, is often ignored. Inspired by the fact that radiologists are often trained to compare with normal tissues when identifying tumor regions, in this paper, we propose a novel brain tumor segmentation framework by adopting normal brain images as reference to compare with tumor brain images in the learned feature space. In this way, tumor-related features can be highlighted and enhanced for accurate tumor segmentation. Considering that the routine tumor brain images are multimodal while the normal brain images are often monomodal, a new contrastive learning based feature comparison module is proposed to solve incomparable issue between features learned from multimodal and monomodal images. In the experiments, both in-house and public (BraTS2019) multimodal tumor brain image datasets are used to evaluate our proposed framework, demonstrating better performance compared to the state-of-the-art methods in terms of Dice score, sensitivity, and Hausdorff distance. Code: https://github.com/hbliu98/CLFC-Brain-Tumor-Segmentation.

Keywords: Brain tumor segmentation · Normal brain images · Feature comparison · Contrastive learning

1 Introduction

Brain tumor segmentation using multimodal magnetic resonance (MR) images is an essential task for subsequent diagnosis and treatment. In the past few years,

L. Wang et al. (Eds.): MICCAI 2022, LNCS 13435, pp. 118–127, 2022.
https://doi.org/10.1007/978-3-031-16443-9_12

many deep learning (DL) based segmentation methods have been proposed and achieved great success [6,11,13,20]. For example, Dong et al. [6] proposed a 2D U-Net [15] for end-to-end brain tumor segmentation, where the soft Dice loss is introduced to handle unbalanced training samples. To utilize volumetric information and multi-scale contextual information, Kamnitsas et al. [11] proposed a 3D convolutional neural network (CNN) with dual-pathway architecture called DeepMedic, to extract features from tumor brain images at multiple scales simultaneously. Wang et al. [20] further integrated Transformers [17] into 3D U-Net to build long-range dependency and learn global semantic features, based on which the segmentation performance can be enhanced.

Although existing DL-based methods have shown promising results, most of them focused on improving the learning capacity of tumor-related features by elaborating deep network's internal structure, while other valuable related information, such as normal brain appearance, is often ignored [18,19]. It is known that radiologists are often trained to compare with normal tissues when identifying tumor regions. Following this observation, the anomaly detection has been introduced for tumor segmentation [1,3]. These methods compare pathological images with their respective normal appearance images reconstructed by the autoencoder [16]. In this way, pathological regions can be highlighted and easily segmented. Note that existing anomaly detection based methods usually work with monomodality [2]. However, in the context of brain tumor segmentation, routine tumor brain images are multimodal, e.g., T1, T1 contrast-enhanced (T1c), T2, and FLAIR MR images, while normal brain images, which are used to train the autoencoder, are often monomodal, e.g., T1 MR images. Comparing multimodal images with monomodal images is difficult, therefore anomaly detection based methods cannot be directly applied to multimodal brain tumor segmentation.

In this paper, we propose a novel multimodal brain tumor segmentation framework, where monomodal normal brain images are adopted as reference and compared with multimodal tumor brain images in the feature space to impel the segmentation performance. To solve the incomparable issue between features learned from different modalities, a new Contrastive Learning based Feature Comparison (CLFC) module is proposed to align the features learned from monomodal normal brain images (normal brain features) to the features learned from multimodal tumor brain images (tumor brain features) at normal brain regions, i.e., non-tumor regions. In this way, tumor regions in tumor brain features, i.e., tumor-related features, can be effectively highlighted and enhanced using normal brain features as reference. Our proposed framework is evaluated using in-house and public (BraTS2019) tumor brain image datasets. Experimental results show that our proposed framework outperforms the state-of-the-art methods in terms of Dice score, sensitivity, and Hausdorff distance in both datasets. The contributions of our work can be summarized as follows:

– We propose a novel deep framework for multimodal brain tumor segmentation, where external information, i.e., monomodal normal brain images, is utilized as reference to impel the segmentation performance.

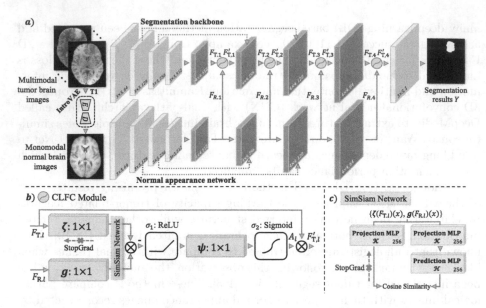

Fig. 1. a) The proposed framework, composed of a segmentation backbone and a normal appearance network; b) Structure of CLFC module; and c) Structure of SimSiam. Kernel sizes and output feature channels are marked at the bottom of each layer.

- A contrastive learning based feature comparison (CLFC) module is designed to solve the incomparable issue between features learned from multimodal tumor brain images and monomodal normal brain images.

2   Method

Our proposed framework is composed of two sub-networks: 1) the segmentation backbone and 2) the normal appearance network, as shown in Fig. 1a. The input of the segmentation backbone is multimodal tumor brain images, while the normal appearance network takes monomodal (T1) normal brain images as the input. The monomodal normal brain images are produced by introspective variational autoencoder (IntroVAE) [8] from the T1 modality contained in the input multimodal tumor brain images. The segmentation backbone and the normal appearance network are of encoder and decoder structures, where each encoder is composed of four convolution layers followed by batch normalization, ReLU and maxpooling layer, and each decoder has four transposed convolution layers to perform upsampling. At the end of the segmentation backbone, an extra convolution layer is designed to produce final segmentation results. In both sub-networks, at each level l of both the decoders, the tumor brain features $F_{T,l}$ learned by the segmentation backbone and the normal brain features $F_{R,l}$ learned by the normal appearance network are aligned and compared by the CLFC modules to produce $F'_{T,l}$, which contains enhanced tumor-related features, making the final segmentation to a high accuracy level. Considering the high computational efficiency of 2D convolution and also the advantage of 3D spatial information, 2.5D slices containing the slice to be segmented and its K adjacent slices (i.e., $2K + 1$ slices) are adopted as the input of our framework.

2.1 The Normal Appearance Network

As aforementioned, the input monomodal normal brain images is set to T1 MR images, which is a commonly adopted modality for normal brain imaging in the context of clinical routine. Instead of using normal brain images from healthy subjects, the monomodal normal brain images are reconstructed from the T1 modality contained in the input multimodal tumor brain images using IntroVAE. In this way, except tumor regions, the reconstructed normal brain images have similar anatomical structure as the original tumor brain images [12,21].

The IntroVAE. The IntroVAE extends traditional VAE to an adversarial learning framework, by which the reconstructed images are of detailed structures. It is composed of an encoder for projecting input brain images to a latent distribution and a decoder for producing reconstructed brain images based on the learned distribution. In the training phase, T1 normal brain MR images from healthy subjects are used to train the IntroVAE, by which low-dimensional manifold representing normal brain appearance can be obtained. In the inference phase, the trained model projects the input tumor brain image to a certain point on the manifold, which represents the closest normal brain appearance. Finally, the corresponding normal brain image can be reconstructed from the point. For more details of the IntroVAE, please refer to the original paper [8].

Based on the reconstructed monomodal normal brain images, normal brain features can be learned in the normal appearance network. At each level of the decoder in the normal appearance network, the learned normal brain features $F_{R,l}$, $l = 1, ..., 4$ are sent to the segmentation backbone as reference to enhance the segmentation performance (discussed below).

2.2 The Segmentation Backbone

The segmentation backbone takes concatenated multimodal tumor brain images as input, and outputs tumor segmentation results. To improve segmentation results, at each decoding level of the segmentation backbone, the learned tumor brain features $F_{T,l}$ are compared with the learned normal brain features $F_{R,l}$, by which tumor regions in $F_{T,l}$, i.e., those tumor-related features, can be highlighted and enhanced. Ideally, for feature vectors $F_{T,l}(x)$ at position x, which are at normal/tumor regions, $F_{R,l}(x)$ should have consistent/inconsistent features at the corresponding positions. In this way, tumor regions in $F_{T,l}$ can then be effectively highlighted according to the feature consistency. Unfortunately, in the context of multimodal brain tumor segmentation, $F_{T,l}$ and $F_{R,l}$ are incomparable (multimodal vs. monomodal).

To tackle the above issue, we propose a new contrastive learning based feature comparison (CLFC) module as shown in Fig. 1b. The CLFC module is composed of two main steps. The *first* step is feature alignment, where two 1×1 convolution layers ζ and g are adopted to align $F_{T,l}$ and $F_{R,l}$ at normal regions. As a result, feature vectors at normal regions are consistent in $\zeta(F_{T,l})$ and $g(F_{R,l})$. To achieve effective feature alignment, a contrastive learning method called the simple Siamese network (SimSiam) [4] is adopted (see Fig. 1c). The SimSiam network is composed of an encoder $\mathcal{F}$, a projection MLP $\mathcal{K}$, and a prediction MLP $\mathcal{H}$. It takes each positive sample pair $\{I_1, I_2\}$ (samples of the same class) as input and maximizes the cosine similarity of output vectors p_1 and v_2:

$$D(I_1, I_2) = -\frac{p_1}{||p_1||_2} \cdot \frac{v_2}{||v_2||_2}, \tag{1}$$

where $p_1 = \mathcal{H}(\mathcal{K}(\mathcal{F}(I_1)))$, $v_2 = \mathcal{K}(\mathcal{F}(I_2))$ and $|| \cdot ||_2$ is the L2-norm. In our framework, feature vector pairs $\{F_{T,l}(x), F_{R,l}(x)\}$, $x \in \Omega_{NR}$, where Ω_{NR} denotes normal regions, are defined as positive samples. Note that Ω_{NR} is determined according to the manually labeled tumor mask in the training dataset. In the CLFC module, ζ and g are regarded as the encoder $\mathcal{F}$ of the SimSiam network, so $\{\zeta(F_{T,l})(x), g(F_{R,l})(x)\}$ are sent to the projection and prediction MLPs to compute the cosine similarity. The loss of the SimSiam network is defined as:

$$\mathcal{L}_{Sim}(F_{T,l}, F_{R,l}) = \frac{1}{2} \sum_{x \in \Omega_{NR}} D(F_{T,l}(x), F_{R,l}(x)) + D(F_{R,l}(x), F_{T,l}(x)). \tag{2}$$

It is worth noting that the SimSiam network is used in the training phase but removed in the inference phase. Moreover, to make the segmentation backbone focus on tumor segmentation rather than feature alignment, a stop gradient operation is also applied before ζ as shown in Fig. 1b.

After the feature alignment step, the *second* step is feature comparison. Since feature vectors in the aligned $F_{T,l}$ and $F_{R,l}$, i.e., $\zeta(F_{T,l})$ and $g(F_{R,l})$, are consistent at normal regions, tumor regions in $\zeta(F_{T,l})$ can be easily identified using $g(F_{R,l})$ as reference. Specifically, we measure the consistency between $\zeta(F_{T,l})$ and $g(F_{R,l})$ and identify the regions of low feature consistency, i.e., $\zeta(F_{T,l})$ at tumor regions vs. $g(F_{R,l})$ at normal regions, to produce the attention map A_l. By using A_l as the mask, tumor-related features in $F_{T,l}$ can be enhanced. The whole procedure of the CLFC module can be summarized as:

$$A_l = \sigma_2(\psi(\sigma_1(-\zeta(F_{T,l}) \otimes g(F_{R,l})))),$$
$$F'_{T,l} = A_l \cdot F_{T,l}, \tag{3}$$

where σ_1 is ReLU, σ_2 is sigmoid activation function, ψ is the 1×1 convolution layer, and $F'_{T,l}$ is the output of the CLFC module containing enhanced tumor-related features. The final segmentation results Y are produced from $F'_{T,4}$ by the last convolution layer in the segmentation backbone. The Dice loss [5] is adopted for Y, and the final loss function of our framework is $\mathcal{L} = \mathcal{L}_{Dice} + \mathcal{L}_{Sim}$.

3   Experiments

Both in-house and public datasets are used to evaluate our proposed framework. Specifically, the in-house dataset contains multimodal tumor brain MR images of T1c, B0, mean diffusivity (MD), and fractional anisotropy (FA) modalities from 104 glioblastoma patients. The public dataset is BraTS2019, which includes tumor brain T1, T1c, T2 and FLAIR brain MR images of 335 glioma patients. Manually labeled tumor masks are available for each patient in both datasets. All images are aligned with MNI152 [7] using affine transformation and normalized by histogram matching. Besides our framework, the state-of-the-art segmentation method nnU-Net [9] is also evaluated. Moreover, ablation experiments of our framework are conducted. Specifically, the Baseline-1 uses only the segmentation backbone of our framework (no normal appearance network), and the Baseline-2 is similar to our framework, but it has no contrastive learning (no SimSiam network) in the CLFC module during training. For both datasets, five-fold patient-wise cross validation is adopted to evaluate each method. The input of all methods under evaluation is 2.5D slices with $K = 2$, i.e., five slices. The batch size is set to four, and the maximal number of epochs is 300. The accuracy of tumor segmentation results is quantified using patient-wise Dice score, sensitivity, and 95% Hausdorff distance (HD95). All methods are implemented using PyTorch and trained with RTX 3090 GPU.

It is worth noting that the IntroVAE is separately trained with public dataset IXI [10] containing 581 normal brain T1 MR images. Specifically, the encoder and decoder of IntroVAE are trained iteratively with the learning rates of 1×10^{-4} and 5×10^{-3}, respectively. The batch size and number of epochs are set to 120 and 200. Other hyper-parameters, like weight terms in the loss function, keep the same as the original paper. After integration into our framework, the parameters of the IntroVAE is fixed during the training of our framework. Moreover, for in-house dataset, which has no T1 modality, we adopt tumor brain T1c MR images as input of IntroVAE to get the reconstructed normal brain T1 MR images, as T1c is T1 with contrast agent in the vessel and both modalities exhibit similar appearance of gray matter and white matter at normal regions [14].

3.1   Evaluation of Segmentation Results

Figure 2 shows some examples of the tumor segmentation results using each method under evaluation. It is clear that as benefiting from feature comparison using normal brain images as reference, our framework can detect subtle normal and tumor regions, which are hard to distinguish using tumor brain images alone, especially in the regions marked in red circles in Fig. 2.

Details of the evaluation results are shown in Table 1, and our framework outperforms all the other methods under evaluation in terms of Dice score, sensitivity, and HD95 in both datasets. The patient-wise Wilcoxon signed rank test is adopted to compare the Dice scores from different methods. For the in-house/public dataset, the p values are 0.0185/0.0076 (Ours vs. nnU-Net), $1.0320 \times 10^{-4}/2.7976 \times 10^{-7}$ (Ours vs. Baseline-1), and $7.7959 \times 10^{-4}/4.9657 \times 10^{-4}$

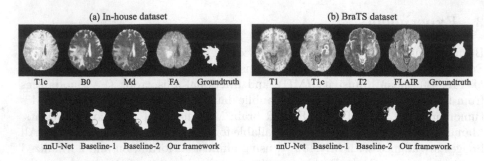

Fig. 2. Examples of the segmentation results using all methods under evaluation. Our framework can detect subtle normal and tumor regions, which cannot be recognized by the other methods (as marked in red circles). (Color figure online)

(Ours vs. Baseline-2), respectively. Baseline-2 achieves better segmentation accuracy than Baseline-1, which shows that the normal appearance network plays a positive role in tumor segmentation. But without the contrastive learning, $F_{T,l}$ and $F_{R,l}$ are difficult to be compared with each other, and the improvement using Baseline-2 is limited. Our framework adopts the contrastive learning for feature alignment in the CLFC module, by which $F_{T,l}$ and $F_{R,l}$ can be well aligned and the segmentation performance is significantly enhanced.

Table 1. Evaluation results of tumor segmentation.

Method	In-house dataset			BraTS dataset		
	Dice (%)	Sensitivity (%)	HD95 (mm)	Dice (%)	Sensitivity (%)	HD95 (%)
nnU-Net	87.44 ± 7.35	85.99 ± 12.28	6.07 ± 3.53	90.53 ± 7.24	89.51 ± 9.82	8.05 ± 17.22
Baseline-1	87.90 ± 3.27	88.86 ± 5.36	6.47 ± 5.36	90.54 ± 7.71	90.00 ± 11.56	4.63 ± 4.96
Baseline-2	88.66 ± 3.61	87.60 ± 6.28	4.74 ± 2.03	91.18 ± 4.90	90.79 ± 8.95	4.23 ± 4.21
Ours	$\mathbf{89.70 \pm 3.68}$	$\mathbf{90.85 \pm 5.50}$	$\mathbf{4.29 \pm 1.81}$	$\mathbf{91.86 \pm 4.63}$	$\mathbf{91.90 \pm 8.02}$	$\mathbf{3.84 \pm 3.04}$

3.2   Evaluation of Contrastive Learning Based Feature Comparison

In the CLFC module, the contrastive learning plays an important role. To give an intuitive visualization of the effect using contrastive learning, distributions of feature vectors in $\zeta(F_{T,l})$ and $g(F_{R,l})$ are shown in Fig. 3.

Specifically, feature vectors are divided into three types: tumor regions in $\zeta(F_{T,l})$, normal regions in $\zeta(F_{T,l})$, and normal regions in $g(F_{R,l})$. All feature vectors are projected onto a 2D plane using the PCA based dimension reduction. Clearly, as compared with Baseline-2 (without contrastive learning), in our framework (with contrastive learning), feature vectors at normal regions in $\zeta(F_{T,l})$ and $g(F_{R,l})$ can be more effectively aligned and are more different from the feature vectors at tumor regions in $\zeta(F_{T,l})$. As a result, tumor regions can be easily identified through the feature comparison. Figure 4 shows some examples of the attention maps A_l produced at each decoding level of Baseline-2 and

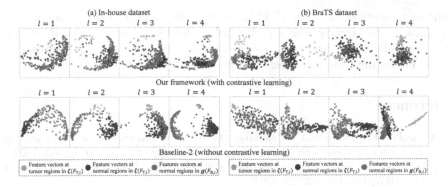

Fig. 3. Distributions of feature vectors in our framework and Baseline-2.

our framework. It is clear that the attention maps produced in our framework has more concentrated tumor regions, and the segmentation results are more consistent with the ground truth than Baseline-2.

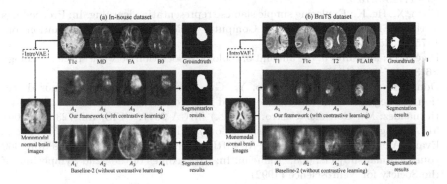

Fig. 4. Examples of the attention maps A_l in our framework and Baseline-2.

4   Conclusion

We proposed a novel multimodal brain tumor segmentation framework, where external information of normal brain appearance was used as reference to highlight and enhance the tumor-related features. Moreover, a contrastive learning based feature comparison (CLFC) module was proposed to address the incomparable issue between features learned from multimodal tumor brain images and monomodal normal brain images, based on which high-quality attention maps A_l as well as tumor-related features $F'_{T,l}$ can be produced for better segmentation results. Both in-house and public BraTS2019 datasets were used to evaluate our framework. The experimental results showed that our framework outperforms the state-of-the-art methods with statistical significance, and the proposed normal appearance network and the proposed CLFC module both play effective role in the segmentation. Since the CLFC module is effective for binary segmentation, our framework currently works with whole tumor segmentation. In the future work, we will extend our framework to segmentation of tumor sub-regions, e.g., edema, enhancing tumor, necrotic and non-enhancing tumor core.

References

1. Astaraki, M., Toma-Dasu, I., Smedby, Ö., Wang, C.: Normal appearance autoencoder for lung cancer detection and segmentation. In: Shen, D., et al. (eds.) MICCAI 2019. LNCS, vol. 11769, pp. 249–256. Springer, Cham (2019). https://doi.org/10.1007/978-3-030-32226-7_28
2. Baur, C., Denner, S., Wiestler, B., Navab, N., Albarqouni, S.: Autoencoders for unsupervised anomaly segmentation in brain MR images: a comparative study. Med. Image Anal., 101952 (2021)
3. Baur, C., Wiestler, B., Albarqouni, S., Navab, N.: Deep autoencoding models for unsupervised anomaly segmentation in brain MR images. In: Crimi, A., Bakas, S., Kuijf, H., Keyvan, F., Reyes, M., van Walsum, T. (eds.) BrainLes 2018. LNCS, vol. 11383, pp. 161–169. Springer, Cham (2019). https://doi.org/10.1007/978-3-030-11723-8_16
4. Chen, X., He, K.: Exploring simple siamese representation learning. In: Proceedings of the IEEE/CVF Conference on Computer Vision and Pattern Recognition, pp. 15750–15758 (2021)
5. Dice, L.R.: Measures of the amount of ecologic association between species. Ecology **26**(3), 297–302 (1945)
6. Dong, H., Yang, G., Liu, F., Mo, Y., Guo, Y.: Automatic brain tumor detection and segmentation using U-Net based fully convolutional networks. In: Valdés Hernández, M., González-Castro, V. (eds.) MIUA 2017. CCIS, vol. 723, pp. 506–517. Springer, Cham (2017). https://doi.org/10.1007/978-3-319-60964-5_44
7. Evans, A., Collins, D., Milner, B.: An MRI-based stereotactic brain atlas from 300 young normal subjects, 408, Anaheim. In: Proceedings of the 22nd Symposium of the Society for Neuroscience (1992)
8. Huang, H., He, R., Sun, Z., Tan, T., et al.: Introvae: introspective variational autoencoders for photographic image synthesis. In: Advances in Neural Information Processing Systems 31 (2018)
9. Isensee, F., Jaeger, P.F., Kohl, S.A., Petersen, J., Maier-Hein, K.H.: nnU-Net: a self-configuring method for deep learning-based biomedical image segmentation. Nat. Methods **18**(2), 203–211 (2021)
10. IXI: Information extraction from images. www.brain-development.org
11. Kamnitsas, K., et al.: Efficient multi-scale 3D CNN with fully connected CRF for accurate brain lesion segmentation. Med. Image Anal. **36**, 61–78 (2017)
12. Luo, Y., et al.: Adaptive rectification based adversarial network with spectrum constraint for high-quality pet image synthesis. Med. Image Anal. **77**, 102335 (2022)
13. Nie, D., Wang, L., Adeli, E., Lao, C., Lin, W., Shen, D.: 3-D fully convolutional networks for multimodal isointense infant brain image segmentation. IEEE Trans. Cybern. **49**(3), 1123–1136 (2019)
14. Radue, E.W., Weigel, M., Wiest, R., Urbach, H.: Introduction to magnetic resonance imaging for neurologists. Continuum Lifelong Learn. Neurol. **22**(5), 1379–1398 (2016)

15. Ronneberger, O., Fischer, P., Brox, T.: U-Net: convolutional networks for biomedical image segmentation. In: Navab, N., Hornegger, J., Wells, W.M., Frangi, A.F. (eds.) MICCAI 2015. LNCS, vol. 9351, pp. 234–241. Springer, Cham (2015). https://doi.org/10.1007/978-3-319-24574-4_28

16. Siddiquee, M.M.R., et al.: Learning fixed points in generative adversarial networks: from image-to-image translation to disease detection and localization. In: Proceedings of the IEEE/CVF International Conference on Computer Vision, pp. 191–200 (2019)

17. Vaswani, A., et al.: Attention is all you need. In: Advances in Neural Information Processing Systems 30 (2017)

18. Wang, K., et al.: Semi-supervised medical image segmentation via a tripled-uncertainty guided mean teacher model with contrastive learning. Med. Image Anal. **79**, 102447 (2022)

19. Wang, L., Shi, F., Lin, W., Gilmore, J.H., Shen, D.: Automatic segmentation of neonatal images using convex optimization and coupled level sets. NeuroImage **58**(3), 805–817 (2011)

20. Wang, W., Chen, C., Ding, M., Yu, H., Zha, S., Li, J.: TransBTS: multimodal brain tumor segmentation using transformer. In: de Bruijne, M., et al. (eds.) MICCAI 2021. LNCS, vol. 12901, pp. 109–119. Springer, Cham (2021). https://doi.org/10.1007/978-3-030-87193-2_11

21. Xiang, L., et al.: Deep embedding convolutional neural network for synthesizing CT image from T1-weighted MR image. Med. Image Anal. **47**, 31–44 (2018)

Meta-hallucinator: Towards Few-Shot Cross-Modality Cardiac Image Segmentation

Ziyuan Zhao[1,2,3], Fangcheng Zhou[2,4], Zeng Zeng[2,3(✉)], Cuntai Guan[1(✉)], and S. Kevin Zhou[5,6]

[1] Nanyang Technological University, Singapore, Singapore
ctguan@ntu.edu.sg
[2] Institute for Infocomm Research (I²R), A*STAR, Singapore, Singapore
zengz@i2r.a-star.edu.sg
[3] Artificial Intelligence, Analytics And Informatics (AI³), A*STAR, Singapore, Singapore
[4] National University of Singapore, Singapore, Singapore
[5] Center for Medical Imaging, Robotics, Analytic Computing and Learning (MIRACLE), School of Biomedical Engineering and Suzhou Institute for Advanced Research, University of Science and Technology of China, Suzhou, China
[6] Key Laboratory of Intelligent Information Processing of Chinese Academy of Sciences (CAS), Institute of Computing Technology, CAS, Beijing, China

Abstract. Domain shift and label scarcity heavily limit deep learning applications to various medical image analysis tasks. Unsupervised domain adaptation (UDA) techniques have recently achieved promising cross-modality medical image segmentation by transferring knowledge from a label-rich source domain to an unlabeled target domain. However, it is also difficult to collect annotations from the source domain in many clinical applications, rendering most prior works suboptimal with the label-scarce source domain, particularly for few-shot scenarios, where only a few source labels are accessible. To achieve efficient few-shot cross-modality segmentation, we propose a novel transformation-consistent meta-hallucination framework, meta-hallucinator, with the goal of learning to diversify data distributions and generate useful examples for enhancing cross-modality performance. In our framework, hallucination and segmentation models are jointly trained with the gradient-based meta-learning strategy to synthesize examples that lead to good segmentation performance on the target domain. To further facilitate data hallucination and cross-domain knowledge transfer, we develop a self-ensembling model with a hallucination-consistent property. Our meta-hallucinator can seamlessly collaborate with the meta-segmenter for learning to hallucinate with mutual benefits from a combined view of meta-learning and self-ensembling learning. Extensive studies on MM-WHS 2017 dataset for cross-modality cardiac segmentation demonstrate

Supplementary Information The online version contains supplementary material available at https://doi.org/10.1007/978-3-031-16443-9_13.

that our method performs favorably against various approaches by a lot in the few-shot UDA scenario.

Keywords: Domain adaptation · Meta-learning · Semi-supervised learning · Segmentation

1 Introduction

Deep learning has made tremendous advancements in recent years, achieving promising performance in a wide range of medical imaging applications, such as segmentation [15,19,31]. However, the clinical deployment of well-trained models to unseen domains remains a severe problem due to the distribution shifts across different imaging protocols, patient populations, and even modalities. While it is a simple but effective approach to fine-tune models with additional target labels for domain adaptation, this would inevitably increase annotation time and cost. In medical image segmentation, it is known that expert-level pixel-wise annotations are usually difficult to acquire and even infeasible for some applications. In this regard, considerable efforts have been devoted in unsupervised domain adaptation (UDA), including feature/pixel-level adversarial learning [4,6,23,32], self-training [14,34], and disentangled representation learning [16,20,24]. Current UDA methods mainly focus on leveraging source labeled and target unlabeled data for domain alignment. Source annotations, however, are also not so easy to access due to expert requirements and privacy problems. Therefore, it is essential to develop a UDA model against the low source pixel-annotation regime. For label-efficient UDA, Zhao *et al.* [29] proposed an MT-UDA framework, advancing self-ensembling learning in a dual-teacher manner for enforcing dual-domain consistency. In MT-UDA, rich synthetic data was generated to diversify the training distributions for cross-modality medical image segmentation, thereby requiring an extra domain generation step in advance. In addition, images generated from independent networks have a limited potential to capture complex structural variations across domains.

On the other hand, many not-so-supervised methods, including self-supervised learning [7,25], semi-supervised learning [1,11,12,30], and few-shot learning [17, 21] have been developed to reduce the dependence on large scale labeled datasets for label-efficient medical image segmentation. However, these methods have not been extensively investigated for either extremely low labeled data regime, *e.g.*, one-shot scenarios or the severe domain shift phenomena, *e.g.*, cross-modality scenarios. Recent works suggest that atlas-based registration and augmentation techniques advance the development of few-shot segmentation [3,27] and pixel-level domain adaptation [10,18]. By approximating styles/deformations between different images, these methods can generate the augmented images with plausible distributions to increase the training data and improve the model generalizability. However, image registration typically increases computational complexity, while inaccurate registrations across modalities can negatively impact follow-up segmentation performance, especially with limited annotations. In this regard, we

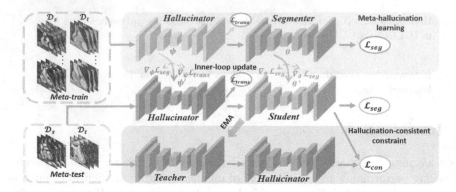

Fig. 1. Overview of our transformation-consistent meta-hallucination framework. In meta-training, hallucinator $\mathcal{G}$ and segmenter $\mathcal{F}$ are optimized together with collaborative objectives. In meta-testing, the transformations generated by $\mathcal{G}$ are used for hallucination-consistent self-ensembling learning to boost cross-modality performance.

pose a natural question: *How can we generate useful samples to quickly and reliably train a good cross-modality segmentation model with only a few source labels?* Recently, model-agnostic meta-learning ("learning to learn") [5] with the goal of improving the learning model itself via the gradient descent process is flexible and independent of any model, leading to broad applications in few-shot learning [9,26] and domain generalization [2,8,13]. Motivated by these observations, we argue that meta-learning can also enable the generator/hallucinator to "learn to hallucinate" meaningful images and obtain better segmentation models under few-shot UDA settings. Therefore, we aim to build a meta-hallucinator for useful sample generation to advance model generalizability on the target domain using limited source annotations.

In this work, we propose a novel transformation-consistent meta-hallucination scheme for unsupervised domain adaptation under source label scarcity. More specifically, we introduce a meta-learning episodic training strategy to optimize both the hallucination and segmentation models by explicitly simulating structural variances and domain shifts in the training process. Both the hallucination and segmentation models are trained concurrently in a collaborative manner to consistently improve few-shot cross-modality segmentation performance. The hallucination model generates helpful samples for segmentation, whereas the segmentation model leverages transformation-consistent constraints and segmentation objectives to facilitate the hallucination process. We extensively investigate the proposed method with the application of cross-modality cardiac substructure segmentation using the public MM-WHS 2017 dataset. Experimental results and analysis have demonstrated the effectiveness of meta-hallucinator against domain shift and label scarcity in the few-shot UDA scenario.

2 Method

Let there be two domains: source $\mathcal{D}_s$ and target $\mathcal{D}_t$, sharing the joint input and label space $\mathcal{X} \times \mathcal{Y}$. Source domain contains N labeled samples $\{(\mathbf{x}_i^s, y_i^s)\}_{i=1}^N$ and M unlabeled samples $\{(\mathbf{x}_i^s)\}_{i=1}^M$, where N is much less than M, while target domain includes P unlabeled samples $\{(\mathbf{x}_i^t)\}_{i=1}^P$. We aim to develop a segmentation model (segmenter) $\mathcal{F}_\theta : \mathcal{X} \to \mathcal{Y}$ by leveraging available data and labels so that it can adapt well to the target domain. The overview of the proposed transformation-consistent meta-hallucination framework is presented in Fig. 1, which we will discuss in detail in this section.

2.1 Gradient-Based Meta-hallucination Learning

In each iteration of gradient-based meta-learning [5], the training data is randomly split into two subsets, *i.e.*, meta-train set $\mathcal{D}_{tr}$ and meta-test set $\mathcal{D}_{te}$ to simulate various tasks, *e.g.*, domain shift or few-shot scenarios, for episodic training to promote robust optimization. Specifically, each episode includes a meta-train step and a meta-test step. In meta-training, the gradient of a meta-train loss $\mathcal{L}_{meta-train}$ on $\mathcal{D}_{tr}$ is first back-propagated to update the model parameters $\theta \to \theta'$. During the meta-test stage, the resulting model $\mathcal{F}_{\theta'}$ is further used to explore $\mathcal{D}_{te}$ via a meta-test loss $\mathcal{L}_{meta-test}$ for fast optimization towards the original parameters θ. Intuitively, such meta-learning schemes not only learn the task on $\mathcal{D}_{tr}$, but also learn how to generalize on $\mathcal{D}_{te}$ for fast adaptation.

In label-scarce domain shift scenarios, we are encouraged to hallucinate useful samples for diversifying training distributions to deal with label scarcity and domain shift. To this end, we introduce a "hallucinator" module $\mathcal{G}_\Psi$ to augment the training set. The objective of the hallucinator is to narrow the domain gap at the image level and generate useful samples for boosting the segmentation performance. We advance the hallucinator into the meta-learning process and promote it to learn how to hallucinate useful samples for the following segmentation model. Specially, in a meta-train step, the parameters Ψ and θ of the hallucinator $\mathcal{G}_\Psi$ and the segmenter $\mathcal{F}_\theta$, respectively, are updated with the meta-train set $\mathcal{D}_{tr}$ via an inner-loop update, defined as:

$$\psi' \leftarrow \psi - \alpha \nabla_\psi \mathcal{L}_{meta-train}(\psi, \theta);$$
$$\theta' \leftarrow \theta - \alpha \nabla_\theta \mathcal{L}_{meta-train}(\psi, \theta), \tag{1}$$

where α denotes the learning rate of the hyperparameters. For the meta-train loss, the segmenter is optimized using the segmentation loss $\mathcal{L}_{seg}$ on the enlarged dataset, whereas the hallucinator objective is to minimize the transformation loss $\mathcal{L}_{trans}$ between source and target images. It is noted that the gradient of the segmentation loss is back-propagated to both hallucinator parameters Ψ and segmenter parameters θ. Therefore, the total meta-train objective is defined as:

$$\mathcal{L}_{meta-train} = \mathcal{L}_{seg} + \lambda_{trans}\mathcal{L}_{trans}, \tag{2}$$

where λ_{trans} is the weighting trade-off parameter. For an input pair consisting of a moving source image and a fixed target image $\{x_i^s, x_i^t\}$, the hallucinator aims to generate a moved target-like image $x_i^{s\rightarrow t}$. We promote fast and robust optimization of both hallucinator and segmenter by sampling tasks of different input pairs for meta-training and meta-testing to simulate structural variances and distribution shifts across domains.

2.2 Hallucination-Consistent Self-ensembling Learning

To effectively leverage the rich knowledge hidden in the unlabeled data, we take advantage of the mean-teacher model based on self-ensembling [22]. Specially, we construct a teacher $\mathcal{F}^{tea}$ with the same architecture as the segmenter and update it with an exponential moving average (EMA) of the segmenter parameters θ at different training steps, i.e., $\theta_t^{tea} = \beta\theta_{t-1}^{tea} + (1 - \beta)\theta_t$, where t and β represent the current step and the EMA smoothing rate, respectively. With a larger β, the teacher model is less reliant on the student model parameters. In general self-ensembling learning, the predictions of the student and teacher models with inputs under different perturbations, such as noises are encouraged to be consistent for model regularization, i.e., $\mathcal{F}^{tea}(x_i; \theta_t^{tea}, \xi') = \mathcal{F}(x_i; \theta_t, \xi)$, where ξ' and ξ represent different perturbations. In contrast to the geometric transformation-invariant property in the context of classification tasks, segmentation is desired to be transformation equivariant at the spatial level. In other words, if the input is transformed with a function f, the output should be transformed in the same manner. Several previous studies [12,28] have demonstrated that the transformation consistency is beneficial for enhancing the regularization of self-ensembling models via various transformation operations, such as rotation. In light of these, we introduce a hallucination-consistent self-ensembling scheme to further promote unsupervised regularization. We apply the same spatial transformations produced by the hallucinator to the student inputs and the teacher outputs, and enable the alignment between their final outputs, i.e., $\mathcal{G}_{\Psi}(\mathcal{F}^{tea}(x_i; \theta_t^{tea})) = \mathcal{F}(\mathcal{G}_{\Psi}(x_i); \theta_t)$. The student model is regularized by minimizing the difference between the outputs of the student and teacher models with a mean square error (MSE) loss. Then, the hallucination-consistent loss is defined as:

$$\mathcal{L}_{con} = \frac{1}{N}\sum_{i=1}^{N}\left\|\mathcal{G}_{\Psi}((\mathcal{F}^{tea}(x_i; \theta_t^{tea}, \xi')) - \mathcal{F}(\mathcal{G}_{\Psi}(x_i); \theta_t, \xi))\right\|^2, \tag{3}$$

where N denotes the number of samples. Different from stochastic transformations, such as random rotation, our hallucination process is learned via meta-learning, producing more meaningful target-like samples in spatial and appearance for domain adaptation. In addition, the hallucination consistency can be used to regularize the meta-optimization of the hallucinator. Note that we only impose the hallucination-consistent loss in the meta-test step since we expect such regularization on unseen data for robust adaptation, thereby improving the network generalization capacity. Then, the meta-test loss is defined as:

$$\mathcal{L}_{meta-test} = \mathcal{L}_{seg} + +\lambda_{con}\mathcal{L}_{con} + \lambda_{trans}\mathcal{L}_{trans}, \tag{4}$$

where λ_{con} is to control the strength of the unsupervised consistency loss. Finally, the total objective of meta-learning is defined as:

$$\underset{\psi,\theta}{\operatorname{argmin}} \ \mathcal{L}_{\text{meta-train}} \left(\mathcal{D}_{\text{tr}} \ ; \psi, \theta\right) + \mathcal{L}_{\text{meta-test}} \left(\mathcal{D}_{\text{te}} \ ; \psi', \theta'\right). \tag{5}$$

3 Experiments and Results

Dataset and Evaluation Metrics. In light of our emphasis on cross-modality segmentation with distinct distribution shifts, we employ the public available Multi-Modality Whole Heart Segmentation (MM-WHS) 2017 dataset [33] to evaluate our meta-hallucinator framework. The dataset contains unpaired 20 MR and 20 CT scans with segmentation maps corresponding to different cardiac substructures. For unsupervised domain adaptation, MR and CT are employed as source $\mathcal{D}_s$ and target $\mathcal{D}_t$, respectively. Following [4], the volumes in each domain are randomly divided into a training set (16 scans) and a testing set (4 scans). For the study on label-scarce scenarios, experiments are conducted with 1-shot and 4-shots source labels. We repeat 4 times with different samples for one-shot scenarios to avoid randomness. For pre-processing, each volume is resampled with unit spacing, and the slices centered on the heart region in the coronal view are cropped to 256×256 and then normalized with z-score into zero mean and unit standard deviation. Four substructures of interest are used for evaluation, *i.e.*, ascending aorta (AA), left atrium blood cavity (LAC), left ventricle blood cavity (LVC), and myocardium of the left ventricle (MYO). Two commonly used metrics for segmentation, *i.e.*, Dice score (Dice) and Average Surface Distance (ASD) [29] are employed to evaluate different methods. Both metrics are reported with the mean performance and the cross-subject variations.

Implementation Details. We employ the 2D U-Net [19] as the segmentation model due to the large variation on slice thickness cross domains. For the hallucinator, we consider image-and-spatial transformer networks (ISTNs) [18], including a CycleGAN-like model [32] for style translation and a spatial transformer network for image registration. Considering memory limitations, we only involve the spatial transformer network in the meta-learning process. More Specifically, we first follow CycleGAN [32] to achieve unpaired image translation for image adaptation. Since limited labels are provided in the source domain, we transform target images to source-like images for training and testing. Then, the pairs of source images and source-like images are fed into our scheme for augmentation and segmentation. For segmentation loss, we use the combination of Dice loss and Cross-entropy loss [29], while the transformation loss involved in meta-learning is based on MSE loss between source images and source-like images [18]. We train the whole framework for 150 epochs using Adam optimizer. The batch size is set as 32, including 8 labeled data, 8 augmented data, and 16 unlabeled data. The number of pairs for meta-train and meta-test are set as 16 and 8, respectively. The learning rate for inner-loop update is set as 0.001. The learning rate for meta-optimization is linearly warmed up from 0 to 0.005 over the first 30 epochs to avoid volatility in early iterations. The EMA decay rate β is set as 0.99, and hyperparameters λ_{con} and λ_{trans} are ramped up individually with the function $\lambda(t) = 10 \times \exp\{-5\,(1 - t/150)^2\}$ (t denotes the current iteration). We apply data augmentations, like random rotation, and extract the largest connected component for each substructure in the 3D mesh for post-processing in all experiments.

Table 1. Segmentation performance of different approaches.

Method	Dice (%) ↑					ASD (voxel) ↓				
	AA	LAC	LVC	MYO	Average	AA	LAC	LVC	MYO	Average
4-shots										
Supervised-only	$85.0_{9.2}$	$87.1_{3.7}$	$75.2_{11.6}$	$63.0_{20.2}$	$77.6_{11.2}$	$2.2_{0.7}$	$3.3_{1.3}$	$5.1_{3.0}$	$5.1_{3.8}$	$3.9_{2.2}$
W/o adaptation	$18.9_{19.7}$	$4.6_{4.6}$	$20.7_{17.0}$	$11.6_{12.5}$	$14.0_{13.5}$	$48.2_{34.0}$	$30.8_{15.0}$	$35.2_{12.7}$	$40.6_{22.6}$	$38.7_{21.1}$
ADDA [23]	$35.5_{24.3}$	$4.2_{4.2}$	$2.1_{3.6}$	$36.9_{3.9}$	$19.7_{9.0}$	$30.9_{38.0}$	$47.9_{31.2}$	$57.4_{21.7}$	$11.8_{3.5}$	$37_{23.6}$
CycleGAN [32]	$43.7_{27.7}$	$49.8_{12.2}$	$43.2_{26.9}$	$23.1_{24.7}$	$40.0_{22.9}$	$25.4_{18.3}$	$12.9_{4.7}$	$17.9_{19.0}$	$36.4_{31.5}$	$23.1_{18.4}$
SIFA [4]	$42.3_{17.4}$	$61.0_{6.6}$	$46.4_{21.1}$	$42.0_{20.2}$	$47.9_{16.3}$	$10.2_{3.4}$	$8.0_{2.8}$	$10.5_{5.6}$	$8.2_{4.3}$	$9.2_{4.0}$
MT [22]	$59.0_{1.8}$	$59.3_{25.8}$	$45.3_{19.2}$	$35.9_{19.2}$	$49.9_{16.5}$	$6.8_{1.0}$	$6.6_{3.6}$	$10.3_{6.8}$	$9.9_{6.2}$	$8.4_{4.4}$
TCSM [12]	$65.3_{3.1}$	$62.7_{16.9}$	$50.9_{13.0}$	$38.3_{7.8}$	$54.3_{10.2}$	$5.6_{0.6}$	$6.2_{2.3}$	$9.6_{6.6}$	$8.2_{2.8}$	$7.4_{3.1}$
ISTN [18]	$34.0_{12.0}$	$61.0_{14.6}$	$47.1_{17.3}$	$32.9_{13.6}$	$43.8_{14.4}$	$10.0_{1.6}$	$5.4_{1.3}$	$9.7_{4.2}$	$10.7_{4.2}$	$9.0_{2.8}$
VoxelMorph [3]	$57.6_{7.2}$	$67.2_{12.9}$	$41.1_{21.0}$	$35.7_{9.3}$	$50.4_{12.6}$	$6.6_{1.1}$	$7.2_{2.7}$	$9.9_{6.3}$	$9.8_{2.8}$	$8.4_{3.2}$
MT-UDA [29]	$67.2_{6.6}$	$\mathbf{80.0_{4.1}}$	$72.1_{8.4}$	$56.2_{11.8}$	$68.9_{7.8}$	$6.3_{2.5}$	$\mathbf{4.1_{1.0}}$	$5.7_{2.6}$	$6.8_{2.5}$	$5.7_{2.2}$
Ours	$\mathbf{75.6_{8.3}}$	$75.1_{11.6}$	$\mathbf{82.3_{4.6}}$	$\mathbf{69.6_{6.8}}$	$\mathbf{75.6_{11.3}}$	$\mathbf{4.8_{2.9}}$	$5.1_{2.5}$	$\mathbf{4.3_{1.7}}$	$\mathbf{4.9_{0.9}}$	$\mathbf{4.8_{2.0}}$
1-shot										
ADDA [23]	$17.3_{12.4}$	$12.7_{7.1}$	$15.7_{12.1}$	$15.2_{11.7}$	$15.2_{10.8}$	$47.1_{12.8}$	$34.5_{4.7}$	$40.5_{12.0}$	$37.3_{10.4}$	$39.9_{10.0}$
CycleGAN [32]	$8.9_{6.3}$	$10.0_{11.7}$	$14.2_{13.2}$	$7.1_{6.9}$	$10.1_{9.5}$	$28.5_{2.8}$	$31.7_{7.1}$	$22.0_{6.7}$	$21.7_{8.7}$	$26.0_{6.3}$
SIFA [4]	$15.3_{12.0}$	$26.3_{21.5}$	$16.8_{12.5}$	$13.0_{10.3}$	$17.9_{14.0}$	$37.6_{15.4}$	$25.3_{21.4}$	$21.7_{12.7}$	$18.5_{9.3}$	$25.8_{14.7}$
MT [22]	$20.1_{16.2}$	$18.2_{9.4}$	$24.1_{13.9}$	$21.1_{5.5}$	$21.0_{11.3}$	$41.8_{22.0}$	$25.7_{6.0}$	$24.5_{9.5}$	$26_{6.8}$	$29.5_{11.1}$
TCSM [12]	$32.7_{15.8}$	$30.8_{9.3}$	$37.7_{13.9}$	$20.1_{5.3}$	$30.3_{11.0}$	$28.0_{11.2}$	$31.9_{12.2}$	$23.3_{11.0}$	$23.1_{7.1}$	$26.6_{9.0}$
ISTN [18]	$24.5_{10.0}$	$21.5_{5.9}$	$26.6_{15.3}$	$18.5_{11.7}$	$22.8_{10.7}$	$32.2_{5.8}$	$46.8_{12.4}$	$25.6_{10.6}$	$27.8_{8.5}$	$33.1_{9.3}$
VoxelMorph [3]	$18.9_{6.1}$	$25.7_{5.8}$	$28.6_{11.3}$	$23.4_{7.6}$	$24.2_{7.7}$	$45.9_{9.4}$	$28.8_{4.9}$	$21.8_{5.0}$	$21.8_{5.2}$	$29.6_{6.1}$
MT-UDA [29]	$37.6_{11.3}$	$\mathbf{43.6_{11.1}}$	$47.5_{15.2}$	$36.0_{5.7}$	$41.2_{10.8}$	$26.8_{11.4}$	$\mathbf{23.5_{12.2}}$	$16.8_{4.7}$	$16.7_{3.0}$	$21.0_{7.8}$
Ours	$\mathbf{64.4_{10.3}}$	$30.9_{10.1}$	$\mathbf{59.1_{6.6}}$	$\mathbf{52.9_{5.0}}$	$\mathbf{51.8_{8.0}}$	$\mathbf{6.3_{1.7}}$	$33.6_{26.5}$	$\mathbf{8.5_{1.7}}$	$\mathbf{7.9_{1.7}}$	$\mathbf{14.1_{7.9}}$

Comparisons of Different Methods. We implement several well-established UDA methods, *i.e.*, a feature adaptation method (**ADDA**) [23], an image adaptation method (**CycleGAN**) [32], and a synergistic image and feature adaptation method (**SIFA**) [4], two recent popular SSL methods, *i.e.*, **MT** [22] and **TCSM** [12], and two representative augmentation (Aug) methods via registration, **ISTN** [18] and **VoxelMorph** [3]. It is noted that we use CycleGAN to close the domain gap at the image level for SSL and Aug methods. Besides, we implement the state-of-the-art few-shot UDA method, **MT-UDA** [29] for comparison. Following previous practices [4,29], we conduct experiments with the **lower "W/o adaptation" baseline** (*i.e.*, directly applying the model trained with source labels to target domain) and the **upper "Supervised-only" baseline** (*i.e.*, training and testing on the target domain).

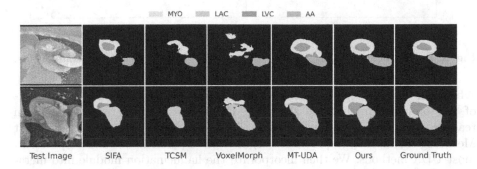

Fig. 2. Visualization of segmentation results generated by different methods.

The results are presented in Table 1. We can see that there is a significant performance gap between the upper and lower bounds due to the domain shifts. Overall, various UDA methods show unsatisfactory adaptation performance compared to the "W/o adaptation" baseline with limited source labels. It is observed that SSL methods, *i.e.*, MT and TCSM can help relax the dependence on source labels by leveraging unlabeled data, while Aug methods such as ISTN and VoxelMorph can also improve the segmentation performance by generating augmented samples. These results suggest that SSL and Aug methods can help unsupervised domain adaptation under source label scarcity. Notably, our method achieves better performance than the UDA, SSL, and Aug methods by a large margin, and outperforms MT-UDA by 6.7% on Dice and 0.9 mm on ASD, showing the effectiveness of our transform-consistent meta-hallucination scheme for few-shot UDA. With fewer source labels (1-shot), our method shows larger performance improvements than other methods, demonstrating that meta-hallucinator is beneficial in label-scarce adaptation scenarios. Moreover, we present the qualitative results of different methods trained on four source labels in Fig. 2 (due to page limit, we only show the best methods in UDA (SIFA), SSL (TCSM) and Aug (VoxelMorph), as well as MT-UDA. More visual comparisons are shown in Appendix). It is observed that our method produces fewer false positives and segments cardiac substructures with smoother boundaries.

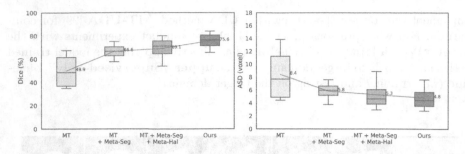

Fig. 3. Boxplot of ablation results (Dice [%] and ASD [voxel]) on different components.

Ablation Study. Here we conduct an ablation analysis on key components of the proposed method, as shown in Fig. 3. We start by advancing mean teacher (MT) into meta-learning with $\mathcal{L}_{Seg}$, *i.e.*, Meta-Seg, emphasizing that Meta-Seg significantly improves the segmentation performance and outperforms most UDA methods. We then incorporate the hallucination module into meta-learning for data augmentation, referred to as Meta-Hal, which yields higher Dice and ASD than Meta-Seg, demonstrating the effectiveness of the meta-hallucination scheme. Finally, by adding hallucination-consistent constraints to enhance the regularization effects for self-ensembling training, consistent performance improvements are obtained with our method.

4 Conclusions

In this work, we propose a novel transformation-consistent meta-hallucination framework for improving few-shot unsupervised domain adaptation in cross-modality cardiac segmentation. We integrate both the hallucination and segmentation models into meta-learning for enhancing the collaboration between the hallucinator and the segmenter and generating helpful samples, thereby improving the cross-modality adaptation performance to the utmost extent. We further introduce the hallucination-consistent constraint to regularize self-ensembling learning simultaneously. Extensive experiments demonstrate the effectiveness of the proposed meta-hallucinator. Our meta-hallucinator can be integrated into different models in a plug-and-play manner and easily extended to various segmentation tasks suffering from domain shifts or label scarcity.

References

1. Bai, W., et al.: Semi-supervised learning for network-based cardiac MR image segmentation. In: Descoteaux, M., Maier-Hein, L., Franz, A., Jannin, P., Collins, D.L., Duchesne, S. (eds.) MICCAI 2017. LNCS, vol. 10434, pp. 253–260. Springer, Cham (2017). https://doi.org/10.1007/978-3-319-66185-8_29
2. Balaji, Y., Sankaranarayanan, S., Chellappa, R.: Metareg: towards domain generalization using meta-regularization. In: Advances in Neural Information Processing Systems, vol. 31 (2018)

3. Balakrishnan, G., Zhao, A., Sabuncu, M.R., Guttag, J., Dalca, A.V.: VoxelMorph: a learning framework for deformable medical image registration. IEEE Trans. Med. Imaging **38**(8), 1788–1800 (2019)
4. Chen, C., Dou, Q., Chen, H., Qin, J., Heng, P.A.: Unsupervised bidirectional cross-modality adaptation via deeply synergistic image and feature alignment for medical image segmentation. IEEE Trans. Med. Imaging **39**(7), 2494–2505 (2020)
5. Finn, C., Abbeel, P., Levine, S.: Model-agnostic meta-learning for fast adaptation of deep networks. In: ICML, pp. 1126–1135. PMLR (2017)
6. Hoffman, J., et al.: CyCADA: cycle-consistent adversarial domain adaptation. In: ICML, pp. 1989–1998. PMLR (2018)
7. Hu, X., Zeng, D., Xu, X., Shi, Y.: Semi-supervised contrastive learning for label-efficient medical image segmentation. In: de Bruijne, M., et al. (eds.) MICCAI 2021. LNCS, vol. 12902, pp. 481–490. Springer, Cham (2021). https://doi.org/10.1007/978-3-030-87196-3_45
8. Khandelwal, P., Yushkevich, P.: Domain generalizer: a few-shot meta learning framework for domain generalization in medical imaging. In: Albarqouni, S., et al. (eds.) DART/DCL -2020. LNCS, vol. 12444, pp. 73–84. Springer, Cham (2020). https://doi.org/10.1007/978-3-030-60548-3_8
9. Kiyasseh, D., Swiston, A., Chen, R., Chen, A.: Segmentation of left atrial MR images via self-supervised semi-supervised meta-learning. In: de Bruijne, M., et al. (eds.) MICCAI 2021. LNCS, vol. 12902, pp. 13–24. Springer, Cham (2021). https://doi.org/10.1007/978-3-030-87196-3_2
10. Lee, M.C.H., Oktay, O., Schuh, A., Schaap, M., Glocker, B.: Image-and-spatial transformer networks for structure-guided image registration. In: Shen, D., et al. (eds.) MICCAI 2019. LNCS, vol. 11765, pp. 337–345. Springer, Cham (2019). https://doi.org/10.1007/978-3-030-32245-8_38
11. Li, S., Zhao, Z., Xu, K., Zeng, Z., Guan, C.: Hierarchical consistency regularized mean teacher for semi-supervised 3D left atrium segmentation. In: 2021 43rd Annual International Conference of the IEEE Engineering in Medicine & Biology Society (EMBC), pp. 3395–3398. IEEE (2021)
12. Li, X., Yu, L., Chen, H., Fu, C.W., Xing, L., Heng, P.A.: Transformation-consistent self-ensembling model for semisupervised medical image segmentation. IEEE Trans. Neural Netw. Learn. Syst. **32**, 523–534 (2020)
13. Liu, Q., Dou, Q., Heng, P.-A.: Shape-aware meta-learning for generalizing prostate MRI segmentation to unseen domains. In: Martel, A.L., et al. (eds.) MICCAI 2020. LNCS, vol. 12262, pp. 475–485. Springer, Cham (2020). https://doi.org/10.1007/978-3-030-59713-9_46
14. Liu, X., et al.: Generative self-training for cross-domain unsupervised tagged-to-cine MRI synthesis. In: de Bruijne, M., et al. (eds.) MICCAI 2021. LNCS, vol. 12903, pp. 138–148. Springer, Cham (2021). https://doi.org/10.1007/978-3-030-87199-4_13
15. Long, J., Shelhamer, E., Darrell, T.: Fully convolutional networks for semantic segmentation. In: Proceedings of the IEEE Conference on Computer Vision and Pattern Recognition, pp. 3431–3440 (2015)
16. Lyu, Y., Liao, H., Zhu, H., Zhou, S.K.: A^{3}DSegNet: anatomy-aware artifact disentanglement and segmentation network for unpaired segmentation, artifact reduction, and modality translation. In: Feragen, A., Sommer, S., Schnabel, J., Nielsen, M. (eds.) IPMI 2021. LNCS, vol. 12729, pp. 360–372. Springer, Cham (2021). https://doi.org/10.1007/978-3-030-78191-0_28

17. Ouyang, C., Biffi, C., Chen, C., Kart, T., Qiu, H., Rueckert, D.: Self-supervision with superpixels: training few-shot medical image segmentation without annotation. In: Vedaldi, A., Bischof, H., Brox, T., Frahm, J.-M. (eds.) ECCV 2020. LNCS, vol. 12374, pp. 762–780. Springer, Cham (2020). https://doi.org/10.1007/978-3-030-58526-6_45

18. Robinson, R., et al.: Image-level harmonization of multi-site data using image-and-spatial transformer networks. In: Martel, A.L., et al. (eds.) MICCAI 2020. LNCS, vol. 12267, pp. 710–719. Springer, Cham (2020). https://doi.org/10.1007/978-3-030-59728-3_69

19. Ronneberger, O., Fischer, P., Brox, T.: U-Net: convolutional networks for biomedical image segmentation. In: Navab, N., Hornegger, J., Wells, W.M., Frangi, A.F. (eds.) MICCAI 2015. LNCS, vol. 9351, pp. 234–241. Springer, Cham (2015). https://doi.org/10.1007/978-3-319-24574-4_28

20. Shin, S.Y., Lee, S., Summers, R.M.: Unsupervised domain adaptation for small bowel segmentation using disentangled representation. In: de Bruijne, M., et al. (eds.) MICCAI 2021. LNCS, vol. 12903, pp. 282–292. Springer, Cham (2021). https://doi.org/10.1007/978-3-030-87199-4_27

21. Tang, H., Liu, X., Sun, S., Yan, X., Xie, X.: Recurrent mask refinement for few-shot medical image segmentation. In: Proceedings of the IEEE/CVF International Conference on Computer Vision, pp. 3918–3928 (2021)

22. Tarvainen, A., Valpola, H.: Mean teachers are better role models: weight-averaged consistency targets improve semi-supervised deep learning results. arXiv preprint arXiv:1703.01780 (2017)

23. Tzeng, E., Hoffman, J., Saenko, K., Darrell, T.: Adversarial discriminative domain adaptation. In: Proceedings of the IEEE Conference on Computer Vision and Pattern Recognition, pp. 7167–7176 (2017)

24. Yang, J., Dvornek, N.C., Zhang, F., Chapiro, J., Lin, M.D., Duncan, J.S.: Unsupervised domain adaptation via disentangled representations: application to cross-modality liver segmentation. In: Shen, D., et al. (eds.) MICCAI 2019. LNCS, vol. 11765, pp. 255–263. Springer, Cham (2019). https://doi.org/10.1007/978-3-030-32245-8_29

25. Zeng, Z., Xulei, Y., Qiyun, Y., Meng, Y., Le, Z.: SeSe-Net: self-supervised deep learning for segmentation. Pattern Recogn. Lett. **128**, 23–29 (2019)

26. Zhang, R., Che, T., Ghahramani, Z., Bengio, Y., Song, Y.: MetaGAN: an adversarial approach to few-shot learning. In: Advances in Neural Information Processing Systems, vol. 31 (2018)

27. Zhao, A., Balakrishnan, G., Durand, F., Guttag, J.V., Dalca, A.V.: Data augmentation using learned transformations for one-shot medical image segmentation. In: Proceedings of the IEEE/CVF Conference on Computer Vision and Pattern Recognition, pp. 8543–8553 (2019)

28. Zhao, N., Chua, T.S., Lee, G.H.: SESS: self-ensembling semi-supervised 3D object detection. In: Proceedings of the IEEE/CVF Conference on Computer Vision and Pattern Recognition, pp. 11079–11087 (2020)

29. Zhao, Z., Xu, K., Li, S., Zeng, Z., Guan, C.: MT-UDA: towards unsupervised cross-modality medical image segmentation with limited source labels. In: de Bruijne, M., et al. (eds.) MICCAI 2021. LNCS, vol. 12901, pp. 293–303. Springer, Cham (2021). https://doi.org/10.1007/978-3-030-87193-2_28

30. Zhao, Z., Zeng, Z., Xu, K., Chen, C., Guan, C.: DSAL: deeply supervised active learning from strong and weak labelers for biomedical image segmentation. IEEE J. Biomed. Health Inform. **25**, 3744–3751 (2021)

31. Zhou, S.K., et al.: A review of deep learning in medical imaging: imaging traits, technology trends, case studies with progress highlights, and future promises. Proc. IEEE **109**, 820–838 (2021)
32. Zhu, J.Y., Park, T., Isola, P., Efros, A.A.: Unpaired image-to-image translation using cycle-consistent adversarial networks. In: Proceedings of the IEEE International Conference on Computer Vision, pp. 2223–2232 (2017)
33. Zhuang, X., Shen, J.: Multi-scale patch and multi-modality atlases for whole heart segmentation of MRI. Med. Image Anal. **31**, 77–87 (2016)
34. Zou, Y., Yu, Z., Liu, X., Kumar, B., Wang, J.: Confidence regularized self-training. In: Proceedings of the IEEE/CVF International Conference on Computer Vision, pp. 5982–5991 (2019)

NestedFormer: Nested Modality-Aware Transformer for Brain Tumor Segmentation

Zhaohu Xing[1], Lequan Yu[2], Liang Wan[1(✉)], Tong Han[3], and Lei Zhu[4,5]

[1] Medical College of Tianjin University, Tianjin, China
{xingzhaohu,lwan}@tju.edu.cn
[2] The University of Hong Kong, Hong Kong, China
[3] Brain Medical Center of Tianjin University, Huanhu Hospital, Tianjin, China
[4] The Hong Kong University of Science and Technology (Guangzhou),
Guangzhou, China
[5] The Hong Kong University of Science and Technology, Hong Kong, China

Abstract. Multi-modal MR imaging is routinely used in clinical practice to diagnose and investigate brain tumors by providing rich complementary information. Previous multi-modal MRI segmentation methods usually perform modal fusion by concatenating multi-modal MRIs at an early/middle stage of the network, which hardly explores non-linear dependencies between modalities. In this work, we propose a novel Nested Modality-Aware Transformer (NestedFormer) to explicitly explore the intra-modality and inter-modality relationships of multi-modal MRIs for brain tumor segmentation. Built on the transformer-based multi-encoder and single-decoder structure, we perform nested multi-modal fusion for high-level representations of different modalities and apply modality-sensitive gating (MSG) at lower scales for more effective skip connections. Specifically, the multi-modal fusion is conducted in our proposed Nested Modality-aware Feature Aggregation (NMaFA) module, which enhances long-term dependencies within individual modalities via a tri-orientated spatial-attention transformer, and further complements key contextual information among modalities via a cross-modality attention transformer. Extensive experiments on BraTS2020 benchmark and a private meningiomas segmentation (MeniSeg) dataset show that the Nested-Former clearly outperforms the state-of-the-arts. The code is available at https://github.com/920232796/NestedFormer.

Keywords: Multi-modal MRI · Brain tumor segmentation · Nested Modality-aware Feature Aggregation · Modality-sensitive gating

1 Introduction

Brain tumor is one of the most common cancers in the world [3], in which gliomas are the most common malignant brain tumors with different levels of aggres-

Supplementary Information The online version contains supplementary material available at https://doi.org/10.1007/978-3-031-16443-9_14.

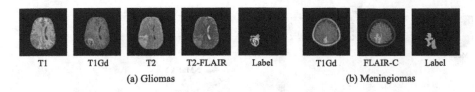

T1   T1Gd   T2   T2-FLAIR   Label   T1Gd   FLAIR-C   Label

(a) Gliomas   (b) Meningiomas

Fig. 1. Multi-modal MRIs for (a) Gliomas; and (b) Meningiomas.

siveness and meningiomas are the most prevalent primary intracranial tumors in adults [21]. Multi-modal magnetic resonance imaging (MRI) is routinely used in the clinic by providing rich complementary information for analyzing brain tumors. Specifically, for gliomas, the commonly used MRI sequences are T1-weighted (T1), post-contrast T1-weighted (T1Gd), T2-weighted (T2) and T2 Fluid Attenuation Inversion Recovery (T2-FLAIR) images; see Fig. 1(a), each with varying roles in distinguishing tumor, peritumoral edema and tumor core [1,2,18]. For meningiomas, they have different characteristic appearances on T1Gd [15] and contrast-enhanced T2-FLAIR (shorted for FLAIR-C) MRI images; see Fig. 1(b). Thus, automatic segmentation of brain tumor structures from multi-modal MRIs is important for clinical diagnosis and treatment planning.

In recent years, convolutional neural networks (CNNs) have achieved promising successes in brain tumor segmentation. The main stream models are built upon the encoder-decoder architecture [22] with skip connections, including S3D-UNet [5], SegResNet [20], HPU-Net [13], etc. Recent works [9,25,28] also explore transformer [24] to model long-range dependencies within images. For instance, TransBTS [25] utilizes 3D-CNN to extract local spatial features, and applies transformer to model global dependencies on high-level features. UNETR [9] uses the ViT transformer as the encoder to learn contextual information, which is merged with the CNN-based decoder via skip connections at multiple resolutions. However, the transformer in these methods is used to enhance the encode path without specific design for multi-modal fusion.

To utilize the multi-modal information, most of existing methods adopt an early-fusion strategy, in which multi-modal images are concatenated as the network input. However, this strategy can hardly explore non-linear relationships between different modalities. To alleviate this problem, recent works follow a layer-fusion strategy [7,27,29], where the modality-specific features extracted by different encoders are fused in the middle layers of the network and share the same decoder. In HyeprDense-Net [7], each modality has a separated stream and dense connections are introduced between layers within the same stream and also across different streams. MAML [27] embeds multi-modal images by different modality-specific FCNs and then applies a modality-aware module to regress attention maps in order to fuse the modality-specific features. Nevertheless, these multi-modal fusion methods do not build the long-range spatial dependencies within and cross modalities, so that they cannot fully utilize the complementary information of different modalities.

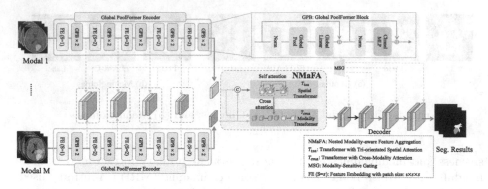

Fig. 2. An overview of the proposed NestedFormer. We design a Nested Modality-aware Feature Aggregation (NMaFA) module to model both the intra- and inter-modality features for multi-modal fusion.

In this paper, we propose a novel nested modality-aware transformer, called NestedFormer, for effective and robust multi-modal brain tumor segmentation. We first design an effective Global Poolformer to extract discriminative volumetric spatial features, with more emphasis on global dependencies, from different MRI modalities. To better extract the complementary features and enable any number of modalities for fusion, we propose a novel Nested Modality-aware Feature Aggregation (NMaFA) module. It explicitly considers both single-modality spatial coherence and cross-modality coherence, and leverages nested transformers to establish the intra- and inter-modality long-range dependencies, resulting in more effective feature representation. Moreover, we design a computationally efficient Tri-orientated Spatial Attention (TSA) paradigm to accelerate the 3D-spatial-coherence calculation. To improve feature reuse effect in the decoding, a novel modality-sensitive gating (MSG) module is developed to dynamically filter modality-aware low-resolution features for effective skip connections. Extensive experiments on BraTS2020 benchmark and a privately collected meningiomas segmentation dataset (MeniSeg) show that our model clearly ourperforms the state-of-the-art methods.

2   Method

Figure 2 illustrates the overview of the proposed NestedFormer, which consists of three components: 1) multiple encoders to obtain multi-scale representations of different modalities, 2) a NMaFA fusion module to explore correlated features within and between multi-modal high-level embeddings, and 3) a gating strategy to selectively transfer modality-sensitive low-resolution features to the decoder.

2.1   Global Poolformer Encoder

Recent works show that transformer is more conducive to modeling global information than CNNs. To better extract local context information for each modality, we extend the Poolformer [26] as the modality-specific encoder. As discussed in [26], replacing the computation-intensive attention module in Transformer with average pooling can achieve superior performance than recent transformer and MLP-like models. Therefore, to enhance global information, we design Global Poolformer Block (GPB), which leverages global pooling instead of average pooling in Poolformer, followed by a fully connected layer. As shown in Fig. 2, given the input feature embedding X, a GPB block consists of a learnable global pooling (GP) and a MLP sub-block. The output Z is computed as,

$$
\begin{aligned}
Y &= GP(LN(X))W_g + X, \\
Z &= MLP(LN(Y)) + Y,
\end{aligned}
\tag{1}
$$

where $LN(*)$ denotes the layer normalization and W_g is the learnable parameter in the FC layer. Our Global Poolformer encoder contains five groups of one feature embedding (FE) layer and two GPB blocks. Each FE layer is a 3D-convolution, while the first FE layer has a convolution patch size of $1 \times 1 \times 1$ and the rest layers have a patch size of $2 \times 2 \times 2$ and a stride of 2. The encoders gradually encode each modality image into high-level feature $F_{L,i} \subset \mathbb{R}^{d \times w \times h \times C}, i \in [1, M]$, where $(d, w, h) = (\frac{D}{16}, \frac{W}{16}, \frac{H}{16})$ are $1/16$ of input spatial resolutions H, W and depth dimension D; M is the number of modal images, the channel dimension C and the layer number L are set as $C = 128$, $L = 5$.

2.2   Nested Modality-Aware Feature Aggregation

Given high-level features $F_{L,i}, i \in [1, M]$, NMaFA leverages a spatial-attention based transformer T_{tsa} and a cross-modality attention based transformer T_{cma} in a nested manner; see Fig. 3. First, transformer T_{tsa} utilizes the self-attention scheme to compute the long-range correlation between different patches in the space within each modality. Specifically, $F_{L,i}$ is *concatenated in the channel dimension* to obtain high-level embedding $F_s \in \mathbb{R}^{d \times w \times h \times MC}$. In this work, each location of F_s is considered as one "patch". Then a patch embedding layer maps F_s to a token sequence $\hat{F}_s \in \mathbb{R}^{dwh \times C}$. T_{tsa} takes $\hat{F}_s$ and the position encoding [24] as the input, and outputs spatially-enhanced feature $\tilde{F}_s \in \mathbb{R}^{dwh \times C}$.

Second, transformer T_{cma} utilizes the cross-attention scheme to further compute the global relation among different modalities to achieve inter-modality fusion. To this end, $F_{L,i}$ is *concatenated in the spatial dimension* to obtain the flatten sequence $\hat{F}_c \in \mathbb{R}^{MP \times C}$. Here, $P(P = 32)$ denotes the number of dominant tokens learnt via the Token Learner strategy [23], which helps to reduce the computational scope especially when the number of tokens increases greatly along with more modalities. After that, both $\tilde{F}_s$ and $\hat{F}_c$ are fed into T_{cma} to get the modality-enhanced feature embedding $\tilde{F}$.

Also note that our two modules are different from traditional channel-spatial attention networks, which reweigh feature maps channel-wise and spatial-wise. Our NMaFA relies on transformer mechanism and the two transformers are fused in a nested form, rather than serial [12] or parallel [19] fusion.

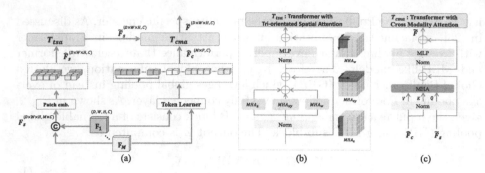

Fig. 3. NMaFA: Nested Modality-aware Feature Aggregation. (a) The overall architecture. (b) The transformer with tri-orientated spatial attention T_{tsa}. (c) The transformer with cross-modality attention T_{cma}.

Transformer with Tri-orientated Spatial Attention. To improve the computational efficiency of spatial attention for volumetric embeddings, inspired by [10,16], we leverage axial-wise attention MHA_z, plane-wise attention MHA_{xy}, and window-wise attention MHA_w; see Fig. 3(b). Concretely, MHA_z models the long-range relationship among feature tokens along the vertical direction; MHA_{xy} models the long-range relationship within each slice; MHA_w uses sliding windows to model the relationship across local 3D-windows. We employ axial and planar learnable absolute position encodings [24] for MHA_z and MHA_{xy}, respectively, and use relative position encoding for window-wise attention MHA_w [16]. The resultant attention is computed as follows,

$$MHA_{tsa}(z) = MHA_z(z) + MHA_{xy}(z) + MHA_w(z), \tag{2}$$

where $z \in \mathbb{R}^{N \times C}$ denotes the embedding tokens with sequence length N and embedding dimension C after layer normalization, $N = d \times w \times h$. By this way, the model not only enhances feature extraction of local important regions, but also calculates global feature dependencies with less computation.

Transformer with Cross-Modality Attention. By concatenating features *in the channel dimension*, T_{tsa} mainly enhances the dependencies within each modality and yields $\tilde{F}_s$, although the inter-modality integration also takes place via patch embedding. To explicitly explore relationship among modalities, we concatenate the feature tokens of different modalities *along the spatial dimension*, yielding $\hat{F}_c \in \mathbb{R}^{MP \times C}$; and then use a cross-attention transformer T_{cma} to enhance the modality dependency information into $\tilde{F}_s$; see Fig. 3(c). The input triplet of (Query, Key, Value) to the cross-attention is computed as

$$Q = \tilde{F}_s W_q, \quad K = \hat{F}_c W_k, \quad V = \hat{F}_c W_v, \tag{3}$$

where $W_q, W_k, W_v \in \mathbb{R}^{C \times d}$ are the weight matrices, $d = 128$ is the dimension of Q, K, V. The cross attention CA is then formulated as

$$CA(\tilde{F}_s, \hat{F}_c) = \hat{F}_s + SoftMax(\frac{QK^T}{\sqrt{d}})V. \tag{4}$$

The resultant token sequence $\tilde{F}$ from T_{cma} fuses and enhances the input features with increasing receptive fields and the cross-modal global relevance.

2.3 Modality-Sensitive Gating

In feature decoding, we first fold the tokens $\tilde{F}$ back to a high-level 4D feature map $R_L \in \mathbb{R}^{d \times w \times h \times C}$. R_L is progressively processed in a regular bottom-up style with a 3D convolution and $2\times$ upsampling operation to recover a full resolution feature map $R_1 \in \mathbb{R}^{D \times W \times H \times N_c}$ for segmentation, where N_c is the number of segments. Note that the encoder features are multi-modal. Hence, we design a modality-sensitive gating strategy in skip connection, to filter the encoder features $\{F_{l,i}, l \in [1,4], i \in [1,M]\}$ according to the modality importance. To be specific, for the l-th layer, an modality importance map $I_l \in \mathbb{R}^{\frac{D}{2^{l-1}} \times \frac{W}{2^{l-1}} \times \frac{H}{2^{l-1}} \times M}$ is learnt from $\tilde{F}$ that is the output of NMaFA, as follows,

$$I_l = \sigma(U_{L-l}^{2\times}(FC(\tilde{F}))), \tag{5}$$

where $FC(*)$ is a $1 \times 1 \times 1$ full connection layer, $U_l^{2\times}$ denotes upsampling l times, and $\sigma(*)$ is the sigmoid function. Denote $\odot$ as element-wise multiplication. Then the filtered encoder feature $\mathcal{F}_l$ is formulated as

$$\mathcal{F}_l = \sum_i (I_{l,i} \odot F_{l,i}). \tag{6}$$

3 Experiment

3.1 Implementation Details

Our NestedFormer was implemented in PyTorch1.7.0 on a NVIDIA GTX 3090 GPU. The parameters were initialized via Xavier [8]. The loss function was a combination of soft dice loss and cross-entropy loss and we adopted the AdamW optimizer [17] with a weight decay of 10^{-5}. The learning rate was empirically set as 10^{-4}. We adopted two T_{tsa} sequentially and just one T_{cma}. In MHA_w, the window-size was set as (2, 2, 2) for BraTS2020 and (2, 4, 4) for MeniSeg.

3.2 Datasets and Evaluation Metrics

For evaluation, we use a public brain tumor segmentation dataset **BraTS2020** [18] and a private 3D meningioma segmentation dataset (**MeniSeg**) collected from Brain Medical Center of Tianjin University, Tianjin Huanhu hospital. Dice score and 95% Hausdorff Distance (HD95) are adopted for quantitative comparison.

BraTS2020 Dataset. The BraTS2020 training dataset contains 369 aligned four-modal MRI data (i.e., T1, T1Gd, T2, T2-FLAIR), with expert segmentation masks (i.e., GD-enhancing tumor, peritumoral edema, and tumor core). Each

Table 1. Quantitative comparison on BraTS 2020 dataset.

Methods	Param (M)	FLOPs (G)	WT		TC		ET		Ave	
			Dice↑	HD95↓	Dice↑	HD95↓	Dice↑	HD95↓	Dice↑	HD95↓
3D-UNet [6]	5.75	1449.59	0.882	5.113	0.830	6.604	0.782	6.715	0.831	6.144
SegResNet [20]	18.79	185.23	0.903	4.578	0.845	5.667	0.796	7.064	0.848	5.763
MAML [27]	5.76	577.65	0.914	4.804	0.854	5.594	0.796	**5.221**	0.855	5.206
nnUNet [11]	5.75	1449.59	0.907	6.94	0.848	5.069	**0.814**	5.851	0.856	5.953
SwinUNet(2D) [4]	27.17	357.49	0.872	6.752	0.809	8.071	0.744	10.644	0.808	8.489
TransBTS [25]	32.99	333	0.910	**4.141**	0.855	5.894	0.791	5.463	0.852	5.166
UNETR [9]	92.58	41.19	0.899	4.314	0.842	5.843	0.788	5.598	0.843	5.251
NestedFormer	10.48	71.77	**0.920**	4.567	**0.864**	**5.316**	0.800	5.269	**0.861**	**5.051**

modality has a $155 \times 240 \times 240$ volume and is already resampled and co-registered. The segmentation task aims to segment the whole tumor (WT), enhancing tumor (ET), and tumor core (TC) regions. Following the recent work [14], we randomly divide the dataset into training (315), validation (17) and test (37).

Meningioma Dataset. The MeniSeg dataset contains 110 annotated two-modal MRIs (i.e., T1Gd and FLAIR-C) from the meningiomas patients, who had undergone tumor resection between March 2016 and March 2021. MRI scans were performed with four 3.0T MRI scanners (Skyra, Trio, Avanto, Prisma from Siemens). Two radiologists annotated meningioma tumor and edema masks on T1Gd and FLAIR-C MRIs, and the third high-experienced radiologist made examination. Each modality data has a volume of $32 \times 256 \times 256$, and is aligned into the same space and sampled to volume sizes of [32, 192, 192] for training. Two-fold cross-validation is conducted for all the compared methods.

3.3   Comparison with SOTA Methods

We compare our network against seven SOTA segmentation methods, including three CNN-based methods (3D-UNet [6], SegResNet [20], MAML [27], nnUNet [11]), and three transformer-based methods (SwinUNet(2D) [4], Trans-BTS [25], and UNETR [9]). For a fair comparison, we utilize the public implementations of compared methods to re-train their networks for generating their best segmentation results. Considering the computation power, all the methods are trained for at most 300 epochs on BraTS2020 and 200 epochs on MeniSeg.

BraTS2020. Table 1 reports the Dice and HD95 scores on three regions (WT, TC, and ET) as well as the averaged scores of all the methods on BraTS2020. Apparently, our NestedFormer achieves the largest Dice score on WT, the largest Dice score on TC, the smallest HD95 scores on TC, and our method also ranks second in Dice score on ET, and second in HD95 score on WT and ET. More importantly, our method has the best quantitative performance with averaging Dice and HD95 scores to be 0.861 and 5.051. It is noted that HD95 is for the distance difference between two sets of points, which is more sensitive than Dice [25]. Hence, Dice is often used as the main metric and HD95 as the reference.

Table 2. Quantitative comparison on MeniSeg dataset.

Methods	Tumor		Edema		Ave	
	Dice↑	HD95↓	Dice↑	HD95↓	Dice↑	HD95↓
3D-UNet [6]	0.799	5.099	0.676	9.655	0.737	7.377
SegResNet [20]	0.813	2.970	0.665	10.438	0.739	6.704
MAML [27]	0.819	2.112	0.682	9.158	0.750	5.635
SwinUNet(2D) [4]	0.807	1.817	0.679	7.986	0.743	4.901
TransBTS [25]	0.809	**1.742**	0.679	6.388	0.744	**4.065**
UNETR [9]	0.818	3.279	0.693	7.837	0.755	5.813
NestedFormer	**0.834**	2.647	**0.695**	**6.173**	**0.765**	4.410

Table 3. Ablation study for different modules on MeniSeg.

Methods	Encoder			Fusion			Dice		
	CNN	PB	GPB	T_{tsa}	T_{cma}	MSG	Tmuor	Edema	Ave
baseline1	✓						0.805	0.675	0.74
baseline2		✓					0.810	0.679	0.75
baseline3		✓		✓			0.816	0.688	0.752
baseline4		✓		✓	✓		0.825	**0.699**	0.762
baseline5	✓			✓	✓	✓	0.823	0.697	0.76
NestedFormer			✓	✓	✓	✓	**0.834**	0.695	**0.765**

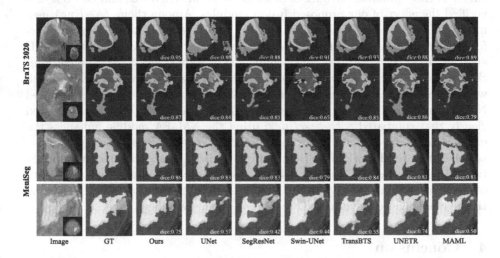

Fig. 4. The visual comparison results on BraTS2020 and MeniSeg dataset.

We also experimented with two-fold cross-validation for UNETR, TransBTS and our method, while our method outperforms the two methods in WT and TC, and is quite close to the best result in ET. As for model complexity, our model has 10.48M parameters and 71.77G FLOPs which is a moderate size model.

MeniSeg. In Table 2, we list Dice and HD95 scores of our network and compared methods on tumor and edema regions on the MeniSeg dataset, as well as the average metrics. Among all the compared methods, MAML has the largest Dice score of 0.819 at the tumor segmentation, while UNETR has the largest Dice score of 0.693 at the edema segmentation, and average Dice score of 0.755. In comparison, our method has a 1.5% Dice improvement in meningioma tumor, 0.2% Dice improvement in edema, and 1.0% average Dice improvement. Regarding HD95, our method achieves the 4th smallest score of 2.647 on the tumor segmentation, and the smallest score of 6.173 on the edema segmentation.

Visual Comparisons on BraTS2020 and MeniSeg. Figure 4 visually compares the segmentation results predicted by our network and SOTA methods on BraTS2020 and MeniSeg. From these visualization results, we can find that our method can more accurately segment brain tumor and peritumoral edema regions than all the compared methods. The reason behind is that our method is able to better fuse multi-modal MRIs by explicitly exploring the intra-modality and the inter-modality relationships among multiple modalities.

3.4 Ablation Study

We conduct ablation studies on the MeniSeg dataset to evaluate the contributions of main modules in our method; see Table 3. We not only compared the effects of three different encoder backbones based on CNN, PB, and GP, but also verified the effect of our proposed fusion modules. Among them, *baseline*1 uses multiple U-Net encoders to extract features of different modal images, and performs feature fusion by concatenation. *baseline*2-*baseline*4 uses multiple GPB encoders to extract features, and conducts skip connection via simple convolution, w/o T_{tsa} and T_{cma} (see Fig. 3), respectively. *baseline*5 replaces GPB block with the original PoolFormer block (PB) in the encoder, using the proposed NMaFA module (including T_{tsa} and T_{cma}) as well as MSG. It can be observed clearly that compared with *baseline*2, using the NMaFA module enhances the extraction of long-distance dependency information and effectively improves the segmentation results, while GPB outperforms PB by considering global information. Moreover, the MSG module is added to increase the feature reuse capability of skip connections, which further improves the segmentation effect, achieving the best average segmentation Dice (0.765) on the MeniSeg dataset.

4   Conclusion

We propose a novel multi-modal segmentation framework, dubbed as Nested-Former. This architecture extracts the features of M modalities by using multiple Global Poolformer Encoders. Then, the high-level features are effectively fused by the NMaFA module, and the low-level features are selected by the modality-sensitive gate (MSG) module. Through these proposed modules, the network effectively extracts and hierarchically fuses features from different modalities. The effectiveness of our proposed NestedFormer is validated on BraTS2020 and MeniSeg datasets. Our framework are modality-agnostic and can be extended to other multimodal medical data. In the future work, we will explore more efficient feature fusion on low-levels to further improve the segmentation performance.

Acknowledgments. This work was supported by the grant from Tianjin Natural Science Foundation (Grant No. 20JCYBJC00960) and HKU Seed Fund for Basic Research (Project No. 202111159073).

References

1. Bakas, S., et al.: Advancing the cancer genome atlas glioma MRI collections with expert segmentation labels and radiomic features. Sci. Data **4**(1), 1–13 (2017)
2. Bakas, S., et al.: Identifying the best machine learning algorithms for brain tumor segmentation, progression assessment, and overall survival prediction in the brats challenge. arXiv preprint arXiv:1811.02629 (2018)
3. Bray, F., Ferlay, J., Soerjomataram, I., Siegel, R., Torre, L., Jemal, A.: Global cancer statistics 2018: Globocan estimates of incidence and mortality worldwide for 36 cancers in 185 countries. CA Cancer J. Clin. **68**(6), 394–424 (2018)
4. Cao, H., et al.: Swin-unet: Unet-like pure transformer for medical image segmentation. arXiv preprint arXiv:2105.05537 (2021)
5. Chen, W., Liu, B., Peng, S., Sun, J., Qiao, X.: S3D-UNet: separable 3D U-Net for brain tumor segmentation. In: Crimi, A., Bakas, S., Kuijf, H., Keyvan, F., Reyes, M., van Walsum, T. (eds.) BrainLes 2018. LNCS, vol. 11384, pp. 358–368. Springer, Cham (2019). https://doi.org/10.1007/978-3-030-11726-9_32
6. Çiçek, Ö., Abdulkadir, A., Lienkamp, S.S., Brox, T., Ronneberger, O.: 3d u-net: learning dense volumetric segmentation from sparse annotation. In: International Conference on Medical Image Computing and Computer-Assisted Intervention, pp. 424 432. Springer (2016)
7. Dolz, J., Gopinath, K., Yuan, J., Lombaert, H., Desrosiers, C., Ayed, I.B.: Hyperdense-net: a hyper-densely connected CNN for multi-modal image segmentation. IEEE Trans. Med. Imaging **38**(5), 1116–1126 (2019)
8. Glorot, X., Bengio, Y.: Understanding the difficulty of training deep feedforward neural networks. In: Proceedings of theTthirteenth International Conference on Artificial Intelligence and Statistics, pp. 249–256. JMLR Workshop and Conference Proceedings (2010)
9. Hatamizadeh, A., et al.: Unetr: transformers for 3d medical image segmentation. In: Proceedings of the IEEE/CVF Winter Conference on Applications of Computer Vision, pp. 574–584 (2022)
10. Ho, J., Kalchbrenner, N., Weissenborn, D., Salimans, T.: Axial attention in multi-dimensional transformers. arXiv preprint arXiv:1912.12180 (2019)
11. Isensee, F., Jaeger, P.F., Kohl, S.A., Petersen, J., Maier-Hein, K.H.: nnu-net: a self-configuring method for deep learning-based biomedical image segmentation. Nat. Methods **18**(2), 203–211 (2021)
12. Khanh, T.L.B., et al.: Enhancing u-net with spatial-channel attention gate for abnormal tissue segmentation in medical imaging. Appl. Sci. **10**(17), 5729 (2020)
13. Kong, X., Sun, G., Wu, Q., Liu, J., Lin, F.: Hybrid pyramid U-Net model for brain tumor segmentation. In: Shi, Z., Mercier-Laurent, E., Li, J. (eds.) IIP 2018. IAICT, vol. 538, pp. 346–355. Springer, Cham (2018). https://doi.org/10.1007/978-3-030-00828-4_35
14. Larrazabal, A.J., Martínez, C., Dolz, J., Ferrante, E.: Orthogonal ensemble networks for biomedical image segmentation. In: de Bruijne, M., et al. (eds.) MICCAI 2021. LNCS, vol. 12903, pp. 594–603. Springer, Cham (2021). https://doi.org/10.1007/978-3-030-87199-4_56
15. Li, X., Lu, Y., Xiong, J., Wang, D., She, D., Kuai, X., Geng, D., Yin, B.: Presurgical differentiation between malignant haemangiopericytoma and angiomatous meningioma by a radiomics approach based on texture analysis. J. Neuroradiol. **46**(5), 281–287 (2019)

16. Liu, Z., et al.: Swin transformer: hierarchical vision transformer using shifted windows. arXiv preprint arXiv:2103.14030 (2021)
17. Loshchilov, I., Hutter, F.: Decoupled weight decay regularization. arXiv preprint arXiv:1711.05101 (2017)
18. Menze, B.H., et al.: The multimodal brain tumor image segmentation benchmark (brats). IEEE Trans. Med. Imaging **34**(10), 1993–2024 (2014)
19. Mou, L., et al.: CS-Net: channel and spatial attention network for curvilinear structure segmentation. In: Shen, D., et al. (eds.) MICCAI 2019. LNCS, vol. 11764, pp. 721–730. Springer, Cham (2019). https://doi.org/10.1007/978-3-030-32239-7_80
20. Myronenko, A.: 3D MRI brain tumor segmentation using autoencoder regularization. In: International MICCAI Brainlesion Workshop, pp. 311–320. Springer (2018)
21. Ostrom, Q.T., Patil, N., Cioffi, G., Waite, K., Kruchko, C., Barnholtz-Sloan, J.S.: Cbtrus statistical report: primary brain and central nervous system tumors diagnosed in the united states in 2013–2017. Neuro Oncol. **22**(iv), 1–96 (2020)
22. Ronneberger, O., Fischer, P., Brox, T.: U-Net: convolutional networks for biomedical image segmentation. In: Navab, N., Hornegger, J., Wells, W.M., Frangi, A.F. (eds.) MICCAI 2015. LNCS, vol. 9351, pp. 234–241. Springer, Cham (2015). https://doi.org/10.1007/978-3-319-24574-4_28
23. Ryoo, M.S., Piergiovanni, A., Arnab, A., Dehghani, M., Angelova, A.: Tokenlearner: what can 8 learned tokens do for images and videos? arXiv preprint arXiv:2106.11297 (2021)
24. Vaswani, A., et al.: Attention is all you need. In: Advances in Neural Information Processing Systems, pp. 5998–6008 (2017)
25. Wang, W., Chen, C., Ding, M., Yu, H., Zha, S., Li, J.: TransBTS: multimodal brain tumor segmentation using transformer. In: de Bruijne, M., et al. (eds.) MICCAI 2021. LNCS, vol. 12901, pp. 109–119. Springer, Cham (2021). https://doi.org/10.1007/978-3-030-87193-2_11
26. Yu, W., et al.: Metaformer is actually what you need for vision. arXiv preprint arXiv:2111.11418 (2021)
27. Zhang, Y., et al.: Modality-Aware Mutual Learning for Multi-modal Medical Image Segmentation. In: de Bruijne, M., et al. (eds.) MICCAI 2021. LNCS, vol. 12901, pp. 589–599. Springer, Cham (2021). https://doi.org/10.1007/978-3-030-87193-2_56
28. Zhang, Y., Pei, Y., Zha, H.: Learning dual transformer network for diffeomorphic registration. In: de Bruijne, M., et al. (eds.) MICCAI 2021. LNCS, vol. 12904, pp. 129–138. Springer, Cham (2021). https://doi.org/10.1007/978-3-030-87202-1_13
29. Zhou, T., Canu, S., Vera, P., Ruan, S.: 3D medical multi-modal segmentation network guided by multi-source correlation constraint. In: 25th International Conference on Pattern Recognition, pp. 10243–10250. IEEE (2020)

MaxStyle: Adversarial Style Composition for Robust Medical Image Segmentation

Chen Chen[1(✉)], Zeju Li[1], Cheng Ouyang[1], Matthew Sinclair[1,2],
Wenjia Bai[1,3,4], and Daniel Rueckert[1,5]

[1] BioMedIA Group, Department of Computing, Imperial College London,
London, UK
chen.chen15@imperial.ac.uk
[2] HeartFlow, Mountain View, USA
[3] Data Science Institute, Imperial College London, London, UK
[4] Department of Brain Sciences, Imperial College London, London, UK
[5] Klinikum rechts der Isar, Technical University of Munich, Munich, Germany

Abstract. Convolutional neural networks (CNNs) have achieved remarkable segmentation accuracy on benchmark datasets where training and test sets are from the same domain, yet their performance can degrade significantly on unseen domains, which hinders the deployment of CNNs in many clinical scenarios. Most existing works improve model out-of-domain (OOD) robustness by collecting multi-domain datasets for training, which is expensive and may not always be feasible due to privacy and logistical issues. In this work, we focus on improving model robustness using a *single-domain* dataset only. We propose a novel data augmentation framework called MaxStyle, which maximizes the effectiveness of style augmentation for model OOD performance. It attaches an auxiliary style-augmented image decoder to a segmentation network for robust feature learning and data augmentation. Importantly, MaxStyle augments data with improved image style diversity and hardness, by expanding the style space with noise and searching for the worst-case style composition of latent features via adversarial training. With extensive experiments on multiple public cardiac and prostate MR datasets, we demonstrate that MaxStyle leads to significantly improved out-of-distribution robustness against unseen corruptions as well as common distribution shifts across *multiple, different, unseen* sites and *unknown* image sequences under both low- and high-training data settings. The code can be found at https://github.com/cherise215/MaxStyle.

1 Introduction

Convolutional neural networks have demonstrated remarkable segmentation accuracy on images that come from the same domain (e.g. images from the

---

Supplementary Information The online version contains supplementary material available at https://doi.org/10.1007/978-3-031-16443-9_15.

L. Wang et al. (Eds.): MICCAI 2022, LNCS 13435, pp. 151–161, 2022.
https://doi.org/10.1007/978-3-031-16443-9_15

same scanner at the same site), yet they often generalize poorly to unseen out-of-domain (OOD) datasets, which hinders the deployment of trained models in real-world medical applications. The performance degradation is mainly caused by the distribution mismatch between training and test data, often originating from the change of vendors, imaging protocols across different hospitals as well as imperfect acquisition processes which affect image quality. To alleviate this problem, a group of works focus on learning domain-invariant features with multi-domain datasets [1–4]. However, collecting and labeling such large datasets can be extraordinarily time-consuming, expensive, and not always feasible due to data privacy and ethical issues. In this work, we consider a more realistic yet more challenging setting: domain generalization from a single source only.

A straightforward solution to improve domain robustness is data augmentation, which transforms and/or perturbs source domain data to resemble unseen data shifts [5–8]. Existing works mainly consider directly perturbing the input space, which in general requires domain knowledge and expertise to design the perturbation function [7]. Very few approaches investigate feature augmentation possibly due to the risk of perturbing semantic content of images. Perturbing features might change semantic attributes such as anatomical structures, which are essential for precise segmentation. An exception is MixStyle [9], which conducts content-preserving feature augmentation by randomly interpolating two samples' feature style statistics. However, there are several limitations with MixStyle: 1) difficulty in visualizing and interpreting augmented features in the high-dimensional space, 2) limited diversity of augmented styles due to the linear interpolation mechanism, 3) sub-optimal effectiveness for model robustness as it does not take the network's vulnerability into account at training.

In this work, we propose a novel data augmentation architecture, where an auxiliary image decoder is attached to a segmentation network to perform self-supervised image reconstruction and style augmentation. Such a design not only improves the interpretability of style augmented examples but also forces the network to learn auxiliary reconstructive features for improved out-of-domain robustness (Sect. 3.2). We further propose an enhanced style augmentation method: **MaxStyle**, which maximizes the effectiveness of style augmentation for model OOD performance (Sect. 3.3). Different from MixStyle, we expand the style space with additional noise and search for the worst-case style composition for the segmentor via adversarial training. Through extensive experiments, we show that MaxStyle achieves competitive robustness against corruptions and improves robustness against common distribution shifts across different sites and different image sequences (Sect. 4.3).

2   Related Work

Style Transfer: Style transfer modifies the visual style of an image while preserving its semantic content. Previous works on style transfer mainly focus on designing advanced generative adversarial networks (GAN) for improved image diversity and fidelity under the assumption that a large-scale dataset with diverse image styles is available [10,11]. Recent findings show that domain shift is closely related to image style changes across different domains [9,12] and can be alleviated by increasing the diversity of training image styles [9,13–17]. One such successful example is MixStyle [9], which generates 'novel' styles via simply linearly mixing style statistics from two arbitrary training instances from the same domain at feature level. DSU [15] augments feature styles with random noise to account for potential style shifts. Compared with MixStyle and DSU, our proposed method, MaxStyle does not only span a *larger* area in the feature style space but also covers *harder* cases.

Adversarial Data Augmentation: Adversarial data augmentation focuses on generating data samples that fool the network to predict the wrong class [18,19]. Such an approach can improve robustness against adversarial and natural corruptions [18,20], and has shown promise for improved model generalization [7,8,17,21–23]. Most methods apply adversarial perturbation to the input space to improve unseen domain performance, e.g., adversarial noise [22–24], and adversarial bias field [7]. Fewer explore feature space perturbation. Huang et al. proposed to adversarially drop dominant features (RSC) during training as a way to improve cross-domain performance [25]. Chen et al. proposed to project masked latent features back to input space in an adversarial manner for improved interpretability and effectiveness [8]. Our method follows this feature-to-input space data augmentation mechanism, but restricts the adversarial perturbation at the style level, generating *style-varying, shape-preserving* hard examples.

3   Method

3.1   Preliminaries: MixStyle

As mentioned before, MixStyle [9] is a feature augmentation method for improving model generalization and robustness, which perturbs feature style statistics to regularize network training. Formally, let $\mathbf{f}_i, \mathbf{f}_j \in \mathbb{R}^{c \times h \times w}$ be c-dimensional feature maps extracted at a certain CNN layer for image $\boldsymbol{x}_i$ and image $\boldsymbol{x}_j$, respectively. MixStyle performs style augmentation for $\mathbf{f}_i$ by first normalizing it with its channel-wise means and standard deviations $\mu(\mathbf{f}_i), \sigma(\mathbf{f}_i) \in \mathbb{R}^c$: $\overline{\mathbf{f}}_i = \frac{\mathbf{f}_i - \mu(\mathbf{f}_i)}{\sigma(\mathbf{f}_i)}$ and then transforming the feature with a linear combination of style statistics $i : \{\sigma(\mathbf{f}_i), \mu(\mathbf{f}_i)\}$ and $j : \{\sigma(\mathbf{f}_j), \mu(\mathbf{f}_j)\}$. This process can be defined as:

$$\mathrm{MixStyle}(\mathbf{f}_i) = \boldsymbol{\gamma}_{mix} \odot \overline{\mathbf{f}}_i + \boldsymbol{\beta}_{mix},$$
$$\boldsymbol{\gamma}_{mix} = \lambda_{mix}\sigma(\mathbf{f}_i) + (1 - \lambda_{mix})\sigma(\mathbf{f}_j), \quad \boldsymbol{\beta}_{mix} = \lambda_{mix}\mu(\mathbf{f}_i) + (1 - \lambda_{mix})\mu(\mathbf{f}_j), \quad (1)$$

where $\odot$ denotes element-wise multiplication; λ_{mix} is a coefficient controlling the level of interpolation, randomly sampled from $[0, 1]$. Acting as a plug-and-play module, MixStyle can be inserted between any CNN layers. It is originally designed as an explicit regularization method for a standard encoder-decoder structure, which perturbs shallow features (e.g., features from the first three convolutional blocks) in an image encoder E_θ parametersied by θ, see Fig. 1.

A segmentation decoder D_{ϕ_s} then performs prediction with the perturbed high-level representation $\hat{z}$. The whole network is optimized to minimize the segmentation loss $\mathcal{L}_{seg}(D_{\phi_s}(\hat{z}), \mathbf{y})$, supervised by the ground-truth $\mathbf{y}$.

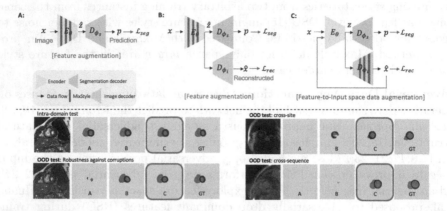

Fig. 1. A) the original use of MixStyle as feature augmentation-based regularization method with a standard encoder-decoder structure. B) MixStyle applied to regularize a dual-branch network with an auxiliary image decoder D_{ϕ_i} attached for image reconstruction. C) We propose to apply MixStyle in the auxiliary decoder D_{ϕ_i} to generate stylized images for feature-to-input space data augmentation instead (MixStyle-DA), which leads to improved model robustness across different OOD test data, compared to A, B (sebottom). GT: ground-truth.

3.2 Robust Feature Learning and Improved Interpretability with Auxiliary Image Decoder

In this work, for improved OOD robustness and better interpretability of style augmentation, we propose to adapt the standard encoder-decoder to a dual-branch network with an auxiliary image decoder D_{ϕ_i} attached, see Fig. 1B, C. The network is supervised with an additional image reconstruction loss $\mathcal{L}_{rec}(\hat{x}, x)$, allowing itself to exploit both complementary image content features and task-specific shape features for the segmentation task [8]. We further propose to insert the style augmentation layers in the image decoder D_{ϕ_i} (Fig. 1C) rather than in the encoder E_θ (Fig. 1B), allowing to generate diverse stylized images with the same high-level semantic features z for *direct* data augmentation. Such a design also improves the interpretability of feature style augmentation. More importantly, our experimental results show that $C > B > A$ in terms of OOD robustness, see Fig. 1 (bottom) and Fig. S1 (supplementary). The segmentation network trained w/approach C is more robust against corruptions and unseen domain shifts across different sites and different image sequences. We name this new method as 'MixStyle-DA' to differentiate it from the original one. We believe this data augmentation-based approach is preferred for model robustness as it tackles model over-fitting from the root, limited training data.

3.3 MaxStyle

On top of the dual-branch architecture with the data augmentation strategy presented above (Fig. 1C), we propose MaxStyle, a novel method that consists of both style mixing and noise perturbation, improving the diversity and sample hardness. As visualized in Fig. 2(a), we introduce additional style noise to expand the style space and apply adversarial training in order to find optimal linear coefficients as well as to generate adversarial style noise for effective data augmentation. Specifically, given feature $\mathbf{f}_i$ extracted at a certain CNN layer in the image decoder D_{ϕ_i} with $\boldsymbol{x}_i$ as input, MaxStyle augments $\mathbf{f}_i$ via:

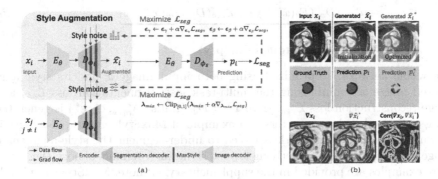

(a)   (b)

Fig. 2. MaxStyle overview. a) MaxStyle reconstructs $\boldsymbol{x}_i$ with augmented feature styles via style mixing and noise perturbation in the image decoder D_{ϕ_i}. Adversarial training is applied, in order to search for 'harder' style composition to fool the segmentation network ($E_\theta \circ D_{\phi_*}$). b) MaxStyle generates a style-optimized image $\hat{\boldsymbol{x}}^*$, which fools the network to under-segment ($\hat{\mathbf{p}}^*$). The anatomical structures remain almost unchanged with *high* correlation (Corr) between two images' gradient fields: $\nabla \boldsymbol{x}$, $\nabla \hat{\boldsymbol{x}}^*$.

$$\text{MaxStyle}(\mathbf{f}_i) = (\boldsymbol{\gamma}_{mix} + \Sigma_\gamma \cdot \boldsymbol{\epsilon}_\gamma) \odot \overline{\mathbf{f}}_i + (\boldsymbol{\beta}_{mix} + \Sigma_\beta \cdot \boldsymbol{\epsilon}_\beta), \qquad (2)$$

where the normalized feature $\overline{\mathbf{f}}_i$ is transformed with mixed styles $\boldsymbol{\gamma}_{mix}, \boldsymbol{\beta}_{mix}$ (Eq. 1) plus additional style noise $\Sigma_\gamma \cdot \boldsymbol{\epsilon}_\gamma, \Sigma_\beta \cdot \boldsymbol{\epsilon}_\beta \in \mathbb{R}^c$ to better explore the space of possible unknown domain shifts. In order to bound the style noise within a reasonable range, similar to [15], we sample the style noises from a re-scaled Gaussian distribution with variance $\Sigma_\gamma, \Sigma_\beta \in \mathbb{R}^c$ estimated from a batch of B instances' style statistics (including $\boldsymbol{x}_i$): $\Sigma_\gamma = \sigma^2(\{\sigma(\mathbf{f}_j)\}_{j=1...i,...B})$, $\Sigma_\beta = \sigma^2(\{\mu(\mathbf{f}_j)\}_{j=1...i,...B})$, $\boldsymbol{\epsilon}_\gamma, \boldsymbol{\epsilon}_\beta \sim \mathcal{N}(\mathbf{0}, \mathbf{1})^1$.

Style Optimization via Adversarial Training: The generated noise $\boldsymbol{\epsilon}_\gamma, \boldsymbol{\epsilon}_\beta$ and the style mixing coefficient λ_{mix} are then updated by maximizing the segmentation loss $\mathcal{L}_{seg}$ so that the synthesized image $\hat{\boldsymbol{x}} = D_{\phi_i}(E_\theta(\boldsymbol{x}); \boldsymbol{\epsilon}_\gamma, \boldsymbol{\epsilon}_\beta, \lambda_{mix})^2$

---

[1] The re-parameterization trick is applied here for ease of follow-up optimization.

[2] For simplicity, we omit non-learnable parameters such as the sampling operator to choose instance $\boldsymbol{x}_j$ from a batch for style mixing.

could fool the network to produce an incorrect prediction $\hat{p} = D_{\phi_s}(E_\theta(\hat{x}))$. Gradient ascent is employed to update the underlying style-related parameters:

$$\epsilon_\gamma \leftarrow \epsilon_\gamma + \alpha \nabla_{\epsilon_\gamma} \mathcal{L}_{seg}(\hat{p}, y), \quad \epsilon_\beta \leftarrow \epsilon_\beta + \alpha \nabla_{\epsilon_\beta} \mathcal{L}_{seg}(\hat{p}, y), \tag{3}$$

$$\lambda_{mix} \leftarrow \text{Clip}_{[0,1]}(\lambda_{mix} + \alpha \nabla_{\lambda_{mix}} \mathcal{L}_{seg}(\hat{p}, y)). \tag{4}$$

Here α denotes the step size. We clip the value of λ_{mix} to ensure it lies in $[0, 1]$. To summarize, training with MaxStyle augmentation can be then written as minimax optimization using the training set $\mathcal{D}$:

$$\min_{\theta, \phi_i, \phi_s} \mathbb{E}_{x, y \sim \mathcal{D}} \underbrace{\mathcal{L}_{seg}(D_{\phi_s}(E_\theta(x)), y) + \mathcal{L}_{rec}(D_{\phi_i}(E_\theta(x)), x)}_{x \text{ as input for network optimization}}$$

$$+ \underbrace{\mathcal{L}_{seg}(D_{\phi_s}(E_\theta(\hat{x}^*)), y) + \mathcal{L}_{rec}(D_{\phi_i}(E_\theta(\hat{x}^*)), x)}_{\hat{x}^* \text{ as input for network optimization}} \tag{5}$$

$$s.t.\ \lambda_{mix}^*, \epsilon_\gamma^*, \epsilon_\beta^* = \arg\max \mathcal{L}_{seg}(\hat{p}, y) \ \textit{(adversarial style optimization)}.$$

The whole network is optimized using both input image x and its style augmented image $\hat{x}^*$, minimizing the multi-task loss: segmentation loss $\mathcal{L}_{seg}$ and image reconstruction loss $\mathcal{L}_{rec}$. Here $\hat{x}^* = D_{\phi_i}(E_\theta(x); \lambda_{mix}^*, \epsilon_\gamma^*, \epsilon_\beta^*)$ is generated using optimized style parameters. An example of MaxStyle generated images is shown in Fig. 2b. The network is fooled to under-segment the style-augmented image $\hat{x}^*$ although the underlying target structures remain almost the same. More examples are provided in the supplementary. Of note, MaxStyle is suitable for general segmentation networks thanks to its plug-and-play nature and the attached decoder can be removed at test time.

4   Experiments and Results

4.1   Data: Cardiac MR Segmentation Datasets

a) Single source domain: The public ACDC dataset [26] is used for network training and intra-domain test, which contains 100 subjects (bSSFP sequences) collected from a **single** site. We randomly split the dataset into 70/10/20 for training/validation/intra-domain test.To simulate common clinical scenarios where the training data is far less than testing data, we randomly selected 10 subjects (10 out of 70) for training and repeated experiments three times. We also validated our method using all 70 training subjects following [8].

b) Multiple unseen test domains: We collated several *public* datasets where each capture different types of distribution shift or quality degradation for comprehensive OOD performance evaluation. They are 1) *ACDC-C* for robustness against imaging artefacts evaluation. Following [8], we synthesized a corrupted version of the ACDC test set where each test image has been randomly corrupted by one of **four, different** types of MR artefacts, three times for each artefact type, using TorchIO toolkit [27]; *ACDC-C* contains 240 ($20 \times 3 \times 4$) subjects in total; 2) a large *cross-site* test set with real-world distribution variations. It consists of 195 subjects (bSSFP sequences) across **four, different** sites, which are from two public challenge datasets: M&Ms [28], MSCMRSeg [29]; 3) *cross-sequence* test set, which consists of 45 LGE sequences from MSCMRSeg [29]. Detailed information can be found in the Supplementary. We employed the image pre-processing and standard data augmentation (incl. common photo-metric and geometric image transformations) described in [8] as our default setting.

4.2 Implementation and Experiment Set-Up

We adopted the dual-branch network presented in [8] as our backbone, which has demonstrated state-of-the-art robustness on the cardiac segmentation task. We applied MaxStyle layers to shallow features in D_{ϕ_i} for optimal performance, i.e. inserted after each of the last three convolutional blocks. Each layer is activated at a probability of 0.5 and batch features at each layer are randomly shuffled for style mixing to trade-off diversity and strength as suggested by [9]. For the adversarial style optimization, we randomly sample λ_{mix} from the uniform distribution $\mathcal{U}_{[0,1]}$ and $\epsilon_{\gamma,\beta}$ from $\mathcal{N}(\mathbf{0},\mathbf{1})$. Adam optimizer was adopted for both style optimization and network optimization. We empirically set $\alpha = 0.1$ with 5 iterations, yielding sufficient improvement. For network optimization, we set learning rate $= 1e^{-4}$ and batch size $= 20$, following [8]. For cardiac small training set, we trained the network for 1,500 epochs to ensure convergence. For larger training data, 600 epochs were sufficient. We employed mean squared error loss for $\mathcal{L}_{rec}$ and cross-entropy loss for $\mathcal{L}_{seg}$. We compare our method with the *baseline* (w/o MaxStyle augmentation) and competitive random and adversarial augmentation methods using the same network and the same multi-task loss, including: a) input space augmentation: RandConv [6], adversarial noise (Adv Noise) [24], and adversarial bias field (Adv Bias) [7]; b) feature augmentation: RSC [25], MixStyle [9], DSU [15]; c) feature-to-input space data augmentation via decoder D_{ϕ_i}: latent space masking (LSM) [8]. All methods were implemented in PyTorch with their recommended set-ups.

4.3 Results

Results of models trained with cardiac low data regime and high data regime are provided Table 1. We also plot the segmentation results in Fig. 3 for visual comparison. From the results, we see that MaxStyle outperforms *all* baseline methods across *all* OOD test sets with different types of domain shifts, providing the largest improvements in both low- and high-data training regimes (+25%, +11%). In particular, MaxStyle significantly improves the segmentation performance on the most challenging cross-sequence dataset with the largest domain shift. By generating images with challenging styles during training, MaxStyle forces the model to be more shape-biased rather than texture/style-biased, which is beneficial for model robustness [30]. To provide additional insights of MaxStyle, we also provide t-SNE visualization of latent feature embeddings in the supplementary. We also found that the model trained w/MaxStyle using 10 ACDC subjects even achieves higher OOD performance than the *baseline* model trained w/70 subjects (0.7420 vs 0.7287), suggesting MaxStyle's superior data efficiency. In addition to cardiac segmentation, we also validate our method on public prostate segmentation datasets [2,3,31–34] to test the generality of our method for medical image segmentation. MaxStyle also achieves the top OOD performance against the competitive methods across *six, different, unseen* test sites. Datasets details and results are shown in the supplementary.

Table 1. Cardiac segmentation performance across multiple unseen test sets. IID: *ACDC* test set performance. OOD: average performance across unseen OOD test sets: *ACDC-C, cross-site,* and *cross-sequence.* Reported values are average Dice scores across different test sets.

Method	Low-data regime (10 training subjects)					High-data regime (70 training subjects)				
	IID	OOD	ACDC-C	Cross-site	Cross-sequence	IID	OOD	ACDC-C	Cross-site	Cross-sequence
baseline	0.8108	0.5925 (-)	0.6788	0.6741	0.4244	0.8820	0.7287 (-)	0.7788	0.8099	0.5974
+RandConv	0.8027	0.6033 (+2%)	0.6872	0.6717	0.4510	0.8794	0.7371 (+1%)	0.7967	0.7747	0.6399
+Adv Noise	0.8080	0.6104 (+3%)	0.6868	0.6804	0.4641	0.8852	0.7291 (+0%)	0.7808	0.8095	0.5970
+Adv Bias	0.8114	0.6594 (+11%)	0.6937	0.7517	0.5327	0.8865	0.7450 (+2%)	0.7839	0.8250	0.6262
+RSC	0.8169	0.6252 (+6%)	0.7010	0.7015	0.4729	0.8844	0.7379 (+1%)	0.7757	0.8115	0.6267
+MixStyle	0.8024	0.6263 (+6%)	0.7002	0.6984	0.4804	0.8788	0.7651 (+5%)	0.7961	0.8130	0.6863
+DSU	0.8127	0.6088 (+3%)	0.6972	0.7008	0.4285	0.8784	0.7553 (+4%)	0.8031	0.8126	0.6503
+LSM	0.7899	0.6462 (+9%)	0.7221	0.7015	0.5149	0.8754	0.7579 (+1%)	0.8213	0.8025	0.6500
+MaxStyle	0.8104~	**0.7420** (+25%)‡	**0.7745**	**0.7645**	**0.6869**	0.8727†	**0.8125** (+11%)‡	**0.8408**	**0.8294**	**0.7673**

~:p-value> 0.5
† p-value> $1e^{-4}$
‡ p-value≪ $1e^{-4}$ (compared to *baseline* results)

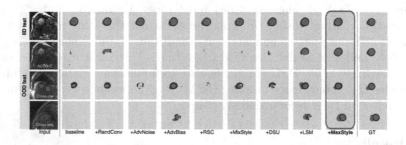

Fig. 3. Qualitative results under the cardiac low-data regime. GT: ground-truth.

Ablation Study: We highlight the importance of increased style diversity via composing style mixing and style noise as well as improved style hardness with adversarial training in the ablation study. Results are shown in Table 2. It can be observed that removing each component leads to performance degradation.

Table 2. The effect of expanding style search space (style noise and style mixing) and applying adversarial style optimization (AdvOpt) for domain generalization. Reported values are mean (std) of Dice scores across all *ten* cardiac test (sub)sets. Experiments were performed under the cardiac high-data training setting.

w/o Style noise	w/o Style mixing	w/o AdvOpt	w/o AdvOpt and Style mixing	w/o AdvOpt and Style noise (MixStyle-DA)	MaxStyle
0.8058 (0.0467)	0.8295 (0.0395)	0.8188 (0.0478)	0.8121 (0.0516)	0.7981 (0.0652)	**0.8321 (0.0339)**

5 Discussion and Conclusion

We introduced MaxStyle, which is a powerful plug-in data augmentation module for single domain generalization. It attaches an auxiliary style-augmented image decoder to a segmentation network for image reconstruction and style augmentation. MaxStyle is capable of generating stylized images with improved diversity and hardness, forcing the network to learn robust features for domain generalization. From the causal perspective [35], the image decoder with MaxStyle can be viewed as a latent data generation process with multi-scale style interventions, conditioned on the extracted high-level causal variables (e.g., content and style). By explicitly feeding the network with images of diverse 'hard' styles, the spurious correlation between image style and labels are weakened. The network thus gains the ability to resist unseen distributional shifts. We also highlighted the importance of combining both reconstructive features and discriminant features for robust segmentation (Sect. 3.2), which is in line with classical robust subspace methods [36]. We validated MaxStyle's efficacy on both cardiac datasets and prostate datasets across various unseen test sets, demonstrating its superiority over competitive methods under both low- and high-data training regimes. It is worth mentioning here that our method does not necessarily sacrifice the IID performance. On the prostate segmentation task, our method can significantly improve the IID performance compared to the baseline method (0.8597 vs 0.8277, see Table S3). For the cardiac segmentation task, the IID performance slightly drops (e.g., 0.8104 vs 0.8108). We hypothesize that the IID performance degradation is task and dataset-dependent. Like other adversarial data augmentation methods [18,24], the main limitation of MaxStyle is that the improved robustness comes at the cost of longer training time due to the adversarial optimization procedure, e.g. increased by a factor of ~1.8 in our case. We hope that MaxStyle will enable more data-efficient, robust and reliable deep models. Integrating MaxStyle into composite data augmentation frameworks [37] would be an interesting direction.

Acknowledgment. This work was supported by two EPSRC Programme Grants (EP/P001009/1, EP/W01842X/1) and the UKRI Innovate UK Grant (No.104691).

References

1. Tao, Q., et al.: Deep learning-based method for fully automatic quantification of left ventricle function from cine MR images: a multivendor, multicenter study. Radiology **290**(1), 180–513 (2019)
2. Liu, Q., et al.: Ms-net: multi-site network for improving prostate segmentation with heterogeneous MRI data. In: TMI (2020)
3. Liu, Q., Dou, Q., Heng, P.-A.: Shape-aware meta-learning for generalizing prostate MRI segmentation to unseen domains. In: Martel, A.L., et al. (eds.) MICCAI 2020. LNCS, vol. 12262, pp. 475–485. Springer, Cham (2020). https://doi.org/10.1007/978-3-030-59713-9_46
4. Dou, Q., et al.: Domain generalization via model-agnostic learning of semantic features. In: NeurIPS 2019, pp. 6447–6458 (2019)
5. Wang, J., et al.: Generalizing to unseen domains: a survey on domain generalization. In: Zhou, Z. (ed.) IJCAI 2021, pp. 4627–4635 (2021)
6. Xu, Z., et al.: Robust and generalizable visual representation learning via random convolutions. In: ICLR 2021 (2021)
7. Chen, C., et al.: Realistic adversarial data augmentation for MR image segmentation. In: Martel, A.L., et al. (eds.) MICCAI 2020. LNCS, vol. 12261, pp. 667–677. Springer, Cham (2020). https://doi.org/10.1007/978-3-030-59710-8_65
8. Chen, C., et al.: Cooperative training and latent space data augmentation for robust medical image segmentation. In: de Bruijne, M., et al. (eds.) MICCAI 2021. LNCS, vol. 12903, pp. 149–159. Springer, Cham (2021). https://doi.org/10.1007/978-3-030-87199-4_14
9. Zhou, K., et al.: Domain generalization with mixstyle. In: ICLR 2021 (2021)
10. Karras, T., Laine, S., Aila, T.: A style-based generator architecture for generative adversarial networks. TPAMI **43**(12), 4217–4228 (2021)
11. Huang, X., Liu, M.-Y., Belongie, S., Kautz, J.: Multimodal unsupervised image-to-image translation. In: Ferrari, V., Hebert, M., Sminchisescu, C., Weiss, Y. (eds.) ECCV 2018. LNCS, vol. 11207, pp. 179–196. Springer, Cham (2018). https://doi.org/10.1007/978-3-030-01219-9_11
12. Li, Y., et al.: Adaptive batch normalization for practical domain adaptation. Pattern Recogn. **80**, 109–117 (2018)
13. Jackson, P.T.G., et al.: Style augmentation: data augmentation via style randomization. In: CVPR Workshops 2019, pp. 83–92 (2019)
14. Yamashita, R., et al.: Learning domain-agnostic visual representation for computational pathology using medically-irrelevant style transfer augmentation. In: IEEE Transactions on Medical Imaging (2021)
15. Li, X., et al.: Uncertainty modeling for out-of-distribution generalization. In: International Conference on Learning Representations (2022)
16. Wagner, S.J., et al.: Structure-preserving multi-domain stain color augmentation using style-transfer with disentangled representations. In: de Bruijne, M., et al. (eds.) MICCAI 2021. LNCS, vol. 12908, pp. 257–266. Springer, Cham (2021). https://doi.org/10.1007/978-3-030-87237-3_25
17. Zhong, Z., et al.: Adversarial style augmentation for domain generalized urban-scene segmentation. Under Review (2021)
18. Madry, A., et al.: Towards deep learning models resistant to adversarial attacks. In: International Conference on Learning Representations (2017)
19. Goodfellow, I.J., Shlens, J., Szegedy, C.: Explaining and harnessing adversarial examples. In: ICLR 2015 (2015)

20. Gilmer, J., et al.: Adversarial examples are a natural consequence of test error in noise. ICML **97**(2019), 2280–2289 (2019)
21. Xie, C., et al.: Adversarial examples improve image recognition. In: CVPR 2020, pp. 816–825 (2020)
22. Volpi, R., et al.: Generalizing to unseen domains via adversarial data augmentation. In: NeurIPS 2018, pp. 5339–5349 (2018)
23. Qiao, F., Zhao, L., Peng, X.: Learning to learn single domain generalization. In: CVPR 2020, pp. 12556–12565 (2020)
24. Miyato, T., et al.: Virtual adversarial training: a regularization method for supervised and semi-supervised learning. In: TPAMI (2018)
25. Huang, Z., Wang, H., Xing, E.P., Huang, D.: Self-challenging improves cross-domain generalization. In: Vedaldi, A., Bischof, H., Brox, T., Frahm, J.-M. (eds.) ECCV 2020. LNCS, vol. 12347, pp. 124–140. Springer, Cham (2020). https://doi.org/10.1007/978-3-030-58536-5_8
26. Bernard, O., et al.: Deep learning techniques for automatic MRI cardiac Multi-Structures segmentation and diagnosis: is the problem solved? TMI **37**(11), 2514–2525 (2018)
27. Pérez-García, F., Sparks, R., Ourselin, S.: Torchio: a python library for efficient loading, preprocessing, augmentation and patch-based sampling of medical images in deep learning. In: Computer Methods and Programs in Biomedicine, p. 106236 (2021)
28. Campello, V.M., et al.: Multi-Centre. The M&Ms Challenge. In: IEEE Transactions on Medical Imaging, Multi-vendor and Multi-disease Cardiac Segmentation (2021)
29. Zhuang, X., et al.: Cardiac segmentation on late gadolinium enhancement MRI: a benchmark study from Multi-sequence cardiac MR segmentation challenge. arXiv: 2006.12434 (2020)
30. Geirhos, R., et al.: ImageNet-trained CNNs are biased towards texture; increasing shape bias improves accuracy and robustness. In: International Conference on Learning Representations, pp. 1–20 (2018)
31. Antonelli, M., et al.: The medical segmentation decathlon. arXiv:2106.05735 (2021)
32. B.N. et al.: NCI-ISBI 2013 challenge: automated segmentation of prostate structures. https://www.cancerimagingarchive.net (2015)
33. Lemaitre, G., et al.: Computer-Aided detection and diagnosis for prostate cancer based on mono and multi-parametric MRI: a review. Comput. Biol. Med. **60**, 8–31 (2015)
34. Litjens, G., et al.: Evaluation of prostate segmentation algorithms for MRI: the PROMISE12 challenge. Med. Image Anal. **18**(2), 359–373 (2014)
35. Castro, D.C., Walker, I., Glocker, B.: Causality matters in medical imaging. Nat. Commun. **11**(1), 3673 (2020)
36. Fidler, S., Skocaj, D., Leonardis, A.: Combining reconstructive and discriminative subspace methods for robust classification and regression by subsampling. TPAMI **28**(3), 337–350 (2006)
37. Hendrycks, D., et al.: Augmix: a simple data processing method to improve robustness and uncertainty. In: ICLR (2020)

A Robust Volumetric Transformer
for Accurate 3D Tumor Segmentation

Himashi Peiris[1(✉)], Munawar Hayat[3], Zhaolin Chen[1,2], Gary Egan[2],
and Mehrtash Harandi[1]

[1] Department of Electrical and Computer Systems Engineering,
Faculty of Engineering, Monash University, Melbourne, Australia
{Edirisinghe.Peiris,Zhaolin.Chen,Mehrtash.Harandi}@monash.edu
[2] Monash Biomedical Imaging (MBI), Monash University, Melbourne, Australia
Gary.Egan@monash.edu
[3] Department of Data Science and AI, Faculty of IT, Monash University,
Melbourne, Australia
Munawar.Hayat@monash.edu

Abstract. We propose a Transformer architecture for volumetric segmentation, a challenging task that requires keeping a complex balance in
encoding local and global spatial cues, and preserving information along
all axes of the volume. Encoder of the proposed design benefits from
self-attention mechanism to simultaneously encode local and global cues,
while the decoder employs a parallel self and cross attention formulation
to capture fine details for boundary refinement. Empirically, we show
that the proposed design choices result in a computationally efficient
model, with competitive and promising results on the Medical Segmentation Decathlon (MSD) brain tumor segmentation (BraTS) Task. We
further show that the representations learned by our model are robust
against data corruptions. Our code implementation is publicly available.

Keywords: Pure volumetric transformer · Tumor segmentation

1 Introduction

Inspired by the strong empirical results of the transformer based models in computer vision [5,8,15], their promising generalization and robustness characteristics [21], and their flexibility to model long range interactions, we propose a volumetric transformer architecture for segmentation of 3D medical image modalities
(*e.g.*, MRI, CT), called VT-UNet. Earlier efforts to develop transformer based
segmentation models for 3D medical scans have been shown to outperform state-
of-the-art CNN counterparts [4]. However, these methods divide 3D volumes
into 2D slices and process 2D slices as inputs [4,6]. As such, considerable and
potentially critical volumetric information, essential to encapsulating inter-slice

---

Supplementary Information The online version contains supplementary material
available at https://doi.org/10.1007/978-3-031-16443-9_16.

dependencies, is lost. While some hybrid approaches (using both convolutional blocks and Transformer layers) keep the 3D volumetric data intact [9,25,29], the design of purely transformer based architecture, capable of keeping intact the volumetric data at input, is yet unexplored in the literature. Our work takes the first step in this direction, and proposes a model, which not only achieves better segmentation performance, but also demonstrates better robustness against data artefacts. The Transformer models have highly dynamic and flexible receptive field and are able to capture long-range interactions, yet designing a Transformer based UNet architecture for volumet-

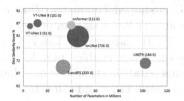

Fig. 1. The model size vs Dice Similarity Coefficient (DSC) is shown in this plot. Circle size indicates Computational Complexity by FLOPs. VT-UNet achieves highest DSC compared to SOTA methods while maintaining a smaller model size and low computational complexity.

ric segmentation remains a challenging task. This is because: (**1**) Encapsulating voxel information and capturing the connections between arbitrary positions in the volumetric sequence is not straightforward. Compared with Transformer based approaches for 2D image segmentation [4], the data in each slice of the volume is connected to three views and discarding either of them can be detrimental. (**2**) Preserving spatial information in a volume is a daunting task. Even for 2D images, while breaking the image into patches and projecting patches into tokens as introduced in Vision Transformer (ViT), local structural cues can be lost, as shown in Tokens-to-token ViT [27]. Effectively encoding the local cues while simultaneously capturing global interactions along multiple axes of a volume is therefore a challenging task. (**3**) Due to the quadratic complexity of the self-attention, and large size of 3D volume tensor inputs, designing a Transformer based segmentation model, which is computationally efficient, requires careful design considerations. Our proposed VT-UNet model effectively tackles the above design challenges by proposing a number of modules. In our UNet based architecture, we develop two types of Transformer blocks. First, our blocks in the encoder which directly work on the 3D volumes, in a hierarchical manner, to jointly capture the local and global information, similar in spirit to the Swin Transformer blocks [14]. Secondly, for the decoder, we introduce parallel cross-attention and self-attention in the expansive path, which creates a bridge between queries from the decoder and keys & values from the encoder. By this parallelization of the cross-attention and self-attention, we aim to preserve the full global context during the decoding process, which is important for the task of segmentation. Since VT-UNet is free from convolutions and combines attention outputs from two modules during the decoding, the order of the sequence is important to get accurate predictions. Inspired by [24], apart from applying relative positional encoding while computing attention in each Transformer block, we augment the decoding process and inject the complementary information extracted from Fourier feature positions of the tokens in the sequence. In summary, our major contributions are, (**1**) We reformulate volumetric tumor segmentation from a sequence-to-sequence perspective, and propose a

UNet shaped *Volumetric Transformer* for multi-modal medical image segmentation. **(2)** We design an encoder block with two consecutive self attention layers to jointly capture local and global contextual cues. Further, we design a decoder block which enables parallel (shifted) window based self and cross attention. This parallelization uses one shared projection of the *queries* and independently computes cross and self attention. To further enhance our features in the decoding, we propose a convex combination approach along with Fourier positional encoding. **(3)** Incorporating our proposed design choices, we substantially limit the model parameters while maintaining lower FLOPs compared to existing approaches (see Fig. 1). **(4)** We conduct extensive evaluations and show that our design achieves state-of-the-art volumetric segmentation results, alongwith enhanced robustness to data artefacts.

2   Methodology

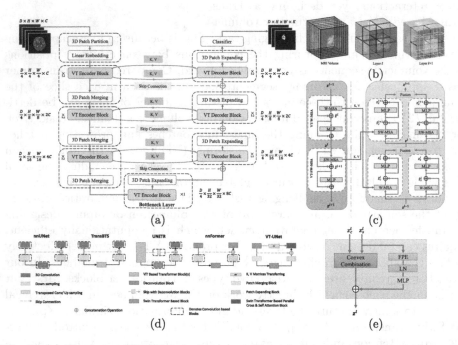

Fig. 2. (a) Illustrates VT-UNet Architecture. Here, k denotes the number of classes. (b) shows visualization of Volumetric Shifted Windows. Consider an MRI volume of size $D \times H \times W$ with $D = H = W = 8$ for the sake of illustration. Further, let the window size for partitioning the volume be $P \times M \times M$ with $P = M = 4$. Here, layer l adopts the regular window partition in the first step of Volumetric Transformer(VT) block which results in $2 \times 2 \times 2 = 8$ windows. Inside layer $l + 1$, volumetric windows are shifted by $(\frac{P}{2}, \frac{M}{2}, \frac{M}{2}) = (2, 2, 2)$ tokens. This results in $3 \times 3 \times 3 = 27$ windows. (c) shows VT Encoder-Decoder Structure. (d) Encoder-Decoder structural comparison with other SOTA methods. The proposed VT-UNet architecture has no convolution modules and is purely based on Transformer blocks. (e) Illustrates the structure of the Fusion Module.

We denote vectors and matrices in bold lower-case $\mathbf{x}$ and bold upper-case $\mathbf{X}$, respectively. Let $\mathbb{X} = \{\mathbf{x}_1, \mathbf{x}_2, \cdots, \mathbf{x}_\tau\}, \mathbf{x}_i \in \mathbb{R}^C$ be a sequence representing a signal of interest (e.g., an MRI volume). We call each $\mathbf{x}_i$ a token. We assume that tokens, in their original form, might not be optimal for defining the span. Therefore, in Self-Attention (SA), we define the span by learning a linear mapping from the input tokens. This we can show with $\mathbb{R}^{\tau \times C_v} \ni \mathbf{V} = \mathbf{X}\mathbf{W_V}$, where we stack tokens $\mathbf{x}_i$s into the rows of $\mathbf{X}$ (i.e., $\mathbf{X} = [\mathbf{x}_1|\mathbf{x}_2|\cdots\mathbf{x}_\tau]^\top$). Following previous work by Hu et al. [10], we use a slight modification of the self-attention [24] (see [13]) in our task as follows:

$$SA(\mathbf{Q}, \mathbf{K}, \mathbf{V}) = \text{SoftMax}\Big(\mathbf{Q}\mathbf{K}^\top / \sqrt{C} + \mathbf{B}\Big)\mathbf{V}, \qquad (1)$$

where $\mathbb{R}^{\tau \times \tau} \ni \mathbf{B}$ is trainable and acts as a relative positional bias across tokens in the volume with $\mathbf{V} = \mathbf{X}\mathbf{W_V}$, $\mathbf{K} = \mathbf{X}\mathbf{W_K}$, and $\mathbf{Q} = \mathbf{X}\mathbf{W_Q}$. In practice, computing SA for multiple attention heads several times in parallel is called Multi-head Self-Attention (MSA). Equation (1) is the basic building block of our Volumetric Transformer Window based Multi-head Self-Attention (VT-W-MSA), and the Volumetric Transformer Shifted Window based Multi-head Self-Attention (VT-SW-MSA), discussed next.

Overview of VT-UNet. Figure 2 shows the conceptual diagram of the proposed volumetric transformer network or **VT-UNet** for short. The input to our model is a 3D volume of size $D \times H \times W \times C$. The output is a $D \times H \times W \times K$ dimensional volume, representing the presence/absence of voxel-level class labels (K is the number of classes). Below, we discuss architectural form of the VT-UNet modules and explain the functionality and rationals behind our design in detail.

The VT Encoder. The VT encoder consists of 3D Patch Partitioning layer together with Linear Embedding layer, and 3D Patch merging layer followed by two successive VT encoder blocks.

3D Patch Partitioning. Transformer-based models work with a sequence of tokens. The very first block of VT-UNet accepts a $D \times H \times W \times C$ dimensional medical volume (e.g., MRI) and creates a set of tokens by splitting the 3D volume into non-overlapping 3D patches (see Fig. 2 (b)). The size of partitioning kernel is $P \times M \times M$, resulting in describing the volume by $\tau = \lfloor D/P \rfloor \times \lfloor H/M \rfloor \times \lfloor W/M \rfloor$ tokens. The 3D patch partitioning is followed by a linear embedding to map each token with dimensionality $P \times M \times M$ to a C dimensional vector. Typical values for M, P and C according to our experiments are 4, 4, and 72, respectively.

VT Encoder Block. In ViT, tokens carry significant spatial information due to the way they are constructed. The importance of performing SA by windowing in ViT has been shown in several recent studies, most notably in Swin Transformer [14]. Following a similar principal in the design of Swin Transformers, albeit for volumetric data, we propose 3D windowing operations in our VT

Encoder Blocks (**VT-Enc-Blks**). In particular, we propose two types of windowing, namely regular window and shifted window, which we show by **VT-W-MSA** and **VT-SW-MSA** for simplicity, respectively. Figure 2b provides the design specifics of **VT-W-MSA** and **VT-SW-MSA**, while Fig. 2 (b) illustrates the windowing operation. Both VT-W-MSA and VT-SW-MSA employ attention layers with windowing, followed by a 2-layer Multi Layer Perceptron (MLP) with Gaussian Error Linear Unit (GELU) non-linearity in between. A Layer Normalization (LN) is applied before every MSA and MLP, and a residual connection is applied after each module. The windowing enables us to inject inductive bias in modeling long range dependencies between tokens. In both VT-W-MSA and VT-SW-MSA, attention across tokens within a window helps representation learning. In the VT-W-MSA, we split the volume evenly into smaller non-overlapping windows as illustrated in Fig. 2 (b). Since tokens in adjacent windows cannot see each other with VT-W-MSA, we make use of a shifted window in VT-SW-MSA (see the right most panel Fig. 2 (b)) which bridges tokens in adjacent windows of VT-W-MSA. The windowing is inspired by the Swin Transformer [14] and can be understood as generalization to volumetric data. Note that the windowing operation in our work resembles [15] that extends the benefits of windowing beyond images to videos. Putting everything together, the VT-Enc-Blk realizes the following functionality:

$$\hat{z}^l = \text{VT-W-MSA}\left(\text{LN}\left(z^{l-1}\right)\right) + z^{l-1}, \qquad \hat{z}^{l+1} = \text{VT-SW-MSA}\left(\text{LN}\left(z^l\right)\right) + z^l,$$
$$z^l = \text{MLP}\left(\text{LN}\left(\hat{z}^l\right)\right) + \hat{z}^l, \qquad\qquad z^{l+1} = \text{MLP}\left(\text{LN}\left(\hat{z}^{l+1}\right)\right) + \hat{z}^{l+1}, \quad (2)$$

where $\hat{z}^l$ and z^l denote the output features of the VT-W-MSA module and the MLP module for block l, respectively.

3D Patch Merging. We make use of 3D patch merging blocks to generate feature hierarchies in the encoder of VT-UNet. Having such hierarchies is essential to generate finer details in the output for the dense prediction tasks [7,14].

After every VT-Enc-Blk, we merge adjacent tokens along the spatial axes in a non-overlapping manner to produce new tokens. In doing so, we first concatenate features of each group of 2×2 neighboring tokens. The resulting vector is projected via a linear mapping to a space where the channel dimensionality of the tokens is doubled (see Fig. 2). The benefit of patch merging is not limited to feature hierarchies. The computational complexity of SA is quadratic in the number of tokens [14,15]. As such, patch merging reduces the FLOPs count of the VT-UNet by a factor of 16 after each VT-Enc-Blk. To give the reader a better idea and as we will discuss in Sect. 4, the tiny VT-UNet model uses only 6.7% FLOPs in comparison to its fully volumetric CNN counterpart [16] while achieving a similar performance (slightly better indeed)! Please note that the patch merging block is not used in the bottleneck stage.

The VT Decoder. After bottleneck layer which consists of a VT-Enc-Blk together with 3D Patch Expanding layer, the VT decoder starts with successive VT Decoder Blocks (**VT-Dec-Blks**), 3D patch expanding layers and a classifier at the end to produce the final predictions. There are some fundamental design differences between VT-Enc-Blk and VT-Dec-Blk which we will discuss next.

3D Patch Expanding. This functionality is used to somehow revert the effect of patch merging. In other words and in order to construct the output with the same spatial-resolution as the input, we need to create new tokens in the decoder. For the sake of discussion, consider the patch expanding after the bottleneck layer (see the middle part of Fig. 2). The input tokens to the patch expanding are of dimensionality $8C$. In the patch expanding, we first increase the dimensionality of the input tokens by a factor of two using a linear mapping. Following a reshaping, we can obtain 2×2 tokens with dimensionality $4C$ from the resulting vector of dimensionality $2 \times 8C$. This, we will reshape along the spatial axes and hence for $D/4 \times H/32 \times W/32 \times 8C$, we create $D/4 \times H/16 \times W/16 \times 4C$ tokens.

VT Decoder Block. The UNet [20] and its variants [18,30] make use of lateral connections between the encoder and the decoder to produce fine-detailed predictions. This is because the spatial information is lost, at the expense of attaining higher levels of semantics, as the input passes through the encoder. The lateral connections in the UNet makes it possible to have the best of both worlds, spatial information from lower layers and semantic information from upper layers (along the computational graph). Having this in mind, we propose a hybrid form of SA at the decoder side (see Fig. 2b for an illustration). Each VT-Dec-Blk receives the generated tokens of its previous VT-Dec-Blk along with the key ($\mathbf{K}_E$) and value ($\mathbf{V}_E$) tokens from the VT-Enc-Blk sitting at the same stage of VT-UNet, see Fig. 2a. Recall that a VT-Enc-Blk has two SA blocks with regular and shifted windowing operations. VT-Dec-Blk enjoys similar windowing operations but makes use of four SA blocks grouped into SA module and Cross Attention (CA) module. The functionality can be described as:

$$\mathrm{SA}_r = \mathrm{SA}(\mathbf{Q}_D, \mathbf{K}_D, \mathbf{V}_D), \qquad \mathrm{CA}_l = \mathrm{SA}(\mathbf{Q}_D, \mathbf{K}_E, \mathbf{V}_E). \qquad (3)$$

Here, r and l, denote right and left branches of the decoder module. The right branch of the SA acts on tokens generated by the previous VT-Dec-Blk according to Eq. (3). We emphasize on the flow of information from the decoder by the subscript D therein. The left branch of the CA, however, uses the queries generated by the decoder along with the keys and values obtained from the VT-Enc-Blk at the same level in the computation graph. The idea here is to use the basis spanned by the encoder (which is identified by values) along with keys to benefit from spatial information harvested by the encoder. These blocks, also use the regular and shifted windowing to inject more inductive bias into the model. Note that the values and keys from the SA with the same windowing operation should be combined, hence the criss-cross connection form in Fig. 2 (c).

Remark 1. One may ask why values and keys are considered from the encoder. We indeed studied other possibilities such as employing queries and keys from the encoder while generating values by the decoder. Empirically, the form described in Eq. (3) is observed to deliver better and more robust outcomes and hence our choice in VT-UNet.[1]

Fusion Module. As illustrated in Fig. 2 (e), tokens generated from the CA module and MSA module are combined together and fed to the next VT-Dec-Blk, $\mathbf{z}^l$ is calculated using by a linear function as:

$$\mathbf{z}^l = \alpha \; \hat{\mathbf{z}}_c^l + (1 - \alpha) \; \hat{\mathbf{z}}_s^l + \mathcal{F}(\hat{\mathbf{z}}_s^l), \tag{4}$$

where $\mathcal{F}(\cdot)$ denotes <u>F</u>ourier Feature <u>P</u>ositional <u>E</u>ncoding (FPE) and α controls the contribution from each CA and MSA module. Aiming for simplicity, in fusing tokens generated by the CA and MSA, we use a linear combination with $\alpha = 0.5$[2].

Classifier Layer. After the final 3D patch expanding layer in the decoder, we introduce a classifier layer which includes a 3D convolutional layer to map deep C dimensional features to K segmentation classes.

A Note on Computational Complexity. The computational complexity of the SA described in Eq. (1) is dictated by computations required for obtaining $\mathbf{Q}, \mathbf{K}, \mathbf{V}$, computing $\mathbf{QK}^\top$ and obtaining the resulting tokens by applying the output of the Softmax (which is a $\tau \times \tau$ matrix) to $\mathbf{V}$. This adds up to $\mathcal{O}(3\tau C^2 + 2\tau^2 C)$, where C and τ are the dimensionality and the number of tokens, respectively. Windowing will reduce the computational load of the SA according to $\mathcal{O}(3\tau C^2 + 2\tau\kappa C)$ where we have assumed that tokens are grouped into κ windows and SA is applied within each window. In our problem, where tokens are generated from volumetric data, $\tau \gg \kappa$ and hence windowing not only helps in having better discriminatory power, but also it helps in reducing the computational load.[3]

---

[1] We empirically observed that employing keys and values from the encoder in CA yields faster convergence of VT-UNet. This, we conjecture, is due to having extra connections from the decoder to encoder during the back-propagation which might facilitate gradient flow.

[2] Breaking the Symmetry: This results in a symmetry, meaning that swapping $\hat{\mathbf{z}}_c^l$ and $\hat{\mathbf{z}}_s^l$ does not change the output. To break this symmetry and also better encapsulate object-aware representations that are critical for anatomical pixel-wise segmentation, we supplement the tokens generated from MSA by a the 3D FPE. The 3D FPE employs sine and cosine functions with different frequencies [24] to yield a unique encoding scheme for each token. The main idea is to use a sine/cosine function with a high frequency and modulate it across the dimensionality of the tokens while changing the frequency according to the location of the token within the 3D volume.

[3] For the sake of simplicity and explaining the key message, we have made several assumptions in our derivation. First, we have assumed $C_k = C_v = C$. We also did not include the FLOPs needed to compute the softmax. Also, in practice, one uses a multi-head SA, where the computation is break down across several parallel head working on lower dimensional spaces (*e.g.*, on for $\mathbf{V}$, we use C/h dimensional spaces where h is the number of heads). This will reduce the computational load accordingly. That said, the general conclusion provided here is valid.

3 Related Work

Vision Transformers have shown superior empirical results for different computer vision tasks [2,15,23], with promising characteristics. For example and as compared with the CNNs, they are less biased towards texture [17], and show better generalization and robustness [17,21]. Transformers have also been recently investigated for image segmentation [4,6,28]. TransUNet [6] is the first Transformer based approach for medical image segmentation. It adapts a UNet structure, and replaces the bottleneck layer with ViT [8] where patch embedding is applied on a feature map generated from CNN encoder (where input is a 2D slice of 3D volume). Unlike these hybrid approaches (using both convolutions and self-attention), Cao et al. [4] proposed Swin-UNet, a purely transformer based network for medical image segmentation. It inherits swin-transformer blocks [14] and shows better segmentation results over TransUNet [6]. The 3D version of TransUnet [6], called TransBTS [25] has a CNN encoder-decoder design and a Transformer as the bottleneck layer. Zhou et al. [29] proposed nnFormer with 3D Swin Transformer based blocks as encoder and decoder with interleaved stem of convolutions. A model which employs a transformer as the encoder and directly connects intermediate encoder outputs to the decoder via skip connections is proposed in [9]. The encoder-decoder structural comparison of SOTA methods are shown in Fig. 2 (d). The aforementioned transformer based approaches for 3D medical image segmentation have shown their promises, by achieving better performances compared with their CNN counterparts. Our proposed model, on the other hand, processes the volumetric data in its entirety, thus fully encoding the interactions between slices. Moreover, our proposed model is built purely based on Transformers and introduces lateral connections to perform CA along-with SA in the encoder-decoder design. These design elements contributed in achieving better segmentation performance, along-with enhanced robustness.

4 Experiments

Table 1. Segmentation results on MSD BraTS dataset.

Method	Average HD95 ↓	Average DSC ↑	WT HD95 ↓	WT DSC ↑	ET HD95 ↓	ET DSC ↑	TC HD95 ↓	TC DSC ↑
UNet [20]	10.19	66.4	9.21	76.6	11.12	56.1	10.24	66.5
AttUNet [18]	9.97	66.4	9.00	76.7	10.45	54.3	10.46	68.3
nnUNet [11]	4.60	81.9	3.64	91.9	4.06	80.97	4.91	85.35
SETR NUP [28]	13.78	63.7	14.419	69.7	11.72	54.4	15.19	66.9
SETR PUP [28]	14.01	63.8	15.245	69.6	11.76	54.9	15.023	67.0
SETR MLA [28]	13.49	63.9	15.503	69.8	10.24	55.4	14.72	66.5
TransUNet [6]	12.98	64.4	14.03	70.6	10.42	54.2	14.5	68.4
TransBTS [25]	9.65	69.6	10.03	77.9	9.97	57.4	8.95	73.5
CoTr [26]	9.70	68.3	9.20	74.6	9.45	55.7	10.45	74.8
UNETR [9]	8.82	71.1	8.27	78.9	9.35	58.5	8.85	76.1
nnFormer [29]	4.05	86.4	3.80	91.3	3.87	81.8	4.49	86.0
VT-UNet-S	3.84	85.9	4.01	90.8	2.91	81.8	4.60	85.0
VT-UNet-B	3.43	87.1	3.51	91.9	2.68	82.2	4.10	87.2

Table 2. Robustness analysis.

	Artefact	Avg. HD95 ↓	Avg. DSC ↑
nnFormer	Clean	4.05	86.4
	Motion	4.81	84.3
	Ghost	4.30	84.5
	Spike	4.63	84.9
VT-UNet	Clean	3.43	87.1
	Motion	3.87	85.8
	Ghost	3.69	86.0
	Spike	3.50	86.6

Table 3. Ablation study.

VT-UNet-B	Avg. HD95 ↓	Avg. DSC ↑
w/o FPE	85.35	4.33
w/o FPE & CA	83.58	6.18

Implementation Details. We use 484 MRI scans from MSD BraTS task [1]. Following [9,29], we divide 484 scans into 80%, 15% and 5% for training, validation and testing sets, respectively. We use PyTorch [19], with a single Nvidia A40 GPU. The weights of Swin-T [14] pre-trained on ImageNet-22K are used to initialize the model. For training, we employ AdamW optimizer with a learning rate of $1e^{-4}$ for 1000 epochs and a batch size of 4. We used rotating, adding noise, blurring and adding gamma as data augmentation techniques.

Experimental Results. Table 1 compares VT-UNet with recent transformer based approaches and SOTA CNN based methods. We introduce variants of the VT-UNet, by changing the number of embedded dimensions used for model training. Our variants are: **(a)** Small *VT-UNet-S* with $C = 48$ **(b)** Base *VT-UNet-B* with $C = 72$. We use Dice Sørensen coefficient (DSC) and Hausdorff Distance (HD) as evaluation metrics, and separately compute them for three classes: (1) Enhancing Tumor (ET), (2) the Tumor Core (TC) (addition of ET, NET and NCR), and (3) the

Table 4. Qualitative Segmentation Results. *Row-1*: Yellow, Green and Red represent Peritumoral Edema (ED), Enhancing Tumor (ET) and Non Enhancing Tumor (NET)/ Necrotic Tumor (NCR), respectively. *Row-2*: volumetric tumor prediction. *Row-3*: segmentation boundaries.

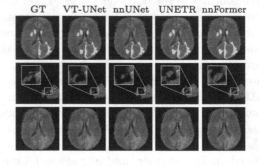

Whole Tumor (WT) (addition of ED to TC), following similar evaluation strategy as in [9,11,29]. Our quantitative results in Table 1 suggest that VT-UNet achieves best overall performance in DSC and HD. Table 4 shows qualitative segmentation results on unseen patient data. We can observe that our model can accurately segment the structure and delineates boundaries of tumor. We believe that, capturing long-range dependencies across adjacent slices plays a vital role in our model's performance. Our empirical results in Table 3 reveal the importance of introducing Parallel CA and SA together with FPE in VT-Dec-Blks along with convex combination. We can notice that all of these components contribute towards model's performance.

Robustness Analysis. Factors such as patient's movement and acquisition conditions can introduce noise to MRI. Here, we investigate the robustness of VT-UNet, by synthetically introducing artefacts to MR images at inference time. These include **(1)** Motion artefacts [22]. **(2)** Ghosting artefacts [3]. **(3)** Spike artefacts (Herringbone artefact) [12]. Table 2 compares the robustness of our method with nnFormer [29]. The results suggest that VT-UNet performs more reliably in the presence of these nuisances. Our findings are consistent with existing works on RGB images, where Transformer based models have shown better robustness against occlusions [17], natural and adversarial perturbations [17,21], owing to their highly dynamic and flexible receptive field.

5 Conclusion

This paper presents a volumetric transformer network for medical image segmentation, that is computationally efficient to handle large-sized 3D volumes, and learns representations that are robust against artefacts. Our results show that the proposed model achieves consistent improvements over existing state-of-the-art methods in volumetric segmentation. We believe our work can assist better clinical diagnosis and treatment planning.

References

1. Antonelli, M., et al.: The medical segmentation decathlon. arXiv preprint arXiv:2106.05735 (2021)
2. Arnab, A., Dehghani, M., Heigold, G., Sun, C., Lučić, M., Schmid, C.: ViViT: a video vision transformer. arXiv preprint arXiv:2103.15691 (2021)
3. Axel, L., Summers, R., Kressel, H., Charles, C.: Respiratory effects in two-dimensional Fourier transform MR imaging. Radiology 160(3), 795–801 (1986)
4. Cao, H., et al.: Swin-Unet: Unet-like pure transformer for medical image segmentation. arXiv preprint arXiv:2105.05537 (2021)
5. Carion, N., Massa, F., Synnaeve, G., Usunier, N., Kirillov, A., Zagoruyko, S.: End-to-end object detection with transformers. In: Vedaldi, A., Bischof, H., Brox, T., Frahm, J.-M. (eds.) ECCV 2020. LNCS, vol. 12346, pp. 213–229. Springer, Cham (2020). https://doi.org/10.1007/978-3-030-58452-8_13
6. Chen, J., et al.: TransUNet: transformers make strong encoders for medical image segmentation. arXiv preprint arXiv:2102.04306 (2021)
7. Chen, L.C., Papandreou, G., Kokkinos, I., Murphy, K., Yuille, A.L.: DeepLab: semantic image segmentation with deep convolutional nets, atrous convolution, and fully connected CRFs. IEEE Trans. Pattern Anal. Mach. Intell. 40(4), 834–848 (2017)
8. Dosovitskiy, A., et al.: An image is worth 16x16 words: transformers for image recognition at scale. arXiv preprint arXiv:2010.11929 (2020)
9. Hatamizadeh, A., et al.: UNETR: transformers for 3D medical image segmentation. In: Proceedings of the IEEE/CVF Winter Conference on Applications of Computer Vision, pp. 574–584 (2022)
10. Hu, H., Zhang, Z., Xie, Z., Lin, S.: Local relation networks for image recognition. In: Proceedings of the IEEE/CVF International Conference on Computer Vision, pp. 3464–3473 (2019)
11. Isensee, F., Jäger, P.F., Full, P.M., Vollmuth, P., Maier-Hein, K.H.: nnU-Net for brain tumor segmentation. In: Crimi, A., Bakas, S. (eds.) BrainLes 2020. LNCS, vol. 12659, pp. 118–132. Springer, Cham (2021). https://doi.org/10.1007/978-3-030-72087-2_11
12. Jin, K.H., Um, J.Y., Lee, D., Lee, J., Park, S.H., Ye, J.C.: MRI artifact correction using sparse+ low-rank decomposition of annihilating filter-based Hankel matrix. Magn. Reson. Med. 78(1), 327–340 (2017)

13. Khan, S., Naseer, M., Hayat, M., Zamir, S.W., Khan, F.S., Shah, M.: Transformers in vision: a survey. arXiv preprint arXiv:2101.01169 (2021)
14. Liu, Z., et al.: Swin Transformer: hierarchical vision transformer using shifted windows. arXiv preprint arXiv:2103.14030 (2021)
15. Liu, Z., et al.: Video Swin Transformer. arXiv preprint arXiv:2106.13230 (2021)
16. Milletari, F., Navab, N., Ahmadi, S.A.: V-Net: fully convolutional neural networks for volumetric medical image segmentation. In: 2016 Fourth International Conference on 3D Vision (3DV), pp. 565–571. IEEE (2016)
17. Naseer, M., Ranasinghe, K., Khan, S., Hayat, M., Khan, F.S., Yang, M.H.: Intriguing properties of vision transformers. arXiv preprint arXiv:2105.10497 (2021)
18. Oktay, O., et al.: Attention U-Net: learning where to look for the pancreas. arXiv preprint arXiv:1804.03999 (2018)
19. Paszke, A., et al.: Automatic differentiation in PyTorch (2017)
20. Ronneberger, O., Fischer, P., Brox, T.: U-Net: convolutional networks for biomedical image segmentation. In: Navab, N., Hornegger, J., Wells, W.M., Frangi, A.F. (eds.) MICCAI 2015. LNCS, vol. 9351, pp. 234–241. Springer, Cham (2015). https://doi.org/10.1007/978-3-319-24574-4_28
21. Shao, R., Shi, Z., Yi, J., Chen, P.Y., Hsieh, C.J.: On the adversarial robustness of visual transformers. arXiv preprint arXiv:2103.15670 (2021)
22. Shaw, R., Sudre, C., Ourselin, S., Cardoso, M.J.: MRI k-space motion artefact augmentation: model robustness and task-specific uncertainty. In: International Conference on Medical Imaging with Deep Learning-Full Paper Track (2018)
23. Touvron, H., Cord, M., Douze, M., Massa, F., Sablayrolles, A., Jégou, H.: Training data-efficient image transformers & distillation through attention. In: International Conference on Machine Learning, pp. 10347–10357. PMLR (2021)
24. Vaswani, A., et al.: Attention is all you need. In: Advances in Neural Information Processing Systems, pp. 5998–6008 (2017)
25. Wang, W., Chen, C., Ding, M., Yu, H., Zha, S., Li, J.: TransBTS: multimodal brain tumor segmentation using transformer. In: de Bruijne, M., et al. (eds.) MICCAI 2021. LNCS, vol. 12901, pp. 109–119. Springer, Cham (2021). https://doi.org/10.1007/978-3-030-87193-2_11
26. Xie, Y., Zhang, J., Shen, C., Xia, Y.: CoTr: efficiently bridging CNN and transformer for 3D medical image segmentation. arXiv preprint arXiv:2103.03024 (2021)
27. Yuan, L., et al.: Tokens-to-token ViT: training vision transformers from scratch on ImageNet. arXiv preprint arXiv:2101.11986 (2021)
28. Zheng, S., et al.: Rethinking semantic segmentation from a sequence-to-sequence perspective with transformers. In: Proceedings of the IEEE/CVF Conference on Computer Vision and Pattern Recognition, pp. 6881–6890 (2021)
29. Zhou, H.Y., Guo, J., Zhang, Y., Yu, L., Wang, L., Yu, Y.: nnFormer: interleaved transformer for volumetric segmentation. arXiv preprint arXiv:2109.03201 (2021)
30. Zhou, Z., Rahman Siddiquee, M.M., Tajbakhsh, N., Liang, J.: UNet++: a nested U-Net architecture for medical image segmentation. In: Stoyanov, D., et al. (eds.) DLMIA/ML-CDS -2018. LNCS, vol. 11045, pp. 3–11. Springer, Cham (2018). https://doi.org/10.1007/978-3-030-00889-5_1

Usable Region Estimate for Assessing Practical Usability of Medical Image Segmentation Models

Yizhe Zhang[1]([✉]), Suraj Mishra[2], Peixian Liang[2], Hao Zheng[2],
and Danny Z. Chen[2]

[1] School of Computer Science and Engineering, Nanjing University of Science and
Technology, Nanjing 210094, Jiangsu, China
yizhe.zhang.cs@gmail.com
[2] Department of Computer Science and Engineering, University of Notre Dame,
Notre Dame, IN 46556, USA

Abstract. We aim to quantitatively measure the practical usability of medical image segmentation models: to what extent, how often, and on which samples a model's predictions can be used/trusted. We first propose a measure, Correctness-Confidence Rank Correlation (CCRC), to capture how predictions' confidence estimates correlate with their correctness scores in rank. A model with a high value of CCRC means its prediction confidences reliably suggest which samples' predictions are more likely to be correct. Since CCRC does not capture the actual prediction correctness, it alone is insufficient to indicate whether a prediction model is both accurate and reliable to use in practice. Therefore, we further propose another method, Usable Region Estimate (URE), which simultaneously quantifies predictions' correctness and reliability of confidence assessments in one estimate. URE provides concrete information on to what extent a model's predictions are usable. In addition, the sizes of usable regions (UR) can be utilized to compare models: A model with a larger UR can be taken as a more usable and hence better model. Experiments on six datasets validate that the proposed evaluation methods perform well, providing a concrete and concise measure for the practical usability of medical image segmentation models. Code is made available at https://github.com/yizhezhang2000/ure.

Keywords: Medical AI models · Model evaluation · Practical usability · Unified measure · Medical image segmentation

1 Introduction

Deep learning (DL) methodologies brought performances of medical image segmentation, detection, and classification to new heights. Many new medical AI

Supplementary Information The online version contains supplementary material available at https://doi.org/10.1007/978-3-031-16443-9_17.

L. Wang et al. (Eds.): MICCAI 2022, LNCS 13435, pp. 173–182, 2022.
https://doi.org/10.1007/978-3-031-16443-9_17

Conventional Eval.: "This model achieves 0.87 ($\pm$0.18) F_1 score and 0.19 ECE (Expected Calibration Error) on the test data."

Doctor: "OK. If future samples are from the same/similar distribution of these test samples, then how often and to what extent the model's predictions are usable in practice?"

URE (proposed).: "Predictions with confidences no smaller than 0.76 satisfy the pre-defined clinical correctness requirement, and 83% of the time, predictions are in this usable region."

Doctor: "Good! I will use this model following the 0.76 confidence guideline." (OR: "Not good enough. I would like the predictions being usable 95% of the time. Please continue to improve.")

Fig. 1. A dialog example to show the merit of clinically oriented model evaluations. Our estimate simultaneously evaluates both predictions' correctness and confidence estimates of a given prediction model, providing a quantitative measure for how usable the model is and concrete guidelines for using the model.

prediction models achieved impressive accuracy (correctness) on test sets. In the meantime, practitioners still face reliability/trustworthiness issues when deploying AI models to real clinical scenarios. Conventional model evaluation measures (e.g., average F_1 score, average Dice Coefficient) mainly focus on prediction correctness. Although a model with good correctness scores is desirable, high correctness alone is not sufficient for being highly usable in clinical practice. As each model inevitably would make some wrong predictions, a highly usable/reliable model should also have its prediction confidence estimates well aligned with the prediction correctness so that the model can alert experts for further input when its confidences for some samples' predictions are low.

Improving the calibration and reliability of confidence estimates is an active research area. Guo et al. [10] found that a DL model is often overconfident in its predictions; they examined a range of methods to calibrate model output confidences. For example, temperature scaling gives the best empirical calibration results due to the strong non-linearity nature; it requires additional labeled data for calibration and does not change the rank order of the original uncalibrated confidence estimates. In [12,15], it was proposed to use model ensembles to improve the reliability of prediction confidence estimates. Dropout-based and Bayesian-based methods [9,13] are also popular choices for better uncertainty/confidence estimates. On the evaluations side, Expected Calibration Error (ECE) [10,15,16] is widely used for evaluating calibration of prediction confidences. ECE computes the absolute difference between the correctness score and confidence score for each prediction (or each bin of predictions) and takes the average of the computed absolute differences as output. Low ECE values indicate that confidences are better matched with the actual correctness scores.

Similarly, Brier score (Br) [3] measures the mean square difference between each probabilistic prediction and its corresponding ground truth label. ECE and Br quantify how well a model's confidence estimates (CE) match with correctness scores (CS) in value but do not measure alignment/matching in rank.

More clinically oriented model evaluations drive more clinically oriented model developments. Prediction correctness (e.g., F_1 score, Dice coefficient) and goodness of alignment between confidence estimates and correctness scores capture two key aspects of a clinically usable medical AI model. But, neither correctness scores nor confidence estimation reliability scores alone provide a unified view on practical usability of a model. For example, a 0.87 average F_1 score and 0.19 ECE score inform little on how one could utilize such a model in practice, and a 0.02 improvement in F_1 score (sample-average) and 0.01 increase in ECE do not offer concrete ideas on how it affects the model's practical usability. There is a strong need to develop a clinically oriented model evaluation method that can inform users to what extent the model's predictions are usable and how often the model's predictions can be used in practice (see Fig. 1).

In this paper, we first propose a new measure, Correctness-Confidence Rank Correlation (CCRC), to assess the goodness of alignment between correctness scores and confidence estimates from the rank perspective. A high CCRC value indicates a sample's prediction with a higher confidence is more likely to be correct (and vice versa). Furthermore, we design a novel estimate named Usable Region Estimate (URE), which quantifies prediction correctness and reliability of confidence estimates in one statistically sound and unified measure, where only **high correctness scores** together with **properly ordered confidence estimates** would yield good evaluation results (see Fig. 2). URE measures to what extent in the confidence space the predictions are usable (with high quality), giving a concrete guideline for using the model in practice and allowing direct comparison between models via comparing the sizes of usable regions.

Our work contributes to clinically oriented and more comprehensive model evaluations in the following aspects. (1) We identify limitations of previous confidence evaluation measures (e.g., ECE), and propose CCRC that takes rank information into account in measuring reliability of confidence estimates. (2) We identify that known methods which separately evaluate prediction correctness and reliability of confidence estimates do not offer a unified measure for a model's practical usability. (3) We propose a novel statistically sound Usable Region Estimate that unifies evaluations of prediction correctness and reliability of confidence estimates in one measure. URE can quantitatively measure a model's usability and give concrete recommendations on how to utilize the model.

2   Methodology

2.1   Preliminaries

Correctness of Predictions. Given predictions $\hat{y}_1, \hat{y}_2, \ldots, \hat{y}_n$ and ground truth labels $y_1, y_2, \ldots, y_n$, a correctness metric ϕ (e.g., F_1 score) applies to each pair of prediction and ground truth label to obtain a correctness score $s_i = \phi(\hat{y}_i, y_i)$. An overall correctness score is defined by $\sum_{i=1}^{n} \frac{s_i}{n}$.

Reliability of Confidence Estimates. <u>NLL</u>: Given an medical AI model $\hat{\pi}(Y|X)$ and n test samples, NLL is defined as $L_{NLL} = \sum_{i=1}^{n} -\log(\hat{\pi}(y_i|x_i))$. When $\hat{\pi}(y_i|x_i)$ is identical to the ground truth conditional distribution $\pi(Y|X)$, the value of L_{NLL} is minimized. Since an overly confident model tends to give a low NLL value, the NLL value is not a reliable indicator for the goodness of alignment between confidence estimates and predictions' correctness. <u>ECE</u>: For a given test set of n samples, one obtains the model confidence of each sample, $conf_i$, and the accuracy/correctness score s_i, for $i = 1, 2, \ldots, n$. One can compare them at the per-sample level or the bin level. For per-sample level comparison, $ECE = \sum_{i=1}^{n} \frac{1}{n}|s_i - conf_i|$. <u>Brier score (Br)</u>: Suppose a model prediction $\hat{y}_i$ is in a probabilistic form; Br measures the distance between the prediction and ground truth: $Br = \sum_{i=1}^{n} \frac{1}{n}(\hat{y}_i - y_i)^2$.

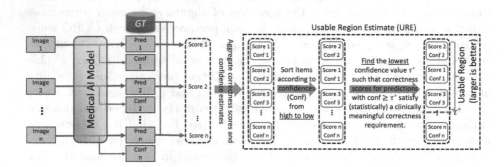

Fig. 2. Illustrating the proposed Usable Region Estimate (URE).

2.2 Correctness-Confidence Rank Correlation (CCRC)

Given a correctness score s_i and a model confidence $conf_i$ for each test sample, $i = 1, 2, \ldots, n$, we use Spearman's rank correlation coefficient [19] to measure the correlation between these two lists of numbers. The coefficient value obtained is between 0 and 1 (inclusive), with 1 for a perfect alignment in rank between confidence estimates and correctness scores. A high CCRC is critical for model deployment as the confidence estimate is informative on which sample's prediction is *more* likely to be correct. Being high in CCRC is a necessary condition for a medical AI model for being highly usable in practice.

2.3 Usable Region Estimate (URE)

Given a correctness score s_i and a model confidence $conf_i$ for each test sample, $i = 1, 2, \ldots, n$, we compute the mean of the correctness scores, μ_τ, for all the predictions whose confidences are no smaller than a confidence level τ, $count_\tau = \sum_{i=1}^{n} \mathbb{1}_{\{conf_i \geq \tau\}}$, and $\mu_\tau = \sum_{i=1}^{n} \frac{s_i \mathbb{1}_{\{conf_i \geq \tau\}}}{count_\tau}$.

URE aims to find the lowest possible value τ^* such that for all the predictions with confidence estimates no smaller than τ^*, the lower end of the 95% CI (confidence interval) of μ_τ is no less than a correctness score requirement (clinically usable segmentation accuracy). We utilize the classic bootstrapping techniques [8] for computing CI, as bootstrapping does not assume the data being in a particular distribution (e.g., normal distribution). The obtained τ^* gives a practical guideline that, using the tested model, any prediction with a confidence value $\geq \tau^*$ is considered being usable, and predictions with confidences lower than τ^* need human attention.[1] To compute the size of the usable region of predictions, we further compute p^*, which is the ratio between the number of predictions with confidences $\geq \tau^*$ and the total number of predictions. A model with a larger value of p^* means that this model has a larger **usable region** of predictions and is considered to be a better model compared to one with a smaller value of p^*. One can efficiently find τ^* by sorting the confidences and iterating through the sorted list and checking the correctness scores. The Python code for URE is given in Listing 1.1. A usability diagram can be created by computing UR using a range of correctness requirements.

```python
def URE(scores, conf, requirement=0.9):
    index = np.argsort(-conf); tau_s=1.0; p_s=0
    for id in range(len(index)):
        pool=scores[index[:id+1]]; statistics=[];
        for b in range(1,100): # Bootstrapping for CI
            poolB=np.random.choice(pool, len(pool));
            stat=np.mean(poolB); # can be of other statistics
            statistics.append(stat);
        ordered = np.sort(statistics)
        lower = np.percentile(ordered, 2.5)
        if lower>=requirement:
            tau_s=conf[index[id]];p_s=len(pool)/len(scores);
    return tau_s, p_s
```

Listing 1.1. Usable Region Estimate (URE): Finding the "Usable Region" in which predictions with confidences $\geq \tau^*$ satisfy a clinically meaningful correctness constraint.

Below we give analyses and technical discussions for the proposed URE.

Clinically Usable Segmentation Accuracy (CUSA) and UR Diagrams.
Setting the CUSA is crucial for developing medical AI models (for determining whether a model is good enough). CUSA is task-dependent and is expected to be determined by medical experts. Moreover, practitioners can use a range of accuracy levels to generate UR diagrams (see Fig. 3) for model evaluation. A model may perform better on a portion of samples, but worse on another portion of samples, and UR diagrams can capture such phenomena when comparing models. Mixed results in UR diagrams can be viewed as the superiority of a model over other models is unstable (unconvincing). Today, segmentation accuracy improvement is often small in the average correctness score. A model with a

---

[1] Assume the test set for computing UR is representative of the true data distribution.

better average score may perform worse on a significant portion of test samples. Using the UR diagram, a model is considered superior only when it consistently gives higher numbers in URE across a range of accuracy levels. This helps avoid drawing premature decisions on which model is superior and helps the Medical AI field develop better models in a more rigorous way.

Statistics of Interest. For the general purpose, "mean" statistics was set as default to estimate the expectation of the correctness scores. For a specific application with a vital requirement on prediction correctness, one may choose to use "2 percentile" or "5 percentile" instead of "mean" to compute the correctness scores' statistics (in line 7 of the code snippet in Listing 1.1).

Stability across Accuracy Levels. For a test set containing fewer samples, large changes on URs may occur across different accuracy levels. A larger test set leads to smoother UR changes. The smoothness of change also depends on the segmentation model itself: if there is a sharp drop in the correctness scores across two accuracy levels, the UR diagram will reflect it. Overall, using a large enough test set (e.g., >1000 samples), the change of UR should be smooth between different accuracy levels. This is guaranteed by the bootstrapping technique used in the URE algorithm and the central limit theorem.

The Role of URE. URE evaluates prediction correctness and confidence estimates simultaneously in a unified measure. It is fundamentally different from *individually* using average correctness scores and CCRC/ECE for model evaluation. For example, when comparing two models, suppose one gives a higher average correctness score but a worse CCRC/ECE result and the other gives a lower average correctness score but a better CCRC/ECE result. It is then *inconclusive* which model can be taken as a more usable one since correctness scores and CCRC/ECE evaluate two key aspects of a model's performance which are both critical for the model usability in practice. In contrast, the proposed URE, an estimate that simultaneously addresses prediction correctness and reliability of confidence estimates, allows to compare the two models in one measure and the one yields a higher value of p^* (with a larger usable region of predictions) should be considered as a more practically usable model.

Other Implications. New research [17,21] suggested that correctness of predictions and reliability of confidences should be both included in evaluations of medical AI models. URE fits this emerging need well. More accurate predictions with more reliable confidence estimates yield larger UR. Thus using URE will help encourage development of accurate and well-calibrated deep learning models. Furthermore, a better-constructed test set would lead to a more accurate UR estimate. Using URE would encourage practitioners to construct a more representative test set, which can benefit clinical practice.

3   Experiments

Our experiments use six public medical image segmentation datasets: (1) BUSI for breast ultrasound image segmentation [1]; (2) ISIC for skin lesion segmentation [7]; (3) GLAS for gland segmentation in histological images [20]; (4) BCSS for breast cancer semantic segmentation in histological images [2]. (5) MoNuSeg for cell segmentation in histological images [11]. (6) PhC-C2DL-PSC [14] for cell segmentation in videos. U-Net [18], DCN [4], DCAN [5], and TransUNet [6] are the models tested in the experiments. A per-sample correctness score is computed for each test sample using the metric specified for every dataset (e.g., F_1 score). Computing per-sample confidence estimation follows a common practice, which takes the max value of prediction logits (after softmax) for each pixel and then takes the average confidence across all the pixels for each test sample. For fair and straightforward comparisons among the networks being evaluated here, we rely on the confidence estimates from the network itself without consulting any external algorithms or models.

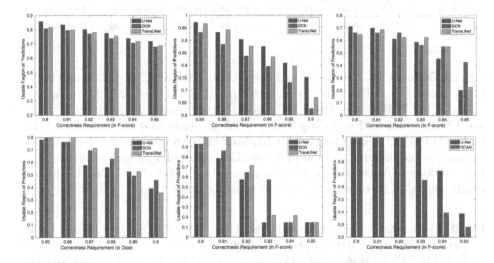

Fig. 3. Usability diagram: the sizes of usable regions for a range of different correctness requirements. Top-Left: BUSI [1]. Top-Mid: ISIC [7]. Top-Right: GLAS [20]. Bottom-Left: BCSS [2]. Bottom-Mid: MoNuSeg [11]. Bottom-Right: PhC-C2DL-PSC [14].

3.1   CCRC and URE Provide New Insights for Model Evaluation

BUSI Dataset. U-Net, DCN, and TransUNet achieve 0.721, 0.690, and 0.716 in F_1 score ($\uparrow$), 0.113, 0.116, and 0.113 in ECE ($\downarrow$), and 0.854, 0.865, and 0.867 in CCRC ($\uparrow$), respectively. According to F_1 score, U-Net performs better than TransUNet with a small margin. ECE suggests that TransUNet and U-Net give the same level of calibration on confidence estimates, but CCRC suggests that TransUNet's confidence estimates are considerably better aligned with the actual correctness scores. We show the sizes of usable regions estimated by the proposed URE for these networks in Fig. 3 (Top-Left). U-Net provides larger usable regions across all the correctness requirement settings.

ISIC Dataset. U-Net, DCN, and TransUNet achieve 0.849, 0.834, and 0.853 in F_1 score (↑), 0.073, 0.094, and 0.082 in ECE (↓), and 0.731, 0.709, and 0.659 (↑) in CCRC, respectively. According to F_1 score, TransUnet performs the best. However, ECE and CCRC both suggest that U-Net performs the best on the reliability of confidence estimates. In Fig. 3 (Top-Mid), we observe that U-Net is the most usable model on ISIC for most the correctness requirement settings.

Table 1. Testing the robustness of the estimated UR: the frequency (%) of new samples violating the correctness requirement associated with the UR (the lower the better).

| Dataset | URE (full) | | | w/o using bootstrapping | | |
	U-Net	DCN	TransUNet	U-Net	DCN	TransUNet
BUSI	9.4 (±2.2)	10.9 (±3.2)	12.3 (±3.1)	48.8 (±5.6)	57.3 (±1.4)	50.4 (±5.2)
ISIC	8.5 (±3.0)	8.8 (±3.6)	10.6 (±2.5)	54.3 (±4.5)	51.2 (±4.9)	48.6 (±5.7)

GLAS Dataset. U-Net, DCN, and TransUNet achieve 0.747, 0.773, and 0.758 in F_1 score (↑), 0.214, 0.183, and 0.197 in ECE (↓), and 0.622, 0.683, and 0.641 (↑) in CCRC, respectively. DCN gives better average F_1 score, ECE score, and CCRC score. In Fig. 3 (Top-Right), URE suggests that U-Net provides the largest UR under a lower correctness requirement (0.8 in F_1 score). For a higher correctness requirement (0.85 in F_1 score), DCN yields larger usable regions with its predictions. TransUNet performs considerably better than U-Net for higher requirements on correctness (0.82 in F_1 score or higher).

BCSS Dataset. U-Net, DCN, and TransUNet achieve 0.832, 0.824, and 0.835 in Dice coefficient (↑), 0.053, 0.064, and 0.063 in ECE (↓), and 0.751, 0.809, and 0.700 (↑) in CCRC, respectively. Comparing to U-Net, TransUNet gives slightly better Dice coefficients but worse ECE and CCRC scores. In Fig. 3 (Bottom-Left), URE suggests that TransUNet provides the largest UR of predictions for most the correctness requirement settings.

MoNuSeg Dataset. UNet, DCN, and TransUNet attain 0.819, 0.803, and 0.828 in F_1 score (↑), 0.074, 0.075, and 0.072 in ECE (↓), and 0.718, 0.815, and 0.635 (↑) in CCRC, respectively. TransUNet performs better in the average correctness (F_1 score), but worse in confidence estimates (ECE and CCRC). Simultaneously measuring these two aspects, URE suggests that TransUNet is a more usable model for most the correctness requirements (see Fig. 3 (Bottom-Mid)).

PhC-C2DL-PSC Dataset. U-Net and DCAN achieve 0.937 and 0.924 in F_1 score ($\uparrow$), 0.023 and 0.038 in ECE ($\downarrow$), and 0.701 and 0.754 in CCRC ($\uparrow$), respectively. U-Net gives a better average F_1 score, better ECE score, but worse CCRC score. In Fig. 3 (Bottom-Right), we show the sizes of usable regions for these two networks. Both the networks give nearly perfect usable regions when the correctness requirement is lower. For a higher correctness requirement, U-Net yields considerably larger usable regions with its predictions.

URE evaluates the prediction correctness and reliability of confidence estimates simultaneously in a unified measure. As validated above, URE provides performance evaluation insights and information that conventional overall correctness scores and confidence calibration errors cannot fully capture.

3.2 Estimated Usable Regions on New Unseen Samples

We examine the robustness of the URs on new unseen samples. For a given dataset, we randomly split the original test set into two halves, and use the 1st half to estimate UR (computing τ^*) and the 2nd half to test whether the new samples satisfy the correctness requirement used in computing the τ^*. We apply the above estimate-and-test procedure to a model 100 times, each time with a different random data split. We then report the frequency when the correctness of new samples violates the correctness requirement associated with the estimated UR. A low frequency indicates the estimated UR is robust to new samples. We repeat the whole procedure 20 times for reporting the mean and standard derivation for this frequency. Table 1 (Left) shows that the estimated URs deliver fairly robust estimates, suggesting that they are informative in guiding model deployment. By comparing to the right part of Table 1, we validate the necessity of using bootstrapping for checking satisfaction of correctness in URE.

4 Conclusions

In this paper, we investigated the importance of measuring practical usability in evaluating medical AI segmentation models. We proposed CCRC for measuring the goodness of alignment between confidence estimates and correctness scores in rank, and developed a novel URE to quantitatively assess usable regions of predictions for AI models. Experiments validated that CCRC and URE are useful measures, bringing new insights for evaluating, comparing, and selecting models.

Acknowledgement. This work was supported in part by the Natural Science Foundation of Jiangsu Province (Grant BK20220949) and NSF Grant CCF-1617735.

References

1. Al-Dhabyani, W., Gomaa, M., Khaled, H., Fahmy, A.: Dataset of breast ultrasound images. Data Brief **28**, 104863 (2020)
2. Amgad, M., et al.: Structured crowdsourcing enables convolutional segmentation of histology images. Bioinformatics **35**(18), 3461–3467 (2019)

3. Brier, G.W., et al.: Verification of forecasts expressed in terms of probability. Mon. Weather Rev. **78**(1), 1–3 (1950)
4. Chen, H., Qi, X.J., Cheng, J.Z., Heng, P.A.: Deep contextual networks for neuronal structure segmentation. In: 30th AAAI Conference on Artificial Intelligence (2016)
5. Chen, H., Qi, X., Yu, L., Heng, P.A.: DCAN: deep contour-aware networks for accurate gland segmentation. In: Proceedings of the IEEE Conference on Computer Vision and Pattern Recognition, pp. 2487–2496 (2016)
6. Chen, J., et al.: TransUNet: transformers make strong encoders for medical image segmentation. arXiv preprint arXiv:2102.04306 (2021)
7. Codella, N.C., et al.: Skin lesion analysis toward melanoma detection: a challenge at the 2017 international symposium on biomedical imaging (ISBI), hosted by the international skin imaging collaboration (ISIC). In: 2018 IEEE 15th International Symposium on Biomedical Imaging (ISBI 2018), pp. 168–172. IEEE (2018)
8. Efron, B., Tibshirani, R.J.: An Introduction to the Bootstrap. CRC Press, Boca Raton (1994)
9. Gal, Y., Ghahramani, Z.: Dropout as a Bayesian approximation: representing model uncertainty in deep learning. In: International Conference on Machine Learning, pp. 1050–1059. PMLR (2016)
10. Guo, C., Pleiss, G., Sun, Y., Weinberger, K.Q.: On calibration of modern neural networks. In: International Conference on Machine Learning, pp. 1321–1330. PMLR (2017)
11. Kumar, N., et al.: A multi-organ nucleus segmentation challenge. IEEE Trans. Med. Imaging **39**(5), 1380–1391 (2019)
12. Lakshminarayanan, B., Pritzel, A., Blundell, C.: Simple and scalable predictive uncertainty estimation using deep ensembles. arXiv preprint arXiv:1612.01474 (2016)
13. Maddox, W.J., Izmailov, P., Garipov, T., Vetrov, D.P., Wilson, A.G.: A simple baseline for Bayesian uncertainty in deep learning. In: Advances in Neural Information Processing Systems, vol. 32 (2019)
14. Maška, M., et al.: A benchmark for comparison of cell tracking algorithms. Bioinformatics **30**(11), 1609–1617 (2014)
15. Mehrtash, A., Wells, W.M., Tempany, C.M., Abolmaesumi, P., Kapur, T.: Confidence calibration and predictive uncertainty estimation for deep medical image segmentation. IEEE Trans. Med. Imaging **39**(12), 3868–3878 (2020)
16. Popordanoska, T., Bertels, J., Vandermeulen, D., Maes, F., Blaschko, M.B.: On the relationship between calibrated predictors and unbiased volume estimation. In: de Bruijne, M., et al. (eds.) MICCAI 2021. LNCS, vol. 12901, pp. 678–688. Springer, Cham (2021). https://doi.org/10.1007/978-3-030-87193-2_64
17. Rajpurkar, P., Chen, E., Banerjee, O., Topol, E.J.: AI in health and medicine. Nat. Med. **28**, 31–38 (2022)
18. Ronneberger, O., Fischer, P., Brox, T.: U-Net: convolutional networks for biomedical image segmentation. In: Navab, N., Hornegger, J., Wells, W.M., Frangi, A.F. (eds.) MICCAI 2015. LNCS, vol. 9351, pp. 234–241. Springer, Cham (2015). https://doi.org/10.1007/978-3-319-24574-4_28
19. Schober, P., Boer, C., Schwarte, L.A.: Correlation coefficients: appropriate use and interpretation. Anesth. Analg. **126**(5), 1763–1768 (2018)
20. Sirinukunwattana, K., et al.: Gland segmentation in colon histology images: the GlaS challenge contest. Med. Image Anal. **35**, 489–502 (2017)
21. Wu, E., Wu, K., Daneshjou, R., Ouyang, D., Ho, D.E., Zou, J.: How medical AI devices are evaluated: limitations and recommendations from an analysis of FDA approvals. Nat. Med. **27**(4), 582–584 (2021)

Modality-Adaptive Feature Interaction for Brain Tumor Segmentation with Missing Modalities

Zechen Zhao[1], Heran Yang[1,2], and Jian Sun[1,2,3(✉)]

[1] Xi'an Jiaotong University, Xi'an, China
zzc496618063@stu.xjtu.edu.cn, {hryang,jiansun}@xjtu.edu.cn
[2] Pazhou Laboratory (Huangpu), Guangzhou, China
[3] Peng Cheng Laboratory, Shenzhen, China

Abstract. Multi-modal Magnetic Resonance Imaging (MRI) plays a crucial role in brain tumor segmentation. However, missing modality is a common phenomenon in clinical practice, leading to performance degradation in tumor segmentation. Considering that there exist complementary information among modalities, feature interaction among modalities is important for tumor segmentation. In this work, we propose Modality-adaptive Feature Interaction (MFI) with multi-modal code to adaptively interact features among modalities in different modality missing situations. MFI is a simple yet effective unit, based on graph structure and attention mechanism, to learn and interact complementary features between graph nodes (modalities). Meanwhile, the proposed multi-modal code, indicating whether each modality is missing or not, guides MFI to learn adaptive complementary information between nodes in different missing situations. Applying MFI with multi-modal code in different stages of a U-shaped architecture, we design a novel network U-Net-MFI to interact multi-modal features hierarchically and adaptively for brain tumor segmentation with missing modality(ies). Experiments show that our model outperforms the current state-of-the-art methods for brain tumor segmentation with missing modalities.

Keywords: Brain tumor segmentation · Missing modalities · Graph · Multi-modal feature interaction

1 Introduction

Brain tumor is a dangerous disease for human health, and accurate segmentation of brain tumor plays an important role in clinical assessment and treatment planning. Multiple Magnetic Resonance Imaging (MRI) images, such as Fluid Attenuation Inversion Recovery (FLAIR), contrast enhanced T1-weighted (T1c), T1-weighted (T1) and T2-weighted (T2), are usually acquired to identify

© The Author(s), under exclusive license to Springer Nature Switzerland AG 2022
L. Wang et al. (Eds.): MICCAI 2022, LNCS 13435, pp. 183–192, 2022.
https://doi.org/10.1007/978-3-031-16443-9_18

the tumor regions since they can provide complementary information [10] for tumor segmentation. In clinical practice, the phenomenon of missing one or more modalities happens possibly because of different imaging protocols [17,22] or image corruption [9,14]. Therefore, designing learning methods robust to missing modality(ies) is necessary.

Previous medical image segmentation methods for missing modality(ies) can be categorized into three approaches, including 1) training a separate model for each subset of modalities [18,21], 2) synthesizing missing modality image(s) and then performing segmentation using the complete modalities [8,13], 3) learning a single segmentation model for different observed modalities [1–4]. The first approach needs to train a large number of models, each of which is for one missing modality scenario, and this strategy is resource-intensive and time-consuming. The second approach requires the training of a synthesizing network, and the quality of the generated image has a great impact on the final segmentation result. The third approach trains a single model (may consist of several modality-specific encoders and decoders for segmentation) for all possible missing modality(ies) scenarios. Recently, there are several methods belonging to the third category of approach. The HeMIS [4] concatenates the first and second moments of features across individual modalities as the fused features for segmentation. Dorent et al. [3] proposed the Multi-modal Variational Auto-Encoder (MVAE) to embed all the observed modalities into a shared latent representation for tumor segmentation. RobustSeg [1] decomposes the features into modality-invariant and modality-specific features, and the invariant features are taken to generate the final segmentation result based on a gating strategy. The current state-of-the-art method RFNet [2] is based on a Region-aware Fusion Module (RFM) by aggregating multi-modal features of different regions adaptively, to model the relations of modalities and tumor regions. The key challenge in incomplete multi-modality image segmentation is how to effectively fuse the multi-modal image features while being robust to the arbitrary missing modality(ies). These prior works either focus on learning shared/invariant features or the relations between modalities and tumor regions, which may not fully consider the complementary information among multi-modality images for segmentation and the model's adaptability to different missing modality(ies).

This work designs an adaptive feature interaction strategy that interacts multi-modal features for multi-modal segmentation adaptive to different missing modality(ies). Specifically, we propose Modality-adaptive Feature Interaction (MFI), which is based on graph structure [7] that models each modality as a graph node and adaptively interacts node features based on attention mechanism [15]. To adaptively learn the complementary features among modalities (i.e., graph nodes), we introduce a multi-modal code to represent if different modalities are observed or not, to guide the learning process. We plug MFI guided by multi-modal code into different stages of a U-shaped architecture [12] to hierarchically interact multi-modal features. Experiments show that our method achieves the state-of-the-art results in incomplete multi-modal brain tumor segmentation on BraTS2018 dataset.

2 Method

In this section, we present network design and details on the *Modality-adaptive Feature Interaction* (MFI) module which is the key module in our network. For brain tumor segmentation, the multi-modal MRI images are in sequences $\mathcal{M} = \{\text{FLAIR}, \text{T1c}, \text{T1}, \text{T2}\}$. As shown in Fig. 1, our network, dubbed as **U-Net-MFI**, is a U-shape [12] net with four stages in both encoder and decoder. There are shortcut connections between corresponding stages with the same resolution in the encoder and decoder by element-wise summation. Different to the conventional U-Net, we interact multi-modal features in different stages hierarchically using MFI modules. To enforce the network to be adaptive to different missing modalities scenarios, we introduce a binary vector of *multi-modal code* to represent if different modalities are observed or not, to guide the feature interaction in MFI. The inputs of our network are multi-modal images and network parameters excluding the MFI modules are not shared for different modalities, same as [2–4] that define different encoders/decoders for different modalities. If a modality image is not provided, its corresponding input is replaced by a zero tensor in the same size as the input image. Please refer to Fig. 1 for the design of transforms with filter sizes.

Based on this architecture, features of different modalities are hierarchically extracted and interact with each other by MFI module guided by multi-modal code for adaptation to different missing modalities. At the end of the decoder, all modality features are concatenated to generate the segmentation map. In the multi-scale stages of decoder before the final stage, we also attach segmentation loss for multi-scale training. We next present the details on MFI module.

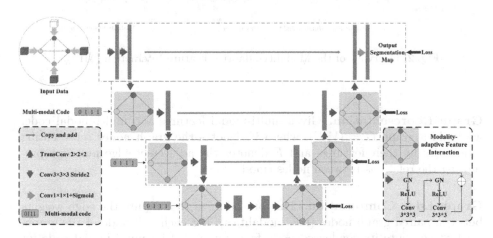

Fig. 1. The architecture of our network (U-Net-MFI).

2.1 Modality-Adaptive Feature Interaction

Different modality contributes differently for identifying different tumor regions [22]. For example, FLAIR is the principle modality for showing whole brain tumor. However, observing more modalities will further improve the accuracy of segmentation. This fact means that different modalities provide complementary information/features for tumor region segmentation. Further considering the different possible missing modalities, the feature interaction should consider to model complementary features among modalities and also be adaptive to different missing modality(ies). Inspired by this motivation, the proposed MFI is designed for multi-modal feature interaction guided by multi-modal code to adaptively learn to interact features among modalities, as shown in Fig. 2.

In the feature space of a certain stage of U-Net-MFI, the input of MFI is multi-modal features $\{f_1, ..., f_N\}$ for N modalities, with $f_i \in \mathbb{R}^{C \times H \times W \times L}$, where C is the number of channels, and W, H, L denote the width, height and slice dimension for the 3D volume features in the current stage. With these modality features and multi-modal code h as inputs, we first build a graph with each node for one modality, then design message passing to learn to incorporate features from other modalities. The proposed multi-modal code h, indicating if different modalities are observed or not, will guide the MFI learning process. We will discuss the details in the following paragraphs.

Modality-adaptive Feature Interaction (MFI)

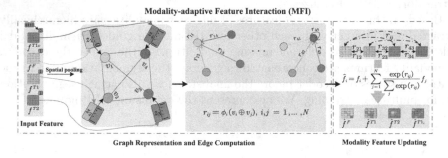

Graph Representation and Edge Computation   Modality Feature Updating

Fig. 2. Overview of the Modality-adaptive Feature Interaction (MFI).

Graph Representation. Given multi-modal features and multi-modal code, we build a graph $G = (V, E)$, where V denotes the graph nodes representing modalities by their features, and E denotes the adjacency edge matrix representing the relations between nodes (modalities).

Graph Edge Computation. This procedure is to compute the edge weights between paired graph nodes (i.e., modalities). For each i-th node of the graph, we first compute its condensed node features guided by multi-modal code for edge weight computation, the condensed node feature is defined as

$$v_i = \psi_{sp}(f_i) \oplus (h_i \mathbf{1}), \quad i = 1, \cdots, N \tag{1}$$

where $\oplus$ denotes the concatenation operation while ψ_{sp} denotes the spatial pooling operation and $\psi_{sp}(f_i) \in \mathbb{R}^C$. $h = \{h_1, ..., h_N\}$ is the *multi-modal code*, and

$h_i \in \{0,1\}$ is i-th element of h indicating whether the i-th modality is missing or not. $\mathbf{1} \in R^C$ is a all one-valued vector and $h_i \mathbf{1} \in \mathbb{R}^C$. In this way, $v_i \in \mathbb{R}^{2 \times C}$ contains both modality feature and modality availability information.

We next present how to compute the edge weights among modalities for feature interaction. We denote the edge weight from j-th node to i-th node as r_{ij}, which is computed as:

$$r_{ij} = \phi_i(v_i \oplus v_j), \quad i, j = 1, \cdots, N \tag{2}$$

where ϕ_i is a linear layer followed by Leaky Rectified Linear Unit [20]. By this, the condensed node features v_i and v_j are fed to ϕ_i to learn $r_{ij} \in \mathbb{R}^C$. Since v_i and v_j contain both modality features and the multi-modal code information, the r_{ij} is therefore dependent on the availability information of modalities i, j.

Modality Feature Updating. Based on the edge weights, this procedure is to interact modality features f_i $(i = 1 \cdots N)$ of nodes to derive the updated feature $\hat{f}_i$ for each node, i.e., modality. Taking the i-th node as an example, we feed the $R_i = \{r_{i1}, ..., r_{iN}\}$ to a softmax layer to get the normalized attention scores, then the output feature $\{\hat{f}_i\}$ is computed as:

$$\hat{f}_i = f_i + \sum_{j=1}^{N} \frac{\exp(r_{ij})}{\sum_j \exp(r_{ij})} f_j, \quad i = 1, \cdots, N \tag{3}$$

Note that the updated feature $\hat{f}_i$ is the addition of input feature f_i with the "borrowed" features from other modalities based on residual skip connection [5]. This implements our idea of feature interaction among modalities.

In summary, MFI is designed to learn complementary features from multi-modal image features guided by multi-modal code. Plugging MFI modules into the hierarchical multi-modal feature spaces of our U-Net-MFI help to deeply capture these complementary features for segmentation.

2.2   Network Details and Training Loss

As shown in Fig. 1, the network is a U-shaped architecture [12] embedded with our proposed MFI. In each stage of Encoder, we use two $3 \times 3 \times 3$ convolutions with Group Normalization (GN) [19] (group size is 8) and ReLU activation, followed by additive identity skip connection. Then a convolution layer with a $3 \times 3 \times 3$ filter with stride of 2 is used to reduce the resolution of the feature maps by factor of 2 in each dimension and simultaneously increase the number of feature channels by factor of 2. Each decoder stage begins with transposed convolution in kernel size of $2 \times 2 \times 2$ with stride of 2 to increase the resolution of feature maps and reduce the number of channels. Then the up-sampled features are added with the features from the corresponding stage of the encoder followed by MFI. At the end of the decoder, a $1 \times 1 \times 1$ convolution is used to decrease the number of output channels to 3, followed by a sigmoid function. In the multi-scale stages of decoder before the final stage, we apply $3 \times 3 \times 3$ convolution and up-sampling after MFI to generate multi-scale segmentation for multi-scale loss.

The multi-scale training loss is based on the combination of cross-entropy loss L_{CE} and the soft Dice loss L_{Dice}, employed in each stage (four stages in decoder of U-Net-MFI) of the decoder as deep supervision [16], and the loss is

$$L(\hat{y}, y) = \sum_{s=1}^{4} \lambda L_{CE}(\hat{y}^s, y^s) + L_{Dice}(\hat{y}^s, y^s) \tag{4}$$

where y^s and $\hat{y}^s$ respectively denote the ground truth and segmentation probabilities in stage s. λ is a trade-off parameter, empirically set to 0.5 in experiments.

3   Experiments

Dataset. We evaluate our method on the Multi-modal Brain Tumor Segmentation Challenge 2018 (BraTS2018) dataset [10], which contains the scans of 285 patients with 210 high grade glioma and 75 low grade glioma. Each subject in the dataset contains four MRI contrasts (FLAIR, T1c, T1, T2), and is pre-processed by the organisers. Scans have been re-sampled to $1\,\mathrm{mm}^3$ isotropic resolution, skull stripped, and co-registered to the same anatomical template. Following the challenge, there are three segmentation classes, including whole tumor ("complete"), core tumor ("core") and enhancing tumor ("enhancing"). The ground truth was obtained by expert manual annotation.

Implementation Details. As a pre-processing procedure, we use intensity normalization for each MRI modality from each patient. Before feeding the data into the deep learning network, we replace each missing modality with a same sized zero tensor and randomly crop 3D volumes of $128 \times 128 \times 128$ voxels from the data with $240 \times 240 \times 155$ voxels due to memory limit. The spatial resolution of feature space at stage 1 of encoder is $128 \times 128 \times 128$ and the resolutions of the latter stages of encoder are half that of the previous one in each dimension. Our network is implemented by PyTorch [11] and is optimized using the Adam optimizer [6] with batch size of 1. The initial learning rate is 5×10^{-5} and we decay it using "poly" learning rate policy where the initial learning rate is multiplied by $(1 - \frac{epoch}{max\_epoch})^{0.9}$. In inference, we segment the $120 \times 120 \times 120$ patches sliding on the test image, and the neighboring patches have 50% overlap. Then the final segmentation map is obtained by averaging the predictions of these patches. Our experiment is performed on one Nvidia GeForce RTX 3090 GPU.

Model Comparison. To evaluate the performance of our model, we compare it to four state-of-the-art methods, HeMIS [4], U-HVED [3], RobustSeg [1] and RFNet [2], using 3-fold cross-validation with the same split lists as [3] on BRATS2018. As shown in Table 1, we use the Dice score as metric and our method achieves preferable results. This highlights the effectiveness of our designed network with

Table 1. Comparison of different methods (Dice %), including HeMIS [4], U-HVED [3], RobustSeg [1] and RFNet [2], on BraTS2018. Available modalities are denoted by ●, and missing modalities are denoted by o.

Modalities				Complete					Core					Enhancing				
F	T1	T1c	T2	HeMIS	U-HVED	RobustSeg	RFNet	Ours	HeMIS	U-HVED	RobustSeg	RFNet	Ours	HeMIS	U-HVED	RobustSeg	RFNet	Ours
o	o	o	●	38.60	80.90	82.24	84.30	**85.63**	19.50	54.10	57.49	**67.62**	67.48	0.00	30.80	28.97	**40.71**	37.42
o	o	●	o	2.60	62.40	73.31	74.93	**75.24**	6.50	66.70	76.83	**80.99**	80.90	11.10	65.50	67.07	69.43	**76.19**
o	●	o	o	0.00	52.40	70.11	**74.68**	74.23	0.00	37.20	47.90	**64.42**	63.13	0.00	13.70	17.29	**34.43**	32.28
●	o	o	o	55.20	82.10	85.69	**86.46**	86.42	16.20	50.40	53.57	**64.89**	63.72	6.60	24.80	25.69	33.92	**37.63**
o	o	●	●	48.20	82.70	85.19	86.39	**86.52**	45.80	73.70	80.20	83.27	**84.03**	55.80	70.20	69.71	73.01	**77.56**
o	●	o	●	15.40	66.80	77.18	78.59	**78.65**	30.40	69.70	78.72	**82.22**	81.92	42.60	67.00	69.06	70.73	**76.78**
●	●	o	o	71.10	84.30	88.24	88.78	**89.52**	11.90	55.30	60.68	71.59	**72.74**	1.20	24.20	32.13	39.68	**41.34**
o	●	●	o	47.30	82.20	84.78	86.15	**86.56**	17.20	57.20	62.19	70.89	**71.41**	0.60	30.70	32.01	**41.42**	41.37
●	o	o	●	74.80	87.50	88.28	89.12	**90.04**	17.70	59.70	61.16	70.82	**71.28**	0.80	34.60	33.84	**43.77**	42.52
●	o	●	o	68.40	85.50	88.51	89.17	**89.99**	41.40	72.90	80.62	82.94	**84.65**	53.80	70.30	70.30	72.84	**78.79**
●	●	●	o	70.20	86.20	88.73	89.71	**90.11**	48.80	74.20	81.06	83.77	**85.16**	60.90	70.10	70.78	73.17	**79.81**
●	●	o	●	75.20	88.00	88.81	89.68	**90.33**	18.70	61.50	64.38	73.09	**73.95**	1.0	34.10	36.41	**44.79**	43.42
●	o	●	●	75.60	88.60	89.27	90.06	**90.80**	54.90	75.60	80.72	83.54	**85.51**	60.50	71.20	70.88	73.13	**79.41**
o	●	●	●	44.20	83.30	86.01	86.78	**86.94**	46.60	75.30	80.33	83.97	**84.13**	55.10	71.10	70.10	72.56	**77.91**
●	●	●	●	73.80	88.80	89.45	90.26	**90.88**	55.30	76.40	80.86	84.02	**85.63**	61.10	71.70	71.13	73.21	**79.98**
Means				50.70	80.10	84.39	85.67	**86.12**	28.70	64.00	69.78	76.53	**77.04**	27.40	50.00	51.02	57.12	**60.16**

adaptive feature interaction. For instance, our proposed U-Net-MFI outperforms the current SoTA method, *i.e.*, RFNet [2] in most missing-modality cases, including 13 out of 15 cases for the whole tumor, 10 out of 15 cases for the core tumor, 10 out of 15 cases for the enhancing tumor. Compared with the RFNet, for the whole tumor and the tumor core regions, we improve the average Dice scores by 0.45% and 0.51% respectively. In particular, without using any post-processing for enhancing tumor region [2,3], we outperform RFNet in the average Dice score by 3.04%. This result shows that our method has higher improvement in enhancing tumor region, which is the most challenging region among the three types of tumor regions. This demonstrates that enhancing tumor region may need more complementary features from multiple modality images.

Table 2. Ablation study on key components of our method (Dice %).

Methods	Complete	Core	Enhancing	Average
Baseline	81.57	68.07	49.93	66.52
+Mean-wise interaction	85.85	75.81	59.27	73.64
+MFI (w/o multi-modal code)	86.34	75.29	59.19	73.61
+MFI (w/ multi-modal code)	86.98	76.90	59.53	74.47

Ablation Study. In this section, we investigate the effectiveness of MFI and multi-modal code, which are two key components of our method, using one of the three-fold data. We first set up a baseline network ("Baseline") that do not use any feature interaction in our network. Then we add the two modules gradually. We compare the performance of these networks on the Dice score, averaging over the 15 possible situations of input modalities. As shown in Table 2, employing MFI without multi-modal code increases the average Dice scores of three tumor regions by 4.77%, 7.22% and 9.26% respectively, compared with "Baseline". Moreover, applying MFI with multi-modal code increases the results over "Baseline" by 5.41%, 8.83% and 9.60%, implying that multi-modal code h additionally increase the results by 0.64%, 1.61% and 0.34% over MFI without multi-modal code. We also apply mean-wise interaction, $i.e.$, $\hat{f}_i = f_i + \frac{\sum_{j=1}^N f_j}{N}$, to substitute Eq. 3 for comparison. As shown in Table 2, the segmentation accuracy of model with mean-wise interaction achieves inferior results than our method, but superior to the baseline method. The above results justify the effectiveness of feature interaction using MFI guided by multi-modal code for adaptation to different missing modality(ies) scenarios.

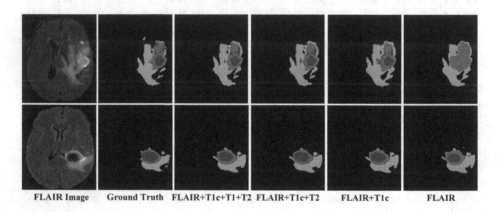

FLAIR Image   Ground Truth   FLAIR+T1c+T1+T2 FLAIR+T1c+T2   FLAIR+T1c   FLAIR

Fig. 3. Examples of segmentation results. Green: whole tumor; Red: core tumor; Blue: enhancing tumor. The first and second columns show input subjects and g.t. labels. The observed modalities are shown under the segmentation results in columns 3–6. (Color figure online)

Visualization of Results. In Fig. 3, we visualize the segmentation results of our method for two examples on BraTS 2018 dataset. The first and second columns show the subjects (only FLAIR image is shown due to space limit) and ground truth labels respectively. The following columns (the input modalities are shown under each column) show the segmentation results with increasingly number of missing modalities. For the second example, even when there is only one available modality (FLAIR), we can still get a decent segmentation result.

4   Conclusion

In this work, we design a novel multi-modal brain tumor segmentation network, which applies Modality-adaptive Feature Interaction with the multi-modal code in multiple stages of U-shaped network, to interact multi-modal features hierarchically and adaptively. Our model outperforms the state-of-the-art approach on

BraTS2018 dataset under both full modalities and various missing modality(ies) situations. The limitations of our method is that the four contrast MR images are required to be pre-aligned (BraTS2018 dataset contains the well aligned multi-contrast MR images for each subject). In the future, we plan to conduct spatial alignment and feature interaction in feature space by attention mechanism at the same time.

Acknowledgments. This work was supported by National Key R&D Program of China (2021YFA1003002), National Natural Science Foundation of China (12125104, 12090021, 12026605, 11971373, 12026603).

References

1. Chen, C., Dou, Q., Jin, Y., Chen, H., Qin, J., Heng, P.-A.: Robust multimodal brain tumor segmentation via feature disentanglement and gated fusion. In: Shen, D., et al. (eds.) MICCAI 2019. LNCS, vol. 11766, pp. 447–456. Springer, Cham (2019). https://doi.org/10.1007/978-3-030-32248-9_50
2. Ding, Y., Yu, X., Yang, Y.: RFNet: region-aware fusion network for incomplete multi-modal brain tumor segmentation. In: Proceedings of the IEEE/CVF International Conference on Computer Vision, pp. 3975–3984 (2021)
3. Dorent, R., Joutard, S., Modat, M., Ourselin, S., Vercauteren, T.: Hetero-modal variational encoder-decoder for joint modality completion and segmentation. In: Shen, D., et al. (eds.) MICCAI 2019. LNCS, vol. 11765, pp. 74–82. Springer, Cham (2019). https://doi.org/10.1007/978-3-030-32245-8_9
4. Havaei, M., Guizard, N., Chapados, N., Bengio, Y.: HeMIS: hetero-modal image segmentation. In: Ourselin, S., Joskowicz, L., Sabuncu, M.R., Unal, G., Wells, W. (eds.) MICCAI 2016. LNCS, vol. 9901, pp. 469–477. Springer, Cham (2016). https://doi.org/10.1007/978-3-319-46723-8_54
5. He, K., Zhang, X., Ren, S., Sun, J.: Deep residual learning for image recognition. In: Proceedings of the IEEE Conference on Computer Vision and Pattern Recognition, pp. 770–778 (2016)
6. Kingma, D.P., Ba, J.: Adam: a method for stochastic optimization. In: ICLR (2015)
7. Kipf, T.N., Welling, M.: Semi-supervised classification with graph convolutional networks. In: ICLR (2016)
8. Li, R., et al.: Deep learning based imaging data completion for improved brain disease diagnosis. In: Golland, P., Hata, N., Barillot, C., Hornegger, J., Howe, R. (eds.) MICCAI 2014. LNCS, vol. 8675, pp. 305–312. Springer, Cham (2014). https://doi.org/10.1007/978-3-319-10443-0_39
9. Liu, Y., et al.: Incomplete multi-modal representation learning for Alzheimer's disease diagnosis. Med. Image Anal. **69**, 101953 (2021)
10. Menze, B.H., et al.: The multimodal brain tumor image segmentation benchmark (BRATS). IEEE Trans. Med. Imaging **34**(10), 1993–2024 (2014)
11. Paszke, A., et al.: Automatic differentiation in PyTorch. In: NIPS Workshop (2017)
12. Ronneberger, O., Fischer, P., Brox, T.: U-Net: convolutional networks for biomedical image segmentation. In: Navab, N., Hornegger, J., Wells, W.M., Frangi, A.F. (eds.) MICCAI 2015. LNCS, vol. 9351, pp. 234–241. Springer, Cham (2015). https://doi.org/10.1007/978-3-319-24574-4_28
13. Shen, L., et al.: Multi-domain image completion for random missing input data. IEEE Trans. Med. Imaging **40**(4), 1113–1122 (2020)

14. Tran, L., Liu, X., Zhou, J., Jin, R.: Missing modalities imputation via cascaded residual autoencoder. In: Proceedings of the IEEE Conference on Computer Vision and Pattern Recognition, pp. 1405–1414 (2017)
15. Vaswani, A., et al.: Attention is all you need. In: Advances in Neural Information Processing Systems, vol. 30, pp. 5998–6008 (2017)
16. Wang, L., Lee, C.Y., Tu, Z., Lazebnik, S.: Training deeper convolutional networks with deep supervision. arXiv:1505.02496 (2015)
17. Wang, S., et al.: LT-Net: label transfer by learning reversible voxel-wise correspondence for one-shot medical image segmentation. In: Proceedings of the IEEE/CVF Conference on Computer Vision and Pattern Recognition, pp. 9162–9171 (2020)
18. Wang, Y., et al.: ACN: adversarial co-training network for brain tumor segmentation with missing modalities. In: de Bruijne, M., et al. (eds.) MICCAI 2021. LNCS, vol. 12907, pp. 410–420. Springer, Cham (2021). https://doi.org/10.1007/978-3-030-87234-2_39
19. Wu, Y., He, K.: Group normalization. In: Proceedings of the European Conference on Computer Vision (ECCV), pp. 3–19 (2018)
20. Xu, B., Wang, N., Chen, T., Li, M.: Empirical evaluation of rectified activations in convolutional network. arXiv:1505.00853 (2015)
21. Zhang, Y., et al.: Modality-aware mutual learning for multi-modal medical image segmentation. In: de Bruijne, M., et al. (eds.) MICCAI 2021. LNCS, vol. 12901, pp. 589–599. Springer, Cham (2021). https://doi.org/10.1007/978-3-030-87193-2_56
22. Zhou, T., Canu, S., Vera, P., Ruan, S.: Brain tumor segmentation with missing modalities via latent multi-source correlation representation. In: Martel, A.L., et al. (eds.) MICCAI 2020. LNCS, vol. 12264, pp. 533–541. Springer, Cham (2020). https://doi.org/10.1007/978-3-030-59719-1_52

Position-Prior Clustering-Based Self-attention Module for Knee Cartilage Segmentation

Dong Liang[1], Jun Liu[1], Kuanquan Wang[1(✉)], Gongning Luo[1(✉)], Wei Wang[1], and Shuo Li[2]

[1] Harbin Institute of Technology, Harbin, China
{wangkq,luogongning}@hit.edu.cn
[2] University of Western Ontario, London, Canada

Abstract. The morphological changes in knee cartilage (especially femoral and tibial cartilages) are closely related to the progression of knee osteoarthritis, which is expressed by magnetic resonance (MR) images and assessed on the cartilage segmentation results. Thus, it is necessary to propose an effective automatic cartilage segmentation model for longitudinal research on osteoarthritis. In this research, to relieve the problem of inaccurate discontinuous segmentation caused by the limited receptive field in convolutional neural networks, we proposed a novel position-prior clustering-based self-attention module (PCAM). In PCAM, long-range dependency between each class center and feature point is captured by self-attention allowing contextual information re-allocated to strengthen the relative features and ensure the continuity of segmentation result. The clutsering-based method is used to estimate class centers, which fosters intra-class consistency and further improves the accuracy of segmentation results. The position-prior excludes the false positives from side-output and makes center estimation more precise. Sufficient experiments are conducted on OAI-ZIB dataset. The experimental results show that the segmentation performance of combination of segmentation network and PCAM obtains an evident improvement compared to original model, which proves the potential application of PCAM in medical segmentation tasks. The source code is publicly available from link: https://github.com/PerceptionComputingLab/PCAMNet

Keywords: Self-attention · Position-prior · Knee cartilage segmentation

1 Introduction

Osteoarthritis (OA) is a kind of chronic degenerative articular disease that causes disability gradually. Knee cartilage is the soft tissue adhering to the end of the bone surface, whose changes in morphological structure are associated with the progression of OA [1]. Compared to other imaging techniques, magnetic

L. Wang et al. (Eds.): MICCAI 2022, LNCS 13435, pp. 193–202, 2022.
https://doi.org/10.1007/978-3-031-16443-9_19

resonance imaging (MRI) shows a higher level of specificity and sensitivity to obtain the biomedical markers of knee cartilage [2]. However, manual cartilage segmentation from MRI has demanded more relative knowledge from specialists. The manual segmentation is tedious, time-consuming and brings inter-/intra-observer variations. Thus, there is a demand to design an effective automatic cartilage segmentation method for the longitudinal analysis.

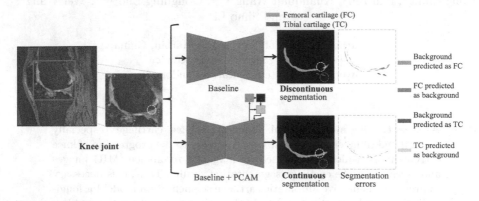

Fig. 1. The knee cartilage segmentation results from the primary baseline [7] and the same model [7] modified by PCAM. In white and red circle, the local contextual information of foreground is similar to the background nearby, which causes the segmentation errors and leads to discontinuous segmentation result. The proposed PCAM can relieve the problem and create a more accurate and continuous segmentation result. (Color figure online)

With the development of deep learning, convolutional neural network(e.g. U-Net [6], V-Net [7])has achieved state-of-the-art segmentation results. Although the 3D deep learning model V-Net has exhibited a superior performance on medical image segmentation tasks, directly applying the primary V-Net on knee MR data may generate low accuracy results. As shown in Fig. 1, the articular structure is complex and the features of tissues around the knee cartilage in MR image are similar to each other, which is difficult to extract continuous knee cartilage accurately from the whole volumetric data. To reduce disturbance brought by the complex background, Ambellan et al. [4] proposed a coarse-to-fine scheme with a 2D CNN for coarse segmentation and a 3D CNN for fine segmentation followed by a statistical shape adjustment. Similarly, Gatti et al. [3] adopted 2D U-Net and 3D U-Net in parallel for coarse segmentation and an additional 3D U-Net for fine segmentation. The coarse-to-fine architecture consists of several sub-networks which brings a huge computational burdens and the input of the behind sub-networks is entirely reliant on the output of the preceding sub-networks. Tan et al. [5] presented a collaborative multi-agent learning framework in an end-to-end scheme that is still limited to the GPU memory. Within 3D MR dataset, the morphological feature of knee cartilage varies greatly. To capture the contextual information of objects with different scales, Zhao et al. [9]

presented a pyramid network using multiple dilated convolution kernels. However, this non-adaptive approach ignored the long-range dependencies and did not distinguish the surrounding pixels from different categories leading to discontinuous segmented results. Sinha et al. [8] pointed out that the self-attention mechanism exhibits good performance on modeling long-range contextual dependency with a high level of flexibility. Nevertheless, when the feature map is large, this attention module is prone to a heavy computational burden.

To overcome the shortcomings mentioned above, we proposed a novel position-prior and clustering-based self-attention module (PCAM) in CNN for automatic knee cartilage segmentation in MR data. The main contributions of our research are concluded as follows: (a) It is the first time that we applied clustering-based self-attention module on knee cartilage segmentation tasks. And the proposed PCAM can be plugged in network flexibly with less GPU memory consumption and computational burdens. (b) We proposed a position-prior module that excludes false positives on the boundary area of knee cartilage from coarse mask to improve the accuracy of feature clustering in PCAM. (c) The presented PCAM captures long-range contextual information to achieve continuous and accurate knee cartilage segmentation. (d) The segmentation models combining proposed PCAM have obtained performance improvement on the OAI-ZIB dataset.

2   Method

General CNN focuses on the local receptive field while neglecting the long-range dependencies that introduces intra-class inconsistency and discontinuous segmentation [8]. The attention module can capture various scale context dependencies and strengthen the relevant feature information to obtain a more accurate result in segmentation tasks. Taking tibial cartilage segmentation as an example, the overview of the network architecture is shown in Fig. 2. The blue box is the side-output function to produced coarse mask of the knee cartilage that is modified by morphological operations. The feature map together with the modified side-output is then fed back to the clustering-based self-attention module to produce an enhanced feature map.

The PCAM is divided into three parts: position-prior module, clustering-based module and self-attention module, which are illustrated in Fig. 3. As it is impractical to use true label mask for calculating each class center in corresponding feature map, the output of the segmentation network is applied for class center approximation [10]. However, the knee cartilage only occupies a small area in a large sized MR image, several false positive points in the coarse segmentation result can make the estimated results deviate from the true class centers evidently. Therefore, we adopt modified result with the help of morphological operations to generate precise position prior in this research.

Position-Prior Module. The position-prior module is designed for excluding the false positives of side-output so as to improve the accuracy of feature

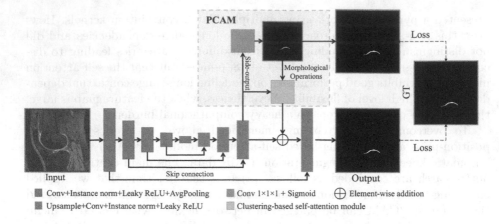

Fig. 2. The segmentation network architecture with PCAM.

clustering. Figure 1 illustrates a common case that V-Net obtains precise segmentation results within the knee cartilage as well as background while fails on the area around the boundary of segmented knee cartilage. In boundary area, the feature distribution is ambiguous because of the low contrast with adjacent tissues and the poor quality of imaging technique. Thus, we divided the predicted probability map into three parts: $M^{boundary}$ (the area around boundary the of predicted knee cartilage); $M^{foreground}$ (the area within predicted knee cartilage); $M^{background}$ (the area except for $M^{boundary}$ and $M^{foreground}$). Areas $M^{foreground}$ and $M^{background}$ are defined as follows:

$$M^{foreground} = \{(x, y, z) | (B)_{(x,y,z)} \subseteq \sigma(F)\} \tag{1}$$

$$M^{background} = \{(x, y, z) | (B)_{(x,y,z)} \subseteq 1 - \sigma(F)\} \tag{2}$$

where F is the feature map and $\sigma(\cdot)$ is the side-output function to generate a predicted probability map; (x, y, z) is the position in $\sigma(F)$; B represents the structure element while function $(B)_{(x,y,z)}$ is the set centered on (x, y, z) containing all elements of B. As shown in Fig. 3, we regard the modified side-output $M^{foreground}$ and $M^{background}$ as position prior, which are then used to assess class centers of foreground and background in feature map, respectively.

Clustering-Based Module. The clustering-based module can weaken the influence of segmentation error by averaging all features belonging to the same class in predicted probability map. In addition, the class center contains abundant contextual information as an aggregation of features within each class. Thus, the clustering method is applied in PCAM to compute the similarity between every position in feature map and each class center so as to construct affinity map. The process of clustering is shown in Fig. 3. Given the feature map $F \in \mathbb{R}^{C \times H \times W \times S}$, foreground mask $M^{foreground} \in \mathbb{R}^{1 \times H \times W \times S}$ and background mask $M^{background} \in \mathbb{R}^{1 \times H \times W \times S}$, F is reshaped to $C \times HWS$ and mask M^{class}

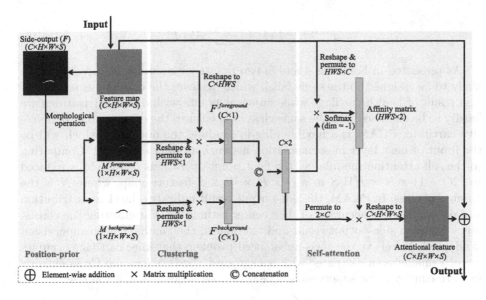

Fig. 3. The details of position-prior and clustering-based attention module.

is adjusted to shape $HWS \times 1$, where C, H, W, S represent channel, height, width and slice, respectively. The class center of each class is calculated as follows:

$$F^{class} = \frac{\sum_{i=1}^{HWS} M^{class}(i) \cdot F(i)}{\sum_{i=1}^{HWS} M^{class}(i)} \tag{3}$$

where the *class* could be one of the elements from set $\{foreground, background\}$, $F(i) \in \mathbb{R}^{C \times 1}$ is the feature vector in position i of feature map F.

Self-attention Module Self-attention module can capture the long-range contextual information to ensure the continuous segmentation result and improve the accuracy. As shown in Fig. 3, the feature map F is firstly reshaped and permuted to $\mathbb{R}^{HWS \times C}$ and class centers are concatenated along the last dimension. The matrix multiplication is then executed between feature map $F \in \mathbb{R}^{HWS \times C}$ and class center $F^{class} \in \mathbb{R}^{C \times 1}$. The results are normalized to generate affinity map A as follows:

$$A_j^{class} = \frac{exp(F^{class} \cdot F(j))}{\sum_i^{classes} exp(F^i \cdot F(j))} \tag{4}$$

where A_j^{class} denotes the similarity between the feature vectors in position j and clustering center of *class*. The class set $classes = \{foreground, background\}$, which also can be extended to multi-class. Affinity map is then multiplied by the transposed feature vectors of class centers to obtain the attention feature map that is further element-wisely added to feature map F. The generation of novel feature map is formulated as follows:

$$F_j^{atten} = F(j) + \sum_{i}^{classes} A_j^i \cdot F^i \tag{5}$$

As presented in Formulas 4 and 5, two points with similar features are more likely to be assigned to the same label, which achieves the continuous segmentation result of health cartilage while adjacent points with dissimilar features are hardly to be classified into the same class that ensures the discontinuity of defective cartilage. PCAM is a flexible plug-in module, the output of which will be the input of next layer in segmentation network as shown in Fig. 2. Comparing to the self-attention module [8], the float point operations of PCAM is reduced to $(2C - 1) \times N \times HWS$ in a $C \times H \times W \times S$ feature map, where N is the number of class. In PCAM, the side-output result indicates the class distribution and determines the accuracy of class center estimation. For ensuring the consistency between side-output result and true mask, the auxiliary deep supervision is adopted. To relieve the class-imbalanced problem that knee cartilage occupies a much smaller area compared with background, the Dice loss and Cross-entropy loss are employed for supervision. The total loss is described as follows:

$$Loss_{Total} = Loss_{Dice} + Loss_{Cross-entropy} \tag{6}$$

3 Experiment

Materials and Evaluation Metrics. The proposed method is validated on the OAI-ZIB Dataset[1]. This public dataset includes 507 3D DESS MR data with 81120 slices. The pixel spacing is 0.3645 mm $\times$ 0.3645 mm and the slice thickness is 0.7 mm for all volumetric data. For each volumetric data, it contains 160 slices in 384 $\times$ 384. The 2-fold cross validation approach is applied in experiment for evaluating the performance of methods. To verify the effectiveness of PCAM, the ablation study is executed as well. Dice similarity coefficient (DSC), average symmetric surface distance (ASSD) and volumetric overlap error (VOE) are adopted for the comparison between the predicted results and ground truth.

Implementation Details. In training phase, the batch size is set to 4 and the initial learning rate is set to 0.01 by Adam optimizer with 0.95 decay rate for every epoch. The training phase would be stopped when the improvement of DSC on validation set is no more than 0.0001 in continuous 10 epochs. The structure element is a 3 $\times$ 3 mask with all elements set to 1. To protect the structure of cartilage, the morphological operation is executed only once. For data augmentation, the elastic deformation, random rotation, random-/center-cropping and random clip contribute to spatial transformation. Moreover, the gamma transformation, Gaussian noise and contrast adjustment are applied to enrich the gray distribution of training data. For mini-batch training in segmentation network, the batch normalization and ReLU are substituted by instance normalization and Leaky ReLU, respectively. The networks are trained and tested on NVIDIA 2080Ti with 11 GB video memory.

---

[1] https://nda.nih.gov/oai/.

Experimental Results. As the cascaded model (e.g. [3]– [5], [11]) is composed of single segmentation models, the segmentation performance of cascaded model depends on its sub-networks with complex computational process and huge computation burdens. In this experiment, several classical segmentation networks with different schemes are evaluated. First of all, the baseline model is derived from V-Net [7] without any modification. The second segmentation model is devised with the joint learning by the generative adversarial network (GAN) whose architecture and training process are adjusted on the basis of [5]. Because the memory usage of attention module in [10] exceeds the limitation of GPU, the third model, that is the combination of baseline model and the attention module [10], is re-designed with the help of side-output and auxiliary supervision. The fourth model is the baseline with proposed PCAM plugged between the third and the fourth upsampling layer as Fig. 2. The fifth model is the primary nnU-Net [11], which obtained the best segmentation results on several medical image segmentation challenges. In the last model, the nnU-Net is combined with the proposed PCAM in the same location as the fourth model.

Table 1. Quantitative comparisons among segmentation methods with evaluation metrics (DSC, VOE and ASSD) by mean and std values.

Model	Femoral cartilage			Tibial cartilage		
	DSC (%)	VOE (%)	ASSD (mm)	DSC (%)	VOE (%)	ASSD (mm)
Baseline [7]	87.71	21.79	0.2259	84.11	27.23	0.2287
	2.77	4.28	0.1020	3.93	5.79	0.1067
Tan et al. [5]	87.86	21.53	0.2390	84.08	27.25	0.2643
	2.84	4.40	0.1221	4.04	5.94	0.1402
Zhang et al. [10]	87.91	21.45	0.2235	84.57	26.52	0.2374
	2.82	4.38	0.1121	4.16	6.20	0.0907
Baseline [7]+PCAM	88.45	20.62	**0.2058**	85.15	25.61	0.2172
	2.57	4.04	0.0900	4.19	6.31	0.1072
nnU-Net [11]	89.03	19.21	0.2551	86.0	24.29	**0.2117**
	2.73	4.37	0.3206	4.52	6.73	0.1074
nnU-Net [11]+PCAM	**89.35**	**19.14**	0.2389	**86.14**	**24.08**	0.2165
	2.69	4.32	0.3196	4.43	6.63	0.1242

The experimental results are demonstrated on Table 1. For adversarial learning scheme, the results are imporved on femoral cartilage segmentation but failed on the tibial cartilage. The structure characteristics that tibial cartilage occupies much fewer pixels may cause the adversarial learning scheme ineffective. For attention module in the third model, it captures the long-range contextual information that is more suitable for elongated cartilage segmentation. Furthermore, the segmentation performance of both baseline model and nnU-Net are improved on condition that the PCAM is plugged in the network which proves the effectiveness of the proposed self-attention module. Comparing to the primary baseline model, the combination of baseline model and PCAM can achieve the continuous and accurate segmentation, which is shown in Fig. 1. To quantify the continuity correctness of segmentation results, 0-dimension Betti number error is calculated slice-by-slice [12]. Under this metric, the average continuity errors of nnU-Net+PCAM model are $0.1323(\pm0.43)$ in femoral cartilage and $0.1267(\pm0.48)$ in tibial cartilage compared to nnU-Net with $0.1792(\pm0.52)$ in femoral cartilage and $0.1358(\pm0.49)$ in tibial cartilage. T-test was conducted between the methods with and without PCAM on segmentation continuity. We obtained p-value<0.01 on both femoral cartilage and tibial cartilage segmentation tasks, indicating that PCAM improves the segmentation continuity significantly.

The decoder of the baseline model contains abundant semantic and spatial information, which is suitable to insert PCAM. However, there are three locations among four upsampling layers of decoder that the PCAM can be plugged in. To find out which upsampling layer of the decoder is optimized, the three locations are remarked as $1, 2, 3$ from low resolution to high resolution and plugged with PCAM, respectively. The experimental results are shown in Table 2. It can be seen that the PCAM plugged between the third and the fourth upsampling layer with the highest resolution has obtained the best segmentation results.

Table 2. Ablation experiment on *PCAM*. *PCAM Net^i* represents that the model is plugged with PCAM behind the *ith* upsampling layer (PCAMNet=Baseline+PCAM). The performance is evaluated on three metrics by mean and std values.

	Femoral cartilage			Tibial Cartilage		
	DSC (%)	VOE (%)	ASSD (mm)	DSC (%)	VOE (%)	ASSD (mm)
PCAMNet^1	87.78	21.67	0.2555	84.57	26.51	0.2374
	2.87	4.47	0.1853	4.16	6.20	0.0907
PCAMNet^2	87.90	21.49	0.2438	84.65	26.39	0.2518
	2.71	4.20	0.1451	4.14	6.15	0.1746
PCAMNet^3	**88.45**	**20.62**	**0.2058**	**85.15**	**25.61**	**0.2172**
	2.57	4.04	0.0900	4.19	6.31	0.1072

In PCAM, the side-output is used to calculate the center of each class. But the original side-output S_{pred} contains many false positives on the area around the boundary of segmented knee cartilage as shown in Fig. 1. In order to filter out segmentation mistakes, the Erode operations (achieved by $maxpooling$) are applied in this research. The modified side-output S'_{pred} has higher precision results that are shown in Table 3.

Table 3. Quantitative comparion by mean and std precision between original side-output S_{pred} results and side-output results S'_{pred} modified by morphological operations. ($Precision = \frac{Truepositives}{TruePositives + Falsepositives} \times 100\%$)

	Precision(%)			
	Femoral cartilage		Tibial cartilage	
	Foreground	Background	Foreground	Background
S_{pred}	87.56	99.83	83.32	99.95
	4.18	0.05	7.51	0.02
S'_{pred}	**98.13**	**99.98**	**94.77**	**99.99**
	1.75	0.02	5.98	0.01

4   Conclusion

In this research, we proposed a novel self-attention module PCAM for ensuring accurate continuous knee cartilage segmentation. The proposed PCAM captures long-range contextual information to relieve limited receptive field brought by convolution filters in neural networks. Besides, the proposed PCAM brings less computational burdens and is flexible to be plugged in any encoder-decoder structured segmentation neural networks. The experimental results show that the proposed method achieves an accurate segmentation both on femoral and tibial cartilage in 3D MR data and has a potential application in future.

Acknowledgements. This work was supported by the National Natural Science Foundation of China under Grant 62001144 and Grant 62001141, and by Science and Technology Innovation Committee of Shenzhen Municipality under Grant JCYJ20210324131800002 and RCBS20210609103820029.

References

1. Tack, A., Zachow, S.: Accurate automated volumetry of cartilage of knee using convolutional neural networks: data from the osteoarthritis initiative. In: 16th International Symposium on Biomedical Imaging, Venice, pp. 40–43. IEEE (2019)
2. Marinetti, A., et al.: Morphological MRI of knee cartilage: repeatability and reproducibility of damage evaluation and correlation with gross pathology examination. Eur. Radiol. **30**(6), 3226–3235 (2019). https://doi.org/10.1007/s00330-019-06627-5
3. Gatti, A.A., Maly, M.R.: Automatic knee cartilage and bone segmentation using multi-stage convolutional neural networks: data from the osteoarthritis initiative. Magn. Reson. Mater. Phys., Biol. Med. **34**(6), 859–875 (2021). https://doi.org/10.1007/s10334-021-00934-z
4. Ambellan, F., Tack., Ehlke, M., Zachow, S.: Automated segmentation of knee bone and cartilage combining statistical shape knowledge and convolutional neural networks: data from the osteoarthritis initiative. Med. Image Anal. **52**, 109–118 (2019)

5. Tan, C., Yan, Z., Zhang, S., Li, K., Metaxas, D.N.: Collaborative multi-agent learning for MR knee articular cartilage segmentation. In: Shen, D., et al. (eds.) MICCAI 2019. LNCS, vol. 11765, pp. 282–290. Springer, Cham (2019). https://doi.org/10.1007/978-3-030-32245-8_32

6. Ronneberger, O., Fischer, P., Brox, T.: U-Net: convolutional networks for biomedical image segmentation. In: Navab, N., Hornegger, J., Wells, W.M., Frangi, A.F. (eds.) MICCAI 2015. LNCS, vol. 9351, pp. 234–241. Springer, Cham (2015). https://doi.org/10.1007/978-3-319-24574-4_28

7. Milletari, F., Navab, N., Ahmadi, S.-A.: V-Net: fully convolutional neural networks for volumetirc medical image segmentation. In: 4th International Conference on 3D Vision, Stanford, pp. 565–571. IEEE (2016)

8. Sinha, A., Dolz, J.: Multi-scale self-guided attention for medical image segmentation. IEEE J. Biomed. Health Inform. $25(1)$, 121–130 (2021)

9. Zhao, H.-S., Shi, J.-P., Qi, X.-J., Wang, X.-G., Jia, J.-Y.: Pyramid scene parsing network. In: IEEE Conference on Computer Vision and Pattern Recognition, Honolulu, pp. 6230–6239. IEEE (2017)

10. Zhang, F., et al.:ACFNet: attentional class feature network for semantic segmentation. In: International Conference on Computer Vision, Seoul, pp. 6797–6806 IEEE (2019)

11. Isensee, F., Jaeger, P.-F., Kohl, S.-A.: nnU-Net: a self-configuring method for deep learning-based biomedical image segmentation. Nature Method **18**, 203–211 (2021)

12. Otter, N., Porter, M.A., Tillmann, U., Grindrod, P., Harrington, H.A.: A roadmap for the computation of persistent homology. EPJ Data Sci. **6**(1), 1–38 (2017). https://doi.org/10.1140/epjds/s13688-017-0109-5

Attentive Symmetric Autoencoder
for Brain MRI Segmentation

Junjia Huang[1,2], Haofeng Li[1(✉)], Guanbin Li[1,2(✉)], and Xiang Wan[1,3]

[1] Shenzhen Research Institute of Big Data, The Chinese University of Hong Kong,
Shenzhen, China
lhaof@foxmail.com
[2] School of Computer Science and Engineering,
Sun Yat-sen University, Guangzhou, China
liguanbin@mail.sysu.edu.cn
[3] Pazhou Lab, Guangzhou 510330, China

Abstract. Self-supervised learning methods based on image patch reconstruction have witnessed great success in training auto-encoders, whose pre-trained weights can be transferred to fine-tune other downstream tasks of image understanding. However, existing methods seldom study the various importance of reconstructed patches and the symmetry of anatomical structures, when they are applied to 3D medical images. In this paper we propose a novel Attentive Symmetric Auto-encoder (ASA) based on Vision Transformer (ViT) for 3D brain MRI segmentation tasks. We conjecture that forcing the auto-encoder to recover informative image regions can harvest more discriminative representations, than to recover smooth image patches. Then we adopt a gradient based metric to estimate the importance of each image patch. In the pre-training stage, the proposed auto-encoder pays more attention to reconstruct the informative patches according to the gradient metrics. Moreover, we resort to the prior of brain structures and develop a Symmetric Position Encoding (SPE) method to better exploit the correlations between long-range but spatially symmetric regions to obtain effective features. Experimental results show that our proposed attentive symmetric auto-encoder outperforms the state-of-the-art self-supervised learning methods and medical image segmentation models on three brain MRI segmentation benchmarks.

This work is supported in part by Chinese Key-Area Research and Development Program of Guangdong Province (2020B0101350001), in part by the National Natural Science Foundation of China (No.62102267), in part by the Guangdong Basic and Applied Basic Research Foundation (No.2020B1515020048), in part by the National Natural Science Foundation of China (No.61976250), in part by the Guangzhou Science and technology project (No.202102020633), and the Guangdong Provincial Key Laboratory of Big Data Computing, The Chinese University of Hong Kong, Shenzhen. Haofeng Li and Guanbin Li are corresponding authors.

Supplementary Information The online version contains supplementary material available at https://doi.org/10.1007/978-3-031-16443-9_20.

Keywords: Masked autoencoder · Self-supervised learning · Brain MRI segmentation · Position encoding

1   Introduction

Accurate segmentation of brain lesion, tumour or tissue for Magnetic Resonance Imaging (MRI) data is essential for building a computer-aided diagnosis (CAD) system, and helps medical experts improve diagnosis and treatment planning. It is necessary to develop an automatic segmentation tool for brain MRI.

Deep convolutional neural networks (DCNNs) have achieved success in brain MRI segmentation [1,21,30], but their local receptive fields fail to capture long-range spatial dependencies. Recently, transformer-based models [2,14] have drawn extensive attention and shown the state-of-the-art results on 3D image segmentation [7,27,29]. These methods collect dense correlations between long-range voxels for representation learning, but they require numerous voxel-level annotations that is scarce in brain medical image. Self-supervised learning (SSL) [22,23,25] uses unlabeled data to pre-train a model that can be fine-tuned to improve the results on downstream tasks. Recently, reconstruction-based SSL methods [8,28], which pre-train transformers for patch-level recovering with natural images. If these methods are applied to 3D medical images, they may fail to model the prior of a brain because they treat all recovered patches equally. Some recent work [24] pre-trains transformers for medical images but it neglects the symmetry of brain structures and the different importance of brain regions.

Motivated by the above observations, we consider a novel transformer-based SSL framework for brain MRI segmentation. Despite individual variations, the structure of brain tissues is relatively stable while lesions have their particular textures and appearance. During the SSL, reconstructing a smooth brain region is not challenging and may cause over-fitting. On the contrary, synthesizing an informative image patch is more difficult, which requires mining the intrinsic representations of anatomical structures. In this work, we propose an attentive reconstruction loss weighting different image regions with their informativeness that is measured by a handcrafted gradient-based score. Moreover, symmetry is an essential prior of brain structure. As transformers encode the coordinates of image patches for computing correlations between different positions, we introduce the symmetry to design a new position encoding method which returns the same code for two distant but symmetrical positions. Transformers with the encoding can enhance the visual features by emphasizing the correlations between contralateral brain regions. Finally, we integrate the proposed loss and encoding with a masked autoencoder to build our proposed SSL framework. Our contributions are summarized as: (1) a novel attentive reconstruction loss function (2) a new symmetric position encoding method, and (3) an SSL framework attentive symmetric autoencoder for brain MRI segmentation. (4) Experimental results show that our method outperforms the state-of-the-art SSL methods and medical image segmentation models on three public benchmarks.

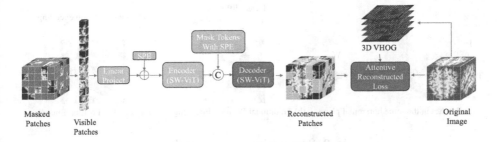

Fig. 1. The architecture of Attentive Symmetric Autoencoder. SPE means symmetric position encoding.

2 Methodology

2.1 Attentive Symmetric Autoencoder

We propose a novel Attentive Symmetric Autoencoder (ASA) that can be trained to obtain generalizable model weights for adapting brain MRI segmentation tasks. As shown in Fig. 1, the proposed ASA consists of a pair of encoder and decoder with symmetric position encoding (SPE) and an attentive reconstruction loss. During the self-supervised training of ASA, the input 3D image is divided into regular non-overlapping image patches (of size $s \times s \times s$). $P\%$ of these image patches are randomly masked and only the unmasked patches are visible. After a linear projection, each visible patch is embedded into a feature vector, which is added with its Symmetric Position Encoding (SPE) to produce the encoder input. The encoder outputs the same number of vectors as its input. Mask Tokens are the same learnable vector added with different SPEs. Each mask token corresponds to a masked image patch. The encoder output is concatenated with the mask tokens to form the decoder input. The decoder reconstructs all the image patches and only the masked ones are used to compute the proposed loss.

Attentive Reconstruction Loss. Considering that learning to recover flatten regions is less helpful for encouraging the model to harvest discriminative representations. We develop an attentive reconstruction loss function that emphasizes the informative regions of brain MRI. To estimate the information of an image patch, we adopt a gradient based metric for 3D images. Inspired by 3D VHOG [3], we calculate the gradient vector $g = (g_x, g_y, g_z)$ for each voxel by applying the filter mask of [-1, 0, 1]. In spherical coordinates, we use two scalars θ and ϕ to represent the orientation of a voxel. θ and ϕ can be calculated as:

$$\theta = cos^{-1}\left(\frac{g_z}{\sqrt{g_x^2 + g_y^2 + g_z^2}}\right), \phi = |atan2(g_y, g_x)|. \tag{1}$$

For each image patch we build a 2D histogram G and the number of bins is $b \times b$. To compute the values G, we traverse each voxel in the image patch. Let θ, ϕ

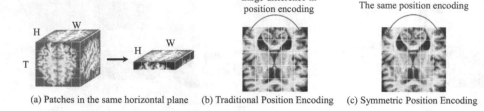

(a) Patches in the same horizontal plane   (b) Traditional Position Encoding   (c) Symmetric Position Encoding

Fig. 2. Symmetric position encoding.

denote the orientation of the current voxel. We first determine the bin indexes of the voxel as $r = \lfloor \theta/(\pi/b) \rfloor, c = \lfloor \phi/(\pi/b) \rfloor$. And then we accumulate $\|g\|$ (the gradient magnitude of the current voxel) to the corresponding bin (r, c) of the 2D histogram G. After processing all voxels in an image patch, L_2 norm is performed on G, the histogram of the patch. We calculate the mean of G as $\bar{G}$ for each image patch. G is normalized among all N masked image regions to characterize the relative importance p_i as $p_i = \frac{\bar{G}_i}{\sum_{i=1}^{N}(\bar{G}_i)}$.

Our proposed loss function adopts mean squared error (MSE) to measure the pixel-level difference between the recovered image areas and the original ones, and pays more attention to the informative brain regions using the gradient-based weight p_i. The overall loss can be formulated as Eq. (2):

$$L(X, Y) = \sum_{i=1}^{N} \left(p_i \cdot \sum_{j}^{M} (X_{ij} - Y_{ij})^2 / M \right). \tag{2}$$

where X, Y are the reconstructed and the original images. N is the number of masked image patches in an image. M is the voxel number in an image patch. X_{ij} denotes the j-th voxel of the i-th patches in the image X.

Symmetric Position Encoding. We observe the left-right symmetry of brain structures, and propose a Symmetric Position Encoding (SPE) method. The proposed method narrow the encoding difference of two symmetric image positions, and can encourage the model to harvest better features from these two correlated regions. For the patches in the same horizontal plane (Fig. 2(a)), the vanilla position encoding [26] of the top left is largely different from that of the top right (Fig. 2(b)), even though these regions have similar contents. However, using our proposed SPE, the leftmost and the rightmost positions (in the same row) can share the same encoding. Let $T \times H \times W$, (t, h, w) denote the patch number of an image and the coordinate of an image patch. The symmetric position encoding is computed as Eq. (3):

$$Pos = (T^2 \cdot t + H \cdot h - |W/2 - w| + W/2)/(10000^{2i/D}),$$
$$PE(t, h, w, 2i) = \sin(Pos), PE(t, h, w, 2i + 1) = \cos(Pos), 1 \le i \le \lfloor D/2 \rfloor, \tag{3}$$

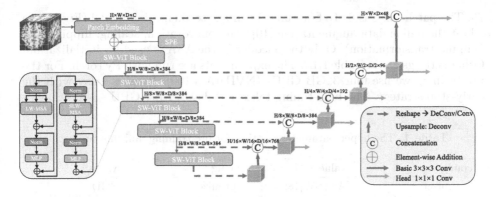

Fig. 3. The architecture of network in downstream tasks.

where D is the dimension number of the SPE vector and is set to the channel number of image patch embeddings. $PE(\cdot)$ returns the $2i$-th/$(2i+1)$-th element of the SPE vector for a patch at (t, h, w). As Fig. 1 shows, the SPE method is used for twice, one for patch embeddings, the other for mask tokens.

2.2 Network Architecture

The proposed ASA model is to provide pre-trained model weights for the downstream task, brain MRI segmentation. Here we describe the architecture of the ASA model and the image segmentation model. The encoder and the decoder of the ASA are based on Vision Transformer (ViT) [2]. The standard ViT [2] uses vanilla self-attention (SA), which leads to high computational cost, especially when processing 3D images. For efficiency, we develop Linear Window-based Multi-head Self-attention (LW-MSA) and Shifted Linear Window-based Multi-head Self-attention(SLW-MSA). Inspired by SwinT [16], we flatten 3D patches into a sequence of patch embeddings, and split the sequence into windows of size S. LW-MSA computes self-attention within each 1D window. SLW-MSA shifts the sequence by $\lfloor \frac{S}{2} \rfloor$ before computing a LW-MSA module, and shifts the sequence by $\lfloor \frac{S}{2} \rfloor$ reversely after the LW-MSA module. LW-MSA and SLW-MSA are computed on a patch level since we convert each image patch to a feature vector via a patch embedding layer at the very beginning. LW-MSA and SLW-MSA are stacked alternately to extract cross-windows features and to build a shifted-window ViT (SW-ViT) for our ASA model. For brain MRI segmentation, we build a U-net with the ASA encoder as the backbone, as shown in Fig. 3.

3 Experiments and Results

Implementation Details. To pre-train the ASA model, we use center-cropping augmentation, Xavier uniform initializer [5] for SW-ViT blocks and set the hyper-parameters following [8] (see Table 1(a)). We follow MAE [8] and set P to

75. The patch size s is 8. To fine-tune the image segmentation model (Fig. 3), we adopt the online data augmentation [10] (random rotation, scaling, flipping and Gamma transformation). Only the encoder of the ASA is used for initialization. Other settings are in Table 1(b). The experiments are run with PyTorch. For the pre-training we use four 32GB GPUs (NVIDIA V100). It takes 1 day with the early-stop strategy. The fune-tuning takes 1–2 days with 1 GPU.

Table 1. The hyper-parameters setting for pre-training and fine-tuning.

config	value	config	value
optimizer	AdamW[18]	optimizer	SGD
optimizer momentum	$\beta_1, \beta_2 = 0.9, 0.95$	optimizer momentum	0.99
weight decay	0.05	weight decay	3e-05
learning rate schedule	cosine decay[17]	initial learning rate	0.01
warmup epochs[6]	1e-6	batch size	2
base learning rate	1.5e-4	num_epoch	1000
batch size	96	loss	Dice and CE loss

<table>
<tr><td>(a) Pre-training setting.</td><td>(b) Fine-tuning setting.</td></tr>
</table>

Datasets. For pre-training our ASA model, we adopt T1 MRI from 2 public datasets, including 9952 cases from Alzheimer's Disease Neuroimaging Initiative(ADNI) dataset[1] [11] and 2041 cases from Open Access Series of Imaging Studies(OASIS) dataset[2] [13]. We convert the data into Brain Imaging Data Structure (BIDS), affinely align the T1 images to the MNI space via Clinica platform [4], strip the brain skull from these images with ROBEX [9] and crop a $128 \times 128 \times 128$ region at their center.

For downstream task, we adopt 3 brain MRI segmentation benchmarks:

Brain Tumor Segmentation (BraTS) 2021 Dataset[3] [19] has 1251 subjects. Each subject has 4 aligned MRI modalities: T1, T1Gd, T2 and T2-FLAIR. The annotations consist of GD-enhancing tumor (ET), peritumoral edematous (ED) and necrotic tumor core (NCR), which are combined into 3 nested subregions: Whole Tumor (WT), Tumor Core (TC), Enhancing Tumor (ET). Following [29], we set the ratio of training/validation/test as 7:1:2.

Internet Brain Segmentation Repository (IBSR) Dataset[4] [20] has 18 T1-weighted MRI volumes of 4 healthy females and 14 healthy males. The ground truth (GT) has 3 categories: Cerebrospinal Fluid (CSF), Gray Matter (GM), White Matter (WM). We adopt 12 cases for training and 6 cases for testing.

---

[1] http://adni.loni.usc.edu/.
[2] https://www.oasis-brains.org/.
[3] http://www.braintumorsegmentation.org/.
[4] https://www.nitrc.org/projects/ibsr/.

Table 2. Comparison on BraTS 2021 dataset. The first group are several competing methods. The best performance is in **bold**.

Task	BraTS 2021					
Metric	Dice (%)↑			HD95 (mm)↓		
Anatomy	WT	TC	ET	WT	TC	ET
nnFormer [29]	91.46	87.42	82.22	10.15	9.59	16.78
TransBTS [27]	92.06	88.20	79.46	4.98	4.86	16.32
UNETR [7]	92.12	88.32	79.61	4.91	4.67	16.32
3D-RPL [23]	93.92	90.13	85.92	3.74	3.98	13.71
3D-Jig [23]	93.87	90.14	86.01	3.85	3.94	11.79
Ours	**94.03**	**90.29**	**86.76**	**3.61**	**3.78**	**10.25**

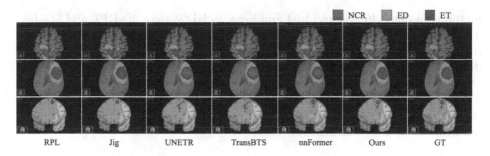

NCR ED ET

RPL   Jig   UNETR   TransBTS   nnFormer   Ours   GT

Fig. 4. Visualization of segmentation results on BraTS 2021 dataset.

White Matter Hyperintensities (WMH) Dataset[5] [12] involves 60 T1 images with pixel-level labels of White Matter Hyperintensities(WMH). We process data as [15] and use 36 cases for training and the rest for testing.
Evaluation Metric. We calculated Dice coefficient scores (Dice) and 95% Hausdorff Distance (HD95) to evaluate the segmentation results in our experiments.

Comparison with the State-of-the-Art. We compare our method with existing 3D transformer-based models (nnFormer [29], TransBTS [27], UNETR [7]) and 3D self-supervised methods (Relative 3D patch location(3D-RPL) [23], 3D Jigsaw puzzle Solving (3D-Jig) [23]) on 3 brain MRI segmentation tasks.

As Table 2 shows, on Brats 2021 dataset our method achieves the Dice scores of 94.03%, 90.29%, 86.76% and the HD95 of 3.61 mm, 3.78 mm and 10.25 mm on WT, TC, ET. Compared to transformer-based methods, our method achieves significantly better performance with both metrics. Specifically, our approach outperforms TransBTS [27] and nnFormer [29] by more than 7% and 4% Dice on ET respectively. Besides, our method shows more competitive results than other SSL methods using the same image segmentation network. For ET category, our

---
[5] https://wmh.isi.uu.nl/data/.

Table 3. Comparison on IBSR dataset and WMH dataset.

Task	IBSR						WMH	
Metric	Dice (%)↑			HD95 (mm)↓			Dice (%)↑	HD95 (mm)↓
Anatomy	CSF	GM	WM	CSF	GM	WM	WMH	WMH
nnFormer [29]	87.31	93.81	92.12	1.52	**1.52**	**1.21**	78.04	2.81
TransBTS [27]	81.42	93.91	92.17	7.84	1.54	1.40	78.81	2.91
UNETR [7]	86.75	93.49	91.86	1.64	1.74	1.48	77.99	3.53
3D-RPL [23]	86.63	93.85	**92.50**	1.83	1.54	1.29	78.63	3.06
3D-Jig [23]	86.93	93.57	92.11	2.00	1.74	1.44	77.86	3.36
Ours	**87.63**	**93.91**	92.44	**1.46**	1.54	1.33	**78.99**	**2.73**

method obtains 3.46 mm and 1.54 mm lower in HD95 than 3D-RPL and 3D-Jig. The visual comparisons are shown in Fig. 4. Our method does predict the ET region (blue) more accurately. As Table 3 shows, on IBSR dataset our method displays the highest Dice on CSF & GM, and obtains the lowest HD95 on CSF. On WMH dataset, the proposed method performs the best on both metrics. These results show that the model weights pre-trained by our method can be transferred to a wide range of datasets and help achieve the state-of-the-art performance.

Table 4. Ablation study on BraTS 2021. SSL denotes the 3D Masked Autoencoder (MAE) SSL method. A-SSL is the MAE method with our AR-Loss.

Metric	Dice (%)↑			HD95 (mm)↓		
Anatomy	WT	TC	ET	WT	TC	ET
Baseline	93.75	89.76	84.98	3.93	4.09	13.93
w/ SSL	94.02	90.28	86.25	4.01	4.06	13.44
w/ A-SSL	93.95	90.24	86.38	3.84	3.79	11.69
w/ SPE	93.90	90.15	85.86	3.69	3.82	11.59
w/ SPE&SSL	93.85	90.04	**86.83**	3.64	3.84	11.59
w/ ASA (ours)	**94.03**	**90.29**	86.76	**3.61**	**3.78**	**10.25**

Ablation Analysis. We verify the strength of the attentive reconstuction loss (AR-Loss), the SPE, and our overall ASA framework on BraTS 2021, as shown in Table 4. 'Baseline' denotes the SW-ViT based segmentation network (see Fig. 3) trained from scratch. 'w/ SSL' denotes training the segmentation network with the model weights pre-trained by a 3D Masked Autoencoder (MAE) SSL method [8]. 'A-SSL' denotes the MAE method with the proposed AR-Loss. As shown as the first half of Table 4, the A-SSL method produces more accurate segmentation results than the competitor SSL at HD95 metric. Especially, on ET the HD95 of using A-SSL is nearly 2 mm lower than that of using the SSL method. Note that HD95 measures the distance between the point sets of two boundaries. The above results show that the proposed AR-Loss can encourage the encoder to learn better representations for boundary information. 'w/ SPE' denotes applying the SPE to the train-from-scratch Baseline. 'w/ SPE' obtains 0.9% higher in Dice and 2.3 mm lower in HD95 than 'Baseline'. 'w/ ASA' denotes using our overall method with the loss and SPE. By comparing ASA with A-SSL, the SPE can further slightly improve A-SSL by 1.4 mm HD95 on ET. These results suggest that our proposed encoding can help the ViT-based encoder understand symmetric structures and harvest discriminative features.

4   Conclusion

In this paper, we propose a novel self-supervised learning architecture for 3D medical images. The proposed framework contains two key components, the symmetric position encoding and the attentive reconstruction loss. The encoding can benefit feature learning for symmetric structures and the attentive loss emphasizes informative image regions for reconstruction-based SSL. Both techniques can improve the generalization of trained models. Extensive experiments are conducted on three public brain MRI datasets. The results suggest that our method can achieve competitive performance with the state-of-the-art SSL methods and medical image segmentation models.

References

1. Çiçek, Ö., Abdulkadir, A., Lienkamp, S.S., Brox, T., Ronneberger, O.: 3D U-Net: learning dense volumetric segmentation from sparse annotation. In: Ourselin, S., Joskowicz, L., Sabuncu, M.R., Unal, G., Wells, W. (eds.) MICCAI 2016. LNCS, vol. 9901, pp. 424–432. Springer, Cham (2016). https://doi.org/10.1007/978-3-319-46723-8_49
2. Dosovitskiy, A., et al.: An image is worth 16×16 words: Transformers for image recognition at scale. arXiv preprint arXiv:2010.11929 (2020)
3. Dupre, R., Argyriou, V., Greenhill, D., Tzimiropoulos, G.: A 3d scene analysis framework and descriptors for risk evaluation. In: 2015 International Conference on 3D Vision, pp. 100–108 (2015). https://doi.org/10.1109/3DV.2015.19
4. El-Rifai, O., et al.: Clinica: an open-source software platform for reproducible clinical neuroscience studies. In: MRI Together 2021-A global workshop on Open Science and Reproducible MR Research (2021)
5. Glorot, X., Bengio, Y.: Understanding the difficulty of training deep feedforward neural networks. In: Proceedings of the Thirteenth International Conference on Artificial Intelligence and Statistics, pp. 249–256. JMLR Workshop and Conference Proceedings (2010)
6. Goyal, P., et al.: Accurate, large minibatch sgd: Training imagenet in 1 hour. arXiv preprint arXiv:1706.02677 (2017)
7. Hatamizadeh, A., et al.: Unetr: Transformers for 3d medical image segmentation. In: Proceedings of the IEEE/CVF Winter Conference on Applications of Computer Vision, pp. 574–584 (2022)

8. He, K., Chen, X., Xie, S., Li, Y., Dollár, P., Girshick, R.: Masked autoencoders are scalable vision learners. arXiv preprint arXiv:2111.06377 (2021)
9. Iglesias, J.E., Liu, C.Y., Thompson, P.M., Tu, Z.: Robust brain extraction across datasets and comparison with publicly available methods. IEEE TMI **30**(9), 1617–1634 (2011)
10. Isensee, F., Jaeger, P.F., Kohl, S.A., Petersen, J., Maier-Hein, K.H.: nnu-net: a self-configuring method for deep learning-based biomedical image segmentation. Nat. Methods **18**(2), 203–211 (2021)
11. Jack Jr, C.R., et al.: The alzheimer's disease neuroimaging initiative (adni): Mri methods. J. Magn. Reson. Imaging Offi. J. Int. Soc. Magn. Reson. Med. **27**(4), 685–691 (2008)
12. Kuijf, H.J., et al.: Standardized assessment of automatic segmentation of white matter hyperintensities and results of the wmh segmentation challenge. IEEE TMI **38**(11), 2556–2568 (2019)
13. LaMontagne, P.J., et al.: Oasis-3: longitudinal neuroimaging, clinical, and cognitive dataset for normal aging and alzheimer disease. MedRxiv (2019)
14. Li, H., et al.: View-disentangled transformer for brain lesion detection. In: 2022 IEEE 19th International Symposium on Biomedical Imaging (ISBI), pp. 1–5. IEEE (2022)
15. Li, H., et al.: Fully convolutional network ensembles for white matter hyperintensities segmentation in mr images. Neuroimage **183**, 650–665 (2018)
16. Liu, Z., et al.: Swin transformer: hierarchical vision transformer using shifted windows. In: Proceedings of the IEEE/CVF ICCV, pp. 10012–10022 (2021)
17. Loshchilov, I., Hutter, F.: Sgdr: Stochastic gradient descent with warm restarts. In: ICLR (2017)
18. Loshchilov, I., Hutter, F.: Decoupled weight decay regularization. In: ICLR (2019)
19. Menze, B.H., et al.: The multimodal brain tumor image segmentation benchmark (brats). IEEE TMI **34**(10), 1993–2024 (2014)
20. Rohlfing, T., Brandt, R., Menzel, R., Maurer, C.R., Jr.: Evaluation of atlas selection strategies for atlas-based image segmentation with application to confocal microscopy images of bee brains. Neuroimage **21**(4), 1428–1442 (2004)
21. Ronneberger, O., Fischer, P., Brox, T.: U-Net: convolutional networks for biomedical image segmentation. In: Navab, N., Hornegger, J., Wells, W.M., Frangi, A.F. (eds.) MICCAI 2015. LNCS, vol. 9351, pp. 234–241. Springer, Cham (2015). https://doi.org/10.1007/978-3-319-24574-4_28
22. Taleb, A., Lippert, C., Klein, T., Nabi, M.: Multimodal self-supervised learning for medical image analysis. In: Feragen, A., Sommer, S., Schnabel, J., Nielsen, M. (eds.) IPMI 2021. LNCS, vol. 12729, pp. 661–673. Springer, Cham (2021). https://doi.org/10.1007/978-3-030-78191-0_51
23. Taleb, A., et al.: 3d self-supervised methods for medical imaging. Adv. Neural. Inf. Process. Syst. **33**, 18158–18172 (2020)
24. Tang, Y., et al.: Self-supervised pre-training of swin transformers for 3d medical image analysis. In: Proceedings of the IEEE/CVF Conference on Computer Vision and Pattern Recognition (CVPR), pp. 20730–20740, June 2022
25. Tao, X., Li, Y., Zhou, W., Ma, K., Zheng, Y.: Revisiting Rubik's cube: self-supervised learning with volume-wise transformation for 3D medical image segmentation. In: Martel, A.L., et al. (eds.) MICCAI 2020. LNCS, vol. 12264, pp. 238–248. Springer, Cham (2020). https://doi.org/10.1007/978-3-030-59719-1_24
26. Vaswani, A., et al.: Attention is all you need. In: Advances in Neural Information Processing Systems, vol. 30 (2017)

27. Wang, W., Chen, C., Ding, M., Yu, H., Zha, S., Li, J.: TransBTS: multimodal brain tumor segmentation using transformer. In: de Bruijne, M., et al. (eds.) MICCAI 2021. LNCS, vol. 12901, pp. 109–119. Springer, Cham (2021). https://doi.org/10.1007/978-3-030-87193-2_11

28. Wei, C., Fan, H., Xie, S., Wu, C.Y., Yuille, A., Feichtenhofer, C.: Masked feature prediction for self-supervised visual pre-training. In: Proceedings of the IEEE/CVF Conference on Computer Vision and Pattern Recognition, pp. 14668–14678 (2022)

29. Zhou, H.Y., Guo, J., Zhang, Y., Yu, L., Wang, L., Yu, Y.: nnformer: interleaved transformer for volumetric segmentation. arXiv preprint arXiv:2109.03201 (2021)

30. Zhou, Z., Rahman Siddiquee, M.M., Tajbakhsh, N., Liang, J.: UNet++: a nested U-Net architecture for medical image segmentation. In: Stoyanov, D., et al. (eds.) DLMIA/ML-CDS -2018. LNCS, vol. 11045, pp. 3–11. Springer, Cham (2018). https://doi.org/10.1007/978-3-030-00889-5_1

Denoising for Relaxing: Unsupervised Domain Adaptive Fundus Image Segmentation Without Source Data

Zhe Xu[1], Donghuan Lu[2], Yixin Wang[3], Jie Luo[4], Dong Wei[2], Yefeng Zheng[2], and Raymond Kai-yu Tong[1(✉)]

[1] Department of Biomedical Engineering,
The Chinese University of Hong Kong, Hong Kong, China
kytong@cuhk.edu.hk
[2] Tencent Healthcare Co., Jarvis Lab, Shenzhen, China
caleblu@tencent.com
[3] Department of Bioengineering, Stanford University, Stanford, CA, USA
[4] Brigham and Women's Hospital, Harvard Medical School, Boston, MA, USA

Abstract. Recently, unsupervised domain adaptation (UDA) has been actively explored for multi-site fundus image segmentation with domain discrepancy. Despite relaxing the requirement of target labels, typical UDA still requires the labeled source data to achieve distribution alignment during adaptation. Unfortunately, due to privacy concerns, the vendor side often cannot provide the source data to the targeted client side in clinical practice, making the adaptation more challenging. To address this, in this work, we present a novel uncertainty-rectified denoising-for-relaxing (U-D4R) framework, aiming at completely relaxing the source data and effectively adapting the pretrained source model to the target domain. Considering the unreliable source model predictions on the target domain, we first present an adaptive class-dependent threshold strategy as the coarse denoising process to generate the pseudo labels. Then, the uncertainty-rectified label soft correction is introduced for fine denoising by taking advantage of estimating the joint distribution matrix between the observed and latent labels. Extensive experiments on cross-domain fundus image segmentation showed that our approach significantly outperforms the state-of-the-art source-free methods and encouragingly achieves comparable or even better performances over the leading source-dependent methods.

Keywords: Domain adaptation · Label denoising · Fundus image

1 Introduction

Accurately segmenting the optic cup and disc from fundus images is a critical step for early glaucoma screening and diagnosis [7]. Although deep neural networks have greatly advanced automatic cup and disc segmentation [7,27],

L. Wang et al. (Eds.): MICCAI 2022, LNCS 13435, pp. 214–224, 2022.
https://doi.org/10.1007/978-3-031-16443-9_21

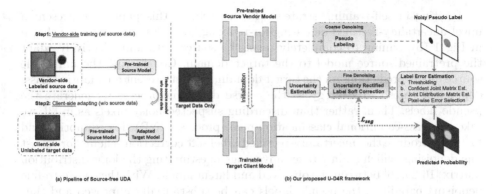

Fig. 1. (a) Source-free UDA where the client side can only access to the pretrained source vendor model instead of the source data due to privacy concerns. (b) Overview of the proposed uncertainty-rectified denoising-for-relaxing (U-D4R) framework for source-free unsupervised domain adaptive fundus segmentation.

they are vulnerable to cross-domain discrepancy [23], e.g., data acquired from different clinical centers or different imaging protocols. Being well trained on the source domain, the model often cannot generalize well in the target domain [22], yet manually labeling the new target domain data is time-consuming or even infeasible. Recently, unsupervised domain adaptation (UDA) has shown promising performance to combat the domain shift problem [11,22]. However, despite relaxing the labels of target data, typical UDA still requires the labeled source data to achieve distribution alignment during adaptation training. Unfortunately, due to privacy concerns, the vendor side often cannot provide (trade) the in-house source data to the targeted client side in clinical practice, making the typical adaptation paradigm infeasible. Such dilemma further poses a more challenging yet practical scenario (Fig. 1 (a)): how to effectively adapt the pretrained source vendor model to the client-side target domain without access to the source data?

Related Work. Being an underexplored topic itself, limited efforts have been made on adapting a model in the absence of source data. SRDA [2] proposed to minimize the class-ratio-regularized entropy with additionally pre-training a class-ratio predictor in the source domain. SDA-Net [10] and DAE [12] proposed to pre-train image reconstruction or label denoising auto-encoders in the source domain for target adaptation. However, these methods are only partially source-free since they need to pre-train other functional branches with source data. Furthermore, TENT [21] proposed to modulate batch normalization and minimize the entropy of model's prediction to achieve test-time adaptation. The self-training based DPL [3] completely relaxes the source data by deriving pixel-level uncertainty and class-level prototypes to select reliable pixels in pseudo labels for loss calculation. Yet, its performance heavily relies on the quality of target prototypes generated from the source model, and directly discarding numerous suspected noisy labels will potentially change the class distribution.

Rooting in self-training strategy [14] as well, in this paper, we present a novel uncertainty-rectified denoising-for-relaxing (U-D4R) framework, as shown in Fig. 1 (b), aiming to completely relax the source data and effectively adapt the pretrained source model to the target domain. Considering the unreliable source model predictions on the target domain, we first present an adaptive class-dependent threshold strategy as the coarse denoising process to generate the pseudo labels. Then, rather than discarding suspected noisy pixels as in [3], we take the class-conditional classification noise process (CNP) [1] assumption and further introduce the uncertainty-rectified label soft correction scheme for robust fine denoising, which mainly takes advantage of estimating the joint distribution matrix [16,25,26] between the observed and latent labels. With the coarse-to-fine denoising paradigm, the pseudo labels can be substantially improved and thus provide effective adaptation guidance. Extensive experiments on cross-domain fundus image segmentation showed that our approach not only outperforms the state-of-the-art source-free methods [2,3,21] with statistical significance, but also encouragingly achieves comparable or even better performances over the leading source-dependent methods [20,22]. A rigorous ablation study is also performed to verify the effectiveness of each component.

2   Methods

Fig. 1 (b) illustrates our uncertainty-rectified denoising-for-relaxing (U-D4R) framework for source-free unsupervised domain adaptive fundus segmentation. In this section, we first introduce the pseudo labeling strategy with adaptive class-dependent thresholds in Sect. 2.1 and then elaborate the uncertainty-rectified label soft correction in Sect. 2.2.

2.1   Pseudo Labeling with Class-Dependent Thresholds

We denote the vendor-side labeled source data as $\mathcal{D}_s = \{\mathbf{X}_{s(i)}, \mathbf{Y}_{s(i)}\}_{i=1}^{M}$ and the client-side unlabeled target data as $\mathcal{D}_t = \{\mathbf{X}_{t(i)}\}_{i=1}^{N}$, where M and N are the number of samples. For our fundus image segmentation problem, each 2D RGB image can be represented as $\mathbf{X}_i \in \mathbb{R}^{H \times W \times 3}$, while its label with C classes as $\mathbf{Y}_i \in \{0,1\}^{H \times W \times C}$. Specifically, $C = 2$ since we have two segmentation targets, i.e., optic cup and disc. In vanilla UDA problem [11,22], $\mathcal{D}_s$ should be accessible during adaptation training, otherwise it cannot align the different domain distributions. Yet, in the more challenging source-free setting, only a pretrained source model $f^s : \mathbf{X}_s \rightarrow \mathbf{Y}_s$ is provided by the vendor side due to privacy concerns, while the client side has to self-adapt f^s as $f^{s \rightarrow t}$ to perform well on their target distribution only with the help of their possessed $\mathcal{D}_t$. Note that such f^s can be obtained via typical supervised learning with standard cross-entropy loss on the source domain. To achieve source-free adaptation, one intuitive solution is performing pseudo labeling for $\mathcal{D}_t$, which has been well demonstrated in semi-supervised learning (SSL) [14]. As an adaptation, in the source-free UDA setting, we utilize f^s to generate pseudo labels $\{\bar{\mathbf{Y}}_t\}$ for $\mathcal{D}_t$, and then optimize the f^s-initialized model to obtain $f^{s \rightarrow t}$ via typical supervised learning.

However, directly applying the argmax operation on the predicted probabilities may lead to unreliable pseudo labels due to the domain shift between
source and target data, while the recent strategy is to empirically set a class-
agnostic threshold γ to select high-confidence labels [3,18]. Although such strategy is effective in filtering out most noisy labels, it potentially poses a dilemma
that the model becomes biased towards the majority classes and neglects the
minority classes. To avoid the imbalanced selection, we propose to compute the
class-dependent threshold as:

$$\hat{\gamma}^c = \alpha \cdot \max(\mathbf{P}^c), \tag{1}$$

where $\mathbf{P}^c$ indicates the predicted probability of class $c \in C$, which is produced
by f^s. The threshold parameter α is set as 0.75, empirically referring to the
previous class-agnostic threshold [22]. Then, for each class $c \in C$, we select the
pixel-wise pseudo label whose confidence is greater than $\hat{\gamma}^c$, i.e., $\bar{y}_t^c = \mathbb{1}\left[p^c > \hat{\gamma}^c\right]$,
where $\mathbb{1}(\cdot)$ is the indicator function.

2.2 Label Self-correction Towards Effective Adaptation

Despite improved pseudo label quality, the simple threshold strategy can only be
regarded as a coarse denoising process. We found that the discrepancy between
source and target domains often leads to significant over-confident wrong predictions, thus heavy noises inevitably exist in the pseudo labels. If not treated properly, the noises will still mislead the network training [3,15,26]. Therefore, further
introducing fine denoising process is necessary. Unlike [3] that takes advantage
of uncertainties and class prototypes derived from f^s to further discard unreliable pixels for final loss calculation, we resort to leveraging the predictions of
$f^{s\to t}$ to serve as a third party to finely correct the pseudo labels. As a start, we
take the class-conditional classification noise process (CNP) [1] assumption into
consideration, which has been proven effective in previous works on noisy labels
[16,25,29]. Here, we assume that every pixel $x_{t(w,h)} \in \mathbf{X}_t$ exists a true (latent)
label $y^*_{t(w,h)}$, while every label in class $j \in C$ may be independently mislabeled
as class $i \in C$ with probability $p\left(\bar{y}_t = i \mid y^*_t = j\right)$.

Class-Conditional Label Error Estimation. Based on the CNP assumption, previous works [6,16] reveal that the confident joint matrix, i.e., integrating a threshold strategy into vanilla confusion matrix [5], increases robustness
to class-imbalanced and over-confident predicted probabilities, while its normalized statistics, named joint distribution matrix here, are effective in estimating
aleatoric uncertainties from latent label noises [16,26]. Here, the target model
$f^{s\to t}$ continuously provides the reference predicted probabilities $\hat{\mathbf{P}}_t$ during training. Ideally, $\hat{\mathbf{P}}_t$ becomes more accurate as the adaptation training goes. Then,
we denote the softmax probability $\hat{p}_t^i(x_t)$ that a pixel x_t belongs to its pseudo
label $\bar{y}_t = i$ as its self-confidence. Intuitively, if x_t with label $\bar{y}_t = i$ satisfies
$\hat{p}_t^j(x_t) \geq \gamma^j$, the true (latent) label y^*_t of x_t can be suspected to be j instead of
i. Here, to alleviate potential over-confident errors due to domain shift, we use
the mean self-confidence [16] instead of the maximum criterion in Eq. 1 as the

per-class threshold $\gamma^j = \text{mean}(\hat{\mathbf{P}}_t^j)$ for each target image $\mathbf{X}_t$. We formulate the confident joint matrix $\mathbf{CJ}_{\bar{y}_t, y_t^*} \in \mathbb{R}^{C \times C}$ for each target image $\mathbf{X}_t$ as:

$$\mathbf{CJ}_{\bar{y}_t, y_t^*}[i][j] = \left| \hat{\mathbf{X}}_{t(\bar{y}_t^i, y_t^{j*})} \right| = \left| \left\{ x_t \in \mathbf{X}_{t(\bar{y}_t^i)} : \hat{p}_t^j(x_t) \geq \gamma^j, j = \underset{c \in C : \hat{p}_t^c(x_t) \geq \gamma^c}{\arg\max} \hat{p}_t^c(x_t) \right\} \right|. \tag{2}$$

Since many pixels are filtered out, we need to calibrate $\mathbf{CJ}_{\bar{y}_t, y_t^*}$ as:

$$\tilde{\mathbf{CJ}}_{\bar{y}_t, y_t^*}[i][j] = \frac{\mathbf{CJ}_{\bar{y}_t, y_t^*}[i][j]}{\sum_{j \in C} \mathbf{CJ}_{\bar{y}_t, y_t^*}[i][j]} \cdot \left| \mathbf{X}_{t(\bar{y}_t = i)} \right|, \tag{3}$$

where $\left| \mathbf{X}_{t(\bar{y}_t = i)} \right|$ is the total number of pixels with label $\bar{y}_t = i$. Then, we further normalize $\tilde{\mathbf{CJ}}_{\bar{y}_t, y_t^*}$ to obtain the joint distribution matrix $\hat{\mathbf{JD}}_{\bar{y}_t, y_t^*}$ as:

$$\hat{\mathbf{JD}}_{\bar{y}_t, y_t^*}[i][j] = \frac{\tilde{\mathbf{CJ}}_{\bar{y}_t, y_t^*}[i][j]}{\sum_{i \in C, j \in C} \tilde{\mathbf{CJ}}_{\bar{y}_t, y_t^*}[i][j]}. \tag{4}$$

This joint distribution has shown more effective and robust in reflecting the distribution of noisy labels and true labels in the real world [6,16,26]. The off-diagonal of $\hat{\mathbf{JD}}_{\bar{y}_t, y_t^*}$ represents how likely these classes are mislabeled in the image, and low confidence can be a heuristic likelihood of being a mislabeled pixel [16]. Thus, for each class $i \in C$, we select $n \cdot \sum_{j \in C : j \neq i} \left(\hat{\mathbf{JD}}_{\bar{y}_t, y_t^*}[i][j] \right)$ pixels with the lowest self-confidence $\hat{p}_t^i(x_t)$ as the mislabeled ones, where n denotes the number of pixels in image $\mathbf{X}_t$. By this way, we can obtain the binary estimated error map $\mathbf{E}^c$ for each class c, where "1" indicates the pixel is identified as an error and "0" for the correct one.

Uncertainty-Rectified Label Soft Correction. Due to the domain shift, ambiguous boundaries and complex organ morphology in multi-site fundus images, disambiguating model error from intrinsic label noise to perfectly estimate such joint distribution is hard to achieve, so that the estimated error map may also contain noises. Therefore, rather than direct correction with fixed human empiricism [26], we introduce the epistemic uncertainty of the target model $f^{s \to t}$ to rectify the label correction process [25]. We denote $\dot{\mathbf{Y}}_t^c$ as the soft corrected label for class c. Intuitively, a more certain pixel should be allocated a more confident correction weight, thus the soft correction process can be formulated as:

$$\dot{\mathbf{Y}}_t^c = \bar{\mathbf{Y}}_t^c + \mathbf{E}^c \cdot (-1)^{\bar{\mathbf{Y}}_t^c} \cdot (1 - \mathbf{U}^c), \tag{5}$$

where $\mathbf{U}^c$ is the uncertainty map for class-c prediction. Specifically, we adopt the Monte Carlo dropout [13] for Bayesian approximation due to its superior robustness against training with noisy labels [9]. Briefly, given an input $\mathbf{X}_t$, K stochastic forward passes through $f^{s \to t}$ are performed with random dropout at inference. Then, we compute the standard deviation of the K softmax predictions as the uncertainty map $\mathbf{U}$.

Loss Function. Taking advantage of the above coarse-to-fine denoising process, the pseudo label quality is substantially improved, thus effective adaptation guidance can be promisingly achieved. Specifically, leveraging the target data and the soft corrected labels, we can adapt the source model f^s towards target distribution by optimizing a standard cross-entropy loss:

$$\mathcal{L}_{seg} = -\sum_v \left[\dot{y}_{t(v)} \cdot \log\left(p_{t(v)}\right) + \left(1 - \dot{y}_{t(v)}\right) \cdot \log\left(1 - p_{t(v)}\right) \right]. \tag{6}$$

3 Experiments

3.1 Datasets

We evaluate the proposed method on the optic disc and cup segmentation from multi-site retinal fundus images. Specifically, we utilize the Drishti-GS [19] dataset (termed as D1) as the source domain to obtain the pretrained source model, while the RIM-ONE-r3 dataset [8] (termed as D2) and the REFUGE dataset [17] (termed as D3) as the target domains. Following their official data split settings, the source domain is divided into 50/51 images for training and testing, respectively, while the two target domain datasets are split to 99/60 (for D2) and 320/80 (for D3) images for training and testing, respectively. Each image is cropped into the 512×512 optic disc region as the network input. Note that since labeled source data is scarce in clinical practice, we utilize the smallest dataset, i.e., Drishti-GS [19], as the source domain data, causing more practical yet more challenging scenario than the previous source-dependent work [22] that employed the largest REFUGE dataset [17] as the labeled source domain.

3.2 Implementation Details and Evaluation Metrics

The framework is implemented with PyTorch 0.4.1 using an NVIDIA Titan X GPU. Following [22], we adopt the same MobileNetV2 adapted DeepLabv3+ [4] as our segmentation backbone without the boundary branch. For Monte Carlo dropout-based uncertainty estimation, we perform 10 stochastic forward passes with a dropout rate of 0.5. We train the target model for 2 epochs (following [3]) with the batch size of 6. Adam optimizer is adopted with the momentum of 0.9 and 0.99, and the learning rate is set to 0.002. Weak data augmentations, including Gaussian noise, random erasing and contrast adjustment, are applied. We adopt two well-known metrics for a comprehensive evaluation, including Dice score and average symmetric surface distance (ASSD). Higher Dice score and lower ASSD indicate more appealing segmentation.

Table 1. Quantitative results of different methods. Standard deviations (over all subjects) are shown in parentheses. ∗ indicates $p \leq 0.05$ from a two-sided paired t-test when comparing the mean of ours with the best-performing baseline under the source-free setting. The best mean results under the source-free setting are shown in bold.

Methods	Source Data	Dice [%] ↑		ASSD [pixel] ↓	
	D1: Drishti-GS [19]	Optic cup	Optic disc	Optic cup	Optic disc
D2: RIM-ONE-r3 [8] (Target)					
No adaptation	–	57.33 (35.36)	89.15 (14.07)	10.92 (6.99)	9.62 (9.95)
Upper bound	–	78.34 (20.88)	95.88 (2.21)	7.51 (5.89)	3.54 (1.82)
BEAL [22]	✓	70.47 (27.40)	90.69 (6.93)	12.68 (14.24)	8.90 (7.69)
AdvEnt [20]	✓	71.23 (26.61)	91.88 (9.87)	10.40 (10.79)	6.97 (5.27)
SRDA [2]	Partial	62.08 (21.48)	90.62 (15.48)	15.83 (8.51)	8.85 (7.93)
TENT [21]	×	62.89 (19.83)	90.25 (10.41)	14.83 (8.62)	8.28 (6.95)
DPL [3]	×	64.01 (19.88)	91.48 (7.09)	16.39 (7.53)	8.95 (7.13)
U-D4R (ours)	×	**68.59 (13.87)**∗	**93.31 (5.41)**∗	**11.62 (5.66)**∗	**6.63 (6.05)**∗
D3: REFUGE [17] (target)					
No adaptation	–	79.45 (10.74)	89.68 (5.62)	8.95 (4.68)	9.03 (4.04)
Upper bound	–	87.67 (6.60)	94.79 (2.01)	5.78 (3.67)	4.70 (1.55)
BEAL [22]	✓	83.24 (10.85)	91.43 (3.79)	8.24 (6.86)	7.93 (3.85)
AdvEnt [20]	✓	83.66 (11.03)	90.39 (3.92)	8.07 (7.08)	8.68 (3.99)
SRDA [2]	Partial	81.57 (7.63)	86.41 (2.93)	8.94 (4.82)	9.56 (4.13)
TENT [21]	×	83.58 (8.42)	92.04 (2.37)	8.01 (4.26)	8.25 (4.58)
DPL [3]	×	84.02 (7.84)	92.57 (2.81)	7.56 (4.05)	7.83 (3.92)
U-D4R (ours)	×	**85.11 (7.77)**∗	**93.60 (2.41)**∗	**6.83 (4.64)**∗	**7.09 (4.59)**

3.3   Results

Comparison Study. The quantitative evaluation results are presented in Table 1. Besides the "No adaptation" lower bound and the supervised training "upper bound" with labeled target data, we also include top-performing vanilla UDA methods (i.e., BEAL [22] and AdvEnt [11]) and recent source-free UDA approaches (i.e., SRDA [2], TENT [21] and DPL [3]). Notably, BEAL [22] is tailored for fundus images via additional boundary constraints. We denote "Partial" on the accessibility of source data for SRDA because it uses the source data to train an additional class ratio predictor [2], while TENT [21] and DPL [3] completely relax the source data. Despite the overall improvement, SRDA performs worse than "No adaptation" in terms of some metrics, which may be the case of negative transfer [24]. Meanwhile, though the source model is pretrained on the smallest D1 set, we observe that the prototype-based DPL [3] can still achieve competitive results. However, directly discarding the suspected noisy labels may change the class distribution and damage the representative ability of the learned features. As observed, the Dice and ASSD of DPL under the D1-to-D2 scenario are still sub-optimal. Overall, our U-D4R achieves more promising performance. Excitingly, under the D1-to-D3 scenario, our U-D4R even outperforms the two vanilla UDA methods, which may be due to the negative transfer effect [24] that impedes finding invariant latent spaces between source and target domains.

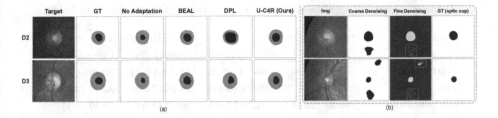

Fig. 2. (a) Exemplar optic cup and disc segmentation results of different methods. (b) Visualized evolution of the pseudo labels for optic cups through the presented coarse-to-fine denoising process.

Figure 2 (a) presents a qualitative comparison. As observed, "No adaptation" is prone to produce under-segmented results for optic cups due to the domain gap, and our U-D4R has comparable performance with BEAL [22], which further demonstrates the effectiveness of our method. Besides, as observed, since the optic disc has clearer boundaries, the domain shift poses more negative effects on cup segmentation. Thus, to better understand our framework, we further visualize two denoising examples of the optic cup in Fig. 2 (b). We can see that most noises can be identified (but some failures also exist) and the pseudo labels are substantially improved with the coarse-to-fine label soft denoising scheme, resulting in more effective guidance for target adaptation training.

Ablation Study and Discussion. As shown in Table 2, we provide a rigorous ablation study for more insights into our method: a) **w/o init**: the target model is not f_s-initialized but with sufficient 100 training epochs; b) **w/o denoising**: the argmax pseudo labels are directly used; c) **w/o cls-denoising**: a fixed-threshold 0.75 [3] is used for all classes (instead of our class-dependent thresholds) during coarse denoising; d) **w/o fine denoising**: only the coarse denoising is performed; e) **w/o uncertainty rect**: the labels are directly corrected without uncertainty rectification. Interestingly, when we train our U-D4R from scratch without pretrained f^s initialization, the target model can still achieve excellent results, where the mean Dice score of optic cup in D2 even reaches 70.78%. This finding shows that our framework has a certain tolerance of severely inaccurate predicted self-confidence at the early training stage and can be potentially adapted to broader applications, e.g., in case of the source model is a black-box [28]. Without any denoising, the target model heavily suffers from inferior training, while each presented denoising component shows its effectiveness in improving the segmentation performance. Especially, without our fine label self-correction process, a significant performance drop can be observed, which emphasizes its necessity. When we elegantly combine these components into our synergistic framework, their better efficacy can be brought into play. Despite the promising performance, our validation only considers the relatively minor domain shifts in practice, i.e., same-modality images from different clinical sites. More challenging cross-modality domain shift is left for our future work.

Table 2. Quantitative ablation study. Best mean results are shown in bold.

Methods	D2: RIM-ONE-r3 [8]		D3: REFUGE [17]	
	$Dice_{cup}$ [%]	$Dice_{disc}$ [%]	$Dice_{cup}$ [%]	$Dice_{disc}$ [%]
U-D4R (ours)	68.59 (13.87)	**93.31 (5.41)**	**85.11 (7.77)**	**93.60 (2.41)**
w/o init	**70.78 (16.39)**	90.43 (6.32)	84.68 (7.90)	93.13 (2.33)
w/o denoising	60.22 (22.58)	84.38 (7.25)	80.33 (8.53)	88.46 (2.47)
w/o cls-denoising	67.42 (14.57)	92.35 (5.83)	85.02 (7.89)	92.50 (2.42)
w/o fine denoising	62.20 (23.72)	87.48 (6.18)	81.26 (7.93)	90.08 (2.24)
w/o uncertainty rect	66.64 (13.97)	92.02 (5.32)	84.42 (7.65)	92.59 (2.43)

4   Conclusion

In this work, we completely relax the source data for unsupervised domain adaptive segmentation via the presented coarse-to-fine label denoising scheme, which is constructed by class-dependent thresholds based pseudo labeling and uncertainty-rectified label soft self-correction. Extensive experiments on cross-domain fundus image segmentation showed that our approach outperforms the state-of-the-art source-free methods and encouragingly achieves comparable or even better performances over the leading source-dependent UDA methods.

Acknowledgement. This research was done with Tencent Healthcare (Shenzhen) Co., LTD and Tencent Jarvis Lab and supported by General Research Fund from Research Grant Council of Hong Kong (No. 14205419) and the Scientific and Technical Innovation 2030-"New Generation Artificial Intelligence" Project (No. 2020AAA0104100).

References

1. Angluin, D., Laird, P.: Learning from noisy examples. Mach. Learn. **2**(4), 343–370 (1988)
2. Bateson, M., Kervadec, H., Dolz, J., Lombaert, H., Ben Ayed, I.: Source-relaxed domain adaptation for image segmentation. In: Martel, A.L., et al. (eds.) MICCAI 2020. LNCS, vol. 12261, pp. 490–499. Springer, Cham (2020). https://doi.org/10.1007/978-3-030-59710-8_48
3. Chen, C., Liu, Q., Jin, Y., Dou, Q., Heng, P.-A.: Source-free domain adaptive fundus image segmentation with denoised pseudo-labeling. In: de Bruijne, M., et al. (eds.) MICCAI 2021. LNCS, vol. 12905, pp. 225–235. Springer, Cham (2021). https://doi.org/10.1007/978-3-030-87240-3_22
4. Chen, L.C., Zhu, Y., Papandreou, G., Schroff, F., Adam, H.: Encoder-decoder with atrous separable convolution for semantic image segmentation. In: Proceedings of the European Conference on Computer Vision, pp. 801–818 (2018)
5. Chen, P., Liao, B.B., Chen, G., Zhang, S.: Understanding and utilizing deep neural networks trained with noisy labels. In: International Conference on Machine Learning, pp. 1062–1070. PMLR (2019)

6. Elkan, C.: The foundations of cost-sensitive learning. In: International Joint Conference on Artificial Intelligence, vol. 17, pp. 973–978. Lawrence Erlbaum Associates Ltd (2001)
7. Fu, H., Cheng, J., Xu, Y., Wong, D.W.K., Liu, J., Cao, X.: Joint optic disc and cup segmentation based on multi-label deep network and polar transformation. IEEE Trans. Med. Imaging **37**(7), 1597–1605 (2018)
8. Fumero, F., Alayón, S., Sanchez, J.L., Sigut, J., Gonzalez-Hernandez, M.: RIM-ONE: An open retinal image database for optic nerve evaluation. In: 24th International Symposium on Computer-Based Medical Systems, pp. 1–6. IEEE (2011)
9. Goel, P., Chen, L.: On the robustness of monte carlo dropout trained with noisy labels. In: Proceedings of the IEEE/CVF Conference on Computer Vision and Pattern Recognition, pp. 2219–2228 (2021)
10. He, Y., Carass, A., Zuo, L., Dewey, B.E., Prince, J.L.: Self domain adapted network. In: Martel, A.L., et al. (eds.) MICCAI 2020. LNCS, vol. 12261, pp. 437–446. Springer, Cham (2020). https://doi.org/10.1007/978-3-030-59710-8_43
11. Kamnitsas, K., et al.: Unsupervised domain adaptation in brain lesion segmentation with adversarial networks. In: International Conference on Information Processing in Medical Imaging, pp. 597–609. Springer (2017)
12. Karani, N., Erdil, E., Chaitanya, K., Konukoglu, E.: Test-time adaptable neural networks for robust medical image segmentation. Med. Image Anal. **68**, 101907 (2021)
13. Kendall, A., Gal, Y.: What uncertainties do we need in Bayesian deep learning for computer vision? arXiv preprint arXiv:1703.04977 (2017)
14. Lee, D.H., et al.: Pseudo-label: The simple and efficient semi-supervised learning method for deep neural networks. In: Workshop on Challenges in Representation Learning (2013)
15. Luo, W., Yang, M.: Semi-supervised semantic segmentation via strong-weak dual-branch network. In: Vedaldi, A., Bischof, H., Brox, T., Frahm, J.-M. (eds.) ECCV 2020. LNCS, vol. 12350, pp. 784–800. Springer, Cham (2020). https://doi.org/10.1007/978-3-030-58558-7_46
16. Northcutt, C., Jiang, L., Chuang, I.: Confident learning: estimating uncertainty in dataset labels. J. Artifi. Intell. Res. **70**, 1373–1411 (2021)
17. Orlando, J.I., et al.: Refuge challenge: A unified framework for evaluating automated methods for glaucoma assessment from fundus photographs. Med. Image Anal. **59**, 101570 (2020)
18. Rizve, M.N., Duarte, K., Rawat, Y.S., Shah, M.: In defense of pseudo-labeling: an uncertainty-aware pseudo-label selection framework for semi-supervised learning. In: The International Conference on Learning Representations (2021)
19. Sivaswamy, J., Krishnadas, S., Chakravarty, A., Joshi, G., Tabish, A.S., et al.: A comprehensive retinal image dataset for the assessment of glaucoma from the optic nerve head analysis. JSM Biomed. Imaging Data Pap. **2**(1), 1004 (2015)
20. Vu, T.H., Jain, H., Bucher, M., Cord, M., Pérez, P.: Advent: Adversarial entropy minimization for domain adaptation in semantic segmentation. In: Proceedings of the IEEE/CVF Conference on Computer Vision and Pattern Recognition, pp. 2517–2526 (2019)
21. Wang, D., Shelhamer, E., Liu, S., Olshausen, B., Darrell, T.: Tent: fully test-time adaptation by entropy minimization. In: The International Conference on Learning Representations (2020)
22. Wang, S., Yu, L., Li, K., Yang, X., Fu, C.-W., Heng, P.-A.: Boundary and entropy-driven adversarial learning for fundus image segmentation. In: Shen, D., et al. (eds.)

MICCAI 2019. LNCS, vol. 11764, pp. 102–110. Springer, Cham (2019). https://doi.org/10.1007/978-3-030-32239-7_12

23. Wang, S., Yu, L., Yang, X., Fu, C.W., Heng, P.A.: Patch-based output space adversarial learning for joint optic disc and cup segmentation. IEEE Trans. Med. Imaging **38**(11), 2485–2495 (2019)

24. Wang, Z., Dai, Z., Póczos, B., Carbonell, J.: Characterizing and avoiding negative transfer. In: Proceedings of the IEEE/CVF Conference on Computer Vision and Pattern Recognition, pp. 11293–11302 (2019)

25. Xu, Z., Lu, D., Luo, J., Wang, Y., Yan, J., Ma, K., Zheng, Y., Tong, R.K.y.: Anti-interference from noisy labels: Mean-teacher-assisted confident learning for medical image segmentation. IEEE Transactions on Medical Imaging (2022)

26. Xu, Z., Lu, D., Wang, Y., Luo, J., Jagadeesan, J., Ma, K., Zheng, Y., Li, X.: Noisy labels are treasure: Mean-teacher-assisted confident learning for hepatic vessel segmentation. In: International Conference on Medical Image Computing and Computer Assisted Intervention. pp. 3–13. Springer (2021)

27. Yu, S., Xiao, D., Frost, S., Kanagasingam, Y.: Robust optic disc and cup segmentation with deep learning for glaucoma detection. Comput. Med. Imaging Graph. **74**, 61–71 (2019)

28. Zhang, H., Zhang, Y., Jia, K., Zhang, L.: Unsupervised domain adaptation of black-box source models. arXiv preprint arXiv:2101.02839 (2021)

29. Zhang, L., Tanno, R., Xu, M.C., Jacob, J., Ciccarelli, O., Barkhof, F., C. Alexander, D.: Disentangling human error from the ground truth in segmentation of medical images. Advances in Neural Information Processing Systems (2020)

Curvature-Enhanced Implicit Function Network for High-quality Tooth Model Generation from CBCT Images

Yu Fang[1,2,5], Zhiming Cui[1], Lei Ma[1], Lanzhuju Mei[1,2,5], Bojun Zhang[3], Yue Zhao[4], Zhihao Jiang[2], Yiqiang Zhan[5], Yongsheng Pan[1], Min Zhu[3], and Dinggang Shen[1,5(✉)]

[1] School of Biomedical Engineering, ShanghaiTech University, Shanghai, China
dgshen@shanghaitech.edu.cn
[2] School of Information Science and Technology, ShanghaiTech University, Shanghai, China
[3] Shanghai Ninth Peoples Hospital, Shanghai Jiao Tong University, Shanghai, China
[4] School of Communication and Information Engineering, Chongqing University of Posts and Telecommunications, Chongqing, China
[5] Shanghai United Imaging Intelligence Co. Ltd., Shanghai, China

Abstract. In digital dentistry, high-quality tooth models are essential for dental diagnosis and treatment. 3D CBCT images and intra-oral scanning models are widely used in dental clinics to obtain tooth models. However, CBCT image is volumetric data often with limited resolution (about 0.3–1.0 mm spacing), while intra-oral scanning model is high-resolution tooth crown surface (about 0.03 mm spacing) without root information. Hence, dentists usually scan and combine these two modalities of data to build high-quality tooth models, which is time-consuming and easily affected by various patient conditions or acquisition artifacts. To address this problem, we propose a learning-based framework to generate high-quality tooth models with both fine-grained tooth crown details and root information only from CBCT images. Specifically, we first introduce a tooth segmentation network to extract individual teeth from CBCT images. Then, we utilize an implicit function network to generate tooth models at arbitrary resolution in a continuous learning space. Moreover, to capture fine-grained crown details, we further explore a curvature enhancement module in our framework. Experimental results show that our proposed framework outperforms other state-of-the-art methods quantitatively and qualitatively, demonstrating the effectiveness of our method and its potential applicability in clinical practice.

1 Introduction

With the development of computer-aided techniques, digital dentistry has been widely used in dental clinics for diagnosis [15], restoration [10], and treatment planning [6,17]. In these systems, the acquisition of high-quality 3D tooth models is essential to assist dentists in extracting [14], implanting [7], or rearranging

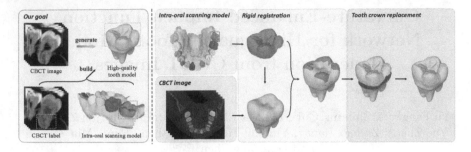

Fig. 1. Left: Overview of generating high-quality tooth model from CBCT images only. Our proposal is to use the tooth crown surfaces obtained from intra-oral scanning models to guide the training process of high-resolution tooth model generation from CBCT images. Right: Overview of high-quality tooth model building from CBCT image and intra-oral scanning model.

teeth [9]. In this regard, segmenting individual teeth from cone-beam computed tomography (CBCT) images [3,4] (the 3D volumetric data of all oral tissues) is a long-standing topic and has achieved promising results. However, due to the imaging techniques and radiation exposure, the spatial resolution in CBCT images is relatively low (about 0.3–1.0 mm spacing), which limits capturing of the tooth crown details. Thus, dentists usually rely on intra-oral scanning models (high-precision tooth crown surface without root information), to analyze occlusion relations of upper and lower jaws [16]. But it is time-consuming to collect both modalities of data, and is easily affected by various patient conditions or acquisition artifacts. In this situation, it is of great significance for developing a framework to generate high-quality tooth models with fine-grained tooth crown details and root information only from CBCT images.

To effectively reconstruct 3D shapes with high resolution, implicit function networks [2,13] have achieved outstanding performance in 3D synthetic datasets for shape recovery, completion, and super-resolution. Their advantage is the ability to handle different objects in a continuous learning space. Unfortunately, most of these methods are designed to capture general shapes, so the predictions tend to be over-smooth, thus ignoring many important geometric details. In the meantime, compared with clean 3D models in the synthetic datasets, 3D models derived from CBCT images or intra-oral scanning models usually introduce more noises from real-world clinical scenarios. Hence, it is extremely challenging to recover high-quality tooth models by segmenting teeth directly from CBCT images, especially on tooth crowns with rich geometric details.

In this study, to tackle the above limitations, we propose a novel curvature-enhanced implicit function network for high-quality tooth model generation from CBCT images. Our key idea is to combine the commonly used CNN-based segmentation network with an implicit function network to generate 3D tooth models with fine-grained geometric details. Specifically, given a 3D CBCT image, we first utilize a segmentation network to segment individual teeth, and represent

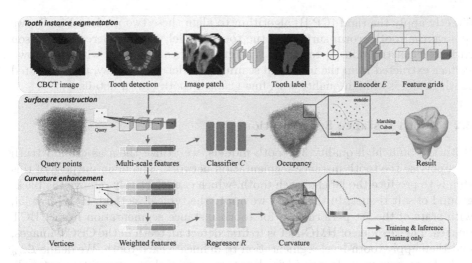

Fig. 2. The overview of our framework. The green arrows indicate the flows for training and inference, and the red arrows are for training only. (Color figure online)

the high-resolution output in the voxel space. Then, we introduce an implicit function network to generate high-quality 3D tooth models at arbitrary resolution in a continuous space. Particularly, to retain tooth crown surfaces with fine-grained details, a curvature enhancement module is proposed to predict local shape properties, which in return guides the implicit function network to reproduce plausible tooth shapes. Note that the ground truth, i.e., high-quality tooth models, in the training stage is built by merging respective surfaces from the CBCT image and intra-oral scanning model as shown in Fig. 1.

2 Method

The overview of our framework to generate high-quality tooth models from CBCT images is shown in Fig. 2, including a *tooth instance segmentation* module, a *surface reconstruction* module, and a *curvature enhancement* module, as detailed below.

2.1 High-quality Tooth Model Building

In this study, we combine the two modalities of data together (i.e., 3D CBCT images and intra-oral scanning models) to build high-quality tooth models, where root surfaces and crown surfaces are produced from the paired CBCT image and the intra-oral scanning model, respectively. As shown in Fig. 1, the ground truth building process is composed of three steps. First, given a CBCT image and its paired intra-oral scanning model, we manually delineate tooth model on the CBCT image, and crown model on the intra-oral scanning model. Then, as the paired tooth model and crown model are scanned from the same patient, we directly apply the rigid ICP [1] algorithm to align these two models. Finally, we remove the tooth crown surface on the tooth model, and use Screened Poisson surface reconstruction [8] to merge the remaining root surface with the crown surface extracted from the intra-oral scanning model. In this way, the generated tooth model is of high quality with fine-grained tooth crown details.

2.2   Tooth Instance Segmentation

With the built high-quality 3D tooth models, we utilize them as ground truth to supervise the tooth instance segmentation network. The main purpose of this step is to produce the label of each tooth, which can effectively remove the background of soft tissues. In this work, we apply the typical method, HMG-Net [5], with state-of-the-art performance for tooth instance segmentation from CBCT images. The key idea of HMG-Net is to first detect all teeth in the CBCT images, and then apply tooth-level segmentation to delineate each tooth. We define $\mathcal{L}_{seg}$ with the cross-entropy loss and Dice loss to supervise the segmentation network.

As shown in Fig. 2, although the method can achieve promising performance on tooth segmentation, details especially on tooth crowns are usually lost due to limited resolution of CBCT images. Hence, we then take the predicted tooth label and the cropped image patch as input to the surface reconstruction module, to generate high-quality tooth models with rich geometric details.

2.3   Surface Reconstruction

In the surface reconstruction module, inspired by the work in 3D model completion [2], we introduce the implicit function strategy with multi-scale encoding and shape decoding, to preserve fine-grained surface details by reconstructing 3D tooth models at arbitrary resolution in a continuous space.

Multi-scale Encoding. With the tooth-level input X (i.e., cropped image patch and tooth label), we first employ an encoder E to extract multi-scale features $\{F_1, F_2, \ldots, F_L\}$ from different convolutional layers. Note that the first feature map F_1 is the input X. In this way, the feature maps at early stages can capture local information, while the feature maps at late stages contain global information. And all the feature maps from different layers preserve 3D volumetric structures aligned with the input data, which is defined as:

$$E(X) := F_1, F_2, \ldots, F_L. \tag{1}$$

Shape Decoding. As multi-scale features are discrete in grids, given a query point $p \in \mathbb{R}^3$ in the continuous space, we can obtain its feature by trilinear interpolation. Moreover, to encode more local neighborhood information for shape decoding, we extract features at the location of the query point p, and additionally at surrounding points in a distance d ($d = 0.05$ in this paper) along the Cartesian axes. At last, we integrate the point-wise features from the multi-scale features $\{F_1, F_2, \ldots, F_L\}$ with different receptive fields, which is denoted as $\{F_1(p), F_2(p), \ldots, F_L(p)\}$. The integrated point-wise features are then fed into a point-wise decoder $\mathcal{D}$, parameterized by a fully connected neural network, to predict the corresponding occupancy value (i.e., inside or outside the surface):

$$\hat{o}_p = \mathcal{D}(F_1(p), F_2(p), \ldots, F_L(p)) \in \{0, 1\}, \tag{2}$$

where $\{0, 1\}$ denotes the query point being outside or inside the surface, respectively. We employ the BCE loss $\mathcal{L}_{occ}$ to supervise the learning process. Note that, to robustly train the network, we sample a number of query points in the continuous space, using the sampling strategy described in Sect. 3.2.

2.4   Curvature Enhancement

Since our method is defined in the continuous space, it is capable of describing a surface at arbitrary resolution. However, it still cannot effectively produce tooth crown surfaces with fine-grained details, for the reason that only the binary occupancy classification on query points cannot faithfully learn the changes of local shape properties (e.g., surface curvature). Thus, we further propose a branch to predict the curvature of each query point.

Specifically, we first extract the vertices $V = \{v_1, v_2, \ldots, v_T; v_t \in \mathbb{R}^3\}$ on the ground truth tooth surface, and compute their corresponding curvature values $\{c_1, c_2, \ldots, c_T; c_t \in \mathbb{R}\}$. For each $v_t \in V$, we extract the features from its K nearest query points $\{p'_1, p'_2, \ldots, p'_K\}$ ($K = 5$ in this paper), and obtain curvature features $F_c(v_t)$ by Inverse Distance Weighting (IDW) [12], which is defined as

$$F_c(v_t) = \sum_{k=1}^{K} \frac{1/D(p'_k, v_t)}{\sum_{k=1}^{K} 1/D(p'_k, v_t)} F(p'_k), \tag{3}$$

where D denotes the Euclidean distance. The curvature features are then fed into the point-wise regressor R, parameterized by a fully connected neural network, to predict the curvature value of v_t. And we use the smooth L1 error $\mathcal{L}_{cur}$ to supervise the curvature enhancement module.

Finally, the overall loss with multiple supervision is computed by:

$$\mathcal{L}_{total} = \mathcal{L}_{seg} + \mathcal{L}_{occ} + \mathcal{L}_{cur}. \tag{4}$$

$\mathcal{L}_{seg}$ and $\mathcal{L}_{occ}$ refer to the loss functions of the tooth segmentation network and the implicit function network, respectively.

3   Experiments

3.1   Dataset and Evaluation Metrics

To evaluate the performance of our proposed method, we collect a dataset with 50 patients in dental clinics. Each subject has a 3D CBCT image (with 0.4 mm spacing) and paired intra-oral scanning model (with 0.03 mm spacing). To build the ground truth of our dataset, 3 experts are first employed to manually delineate the tooth labels and crown labels on CBCT images and intra-oral scanning models, respectively. Then, we merge the two modalities of data to obtain the high-quality tooth models (see details in Sect. 2.1). In this study, 50 samples are randomly divided into 3 subsets, using 20 for training, 10 for validation, and the remaining 20 samples for testing. Note that only CBCT images are fed into the framework to generate high-quality tooth models.

To quantitatively analyze the performance of our method, we report the following four metrics, including Intersection over Union (IoU), Chamfer-L2, Normal Consistency (Normals), and occupancy accuracy (OccAcc). IoU measures the similarity between two volumes, and Chamfer-L2 is the metric to measure a bidirectional distance between two surfaces. Normals is first proposed by OccNet [13] to measure the normal consistency between two surfaces. We define OccAcc as an additional accuracy metric to evaluate the occupancy prediction.

3.2 Sampling Strategies

To approximate the continuous query space, we briefly introduce the sampling strategies during network training and inference stages. In the training stage, the most intuitive way is to sample points around the ground truth tooth surface within a small distance. Specifically, we first sample points on the ground truth tooth surface, and then add random displacements with two Gaussian distributions, where their deviations are $\sigma_1 = 0.02$ and $\sigma_2 = 0.1$, respectively. Thus, the sample points within σ_1 can capture fine-grained surface details, and the sample points within σ_2 can cover the entire geometric space. In the network training, we sample 50K points from each of the two distributions. In the inference stage, since the ground truth tooth model is not available, we uniformly sample points along each axis in the continuous query space. Note that the retrieval resolution is determined by the density of query points on each axis. In our experiments, to obtain the tooth model with rich geometry information, especially on the tooth crowns, we query output with a resolution of 256^3, which is about 4^3 times larger than the image patch cropped from the original 3D CBCT image.

3.3 Implementation Details

Our framework is built on a PyTorch platform with an NVIDIA Tesla V100S GPU. The encoder is composed of four blocks, including one Convolution(Conv, with a $3 \times 3 \times 3$ kernel and a $1 \times 1 \times 1$ padding)-ReLU-Batch Normalization(BN) block and three MaxPooling-Conv-ReLU-Conv-ReLU-BN blocks. And the network architectures of the point-wise decoder C and the point-wise regressor R are the same, which include four fully connected layers with ReLU. We use the Adam optimizer with a learning rate of 0.0001, divided by 10 for every 20 epochs. In the testing stage, with the predicted occupancy of each query point (i.e., inside or outside the surface), we apply the traditional Marching Cubes algorithm [11] to generate high-quality tooth models.

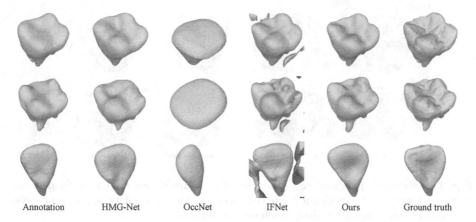

Annotation   HMG-Net   OccNet   IFNet   Ours   Ground truth

Fig. 3. Qualitative results of three typical cases generated by different methods.

Table 1. Quantitative comparison with different methods for high-quality tooth model generation.

Method	OccAcc ↑	IoU ↑	Chamfer-L2 ↓	Normals ↑
HMG-Net	75.95	77.10	6.30e-4	95.09
OccNet	61.41	54.04	2.92e-3	85.08
IF-Net	77.47	65.52	9.34e-3	88.67
Ours	**79.70**	**83.03**	**3.00e-4**	**96.25**

3.4   Comparison with Other Methods

Recently, many methods have been proposed to reconstruct 3D shapes with implicit functions, including the Occupancy Networks (OccNet) [13], and the Implicit Feature Networks (IF-Net) [2]. We also implement the baseline segmentation method (i.e., HMG-Net [5]) with high-quality tooth model supervision.

The quantitative results of different methods are presented in Table 1. It can be found that our approach achieves the best performance across all metrics. Moreover, the accuracy of directly applying the implicit function based methods on tooth surfaces (i.e., OccNet and IF-Net) is relatively low. The main reason is that there is a large domain gap between the input CBCT images and output tooth surfaces. Directly applying surface reconstruction algorithms cannot effectively address the distribution shift problem.

In order to analyze the advantage of our algorithm more comprehensively, we further provide visual comparison of three typical examples in Fig. 3. It can be found that, the tooth instance segmentation method (i.e., HMG-Net) can generate promising tooth shapes, but many details, especially on the tooth crowns, are missed due to limited resolution of CBCT images. For the implicit function based methods, we observe that the tooth surfaces produced by OccNet are too smooth with only global structures. With the multi-scale

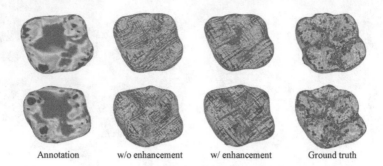

Annotation   w/o enhancement   w/ enhancement   Ground truth

Fig. 4. Investigation on surface curvatures. The figure shows the comparison of our framework without and with curvature enhancement for surface details. Different colors are used to show curvature distribution on the surface.

Table 2. Statistical results for analyzing different components in our framework. "H-IF" denotes the joint learning of HMG-Net and IF-Net.

Method	OccAcc ↑	IoU ↑	Chamfer-L2 ↓	Normals ↑
HMG-Net	75.95	77.10	6.30e-4	95.09
H-IF	78.67	82.74	**2.88e-4**	95.87
Ours	**79.70**	**83.03**	3.00e-4	**96.25**

feature extraction scheme, IF-Net generates better results with fine-grained details. However, many noises are introduced due to large domain gap between CBCT images and tooth models. Notably, our method matches better with the ground truth, where global structures and local details can be successfully retained, indicating the effectiveness of both the image-to-surface tooth model generation scheme and the curvature enhancement in this specific task.

3.5 Ablation Study

To validate the effectiveness of each component in our method, we conduct several ablation experiments by gradually augmenting the baseline network.

Surface Reconstruction. Based on the baseline tooth instance segmentation network (i.e., HMG-Net), we add another branch, surface reconstruction by the implicit function learning, to enhance the network capability on global tooth structures, and denote it as H-IF. The results are shown in Table 2. It can be found that the accuracy is consistently improved in terms of all metrics. Particularly, the surface distance error (i.e., Chamfer-L2) is greatly reduced (6.30$e-4$ vs. 2.88$e-4$), demonstrating that the implicit function strategy can more effectively learn global tooth structures.

Curvature Enhancement. To validate the effectiveness of the curvature enhancement module, we include this module based on H-IF as our final framework. As shown in Table 2, the statistical results are improved, especially on the metric of surface normals (95.87% vs. 96.25%). Note that surface normals are usually sensitive to the geometric details on surfaces. Moreover, we further present the visual comparison of curvature distributions in Fig. 4. The first column is the curvature distributions computed of manual annotations in original CBCT images, where many geometric details, especially on the crowns, are missed due to limited CBCT resolution. The second and third columns are the tooth models produced by our method without and with the curvature enhancement module. It can be seen that, with the curvature enhancement (3rd column), our framework can produce local shape properties more clearly, and the curvature distribution matches with the ground truth more consistently.

4 Conclusion

In this paper, we propose an effective framework to generate high-quality tooth models from CBCT images. Our framework first introduces a tooth instance segmentation network to segment individual teeth coarsely, and then learns to generate high-quality tooth models under implicit function learning. Meanwhile, a curvature enhancement module is further proposed to guide the surface reconstruction. Experimental results on the real-patient dataset demonstrate that our proposed method outperforms other state-of-the-art methods, showing its potential to be applied in dental clinics.

References

1. Besl, P.J., McKay, N.D.: Method for registration of 3-d shapes. In: Sensor Fusion IV: Control Paradigms and Data Structures, vol. 1611, pp. 586–606. International Society for Optics and Photonics (1992)
2. Chibane, J., Alldieck, T., Pons-Moll, G.: Implicit functions in feature space for 3d shape reconstruction and completion. In: Proceedings of the IEEE/CVF Conference on Computer Vision and Pattern Recognition, pp. 6970–6981 (2020)
3. Chung, M., et al.: Pose-aware instance segmentation framework from cone beam CT images for tooth segmentation. Comput. Biol. Med. **120**, 103720 (2020)
4. Cui, Z., Li, C., Wang, W.: Toothnet: automatic tooth instance segmentation and identification from cone beam ct images. In: Proceedings of the IEEE/CVF Conference on Computer Vision and Pattern Recognition, pp. 6368–6377 (2019)
5. Cui, Z., et al.: Hierarchical morphology-guided tooth instance segmentation from CBCT images. In: Feragen, A., Sommer, S., Schnabel, J., Nielsen, M. (eds.) IPMI 2021. LNCS, vol. 12729, pp. 150–162. Springer, Cham (2021). https://doi.org/10.1007/978-3-030-78191-0_12
6. Ewers, R., et al.: Computer-aided navigation in dental implantology: 7 years of clinical experience. J. Oral Maxillof. Surg. **62**(3), 329–334 (2004)
7. Hoffmann, O., Zafiropoulos, G.G.: Tooth-implant connection: a review. J. Oral Implantol. **38**(2), 194–200 (2012)

8. Kazhdan, M., Hoppe, H.: Screened poisson surface reconstruction. ACM Trans. Graph. **32**(3), 1–13 (2013)

9. Li, Y., Jacox, L.A., Little, S.H., Ko, C.C.: Orthodontic tooth movement: the biology and clinical implications. Kaohsiung J. Med. Sci. **34**(4), 207–214 (2018)

10. Liu, C., Guo, J., Gao, J., Yu, H.: Computer-assisted tooth preparation template and predesigned restoration: a digital workflow. Int. J. Computeriz. Dentist. **23**(4), 351–362 (2020)

11. Lorensen, W.E., Cline, H.E.: Marching cubes: a high resolution 3d surface construction algorithm. ACM Siggraph Comput. Graph. **21**(4), 163–169 (1987)

12. Lu, G.Y., Wong, D.W.: An adaptive inverse-distance weighting spatial interpolation technique. Comput. Geosci. **34**(9), 1044–1055 (2008)

13. Mescheder, L., Oechsle, M., Niemeyer, M., Nowozin, S., Geiger, A.: Occupancy networks: learning 3d reconstruction in function space. In: Proceedings of the IEEE/CVF Conference on Computer Vision and Pattern Recognition, pp. 4460–4470 (2019)

14. Orlowska, M., Jozwiak, R., Regulski, P.: Virtual tooth extraction from cone beam computed tomography scans. In: Augustyniak, P., Maniewski, R., Tadeusiewicz, R. (eds.) PCBBE 2017. AISC, vol. 647, pp. 275–285. Springer, Cham (2018). https://doi.org/10.1007/978-3-319-66905-2_24

15. Paredes, V., Gandia, J.L., Cibrián, R.: Digital diagnosis records in orthodontics. An overview. Med. Oral Patol. Oral Cir. Bucal. **11**(1), E88–E93 (2006)

16. Roberta, T., Federico, M., Federica, B., Antonietta, C.M., Sergio, B., Ugo, C.: Study of the potential cytotoxicity of dental impression materials. Toxicol. In vitro **17**(5–6), 657–662 (2003)

17. Van Der Meer, W.J., Vissink, A., Ng, Y.L., Gulabivala, K.: 3d computer aided treatment planning in endodontics. J. Dentist. **45**, 67–72 (2016)

PHTrans: Parallelly Aggregating Global and Local Representations for Medical Image Segmentation

Wentao Liu[1], Tong Tian[2], Weijin Xu[1], Huihua Yang[1,3(✉)], Xipeng Pan[3,4], Songlin Yan[1], and Lemeng Wang[1]

[1] School of Artificial Intelligence, Beijing University of Posts and Telecommunications, Beijing 100876, China
yhh@bupt.edu.cn
[2] State Key Laboratory of Structural Analysis for Industrial Equipment, School of Aeronautics and Astronautics, Dalian University of Technology, Dalian 116024, China
[3] School of Computer Science and Information Security, Guilin University of Electronic Technology, Guilin 541004, China
[4] Department of Radiology, Guangdong Provincial People's Hospital, Guangdong Academy of Medical Sciences, Guangzhou 510080, China

Abstract. The success of Transformer in computer vision has attracted increasing attention in the medical imaging community. Especially for medical image segmentation, many excellent hybrid architectures based on convolutional neural networks (CNNs) and Transformer have been presented and achieve impressive performance. However, most of these methods, which embed modular Transformer into CNNs, struggle to reach their full potential. In this paper, we propose a novel hybrid architecture for medical image segmentation called PHTrans, which parallelly hybridizes Transformer and CNN in main building blocks to produce hierarchical representations from global and local features and adaptively aggregate them, aiming to fully exploit their strengths to obtain better segmentation performance. Specifically, PHTrans follows the U-shaped encoder-decoder design and introduces the parallel hybrid module in deep stages, where convolution blocks and the modified 3D Swin Transformer learn local features and global dependencies separately, then a sequence-to-volume operation unifies the dimensions of the outputs to achieve feature aggregation. Extensive experimental results on both Multi-Atlas Labeling Beyond the Cranial Vault and Automated Cardiac Diagnosis Challeng datasets corroborate its effectiveness, consistently outperforming state-of-the-art methods. The code is available at: https://github.com/lseventeen/PHTrans.

Keywords: Medical image segmentation · Transformer · CNN · Hybrid architecture

L. Wang et al. (Eds.): MICCAI 2022, LNCS 13435, pp. 235–244, 2022.
https://doi.org/10.1007/978-3-031-16443-9_23

1   Introduction

Medical image segmentation aims to extract and quantify regions of interest in biological tissue/organ images, which are essential for disease diagnosis, preoperative planning, and intervention. Benefiting from the excellent representation learning ability of deep learning, convolutional neural networks (CNNs) have achieved tremendous success in medical image analysis. Many excellent network models (e.g., U-Net [14], 3D U-Net [4] and Attention U-Net [12]) have emerged, constantly refreshing the upper limit of performance for various segmentation tasks. In spite of achieving extremely competitive results, CNN-based methods lack the ability to model long-range dependencies due to inherent inductive biases such as locality and translational equivariance. Several researchers have alleviated this problem by increasing the size of the convolution kernel [13], using atrous convolution [20], and embedding self-attention mechanisms [17]. However, it cannot be fundamentally solved provided that the convolution operation remains at the heart of the network architecture.

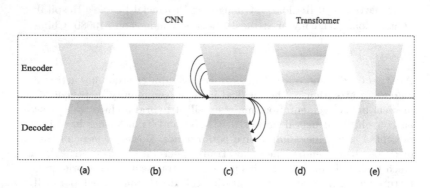

Fig. 1. Comparison of several hybrid architectures of Transformer and CNN. (Color figure online)

Transformer [15], relying purely on attention mechanisms to model global dependencies without any convolution operations, have emerged as an alternative architecture which has delivered better performance than CNNs in computer vision (CV) on the condition of being pre-trained on large-scale datasets. Among them, Vision Transformer (ViT) [5] splits images into a sequence of tokens and models their global relations with stacked Transformer blocks, which have revolutionized the CV field. Swin Transformer [11] can produce hierarchical feature representations with lower computational complexity in shiftable windows, achieving state-of-the-art performance in various CV tasks. However, medical image datasets are much smaller in magnitude than the datasets used by the pre-trained in the mentioned works (e.g., ImageNet-21k and JFT-300M) because medical images are not always available and require professional annotation. As a result, Transformer produces unsatisfactory performance in medical image segmentation. Meanwhile, many hybrid structures derived from the combination of CNN and Transformer have emerged, which offer the advantages of both and have gradually become a compromise solution for medical image segmentation without being pre-trained on large datasets.

We summarize several popular hybrid architectures based on Transformer and CNN in medical image segmentation. These hybrid architectures add Transformer into models with CNN as the backbone, or replace parts of the architecture's components. For example, UNETR [7] and Swin UNETR [2] used the encoder-decoder structure in which the encoder is composed of a cascade of blocks built with self-attention and multilayer perceptron, i.e., Transformer, while the decoder is stacked convolutional layers, see Fig. 1(a). TransBTS [16] and TransUNet [3] introduced a Transformer between the encoder and decoder composed of CNN, see Fig. 1(b). MISSFormer [8] and CoTr [18] bridged all stages from encoder to decoder by Transformer instead of only adjacent stages, which captures the multi-scale global dependency, see Fig. 1(c). In addition, nnFormer [21] interleaved Transformer and convolution blocks into a hybrid model, where convolution encodes precise spatial information and self-attention captures global context, see Fig. 1(d). From Fig. 1, it can be seen that these architectures implement a serial combination of Transformer and CNN from a macro perspective. Nevertheless, in the serial combination, convolution and self-attention cannot run through the entire network architecture, making it difficult to continuously model local and global representations, so it does not fully exploit their potential.

In this paper, we propose a parallel hybrid Transformer (PHTrans) for medical image segmentation where the main building blocks consist of CNN and Swin Transformer to simultaneously aggregate global and local representations, see Fig. 1(e). In PHTrans, we extend the standard Swin Transformer to a 3D version by extracting 3D patches that partition a volume and constructing 3D self-attention mechanisms. Given the hierarchical property of the Swin Transformer can conveniently leverage advanced techniques for dense prediction such as U-Net [11], we followed the successful U-shaped architecture design and introduced a transformation operation of sequence and volume to achieve the parallel combination of Swin Transformer and CNN in a block. In contrast to serial hybrid architecture, PHTrans can independently construct hierarchical local and global representations and fuse them in each stage, fully exploiting the potential of CNN and the Transformer. Extensive experiments demonstrate the superiority of our method against other competing methods on various medical image segmentation tasks.

2   Method

2.1   Overall Architecture

An overview of the PHTrans architecture is illustrated in Fig. 2(a). PHTrans follows the U-shaped encoder and decoder design, which is mainly composed of pure convolution modules and parallel hybrid ones. Our original intention was to construct a completely hybrid architecture composed of Transformer and CNN, but due to the high computational complexity of the self-attention mechanism, Transformer cannot directly receive input with pixels serving as tokens. In our implementation, a cascade of convolution blocks and down-sampling operations are introduced to reduce the spatial size, which progressively extracts low-level

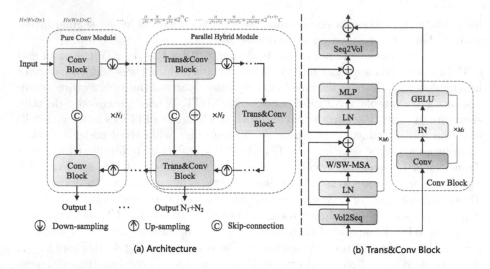

Fig. 2. (a) The architecture of PHTrans; (b) Parallel hybird block consisting of Transformer and convolution (Trans&Conv block).

features with high resolution to obtain fine spatial information. Similarly, these pure convolution modules are deployed in the decoder at the same stage to recover the original image dimension by up-sampling.

Given an input volume $x \in \mathbb{R}^{H \times W \times D}$, where H, W and D denote the height, width, and depth, respectively, we first utilize several pure convolution modules to obtain feature maps $f \in \mathbb{R}^{\frac{H}{2^{N_1}} \times \frac{W}{2^{N_1}} \times \frac{D}{2^{N_1}} \times 2^{N_1}C}$, where N_1 and C denote the number of modules and base channels, respectively. Afterwards, parallel hybrid modules consisting of Transformer and CNN were applied to model the hierarchical representation from local and global feature. The procedure is repeated N_2 times with $\frac{H}{2^{N_1+N_2}} \times \frac{W}{2^{N_1+N_2}} \times \frac{D}{2^{N_1+N_2}}$ as the output resolutions and $2^{N_1+N_2}C$ as the channel number. Corresponding to the encoder, the symmetric decoder is similarly built based on pure convolution modules and parallel hybrid modules, and fuses semantic information from the encoder by skip-connection and addition operations. Furthermore, we use deep supervision at each stage of the decoder during the training, resulting in a total of $N_1 + N_2$ outputs, where joint loss consisting of cross entropy and dice loss is applied. The architecture of PHTrans is straightforward and changeable, where the number of each module can be adjusted according to medical image segmentation tasks, i.e., N_1, N_2, M_1 and M_2. Among them, M_1 and M_2 are the numbers of Swin Transformer blocks and convolution blocks in the parallel hybrid module.

2.2 Parallel Hybrid Module

The parallel hybrid modules are deployed in the deep stages of PHTrans, where the Trans&Conv block, as its heart, achieves hierarchical aggregation of local and global representations by CNN and Swin Transformer.

Trans&Conv Block. The scale-reduced feature maps are fed into Swin Transformer (ST) blocks and convolution (Conv) blocks, respectively. We introduce Volume-to-Sequence (V2S) and Sequence-to-Volume (S2V) operations at the beginning and end of ST blocks, respectively, to implement the transform of volume and sequence, making it concordant with the dimensional space of the output that Conv blocks produce. Specifically, V2S is used to reshape the entire volume (3D image) into a sequence of 3D patches with a window size. S2V is the opposite operation. As shown in Fig. 2(b), a ST block consists of a shifted window based multi-head self attention (MSA) module, followed by a 2-layer MLP with a GELU activation function in between. A LayerNorm (LN) layer is applied before each MSA module and each MLP, and a residual connection is applied after each module [11]. In M_1 successive ST blocks, the MSA with regular and shifted window configurations, i.e., W-MSA and SW-MSA, is alternately embedded into ST blocks to achieve cross-window connections while maintaining the efficient computation of non-overlapping windows.

For medical image segmentation, we modified the standard ST block into a 3D version, which computes self-attention within local 3D windows that are arranged to evenly partition the volume in a non-overlapping manner. Supposing $x \in \mathbb{R}^{H \times W \times S \times C}$ is the input of ST block, it would be first reshaped to $N \times L \times C$, where N and $L = W_h \times W_w \times W_s$ denote the number and dimensionality of 3D windows, respectively. The self-attention in each head is calculated as:

$$Attention(Q, K, V) = SoftMax(\frac{QK^T}{\sqrt{d}} + B)V, \tag{1}$$

where $Q, K, V \in \mathbb{R}^{L \times d}$ are the *query, key* and *value* matrices, d is the *query/key* dimension, and $B \in \mathbb{R}^{L \times L}$ is the relative position bias. We parameterize a smaller-sized bias matrix $\hat{B} \in \mathbb{R}^{(2W_h-1) \times (2W_w-1) \times (2W_s-1)}$, and values in B are taken from $\hat{B}$.

The convolution blocks are repeated M_2 times with a $3 \times 3 \times 3$ convolutional layer, a GELU nonlinearity, and an instance normalization layer as a unit. The configuration of the convolution blocks is simple and flexible enough that any off-the-shelf convolutional network can be applied. Finally, we fuse the outputs of the ST blocks and Conv blocks by an addition operation. The computational procedure of the Trans&Conv block in the encoder can be summarized as follows:

$$y_i = S2V(ST^{M_1}(V2S(x_{i-1}))) + Conv^{M_2}(x_{i-1}), \tag{2}$$

where x_{i-1} is the down-sampling results of the encoder's $i - 1^{th}$ stage. In the decoder, besides skip-connection, we supplement the context information from the encoder with an addition operation. Therefore, the Trans&Conv block in the decoder can be formulated as:

$$z_i = S2V(ST^{M_1}(V2S(x_{i+1} + y_i)) + Conv^{M_2}([x_{i+1}, y_i]), \tag{3}$$

where x_{i+1} is the up-sampling results of the decoder's $i + 1^{th}$ stage and y_i is output of the encoder's i^{th} stage.

Down-sampling and Up-sampling. The down-sampling contains a strided convolution operation and an instance normalization layer, where the channel number is halved and the spatial size is doubled. Similarly, the up-sampling is a strided deconvolution layer followed by an instance normalization layer, which doubles the number of feature map channels and halved the spatial size. The stride is generally set to 2 in all dimensions. However, when 3D medical images are anisotropic, the stride with respect to specific dimensions is set to 1.

3 Experiments

3.1 Dataset

The Multi-Atlas Labeling Beyond the Cranial Vault (BCV) [10] includes 30 cases with 3779 axial abdominal clinical CT images. Similar to [21], the dataset is split into 18 training samples and 12 testing samples. And the average Dice-Similarity coefficient (DSC) and average Hausdorff Distance (HD) are used as evaluation metrics to evaluate our method on 8 abdominal organs (aorta, gallbladder, spleen, left kidney, right kidney, liver, pancreas, spleen, stomach).

The Automated Cardiac Diagnosis Challenge (ACDC) [1] dataset is collected from different patients using MRI scanners. For each patient's MR image, left ventricle (LV), right ventricle (RV), and myocardium (MYO) are labeled. Following [21], 70 samples are divided into the training set, 10 samples are divided into validation set, and 20 samples are divided into test set. The average DSC is used to evaluate our method on this dataset.

3.2 Implementation Details

For a fair comparison, we employed the code framework of nnUNet [9] to evaluate the performance of PHTrans as the same as CoTr [18] and nnFormer [21]. All experiments were performed under the default configuration of nnUNet. In PHTrans, we empirically set the hyper-parameters $[N_1,N_2,M_1,M_2]$ to $[2,4,2,2]$ and adopted the stride strategy of nnU-Net [9] for down-sampling and up-sampling without elaborate design. Moreover, the base number of channels C is 24 and the numbers of heads of multi-head self-attention used in different encoder stages are [3,6,12,24]. For BCV and ACDC datasets, we set the size of 3D windows $[W_h,W_w,W_s]$ to [3,6,6] and [2,8,7] in ST blocks, respectively. In the training stage, we randomly cropped sub-volumes of size $48\times192\times192$ and $16\times256\times224$ from BCV and ACDC datasets as the input, respectively. We implemented PHTrans under PyTorch 1.9 and conducted experiments with a single GeForce RTX 3090 GPU.

3.3 Results

Comparing with State-of-the-Arts. We compared the performance of PHTrans with previous state-of-the-art methods. In addition to the hybrid architectures mentioned in the introduction, it also includes, LeVIT-Unet [19], Swin-Unet [2], and nnU-Net [9]. Furthermore, we reproduced Swin UNETR [6] by

Table 1. Results of segmentation on the BCV dataset (average dice score % and average hausdorff distance in mm). † indicates that the model is pre-trained on ImageNet.

Methods	DSC↑	HD↓	Aot	Gal	Kid (L)	Kid (R)	Liv	Pan	Spl	Sto
Swin-Unet† [2]	79.13	21.55	85.47	66.53	83.28	79.61	94.29	56.58	90.66	76.60
TransUNet† [3]	77.48	31.69	87.23	63.13	81.87	77.02	94.08	55.86	85.08	75.62
LeViT-Unet† [19]	78.53	16.84	87.33	62.23	84.61	80.25	93.11	59.07	88.86	72.76
MISSFormer [8]	81.96	18.20	86.99	68.65	85.21	82.00	94.41	65.67	91.92	80.81
CoTr [18]	86.33	12.63	92.10	**81.47**	85.33	86.41	96.87	80.20	92.21	76.08
nnFormer† [21]	86.45	14.63	89.06	78.19	**87.53**	87.09	95.43	81.92	89.84	82.58
nnU-Net [9]	87.75	9.83	**92.83**	80.66	84.86	89.78	**97.17**	82.00	**92.39**	82.31
UNETR [7]	79.42	29.27	88.92	69.80	81.38	79.71	94.28	58.93	86.14	76.22
Swin UNETR [6]	85.78	17.75	92.78	76.55	85.25	89.12	96.91	77.22	88.70	79.72
PHTrans	**88.55**	**8.68**	92.54	80.89	85.25	**91.30**	97.04	**83.42**	91.20	**86.75**

modifying ViT in the encoding stage of UNETR [7] into a Swin transformer and also evaluated the performance of UNETR and Swin UNETR in the same way and used the same dataset partition as ours. The results of segmentation on the BCV dataset are shown in Table 1. Our PHTrans achieves the best performance with 88.55% (DSC↑) and 8.68 (HD↓) and surpasses the previous best model by 0.8% on average DSC and 1.15 on HD. The representative samples in Fig. 3 demonstrate the success of identifying organ details by PHTrans, e.g., "Stomach" in rows 1, 2, and "Left Kidney" in rows 2. The results of segmentation on the ACDC dataset are presented in Table 2. Similarly, it is obviously apparent that PHTrans is competitive with other state-of-the-art methods by achieving the highest average DSC. It is worth mentioning that Swin-Unet, TransUNet, LeViT, and nnFormer use the pre-trained weights on ImageNet to initialize their networks, while PHTrans was trained on both datasets from scratch. Additionally, we compared the number of parameters and FLOPs to evaluate the model complexity of 3D approaches, i.e., nnformer, CoTr, nnU-Net, UNETR, Swin UNETR and PHTrans, in BCV experiments. As shown in Table 3, PHTrans has few parameters (36.3M), and its FLOPs (187.4G) is significantly lower than CoTr, nnU-Net and Swin UNETR. In summary, the results of PHTrans on BCV and ACDC datasets fully demonstrate its excellent medical image segmentation and generalization ability with preserved moderate model complexity.

Table 2. Results of segmentation on the ACDC dataset. † indicates that the model is pre-trained on ImageNet.

Methods	DSC	RV	MLV	LVC
Swin-Unet† [2]	90.00	88.55	85.62	**95.83**
TransUNet† [3]	89.71	88.86	84.53	95.73
LeViT-Unet† [19]	90.32	89.55	87.64	93.76
MISSFormer [8]	90.86	89.55	88.04	94.99
nnFormer† [21]	91.62	**90.27**	89.23	95.36
nnU-Net [9]	91.36	90.11	88.75	95.23
PHTrans	**91.79**	90.13	**89.48**	95.76

Table 3. Comparison of number of parameters and FLOPs for 3D segmentation models in BCV experiments.

Methods	Params (M)	FLOPs(G)
CoTr [18]	41.9	318.5
nnformer [21]	158.7	171.9
nnU-Net [9]	**30.8**	313.2
UNETR [7]	92.6	**82.6**
Swin UNETR [6]	61.98	394.84
PHTrans	36.3	187.4

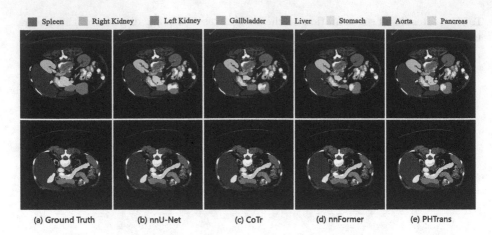

Fig. 3. Qualitative visualizations of the proposed PHTrans and other 3D methods.

Table 4. Ablation study on the architecture.

Methods	DSC↑	HD↓	Aot	Gal	Kid (L)	Kid (R)	Liv	Pan	Spl	Sto
3D Swin-Unet	83.47	20.13	87.46	71.37	86.77	83.53	95.47	72.51	92.19	78.48
3D Swin-Unet+PCM	84.95	19.52	87.36	77.55	85.76	84.48	95.59	79.66	88.93	80.26
PHTrans w/o ST	87.71	14.37	92.22	79.35	85.94	88.08	96.62	83.26	90.63	85.59
PHTrans w/o PCM	86.10	18.29	88.73	76.56	**86.81**	85.56	96.09	78.04	**94.28**	82.77
PHTrans	**88.55**	**8.68**	**92.54**	**80.89**	85.25	**91.30**	**97.04**	**83.42**	91.20	**86.75**

Ablation Study. Using the modified 3D Swin-Unet as the baseline, we progressively integrated the components of PHTrans to explore the influence of different components on the model performance. Table 4 provides quantitative results of the ablation study on the architecture. "+PCM" denotes using stacked pure convolutional modules instead of a strided convolution operation for patch partition, while "w/o PCM denotes the opposite. "w/o ST" means that the parallel hybrid module in PHTrans removes the Swin Transformer blocks, resulting in a similar architecture to nnU-Net. From these results, it can be seen that the performance of 3D Swin-Unet and PHTrans is improved by PCM, which is owed to its ability to capture fine-grained details in the first few stages. Furthermore, the PHTrans brings more significant performance gains compared with the single architecture and outperforms "3D Swin-Unet+PCM" and "PHTrans w/o ST" by 3.6% and 0.84% in average DSC and 10.84 and 5.69 in HD, respectively. The results indicate the effectiveness of using a parallel combining strategy of CNN and Transformer to aggregate global and local representations.

Discussion. In PHTrans, a vanilla Swin Transformer and simple convolutional blocks are applied, which demonstrates that significant performance gains stem from the parallel hybrid architecture design rather than the Transformer and CNN blocks compared to state-of-the-arts. Furthermore, PHTrans is not pre-trained since there is no large enough all-purpose dataset of 3D medical images so far. From the above considerations, in the future, we will elaborately design Transformer and CNN blocks and explore how to pre-train Transformer end-to-end to further improve the segmentation performance.

4   Conclusions

In this paper, we propose a parallel hybrid architecture (PHTrans) based on the Swin Transformer and CNNs for accurate medical image segmentation. Different from other hybrid architectures that embed modular Transformer into CNNs, PHTrans constructs hybrid modules consisting of Swin Transformer and CNN throughout the model, which continuously aggregates hierarchical representations from global and local features to give full play to the superiority of both. Extensive experiments on BCV and ACDC datasets show our method is superior to several state-of-the-art alternatives. As an all-purpose architecture, PHTrans is flexible and can be replaced with off-the-shelf convolution and Transformer blocks, which open up new possibilities for more downstream medical image tasks.

Acknowledgment. This work was supported in part by the National Key R&D Program of China (No.2018AAA0102600) and National Natural Science Foundation of China (No.62002082).

References

1. Bernard, O., et al.: Deep learning techniques for automatic MRI cardiac multi-structures segmentation and diagnosis: is the problem solved? IEEE Trans. Med. Imaging **37**(11), 2514–2525 (2018)
2. Cao, H., et al.: Swin-unet: Unet-like pure transformer for medical image segmentation. arXiv preprint arXiv:2105.05537 (2021)
3. Chen, J., et al.: Transunet: transformers make strong encoders for medical image segmentation. arXiv preprint arXiv:2102.04306 (2021)
4. Cui, Z., et al.: Hierarchical morphology-guided tooth instance segmentation from CBCT images. In: Feragen, A., Sommer, S., Schnabel, J., Nielsen, M. (eds.) IPMI 2021. LNCS, vol. 12729, pp. 150–162. Springer, Cham (2021). https://doi.org/10.1007/978-3-030-78191-0_12
5. Dosovitskiy, A., et al.: An image is worth 16x16 words: transformers for image recognition at scale. In: 9th International Conference on Learning Representations, ICLR 2021, Virtual Event, Austria, 3–7 May 2021. OpenReview.net (2021). https://openreview.net/forum?id=YicbFdNTTy
6. Hatamizadeh, A., Nath, V., Tang, Y., Yang, D., Roth, H., Xu, D.: Swin unetr: swin transformers for semantic segmentation of brain tumors in MRI images. arXiv preprint arXiv:2201.01266 (2022)
7. Hatamizadeh, A., et al.: Unetr: transformers for 3d medical image segmentation. In: Proceedings of the IEEE/CVF Winter Conference on Applications of Computer Vision, pp. 574–584 (2022)

8. Huang, X., Deng, Z., Li, D., Yuan, X.: Missformer: an effective medical image segmentation transformer. arXiv preprint arXiv:2109.07162 (2021)
9. Isensee, F., Jaeger, P.F., Kohl, S.A., Petersen, J., Maier-Hein, K.H.: nnu-net: a self-configuring method for deep learning-based biomedical image segmentation. Nat. Methods **18**(2), 203–211 (2021)
10. Landman, B., Xu, Z., Igelsias, J., Styner, M., Langerak, T., Klein, A.: Miccai multi-atlas labeling beyond the cranial vault-workshop and challenge. In: Proceedings of the MICCAI Multi-Atlas Labeling Beyond Cranial Vault-Workshop Challenge, vol. 5, p. 12 (2015). https://www.synapse.org/#!Synapse:syn3193805/wiki/217789
11. Liu, Z., et al.: Swin transformer: hierarchical vision transformer using shifted windows. In: Proceedings of the IEEE/CVF International Conference on Computer Vision, pp. 10012–10022 (2021)
12. Oktay, O., et al.: Attention u-net: Learning where to look for the pancreas. arXiv preprint arXiv:1804.03999 (2018)
13. Peng, C., Zhang, X., Yu, G., Luo, G., Sun, J.: Large kernel matters-improve semantic segmentation by global convolutional network. In: Proceedings of the IEEE Conference on Computer Vision and Pattern Recognition, pp. 4353–4361 (2017)
14. Ronneberger, O., Fischer, P., Brox, T.: U-Net: convolutional networks for biomedical image segmentation. In: Navab, N., Hornegger, J., Wells, W.M., Frangi, A.F. (eds.) MICCAI 2015. LNCS, vol. 9351, pp. 234–241. Springer, Cham (2015). https://doi.org/10.1007/978-3-319-24574-4_28
15. Vaswani, A., et al.: Attention is all you need. Adv. Neural Inf. Process. Syst. **30** (2017)
16. Wang, W., Chen, C., Ding, M., Yu, H., Zha, S., Li, J.: TransBTS: multimodal brain tumor segmentation using transformer. In: de Bruijne, M., et al. (eds.) MICCAI 2021. LNCS, vol. 12901, pp. 109–119. Springer, Cham (2021). https://doi.org/10.1007/978-3-030-87193-2_11
17. Wang, X., Girshick, R., Gupta, A., He, K.: Non-local neural networks. In: Proceedings of the IEEE Conference on Computer Vision and Pattern Recognition, pp. 7794–7803 (2018)
18. Xie, Y., Zhang, J., Shen, C., Xia, Y.: CoTr: Efficiently bridging CNN and transformer for 3D medical image segmentation. In: de Bruijne, M., et al. (eds.) MICCAI 2021. LNCS, vol. 12903, pp. 171–180. Springer, Cham (2021). https://doi.org/10.1007/978-3-030-87199-4_16
19. Xu, G., Wu, X., Zhang, X., He, X.: Levit-unet: make faster encoders with transformer for medical image segmentation. arXiv preprint arXiv:2107.08623 (2021)
20. Yu, F., Koltun, V.: Multi-scale context aggregation by dilated convolutions. arXiv preprint arXiv:1511.07122 (2015)
21. Zhou, H.Y., Guo, J., Zhang, Y., Yu, L., Wang, L., Yu, Y.: nnformer: interleaved transformer for volumetric segmentation. arXiv preprint arXiv:2109.03201 (2021)

Learning Tumor-Induced Deformations to Improve Tumor-Bearing Brain MR Segmentation

Meng Jia[✉] and Matthew Kyan

Department of Electrical Engineering and Computer Science,
York University, Toronto, Canada
{mjia,mkyan}@eecs.yorku.ca

Abstract. We propose a novel framework that applies atlas-based whole-brain segmentation methods to tumor-bearing MR images. Given a patient brain MR image where the tumor is initially segmented, we use a point-cloud deep learning method to predict a displacement field, which is meant to be the deformation (inverse) caused by the growth (mass-effect and cell-infiltration) of the tumor. It's then used to warp and modify the brain atlas to represent the change so that existing atlas-based healthy-brain segmentation methods could be applied to these pathological images. To show the practicality of our method, we implement a pipeline with nnU-Net MRI tumor initial segmentation and SAMSEG, an atlas-based whole-brain segmentation method. To train and validate the deformation network, we synthesize pathological ground truth by simulating artificial tumors in healthy images with TumorSim. This method is evaluated with both real and synthesized data. These experiments show that segmentation accuracy can be improved by learning tumor-induced deformation before applying standard full brain segmentation. Our code is available at https://github.com/jiameng1010/Brain_MRI_Tumor.

Keywords: Brain MRI segmentation · Point cloud deep learning

1   Introduction

Brain tumors are among the most fatal cancers in the human body because a large portion of them are malignant or borderline-malignant like glioblastomas and astrocytomas [5], and even benign tumors like gliomas and meningiomas may damage vital brain functions by compressing or eroding the surrounding tissue. Magnetic Resonance Imaging (MRI) is the most widely used neural imaging technique in brain cancer diagnosis and treatment. Accurate and robust segmentation of the tumor body, peritumoral edema, brain tissues affected by the tumor [24], and organ-at-risk [1] is essential for diagnosis, survival prediction, treatment planning, and many other medical practices.

Supplementary Information The online version contains supplementary material available at https://doi.org/10.1007/978-3-031-16443-9_24.

Semantic whole Brain MRI segmentation refers to the task that segments the image into regions according to tissue types (WM, GM, and CSF) or anatomical structures (cortex, hippocampus, thalamus, lentiform nucleus). For fine anatomic segmentation, atlas-based generative approaches provide the most reliable and robust results [31]. This strategy relies on aligning a probabilistic atlas to the input images to provide a segmentation prior. However, directly using atlas-based segmentation methods in tumor-bearing patient images is problematic because the tumor's mass-effect may break the spatial information of the brain anatomy represented in the population-based atlas. There are two common solutions to this problem. First, in some atlas-based approaches, brain (including the tumor) segmentation and registration problems are solved jointly [1,11,29]. Scheufele et al. [29] introduce a joint image registration and biophysical inversion framework. Agn et al. [1] combine a generative model for whole-brain segmentation with a spatial regularization model of tumor shape, and *simultaneously* segment the tumor and organs-at-risk. However, given an accurate tumor segmentation, finding a solution with the reaction-diffusion model that grows a tumor in the atlas is still highly complex and challenging because of its inherent non-linearity, ill-posedness, and ill-conditioned nature [30]. A second possible solution is warping the input image (reverting toward a non-pathological state) with the estimated displacement to eliminate the tumor-induced deformation [19]; or warping (grow a tumor in) the brain atlas to reflect the tumor and spatial changes in the patient image [4,7,23,25]. The latter is sometimes referred to as "Seeded-Atlas" [25]. However, these approaches require a good tumor segmentation or even a well-positioned seed manually placed by experts to initialize the tumor growth. Our method follows the same principle with the hope of being more robust to those problems by learning from the synthesized deformation fields.

Recent years have witnessed an influx of discriminative neural-network approaches for brain tumor segmentation [10,14,28,32], and they show remarkable performance in the BraTS [22] challenge. However, joint generative model-based whole-brain segmentation couldn't benefit from these, as they require a generative tumor model, like RBMs and VAEs used in [1,24], to represent the tumor shape distribution.

Due to the difficulties above, we propose to train a regression model that *directly* predicts the tumor-induced deformation so that the probabilistic atlas can be warped and modified to represent the changed geometric structure. Our method is inspired by recent CNN-based volumetric brain image-to-image registration methods [3,8] and occupancy-function-based mesh-to-points liver registration [16]. The proposed method is similar to [19], in terms of trying to find out the inversion of the tumor growth. Like VoxelMorph methods [8], our network is a regression model that directly produces a vector field. The point cloud network method proposed in [2,15] maps a sparse point set to a continuous function in 3D space, and has also been used to learn the occupancy function of deformed livers in open-abdominal liver surgery [16]. We use a similar point cloud network to learn tumor-induced deformations to aid in whole-brain segmentation. This novel solution robustly combines a discriminative tumor segmentation with an atlas-based brain semantic segmentation. A novel training data synthesizing

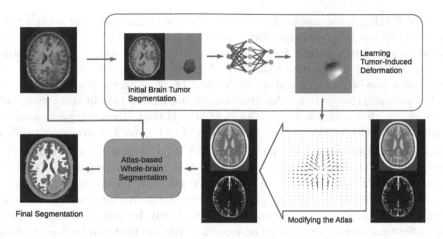

Fig. 1. A graphical abstract of the proposed segmentation method.

method is implemented based on TumorSim [26]. In addition, we introduce new evaluation methods to validate the proposed method.

2 Method

Figure 1 shows a graphical abstract of our method. Given an initial tumor segmentation, we sample a number of points from the boundary (surface) of the tumor segment and feed that into a point-cloud network [15] that predicts tumor growth. It produces a 3D vector field that models the brain's deformation caused by the growth of the tumor (Sect. 2.1). We use TumorSim [26], an open-source package that generates pathological ground truth from healthy brain templates, to generate displacement data for training (Sect. 2.2). Once trained, the network produces the tumor-induced mass-effect deformation as a spatial transformation $\phi : \mathbb{R}^3 \to \mathbb{R}^3$. As indicated in Fig. 1, the population-based probabilistic atlas is modified by warping with the predicted transformation to reflect the geometric change in the patient images. In addition, the initial tumor segment information is overlaid onto the atlas by appending a "tumorous" probability to the labels' prior in the atlas (Sect. 2.3). Finally, the patient image is segmented with a generative whole-brain segmentation method using the modified atlas. This pipeline uses the pre-trained nnU-Net model [13] for initial tumor segmentation and SamSeg [26] for the final brain segmentation.

2.1 Deformation Network

In most glioma models [12,20,26], the tumor behavior consisting of two components: proliferation and infiltration. We decide to learn a combined overall deformation with the network because the purpose is not an accurate tumor characterization, but a transformation that is good enough to improve atlas-based segmentation. Meanwhile, as indicated by [26], there is a correlation between cancer cell infiltration and edema. So we take the tumor core and edema from the initial tumor label and transform that to a point cloud $P : (P_{TC}, P_{ED})$ consists of two groups of points: P_{TC} representing tumor core and P_{ED} representing edema. A similar network and the same form of input as [16] is used, i.e., we associate a two-bit one-hot feature indicating TC/ED for each point.

Let ϕ denote the predicted deformation, we implement two different setups for learning ϕ with a point cloud network $\text{def}(\cdot|P)$: 1) that directly regresses on the displacement field (DISP), $\phi_{DISP} : q \mapsto q + \tau\text{def}(q|P)$ and 2) we represent diffeomorphism with a Stationary Velocity Field (SVF), $\phi_{SVF}^{(1)} : q \mapsto \exp(\tau\text{def}(q|P))$, where the SVF is integrated over $[0,1]$ to obtain the transformation (exponential map of SVF approximated in a certain resolution). When warping atlases, the scalar τ that adjusts the magnitude of transform can be determined either empirically by trials (as a constant) or interactively by a human expert (as a per-patient parameter), providing an easy intervention that can be used to compensate for the incorrect overall size of the initial tumor segment and the scale of predicted deformation, thus optimize the whole-brain segmentation.

2.2   Synthesizing Tumor Images for Training

We use TumorSim [26] to grow artificial tumors in healthy brain models of BrainWeb [6]. The input brain structure is modelled with a labelled tetrahedral volume mesh, where tissue mechanical properties (for biophysical deformation) are associated with each mesh element, and the WM tractography (for cancer cell infiltration) is captured with a diffusion tensor image (DTI). Given an initial tumor seed and the required tumor-growth parameters, the pathological brain is produced by solving the reaction-diffusion system. The initial seeds are sampled from ground-truth tumor masks in the BraTS dataset [22] with randomization and augmentation: 1) seed marks are flipped from one hemisphere to another to double the number of instances; 2) random erosion and dilation are applied to approximate the longitudinal earlier and later stages of the tumor. For each simulation, we draw random tumor-growth parameters: initial mass-effect pressure and tensor infiltration multipliers of GM and WM. In TumorSim, the initial tumor surface pressure direction is randomized by Von-Mises-Fisher distribution to further increase the output variability.

TumorSim produces as output a ground truth tumor segment $T : (T_{TC}, T_{ED})$ and dense deformation fields of mass-effect and infiltration, which are then combined to form the overall deformation $D \in \mathbb{R}^{nx \times ny \times nz \times 3}$. We first turn this GT tumor and edema volume to a triangle mesh using Marching Cubes [21]. The training point clouds P_{TC} and P_{ED} are uniformly sampled from the mesh surfaces ∂T_{TC} and ∂T_{ED}, respectively. Then, a set of query points $Q : \{q_1, q_2, \cdots, q_J\}$ are randomly drawn from the inside and peripheral tumor space. The deformation vectors $D(q_j)$ of those query points are interpolated from D. In the SVF diffeomorphism learning, the displacement field D is first transformed[1] to SVF before interpolation.

---

[1] We use Scaling-and-Squaring implemented in MIRTK (https://mirtk.github.io) to transform between DISP and SVF, specifically, `calculate-logarithmic-map` and `calculate-exponential-map`.

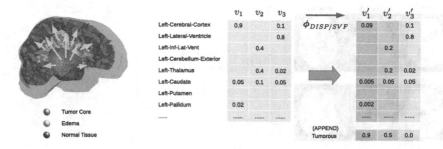

Fig. 2. Modifying the probabilistic atlas.

Let the tuple $\left(P_i, Q_i, \left\{D_i(q_j^{(i)}) | j = [1, \ldots, J]\right\}\right)$ indicate the i-th instance of our training data, the batch loss used to train the network is:

$$\mathcal{L}_B(\theta) = \frac{1}{B} \sum_{i=1}^{B} \sum_{j=1}^{J} \mathcal{L}\left(\text{def}_\theta\left(q_j^{(i)} | P_i\right), D_i\left(q_j^{(i)}\right)\right), \tag{1}$$

where θ indicates the network's trainable parameters and B is batch size. For simplicity and robustness, we use mean absolute error as the per-point loss $\mathcal{L}$ and train the network with the Adam [17] optimizer.

2.3 Modifying the Atlas

After the deformation network is trained, it infers the tumor-induced deformation from a tumor segment. The initial tumor segment and the inferred deformation provide strong clues to help whole-brain segmentation. Like seeded-atlas approaches [4,11,23], we incorporate the tumor shape information into the population-based healthy atlas to reflect the structural changes of a specific patient. The probabilistic atlas used in SAMSEG [1,27] is a tetrahedron mesh where each vertex v_i is associated with a vector α_i that represents the prior of each anatomical label. As shown in Fig. 2, the atlas is modified in two aspects: geometric and probabilistic. First, the atlas A is deformed (by interpolation) so that $A \circ \phi_{DISP/SVF}$ appears the same as if the vertices are transformed by the predicted deformation $\phi_{DISP/SVF} : v_i \mapsto v_i'$ (directly transforming the vertices' coordinates results in bad triangles and thus break the mesh's local deformation property). And second, we append the tumor label to the prior's vector. In the case that transformed vertices v_i' fall in the tumor segment T_{TC} or T_{ED} (indicated with green and yellow in Fig. 2), a "tumorous" probability is appended, and the original label priors are equally rescaled to keep their sum as 1, i.e., $\alpha_i = [(1 - \alpha_T) \alpha_i', \alpha_T]^T$. The hyper-parameters α_{TC} and α_{ED} could be determined based on how we trust the initial tumor segmentation. In our implementation, they are tentatively set as 0.9 and 0.5 so that edema segments can be further refined.

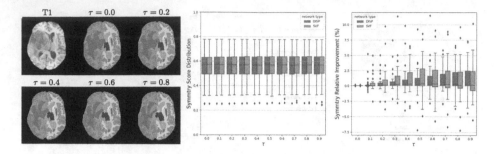

Fig. 3. The measurement of left-right symmetry when using different τ

3   Experiments

Validation of medical image segmentation with comparable metrics is a difficult task because of the lack of reliable labelled data [22,26]. In this study, though, it's imaginable that even if experts are recruited to produce such ground truth, their manual segments might show significant variations and disagree for tumor-bearing scans. We solve this problem with novel evaluation methods. First, we validate the network by showing that the degree of brain inter-hemisphere symmetry is improved when transforming the tumor-bearing images with $\phi_{DISP/SVF}^{-1}$, the inverse of the predicted tumor-growth (Sect. 3.1). Second, we synthesize pathological ground-truth by fusing well-labelled health brain images (MindBoggle-101 [18]) and tumor scans (BraTS [22]), and conducting quantitative evaluations against these (Sect. 3.2). In addition, a regular segmentation test is performed with real tumor images for subjective evaluation (Sect. 3.3).

3.1   Inter-Hemisphere Symmetry Validation

In this proof-of-concept experiment, we naively assume that a healthy brain's left and right cerebral hemispheres are roughly symmetric, and the growth of a tumor tends to break the symmetry. Therefore the degree-of-symmetry of a pathological brain can be increased by inversing the tumor-induced deformation. We randomly take 47 multi-model scans $\{\mathcal{M}_i\}_{i=1,2,\cdots,47}$ from the BraTS2020 training set and produce their deformation ϕ_i. Their segmentations $\mathcal{S}_i$ are produced by SAMSEG using an atlas that has no specific tumor information, i.e., uniform tumor prior over the brain area[2]. A mirrored segmentation $\mathcal{S}_i^*$ is created by flipping $\mathcal{S}_i$ and swapping the left-right labels. We use the weighted mean Dice score [9] of each segment as the symmetry measurement: $\mathrm{Dice}\,(\mathcal{S}_i^*, \mathcal{S}_i)$. Similarly, we compute $\mathrm{Dice}\left(\left(\phi_i^{-1} \circ \mathcal{S}_i\right)^*, \phi_i^{-1} \circ \mathcal{S}_i\right)$ as the symmetry score of when tumor-induced deformation is eliminated by inverse transformation. We run tests with

---

[2] This atlas is generated from the SAMSEG atlas comes with FreeSurfer (https:// surfer.nmr.mgh.harvard.edu). We remove the non-brain labels like skull and optic-chiasm, and re-distribute the label priors because BraTS data are skull-stripped.

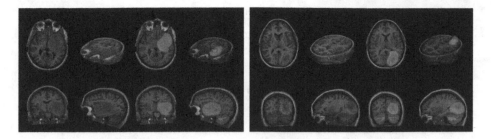

Fig. 4. Examples of the synthesized test images.

different scale factors $0 \leq \tau < 1$. The score's distribution and its relative improvement are shown in Fig. 3. Both DISP or SVF deformations have a clear positive effect on the symmetry comparing with the original brain structure ($\tau = 0$). Meanwhile, we notice that the original brain symmetry score has a large deviation. It is worth mentioning that some tissues are not precisely symmetric, and the segmentation is not 100% accurate. The improvement of symmetry shown in Fig. 3 is a sign that some useful information about the brain geometric change is captured by the deformation network.

3.2   Synthesized Images Evaluation

In Mindboggle-101 [18], OASIS-TRT-20 is a subset of 20 health subjects that the subcortical volumes are manually labelled. We use these images to synthesize pathological images by deforming and smoothly fusing tumor image patches (from BraTS) into them (see Fig. 4). The high-quality labels are deformed accordingly as the ground-truth. Details of this data synthesis are sketched in the supplementary materials. Our DISP and SVF models are tested with a fixed scale factor $\tau = 0.5$, and we compare the performances with SAMSEG and a no-deformation baseline ($\tau = 0.0$) that the atlas is only modified with the pre-segmented tumor (see the table below). Our method, including the baseline, outperforms plain SAMSEG. We speculate that the GMM estimation in SAM-SEG benefits from the accurate tumor prior. Compared with the baseline, we can see the tumor side segmentation, especially the subcortical tissues (indicated as *sub*), is further improved with the learned deformation.

3.3   Test with Real Data

Lastly, we test our method with the real patient scans found in BraTS. The result is in accordance with previous experiments and shows that modifying the atlas with our method has a visible positive effect on the segmentation. Figure 5 show four comparisons ($\tau = 0.7$). The subcortical brain structures, especially tissues that are close or adjacent to the tumor or edema, tend to be better segmented. Note the segmentation outputs are different in DISP/SVF setups. Different τ also plays a role in segmentation in the area close to the lesion

Table 1. The synthesized test set segmentation accuracy (weighted-mean-Dice).

	Tumor side			Non-Tumor side			All sides		
	Sub	Cortex	All	Sub	Cortex	All	Sub	Cortex	All
SAMSEG	0.704	0.794	0.783	0.727	0.814	0.804	0.716	0.805	0.794
Our(τ=0)	0.722	0.827	0.815	**0.789**	0.845	0.838	0.760	0.836	0.827
Our-DISP	**0.741**	**0.833**	**0.821**	0.777	**0.848**	**0.839**	0.760	**0.841**	**0.831**
Our-SVF	0.736	0.829	0.818	0.780	0.847	0.838	**0.760**	0.838	0.828

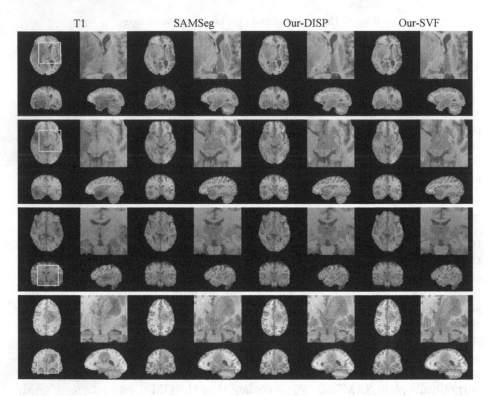

Fig. 5. Results of testing our method in BraTS dataset, compared with SAMSEG. Note, this experiment is not a strict *test* because nnU-Net is trained with the same BraTS dataset. So the initial tumor label is quite accurate and could be instead treated as if it came from a high-quality manual segmentation.

(cf. Sect. 3.2). Therefore, an advocated workflow of using the proposed method in a medical scenario would be to run a batch of segmentations in a tumor-bearing scan with different τ. Experts can pick one to work with or make a further refinement.

4   Conclusion

We propose a novel method that directly estimates tumor-induced deformation (inverse problem of the tumor growth) from one single MR scan. This way, the highly entangled image-to-atlas registration and tumor-growth calibration problems are avoided. Acting as an intermediate process, this learned mass-effect deformation improves anatomical/semantic segmentation of tumor-bearing brain images. Moreover, this framework is not specific to any particular type of tumor or segmentation methods, and offers the flexibilty to work with any existing atlas-based generative approaches.

References

1. Agn, M.: A modality-adaptive method for segmenting brain tumors and organs-at-risk in radiation therapy planning. Med. Image Anal. **54**, 220–237 (2019)
2. Atzmon, M., Maron, H., Lipman, Y.: Point convolutional neural networks by extension operators. arXiv preprint arXiv:1803.10091 (2018)
3. Balakrishnan, G., Zhao, A., Sabuncu, M.R., Guttag, J., Dalca, A.V.: Voxelmorph: a learning framework for deformable medical image registration. IEEE Trans. Med. Imag. **38**(8), 1788–1800 (2019)
4. Bauer, S., Seiler, C., Bardyn, T., Buechler, P., Reyes, M.: Atlas-based segmentation of brain tumor images using a markov random field-based tumor growth model and non-rigid registration. In: 2010 Annual International Conference of the IEEE Engineering in Medicine and Biology, pp. 4080–4083, IEEE (2010)
5. Bauer, S., Wiest, R., Nolte, L.P., Reyes, M.: A survey of MRI-based medical image analysis for brain tumor studies. Phys. Med. & Biol. **58**(13), R97 (2013)
6. Cocosco, C.A., Kollokian, V., Kwan, R.K.S., Pike, G.B., Evans, A.C.: Brainweb: Online interface to a 3D MRI simulated brain database. In: NeuroImage. Citeseer (1997)
7. Cuadra, M.B., Pollo, C., Bardera, A., Cuisenaire, O., Villemure, J.G., Thiran, J.P.: Atlas-based segmentation of pathological MR brain images using a model of lesion growth. IEEE Trans. Med. Imag. **23**(10), 1301–1314 (2004)
8. Dalca, A.V., Balakrishnan, G., Guttag, J., Sabuncu, M.R.: Unsupervised learning of probabilistic diffeomorphic registration for images and surfaces. Med. Image Anal. **57**, 226–236 (2019)
9. Dice, L.R.: Measures of the amount of ecologic association between species. Ecology **26**(3), 297 302 (1945)
10. Gholami, A.: A novel domain adaptation framework for medical image segmentation. In: International MICCAI Brainlesion Workshop, pp. 289–298, Springer (2018)
11. Gooya, A., Pohl, K.M., Bilello, M., Cirillo, L., Biros, G., Melhem, E.R., Davatzikos, C.: GLISTR: glioma image segmentation and registration. IEEE Trans. Med. Imaging **31**(10), 1941–1954 (2012)
12. Harpold, H.L., Alvord, E.C., Jr., Swanson, K.R.: The evolution of mathematical modeling of glioma proliferation and invasion. J. Neuropathol. Exp. Neurol. **66**(1), 1–9 (2007)
13. Isensee, F., Jaeger, P.F., Kohl, S.A., Petersen, J., Maier-Hein, K.H.: nnU-Net: a self-configuring method for deep learning-based biomedical image segmentation. Nat. Methods **18**(2), 203–211 (2021)

14. Isensee, F., Jäger, P.F., Full, P.M., Vollmuth, P., Maier-Hein, K.H.: nnU-Net for brain tumor segmentation. In: International MICCAI Brainlesion Workshop, pp. 118–132, Springer (2020)

15. Jia, M., Kyan, M.: Learning occupancy function from point clouds for surface reconstruction. arXiv preprint arXiv:2010.11378 (2020)

16. Jia, M., Kyan, M.: Improving intraoperative liver registration in image-guided surgery with learning-based reconstruction. In: ICASSP 2021–2021 IEEE International Conference on Acoustics, Speech and Signal Processing (ICASSP), pp. 1230–1234, IEEE (2021)

17. Kingma, D.P., Ba, J.: Adam: A method for stochastic optimization. arXiv preprint arXiv:1412.6980 (2014)

18. Klein, A., Tourville, J.: 101 labeled brain images and a consistent human cortical labeling protocol. Front. Neurosci. **6**, 171 (2012)

19. Kyriacou, S.K., Davatzikos, C., Zinreich, S.J., Bryan, R.N.: Nonlinear elastic registration of brain images with tumor pathology using a biomechanical model [MRI]. IEEE Trans. Med. Imaging **18**(7), 580–592 (1999)

20. Lipková, J., et al.: Personalized radiotherapy design for glioblastoma: integrating mathematical tumor models, multimodal scans, and Bayesian inference. IEEE Trans. Med. Imaging **38**(8), 1875–1884 (2019)

21. Lorensen, W.E., Cline, H.E.: Marching cubes: A high resolution 3D surface construction algorithm. In: ACM SIGGRAPH computer graphics, vol. 21, pp. 163–169, ACM (1987)

22. Menze, B.H., Jakab, A., Bauer, S., Kalpathy-Cramer, J., Farahani, K., Kirby, J., Burren, Y., Porz, N., Slotboom, J., Wiest, R., et al.: The multimodal brain tumor image segmentation benchmark (BRATS). IEEE Trans. Med. Imaging **34**(10), 1993–2024 (2014)

23. Mohamed, A., Zacharaki, E.I., Shen, D., Davatzikos, C.: Deformable registration of brain tumor images via a statistical model of tumor-induced deformation. Med. Image Anal. **10**(5), 752–763 (2006)

24. Pálsson, S., Cerri, S., Poulsen, H.S., Urup, T., Law, I., Van Leemput, K.: Predicting survival of glioblastoma from automatic whole-brain and tumor segmentation of mr images. arXiv preprint arXiv:2109.12334 (2021)

25. Pollo, C., Cuadra, M.B., Cuisenaire, O., Villemure, J.G., Thiran, J.P.: Segmentation of brain structures in presence of a space-occupying lesion. Neuroimage **24**(4), 990–996 (2005)

26. Prastawa, M., Bullitt, E., Gerig, G.: Simulation of brain tumors in MR images for evaluation of segmentation efficacy. Med. Image Anal. **13**(2), 297–311 (2009)

27. Puonti, O., Iglesias, J.E., Van Leemput, K.: Fast and sequence-adaptive whole-brain segmentation using parametric Bayesian modeling. Neuroimage **143**, 235–249 (2016)

28. Rehman, M.U., Cho, S., Kim, J., Chong, K.T.: Brainseg-Net: Brain tumor MR image segmentation via enhanced encoder-decoder network. Diagnostics **11**(2), 169 (2021)

29. Scheufele, K., Mang, A., Gholami, A., Davatzikos, C., Biros, G., Mehl, M.: Coupling brain-tumor biophysical models and diffeomorphic image registration. Comput. Methods Appl. Mech. Eng. **347**, 533–567 (2019)

30. Scheufele, K., Subramanian, S., Biros, G.: Fully automatic calibration of tumor-growth models using a single mpMRI scan. IEEE Trans. Med. Imaging **40**(1), 193–204 (2020)

31. Sederevičius, D., et al.: Reliability and sensitivity of two whole-brain segmentation approaches included in freesurfer-ASEG and SAMSEG. Neuroimage **237**, 118113 (2021)
32. Wang, G., Li, W., Ourselin, S., Vercauteren, T.: Automatic brain tumor segmentation based on cascaded convolutional neural networks with uncertainty estimation. Front. Comput. Neurosci. **13**, 56 (2019)

Contrastive Re-localization and History Distillation in Federated CMR Segmentation

Xiaoming Qi[1], Guanyu Yang[1,3]([✉]), Yuting He[1], Wangyan Liu[2], Ali Islam[5], and Shuo Li[4]

[1] LIST, Key Laboratory of Computer Network and Information Integration (Southeast University), Ministry of Education, Nanjing 210096, China
yang.list@seu.edu.cn
[2] Department of Radiology, The First Affiliated Hospital of Nanjing Medical University, Nanjing, China
[3] Jiangsu Provincial Joint International Research Laboratory of Medical Information Processing, Southeast University, Nanjing 210096, China
[4] Department of Medical Biophysics, University of Western Ontario, London, ON, Canada
[5] St. Joseph's Health Care London, London, ON, Canada

Abstract. Federated learning (FL) has shown value in multi-center multi-sequence cardiac magnetic resonance (CMR) segmentation, due to imbalanced CMR distributions and privacy preservation in clinical practice. However, the larger heterogeneity among multi-center multi-sequence CMR brings challenges to the FL framework: (1) Representation bias in the model fusion. The FL server model, which is generated by an average fusion of heterogeneous client models, is biased to representation close to the mean distribution and away from the long-distance distribution. Hence, the FL has poor representation ability. (2) Optimization stop in the model replacing. The heterogeneous server model replacing client model in FL directly causes the long-distance clients to utilize worse optimization to replace the original optimization. The client has to recover the optimization with the worse initialization, hence it lacks the continuous optimization ability. In this work, a cross-center cross-sequence medical image segmentation FL framework (FedCRLD) is proposed for the first time to facilitate multi-center multi-sequence CMR segmentation. (1) The contrastive re-localization module (CRL) of FedCRLD enables the correct representation from the heterogeneous model by embedding a novel contrastive difference metric of mutual information into a cross-attention localization transformer to transfer client-correlated knowledge from server model without bias. (2) The momentum distillation strategy (MD) of FedCRLD enables continuous optimization by conducting self-training on a dynamically updated client momentum bank to refine optimization by local correct optimization history. FedCRLD is validated on 420 CMR images 6 clients from 2 public datasets scanned by different hospitals, devices and

Supplementary Information The online version contains supplementary material available at https://doi.org/10.1007/978-3-031-16443-9_25.

contrast agents. Our FedCRLD achieves superior performance on multi-center multi-sequence CMR segmentation (average dice 85.96%). https://github.com/JerryQseu/FedCRLD.

Keywords: Federated learning · Contractive learning · Momentum distillation

1   Introduction

Federated learning (FL) [10] shows significant value on multi-center multi-sequence cardiac magnetic resonance (CMR) segmentation. With the growing awareness of privacy preservation, the medical data, which exists in isolated medical centers and reveals patients' situations and medical diagnosis information [1], become more sensitive. The medical data has been protected by the Health Insurance Portability and Accountability Act (HIPPA) [8] and EU General Data Protection Regulation (GDPR) [15]. In clinical practice, multi-sequence CMR is required in different cardiovascular diseases, such as cine-CMR for viable tissues assessment, delayed enhancement-CMR (DE-CMR) for myocardial infarction evaluation, and T2 weighted-CMR (T2-CMR) for no-flow regions evaluation [7,9]. This directly caused imbalanced sequence distributions in different hospitals. Hence, the limitations of imbalanced CMR data distributions in different centers and privacy preservation become extremely urgent in multi-center multi-sequence CMR segmentation [3]. FL, which takes multiple models fusion to achieve the model jointly trained on numerous CMR sequences from different hospitals [14,17] and protects privacy [18,19], is able to handle the imbalanced CMR sequence segmentation with privacy preservation.

Although the existing FL approaches [4,11–14] have success on inter-client tasks, the larger heterogeneity of multi-center multi-sequence CMR makes the segmentation still a challenging task. The different scanning devices and contrast agents in the clinical diagnosis requirements cause the multi-center multi-sequence CMR to have larger heterogeneity than regular studies in FL (Fig. 1). This directly brings challenges to FL multi-center multi-sequence CMR segmentation: (1) Representation bias in the model fusion. The FL server model, which is generated by an average fusion of heterogeneous client models, is biased to representation close to the mean distribution and away from the long-distance distribution (Fig. 1 (1)). Hence, the FL framework has poor representation ability for all multi-center multi-sequence CMR. (2) Optimization stop in the model replacing. The regular FL application can benefit from a fused model replacing, due to the similar client distributions. However, the heterogeneous server model replacing client model in FL directly causes the long-distance clients to utilize worse optimization to replace the original optimization (Fig. 1 (2)). The client have to recover the optimization with the worse initialization, hence it lacks the continuous optimization ability.

To overcome these challenges, we propose a novel FL framework, contrastive re-localization and history distillation in federated CMR segmentation (FedCRLD),

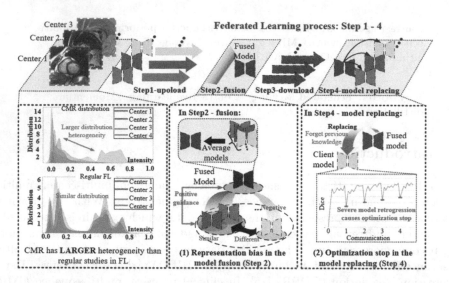

Fig. 1. The CMR has larger heterogeneity than regular FL studies. The large heterogeneity brings challenges: (1) Representation bias in the model fusion. The average on client models with large heterogeneity makes the FL server show negative representation on partial heterogeneous clients. (2) Optimization stop in the model replacing. The server model with heterogeneous guidance replaces the client correct local model for following training. The client optimization suffers model retrogression and has no model refinement in this round of training.

to greatly facilitate multi-center multi-sequence CMR segmentation while preserving privacy. (1) Our FedCRLD enables cross-center cross-sequence medical segmentation possible, by re-localizing server representation to client and distilling history to facilitate CMR segmentation. This framework in line with the scenarios of real-world medical FL will make multi-center multi-sequence CMR joint diagnosis. (2) Our contrastive re-localization module (CRL) establishes a novel contrastive difference metric of mutual information with a cross-attention localization transformer to transfer client-correlated knowledge without negative bias. Hence, CRL enables the correct representation in the model fusion by just transferring only client-correlated representation from the server model for enhancement. (3) Our momentum distillation strategy (MD) builds self-training by distilling the client history momentum version as additional optimization guidance on a dynamically updated momentum bank. It enables continuous optimization in the model fusion by summarizing the client's correct optimization history as one of the guidance for the following optimization.

2   Methodology

Our FedCRLD provides a novel way to achieve multi-center multi-sequence CMR segmentation (Fig. 2). It consists of: (1) CRL introduces a contrastive metric in

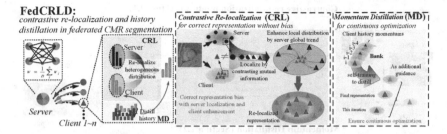

Fig. 2. FedCRLD corrects the representation bias by embedding a contrastive difference metric into cross-attention localization. FedCRLD ensures continuous optimization by self-training on history momentums.

cross-attention representation to correct representation bias and benefit from server (Sect. 2.1). (2) MD distills history momentums for self-training to ensure continual optimization (Sect. 2.2).

The CRL and MD are conducted at the same time during client E epochs training with private data. Then each client uploads the feature perception model to the server for fusion and downloads the server model for the next iteration. The whole process repeats T rounds. In the server model fusion, FedCRLD takes the average model of K client feature perception modules.

2.1   CRL for Representation Bias Correction:

CRL enables the correct representation enhancement from the heterogeneous CMR sequence. It takes two steps to correct the bias: localizing server distribution by mutual information and correlating positive server-client representation by a cross-attention transformer.

Localizing Server Distribution. To localize the server distribution, CRL maximizes mutual information between server distribution D_{server} and client distribution d_{client} by contrastive difference metric (Fig. 2). The mutual information is in the form of Kullback-Leibler (KL) divergence, which is represented as: $L_m = (\sum D_{server} log(\frac{D_{server}}{d_{client}}) + \sum d_{client} log(\frac{d_{client}}{D_{server}}))/2$.

Due to the $KL(p||q) \neq KL(q||p)$, CRL sets up two-way on D_{server} and d_{client}. Through minimizing KL divergence with weight α, the distance between D_{server} and d_{client} is minimized. The representation of server are matched to d_{server} during training. The αL_m is utilized as the term in the loss of client training. Through the maximizing of mutual information, the heterogeneity caused by different centers could be corrected.

Correlating Positive Server-Client Representation. To facilitate client distribution by server distribution, CRL proposes a cross-attention transformer between the localized server distribution (X_g) and client distribution (X_l). The cross-attention transformer consists of two parts (Fig 3): (1) **Self attention** [16] on client (X_l) for further representation. With the X_l (size b, c, d, w, h) as

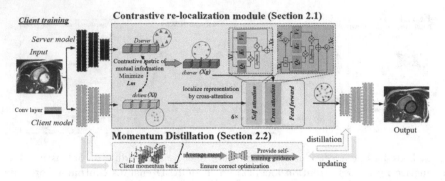

Fig. 3. CRL embeds contrastive difference metric of mutual information into cross-attention representation localization to correct representation bias. MD conducts self-training by history momentum to ensure continual optimization.

the input of self attention, the Q_s, K_s and V_s convert the distribution to $1 \times (c \times d \times w \times h)$ and $(c \times d \times w \times h) \times 1$. Then the further global correlation of X_l is perceived as (X_s). The whole process is represented as follows: $X_s = X_l + softmax(\frac{Q_s K_s^T}{\sqrt{d_k}})V_s$. (2) **Cross attention** takes X_s and X_g as input to transfer the generalization to X_s. The X_s and X_g are concatenated as the input of K_c and V_c. The Q_c, K_c and V_s convert the distribution to $1 \times (c \times d \times w \times h)$ and $(c \times d \times w \times h) \times 1$. Then the global correlation between X_s and X_g is perceived and concatenated to X_s as the final distribution X_c. Finally, the feed forward operation is performed on the X_c. The process is represented as: $X_c = Concat(X_s, f(X_s) + softmax(\frac{Q_c K_c^T}{\sqrt{d_k}})V_c)$

The cross attention transformer has 6 heads. Through the cross attention transformer, the correlation between each position of client distribution and the global position localized server distribution is utilized to improve the client representation. Due to the heterogeneity caused by different sequences, the localized server distribution could not ensure the correspondence of different tissues in different sequence. The cross attention transformer takes local distributions of different tissues as a base and searches the corresponding tissue distributions from the server model for improving local distribution. The heterogeneity caused by different sequences is corrected.

2.2   MD for Continuous Optimization of Each Center

MD strategy ensures continuous optimization on dynamically updated client models. MD builds a self-training in each client by distilling history momentums as one the guidance. The local optimization of each client is correct and trends to convergence. First, MD updates the correct history momentum into the momentum bank. Second, MD distills the momentum bank for continuous optimization guidance (Fig 2 MD).

The history momentum model of each client is a continuously-evolving teacher which consists of exponential-moving-average versions of the client model (Fig 3 MD). After client E epochs training with private data, the latest client model (Mod_s) is used to update the history momentums (Mod_h). This can be represented as $Mod_h = (Mod_s + Mod_h)/2$. During client training, the Mod_h not only provides pseudo-targets (Seg_p) as soft constraints but also provides history perception (d_{mom}) to regularize d_{client} directly. For the segmentation result, the dice and cross entropy is used on client model segmentation (Seg_c) and labels (Lab). The Seg_p is the regularization term. $Loss_{seg} = (CE(Seg_c, Lab) + Dice(Seg_c, Lab)) + \beta CE(Seg_c, Seg_p)$. For the distribution d_{client}, the distance between latest distribution and history distribution is calculate to further guide the perception of client model $Loss_{dis} = MSE(d_{client}, d_{mom})$. Finally, the whole constrain of client loss consists of L_m, $Loss_{seg}$, $Loss_{dis}$ $(Loss = Loss_{seg} + Loss_{dis} + \alpha L_m)$. The parameters of α and β are set 1 in FedCRLD.

3   Experiments and Results

Dataset: To validate the performance of FedCRLD on multi-center multi-sequence CMR, we construct a real-world FL benchmark from publicly available datasets (M&M [2] and Emidec [9]). The subjects of M&M are scanned in three different countries (Spain, Germany and Canada) using four different scanner vendors (Siemens, General Electric, Philips and Canon). The DE-MRI of Emidec is scanned by Siemens scanners acquired at the University Hospital of Dijon (France). The details are illustrated in Table 1. Each client is divided into the training, validation and test sets as 7:1:2. The heterogeneity of CMR is obvious (different centers, devices, and contrast agents).

Table 1. The details of multi-domain CMR in 6 clients.

Client	Center	Device	Contrast agents	Num	Category
A(M&M)	1	Siemens	cine-CMR	95	LV+RV+MYO
B(M&M)	2	Philips	cine-CMR	74	LV+RV+MYO
C(M&M)	3	Philips	cine-CMR	51	LV+RV+MYO
D(M&M)	4	GE	cine-CMR	50	LV+RV+MYO
E(M&M)	5	Canon	cine-CMR	50	LV+RV+MYO
F(Emidec)	6	Siemens	DE-CMR	100	LV+MYO+no reflow+myocardial infarction

Implementation: Our FedCRLD takes 3D U-net [5] as the basic of encoder and decoder. In client training, the network is based on Pytorch with the learning rate of 1×10^{-5}. The network is optimized using Adam with a batch size of 1. The communication is conducted after every $E = 5$ in client training until $T = 100$ in total. In each client training, data augmentation (rotation, translation, scale, and mirror) and maximum connected domain are the post-processing. The best model is selected using the hold-out validation set on each client. The output channel of $ClientF$ is set to 5, and the others are 4. All experiments are performed on a single NVIDIA TITAN RTX GPU. Dice [6] similarity coefficient is the segmentation performance evaluation.

Table 2. FedCRLD has superior segmentation ability in real-world multi-center multi-sequence CMR. The average Dice score(%) reveals the effectiveness of existing methods on real-world CMR segmentation.(L_m means contrastive difference metric of CRL, ca means cross-attention transformer of CRL)

| Methods | Fedavg | FedProx | FedBN | PRRF | FedCRLD | | | | Data | Client |
Clients	[12]	[9]	[10]	[3]	$-L_m$	-ca	-MD	Full	sharing	Ind
A	85.84	86.70	86.98	87.04	86.17	86.37	86.61	**88.06**	85.79	85.91
B	85.39	84.41	85.87	86.11	86.66	85.97	86.01	**87.28**	83.86	81.05
C	89.08	88.81	89.58	88.05	89.07	88.51	90.00	**90.88**	86.43	84.11
D	79.77	82.85	81.91	84.65	84.44	85.05	86.10	**86.96**	79.90	81.33
E	84.42	83.66	84.73	83.85	83.41	83.58	84.29	**86.40**	83.22	84.00
F	18.36	27.26	27.49	53.09	73.12	75.11	74.00	**76.15**	32.48	50.41
AVG	73.81	75.62	76.09	80.47	83.81	84.10	84.50	**85.96**	75.28	77.80

Significant Performance (Table 2): Our FedCRLD has superior performance in multi-center multi-sequence CMR segmentation than the state-of-the-art FL methods (FedAvg [13], FedProx [11], FedBN [12] and PRRF [4]). Fed-CRLD achieves the highest Dice score (85.96%, the visual results are shown in Appendix) by the bias correction of CRL and continuous optimization of MD. The FedProx, FedBN and PRRF, which are designed for the non-IID scenario, have better performance than FedAvg. However, the existing FL methods can hardly handle the heterogeneous CMR segmentation (client D79.77%, E83.66%, and F18.36%). This also reveals that our FedCRLD is able to ensure each client optimization with heterogeneous CMR segmentation.

Our FedCRLD is robust and more effective than traditional training strategies (data-sharing and data-private in 3D U-net). (1) In Data sharing, we make the datasets of 6 clients as a whole to train CMR segmentation by 3D U-net (Table 2). This model is optimized with the learning rate of 1×10^{-5} for 500 epochs. The data-sharing strategy makes the models are biased toward similar data(86.43% vs. 79.90%). This reveals that the FL is effective to the unbalanced data distribution in real-world(10.41% up). (2) In Client Ind, six CMR segmentation models are trained by private data respectively. These models, 3D U-net structures, are all trained with the learning rate of 1×10^{-5} for 500 epochs. FedCRLD has better performance than local training(8.160% up). It represents that CMR data of different clients can benefit from mutual training, especially on difficult tasks (DE-CMR segmentation 76.15% vs. 50.41%).

Ablation Study (Table 2): Each part of FedCRLD is effective in multi-center multi-sequence CMR segmentation (Table 2). (1) In $-L_m$, the contrastive difference metric of CRL is removed. This evaluates that the contrastive difference metric of CRL localizes the server effectively(2.15% up). (2) In -ca, the cross-attention transformer of CRL is remove, and localized d_{server} is directly utilized to replace d_{client}. This evaluates that the cross-attention representation enhances the client representation from the server model further eliminates bias and enhances client representation (1.86% up). (3) In -MD, MD is removed. This evaluates that the self-training on history momentum is able to ensure the continuous optimization in client training (1.46% up).

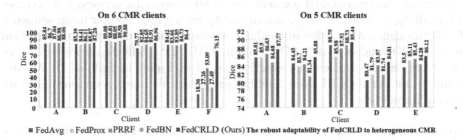

Fig. 4. FedCRLD has robust adaptability to heterogeneous CMR in the real-world. FedCRLD not only achieves the best performance on 6 CMR client testing (especially on heterogeneous client F), but also has superior performance on 5 CMR client testing.

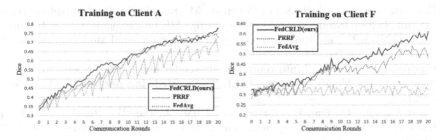

Fig. 5. FedCRLD keeps continuous optimization in FL framework. FedCRLD not only has stable optimization in Client A (close to mean model), but also keeps continuous optimization in Client F (heterogeneous model).

Model Analysis: FedCRLD has a comprehensive representation of heterogeneous CMR. Comparing the performance on cine-CMR (client A, B, C, D, E) with cine-CMR & DE-CMR (Fig. 4), FedCRLD not only keeps robust performance on data heterogeneity of cine-CMR caused by different devices and centers, but also can deal with the heterogeneous CMR caused by different contrast agents (cine-CMR and DE-CMR).

FedCRLD enables continuous optimization. The FedCRLD training processes of client A (cine-CMR) and client F (DE-CMR) indicate that FedCRLD has a stable convergence process than other FL methods (Fig. 5). Due to the FedCRLD being regularized by client history momentums based on the server model, the optimization stop caused by knowledge forgetting will not influence the correct model training.

4   Discussion and Conclusion

We propose a novel FL framework, contrastive re-localization and history distillation in federated CMR segmentation (FedCRLD), to greatly facilitate multi-center multi-sequence CMR segmentation while preserving privacy. It overcomes FL challenges caused by larger heterogeneity among multi-center multi-sequence CMR through: a contrastive re-localization module and a momentum distillation strategy We validate FedCRLD on public CMR data collected from 6 different clients or devices to show the effectiveness of each part of FedCRLD. We also want to use our FedCRLD on MindSpore[1], which is a new deep learning computing framework.

Acknowledgments. This work was supported in part by the CAAI-Huawei Mind-Spore Open Fund, CANN(Compute Architecture for Neural Networks), Ascend AI Processor, and Big Data Computing Center of Southeast University.

References

1. Baumgartner, C.F., Koch, L.M., Pollefeys, M., Konukoglu, E.: An exploration of 2D and 3D deep learning techniques for cardiac MR image segmentation. In: International Workshop on Statistical Atlases and Computational Models of the Heart, pp. 111–119 (2017)
2. Campello, V.M., et al.: Multi-centre, multi-vendor and multi-disease cardiac segmentation: the M&Ms challenge. IEEE Trans. Med. Imaging **40**(12), 3543–3554 (2021)
3. Chen, C., Qin, C., Qiu, H., Tarroni, G., Duan, J., Bai, W., Rueckert, D.: Deep learning for cardiac image segmentation: a review. Front. Cardiovasc. Med. **7**, 25 (2020)
4. Chen, Z., Zhu, M., Yang, C., Yuan, Y.: Personalized retrogress-resilient framework for real-world medical federated learning. In: International Conference on Medical Image Computing and Computer-Assisted Intervention, pp. 347–356 (2021)
5. Çiçek, Ö., Abdulkadir, A., Lienkamp, S.S., Brox, T., Ronneberger, O.: 3D u-Net: learning dense volumetric segmentation from sparse annotation. International conference on medical image computing and computer-assisted intervention, pp. 424–432 (2016)
6. Dice, L.R.: Measures of the amount of ecologic association between species. Ecology **26**(3), 297–302 (1945)
7. Isensee, F., Jaeger, P.F., Full, P.M., Wolf, I., Engelhardt, S., Maier-Hein, K.H.: Automatic cardiac disease assessment on cine-MRI via time-series segmentation and domain specific features. In: International workshop on statistical atlases and computational models of the heart, pp. 120–129 (2017)
8. Kairouz, P., et al.: Advances and open problems in federated learning. Foundations and Trends® in Machine Learning **14**(1–2), 1–210 (2021)
9. Lalande, A., et al.: Emidec: a database usable for the automatic evaluation of myocardial infarction from delayed-enhancement cardiac MRI. Data **5**(4), 89 (2020)

---

[1] https://www.mindspore.cn/.

10. Li, T., Sahu, A.K., Talwalkar, A., Smith, V.: Federated learning: challenges, methods, and future directions. IEEE Signal Process. Mag. **37**(3), 50–60 (2020)
11. Li, T., Sahu, A.K., Zaheer, M., Sanjabi, M., Talwalkar, A., Smith, V.: Federated optimization in heterogeneous networks. Proc. Mach. Learn. Syst. **2**, 429–450 (2020)
12. Li, X., JIANG, M., Zhang, X., Kamp, M., Dou, Q.: FedBN: Federated learning on non-IID features via local batch normalization. In: International Conference on Learning Representations (2020)
13. McMahan, B., Moore, E., Ramage, D., Hampson, S., Arcas, B.A.: Communication-efficient learning of deep networks from decentralized data. Artificial intelligence and statistics, pp. 1273–1282 (2017)
14. Roth, H.R., et al.: Federated whole prostate segmentation in MRI with personalized neural architectures. In: International Conference on Medical Image Computing and Computer-Assisted Intervention, pp. 357–366 (2021)
15. Truong, N., Sun, K., Wang, S., Guitton, F., Guo, Y.: Privacy preservation in federated learning: an insightful survey from the GDPR perspective. Comput. Secur. **110**, 102402 (2021)
16. Vaswani, A., et al.: Attention is all you need. Advances in neural information processing systems, pp. 5998–6008 (2017)
17. Wu, Y., Zeng, D., Wang, Z., Shi, Y., Hu, J.: Federated contrastive learning for volumetric medical image segmentation. In: International Conference on Medical Image Computing and Computer-Assisted Intervention, pp. 367–377 (2021)
18. Xu, J., Glicksberg, B.S., Su, C., Walker, P., Bian, J., Wang, F.: Federated learning for healthcare informatics. J. Healthcare Inform. Res. **5**(1), 1–19 (2021)
19. Yang, D., et al.: Federated semi-supervised learning for COVID region segmentation in chest CT using multi-national data from china, italy, japan. Med. Image Anal. **70**, 101992 (2021)

Contrast-Free Liver Tumor Detection Using Ternary Knowledge Transferred Teacher-Student Deep Reinforcement Learning

Chenchu Xu[1,2], Dong Zhang[3(✉)], Yuhui Song[1], Leonardo Kayat Bittencourt [4], Sree Harsha Tirumani[5], and Shuo Li[6]

[1] Anhui University, Hefei, China
[2] Institute of Artificial Intelligence, Hefei Comprehensive National Science Center, Hefei, China
[3] University of British Columbia, Vancouver, Canada
zhangdong9612@gmail.com
[4] University Hospitals, Case Western Reserve University, Cleveland, USA
[5] UH Cleveland Medical Center, Cleveland, USA
[6] Case Western Reserve University, Cleveland, USA

Abstract. Contrast-free liver tumor detection technology has a significant impact on clinics due to its ability to eliminate contrast agents (CAs) administration in the current tumor diagnosis. In this paper, we proposed a novel ternary knowledge transferred teacher-student DRL (Ts-DRL) as a safe, speedy, and inexpensive contrast-free technology for liver tumor detection. Ts-DRL leverages a teacher network to learn tumor knowledge after CAs administration, and create a pipeline to transfer teacher's knowledge to guide a student network learning of tumor without CAs, thereby realizing contrast-free liver tumor detection. Importantly, Ts-DRL possesses a new ternary knowledge set (actions, rewards, and features of driven actions), which for the first time, allows the teacher network to not only inform the student network what to do, but also teach the student network why to do. Moreover, Ts-DRL possesses a novel progressive hierarchy transferring strategy to progressively adjust the knowledge rationing between teachers and students during training to couple with knowledge smoothly and effectively transferring. Evaluation on 325 patients including different types of tumors from two MR scanners, Ts-DRL significantly improves performance (Dice by at least 7%) when comparing the five most recent state-of-the-art methods. The results proved that our Ts-DRL has greatly promoted the development and deployment of contrast-free liver tumor technology.

Keywords: Contrast-free technology · Deep reinforcement learning · Teacher-student framework

Supplementary Information The online version contains supplementary material available at https://doi.org/10.1007/978-3-031-16443-9_26.

1 Introduction

The contrast-free liver tumor detection technology [1–3] has generated significant clinic impact for diagnosis. Such a novel technology detects liver tumors accurately without contrast agents (CAs) comparable to radiologists' results on contrast-enhanced liver MR images that are clinical "gold standard" based on CA administration (Fig. 1). It avoids the large gap in clinical effect due to CA administration that the tumor is barely visible in non-enhanced liver MR images and the tumor has explicit appearances in contrast-enhanced images [4]. Thus, the technology presents a great future to eliminate the CA administration and prevent CA-associated high-risk [5], non-reproducible, and time-consuming issues [6], thus substantially improving the safety and effectiveness of diagnosis [2].

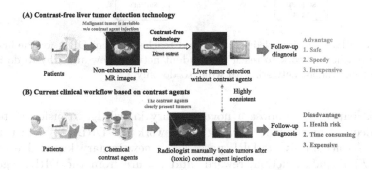

Fig. 1. The contrast-free liver tumor detection technology is a safe, speedy, and inexpensive alternative to clinic contrast agent based liver tumor detection workflow.

The contrast-free liver tumor detection technology's success is because it relies on a powerful framework, i.e., the student-teacher deep reinforcement learning (DRL) framework [1]. This framework leverages DRL to raise the feature space of tumors validly [7] and builds two similar DRL-based networks: a teacher network and a student network learn contrast-enhanced and non-enhanced liver MR images, respectively. Then, the framework builds a pipeline to take learned tumor knowledge (i.e., actions and rewards $\langle \mathcal{A}, \mathcal{R} \rangle$ for DRL agent to detect tumors) from the teacher network to guide the student network detecting the barely-visible tumor in non-enhanced images [1].

However, even the state-of-the-art contrast-free liver tumor detection technology [1] with its teacher-student DRL framework is still suffered in detecting malignant tumors. This issue is because its transferred knowledge is unexplained and often misleads the student network. The transferred knowledge in the existing framework is *binary knowledge $\langle \mathcal{A}, \mathcal{R} \rangle$ from the teacher network that only instructs the student network what to do (i.e., enforce the student network to do the same action as the teacher network), but not teaches the student network why to do (i.e., the cause for this action in this state).* This binary knowledge causes the learning of tumors fully relying on the decision (i.e., actions and resulting rewards) learned from the contrast-enhanced liver images. But when a large margin between the features from the enhanced and non-enhanced images (often occurs in malignant tumors), such knowledge confuses the student network's learning, misleading the tumor knowledge and resulting in a false detection.

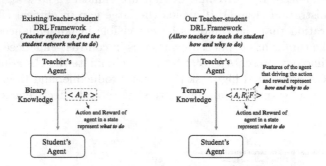

Fig. 2. A novel ternary knowledge transferred teacher-student DRL, for the first time, allows the teacher network to teach the student network how and why to do it, than to enforce feed student network with what to do only.

In this paper, we propose a novel ternary knowledge transferred teacher-student deep reinforcement learning framework (Ts-DRL) to detect tumors without CAs accurately. Ts-DRL possesses a new ternary knowledge set that is $\langle \mathcal{A}, \mathcal{R}, \mathcal{F} \rangle$ including action, reward, and feature from the DRL's agents of teacher network to guide the student network learning. Such ternary knowledge not only leverages $\mathcal{A}, \mathcal{R}$ inform the student network what to do, but also innovatively embeds $\mathcal{F}$ to teach the student the reason and purpose behind the $\mathcal{A}, \mathcal{R}$. Thus, it improves the effectiveness of the teacher-student DRL framework and the accuracy of detection. Moreover, Ts-DRL possesses a novel progressive hierarchy transferring strategy (P-strategy) to strengthen the transfer of ternary knowledge more smoothly and effectively. This strategy progressively adjusts the weight of the ternary knowledge of the teacher network in the student network, and promotes the student network from rigidly relying on the teacher's knowledge to understanding the knowledge. Thus, P-strategy enables ternary knowledge set to be better transferred during training and guarantees the student network that performs better on non-enhanced Liver MR images independently.

The contributions of this paper are: **1)** A novel teacher-student framework-driven DRL approach is proposed to detect liver tumors accurately without CA administration to eliminate CAs-associated health risks and streamline clinical workflows potentially. **2)** A ternary knowledge set is first proposed to improve the performance of DRL-based teacher-student frameworks by allowing teachers to teach the students how and why to do it than to enforce feed students with what to do only. **3)** A p-strategy is creatively proposed to meet the transferring effectiveness and stability of the ternary knowledge by progressively adjusting the knowledge rationing between teachers and students during training. 4) a new state-of-the-art performance in the field of contrast-free liver detection is attained, and design choices and model variants are detailed analyzed.

2   Methodology

Our Ts-DRL consists of two same DRL networks with iterative exploration, and they are a teacher and student networks. Both networks are designed on the soft actor-critic model, including an actor and a critic module. Such a model outputs actions in each exploration by the actor module and independently provides features that correspond to actions by the state value in the critic module. Thus, it best matches our ternary knowledge transferring between teacher and student networks. In the training phase, the teacher network inputs contrast-enhanced liver MR images and builds an agent to self-explore iteratively to find optimal actions and map a bounding box on tumors. Meanwhile, the student network inputs non-enhanced liver MR images, builds an agent, and trains it to map the box on tumors without CAs by introducing the ternary knowledge from the teacher network in each iterative. In the testing phase, the well-trained student network gets rid of the ternary knowledge from the teacher network and independently maps the box on the tumors.

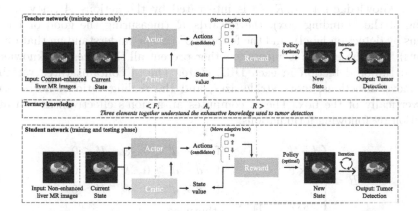

Fig. 3. Ts-DRL consists of two same DRL networks: teacher network and student network. During training, the teacher network learns and transfers the tumor knowledge in contrast-enhanced liver MR images to the student network. The well-trained student network detects tumors accurately and independently without CAs. Each network includes an actor module and a critic module, and they output actions, rewards, and features respectively to build a ternary knowledge set in each iterative.

2.1   Network Architecture

Both teacher and student networks in Ts-DRL are built the same actor and critic modules with ResNet-18 so that they have high consistency to improve the robustness during the ternary knowledge transferring. Furthermore, actor and critic modules enable to cooperate with each other by the actor selecting actions and the critic criticizing the actions to improve the accuracy of actions. The actor module consists of a feature-extractor built with ResNet-18 and a decision-maker built with fully-connected layers. ResNet-18 is deep enough to allow the feature-extractor to collect high-level decision-related features from the input, and its residual blocks also ensure to capture all details. The decision-maker predicts a Gaussian distribution as policy from the collected tumor feature. This policy indicates the possibilities of all candidate actions. The critic module has a similar network architecture as the actor module. The critic module employs four dilated blocks before the feature-extractor (ResNet-18) and predicts state-values with the decision-maker. The four dilated blocks are constructed with convolution layers with kernel sizes [1,3,5,7] and dilated convolution layers with dilated rates [2,4,6,8] to perceive multi-scale low-level features. *Figure see Supplementary Material.*

2.2   Ternary Knowledge Set

Ternary knowledge set $\langle \mathcal{A}, \mathcal{R}, \mathcal{F} \rangle$ is integrated by the actions $\mathcal{A}$ (moving and scaling of the bounding box), the rewards $\mathcal{R}$ (measuring the good or bad of actions by detection label), and the features $\mathcal{F}$ (image understanding that drives actions), and these three elements together present all the necessary knowledge used to tumor detection in each DRL state. The $\mathcal{A}$ contains three variables that move directions, move stride-sizes, and the scaling ratio of horizontally and vertically of the box. The $\mathcal{R}$ computes an action's effectiveness to tumor detection by following formulations:

$$
R = \begin{cases}
d - d', & \text{if} \quad d - d' < 0 \quad \& \quad I' \leq 0 \\
\alpha \times e^{-d'/\beta}, & \text{if} \quad d - d' \geq 0 \quad \& \quad I' \leq 0 \\
0, & \text{if} \quad d - d' < 0 \quad \& \quad I' > 0 \\
e^{(\alpha \times e^{-d'/\beta}) + I'}, & \text{otherwise}
\end{cases} \tag{1}
$$

where d and d' indicate the Euclidean distance of center points between the bounding box and the detection label before and after the action A in a state. I' indicates the overlap rate between the box and the label after the action. α and β are weight hyper-parameters. The $\mathcal{F}$ is a low-level feature of the current state after action taken. Namely, $\mathcal{F} = \{x_i\}_{i=1}^{4}$, where x_i indicates the feature from the i-th dilated block in the critic module to train the main body (feature-extractor and decision-maker).

2.3   P-strategy

P-strategy transfer the $\langle \mathcal{A}, \mathcal{R}, \mathcal{F} \rangle$ in each state to benefit the student network from the teacher network and thus detect liver tumors accurately. For the action $\mathcal{A}$ and the reward $\mathcal{R}$, P-strategy makes the actor module in the student network imitate the teacher network's decision after they observe the bounding boxes' states in the non-enhanced and contrast-enhanced images and distributes the same reward value to the critic module in the student network. Such a direct action imitation and reward distribution as priors and supervision effectively constrain the student network in the right exploration. For the feature $\mathcal{F}$, P-strategy makes the critic module in the student network from fully dependent to completely independent of the low-level features in the teacher network via a progressive-replacing and finally stables the student network. Particularly, assuming the features to train the critic's main-body are $\{X_i\}_{i-1}^4$ in the student network. And the features extracted by the four dilated blocks in teacher network and student network are $\{x_i^t, x_i^s\}_{i=1}^4$ In the beginning, $\{X_i = x_i^t\}_{i=1}^4$, i.e., all the low-level features to train the critic module in the student network are from the teacher network. After warming-up learning, P-strategy introduces one item from the student network's features that is most similar to the teacher network's features into $\{X_i\}_{i=1}^4$. Namely:

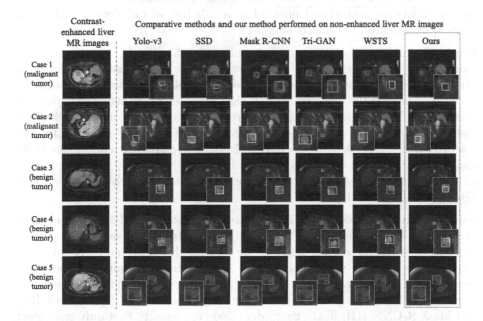

Fig. 4. Our Ts-DRL outperforms all five comparative methods. The green bounding box is the detection results of different methods, and the red bounding box is the detection label. Our Ts-DRL achieves the highest overlap between green and red bounding boxes and has almost no false positives. (Color figure online)

$$X_{i \in [1,4]} \begin{cases} x_i^s, & \text{if } i \text{ is } I \\ x_i^t, & \text{otherwise} \end{cases} \quad where \ \ I = \text{argmin} \left\{ \mathcal{D}(x_j^s, x_j^t) \right\}_{j=1}^4 \tag{2}$$

where I is an index indicates that the student network's feature x_I^s is most similar to the teacher network's feature x_I^t, where $\mathcal{D}(\cdot)$ is the Euclidean distance between two features. As training progresses, P-strategy introduces more features from the student network into $\{X_i\}_{i=1}^4$. After all features are from the student network, i.e., $\{X_i = x_i^s\}_{i=1}^4$, P-strategy cut off all the transferring and make the student network independent from the teacher network.

3   Experiment

3.1   Experimental Setup

Datasets. Ts-DRL is trained and tested on a generalized dataset that consists of 325 liver tumor cases including benign tumors (hemangiomas, 100), malignant tumors (hepatocellular carcinoma, 150), and normal controls (75). Contrast-enhanced and non-enhanced liver MRI images were obtained using 1.5 T MRI scanners (two types: Signa Artist, GE, and Aera, Siemens) using T1-weighted imaging (0.1 mmol/kg dose of gadolinium-based CA). *For more imaging details, please see Supplementary Material.* The tumor detection label is the bounding box of tumors segmentation delineated manually by two radiologists with seven years of experience in liver MR imaging.

Implementation. Ts-DRL randomly selected 4/5 cases for training and the remaining 1/5 for independent testing (patient-wise). Each patient includes an MR stack of images (36–120 slices, the average number of input slices is 62). All codes are based on the PyTorch, and all high-parameters are tuned to obtain the best results. The batch size of experience-replay is 80 (tuned from [40,60,80,100]) and the size of experience pool is 10000. ADAM optimizer [8] and learning rate starts from 1e-4 (0.95 decays, tuned from [1e-3,1e-4,1e-5]). Max-episode is 80,000, and the length of each episode is 35 (tuned from [55, 45, 35, 30, 25]). It required 52 h for training, and 0.54 sec on average for a test image whose size is 256×256 on $4 \times$ Tesla p100 GPUs.

Comparative Methods and Evaluation Metrics. Ts-DRL is compared to five state-of-the-art methods. They are divided into two types: 1) Two recent contrast-free liver detection methods are WSTS ([1] 2021) and Tri-GAN ([3] 2020). Both methods were published in Medical Image Analysis (One of the best journals in medical image analysis) and represent the state-of-the-art in this task. 2) Three standard object detection methods (SSD [9], Yolo-v3 [10], and Mask R-CNN [11]) in the computer vision community that only can detect liver tumors from the non-enhanced liver MR images to meet the tumor detection without CA. Ts-DRL and five comparative methods evaluate their performance by comparing the detection results and the detection label and use four well-recognized metrics [12] that include the dice coefficient, the Recall value, the Precision value, and the Kappa coefficient to measure the comparing.

3.2   Experimental Results

Correct Contrast-Free Liver Tumor Detection. Figure 4 presents that our Ts-DRL enables the correct detection of liver tumors without CAs. The visualization results show that the two types of color bounding boxes (the green bounding box is the detection results, and the red bounding box is the detection label) are highly overlapping, which means the high performance of our Ts-DRL. Note that visualization results are obtained from non-enhanced liver MR images, and tumors may barely be visible in these images (such as case 1) unlike they can be visible in contrast-enhanced liver MR images. Moreover, Ts-DRL obtains a Dice of 69.05%, an Recall of 70.71%, a Precision of 64.63%, and a Kappa of 69.04%, as shown in Table 1. These results proved that our Ts-DRL has great potential to promote the contrast-free liver tumor detection technology applied in clinical and eliminate the contrast-agents injection with its high health risk.

Table 1. Ts-DRL attains a new state-of-the-art performance in the field of contrast-free liver detection (Only b-knowl means Only binary knowledge)

Experiments	Methods	Dice (%)	Recall (%)	Precision (%)	KAP (%)
Comparative methods	Tri-GAN	55.76±24.34	61.26±22.19	59.75±23.87	55.73 ±24.38
	SSD	60.19±22.10	62.21±20.82	61.52±20.64	60.16±22.13
	Yolo-v3	40.95±31.29	58.40±28.77	34.22±29.33	40.92±31.33
	Mask R-CNN	60.54±21.67	67.23±15.74	63.22±18.20	60.52±21.69
	WSTS	62.50±16.49	63.88±15.05	59.71±16.06	62.48±16.50
Ablation experiments	Only student	52.20±22.59	60.58±24.18	58.46±21.02	52.18±22.61
	Only b-knowl	61.82±15.41	63.93±14.73	59.34±16.85	61.80±15.43
	Ts-DRL	**69.05±13.65**	**70.71±12.81**	**64.63±15.00**	**69.04±13.67**

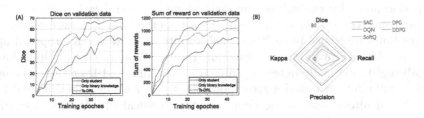

Fig. 5. Superiority of Ts-DRL in ablation experiments

Outperformance of Our Ts-DRL than All Comparative Methods. Figure 4 and Table 1 indicate that Ts-DRL outperforms all five comparative methods in the evaluation of the visualization and all four metrics. In Fig 4, Ts-DRL achieves the highest overlap between green and red bounding boxes, especially has almost no false positives. In Table 1, comparing the results of five comparative methods, our Ts-DRL improved the Dice by 8–29%, the Recall by 3–12%, the Precision by 2–14%, and Kappa by 8–29%. Note that our Ts-DRL also obtains the lowest standard deviation in all metrics, which means our Ts-DRL is better robust than all comparative methods. All the outperformance is because our ternary knowledge set teaches the student network's agent the reason behind the actions and rewards from the teacher's agent, improving the effectiveness and robustness of the student agent.

Superiority of our Components in Ablation Experiments. Figure 5 and Table 1 show that ternary knowledge set with P-strategy is the main reason for Ts-DRL to attain new state-of-the-art performance in this task. Comparing our Ts-DRL and two different versions (1. Only student that only use student network performed on non-enhance liver MR; 2. Only binary knowledge that Ts-DRL removed ternary knowledge set with P-strategy and only used binary knowledge), our Ts-DRL improved a Dice by at least 7%, a Recall by at least 7%, a Precision by at least 15%, and a Kappa by at least 7%. Note that the P-strategy is specially designed to couple with the ternary knowledge set transfer, and it cannot independently perform ablation experiments. Figure 5 proved that the soft actor-critic is the best chosen to realize ternary knowledge set with P-strategy. By comparing different DRL models including DQN [13], SoftQ [14], DPG [15], and DDPG [16], soft actor-critic-based Ts-DRL perform the better Dice (18%), Recall (15%), Precision (13%), and Kappa (18%) than other frameworks at least.

4   Conclusion

In this paper, we proposed a new teacher-student DRL-based contrast-free approach to detect liver tumors accurately. Our approach introduces a novel ternary knowledge set with a P-strategy that is the first to allow the teacher network to teach the student network what to do and why to do, thereby significantly improving the detection performance. Our approach obtains the highest performance by evaluating our method on a generalization dataset (325 patients) and comparing five state-of-the-art methods in four well-recognized metrics (improved Dice by 8–29%). Such results prove that our proposed approach provides accurate tumor detection results comparable to results based on radiologists manually detecting from contrast-enhanced liver tumor images. These results demonstrated a great potential to offer a safe, efficient, and low-cost clinical alternative to the department of medical imaging for a credible liver tumor diagnosis. Thus, our approach presents a great prospect to eliminate Gd-associated health risks, streamline clinical workflows and conserve clinical resources. Our approach will further work in the aspects of robustness by increasing the amount of data and the timeliness by introducing model compression algorithms, thereby increasing clinical impact.

Acknowledgment. This work was supported in part by The National Natural Science Foundation of China (62106001) and The University Synergy Innovation Program of Anhui Province under Grant (GXXT-2021–007).

References

1. Zhang, D., Chen, B., Chong, J., Li, S.: Weakly-supervised teacher-student network for liver tumor segmentation from non-enhanced images. Med. Image Anal. **70**, 102005 (2021)
2. Xu, C., Zhang, D., Chong, J., Chen, B., Li, S.: Synthesis of gadolinium-enhanced liver tumors on nonenhanced liver MR images using pixel-level graph reinforcement learning. Med. Image Anal. **69**, 101976 (2021)
3. Zhao, J., et al.: Tripartite-GAN: synthesizing liver contrast-enhanced MRI to improve tumor detection. Med. Image Anal. **63**, 101667 (2020)
4. Gandhi, S.N., Brown, M.A., Wong, J.G., Aguirre, D.A., Sirlin, C.B.: MR contrast agents for liver imaging: what, when, how. Radiographics **26**(6), 1621–1636 (2006)
5. Schieda, N., et al.: Gadolinium-based contrast agents in kidney disease: a comprehensive review and clinical practice guideline issued by the Canadian association of radiologists. Can. J. Kidney Health Dis. **5**, 2054358118778573 (2018)
6. Xu, C., et al.: Direct delineation of myocardial infarction without contrast agents using a joint motion feature learning architecture. Med. Image Anal. **50**, 82–94 (2018)
7. Zhu, Z., Lin, K., Zhou, J.: Transfer learning in deep reinforcement learning: A survey. arXiv preprint arXiv:2009.07888 (2020)
8. Zhang, Z.: Improved Adam optimizer for deep neural networks. In: 2018 IEEE/ACM 26th International Symposium on Quality of Service (IWQoS), pp. 1–2, IEEE (2018)
9. Liu, W., et al.: SSD: Single Shot MultiBox Detector. In: Leibe, Bastian, Matas, Jiri, Sebe, Nicu, Welling, Max (eds.) ECCV 2016. LNCS, vol. 9905, pp. 21–37. Springer, Cham (2016). https://doi.org/10.1007/978-3-319-46448-0_2
10. Redmon, J., Farhadi, A.: Yolov3: An incremental improvement. arXiv preprint arXiv:1804.02767 (2018)
11. He, K., Gkioxari, G., Dollár, P., Girshick, R.: Mask r-cnn. In: Proceedings of the IEEE international conference on computer vision, pp. 2961–2969 (2017)
12. Zou, K.H., et al.: Statistical validation of image segmentation quality based on a spatial overlap index1: scientific reports. Acad. Radiol. **11**(2), 178–189 (2004)
13. Mnih, V., et al.: Playing atari with deep reinforcement learning. arXiv preprint arXiv:1312.5602 (2013)
14. Haarnoja, T., Tang, H., Abbeel, P., Levine, S.: Reinforcement learning with deep energy-based policies. In: International Conference on Machine Learning, pp. 1352–1361, PMLR (2017)
15. Silver, D., Lever, G., Heess, N., Degris, T., Wierstra, D., Riedmiller, M.: Deterministic policy gradient algorithms. In: International conference on machine learning, pp. 387–395, PMLR (2014)
16. Lillicrap, T.P.: Continuous control with deep reinforcement learning. arXiv preprint arXiv:1509.02971 (2015)

DeepPyramid: Enabling Pyramid View and Deformable Pyramid Reception for Semantic Segmentation in Cataract Surgery Videos

Negin Ghamsarian[1(✉)], Mario Taschwer[2], Raphael Sznitman[1], and Klaus Schoeffmann[2]

[1] Center for AI in Medicine, Faculty of Medicine, University of Bern, Bern, Switzerland
{negin.ghamsarian,raphael.sznitman}@unibe.ch
[2] Department of Information Technology, Alpen-Adria-Universität Klagenfurt, Klagenfurt, Austria
{mt,ks}@itec.aau.at

Abstract. Semantic segmentation in cataract surgery has a wide range of applications contributing to surgical outcome enhancement and clinical risk reduction. However, the varying issues in segmenting the different relevant structures in these surgeries make the designation of a unique network quite challenging. This paper proposes a semantic segmentation network, termed DeepPyramid, that can deal with these challenges using three novelties: (1) a Pyramid View Fusion module which provides a varying-angle global view of the surrounding region centering at each pixel position in the input convolutional feature map; (2) a Deformable Pyramid Reception module which enables a wide deformable receptive field that can adapt to geometric transformations in the object of interest; and (3) a dedicated Pyramid Loss that adaptively supervises multi-scale semantic feature maps. Combined, we show that these modules can effectively boost semantic segmentation performance, especially in the case of transparency, deformability, scalability, and blunt edges in objects. We demonstrate that our approach performs at a state-of-the-art level and outperforms a number of existing methods with a large margin (3.66% overall improvement in intersection over union compared to the best rival approach).

Keywords: Cataract surgery · Semantic segmentation · Surgical data science

1 Introduction

Cataracts are naturally developing opacity that obfuscates sight and is the leading cause of blindness worldwide, with over 100 million people suffering from them. Today,

This work was funded by Haag-Streit Switzerland and the FWF Austrian Science Fund under grant P 31486-N31.

Supplementary Information The online version contains supplementary material available at https://doi.org/10.1007/978-3-031-16443-9_27.

surgery is the most effective way to cure patients by replacing natural eye lenses with artificial ones. More than 10 million cataract surgeries are performed every year, making it one of the most common surgeries globally [19]. With the aging world population growing, the number of patients at risk of complete cataract-caused blindness is sharply increasing [20] and the number of surgeries needed brings unprecedented organizational and logistical challenges.

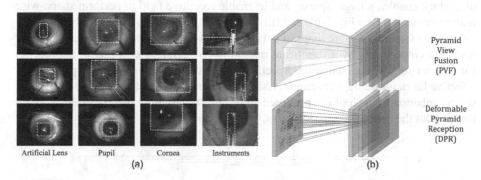

Fig. 1. (a) Challenges in semantic segmentation of relevant objects in cataract surgery. (b) Proposed Pyramid View Fusion and Deformable Pyramid Reception modules.

To help train future surgeons and optimize surgical workflows, automated methods that analyze cataract surgery videos have gained significant traction in the last decade. With the prospect of reducing intra-operative and post-operative complications [5], recent methods have included surgical skill assessment [8,25], remaining surgical time estimation [13], irregularity detection [7] or relevance-based compression [6]. In addition, a reliable relevant-instance-segmentation approach is often a prerequisite for a majority of these applications [17]. In this regard, four different structures are typically of interest: the intraocular lens, the pupil, the cornea, and surgical instruments. Due to the diversity in the appearance of these structures, segmentation methods must overcome several hurdles to perform well on real-world video sequences. Specifically, a semantic segmentation network is required to simultaneously deal with: 1) a transparent artificial lens that undergoes deformations, 2) color, shape, size, and texture variations in the pupil, 3) unclear edges of the cornea, and 4) severe motion blur, reflection distortion, and scale variations in instruments (see Fig. 1-a). This work looks to provide a method to segment these structures despite the mentioned challenges.

Several network architectures for cataract surgery semantic segmentation have been proposed or have been used in the recent past [1,14–16,23]. Many of these methods have been based on the U-Net architecture [18] and aimed at improving accuracy by addressing different limitations from the original architecture. In [14,16], different attention modules were used to guide the network's computational efforts toward the most discriminative features in the input feature map considering the characteristics of the objects of interest. Additionally, fusion modules have been proposed to improve semantic representation via combining several feature maps [1,23]. However, as we show in our experiments, these methods still have difficulties with the aforementioned challenges in cataract video segmentation.

In this work, we propose a novel architecture that is tailored to adaptively capture semantic information despite the challenges typically found in cataract surgery videos. Our proposed network, *DeepPyramid*[1], introduces three key contributions: (i) a Pyramid View Fusion (PVF) module allowing a varying-angle surrounding view of the feature maps for each pixel position, (ii) a Deformable Pyramid Reception (DPR) module, which enables a large, sparse, and learnable receptive field to perform shape-wise feature extraction (see Fig. 1-b), and (iii) a Pyramid Loss, ($\mathcal{PL}$) to explicitly supervise multi-scale semantic feature maps in our network. We show in the experiments that our approach outperforms by a significant margin twelve rival state-of-the-art approaches for cataract surgery segmentation. Specifically, we show that our model is particularly effective for deformable, transparent, and changing scale objects. In addition, we show the contribution of each of the proposed additions, highlighting that the addition of all three yields the observed improvements.

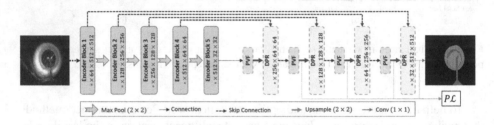

Fig. 2. Overall architecture of the proposed DeepPyramid network. It contains Pyramid View Fusion (PVF), Deformable Pyramid Reception (DPR), and Pyramid Loss ($\mathcal{PL}$).

2   Methodology

Our proposed segmentation strategy aims to explicitly model deformations and context within its architecture. Using a U-Net-based architecture, our proposed model is illustrated in Fig. 2. At its core, the encoder network remains that of a standard VGG16 network. Our approach is to provide useful decoder modules to help alleviate segmentation concerning relevant objects' features in cataract surgery[2]. Specifically, we propose a Pyramid View Fusion (PVF) module and a Deformable Pyramid Reception (DPR) module. These are then trained using a dedicated Pyramid Loss ($\mathcal{PL}$).

Conceptually, the PVF module is inspired by the human visual system and aims to recognize semantic information found in images considering not only the internal

---

[1] The PyTorch implementation of DeepPyramid is publicly available at https://github.com/Negin-Ghamsarian/DeepPyramid_MICCAI2022.

[2] Since changing the encoder network entails pretraining on a large dataset (such as ImageNet), which in turn imposes more computational costs, we only add the proposed modules after the bottleneck. Nevertheless, since these modules are applied to concatenated features coming from the encoder network via skip connections, the encoder features can be effectively guided.

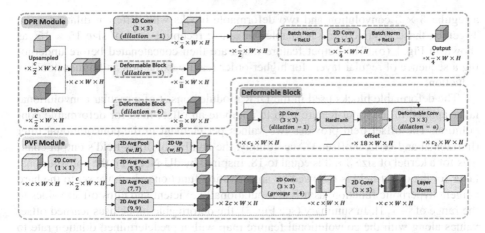

Fig. 3. Detailed architecture of the Deformable Pyramid Reception (DPR) and Pyramid View Fusion (PVF) modules.

object's content but also the relative information between the object and its surrounding area. Thus the role of the PVF is to reinforce the observation of relative information at every distinct pixel position. Specifically, we use average pooling to fuse the multi angle local information for this novel attention mechanism. Conversely, our DPR module hinges on a novel deformable block based on dilated convolutions that can help recognize each pixel position's semantic label based on its cross-dependencies with varying-distance surrounding pixels without imposing additional trainable parameters. Due to the inflexible rectangle shape of the receptive field in regular convolutional layers, the feature extraction procedure cannot be adapted to complex deformable shapes [12]. Our proposed dilated deformable convolutional layers attempt to remedy this explicitly in terms of both scale and shape. We now specify these modules and our loss function in the following subsections.

Pyramid View Fusion (PVF). First, a bottleneck is formed by employing a convolutional layer with a kernel size of one to curb computational complexity. The convolutional feature map is then fed into four parallel branches: a global average pooling layer followed by upsampling and three average pooling layers with progressively larger filter sizes and a common stride of 1. Note that using a one-pixel stride is essential to obtain pixel-wise centralized pyramid views in contrast with region-wise pyramid attention as shown in PSPNet [23]. The output feature maps are then concatenated and fed into a convolutional layer with four groups. This layer is responsible for extracting inter-channel dependencies during dimensionality reduction. A regular convolutional layer is then applied to extract joint intra-channel and inter-channel dependencies before being fed into a layer-normalization function. A summary of this module is illustrated in Fig. 3.

Deformable Pyramid Reception (DPR). As shown in Fig. 3 (top), the fine-grained feature map from the encoder and coarse-grained semantic feature map from the previous layer are first concatenated. These features are then fed into three parallel branches:

a regular 3×3 convolution and two deformable blocks with different dilation rates. Together, these layers cover a learnable but sparse receptive field of size 15×15^3 as shown in Fig. 1 (b). The output feature maps are then concatenated before undergoing a sequence of regular layers for higher-order feature extraction and dimensionality reduction.

The deformable blocks used in the DPR module consist of a regular convolutional layer applied to the input feature map to compute an offset field for deformable convolution. The offset field provides two values per element in the convolutional filter (horizontal and vertical offsets). Accordingly, the number of offset field's output channels for a kernel of size 3×3 is equal to 18. Inspired by dU-Net [22], the convolutional layer for the offset field is followed by an activation function, which we set to the hard tangent hyperbolic function, as it is computationally efficient and clips offset values to the range of $[-1, 1]$. In summary (see Fig. 3), the deformable block uses learned offset values along with the convolutional feature map with a predetermined dilation rate to extract object-adaptive features.

The output feature map (y) for each pixel position (p_o) and the receptive field $(\mathcal{RF})$ for a regular 2D convolution with a 3×3 filter and dilation rate of 1 can be computed by,

$$y(p_o) = \sum_{p_i \in \mathcal{RF}_1} x(p_0 + p_i) w(p_i), \tag{1}$$

where $\mathcal{RF}_1 = \{(-1, -1), (-1, 0), ..., (1, 0), (1, 1)\}$, x denotes the input convolutional feature map, and w refers to the weights of the convolutional kernel. In a dilated 2D convolution with a dilation rate of α, the receptive field can be defined as $\mathcal{RF}_\alpha = \alpha \times \mathcal{RF}_1$. Although the sampling locations in a dilated receptive field have a greater distance to the central pixel, they follow a firm structure. In a deformable dilated convolution with a dilation rate of α, the sampling locations of the receptive field are dependent on the local contextual features. In the proposed deformable block, the sampling location for the ith element of the receptive field and the input pixel p_0 are calculated as,

$$\mathcal{RF}_{def,\alpha}[i, p_0] = \mathcal{RF}_\alpha[i] + f \left(\sum_{p_j \in \mathcal{RF}_1} x(p_0 + p_j) \hat{w}(p_j) \right), \tag{2}$$

where f denotes the activation function, which is the tangent hyperbolic function in our case, and $\hat{w}$ refers to the weights of the offset filter. This learnable receptive field can be adapted to every distinct pixel in the convolutional feature map and allows the convolutional layer to extract stronger informative semantic features when compared to the regular convolution.

Pyramid Loss *(PL)*. To train our network using the PVF and DPR modules, we wish to directly supervise the multi-scale semantic feature maps of the decoder. To enable

---

[3] The structured 3×3 filter covers up to 1 pixel from the central pixel. The deformable filter with *dilation* $= 3$ covers an area of 2 to 4 pixels away from each central pixel. Similarly, the deformable convolution with *dilation* $= 6$ covers an area of 5 to 7 pixels away from each central pixel. Together, these form a sparse filter of size 15×15 pixels.

direct supervision, a depth-wise fully connected layer is formed using a pixel-wise convolution operation. The output feature map presents the semantic segmentation results with the same resolution as the input feature map. To compute the loss for varying-scale outputs, we downscale the ground-truth masks using inter-nearest downsampling for multi-class segmentation and max-pooling for binary segmentation. Our overall loss is then defined as,

$$PL = \mathcal{L}_1 + \alpha\mathcal{L}_2 + \beta\mathcal{L}_4 + \gamma\mathcal{L}_8, \tag{3}$$

where α, β, and γ are predetermined weights in the range of $[0, 1]$ and $\mathcal{L}_i$ denotes the loss of output mask segmentation result with the resolution of $(1/i)$ compared to the input resolution.

3   Experimental Setup

To evaluate the performance of our approach, we make use of data from three datasets. These include the "Cornea" [6] and "Instruments" mask annotations from the CaDIS dataset [9]. In addition, we have collected a separate dataset from which we performed the "Intraocular Lens" and "Pupil" pixel-wise segmentations[4]. The total number of training and test images for the aforementioned objects are 178:84, 3190:459, 141:48, and 141:48, respectively[5]. In the following experiments, all training and test images were split patient-wise to ensure realistic conditions.

We compare the performance of DeepPyramid with thirteen different state-of-the-art segmentation approaches including UNet++ and UNet++/DS [24], CPFNet [4], BARNet [15], PAANet [14], dU-Net[6] [22], MultiResUNet [11], CE-Net [10], RAUNet [16], FED-Net [2] UPerNet [21], PSPNet[7] [23], and U-Net [18][8]. With the exception of the U-Net, MultiResUNet, and dU-Net, which do not use a pretrained backbone, the weights of the backbone for all networks were initialized with ImageNet [3] training weights. The input size of all models is set to $3 \times 512 \times 512$.

For all methods, training is performed using data augmentation. Transformations considered the inherent and statistical features of datasets. For instance, we use motion blur transformation to encourage the network to deal with harsh motion blur regularly occurring in cataract surgery videos. We further use brightness and contrast, shift and scale, and rotate augmentation.

Due to the different depth and connections of the proposed and rival approaches, all networks are trained with three different initial learning rates ($lr \in \{5, 2, 10\} \times 10^{-4}$), and the results with the highest IoU for each network are listed. The learning

---

[4] The customized datasets is publicly released in https://ftp.itec.aau.at/datasets/ovid/DeepPyram/.

[5] Our evaluations are based on binary segmentation per relevant object so that we do not have the imbalance problem. In the case of multi-class classification, methods such as oversampling can mitigate the imbalance problem [17].

[6] Our version of du-Net has the same number of filter-response maps as the U-Net.

[7] To provide a fair comparison, we adopt our improved version of PSPNet, featuring a decoder designed similarly to U-Net (with four sequences of double-convolution blocks).

[8] BARNet, PAANet, and RAUNet are tailored for instrument segmentation in surgical videos. Other methods are state-of-the-art for medical image segmentation.

rate is scheduled to decrease every two epochs with the factor of 0.8. In all evaluations, the networks are trained end-to-end and for 30 epochs. We use a threshold of 0.1 for gradient clipping during training.

The loss function used during training is a weighted sum of binary cross-entropy (BCE) and the logarithm of the soft Dice coefficient. We set $\alpha = 0.75$, $\beta = 0.5$, and $\gamma = 0.25$ in Eq. (3). Additional information on our experimental section can be found in the supplementary materials.

Table 1. Quantitative comparison between the proposed DeepPyramid and state-of-the-art approaches.

Network	IoU%|Dice%				
	Lens	Pupil	Cornea	Instrument	Mean
U-Net [18]	58.19|67.91	85.51|89.36	79.83|86.20	56.12|67.02	69.91|77.62
PSPNet+ [23]	80.56|88.89	93.23|96.45	88.09|93.55	65.37|76.47	81.81|88.84
UPerNet [21]	77.78|86.93	93.34|96.52	86.62|92.67	68.51|78.68	81.56|88.70
FEDNet [2]	78.12|87.38	93.93|96.85	85.73|92.10	65.13|76.11	80.72|88.11
RAUNet [16]	76.40|85.34	89.26|94.26	85.73|92.10	65.13|76.11	79.13|86.95
CE-Net [10]	68.40|80.43	83.59|90.89	83.47|90.85	61.57|74.64	74.25|84.20
MultiResUNet [11]	60.73|71.62	58.36|66.80	73.10|83.40	55.43|66.07	61.90|71.97
dU-Net [22]	59.83|69.46	71.86|79.53	82.39|90.00	61.36|71.55	68.86|77.63
PAANet [14]	74.92|84.83	90.02|94.59	86.75|92.71	64.47|75.24	79.04|86.74
BARNet [15]	67.33|78.85	91.33|95.32	83.98|91.09	66.72|77.14	77.34|85.60
CPFNet [4]	73.56|83.74	90.27|94.83	87.63|93.28	61.16|73.51	78.18|86.34
UNet++/DS [24]	79.50|87.85	95.28|97.53	86.72|92.57	66.05|75.91	81.88|88.46
UNet++ [24]	81.32|89.34	95.66|97.77	85.08|91.72	70.11|79.56	83.04|89.59
DeepPyramid	**85.61 | 91.98**	**96.56|98.24**	**90.24|94.63**	**74.40|83.30**	**86.70|92.03**

4   Experimental Results

Table 1 compares the performance of all evaluated methods. Accordingly, DeepPyramid, Unet++, and PSPNet+ are the top three segmentation methods in terms of IoU for the relevant objects in cataract surgery videos. However, DeepPyramid shows considerable improvements in segmentation accuracy compared to the second-best approach in each class. Specifically, DeepPyramid achieves more than 4% improvement in lens segmentation (85.61% vs. 81.32%) and more than 4% improvement in instrument segmentation (74.40% vs. 70.11%) compared to UNet++. Similarly, DeepPyramid achieves the highest dice coefficient compared to the evaluated approaches for all classes.

Table 2 validates the effectiveness of the proposed modules in an ablation study while also showing the impact on the different segmentation classes. The PVF module appears to enhance the performance for the cornea and instrument segmentation (2.41% and 2.76% improvement in IoU, respectively). This improvement is most likely due to the ability of the PVF module to provide a global view of varying-size sub-regions centered around each spatial position. Such a global view can reinforce semantic representation in the regions corresponding to blunt edges and reflections. Due to scale variance in instruments, the DPR module boosts the segmentation performance for surgical instruments. The addition of the $P\mathcal{L}$ loss results in the improvement in IoU for all the relevant classes, especially the lens (roughly 2% improvement) and instrument (1.64% improvement) classes. The combination of PVF, DPR, and $P\mathcal{L}$ show a marked 4.58% improvement in instrument segmentation and 4.22% improvement in cornea segmentation (based on IoU%). These modules improve the IoU for the lens and pupil by 2.85% and 1.43%, respectively. Overall, the addition of the different proposed components in DeepPyramid lead to considerable improvements in segmentation performance (3.27% improvement in IoU) when compared to the evaluated baselines.

Table 2. Impact of different modules on DeepPyramid's performance (ablation study).

Modules			Params	IoU%/Dice%				
PVF	DPR	$P\mathcal{L}$		Lens	Pupil	Cornea	Instrument	Overall
✗	✗	✗	22.55 M	82.98/90.44	95.13/97.48	86.02/92.28	69.82/79.05	83.49/89.81
✓	✗	✗	22.99 M	83.73/90.79	96.04/97.95	88.43/93.77	72.58/81.84	85.19/91.09
✗	✓	✗	23.17 M	81.85/89.58	95.32/97.59	86.43/92.55	71.57/80.60	83.79/90.08
✓	✓	✗	23.62 M	83.85/90.89	95.70/97.79	89.36/94.29	72.76/82.00	85.42/91.24
✓	✓	✓	23.62 M	**85.84/91.98**	**96.56/98.24**	**90.24/94.77**	**74.40/83.30**	**86.76/92.07**

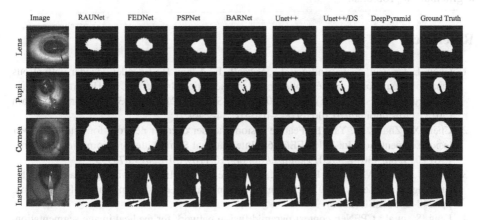

Fig. 4. Qualitative comparisons among DeepPyramid and the six top-performing rival approaches for the relevant objects in cataract surgery videos. The representative images are selected from the test set.

Figure 4 illustrates the qualitative results of our method and evaluated baselines. Specifically, we see the effectiveness DeepPyramid has in segmenting challenging cases. Taking advantage of the pyramid view provided by the PVF module, DeepPyramid can handle reflection and brightness variation in instruments, blunt edges in the cornea, color and texture variation in the pupil, as well as transparency in the lens. Furthermore, powered by deformable pyramid reception, DeepPyramid can tackle scale variations in instruments and blunt edges in the cornea. In particular, we see from Fig. 4 that DeepPyramid shows much less distortion in the region of edges, especially in the case of the cornea. Furthermore, based on these qualitative experiments, DeepPyramid shows much better precision and recall in the narrow regions for segmenting the instruments and other relevant objects in the case of occlusion by the instruments. Further results are shown in the supplementary materials of the paper.

5   Conclusion

In this work, we have proposed a novel network architecture for semantic segmentation in cataract surgery videos. The proposed architecture takes advantage of two modules, namely "Pyramid View Fusion" and "Deformable Pyramid Reception", as well as a dedicated "Pyramid Loss", to simultaneously deal with (i) geometric transformations such as scale variation and deformability, (ii) blur degradation and blunt edges, and (iii) transparency, texture and color variation typically observed in cataract surgery images. We show in our experiments that our approach provides state-of-the-art performances in segmenting key anatomical structures and surgical instruments typical with such surgeries. Beyond this, we demonstrate that our approach outperforms a large number of recent segmentation methods by a considerable margin. The proposed architecture can also be adopted for various other medical image segmentation and general semantic segmentation problems.

References

1. Chen, L.-C., Zhu, Y., Papandreou, G., Schroff, F., Adam, H.: Encoder-decoder with atrous separable convolution for semantic image segmentation. In: Ferrari, V., Hebert, M., Sminchisescu, C., Weiss, Y. (eds.) ECCV 2018. LNCS, vol. 11211, pp. 833–851. Springer, Cham (2018). https://doi.org/10.1007/978-3-030-01234-2_49
2. Chen, X., Zhang, R., Yan, P.: Feature fusion encoder decoder network for automatic liver lesion segmentation. In: 2019 IEEE 16th International Symposium on Biomedical Imaging, ISBI 2019, pp. 430–433 (2019). https://doi.org/10.1109/ISBI.2019.8759555
3. Deng, J., Dong, W., Socher, R., Li, L.J., Li, K., Fei-Fei, L.: ImageNet: a large-scale hierarchical image database. In: 2009 IEEE Conference on Computer Vision and Pattern Recognition, pp. 248–255. IEEE (2009)
4. Feng, S., et al.: CPFNet: context pyramid fusion network for medical image segmentation. IEEE Trans. Med. Imaging 39(10), 3008–3018 (2020). https://doi.org/10.1109/TMI.2020.2983721
5. Ghamsarian, N.: Enabling relevance-based exploration of cataract videos. In: Proceedings of the 2020 International Conference on Multimedia Retrieval, ICMR 2020, pp. 378–382 (2020). https://doi.org/10.1145/3372278.3391937

6. Ghamsarian, N., Amirpourazarian, H., Timmerer, C., Taschwer, M., Schöffmann, K.: Relevance-based compression of cataract surgery videos using convolutional neural networks. In: Proceedings of the 28th ACM International Conference on Multimedia, MM 2020, pp. 3577–3585. Association for Computing Machinery, New York (2020). https://doi.org/10.1145/3394171.3413658

7. Ghamsarian, N., Taschwer, M., Putzgruber-Adamitsch, D., Sarny, S., El-Shabrawi, Y., Schoeffmann, K.: LensID: a CNN-RNN-based framework towards lens irregularity detection in cataract surgery videos. In: de Bruijne, M., et al. (eds.) MICCAI 2021. LNCS, vol. 12908, pp. 76–86. Springer, Cham (2021). https://doi.org/10.1007/978-3-030-87237-3_8

8. Ghamsarian, N., Taschwer, M., Putzgruber-Adamitsch, D., Sarny, S., Schoeffmann, K.: Relevance detection in cataract surgery videos by spatio-temporal action localization (2021)

9. Grammatikopoulou, M., et al.: CaDIS: cataract dataset for image segmentation (2020)

10. Gu, Z., et al.: CE-Net: context encoder network for 2d medical image segmentation. IEEE Trans. Med. Imaging 38(10), 2281–2292 (2019). https://doi.org/10.1109/TMI.2019.2903562

11. Ibtehaz, N., Rahman, M.S.: MultiResUNet: rethinking the U-Net architecture for multimodal biomedical image segmentation. Neural Netw. 121, 74–87 (2020). https://doi.org/10.1016/j.neunet.2019.08.025, https://www.sciencedirect.com/science/article/pii/S0893608019302503

12. Lei, T., Wang, R., Zhang, Y., Wan, Y., Liu, C., Nandi, A.K.: DefED-Net: deformable encoder-decoder network for liver and liver tumor segmentation. IEEE Trans. Radiat. Plasma Med. Sci. 6(1), 68–78 (2022). https://doi.org/10.1109/TRPMS.2021.3059780

13. Marafioti, A., et al.: CataNet: predicting remaining cataract surgery duration. In: de Bruijne, M., et al. (eds.) MICCAI 2021. LNCS, vol. 12904, pp. 426–435. Springer, Cham (2021). https://doi.org/10.1007/978-3-030-87202-1_41

14. Ni, Z.L., et al.: Pyramid attention aggregation network for semantic segmentation of surgical instruments. Proc. AAAI Conf. Artif. Intell. 34(07), 11782–11790 (2020). https://doi.org/10.1609/aaai.v34i07.6850, https://ojs.aaai.org/index.php/AAAI/article/view/6850

15. Ni, Z.L., et al.: BARNet: bilinear attention network with adaptive receptive fields for surgical instrument segmentation. In: Bessiere, C. (ed.) Proceedings of the 29th International Joint Conference on Artificial Intelligence, IJCAI-20, pp. 832–838 (July 2020). https://doi.org/10.24963/ijcai.2020/116

16. Ni, Z.L., et al.: RAUNet: residual attention U-Net for semantic segmentation of cataract surgical instruments. In: Gedeon, T., Wong, K.W., Lee, M. (eds.) Neural Information Processing (2019)

17. Pissas, T., Ravasio, C.S., Da Cruz, L., Bergeles, C.: Effective semantic segmentation in cataract surgery: what matters most? In: de Bruijne, M., et al. (eds.) MICCAI 2021. LNCS, vol. 12904, pp. 509–518. Springer, Cham (2021). https://doi.org/10.1007/978-3-030-87202-1_49

18. Ronneberger, O., Fischer, P., Brox, T.: U-Net: convolutional networks for biomedical image segmentation. In: Navab, N., Hornegger, J., Wells, W.M., Frangi, A.F. (eds.) MICCAI 2015. LNCS, vol. 9351, pp. 234–241. Springer, Cham (2015). https://doi.org/10.1007/978-3-319-24574-4_28

19. Trikha, S., Turnbull, A., Morris, R., Anderson, D., Hossain, P.: The journey to femtosecond laser-assisted cataract surgery: new beginnings or a false dawn? Eye (London, England) 27 (2013). https://doi.org/10.1038/eye.2012.293, https://www.ncbi.nlm.nih.gov/pubmed/23370418

20. Wang, W., Yan, W., Müller, A., He, M.: A global view on output and outcomes of cataract surgery with national indices of socioeconomic development. Invest. Ophthalmol. Vis. Sci. 58, 3669–3676 (2017). https://doi.org/10.1167/iovs.17-21489

21. Xiao, T., Liu, Y., Zhou, B., Jiang, Y., Sun, J.: Unified perceptual parsing for scene under-standing. In: Ferrari, V., Hebert, M., Sminchisescu, C., Weiss, Y. (eds.) ECCV 2018. LNCS, vol. 11209, pp. 432–448. Springer, Cham (2018). https://doi.org/10.1007/978-3-030-01228-1_26

22. Zhang, M., Li, X., Xu, M., Li, Q.: Automated semantic segmentation of red blood cells for sickle cell disease. IEEE J. Biomed. Health Inform. **24**(11), 3095–3102 (2020). https://doi.org/10.1109/JBHI.2020.3000484

23. Zhao, H., Shi, J., Qi, X., Wang, X., Jia, J.: Pyramid scene parsing network. In: Proceedings of the IEEE Conference on Computer Vision and Pattern Recognition (CVPR), July 2017

24. Zhou, Z., Siddiquee, M.M.R., Tajbakhsh, N., Liang, J.: UNet++: redesigning skip connec-tions to exploit multiscale features in image segmentation. IEEE Trans. Med. Imaging **39**(6), 1856–1867 (2020). https://doi.org/10.1109/TMI.2019.2959609

25. Zisimopoulos, O., et al.: DeepPhase: surgical phase recognition in CATARACTS videos. In: Frangi, A.F., Schnabel, J.A., Davatzikos, C., Alberola-López, C., Fichtinger, G. (eds.) MICCAI 2018. LNCS, vol. 11073, pp. 265–272. Springer, Cham (2018). https://doi.org/10.1007/978-3-030-00937-3_31

A Geometry-Constrained Deformable Attention Network for Aortic Segmentation

Weiyuan Lin[1,2], Hui Liu[3], Lin Gu[4,5], and Zhifan Gao[1,2(✉)] ⓘ

[1] School of Biomedical Engineering, Shenzhen Campus of Sun Yat-sen University,
Shenzhen, China
gaozhifan@gmail.com
[2] Sun Yat-sen University, Guangzhou, China
[3] Department of Radiology, Guangdong Provincial People's Hospital, Guangdong
Academy of Medical Sciences, Guangzhou, China
[4] RIKEN AIP, Tokyo, Japan
[5] The University of Tokyo, Tokyo, Japan

Abstract. Morphological segmentation of the aorta is significant for aortic diagnosis, intervention, and prognosis. However, it is difficult for existing methods to achieve the continuity of spatial information and the integrity of morphological extraction, due to the gradually variable and irregular geometry of the aorta in the long-sequence computed tomography (CT). In this paper, we propose a geometry-constrained deformable attention network (GDAN) to learn the aortic common features through interaction with context information of the anatomical space. The deformable attention extractor in our model can adaptively adjust the position and the size of patches to match different shapes of the aorta. The self-attention mechanism is also helpful to explore the long-range dependency in CT sequences and capture more semantic features. The geometry-constrained guider simplifies the morphological representation with a high spatial similarity. The guider imposes strong constraints on geometric boundaries, which changes the sensitivity of gradually variable aortic morphology in the network. Guider can assist the correct extraction of semantic features combining deformable attention extractor. In 204 cases of aortic CT dataset, including 42 normal aorta, 45 coarctation of the aorta, and 107 aortic dissection, our method obtained a mean dice similarity coefficient of 0.943 on the test set (20%), outperforming 6 state-of-the-art methods about aortic segmentation.

Keywords: Aortic segmentation · Computed tomography · Geometric constraint

1 Introduction

Segmentation of the aorta on computed tomography (CT) images is significant for aortic diagnosis, intervention planning and prognosis. Aorta is an indispensable cardiovascular organ, because a diseased aorta with abnormal blood flow

L. Wang et al. (Eds.): MICCAI 2022, LNCS 13435, pp. 287–296, 2022.
https://doi.org/10.1007/978-3-031-16443-9_28

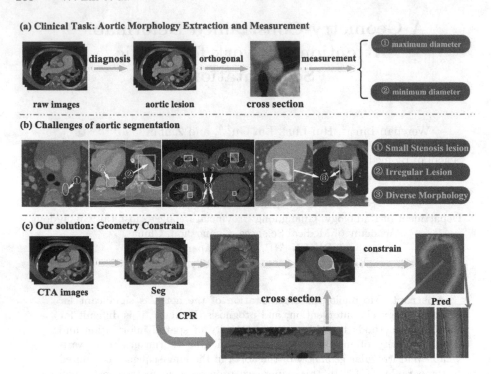

Fig. 1. (a) Aortic segmentation from CT sequences is beneficial to the diagnosis and morphological measurement of clinical aortic disease; (b) Various challenges remain in the aortic segmentation; (c) Our solution: CPR-based geometry-constrained module and deformable self-attention module are designed to guide segmentation.

endangers the patient. According to the Centers for Disease Control and Prevention, on average, aortic diseases account for more than 43,000 deaths annually each year [1]. Diagnosis, intervention and prognosis of aortic diseases rely on morphological information of medical CT images. Clinical guidelines require the accurate acquisition of morphological information for conservative treatment or individualized intervention [1,2], as shown in Fig. 1. Morphological abnormality is the basis for diagnosis on CT images, such as coarctation and tearing [3,4]. The selection of stents in the intervention requires maximum and minimum diameters of the lesion [5]. For prognosis, morphological recovery after treatment is used as predictive factors for curative effects, such as shrinkage of the true and false lumen [6]. Clinicians have to focus on aortic morphology and reconstruct vessels along center points for better measurement, so morphological extraction is beneficial to clinical decision-making. However, traditional CT workstations rely on clinical experience and only support manual or semi-automatic operation [7]. It is time-consuming and laborious.

Challenge in aortic segmentation is the nonlinear relationship between the spatial position and the gradually variable morphological structures in the long-sequence CT images. Although morphological differences between adjacent slices are relatively small, morphological irregularity and aortic complexity will increase gradually with the movement of the scanning position in the CT sequence. For example, the cross-sectional ascending aorta is shaped like an ellipse, while it is irregular in the aortic arch. This phenomenon is caused by inherent physiological characteristics. Moreover, diversity of aortic diseases further aggravates the difficulty of segmentation. For the coarctation of the aorta (CoA), stenosis is small in the whole CT images without obvious features [8]. For the aortic dissection (AD), lesion involves most of descending aorta with irregular beak signs and a small true lumen compressed by false lumen [9]. It is difficult to obtain common representation features in the aorta with different morphology, because the variable anatomy is not conducive to the learning of overall sample distribution and lesion features.

Traditional rule-based methods belong to manual or semi-automatic segmentation, such as multi-scale wavelet analysis. These methods are only suitable for a small number of datasets, and their generalization is poor in the clinic due to parameters readjustment in the face of new data. Deep learning methods learn image features automatically by neural network [10–12]. Multi-view coarse-to-fine methods cause errors accumulation, because they rely on cooperation of each stage [13]. Hypothetical prior algorithms cause false-positive results because of their low sensitivity to local information, when dealing with the complex aorta [14]. Cascade U-net and multi-stage straightening method only consider the correlation between short-range pixels, so the spatial continuity is poor and profile details are lost [15,16]. Long CT sequences and variable geometry are associated with risk of spatial information loss during aortic segmentation. Existing methods are still difficult to solve the challenge of aortic segmentation [17].

This study proposes a geometry-constrained deformable attention network (GDAN) for segmenting the aorta. GDAN consists of a deformable attention extractor and a geometry-constrained guider. Specially, the main purpose of the extractor is to generate variable-size patches to better match the aorta in different structures and enhance original self-attention representation. Self-attention mechanism is beneficial to capture long-range dependency in the long CT sequence. Guider introduces curved planar reformation (CPR) to reorganize voxel data and stretch twisted vessels [18,19], which makes a higher spatial similarity and homogeneity in geometric structure. The main purpose of the guider is to represent the irregular and variable structure as a more regular tubular structure and simplify the nonlinear relationship. Guider learns simplified feature representations to indirectly explore the relationship of the original nonlinear mapping. This process is equivalent to imposing strong constraints on the contour of the aorta. Guider adjust sensitivity of the GDAN to fit the aortic morphology at different locations. Finally, the guider gradually improves the extractor, making it sensitive to gradually variable aortic morphology.

The main contributions in the paper are three-fold: (1) GDAN provides a clinical tool for aortic extraction and diameter measurement. (2) Geometry-constrained guider can simplify and constrain aortic morphology to improve nonlinear mapping in the segmentation. (3) The deformable attention extractor provides a feature extraction mode with patches of adaptive size and position to capture variable aortic structures.

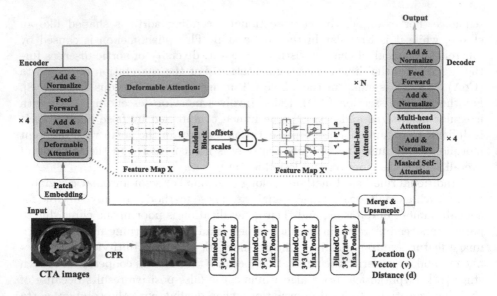

Fig. 2. The architecture of GDAN. Variable patches of the deformable attention extractor captures aortic features in the long sequence. Geometry-constrained guider improves the information flow and corrects the nonlinear feature mapping of the aorta.

2   Geometry-Constrained Deformable Attention Network

2.1   Problem Formulation

The purpose of GDAN is to learn aortic feature representation and map geometric offsets from original representation to CPR. All images and their corresponding labels are defined as x and y, where x_i, y_i are input and ground truth values at the pixel location, respectively. The objective mapping J is defined as $y = J(x)$. The specialized mapping y' is recorded as $y' = H(y)$, where H is representation mapped from y to CPR. The optimized loss function can be solved by minimizing empirical error on the dataset $z = (x, y)$,

$$h = \arg\min_{\Omega^*} \frac{1}{M} \sum_{i=1}^{M} L(J(x_i; \Omega), y_i) = \arg\min_{\Omega^*} \frac{1}{M} \sum_{i=1}^{M} L(H(J(x_i; \Omega)), H(y_i)) \quad (1)$$

where h is target space, M is the number of training samples, L is the loss function and Ω is the model parameters to be learned.

2.2   Deformable Attention Extractor

The rectangular patch embedding with a fixed size and position is not conducive to handling geometric deformation [20], because it causes a complete structure to be divided into different patches and destroys semantic integrity. Considering the variability of aortic morphology and location in the long CT sequences, we improve a long-range deformable attention module to extract local and global contextual information [21].

Given the feature map $x \in R^{H \times W \times C}$, the coordinate of reference point and the size of a fixed patch are generated as $p = (p_x, p_y)$ and $s \times s$. Now, we add an offset $\Delta p = (o_x, o_y)$ and redefine a new scale $s' = (s_x, s_y)$, so the coordinate and the size are updated to $p' = p + \Delta p$ and $s_x \times s_y$. To obtain parameter matrix $(\Delta p, s')$, input and all patches are fed into a residual block M_{RB} for prediction. Offsets and scales are trained as followed:

$$(p + \Delta p), s' = (x + w_{\Delta p} \cdot M_{RB}(x \cdot W_q)), (w_{s'} \cdot M_{RB}(x \cdot W_q)) \qquad (2)$$

where $w_{\Delta p}$ and $w_{s'}$ are learned weights. Although patches with different scales are more flexible to feature extraction, they are not conducive to batches of input in the network. To solve this problem, bilinear interpolation is chosen for sampling deformed regions [22], so features extracted at point p' are:

$$\begin{aligned} F(p') &= \sum_{q_x, q_y} B(p_x, p_y; q_x, q_y) Z(q_x, q_y) = \sum_{q_x, q_y} B(p_x, q_x) B(p_y, q_y) Z(q_x, q_y) \\ &= \sum_{q_x, q_y} \max(0, 1 - |p_x - q_x|) \max(0, 1 - |p_y - q_y|) Z(q_x, q_y) \end{aligned} \qquad (3)$$

where B is the bilinear interpolation function and Z is the embedding output of all positions from the multi-head self-attention block. The final output of deformable attention on q, k, v with position offsets and scales is formulated as:

$$y = ((q \cdot (F(x) \cdot W_k)^T / \sqrt{dim} + F(p')) \cdot (F(x) \cdot W_v)) \qquad (4)$$

Deformable attention extractor is an adaptive feature extraction module. This module introduces residual block bringing more operations of feature concatenation and a wider range of receptive fields, so it is beneficial for the integration of offset and size in the feature maps. Self-attention mechanism is combined to facilitate long-range interaction between local and global features. Deformable structure helps learn strong representations that fit mappings of variable aortic morphology. The overall network is illustrated in Fig. 2. Given an CT image of size $H \times W \times 1$, it is divided into non-overlapped patches of size 4×4. The feature maps at multi-level are proportionally scaled down $(\frac{1}{4}, \frac{1}{8}, \frac{1}{16}, \frac{1}{32})$ to reduce computational cost caused by dot products operation and bilinear interpolation of large high-resolution images.

2.3   Geometry-Constrained Guider

Guider takes geometric input based on CPR as prior knowledges and the output are backpropagated to the extractor.

CPR refers to straightening 3D model along center points, which gets cross-sectional images perpendicular to centerline. The approximate process is to calculate aortic skeleton from the mask by skeletonization algorithm [23], then extract the longest path of the skeleton and set it as centerline of the whole aorta. Reconstructed images can be represented as $C = \{C^m | m = 1, 2, ..., M\}, C^m =$

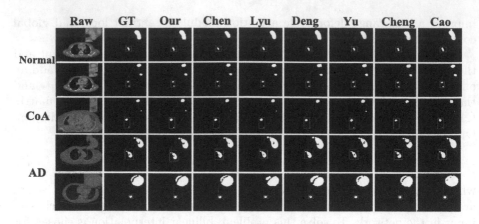

Fig. 3. Representative segmentation results show our GDAN is better than six state-of-the-art methods. Types of the aortic diseases include the normal aorta, coarctation of the aorta (CoA), and aortic dissection (AD). "GT" is the ground truth.

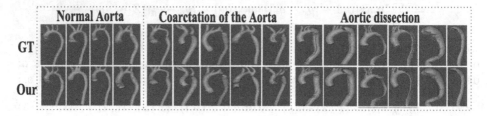

Fig. 4. 3D visualization of ground truth (GT) and our GDAN in patients with normal aorta, coarctation of the aorta and aortic dissection (including true and false lumen).

(l_m, v_m, d_m), where M is the number of center points; and l_m, v_m and d_m are respectively location, tangential vector, and distance to skeleton of mth center point. Guider makes a aortic structure with similarity and homogeneity in the space, while non-aortic region has greater anisotropy. We combine extracted geometric features with images for embedding, and input to the decoder. Error of the output continuously optimizes parameters through the back propagation of the loss function:

$$L_{guider} = \arg\min_{C} \frac{1}{N} \sum_{n=1}^{N} \frac{1}{M} \sum_{m=1}^{M} ||C_{Seg'}^{(n,m)} - C_{Gt'}^{(n,m)}|| \tag{5}$$

where $||C||$ is the Frobenius norm of matrix C. $C_{Seg'}^{(n,m)}$ and $C_{Gt'}^{(n,m)}$ are the predicted results and real labels of mth point of nth sample after CPR, respectively. N is the number of sample. Guider can impose strong constraints on the contour of the aorta, and correct semantic feature mapping to fit nonlinear function.

Table 1. Comparison of our GDAN and the state-of-the-art methods for aortic segmentation. Dice similarity coefficient (DSC), root mean square distance (RMSE) and Hausdorff distance (HD) are used for evaluation. The units of RMSD and HD are mm.

Method	Whole aorta			Ascending aorta			Aortic arch			Descending aorta		
	DSC	RMSD	HD	DSC	RMSD	HD	DSC	RMSD	HD	DSC	RMSD	HD
Chen [16]	0.906	4.65	3.84	0.918	3.67	3.26	0.862	5.25	4.98	0.907	4.61	3.81
Lyu [24]	0.912	3.82	3.76	**0.927**	**3.04**	3.09	0.844	5.88	5.84	0.885	5.02	5.22
Deng [25]	0.907	4.63	3.81	0.906	4.63	3.53	0.835	6.33	5.99	0.902	4.96	4.12
Yu [26]	0.855	5.73	5.28	0.903	4.78	4.04	0.812	6.91	6.37	0.851	5.75	5.36
Cheng [27]	0.863	5.24	4.36	0.887	4.97	4.88	0.845	5.82	5.83	0.857	5.71	5.39
Cao [15]	0.844	5.89	5.67	0.875	5.04	5.13	0.797	7.18	6.45	0.860	5.29	5.06
Our GDAN	**0.943**	**3.56**	**3.52**	0.921	3.12	**3.01**	**0.899**	**4.82**	**4.16**	**0.916**	**3.08**	**3.31**

Table 2. Ablation study on our GDAN. "DAM" is the deformable attention module and "GcM" is geometry-constrained module. "DSC" is Dice similarity coefficient, "ASSD" is average symmetric surface distance and "VOE" is volumetric overlap error. "×" means to remove the module and "√" means to keep the module.

DAM	GcM	DSC	Precision	Recall	ASSD	VOE
×	×	0.890	0.861	0.901	6.13	0.206
√	×	0.914	0.887	**0.918**	4.58	0.153
×	√	0.931	0.935	0.912	3.55	0.131
√	√	**0.943**	**0.946**	0.914	**3.47**	**0.107**

3 Experiment and Result

Dataset. A total of 204 patients are retrospectively enrolled, including 42 normal aorta, 45 CoA, and 107 aortic dissection. All CT images are performed using a 256-slice CT scanner (Philips Brilliance iCT). Average 100 mL of contrast agent (iopamidol 370, BRACCO) is injected into the antecubital vein, followed by 30 mL of saline solution. The bolus tracking technique (BolusPro, Philips Healthcare) is used simultaneously for imaging. Scanning range is from the aortic thoracic inlet to the beginning of the bilateral femoral artery. A radiologist with 6 years of experience in CT manually annotates images in a voxel-wise manner as ground truth and verifies them again.

Train and Test. All dataset is trained and tested on a NVIDIA RTX A6000 48 GB GPU. We use the Adam optimizer to minimise the loss function of our GDAN, with a batch of 16 images per step. Initial learning rate is 0.01 and the decay rate is 0.9. Training and test phase respectively use 80% and 20% dataset.

Aortic Segmentation. We compare with six state-of-the-art methods concerned aortic segmentation. The evaluation metrics are dice similarity coefficient

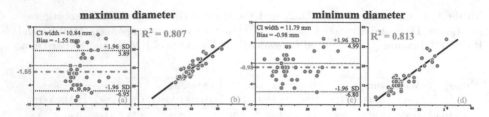

Fig. 5. High agreement between our GDAN and ground truth (GT) manually measured. (a) and (c) are the Bland-Altman analysis. (b) and (d) are Pearson correlation.

(DSC), root mean square distance (RMSE) and Hausdorff distance (HD). Figure 3 and Fig. 4 directly shows segmentation results on the representative images of the normal aorta, CoA and aortic dissection. Aortic areas segmented by our GDAN are highly similar to the ground truth (GT). Table 1 also evidences that our GDAN performs well with a overall DSC of 0.943, root mean square distance of 3.56 and HD of 3.52. They respectively improved 0.032–0.099, 0.30–2.37 and 0.24–2.15, compared with the advanced methods. Moreover, our GDAN mainly improves segmentation in the aortic arch and descending aorta.

Diameter Measurement. In order to verify measurement accuracy of two clinical indicators of maximum and minimum diameter, we automatically calculate their values in the segmented aorta. Bland-Altman analysis and Pearson correlation are used to evaluate the differences between our GDAN and manual measurement by the radiologist. Figure 5 presents high segmentation accuracy. The average bias and the correlation coefficient R^2 of the maximum diameter are -1.55 mm and 0.807, while them of the minimum diameter are -0.98mm and 0.813. The curves demonstrate the superiority of our GDAN.

Ablation Study. We also validate the effectiveness of each component in our GDAN, as shown in Table 2. The DSC and average symmetric surface distance (ASSD) are 0.890 and 6.13 when removing deformable attention module and Geometry-constrained module. Then, the results improve to 0.914 and 4.58 when only adding deformable attention module. When only Geometry-constrained module is added, the results improve to 0.931 and 3.55. The best results are 0.943 and 3.47 if both of them are used. Our modules play a key role in aortic segmentation.

4   Conclusion

This paper proposes an available geometry-constrained deformable attention network (GDAN) for aortic segmentation. The CPR-based geometry combining deformable attention extractor can impose constraints on the aortic boundary to guide the learning of nonlinear mapping. The method is conducted on 204 subjects and compared with six state-of-the-art methods. The results (a overall DSC of 0.943; average bias of -1.55 mm and -0.98 mm) demonstrate that our GDAN can aid in the clinical morphological extraction of the aorta.

Acknowledgement. This work is supported by Shenzhen Science and Technology Program (Grant No. GXWD20201231165807008, 20200825113400001), Guangdong Basic and Applied Basic Research Foundation (2022A1515011384), National Natural Science Foundation of China (62101606), Guangdong Natural Science Funds (2020B1515120061), and Natural Science Foundation of Guangdong Province (Grant No. 2020A1515010650).

References

1. Hiratzka, L.F., et al.: 2010 ACCF/AHA/AATS/ACR/ASA/SCA/SCAI/SIR/STS/ SVM guidelines for the diagnosis and management of patients with thoracic aortic disease. J. Am. Coll. Cardiol. **55**(14), 27–129 (2010)
2. Erbel, R., et al.: 2014 ESC guidelines on the diagnosis and treatment of aortic diseases. Russ. J. Cardiol. **123**(7), 7–72 (2015)
3. Roberts, C.S., Roberts, W.C.: Aortic dissection with the entrance tear in the descending thoracic aorta. Analysis of 40 necropsy patients. Ann. Surg. **213**(4), 356–368 (1991)
4. Zhao, Q., et al.: Predictors of aortic dilation in patients with coarctation of the aorta: evaluation with dual-source computed tomography. BMC Cardiovasc. Disord. **18**(1), 1–7 (2018)
5. Garzón, G., Fernández-Velilla, M., Martí, M., Acitores, I., Ybáñez, F., Riera, L.: Endovascular stent-graft treatment of thoracic aortic disease. Radiographics **25**(suppl 1), S229–S244 (2005)
6. Spinelli, D., et al.: Current evidence in predictors of aortic growth and events in acute type B aortic dissection. J. Vasc. Surg. **68**(6), 1925–1935 (2018)
7. Dugas, A., et al.: Reproducibility of abdominal aortic aneurysm diameter measurement and growth evaluation on axial and multiplanar computed tomography reformations. Cardiovasc. Interv. Radiol. **35**(4), 779–787 (2012)
8. Nance, J.W., Ringel, R.E., Fishman, E.K.: Coarctation of the aorta in adolescents and adults: a review of clinical features and CT imaging. J. Cardiovasc. Comput. Tomogr. **10**(1), 1–12 (2016)
9. Fleischmann, D., et al.: Imaging and surveillance of chronic aortic dissection: a scientific statement from the American heart association. Circ. Cardiovasc. Imaging **15**(3), e000075 (2022)
10. Gao, Z., Liu, X., Qi, S., Wu, W., Hau, W.K., Zhang, H.: Automatic segmentation of coronary tree in CT angiography images. Int. J. Adapt. Control Signal Process. **33**(8), 1239–1247 (2019)
11. Wu, C., et al.: Vessel-GAN: angiographic reconstructions from myocardial CT perfusion with explainable generative adversarial networks. Future Gener. Comput. Syst. **130**, 128–139 (2022)
12. Gao, Z., et al.: Learning physical properties in complex visual scenes: an intelligent machine for perceiving blood flow dynamics from static CT angiography imaging. Neural Netw. **123**, 82–93 (2020)
13. Fantazzini, A., et al.: 3D automatic segmentation of aortic computed tomography angiography combining multi-view 2D convolutional neural networks. Cardiovasc. Eng. Technol. **11**(5), 576–586 (2020)

14. Kovács, T., Cattin, P., Alkadhi, H., Wildermuth, S., Székely, G.: Automatic segmentation of the aortic dissection membrane from 3D CTA images. In: Yang, G.-Z., Jiang, T.Z., Shen, D., Gu, L., Yang, J. (eds.) MIAR 2006. LNCS, vol. 4091, pp. 317–324. Springer, Heidelberg (2006). https://doi.org/10.1007/11812715_40

15. Cao, L., et al.: Fully automatic segmentation of type B aortic dissection from CTA images enabled by deep learning. Eur. J. Radiol. **121**, 108713 (2019)

16. Chen, D., et al.: Multi-stage learning for segmentation of aortic dissections using a prior aortic anatomy simplification. Med. Image Anal. **69**, 101931 (2021)

17. Pepe, A., et al.: Detection, segmentation, simulation and visualization of aortic dissections: a review. Med. Image Anal. **65**, 101773 (2020)

18. Fu, F., et al.: Rapid vessel segmentation and reconstruction of head and neck angiograms using 3D convolutional neural network. Nat. Commun. **11**(1), 1–12 (2020)

19. Raman, R., Napel, S., Beaulieu, C.F., Bain, E.S., Jeffrey Jr., R.B., Rubin, G.D.: Automated generation of curved planar reformations from volume data: method and evaluation. Radiology **223**(1), 275–280 (2002)

20. Liu, Z., et al.: Swin transformer: hierarchical vision transformer using shifted windows. In: Proceedings of the IEEE/CVF International Conference on Computer Vision, pp. 10012–10022(2021)

21. Xia, Z., Pan, X., Song, S., Li, L.E., Huang, G.: Vision transformer with deformable attention. arXiv preprint arXiv:2201.00520 (2022)

22. Kirkland, E.J.: Bilinear interpolation. In: Kirkland, E.J. (ed.) Advanced Computing in Electron Microscopy, pp. 261–263. Springer, Boston (2010). https://doi.org/10.1007/978-1-4419-6533-2_12

23. Lee, T.C., Kashyap, R.L., Chu, C.N.: Building skeleton models via 3-D medial surface axis thinning algorithms. Graph. Models Image Process. **56**(6), 462–478 (1994)

24. Lyu, T., et al.: Dissected aorta segmentation using convolutional neural networks. Comput. Methods Programs Biomed. **211**, 106417 (2021)

25. Deng, X., Zheng, Y., Xu, Y., Xi, X., Li, N., Yin, Y.: Graph cut based automatic aorta segmentation with an adaptive smoothness constraint in 3D abdominal CT images. Neurocomputing **310**, 46–58 (2018)

26. Yu, Y., et al.: A threedimensional deep convolutional neural network for automatic segmentation and diameter measurement of type B aortic dissection. Korean J. Radiol. **22**(2), 168–178 (2021)

27. Cheng, J., Tian, S., Yu, L., Ma, X., Xing, Y.: A deep learning algorithm using contrast-enhanced computed tomography (CT) images for segmentation and rapid automatic detection of aortic dissection. Biomed. Signal Process. Control **62**, 102145 (2020)

ConTrans: Improving Transformer with Convolutional Attention for Medical Image Segmentation

Ailiang Lin, Jiayu Xu, Jinxing Li, and Guangming Lu[✉]

Harbin Institute of Technology, Shenzhen, China
luguangm@hit.edu.cn

Abstract. Over the past few years, convolution neural networks (CNNs) and vision transformers (ViTs) have been two dominant architectures in medical image segmentation. Although CNNs can efficiently capture local representations, they experience difficulty establishing long-distance dependencies. Comparably, ViTs achieve impressive success owing to their powerful global contexts modeling capabilities, but they may not generalize well on insufficient datasets due to the lack of inductive biases inherent to CNNs. To inherit the merits of these two different design paradigms while avoiding their respective limitations, we propose a concurrent structure termed ConTrans, which can couple detailed localization information with global contexts to the maximum extent. ConTrans consists of two parallel encoders, i.e., a Swin Transformer encoder and a CNN encoder. Specifically, the CNN encoder is progressively stacked by the novel Depthwise Attention Block (DAB), with the aim to provide the precise local features we need. Furthermore, a well-designed Spatial-Reduction-Cross-Attention (SRCA) module is embedded in the decoder to form a comprehensive fusion of these two distinct feature representations and eliminate the semantic divergence between them. This allows to obtain accurate semantic information and ensure the up-sampling features with semantic consistency in a hierarchical manner. Extensive experiments across four typical tasks show that ConTrans significantly outperforms state-of-the-art methods on ten famous benchmarks.

Keywords: Medical image segmentation · Transformer · Convolution Neural Network · Cross-attention

1 Introduction

Medical image segmentation is an extremely important part of modern computer-assisted diagnosis (CAD) applications and image-guided surgery systems, where the main goal is to segment objects of interest through pixel-level classification in medical images. Recent years have witnessed the rapid development of deep learning, especially convolution neural networks (CNNs), which have widely adopted and achieved remarkable performances in a range of medical image segmentation tasks such as polyp segmentation, cell segmentation, skin lesion segmentation, pneumonia lesion segmentation, etc. In particular, the typical encoder-decoder based U-shaped architecture [18] and its variants [13, 17, 32] exhibit excellent segmentation potential, which is

L. Wang et al. (Eds.): MICCAI 2022, LNCS 13435, pp. 297–307, 2022.
https://doi.org/10.1007/978-3-031-16443-9_29

largely attributed to the strong local feature encoding capacity of convolution operation. Despite the impressive success, CNNs still struggle to capture long-distance feature dependencies due to the inevitable limitation of the receptive field. Meanwhile, such global interaction is crucial for medical images where the segmentation of objects in different regions depends on contextual information. To alleviate such issues, considerable efforts are devoted to enlarging the receptive fields by introducing effective sampling strategies [6, 13], spatial pyramid enhancement [30] or various attention mechanism [27]. Nevertheless, all these methods still suffer from challenges in efficient dynamic context modeling between arbitrary pixels, which means there is great potential for improvement.

Recently, motivated by the great success of Transformer [24] in natural language processing (NLP), a large amount of literature has emerged to apply the multi-head self-attention (MHSA) mechanism to various visual domain [9, 31]. The Vision Transformer (ViT) [9] is the first image recognition model purely based on Transformer where input images are split into sequences of patches with positional embeddings, leading to a competitive performance on ImageNet but requires pre-training on a large supplementary dataset. To reduce the high computational complexity and introduce the pyramid structure which is very important for dense prediction tasks, a hierarchical Swin Transformer [16] is proposed with Window based MSA (W-MHSA) that only computes self-attention within local windows. Furthermore, SETR [31] achieves state-of-the-art performance in segmentation tasks by using ViT as the backbone. For medical image segmentation, TransUNet [5] effectively boosts the segmentation performance by computing MHSA on the highest-level features of U-Net, while MedT [22] utilizes the modified axial attention to train effectively on small-scale medical datasets. To some extent, the above methods are encouraging but still cannot produce satisfactory results on various medical segmentation tasks for the following reasons: 1) most ViTs ignore the local feature details in non-overlapping patches and lack some of the inductive biases inherent to CNNs; 2) CNNs and ViTs follow different design paradigms, which means neglecting the semantic gap between them will lead to feature redundancy; 3) the contribution of MHSA in the up-sampling process has not been fully explored.

Based on the above discussions, we propose a novel framework named ConTrans in this work, which aims to inherit the merits of self-attention and convolution for effective medical image segmentation. ConTrans is a dual structure with two parallel encoders in feature extraction, i.e., a Swin Transformer encoder and a well-designed CNN encoder. In special, the CNN encoder is progressively stacked by the Depthwise Attention Block (DAB), which is composed of three key modules in sequence, i.e., depthwise convolution, channel attention, and spatial attention. Our DAB can not only provide the crucial inductive biases and local features missing from Transformer encoder, but also reduce the local redundancy. Although convolution and self-attention are generally considered as two distinct techniques for representation learning, we argue that there exists a strong underlying relationship between them. To form a comprehensive integration of these two distinct feature representations, we develop the novel Spatial-Reduction-Cross-Attention (SRCA) module. It first leverages the Spatial-Reduction Attention (SRA) to learn contextual dependencies among the up-sampled features while reducing the resource consumption, then another SRA is applied for feature integration where the

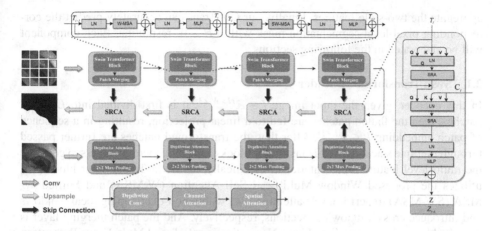

Fig. 1. An overview of the proposed ConTrans framework.

Transformer-based patch embeddings serve as the query inputs, key and value inputs are reasoned from the local features of DAB in the same layer. By embedding SRCA into the decoder in a hierarchical manner, we can eliminate the semantic divergence between local features and global representations in shallow and deep layers, ensuring the fused features with semantic consistency and improving the semantic segmentation quality.

To summarize, the main contributions of this work are as follows: (1) We present the ConTrans, which inherits the merits of Transformer and CNN for medical image segmentation. (2) The proposed Depthwise Attention Block (DAB) can not only provide the required inductive biases and local features but also reduce the local redundancy. (3) A well-designed SRCA is embedded in the decoder to form a comprehensive fusion of local features and global representations while eliminating the semantic divergence between them. (4) Empirical evidence on ten medical datasets demonstrates that our ConTrans consistently outperforms other existing state-of-the-art methods.

2 Method

Figure 1 illustrates an overview of the proposed ConTrans. Unlike the previous encoder-decoder works, ConTrans adopts a novel dual encoding mechanism, i.e. a Swin Transformer encoder and a well-designed CNN encoder, which are responsible for extracting global representations and local features, respectively. Specifically, the CNN encoder is hierarchically stacked by Depthwise Attention Block (DAB), which is composed of depthwise convolution (DWConv), channel attention, and spatial attention. In practice, given the input of medical image $I \in \mathbb{R}^{H \times W \times 3}$, we first feed it into the two parallel encoders to extract hierarchical features at $\{\frac{1}{4}, \frac{1}{8}, \frac{1}{16}, \frac{1}{32}\}$ of the input image resolution, then the two-style hierarchical features are passed to the decoder together via skip connections. Following the typical design of U-Net [18], the decoder performs bilinear up-sampling on the extracted multi-level features. Moreover, a novel Spatial-Reduction-Cross-Attention (SRCA) module is embedded into each decoding stage to consecutively aggregate the two-style features. Finally, our ConTrans can accurately predict the corresponding pixel-level segmentation mask with a $H \times W$ resolution. Each component will be elaborated in the following sections.

2.1 Swin Transformer Encoder

In this encoder, given the input image $I \in \mathbb{R}^{H \times W \times 3}$, it is first divided into $L = \frac{HW}{16}$ patches, then the flatten patches are fed to a linear projection, resulting in a sequence of patch embedding $P \in \mathbb{R}^{L \times D}$. After that, the transformed patches are further passed through multiple stacked Swin Transformer blocks and patch merging layers to generate the multi-level feature representations. Specifically, the Swin Transformer block [16] utilizes the proposed Window Multi-head Self-Attention (W-MSA) and Shifted W-MSA (SW-MSA) to perform self-attention [24] within non-overlapping local windows and introduce cross-window connections, respectively. And the patch merging layer is responsible for down-sampling. Layer Normalization (LN) and Multi Layer Perceptron (MLP) are also applied after each modified self-attention module. As shown in Fig. 1, the Swin Transformer block is formulated as:

$$\hat{T}_i = \text{W-MSA}\left(\text{LN}\left(T_{i-1}\right)\right) + T_{i-1}, \qquad T_i = \text{MLP}\left(\text{LN}\left(\hat{T}_i\right)\right) + \hat{T}_i,$$
$$\hat{T}_{i+1} = \text{SW-MSA}\left(\text{LN}\left(T_i\right)\right) + T_i, \qquad T_{i+1} = \text{MLP}\left(\text{LN}\left(\hat{T}_{i+1}\right)\right) + \hat{T}_{i+1}, \tag{1}$$

where T_{i-1} and $\hat{T}_i$ (specially, $T_0 \leftarrow P$) denote the inputs of the (S)W-MSA module and the MLP module in the i-th block, respectively.

2.2 Depthwise Attention Block

The Swin Transformer encoder can easily capture global dependencies at a low computation cost, but they may not perform well on medical datasets due to the lack of inductive biases inherent to CNNs. To overcome this limitation, we devise the Depthwise Attention Block (DAB) to preserve the missing local details in Transformer while reducing the feature redundancy. Specifically, DAB is a lightweight module including depthwise convolution, channel attention, and spatial attention. As figured in Fig. 1, the CNN encoder is split into 4 stages. Each stage consists of the proposed DAB, followed by a max-pooling with stride 2, which results in a $4\times$ spatial downsampling of the input features. Hence, the DAB can be formulated as:

$$\text{DAB}(C_j) = \text{Att}_{\text{spatial}}(\text{Att}_{\text{channel}}(\text{DWConv}(C_j))), \tag{2}$$

where C_j refers to the input feature map of j-th stage $\{j = 1, 2, 3, 4\}$ and DWConv denotes depthwise convolution with zero paddings. The channel attention and spatial attention are adopted from CBAM [27], with the aim to focus on obtaining the CNN inductive biases we need, and leverage the attention mechanism to reduce feature redundancy for better fusion with global representations.

2.3 Spatial-Reduction-Cross-Attention

Given the two-style feature representations T_l, $C_l \in \mathbb{R}^{\frac{HW}{4^{l+1}} \times C_l}$ in the l-th stage $\{l = 1, 2, 3, 4\}$, how to inherit the merits from them while avoiding their respective limitations remains. To address such difficulties, we propose the Spatial-Reduction-Cross-Attention (SRCA), in which successive spatial-reduction-attention (SRA) are employed to build long-range dependencies during the up-sampling process and fuse feature representations via cross-attention. It should be noted that PVT [26] proposes SRA to reduce the computational overhead in traditional MHSA. As illustrated in Fig. 1, the SRCA module contains several layers, each of which consists of two SRA and an MLP module. Details of the first SRA can be formulated as follows:

$$\hat{\mathbf{Q}}_l = T_l \hat{\mathbf{W}}_l^Q, \hat{\mathbf{K}}_l = T_l \hat{\mathbf{W}}_l^K, \hat{\mathbf{V}}_l = T_l \hat{\mathbf{W}}_l^V, \tag{3}$$

$$\hat{Z}_l = \text{MLP}(\text{SRA}(\hat{\mathbf{Q}}_l, \hat{\mathbf{K}}_l, \hat{\mathbf{V}}_l)), \tag{4}$$

where $\hat{\mathbf{W}}_l^Q \in \mathbb{R}^{C_l \times d_q}, \hat{\mathbf{W}}_l^K \in \mathbb{R}^{C_l \times d_k}$ and $\hat{\mathbf{W}}_l^V \in \mathbb{R}^{C_l \times d_v}$ are projection matrices of l-th stage. $\hat{\mathbf{Q}}_l, \hat{\mathbf{K}}_l, \hat{\mathbf{V}}_l$ denote the query, key and value matrices. After we establish long-range contextual information at the global representations T_l, another SRA is used to integrate the resulting feature $\hat{Z}_l$ with local features through cross-attention mechanism. To this end, we map $\hat{Z}_l$ to query $\mathbf{Q}_l$, while key $\mathbf{K}_l$ and value $\mathbf{V}_l$ are computed from the local features C_l, respectively.

$$\mathbf{Q}_l = \hat{Z}_l \mathbf{W}_l^Q, \mathbf{K}_l = C_l \mathbf{W}_l^K, \mathbf{V}_l = C_l \mathbf{W}_l^V, \tag{5}$$

$$Z = \text{MLP}(\text{SRA}(\mathbf{Q}_l, \mathbf{K}_l, \mathbf{V}_l)), \tag{6}$$

$$Z_l = \text{Conv}_{3 \times 3}(\text{Concat}(Z, Z_{l+1})), \tag{7}$$

where $\mathbf{W}_l^Q \in \mathbb{R}^{C_l \times d_q}, \mathbf{W}_l^K \in \mathbb{R}^{C_l \times d_k}$ and $\mathbf{W}_l^V \in \mathbb{R}^{C_l \times d_v}$ are linear projection parameters too. d_q, d_k, d_v represent the dimension of the query, key and value, respectively. The resulting features Z are concatenated with the up-sampling features Z_{l+1}. Finally, the fused features are passed into a 3×3 convolution to obtain the final output Z_l. Here, SRCA is considered as a bridge between two encoders, computing self-attention for each pixel-pair between local features and global representations. In this way, we can eliminate the semantic divergence and establish effective fusion between the two-style features, significantly improving segmentation performance. By inserting SRCA into each stage of the decoder, we can eliminate the semantic divergence and establish effective fusion between the two-style features, significantly improving segmentation performance. Finally, the output Z_0 of the last SRCA is up-sampled to the original resolution of input image and then a 1×1 convolution takes the up-sampling features to generate the segmentation mask predictions of size $H \times W \times N_{cls}$, where N_{cls} denotes the number of categories.

3 Experiments

3.1 Datasets

To evaluate the effectiveness of our ConTrans, we conduct experiments on four common medical image segmentation tasks including ten datasets. (1) Polyp Segmentation: Kvasir [14] (1,000 samples), CVC-ClinicDB [19] (612 samples), CVC-ColonDB [2] (380 samples), ETIS [25] (196 samples), (2) Skin Lesion Segmentation: ISIC2017 [8] (2,000 samples for training, 150 samples for validation and 600 samples for testing),

ISIC2018 [7,21], (2,594 samples), (3) Pneumonia Lesion Segmentation: COVID-19 [1] (50 samples for training and 50 samples for testing), (4) Cell Segmentation: GLAS [20] (85 samples for training and 80 samples for testing), Bowl2018 [3] (670 samples), Pannuke [12] (7904 samples). For COVID-19 and GLAS datasets, we follow the same split in [11] and [23], respectively. For Pannuke dataset, we follow the official three-fold cross-validation. As for other datasets, we performed five-fold cross-validation on them because the annotations of their public test sets are missing.

3.2 Implementation Details

ConTrans is implemented using the PyTorch framework and trained on an NVIDIA RTX 3090 GPU. In special, the Swin Transformer encoder is initialized with pre-trained weights Swin-Base [16]. All images are resized to 256×256 and augmented with random rotation, horizontal flip, and vertical flip. We use the weighted IoU loss and binary cross-entropy loss to train our network for 300 epochs with a batch size of 16. Besides, we use deep supervision following [10] and apply the multi-scale training strategy $\{0.75, 1, 1.25\}$. In all experiments, we utilize the SGD optimizer with momentum 0.9, weight decay 1e–4, and base learning rate of 1e–2. Early stopping and the "Poly" learning rate decay are also applied. With regard to the evaluation metric, we adopt the Dice Similarity Coefficient (DSC) for a fair comparison.

3.3 Comparisons with State-of-the-Art Methods

Two broad approaches are involved in our comparative experiments as the baseline, i.e., CNN-based and Transformer-based methods. Table 1 and Table 2 present the comparison results of our ConTrans with other state-of-the-art methods. Moreover, we visualize the predictions in Fig. 2.

1) Results on Polyp Segmentation: Polyp segmentation is an essential part of clinical colonoscopy, but the varied appearance and blurred boundaries of polyps make it still a challenging task. As illustrated in Table 1, we can obviously observe that ConTrans achieves the best segmentation performance on four challenging datasets, even outperforms PraNet [10], which is specially designed for polyp segmentation. In particular, our method achieves a significant improvement of 2.93% on the ETIS dataset compared to PraNet. We also visualize the qualitative results in Fig. 2, showing that our proposed method can accurately identify and segment the polyp tissues.

2) Results on Skin Lesion Segmentation: On the ISIC2017 dataset, our ConTrans outperforms the existing baselines and yields the highest score of 87.52%. We extensively conduct experiments on the ISIC2018 dataset, ConTrans still achieves excellent performance with a score of 91.17%, which further proves that our method can effectively capture the boundaries of skin lesions and obtain better edge predictions.

3) Results on Pneumonia Lesion Segmentation: [1] is the first open-access COVID-19 dataset for pneumonia lesion segmentation, but it only contains 100 labeled images, and the objects of pneumonia lesions usually vary widely in texture, size, and position, which makes Transformer-based architectures difficult to generalize well. As illustrated in Table 1, most CNN baselines outperform Transformer

Table 1. Quantitative results on nine datasets compared to other leading methods.

Method	Clinic	Colon	ETIS	Kvasirs	ISIC17	ISIC18	Bowl	COVID	GLAS
UNet [18]	88.59	82.24	80.89	84.32	85.76	88.78	91.35	73.73	87.99
UNet++ [32]	89.30	82.86	80.77	84.95	85.77	88.85	91.38	76.63	89.98
CENet [13]	91.53	83.11	75.03	84.92	86.00	89.53	91.34	74.35	89.02
AttentionUNet [17]	90.57	83.25	79.68	80.25	86.02	88.95	91.46	75.98	87.68
DeepLabv3 [6]	92.46	89.90	89.88	89.19	85.47	89.92	90.86	73.35	87.49
PraNet [10]	94.14	91.94	90.65	91.45	86.53	90.53	87.87	76.94	91.20
MedT [22]	86.14	75.28	69.20	82.69	84.53	87.17	90.50	71.84	82.52
MCTrans [15]	92.30	86.58	83.69	86.20	86.11	90.35	91.28	76.20	90.71
SegFormer [28]	93.45	91.33	89.64	91.47	87.48	90.77	91.17	75.32	89.73
SETR-PUP [31]	90.78	88.55	83.42	88.27	85.25	89.77	85.45	72.56	88.75
TransUNet [5]	93.40	90.81	90.14	91.08	86.47	90.20	91.09	74.36	89.94
Swin-Unet [4]	90.38	86.26	87.80	89.70	86.28	90.43	91.00	74.82	88.94
TransFuse [29]	93.99	91.17	90.39	91.16	87.14	90.83	91.19	77.70	90.79
ConTrans	**94.76**	**92.63**	**93.58**	**92.54**	**87.52**	**91.17**	**92.17**	**78.83**	**92.06**

Table 2. Quantitative results on the Pannuke dataset compared to other leading methods. Ave represents the average score of five tissues.

Method	Neo	Inflam	Conn	Dead	Epi	Ave
UNet [18]	82.85	65.48	62.29	40.11	75.57	65.26
UNet++ [32]	82.03	67.58	62.79	40.79	77.21	66.08
CENet [13]	82.73	68.25	63.15	41.12	77.27	66.50
AttentionUNet [17]	82.74	65.42	62.09	38.60	76.02	64.97
MCTrans [15]	84.22	68.21	65.04	48.30	78.70	68.90
ConTrans	**84.26**	**68.61**	**65.92**	**48.62**	**79.71**	**69.42**

baselines, i.e. PraNet (76.94%) vs SegFormer (75.32%). It can be noted that Trans-Fuse achieves an encouraging score of 77.70% by fusing ViT and ResNet, while our ConTrans surpasses it by 1.13%. We attribute this superiority to CNN's inductive biases and our efficient feature integration strategy.

4) Results on Cell Segmentation: We first conduct comparative experiments on Bowl2018 and GLAS dataset as shown in Table 1. The scores of our ConTrans on both datasets are 92.17% and 92.06%, which clearly outperforms other advanced methods, i.e. AttentionUNet (91.46%) and PraNet (91.20%). In addition, Table. 2 presents our results on the Pannuke dataset, which consists of five tissues: Neoplastic (Neo), Inflammatory (Inflam), Connective (Conn), Dead, and Non-Neoplastic Epithelial (Epi). Compared with the previous state-of-the-art MCTrans [15], which is dedicated to learning the global semantic relationship among multiple categories, our ConTrans achieves consistent improvement on all tissues and promotes the average score (69.42% vs. 68.90%). These experimental results demonstrate the superiority of our method in recognizing multiple segmented objects.

Table 3. Ablation study on four medical image segmentation tasks

DAB	SRCA	Kvasir	ISIC17	COVID	GLAS
✗	✗	89.96	85.22	75.74	90.49
✓	✗	91.06	86.40	76.47	91.15
✓	✓	92.54	87.52	78.83	92.06

3.4   Ablation Studies

We further conduct ablation studies to investigate the influence of various factors on ConTrans. As shown in Table 3, after adding the DAB, the local features inherent to CNNs can effectively provide the missing inductive biases in Transformer, promoting the score by a large margin. Moreover, the segmentation performance is further improved by applying the SRCA to fuse the two-style features, confirming the effectiveness of this feature fusion strategy. It indicates that eliminating the semantic divergence between local features and global representations can make them complement each other and obtain better fusion results.

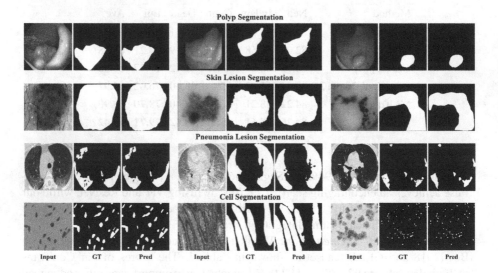

Fig. 2. Visual segmentation results on four tasks. Each row contains input medical image (Input), ground truth (GT) and prediction (Pred) in turn.

4   Conclusions

In this paper, we propose a novel architecture termed ConTrans for accurate segmentation in medical images. Our method not only exploits CNN's capacity in capturing detailed local information, but also leverages Transformer to build long-range dependencies and global contexts. Extensive experiments on four typical medical segmentation tasks across ten public datasets demonstrate that our ConTrans significantly outperforms the previous state-of-the-art methods.

Acknowledgments. This work was supported in part by the NSFC fund (62176077, 61906162), in part by the Guangdong Basic and Applied Basic Research Foundation under Grant 2019Bl515120055, in part by the Shenzhen Key Technical Project under Grant 2020N046, in part by the Shenzhen Fundamental Research Fund under Grant JCYJ20210324132210025, in part by the Medical Biometrics Perception and Analysis Engineering Laboratory, Shenzhen, China, in part by Shenzhen Science and Technology Program (RCBS20200714114910193), and in part by Education Center of Experiments and Innovations at Harbin Institute of Technology, Shenzhen.

References

1. COVID-19 CT segmentation dataset. https://medicalsegmentation.com/covid19/. Accessed 11 Apr 2014
2. Bernal, J., Sánchez, F.J., Fernández-Esparrach, G., Gil, D., Rodríguez, C., Vilariño, F.: WM-DOVA maps for accurate polyp highlighting in colonoscopy: Validation vs. saliency maps from physicians. Comput. Medi. Imaging Graph. **43**, 99–111 (2015)
3. Caicedo, J.C., et al.: Nucleus segmentation across imaging experiments: the 2018 data science bowl. Nat. MethodsD **16**(12), 1247–1253 (2019)
4. Cao, H., et al.: Swin-UNet: UNet-like pure transformer for medical image segmentation. arXiv preprint arXiv:2105.05537 (2021)
5. Chen, J., et al.: Transunet: Transformers make strong encoders for medical image segmentation. arXiv preprint arXiv:2102.04306 (2021)
6. Chen, L.C., Papandreou, G., Schroff, F., Adam, H.: Rethinking atrous convolution for semantic image segmentation. arXiv preprint arXiv:1706.05587 (2017)
7. Codella, N., et al.: Skin lesion analysis toward melanoma detection 2018: a challenge hosted by the international skin imaging collaboration (ISIC). arXiv preprint arXiv:1902.03368 (2019)
8. Codella, N.C., et al.: Skin lesion analysis toward melanoma detection: a challenge at the 2017 international symposium on biomedical imaging (ISBI), hosted by the international skin imaging collaboration (ISIC). In: 2018 IEEE 15th International Symposium on Biomedical Imaging (ISBI 2018), pp. 168–172. IEEE (2018)
9. Dosovitskiy, A., et al.: An image is worth 16x16 words: transformers for image recognition at scale. arXiv preprint arXiv:2010.11929 (2020)
10. Fan, D.P., et al.: PraNet: parallel reverse attention network for polyp segmentation. In: Martel, A.L., et al. (eds.) MICCAI 2020. LNCS, vol. 12266, pp. 263–273. Springer, Cham (2020). https://doi.org/10.1007/978-3-030-59725-2_26
11. Fan, D.P., Zhou, T., Ji, G.P., Zhou, Y., Chen, G., Fu, H., Shen, J., Shao, L.: Inf-Net: automatic covid-19 lung infection segmentation from CT images. IEEE Trans. Med, Imaging **39**(8), 2626–2637 (2020)

12. Gamper, J., Alemi Koohbanani, N., Benet, K., Khuram, A., Rajpoot, N.: PanNuke: an open pan-cancer histology dataset for nuclei instance segmentation and classification. In: Reyes-Aldasoro, C.C., Janowczyk, A., Veta, M., Bankhead, P., Sirinukunwattana, K. (eds.) ECDP 2019. LNCS, vol. 11435, pp. 11–19. Springer, Cham (2019). https://doi.org/10.1007/978-3-030-23937-4_2

13. Gu, Z., et al.: Ce-Net: context encoder network for 2D medical image segmentation. IEEE Trans. Med. Imaging 38(10), 2281–2292 (2019)

14. Jha, D., et al.: Kvasir-SEG: a segmented polyp dataset. In: Ro, Y.M., et al. (eds.) MMM 2020. LNCS, vol. 11962, pp. 451–462. Springer, Cham (2020). https://doi.org/10.1007/978-3-030-37734-2_37

15. Ji, Y., et al.: Multi-compound transformer for accurate biomedical image segmentation. In: de Bruijne, M., et al. (eds.) MICCAI 2021. LNCS, vol. 12901, pp. 326–336. Springer, Cham (2021). https://doi.org/10.1007/978-3-030-87193-2_31

16. Liu, Z., et al.: Swin transformer: hierarchical vision transformer using shifted windows. arXiv preprint arXiv:2103.14030 (2021)

17. Oktay, O., et al.: Attention U-Net: learning where to look for the pancreas. arXiv preprint arXiv:1804.03999 (2018)

18. Ronneberger, O., Fischer, P., Brox, T.: U-Net: convolutional networks for biomedical image segmentation. In: Navab, N., Hornegger, J., Wells, W.M., Frangi, A.F. (eds.) MICCAI 2015. LNCS, vol. 9351, pp. 234–241. Springer, Cham (2015). https://doi.org/10.1007/978-3-319-24574-4_28

19. Silva, J., Histace, A., Romain, O., Dray, X., Granado, B.: Toward embedded detection of polyps in WCE images for early diagnosis of colorectal cancer. Int. J. Comput. Assist. Radiol. Surg. 9(2), 283–293 (2014)

20. Sirinukunwattana, K., et al.: Gland segmentation in colon histology images: the glas challenge contest. Med. Image Anal. 35, 489–502 (2017)

21. Tschandl, P., Rosendahl, C., Kittler, H.: The ham10000 dataset, a large collection of multi-source dermatoscopic images of common pigmented skin lesions. Sci. Data 5(1), 1–9 (2018)

22. Valanarasu, J.M.J., Oza, P., Hacihaliloglu, I., Patel, V.M.: Medical transformer: gated axial-attention for medical image segmentation. arXiv preprint arXiv:2102.10662 (2021)

23. Valanarasu, J.M.J., Sindagi, V.A., Hacihaliloglu, I., Patel, V.M.: KiU-Net: towards accurate segmentation of biomedical images using over-complete representations. In: Martel, A.L., et al. (eds.) MICCAI 2020. LNCS, vol. 12264, pp. 363–373. Springer, Cham (2020). https://doi.org/10.1007/978-3-030-59719-1_36

24. Vaswani, A., et al.: Attention is all you need. arXiv preprint arXiv:1706.03762 (2017)

25. Vázquez, D.: A benchmark for endoluminal scene segmentation of colonoscopy imageA benchmark for endoluminal scene segmentation of colonoscopy images. J. Healthc Eng. 2017 (2017)

26. Wang, W., et al.: Pyramid vision transformer: a versatile backbone for dense prediction without convolutions. In: Proceedings of the IEEE/CVF International Conference on Computer Vision, pp. 568–578 (2021)

27. Woo, S., Park, J., Lee, J.-Y., Kweon, I.S.: CBAM: convolutional block attention module. In: Ferrari, V., Hebert, M., Sminchisescu, C., Weiss, Y. (eds.) ECCV 2018. LNCS, vol. 11211, pp. 3–19. Springer, Cham (2018). https://doi.org/10.1007/978-3-030-01234-2_1

28. Xie, E., Wang, W., Yu, Z., Anandkumar, A., Alvarez, J.M., Luo, P.: SegFormer: simple and efficient design for semantic segmentation with transformers. Adv. Neural Inf. Process. Syst. 34 (2021)

29. Zhang, Y., Liu, H., Hu, Q.: TransFuse: fusing transformers and cnns for medical image segmentation. In: de Bruijne, M., et al. (eds.) MICCAI 2021. LNCS, vol. 12901, pp. 14–24. Springer, Cham (2021). https://doi.org/10.1007/978-3-030-87193-2_2

30. Zhao, H., Shi, J., Qi, X., Wang, X., Jia, J.: Pyramid scene parsing network. In: Proceedings of the IEEE Conference on Computer Vision and Pattern Recognition, pp. 2881–2890 (2017)
31. Zheng, S., et al.: Rethinking semantic segmentation from a sequence-to-sequence perspective with transformers. arXiv preprint arXiv:2012.15840 (2020)
32. Zhou, Z., Rahman Siddiquee, M.M., Tajbakhsh, N., Liang, J.: UNet++: a nested U-Net architecture for medical image segmentation. In: Stoyanov, D., et al. (eds.) DLMIA/ML-CDS - 2018. LNCS, vol. 11045, pp. 3–11. Springer, Cham (2018). https://doi.org/10.1007/978-3-030-00889-5_1

End-to-End Segmentation of Medical Images via Patch-Wise Polygons Prediction

Tal Shaharabany[(⊠)] and Lior Wolf

Tel-Aviv University, Tel Aviv-Yafo, Israel
{shaharabany,wolf}@mail.tau.ac.il

Abstract. The leading medical image segmentation methods represent the output map as a pixel grid. We present an alternative in which the object edges are modeled, per image patch, as a polygon with k vertices that is coupled with per-patch label probabilities. The vertices are optimized by employing a differentiable neural renderer to create a raster image. The delineated region is then compared with the ground truth segmentation. Our method obtains multiple state-of-the-art results for the Gland segmentation dataset (Glas), the Nucleus challenges (MoNuSeg), and multiple polyp segmentation datasets, as well as for non-medical benchmarks, including Cityscapes, CUB, and Vaihingen. Our code for training and reproducing these results is attached as a supplement.

1 Introduction

In recent years, fully convolutional networks based on encoder-decoder architectures [22], in which the encoder is pre-trained on ImageNet, have become the standard tool in the field of medical image segmentation. Several techniques were developed to better leverage the capacity of the architectures [15,31,35]. For example, skip connections between the encoder and the decoder were added to overcome the loss of spatial information and to better propagate the training signal through the network [27]. What all these methods have in common is the representation of the output segmentation map as a binary multi-channel image, where the number of channels is the number of classes C. In our work, we add a second type of output representation, in which, for every patch in the image and for every class, a polygon comprised of k points represents the binary mask. Active contour methods [8,18] which also optimize a polygon, are, unlike our method, iterative, produce a single polygon, lacking to cope with occlusions.

For optimization purposes, we use a neural renderer [19,23] to translate the polygons to a binary raster-graphics mask. The neural renderer provides a gradient for the polygon vertices and enables us to maximize the overlap between the output masks and the ground truth.

Supplementary Information The online version contains supplementary material available at https://doi.org/10.1007/978-3-031-16443-9_30.

L. Wang et al. (Eds.): MICCAI 2022, LNCS 13435, pp. 308–318, 2022.
https://doi.org/10.1007/978-3-031-16443-9_30

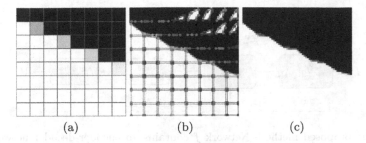

(a)   (b)   (c)

Fig. 1. The process of combining the data of polygons with the grid-based predictions for obtaining a refined segmentation map. (a) The low-resolution segmentation map M is shown for a single object. White pixels denote in-object, black ones are the background. (b) The corresponding polygons R_p. Different polygons are drawn in different colors and are slightly scaled to show them separately. (c) The segmentation map M_o, in the final resolution. The polygons on the top right part of R_p are gated out by M.

Since there are polygons for every patch and every class, whether the object appears there or not, we multiply the binary mask with a low-resolution binary map that indicates the existence of each object.

This method achieves state-of-the-art results on multiple medical microscopy benchmarks, including Gland segmentation (Glas), a challenge from the MIC-CAI'15, the Nucleus challenges (MoNuSeg) and multiple polyp segmentation datasets, as well as on well-established non-medical datasets, including Cityscapes.

2   Related Work

Many versions of the U-net segmentation architecture [27] have been proposed. Improved generalization is obtained by reducing the gap between the encoder and decoder semantic maps [48]. Another contribution adds residual connections for each encoder and decoder block and divides the image into patches with a calculated weighted map for each patch as input to the model [43]. A transformer-based module [40] and an A Feature Enhance Module (SFEM) [25] were proposed for processing skip connection information efficiently. In another contribution, a self-attention mechanism is applied to each feature map in the encoder-decoder block in order to model long-range dependencies in the image, with good computational efficiency [41]. This is extended by adding scalar gates to the attention layers [37]. The same contribution also adds a Local-Global training strategy (LoGo), which further improves performance. Fang et al. proposed a novel selective feature aggregation network with the area and boundary constraints to collect global and local context [47]. Yin et al. learn an external contextual relation across the images in the dataset [44]. Another work employs a coarse low-resolution mask that is learned and refined with an attention mechanism [11]. Hardnet architectures are based on a Harmonic Densely Connected Network and were shown to provide an excellent accuracy/efficiency tradeoff [4,16] without an attention mechanism. Due to performance degradation for

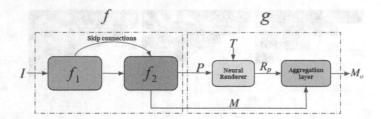

Fig. 2. The proposed method. Network f contains an encoder-decoder network with skip connections between them. Given an input Image I, it generates (1) a low-resolution segmentation probability map M (2) a patch-wise polygon array P. Network g contains (1) a neural renderer that employs a predefined triangulation T to generate the binary polygon segmentation map R_p from P (2) an aggregation layer that rearranges R_p according to the patch location in the original image. The accumulated map is then gated, i.e., multiplied pixel-wise, by the initial probabilities provided by the upsampled version of M to generate the output segmentation map M_o.

small polyps and background-foreground imbalance, a shallow attention module and a probability correction strategy were used [42]. A transformer encoder and three modules of feature aggregators were also used for Polyp detection [10]. Another Polyp detection algorithm [39] utilized Context Enhancement Module (CEM) to increase the receptive field. . Many methods use dilated convolutions [34,45] to preserve the spatial size of the feature map and enlarge the receptive field [5,46]. The receptive field can also be enlarged by using larger kernels [26] and enriched with contextual information, by using kernels of multiple scales [6].

The first deep learning-based snakes predict a correction vector for each point, by considering the small enclosing patch, in order to converge to the object outline [29]. Acuna et al. [1] propose automated and semi-automated methods for object annotation and segmentation based on Graph neural networks and reinforcement learning. Ling et al. [21] used a Graph Convolutional Network (GCN) to fit the polygon's positions to the outline. A different approach predicts a shift map and iteratively changes the positions of the polygon vertices, according to this map [13]. In order to obtain a gradient signal, the authors use a neural renderer that transforms the 2D polygon into a 2D segmentation map. Our method also employs a neural renderer. However, we perform local predictions per patch and recover the vertex locations of the polygon in each patch non-iteratively. This allows us to make finer local predictions and overcome discontinuities caused by occlusions or by the existence of multiple objects.

3   Methods

The input to our method is an RGB image I, of dimensions $3 \times H \times W$. The image is divided into patches of size $s \times s$ and a polygon with k vertices is predicted for each patch. In addition, each patch has an associate scalar that determines the probability of the object being present in the patch.

This mask representation is illustrated in Fig. 1. The low-res probability map in panel (a) assigns one value per grid cell, with higher probabilities denoted by a brighter value. The local polygons in panel (b) determine a local segmentation mask in each grid cell. When the grid cell is completely enclosed by the segmented object, the polygon covers the entire cell. At object boundaries, it takes the form of the local boundary. When out of the object, it can degenerate to a zero-area polygon or an arbitrary one. The final segmentation map, depicted in Fig. 1(c), is obtained by multiplying the low-resolution probability map with the local patch rastered for each polygon.

The learned network we employ, see Fig. 2, is given as $f = f_2 \circ f_1$, where $f_1 : \mathbb{R}^{C \times H \times W} \to \mathbb{R}^{C \times \frac{H}{32} \times \frac{W}{32}}$ is the backbone (encoder) network based on a pretrained imagenet models, and f_2 is the decoder network. The output space of f_2 is a tensor that can be seen as a field with $\frac{H}{s} \times \frac{W}{s}$ vectors $\mathbb{R}^{2k+1}$, each representing a single polygon associated with a single image patch. This representation is the concatenation of the 2D coordinates of the k vertices together with a probability map. The image coordinates are normalized, such that their range is $[-1,1]$.

A fixed triangulation T is obtained by performing Delaunay triangulation on a polygon of k vertices. It is used to convert all polygons to a raster image, assuming that the polygons are locally almost always close to being convex. We set the z-axis to be constant and the perspective to zero, in order to use a 3D renderer for 2D objects.

We denote by $P = \{P^i\}$ the set of all $\frac{H}{s} \times \frac{W}{s}$ polygons, and by M the map of size $\frac{H}{s} \times \frac{W}{s}$ that contains all the patches probabilities. The neural renderer r, given the vertices P and the fixed triangulation T, returns the polygon shape as a local image patch. In this raster image, all pixels inside the polygon defined by P through the triangulation T are assigned a value of one, zero for pixels outside, and in-between values for pixels divided by the polygon:

$$M_p^i = r(P^i, T) \in \mathbb{R}^{s \times s}, \tag{1}$$

The aggregation layer collects all local patches M_p^i, each in its associated patch location, to obtain the map $R_p \in \mathbb{R}^{H \times W}$. By performing nearest neighbor upsampling to the probability map M by a factor of s in each dimension, we obtain a probability map of the same size as R_p. The final mask is the pixel-wise product of these two maps.

$$M_o = \uparrow M \odot R_p, \tag{2}$$

where $\odot$ denotes elementwise multiplication and $\uparrow$ denotes upsampling. This way, the low-resolution mask serves as a gating mechanism that determines whether the polygon exists within an object of a certain class in the output segmentation map. To maximize the overlap between the output mask (before and after incorporating the polygons) and the ground truth segmentation map Y, a soft dice loss [33] is used:

$$\mathcal{L}_{dice}(y, \hat{y}) = 1 - \frac{2TP(y, \hat{y}) + 1}{2TP(y, \hat{y}) + FN(y, \hat{y}) + FP(y, \hat{y}) + 1} \tag{3}$$

Where TP is the true positive between the ground truth y and output mask $\hat{y}$, FN is a false negative and FP is a false positive. In addition, a binary cross-entropy is used for M:

$$\mathcal{L} = \mathcal{L}_{BCE}(\uparrow M, Y) + \mathcal{L}_{dice}(M_o, Y) \tag{4}$$

To avoid an additional parameter, this loss appears without weighting. During inference, we employ only M_o and the second term can be seen as auxiliary.

Architecture. For the backbone f_1, the Harmonic Dense Net [4] is used, in which the receptive field is of size 32×32. The network contains six "HarD" blocks, as described in [4], with 192, 256, 320, 480, 720, 1280 output channels, respectively. The last block contains a dropout layer of 0.1.

The decoder f_2 contains two upsampling blocks to obtain an output resolution of $s = 8$ times less than the original image size. The number of upsampling blocks is defined by s (lower values require more upsampling blocks), and so is the receptive field of f_1. Each block contains two convolutional layers with a kernel size of 3 and zero padding equal to one. In addition, we use batch normalization after the last convolution layer, before the activation function. The first layer's activation function is a ReLU, while the second layer's activation functions are sigmoid for the channel of the low-resolution map and *tanh* for the polygon vertices. Each layer receives a skip connection from the block of the encoder that has the same spatial resolution. We note that our decoder requires considerably fewer learnable parameters compared to a regular decoder and, since only two blocks are used, fewer skip connections from the encoder.

The subsequent network g employs the mesh renderer of [19], with a zero perspective, a camera distance of one, and an output mask of size $s \times s$. This way, the 3D renderer is adapted for the 2D rendering task.

Our decoder employs 17M parameters, while FCN-HarDNet-85 employs a decoder with 22M parameters. In term of FLOPs, for an image size of 256^2, 16.51 GMacs are measured for our model and 25.44 GMacs for the comparable FCN. Peak memory consumption is 486 MB for our model and 506 MB for the FCN. For training our network, we use the ADAM optimizer, with an initial learning rate of 0.0003 and a gamma decay of 0.7, applied every 50 epochs. The batch size is 32 and the parameter of weight decay regularization is set to $5 \cdot 10^{-5}$. A single GeForce RTX 2080 Ti is used for training on all datasets.

4   Experiments

We present our results on three medical segmentation benchmarks and three non-medical segmentation benchmarks and compare our performance with the state-of-the-art methods. The MoNuSeg dataset [20] contains a training set with 30 microscopic images from seven organs with annotations of 21,623 individual nuclei; the test dataset contains 14 similar images. Following previous work, we resized the images to a resolution of 512×512 [37]. An encoder-decoder architecture based on the HarDNet-85 [4] backbone is used. Since the recovered elements are very small, our method employs a map M at 1/4 resolution ($s = 4$), and not lower resolutions, as in other datasets. Under this settings where

Table 1. MoNu and Glas results

Method	Monu		GlaS	
	Dice	IoU	Dice	IoU
FCN [2]	28.84	28.71	-	-
U-Net [27]	79.43	65.99	86.05	75.12
U-Net++ [48]	79.49	66.04	87.36	79.03
Res-UNet [43]	79.49	66.07	-	-
Axial attention [41]	76.83	62.49	-	-
MedT [37]	79.55	66.17	88.85	78.93
UCTransNet [40]	79.87	66.68	89.84	82.24
FCN-Hardnet85 [4]	79.52	66.06	89.37	82.09
Low res Mask (M)	65.82	49.13	85.38	74.91
Ours (M_o)	**80.30**	**67.19**	**91.19**	**84.34**

Table 2. Polyp benchmarks

Method	Kvasir33 [17]		Clinic [3]		Colon [36]		ETIS [30]	
	Dice	IoU	Dice	IoU	Dice	IoU	Dice	IoU
U-Net [27]	81.8	74.6	82.3	75.5	51.2	44.4	39.8	33.5
U-Net++ [48]	82.1	74.3	79.4	72.9	48.3	41.0	40.1	34.4
SFA [12]	72.3	61.1	70.0	60.7	46.9	34.7	29.7	21.7
MSEG [16]	89.7	83.9	90.9	86.4	73.5	66.6	70.0	63.0
DCRNet [44]	88.6	82.5	89.6	84.4	70.4	63.1	55.6	49.6
ACSNet [47]	89.8	83.8	88.2	82.6	71.6	64.9	57.8	50.9
PraNet [11]	89.8	84.0	89.9	84.9	71.2	64.0	62.8	56.7
EU-Net [25]	90.8	85.4	90.2	84.6	75.6	68.1	68.7	60.9
SANet [42]	90.4	84.7	91.6	85.9	75.3	67.0	75.0	65.4
Polyp-PVT [10]	91.7	86.4	93.7	88.9	80.8	72.7	78.7	70.6
FCN-Hardnet85 [4]	90.0	84.9	92.0	86.9	77.3	70.2	76.9	69.5
Low res Mask (M)	87.5	79.3	85.7	76.4	74.8	63.9	68.9	56.8
Ours (M_o)	**91.8**	**86.5**	**93.8**	**89.0**	**80.9**	**73.4**	**79.1**	**71.4**

$s = 4$ instead of $s = 8$, we improved the performances by 1.5%. The Gland segmentation (GlaS) challenge was presented at MICCAI'15 [32]. It involves 85 images for training and 80 for testing. All images are resized to a resolution of 224 × 224, following [40]. Following [11], our algorithm was also tested on four Polyp datasets, including Kvasir-SEG [17], ClinicDB [3], ColonDB [36] and ETIS [30]. The dataset is split into a training set of 1448 images, composed of 900 images from ClinicDB and 548 from Kvasir, and a test set consisting of 100 images from ClinicDB, 64 from Kvasir, 196 from ETIS, and 380 from ColonDB.

For instance segmentation, the Cityscapes dataset [9] contains eight object classes with 5,000 fine-annotated images, divided into 2,975 images for training, 500 for validation, and 1,525 for testing. Following [21], we train and evaluate given the bounding box of the object, such that the network outputs a single mask for the entire instance. The baseline"FCN-HarDNet-85" employs a fully-convolutional network with the same backbone [4] we used for our own method. The Vaihingen [28] dataset consists of 168 aerial images, with a resolution of 256 × 256, from this German city. The task is to segment the central building in a dense and challenging environment, which includes other structures, streets, trees, and cars. The dataset is divided into 100 buildings for training, and the remaining 68 for testing. In addition to the conventional IoU and F1-score metrics, the weighted Coverage (WCov) score is computed as the IoU for each object in the image weighted by the related size of the object from the whole image and the Boundary F-score (BoundF) is the average of the F1-scores computed with thresholds ranging from 1 to 5 pixels [28]. For this dataset, we use an encoder-decoder architecture based on the HarDNet-85 [4] backbone and produce M at $1/8$ resolution ($s = 8$). Finally, we report results on CUB-200-2011 [38]. This dataset contains 5994 images of different birds for training and 5794 for validation. We trained the model for the same setting of cityscapes, with an input image size of 256^2 and evaluating on the original image size.

Results. The results for the MoNu dataset are reported in Table 1. We outperform all baselines, including the latest Axial attention Unet [41] and Medical transformer [37], for both the Dice score and mean-IoU. Our algorithm also

Table 3. Cityscapes segmentation results for two widespread protocols: the top part refers to segmentation results with 15% expansion around the bounding box; the bottom part refers to segmentation results with no expansion around the bounding box

Method	Bike	Bus	Person	Train	Truck	Mcycle	Car	Rider	Mean
Polygon-RNN++ (with BS) [1]	63.06	81.38	72.41	64.28	78.90	62.01	79.08	69.95	71.38
PSP-DeepLab [6]	67.18	83.81	72.62	68.76	80.48	65.94	80.45	70.00	73.66
Polygon-GCN (with PS) [21]	66.55	85.01	72.94	60.99	79.78	63.87	81.09	71.00	72.66
Spline-GCN (with PS) [21]	67.36	**85.43**	73.72	64.40	80.22	64.86	81.88	71.73	73.70
Ours	**68.20**	85.10	**74.52**	74.80	**80.50**	66.20	**82.10**	71.81	**75.40**
Deep contour [13]	68.08	83.02	75.04	74.53	79.55	66.53	81.92	72.03	75.09
FCN-HarDNet-85 [4]	68.26	84.98	74.51	76.60	80.20	66.25	82.36	71.57	75.59
Ours	**69.53**	**85.50**	**75.15**	76.90	**81.20**	**66.96**	**82.69**	**72.18**	**76.26**

Table 4. Vaihingen results.

Method	F1-score	mIoU	WCov	FBound
FCN-UNet [27]	87.40	78.60	81.80	40.20
FCN-ResNet34	91.76	87.20	88.55	75.12
FCN-HarDNet-85 [4]	93.97	88.95	93.60	80.20
DSAC [24]	-	71.10	70.70	36.40
DarNet [7]	93.66	88.20	88.10	75.90
Deep contour [13]	94.80	90.33	93.72	78.72
TDAC [14]	94.26	89.16	90.54	78.12
Ours	**95.15**	**90.92**	**94.36**	**83.89**

Table 5. Results for CUB dataset

Method	Dice	mIoU	FBound
U-Net [27]	93.84	88.80	76.54
Deep contour [13]	93.72	88.35	75.92
FCN-Hardnet85 [4]	94.91	90.51	79.73
Low res Mask (M)	84.45	73.43	39.47
Ours (M_o)	**95.11**	**90.91**	**79.86**

performs better in comparison to the fully convolutional segmentation network with the same backbone Hardnet-85. The improvement from low-resolution mask (M) to the output mask (M_o) is from 49.13% IoU to 67.19%. The full-resolution FCN-Hardnet85 obtains a lower result of 66.06 IoU. A sample visualization is presented in Fig. 3 (first row).

The results for GlaS are also shown in Table 1. Our algorithm outperforms Medical transformer by more than 2% IoU [37]. The improvement from low-resolution mask (M) to the output mask (M_o) is nearly 10% IoU. See Fig. 3 (2nd row) for a sample.

Polyp results are listed in Table 2. Our method outperforms the state of the art on this benchmark EU-Net [25], SANet [42], Polyp-PVT [10]. For Kvasir-SEG and ClinicDB the gap is 0.1% IoU, for ColonDB it is 0.7% IoU, and for ETIS 0.8% IoU. The refinement from low-resolution mask(M) to the output mask(M_o) shows an improvement of 7–15% IoU. Figure 3 (3rd row) depicts an example.

The Cityscapes results for two existing protocols are presented in Table 3. As can be seen, our method outperforms all baseline methods, including active contour methods and encoder-decoder architectures. Furthermore, one can observe the gap in performance between our method and the U-Net-like method that is based on the same backbone architecture ("FCN-HarDNet-85"). This is despite

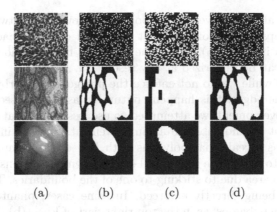

(a)   (b)   (c)   (d)

Fig. 3. Sample results of the proposed method on the Nucleus challenges (MoNuSeg, first row) - first row, the gland segmentation dataset (Glas, second row), and the Kvasir polyp segmentation dataset (last row). (a) Input image. (b) Ground truth segmentation. (c) The low-resolution segmentation map M. (d) The final segmentation map M_o.

the reduction in the number of trainable parameters and the size of the representation (each 8×8 patch is represented by $k + 1$ output floats, instead of 64 floats). Appendix Fig. 4 presents sample results and compares our method to the leading active contour method ("Deep contour") and to the FCN-HarDNet-85 baseline. As can be seen, the active contour method struggles to model objects that deviate considerably from their convex hull, and cannot handle occlusions well. In comparison to the FCN baseline, our method provides more accurate details (consider the bicycle fork in the first row), but the overall shape is the same. The results for the Vaihingen building dataset are presented in Table 4. Our method outperforms both fully connected segmentation methods and active-contour based methods. Appendix Fig. 6 presents typical samples, demonstrating the refinement obtained by the polygons over the initial segmentation mask. The results for CUB are reported in Table 5. We obtain better results than both FCN-Hardnet85 and Deep Contour, on all metrics. Appendix Fig. 5 present the output of our algorithm for samples from CUB-200-2011.

More results are presented in the appendix. Figure 7 and Table 7 present our network's performance for building segmentation and the Glas dataset, respectively, when varying the two parameters. As can be seen, for small patches, for instance 4×4 or 8×8, the number of nodes that optimizes performance is five. As may be expected, for larger patches, the method performs better with a higher number of vertices. We also checked our performance for different Hardnet backbones - HarDNnet39DS, HarDNnet68DS, HarDNnet68, HarDNnet85 (DS stands for depth-wise-separable, see [4]). The results are shown in appendix Table 6 for the building dataset. As expected, performance improves with the increase in the number of parameters in the network. It can also be observed that between the low-resolution masks and the refined mask there is a performance gap of 4% in terms of IoU, in favor of the latter, across all backbones.

Discussion. The ability to work with polygons allows our network to produce segmentation maps at a resolution that is limited only by machine precision. This can be seen in Fig. 1(b), where the polygons have fractional coordinates. These polygons can be rasterized at any resolution.

We note that while we do not enforce the polygons of nearby patches to be compatible at the edges, this happens naturally, as can be seen in Fig. 1(b). Earlier during development, we attempted to use loss terms that encourage such compatibility, but did not observe an improvement. Further inspection of the obtained polygons reveals that polygons with $k > 4$ often employ overlapping vertices in order to match the square cell edges. In empty regions, polygons tend to become of zero area due to sticking to one of the boundaries. These behaviors emerge without being directly enforced. In some cases, phantom objects are detected by single polygons, as in the top right part of Fig. 1(b). These polygons are removed by the gating process that multiplies with M, as can be seen in Fig. 1(c). We do not observe such polygons in the output images.

5   Conclusions

The pixel grid representation is commonly used by deep segmentation networks applied to medical imaging. We present an alternative approach that encodes the local segmentation maps as polygons. A direct comparison to a method that employs the same backbone but without polygonal representation reveals a significant improvement in performance on multiple benchmarks, despite using fewer training parameters.

Acknowledgments. This project has received funding from the European Research Council (ERC) under the European Union's Horizon 2020 research and innovation programme (grant ERC CoG 725974).

References

1. Acuna, D., Ling, H., Kar, A., Fidler, S.: Efficient interactive annotation of segmentation datasets with polygon-rnn++. In: CVPR (2018)
2. Badrinarayanan, V., Kendall, A., Cipolla, R.: SegNet: a deep convolutional encoder-decoder architecture for image segmentation. Trans. Pattern Anal. Mach. Intell. **39**(12) (2017)
3. Bernal, J., Sánchez, F.J., Fernández-Esparrach, G., et al.: WM-DOVA maps for accurate polyp highlighting in colonoscopy: validation vs. saliency maps from physicians. Comput. Med. Imag. Graph. **43** (2015)
4. Chao, P., Kao, C.Y., Ruan, Y.S., Huang, C.H., Lin, Y.L.: HardNet: a low memory traffic network. In: ICCV (2019)
5. Chen, L.C., Papandreou, G., Kokkinos, I., Murphy, K., Yuille, A.L.: Semantic image segmentation with deep convolutional nets and fully connected CRFs. arXiv (2014)
6. Chen, L.C., Papandreou, G., Kokkinos, I., Murphy, K., Yuille, A.L.: Deeplab: Semantic image segmentation with deep convolutional nets, atrous convolution, and fully connected CRFs. Trans. Pattern Anal. Mach. Intell. **40**(4) (2017)

7. Cheng, D., Liao, R., Fidler, S., Urtasun, R.: DarNet: deep active ray network for building segmentation. In: CVPR (2019)
8. Cohen, L.D.: On active contour models and balloons. CVGIP Image Underst. **53**(2) (1991)
9. Cordts, M., et al.: The cityscapes dataset for semantic urban scene understanding. In: CVPR (2016)
10. Dong, B., Wang, W., Fan, D.P., Li, J., Fu, H., Shao, L.: Polyp-PVT: polyp segmentation with pyramid vision transformers. arXiv (2021)
11. Fan, D.P., et al.: PraNet: parallel reverse attention network for polyp segmentation. In: Martel, A.L., et al. (eds.) MICCAI 2020. LNCS, vol. 12266, pp. 263–273. Springer, Cham (2020). https://doi.org/10.1007/978-3-030-59725-2_26
12. Fang, Y., Chen, C., et al.: Selective feature aggregation network with area-boundary constraints for polyp segmentation. In: MICCAI (2019)
13. Gur, S., Shaharabany, T., Wolf, L.: End to end trainable active contours via differentiable rendering. In: ICLR (2019)
14. Hatamizadeh, A., Sengupta, D., Terzopoulos, D.: End-to-end trainable deep active contour models for automated image segmentation: delineating buildings in aerial imagery. In: ECCV (2020)
15. He, K., Zhang, X., Ren, S., Sun, J.: Deep residual learning for image recognition. In: CVPR (2016)
16. Huang, C.H., Wu, H.Y., Lin, Y.L.: HarDNet-MSEG: a simple encoder-decoder polyp segmentation neural network that achieves over 0.9 mean dice and 86 fps. arXiv (2021)
17. Jha, D., et al.: Kvasir-SEG: a segmented polyp dataset. In: Ro, Y.M., et al. (eds.) MMM 2020. LNCS, vol. 11962, pp. 451–462. Springer, Cham (2020). https://doi.org/10.1007/978-3-030-37734-2_37
18. Kass, M., Witkin, A., Terzopoulos, D.: Snakes: active contour models. IJCV (1988)
19. Kato, H., Ushiku, Y., Harada, T.: Neural 3d mesh renderer. In: CVPR (2018)
20. Kumar, N., Verma, R., Anand, D., Zhou, Y., Onder, O.F., et al.: A multi-organ nucleus segmentation challenge. IEEE Trans. Med. Imaging **39**(5), 1380–1391 (2019)
21. Ling, H., Gao, J., Kar, A., Chen, W., Fidler, S.: Fast interactive object annotation with curve-GCN. In: CVPR (2019)
22. Long, J., Shelhamer, E., Darrell, T.: Fully convolutional networks for semantic segmentation. In: CVPR (2015)
23. Loper, M.M., Black, M.J.: OpenDR: an approximate differentiable renderer. In: Fleet, D., Pajdla, T., Schiele, B., Tuytelaars, T. (eds.) ECCV 2014. LNCS, vol. 8695, pp. 154–169. Springer, Cham (2014). https://doi.org/10.1007/978-3-319-10584-0_11
24. Marcos, D., et al.: Learning deep structured active contours end-to-end. In: CVPR (2018)
25. Patel, K., Bur, A.M., Wang, G.: Enhanced u-net: a feature enhancement network for polyp segmentation. In: CRV. IEEE (2021)
26. Peng, C., Zhang, X., Yu, G., Luo, G., Sun, J.: Large kernel matters-improve semantic segmentation by global convolutional network. In: CVPR (2017)
27. Ronneberger, O., Fischer, P., Brox, T.: U-Net: convolutional networks for biomedical image segmentation. In: MICCAI (2015)
28. Rottensteiner, F., et al.: International society for photogrammetry and remote sensing, 2D semantic labeling contest (2022)
29. Rupprecht, C., Huaroc, E., et al.: Deep active contours. arXiv (2016)

30. Silva, J., Histace, A., Romain, O., et al.: Toward embedded detection of polyps in WCE images for early diagnosis of colorectal cancer. Int. J. CARS 9, 283–293 (2014)
31. Simonyan, K., Zisserman, A.: Very deep convolutional networks for large-scale image recognition. arXiv (2014)
32. Sirinukunwattana, K., et al.: Gland segmentation in colon histology images: the GLAS challenge contest. Med. Image Aanal. **35** (2017)
33. Sudre, C.H., Li, W., et al.: Generalised dice overlap as a deep learning loss function for highly unbalanced segmentations. In: DLMIA (2017)
34. Sun, X., Zhang, P., Wang, D., Cao, Y., Liu, B.: Colorectal polyp segmentation by u-net with dilation convolution. In: ICMLA, pp. 851–858. IEEE (2019)
35. Szegedy, C., et al.: Going deeper with convolutions. In: CVPR (2015)
36. Tajbakhsh, N., Gurudu, S.R., Liang, J.: Automated polyp detection in colonoscopy videos using shape and context information. IEEE Trans. Med. Image **35**(2) (2015)
37. Valanarasu, J.M.J., Oza, P., Hacihaliloglu, I., Patel, V.M.: Medical transformer: gated axial-attention for medical image segmentation. arXiv (2021)
38. Wah, C., Branson, S., Welinder, P., Perona, P., Belongie, S.: The Caltech-UCSD birds-200-2011 dataset (2011)
39. Wang, D., et al.: AFP-mask: anchor-free polyp instance segmentation in colonoscopy. IEEE J. Biomed. Health Inform. **26**(7):2995–3006 (2022)
40. Wang, H., Cao, P., Wang, J., Zaiane, O.R.: Uctransnet: rethinking the skip connections in u-net from a channel-wise perspective with transformer. arXiv (2021)
41. Wang, H., et al.: Axial-DeepLab: stand-alone axial-attention for panoptic segmentation. In: Vedaldi, A., Bischof, H., Brox, T., Frahm, J.-M. (eds.) ECCV 2020. LNCS, vol. 12349, pp. 108–126. Springer, Cham (2020). https://doi.org/10.1007/978-3-030-58548-8_7
42. Wei, J., Hu, Y., Zhang, R., Li, Z., Zhou, S.K., Cui, S.: Shallow attention network for polyp segmentation. In: de Bruijne, M., et al. (eds.) MICCAI 2021. LNCS, vol. 12901, pp. 699–708. Springer, Cham (2021). https://doi.org/10.1007/978-3-030-87193-2_66
43. Xiao, X., Lian, S., Luo, Z., Li, S.: Weighted res-UNet for high-quality retina vessel segmentation. In: ITME. IEEE (2018)
44. Yin, Z., Liang, K., Ma, Z., Guo, J.: Duplex contextual relation network for polyp segmentation. arXiv (2021)
45. Yu, F., Koltun, V.: Multi-scale context aggregation by dilated convolutions. arXiv (2015)
46. Zhang, H., et al.: Context encoding for semantic segmentation. In: CVPR (2018)
47. Zhang, R., Li, G., Li, Z., Cui, S., Qian, D., Yu, Y.: Adaptive context selection for polyp segmentation. In: Martel, A.L., et al. (eds.) MICCAI 2020. LNCS, vol. 12266, pp. 253–262. Springer, Cham (2020). https://doi.org/10.1007/978-3-030-59725-2_25
48. Zhou, Z., Rahman Siddiquee, M.M., Tajbakhsh, N., Liang, J.: UNet++: a nested U-Net architecture for medical image segmentation. In: Stoyanov, D., et al. (eds.) DLMIA/ML-CDS -2018. LNCS, vol. 11045, pp. 3–11. Springer, Cham (2018). https://doi.org/10.1007/978-3-030-00889-5_1

Automatic Identification of Segmentation Errors for Radiotherapy Using Geometric Learning

Edward G. A. Henderson[1]([✉]) [ID], Andrew F. Green[1,2] [ID], Marcel van Herk[1,2] [ID], and Eliana M. Vasquez Osorio[1,2] [ID]

[1] The University of Manchester, Oxford Road, Manchester M13 9PL, UK
edward.henderson@postgrad.manchester.ac.uk
[2] Radiotherapy Related Research, The Christie NHS Foundation Trust, Manchester M20 4BX, UK

Abstract. Automatic segmentation of organs-at-risk (OARs) in CT scans using convolutional neural networks (CNNs) is being introduced into the radiotherapy workflow. However, these segmentations still require manual editing and approval by clinicians prior to clinical use, which can be time consuming. The aim of this work was to develop a tool to automatically identify errors in 3D OAR segmentations without a ground truth. Our tool uses a novel architecture combining a CNN and graph neural network (GNN) to leverage the segmentation's appearance and shape. The proposed model was trained using data-efficient learning using a synthetically-generated dataset of segmentations of the parotid gland with realistic contouring errors. The effectiveness of our model was assessed with ablation tests, evaluating the efficacy of different portions of the architecture as well as the use of transfer learning from a custom pretext task. Our best performing model predicted errors on the parotid gland with a precision of 85.0% & 89.7% for internal and external errors respectively, and recall of 66.5% & 68.6%. This offline QA tool could be used in the clinical pathway, potentially decreasing the time clinicians spend correcting contours by detecting regions which require their attention. All our code is publicly available at https://github.com/rrr-uom-projects/contour_auto_QATool.

Keywords: Segmentation error detection · Geometric learning · Data-efficient learning

1 Introduction

The 3D segmentation of organs-at-risk (OARs) in computed tomography (CT) scans is a crucial step in the radiotherapy pathway. However, manual segmentation is time-consuming and prone to inter- and intra-observer variability [1].

Supplementary Information The online version contains supplementary material available at https://doi.org/10.1007/978-3-031-16443-9_31.

L. Wang et al. (Eds.): MICCAI 2022, LNCS 13435, pp. 319–329, 2022.
https://doi.org/10.1007/978-3-031-16443-9_31

Automatic segmentation using convolutional neural networks (CNNs) is now considered state-of-the-art for medical image segmentation [2]. However, every contour needs to be evaluated and approved by clinical staff before use in treatment planning [21]. The aim of this study was to develop a quality assurance (QA) tool to automatically detect segmentation errors without a ground-truth. Such a tool could complement current radiotherapy workflows by drawing the clinicians' attention to regions of a contour which may need updating.

There are already a few methods for automatic QA of anatomical segmentations. Some approaches use statistical models to classify contours requiring further updates or to predict performance metrics. In 2012, Kohlberger et al. applied a support vector machine regression algorithm to predict a volumetric overlap metric for segmentations based on hand-selected shape and appearance features [9]. Chen et al. trained a statistical model based on geometrical attributes of segmentations and their neighbouring structures which could classify contours as correct or incorrect [3]. McCarroll et al. used a bagged tree classification model to flag contours needing attention [11], while Hui et al. developed a QA tool using a hand-crafted set of volumetric features[8].

Some more recent studies used secondary auto-segmentation methods or CNN model ensembling to highlight regions of uncertainty in auto-contouring [12,13,17]. Valindria et al. proposed a reverse classification accuracy method in which a secondary segmentation model was trained from an image and segmentation prediction pair without a ground truth [20]. By evaluating this new model on a database of images with known ground truth segmentations they could predict the Dice similarity coefficient (DSC) for the original prediction. Sander et al. developed a two-step auto-segmentation method for cardiac MR images capable of estimating uncertainties and predicting local segmentation failures [18].

In contrast, our proposed automatic QA method can be applied to any segmentation, automatic or manual, and estimates error on the entire 3D segmentation shape without a gold-standard. In this proof of concept study we predicted errors on a single OAR delineated in head and neck CT scans. We chose the parotid gland as it is suitably complex, with both convex and concave regions, to showcase a wide range of potential segmentation failures.

In this paper we make several contributions. First, we developed a novel automatic error detection system, combining convolutional neural networks (CNNs) and Graph Neural Networks (GNNs). We also performed pre-training and transfer learning to speed up training convergence. Our methodology is data-efficient in the sense that unlimited training data can be generated from a small amount of OAR segmentations and images prior to training.

2   Materials and Method

Our proposed automatic QA tool uses multiple forms of input information: shape information of the segmentation structure and information from the appearance of the surrounding CT scan. To train our method we used a public dataset with segmentations and generated additional segmentations with realistic errors.

2.1   Dataset

For this study we used a publicly-available dataset of 34 head and neck (HN) CT scans resampled to $1 \times 1 \times 2.5$ mm voxel spacing [14]. Each of the 34 HN CTs have segmentations of OARs performed by two doctors. For this study we used one doctor's segmentations (the "oncologist" set of Nikolov et al.) of the left and right parotid glands as the ground truth. For a given patient, the left and right parotid glands are bilateral structures and mostly symmetric. However, between patients the gland shapes vary considerably. All patient data was flipped laterally to create a dataset of 68 left parotid gland ground truth segmentations.

2.2   Generating a Training Dataset

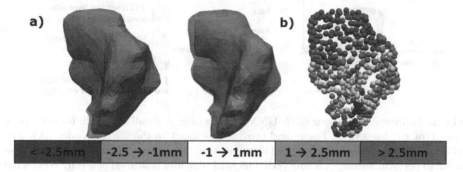

Fig. 1. An example of a perturbed segmentation and classes related to the distance from the ground truth. a) The ground-truth segmentation is shown in orange and the perturbed segmentation is shown in blue. b) The coloured nodes correspond to the signed closest distance to the ground truth. (Color figure online)

To create a large dataset of training samples containing segmentation errors we perturbed the ground truth parotid gland segmentations (Fig. 1a). Practically, we created signed distance transforms of the ground-truth segmentations and added structured noise voxel-wise. The structured noise was created by drawing noise from a normal distribution which was then convolved with a 7.5 mm Gaussian kernel with an amplitude such that the resultant structured noise had a standard deviation of $\sim$1 mm. The perturbed segmentation was extracted using the marching cubes algorithm (at level = 0) to acquire a triangular mesh manifold. A connected components algorithm was applied to eliminate any disconnected spurious segmentations. To simplify the mesh we applied 100 iterations of Taubin smoothing [19] and quadric error metric decimation reducing the mesh size to $\sim$1000 triangles followed by a final 10 iterations of Taubin smoothing. We perturbed each ground-truth segmentation 100 times to generate a dataset of 6800 parotid glands.

Node-wise classes were then determined for each of the perturbed segmentations. The signed nearest distance to the ground truth segmentation was calculated for each node, categorising them into one of five classes (Fig. 1b).

2.3   A Hybrid CNN-GNN Model for Contour Error Prediction

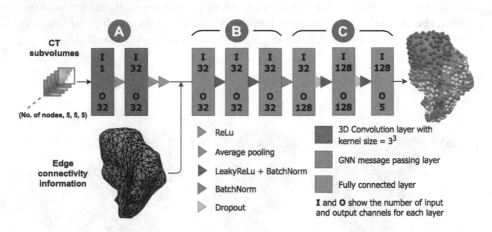

Fig. 2. An illustration of our CNN-GNN architecture. A small 3D patch for each node was extracted from the CT scan and provided as input to the CNN encoder (**A**). The edge connectivity information for the segmentation structure is provided to the GNN in order to perform message passing (**B**). The MLP decoder is formed of 3 fully-connected layers which make the node-wise classification predictions (**C**). The output on the right is an actual prediction.

Our proposed method performs node-wise classification and makes predictions which correspond to the signed distance from the ground-truth segmentation without explicit knowledge of the ground-truth (Fig. 1b). Since we implemented a novel method we investigated several architectures, considering different types of GNN message passing layers to use in the processor and the style of feature extractor to use for the CNN encoder. For the sake of conciseness, in this paper we only describe our best architecture with brief justifications.

We used an encoder, processor and decoder structure inspired by the work of Pfaff et al. on learning mesh-based simulations using GNNs [16]. Figure 2 shows a schematic of our model structure, which consists of a CNN encoder, a GNN processor and a multi-layer perceptron (MLP) decoder (Fig. 2, labels **A**, **B** and **C**, respectively). A $5 \times 5 \times 5$ voxel CT sub-volume centred on each node was provided to the CNN encoder, a feature extractor of two 3D convolutional layers, to produce node-wise representations. The GNN takes these representations and performs message passing between connected nodes, updating each node's representation according to its local neighbourhood. The MLP takes the final representations to predict node-wise classifications. Our hypotheses behind this model design were that the CNN encoder would allow our method to gain context from the CT scan appearance while the GNN processor would leverage the geometrical structure of the input and make node-wise predictions with awareness of the local neighbourhood, similar to active appearance models [4].

For the GNN processor we used the *SplineConv* layer, introduced by Fey et al., to perform message passing [6]. The *SplineConv* is a generalisation of traditional CNN convolution layers for geometric input. Node features in local neighbourhoods are weighted by B-Spline continuous kernel functions, which are in turn parameterised by sets of trainable weights. How node features are aggregated among neighbouring nodes is determined by spatial relations. The relative spatial positions of each node are encoded using 3D Cartesian pseudo-coordinates on the range [0, 1]. In our GNN layers we used the addition aggregation operation as this has been shown to produce the most expressive GNN models [23]. All *SplineConv* layers in our model used spline degree 2, kernel size 5 and were followed by LeakyReLU activation functions and batch normalisation.

2.4   CNN Pre-training

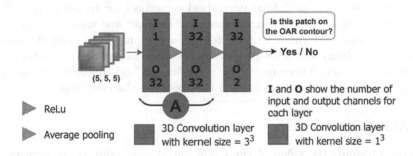

Fig. 3. A schematic of the pre-training task. The CNN encoder, marked A, was connected to a $1 \times 1 \times 1$ convolution layer and an average pooling operation to perform binary classification. The pretext task consisted of classifying whether $5 \times 5 \times 5$ CT patches were sampled from the parotid gland boundary.

Pre-training is a popular transfer learning technique where a model is trained to perform a pretext task in order to learn a set of features which are helpful for the model when applied to the main task[15]. We applied it to initialise the weights of the CNN encoder of our model.

We designed a custom pretext task: to classify 3D CT patches as either "on" or "off" the parotid gland boundary (Fig. 3). These 3D CT image patches were extracted from the same dataset described in Sect. 2.2 live at training time in an equal ratio of "on" and "off" contour.

2.5   Implementation Details

All our models were implemented using PyTorch 1.9.0 and PyG 2.0.1 [5] and operations with triangular meshes performed using Open3D 0.13.0 [25]. Our proposed model has 437,829 parameters. All model training was performed using a 24 GB NVidia GeForce RTX 3090 and AMD Ryzen 9 3950X 16-Core Processor.

In the pre-training phase the encoder was trained on 512 patch samples per epoch, with a batch size of 64, for a maximum of 10000 epochs using binary cross-entropy loss. Our error-prediction models were trained on 25 perturbations per image per epoch, with a batch size of 16, for a maximum of 50 epochs using focal loss with $\gamma = 2$. For both phases we used the AdamW optimizer [10] with an initial learning rate of 0.001, weight decay set to 0.001 and learning rate cosine annealing. Training individual models took $\sim$10 m and inference for a single segmentation takes $\sim$0.03 s.

3   Experiments

We performed a series of experiments in which we ablated different portions of our proposed method. A 5-fold cross-validation was performed for each experiment. In each fold, we used 4800 perturbed structures for training and 600 for validation. Sets of 1400 testing structures were held out and used to evaluate the final classification performance of the fold. Since we used the same dataset for both the pre-training and error-prediction tasks the same cross-validation seeding was used for each. This ensured the same images and contours were used for training, validation and testing in each fold between the two tasks.

3.1   Ablation Tests

First, to determine the value of the CNN encoder we trained an identical set of models with the 3D CT sub-volumes uniformly overwritten with a constant value. This effectively blinded our model to the CT scan appearance. With this ablation the model learned to predict errors from the segmentation shape alone. Second, to evaluate whether the GNN processor was learning useful information from the segmentation's geometric structure, we trained a model without GNN message passing. In this ablation the node-wise representations from the encoder were passed directly to the MLP decoder. Finally we trained an additional set of models from scratch to analyse whether pre-training the CNN encoder, as described in Sect. 2.4, was advantageous.

Confusion matrices were used to evaluate these tests. We calculated each models precision and recall of internal and external segmentation errors (predictions of the edge classes, < -2.5 mm and > 2.5 mm). Errors predicted with the correct sign were considered correct for precision but not recall.

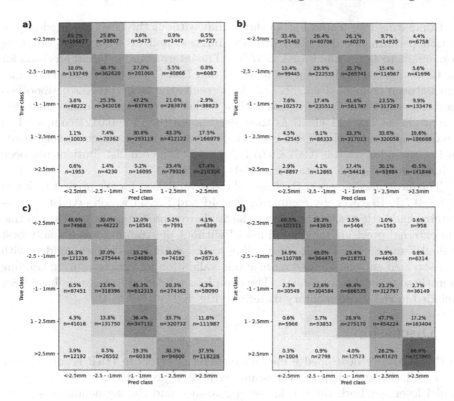

Fig. 4. Confusion matrices of the results of our experiments. **a)** Our proposed full CNN-GNN-MLP method; **b)** CNN encoder ablation; **c)** GNN processor ablation; **d)** pre-training ablation. The percentages in each plot corresponds to the recall. The CNN and GNN ablations lead to a performance reduction in all classes, most significantly for the two extremes. Classification performance was similar with and without pre-training.

4 Results

In Fig. 4 we show confusion matrices of the results of our proposed method (Fig. 4a) alongside the results of the 3 experiments performed (Fig. 4b–d). The percentages included in the confusion matrices show the recall for each class. The internal and external precision for our full model with pre-training was 79.9% and 89.2%.

4.1 Ablation Tests

By comparing Fig. 4b and Fig. 4a we observed that the CNN encoder was necessary for our method to make good predictions, i.e. the CT input was essential. This is unsurprising as the segmentation under consideration could have identical shape to the ground truth but a spatial offset would introduce errors. Information from the imaging is the only way of determining if this is the case.

Comparing Fig. 4c and Fig. 4a shows that without the GNN the models prediction performance decreased with the internal and external recall dropping from 69.2% to 48.6% and 67.4% to 37.9%. The internal and external precision for the GNN-ablated model dropped to 58.2% and 71.6%. The GNN shared information between connected neighbours. As a result, the GNN processor smoothed predictions made in local regions with a higher proportion of nodes allocated similar classes (i.e. ±1 class) as their neighbours (0.974 ± 0.012) compared to the GNN-ablated model (0.894 ± 0.024). The proportion of matching neighbours for the ground truth was 0.967 ± 0.014. A figure of the full distributions has been included in the supplementary materials (Fig. S.1).

The mean accuracy over the five-fold cross-validation for the pretext task was 88.8 ± 0.7%. However, by comparing Fig. 4d and Fig. 4a we observed that pre-training had little positive impact on the final error-prediction rates. The internal and external precision for the full model without pre-training was actually best at 85.0% and 89.7%. Despite converging to the similar loss values, models with a pre-trained encoder showed more stable validation loss curves during training than those trained from scratch. The complete validation loss curves have been included in the supplementary materials (Fig. S.2).

5   Discussion

In this study we developed a method to identify errors in OAR segmentations from radiotherapy planning CT scans without a ground truth. Our proposed model leverages both the CT image appearance and the segmentation's shape with a novel hybrid CNN-GNN-MLP architecture. To the best of our knowledge this is the first time geometric learning with GNNs has been applied to segmentation error prediction. Using GNNs, our method gained greater context of the geometric shape of the segmentation by leveraging the manifold structure directly. In our study, we found that pre-training the CNN part of the model did not improve prediction results, but only enabled a smoother training process.

Several approaches have been published to perform automatic segmentation QA based on statistical models or using multiple segmentation models. However, most of these previous methods predict a global accuracy measure such as DSC [3,9,11,13,17,20]. Our proposed method makes node-wise predictions of errors in distinct locations on the segmentation to help guide clinicians when completing the manual contour QA process.

There have been some previous works to predict uncertainties using model ensembles [12] or directly using new CNN models [18]. However, such approaches require replacing current segmentation systems with new models. Our method can be applied to segmentations generated by any means, allowing QA of existing methods. As most current commercial auto-segmentation packages do not provide uncertainty estimations, our proposed method is particularly relevant [7].

The perturbed segmentations that were generated for the training dataset were not informed by real anatomical shape variations at this time. However, we ensured that the magnitude of the perturbations was consistent with the inter-observer variability for parotid gland contouring, which Brouwer et al. found to range between 2.0 and 2.6 mm (3D SD) [1]. In future, training data could be generated based on observer variation data [22]. Further, our methodology would benefit from validation on sets of real erroneous contours. However, this was out of scope for this paper.

In this study we only performed experiments on one OAR, the parotid gland, in the head and neck anatomy. This is a difficult organ to segment and creates a broad range of error scenarios for our model to learn to recognise [1]. Given our proposed methods performance on this OAR, we are confident our method can be extended to further structures in the future.

The CT resolution and method of sampling sub-volumes for the CNN will indirectly affect the model performance. However, as the nodes of the OAR mesh were not anchored to the CT grid we were able to predict realistic small errors. In future we aim to quantify the impact of the CT resolution and sub-volume characteristics on our methods performance.

Although the range of the class bins were narrow (±2.5 mm), errors of this magnitude are important in modern radiotherapy. For example, very steep dose gradients of $\sim5\%$/mm are seen in proton therapy plans [24]. Small errors in delineations can result in significant changes, making plans unacceptable.

As it stands, our tool could easily be dropped into the clinical pathway, potentially decreasing the time clinicians spend correcting contours by drawing their attention to regions that need updating. It is potentially also fast enough to be applied in an interactive setting, e.g. during the final editing process.

6 Conclusion

We have developed a novel method to predict segmentation errors without a ground-truth. Our model combines shape and appearance features using a hybrid CNN-GNN architecture. This method could provide automatic segmentation QA to improve consistency and outcomes for patients treated with radiotherapy.

Acknowledgements. Marcel van Herk was supported by NIHR Manchester Biomedical Research Centre. This work was also supported by Cancer Research UK via funding to the Cancer Research Manchester Centre [C147/A25254] and by Cancer Research UK RadNet Manchester [C1994/A28701].

References

1. Brouwer, C.L., et al.: 3D variation in delineation of head and neck organs at risk. Radiat. Oncol. **7**(1) (2012). https://doi.org/10.1186/1748-717X-7-32
2. Cardenas, C.E., Yang, J., Anderson, B.M., Court, L.E., Brock, K.B.: Advances in auto-segmentation. Semin. Radiat. Oncol. **29**(3), 185–197 (2019). https://doi.org/10.1016/j.semradonc.2019.02.001
3. Chen, H.C., et al.: Automated contouring error detection based on supervised geometric attribute distribution models for radiation therapy: a general strategy. Med. Phys. **42**(2), 1048–1059 (2015). https://doi.org/10.1118/1.4906197

4. Cootes, T., Edwards, G., Taylor, C.: Active appearance models. IEEE Trans. Pattern Anal. Mach. Intell. **23**(6), 681–685 (2001). https://doi.org/10.1109/34.927467

5. Fey, M., Lenssen, J.E.: Fast graph representation learning with PyTorch geometric. In: ICLR Workshop on Representation Learning on Graphs and Manifolds (2019). https://doi.org/10.48550/arXiv.1903.02428

6. Fey, M., Lenssen, J.E., Weichert, F., Muller, H.: SplineCNN: fast geometric deep learning with continuous B-spline kernels. In: Proceedings of the IEEE Computer Society Conference on Computer Vision and Pattern Recognition, pp. 869–877. IEEE Computer Society, November 2018. https://doi.org/10.1109/CVPR.2018.00097

7. Green, A.F., Aznar, M.C., Muirhead, R., Vasquez Osorio, E.M.: Reading the mind of a machine: hopes and hypes of artificial intelligence for clinical oncology imaging. Clin. Oncol. **34**(3), e130–e134 (2022). https://doi.org/10.1016/j.clon.2021.11.008

8. Hui, C.B., et al.: Quality assurance tool for organ at risk delineation in radiation therapy using a parametric statistical approach. Med. Phys. **45**(5), 2089–2096 (2018). https://doi.org/10.1002/mp.12835

9. Kohlberger, T., Singh, V., Alvino, C., Bahlmann, C., Grady, L.: Evaluating segmentation error without ground truth. In: Ayache, N., Delingette, H., Golland, P., Mori, K. (eds.) MICCAI 2012. LNCS, vol. 7510, pp. 528–536. Springer, Heidelberg (2012). https://doi.org/10.1007/978-3-642-33415-3_65

10. Loshchilov, I., Hutter, F.: Decoupled weight decay regularization. In: International Conference on Learning Representations (ICLR) (2019). https://doi.org/10.48550/arXiv.1711.05101

11. McCarroll, R., et al.: Machine learning for the prediction of physician edits to clinical autocontours in the head-and-neck. Med. Phys. **44**(6), 3160 (2017). https://doi.org/10.1002/mp.12304

12. Mehrtash, A., Wells, W.M., Tempany, C.M., Abolmaesumi, P., Kapur, T.: Confidence calibration and predictive uncertainty estimation for deep medical image segmentation. IEEE Trans. Med. Imaging **39**(12), 3868–3878 (2020). https://doi.org/10.1109/TMI.2020.3006437

13. Men, K., Geng, H., Biswas, T., Liao, Z., Xiao, Y.: Automated quality assurance of OAR contouring for lung cancer based on segmentation with deep active learning. Front. Oncol. **10**, 986 (2020). https://doi.org/10.3389/fonc.2020.00986

14. Nikolov, S., et al.: Deep learning to achieve clinically applicable segmentation of head and neck anatomy for radiotherapy. ArXiv e-prints (2018). https://doi.org/10.48550/arXiv.1809.04430

15. Noroozi, M., Favaro, P.: Unsupervised learning of visual representations by solving Jigsaw puzzles. In: Leibe, B., Matas, J., Sebe, N., Welling, M. (eds.) ECCV 2016. LNCS, vol. 9910, pp. 69–84. Springer, Cham (2016). https://doi.org/10.1007/978-3-319-46466-4_5

16. Pfaff, T., Fortunato, M., Sanchez-Gonzalez, A., Battaglia, P.W.: Learning mesh-based simulation with graph networks. In: 9th International Conference on Learning Representations, ICLR (2021). https://doi.org/10.48550/arXiv.2010.03409

17. Rhee, D.J., et al.: Automatic detection of contouring errors using convolutional neural networks. Med. Phys. **46**(11), 5086–5097 (2019). https://doi.org/10.1002/mp.13814

18. Sander, J., de Vos, B.D., Išgum, I.: Automatic segmentation with detection of local segmentation failures in cardiac MRI. Sci. Rep. **10**(1), 1–19 (2020). https://doi.org/10.1038/s41598-020-77733-4

19. Taubin, G.: Curve and surface smoothing without shrinkage. In: Proceedings of IEEE International Conference on Computer Vision. IEEE Computer Society Press (1995). https://doi.org/10.1109/iccv.1995.466848
20. Valindria, V.V., et al.: Reverse classification accuracy: predicting segmentation performance in the absence of ground truth. IEEE Trans. Med. Imaging **36**(8), 1597–1606 (2017). https://doi.org/10.1109/TMI.2017.2665165
21. Vandewinckele, L., et al.: Overview of artificial intelligence-based applications in radiotherapy: recommendations for implementation and quality assurance (2020). https://doi.org/10.1016/j.radonc.2020.09.008
22. Vasquez Osorio, E.M., Shortall, J., Robbins, J., Van Herk, M.: Contour generation with realistic inter-observer variation. In: 19th International Conference on the use of Computers in Radiation Therapy, pp. 222–223 (2019)
23. Xu, K., Hu, W., Leskovec, J., Jegelka, S.: How powerful are graph neural networks? In: International Conference on Learning Representations (2019). https://openreview.net/forum?id=ryGs6iA5Km
24. Yan, Y., Yang, J., Li, Y., Ding, Y., Kadbi, M., Wang, J.: Impact of geometric distortion on dose deviation for photon and proton treatment plans. J. Appl. Clin. Med. Phys. **23**(3) (2022). https://doi.org/10.1002/acm2.13517
25. Zhou, Q.Y., Park, J., Koltun, V.: Open3D: a modern library for 3D data processing. arXiv:1801.09847 (2018). https://doi.org/10.48550/arXiv.1801.09847

A Novel Knowledge Keeper Network for 7T-Free but 7T-Guided Brain Tissue Segmentation

Jieun Lee[1], Kwanseok Oh[1], Dinggang Shen[3,4], and Heung-Il Suk[1,2(✉)]

[1] Department of Artificial Intelligence, Korea University, Seoul, Republic of Korea
{ljelje415,ksohh,hisuk}@korea.ac.kr
[2] Department of Brain and Cognitive Engineering, Korea University,
Seoul, Republic of Korea
[3] School of Biomedical Engineering, ShanghaiTech University, Shanghai, China
dgshen@shanghaitech.edu.cn
[4] Shanghai United Imaging Intelligence Co., Ltd., Shanghai, China

Abstract. An increase in signal-to-noise ratio (SNR) and susceptibility-induced contrast at higher field strengths, *e.g.*, 7T, is crucial for medical image analysis by providing better insights for the pathophysiology, diagnosis, and treatment of several disease entities. However, it is difficult to obtain 7T images in real clinical practices due to the high cost and low accessibility. In this paper, we propose a novel knowledge keeper network (KKN) to guide brain tissue segmentation by taking advantage of 7T representations without explicitly using 7T images. By extracting features of a 3T input image substantially and then transforming them to 7T features via knowledge distillation (KD), our method achieves deriving 7T-like representations from a given 3T image and exploits them for tissue segmentation. On two independent datasets, we evaluated our method's validity in qualitative and quantitative manners on 7T-like image synthesis and 7T-guided tissue segmentation by comparing with the comparative methods in the literature.

Keywords: Brain tissue segmentation · Knowledge distillation · Medical image synthesis · 7T MRI

1 Introduction

With the advent of a 7T Magnetic Resonance Imaging (MRI) scanner, structural brain images have shown a higher signal-to-noise ratio (SNR), spatial resolution, and tissue contrast [3]. Such accurate anatomical details could help gain important insights regarding the pathophysiology, diagnosis, and treatment of several disease entities [12,16]. However, in real clinical environments, most MR images are acquired at routine 3T MRI because the 7T MRI scanner is more expensive

Supplementary Information The online version contains supplementary material available at https://doi.org/10.1007/978-3-031-16443-9_32.

and hence less available. There have been methods for contrast enhancement [17,18] using the images produced at a lower magnetic field strength, but it is still limited to revealing the significant contrast between tissues similar to 7T.

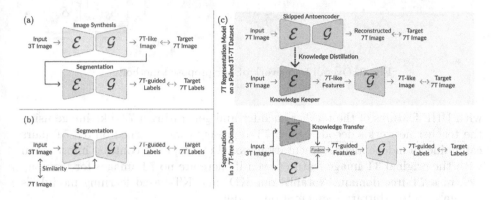

Fig. 1. The conceptual differences between conventional 7T-assisted methods (a, b) and ours (c). $\mathcal{E}$ and $\mathcal{G}$ denote an encoder and decoder, respectively.

To this end, deep-learning methods have recently emerged for tissue segmentation to map a 3T image to its corresponding 7T image in either image-level or feature-level. There are two different approaches in the existing 3T-to-7T mapping methods for tissue segmentation: (1) synthesizing 7T-like image first and then conducting segmentation (Fig. 1(a)); (2) learning feature representations with the guidance of 7T features (Fig. 1(b)). In synthesizing 7T-like image first and then conducting segmentation [2,19,25], it is prone to unexpected deformation or distortion of the inherent features in 3T informative for segmentation while mapping to 7T at the image level. Meanwhile, in learning feature representations with the guidance of 7T features [8,23], it is common to use handcrafted or task-dependent features extracted from 7T images, revealing its limitations in robustness and generality. More importantly, both approaches lack applicability and generalizability on datasets not including 7T images because they require pairs of a 3T image and its corresponding 7T image for training.

To circumvent the limitations of the existing methods mentioned above, in this work, we propose a novel knowledge keeper network (KKN) to estimate hierarchical 7T features from a 3T image via knowledge distillation (KD) [4,10] and guide a tissue segmentation model as a way of knowledge transfer (KT). While KT is a direct way, KD has been used to transfer information indirectly from a well-trained teacher network to a student network. For feature-based distillation of the low-level vision tasks (*e.g.*, segmentation and synthesis), encoder-decoder networks can effectively transfer the knowledge in intermediate layers [6,20,24]. As shown Fig. 1(c), we train a teacher network formulated as an autoencoder with skip connections to extract ultra-high field (UHF) features, *i.e.*, 7T features, and then KKN transforms the 3T image to 7T-like features by matching

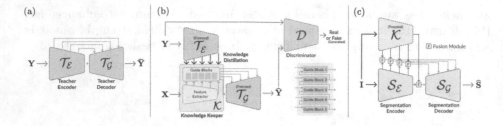

Fig. 2. Overall framework of our proposed method.

with UHF features of the teacher encoder and generating a 7T-like image using the teacher network's decoder. Such 7T representations learned on 3T-7T pairs can guide an independent segmentation model to adapt the contrast information with the original 3T image features even if there are no 7T images for training, *i.e.*, in a 7T-free domain. Notably, our KD- and KT-based learning module is pluggable into arbitrary segmentation models.

The main contributions of this work are as follows: (1) We propose a novel knowledge keeper network (KKN) to extract 7T-like representations from a 3T image that can further guide tissue segmentation to take advantage of the contrast information of 7T without 7T images; (2) We show the generalizability of our proposed method by plugging in and transferring the pre-trained KKN to the existing brain tissue segmentation models on Internet Brain Segmentation Repository (IBSR) dataset; (3) Our proposed method outperforms the existing methods for 7T-like image synthesis and brain tissue segmentation robustly.

2 Proposed Method

This section introduces our proposed method, illustrated in Fig. 2, which consists of a teacher network $\mathcal{T}$, a knowledge keeper network (KKN) $\mathcal{K}$, and a segmentation network $\mathcal{S}$. Through two-stage KD methods, KKN learns to infer 7T-like representations from a 3T image using a paired 3T-7T dataset $\{\mathbf{X}, \mathbf{Y}\}$. And then, the representations are provided to the segmentation network in order to achieve 7T-guided segmentation using only a 3T image $\mathbf{I}$ and its corresponding segmentation mask $\mathbf{S}$ in a 7T-free domain.

2.1 Teacher Network

At the first stage of KD, the teacher network takes a 7T image $\mathbf{Y}$ as input and produces a reconstructed image $\widetilde{\mathbf{Y}}$ to extract ultra-high field (UHF) features, as shown in Fig. 2(a). For the designing of our teacher network, we adopt the architecture of 3D U-Net [7] with a minor modification that uses upsampling followed by $1 \times 1 \times 1$ convolution instead of transposed convolution (details in Supplementary S1).

Since the skipped feature maps provide significantly strong supervision to reconstruct the input image, especially in the lower feature level, the network might be hindered from extracting useful information for knowledge distillation. Thus, we apply two regularization tricks for training. The first regularization is feature-level reconstruction, inspired by multi-level autoencoders of [9]. Similar to image-level reconstruction that the whole image is reproduced through contracting and expanding paths, feature-level reconstruction is a process that reproduces the $(i\text{-}1)$-level feature map from the contracted i-level feature map. We employ feature-level reconstruction at each encoder level by training additional expanding blocks that consist of upsampling followed by convolution. So the teacher network is trained by the sum of traditional image-level reconstruction error and feature-level reconstruction errors as follows:

$$\mathcal{L}_{\text{teacher}} = \lambda_{\text{img}} \| \mathbf{Y} - \widetilde{\mathbf{Y}} \|_1 + \sum_{i=1}^{N-1} \| f_i^{\mathcal{T}} - \widetilde{f}_i^{\mathcal{T}} \|_2, \qquad (1)$$

where $f_i^{\mathcal{T}}$ and $\widetilde{f}_i^{\mathcal{T}}$ denote the output feature maps of the i-th level in the teacher encoder and the reconstructed feature map from $f_{i+1}^{\mathcal{T}}$, respectively. λ_{img} is a weighting hyperparameter, and N is the number of encoder levels. The second regularization is adding random Gaussian noises to feature maps when passing to their corresponding expanding levels. By injecting the noise into respective feature maps, the network learns features that have the ability to reconstruct the image despite slight distortion, thus being robust to variations. With two feature-level regularization tricks, we build a well-trained teacher network for knowledge distillation of the next stage.

2.2   Knowledge Keeper Network

KKN has an intermediate role between 7T and 7T-free domains. In a 7T domain that 7T images are available for, the network learns how to transform the extracted 3T features to 7T in a task of image synthesis. And then, in a 7T-free domain, the network guides the segmentation model to reveal the tissue contrast without using 7T images. So KKN consists of a feature extractor and guide blocks, as shown in Fig. 2(b).

The feature extractor is an encoder structure with densely connected convolution blocks, which are smaller but deeper compared to the teacher encoder so that 3T features are concatenated progressively. The highest feature map is transformed at the guide block and then fed into the lower level guide block. Thus, the guide block of each level except the highest level combines features of both current and higher levels. In the guide block, the current level feature map following convolution block and higher-level feature map following upsampling are concatenated to generate two attention maps using $1 \times 1 \times 1$ convolution. The two attention maps are multiplied with current and higher-level feature maps, respectively, and then two features are added channel-wisely (details in Supplementary S2). Lastly, the transformation of the feature map in the guide block utilizes the relative positions and similarities between global context and the most characteristic features as a position-aware recalibration module inspired by [14].

For training, we use both a pre-trained teacher network and a patch discriminator [11] to generate 7T-like feature maps. Specifically, the feature maps of KKN are optimized to match UHF features that the teacher encoder $\mathcal{T}_{\mathcal{E}}$ extracts from a 7T image. The loss for knowledge distillation is defined as:

$$\mathcal{L}_{\text{distill}} = \sum_{i=1}^{N} (\|f_i^{\mathcal{T}} - f_i^{\mathcal{K}}\|_2 + \|\mu(f_i^{\mathcal{T}}) - \mu(f_i^{\mathcal{K}})\|_2 + \|\sigma(f_i^{\mathcal{T}}) - \sigma(f_i^{\mathcal{K}})\|_2), \quad (2)$$

where $f_i^{\mathcal{K}}$ denotes the i-th level feature map of KKN. The channel-wise mean μ and standard deviation σ are calculated for matching the distribution of feature maps [22] between KKN and the teacher encoder. The teacher decoder $\mathcal{T}_{\mathcal{G}}$ generates an image using the feature maps of the KKN and allows us to compute the synthesis loss $\mathcal{L}_{\text{vox}} = \|\mathbf{Y} - \widehat{\mathbf{Y}}\|_1$. The discriminator $\mathcal{D}$ is adversarially trained for realistic 7T image synthesis with a loss of $\mathcal{L}_{\text{adv}} = \|\mathcal{D}(\widehat{\mathbf{Y}}) - \mathbf{1}\|_2$, where $\mathbf{1}$ denotes a label of real. By replacing the teacher encoder with KKN for synthesizing the 7T-like image $\widehat{\mathbf{Y}}$, the 3T image can be transformed to the 7T-like features. Putting all together, a total loss is composed of the distillation loss, the synthesis loss, and the adversarial loss [15] as follows:

$$\mathcal{L}_{\text{keeper}} = \mathcal{L}_{\text{distill}} + \lambda_{\text{vox}} \mathcal{L}_{\text{voxel}} + \lambda_{\text{adv}} \mathcal{L}_{\text{adv}}, \quad (3)$$

where λ_{vox} and λ_{adv} are hyperparameters. By optimizing the loss, our proposed network is expected to learn 7T-like feature representations that can be generalized to a 7T-free domain.

2.3   Brain Tissue Segmentation Network

We employ tissue segmentation on both paired and unpaired datasets. On a paired dataset, we validate that a decoder can segment the 3T image into 7T-guided labels using only 7T-like features. Since KKN extracts reliable 7T-like representations, additional feature extraction or transformation is not necessary for 7T-guided segmentation. On the other hand, we use a segmentation network with an encoder-decoder architecture on an unpaired dataset. For an unseen image, the segmentation encoder $\mathcal{S}_{\mathcal{E}}$ extracts an input 3T image features, which are transferred into fusion modules, as indicated by blue lines in Fig. 2(c). Simultaneously, 7T-like representations are fused as guidance to adapt the tissue contrast for segmentation, as indicated by purple lines in Fig. 2(c). The fusion module of i-th feature level includes two learnable parameters, α_i and β_i, so that they are used for the weighted sum between guidance and original feature as $\alpha_i f_i^{\mathcal{K}} + \beta_i f_i^{\mathcal{S}}$ where $f_i^{\mathcal{S}}$ denotes the i-th level feature map of the segmentation encoder. By adding the simple fusion modules and pre-trained KKN, any segmentation model that has an encoder and decoder can take advantage of 7T-like representations. While training the model for 7T-guided segmentation, we can optimize categorical cross-entropy loss using only the predicted and ground truth labels without 7T images.

Table 1. Dice Similarity Coefficient (DSC, %) of brain tissue segmentation on a paired 3T-7T dataset across 15 folds (mean ± standard deviation).

	Method		Performance			
	3T-to-7T	Segmentation	CSF	GM	WM	Average
(a)	WATNet [19]	ANTs	64.19 ± 6.45	71.85 ± 6.67	80.04 ± 7.37	72.03 ± 5.54
(b)	Ours	ANTs	72.33 ± 5.81	75.05 ± 6.40	83.49 ± 7.60	76.96 ± 5.49
(c)	Ours	3D U-Net decoder	72.66 ± 4.99	76.08 ± 4.04	86.18 ± 6.62	78.31 ± 4.03

3 Experimental Settings and Results

3.1 Dataset

Paired 3T-7T Dataset. We used 15 pairs of 3T and 7T MR images from 15 adults. All T1-weighted images were acquired using Siemens Magnetom Trio 3T and 7T MRI scanners. Similar to [19,25], we normalized the image intensity to the range of [0, 1] and applied histogram matching on the 3T and 7T images, respectively. For evaluation, we adopted leave-one-subject-out cross-validation that one pair of 3T and 7T images of a subject was used for testing and the remaining pairs used for training and validation. For the segmentation task, we obtained the ground truth of 7T tissue segmentation masks using advanced normalization tools (ANTs) [1].

IBSR Dataset. In order to verify the generalizability of our proposed method, we used the public IBSR[1] dataset on the brain tissue segmentation task. The dataset consists of 18 subjects aged 7 to 71 years and corresponding tissue segmentation masks annotated by the expert. Following [5,21], we applied min-max normalization in the range of [0, 1] and Contrast-Limited Adaptive Histogram Equalization (CLAHE) for the images. We randomly split the dataset into 7 training, 3 validation, and 8 testing samples with the condition that each dataset has a similar age group. Other additional experimental details, including network architectures and training hyperparameters, can be found in Supplementary S3.

3.2 Results and Analysis

Here, we performed quantitative and qualitative evaluations to demonstrate the validity of our proposed method in various metrics. First, for image synthesis analysis, we adopted two commonly used image quality metrics, peak signal-to-noise ratio (PSNR) and structural similarity (SSIM). Second, for the brain tissue segmentation, we used Dice Similarity Coefficient (DSC) of cerebrospinal fluid (CSF), gray matter (GM), and white matter (WM). Lastly, we visualized the synthesized 7T images and tissue-segmented images.

---

[1] https://www.nitrc.org/projects/ibsr.

Table 2. Dice Similarity Coefficient (DSC, %) of brain tissue segmentation on IBSR dataset. "Ours" indicates to plug in our KKN and fusion module to the baseline models.

	Method	CSF	GM	WM	Average
Baseline	3D U-Net [7]	82.09	91.81	88.49	87.46
	VoxResNet [5]	77.70	90.72	86.75	85.06
	U-SegNet [13]	72.03	87.79	84.22	81.35
	RP-Net [21]	74.89	87.29	83.41	81.86
Ours	3D U-Net [7]	82.56 (0.47↑)	93.00 (1.19↑)	89.60 (1.11↑)	88.39 (0.93↑)
	VoxResNet [5]	79.32 (1.62↑)	91.68 (0.96↑)	87.71 (0.96↑)	86.24 (1.18↑)
	U-SegNet [13]	75.15 (3.12↑)	88.88 (1.09↑)	85.52 (1.30↑)	83.18 (1.83↑)
	RP-Net [21]	76.74 (1.85↑)	88.33 (1.04↑)	84.72 (1.31↑)	83.26 (1.40↑)

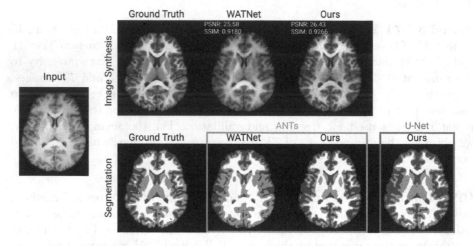

Fig. 3. Illustration of image synthesis (top) and brain tissue segmentation (bottom) results on a paired 3T-7T dataset.

Reliability of 7T-Like Representations. We verified the reliability of 7T-like representations available for image synthesis and segmentation. In Fig. 3, we compared the synthesized images and segmented masks with WATNet [19], which records the state-of-the-art performance in 7T image synthesis. Regarding the image synthesis (top of Fig. 3), we observed that our proposed method not only achieved the better performance of PSNR (+0.85) and SSIM (+0.0086) than WATNet but also provided the synthesized image that has strong tissue contrast compared to the relatively blurred image of WATNet. Through tissue segmentation using ANTs, the difference of anatomical structures between the synthesized images was clearly visible in an orange box of Fig. 3. It can be seen that our method accurately captured the tissue contrast with the dramatically increased DSC of all tissue labels (Table 1(a) and (b)). Moreover, we evaluated the performance of tissue segmentation when training only a 3D U-Net decoder using 7T-like representations from the pre-trained KKN. The segmentation results with 7T-like features along with the 3T input image features

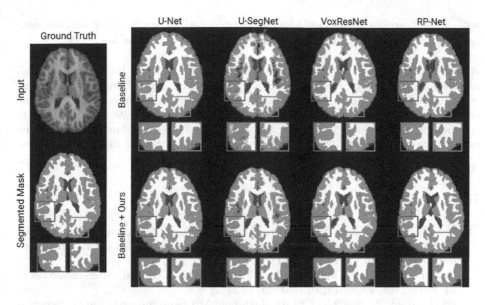

Fig. 4. Comparison of the segmentation outcomes on IBSR dataset. The first row shows the segmentation results of baselines, and the second row shows the segmentation results plugged in our proposed method to baselines (blue box: middle temporal gyrus, orange box: middle occipital gyrus). (Color figure online)

produced the more precise and accurate segmentation qualitatively (a blue box of Fig. 3) and quantitatively (Table 1(c)).

Generalizability on Tissue Segmentation. Our proposed method is compared with four competing methods: 3D U-Net [7], VoxResNet [5], U-SegNet [13], and RP-Net [21]. Table 2 presents that our proposed method plugged into baseline models improved performance in all tissue labels consistently. It means that the integration of the 7T-like features obtained by our proposed KKN and those of the 3T input in the fusion modules helped obtain better tissue-discriminative information. From the visual inspection of the segmentation results in Fig. 4, the baseline methods combined with our proposed KNN could identify tissues in a fine-grained level.

4   Conclusion

In this paper, we proposed a novel knowledge keeper network (KKN), trained via KD and KT, to guide and train a brain tissue segmentation model using 7T-like representations in a 7T-free domain. Our proposed method demonstrated the reliability and generalizability of 7T-like representations on paired 3T-to-7T and 7T-free IBSR datasets with superior performance quantitatively and qualitatively. Thanks to its pluggable property into U-Net-shaped deep networks with minor modifications, we believe its potential to apply other downstream tasks, for which higher field strength, *e.g.*, 7T, images can be beneficial.

Acknowledgements. This work was supported by Institute of Information & communications Technology Planning & Evaluation (IITP) grant funded by the Korea government (MSIT) No. 2022-0-00959 ((Part 2) Few-Shot Learning of Causal Inference in Vision and Language for Decision Making) and by Institute of Information & communications Technology Planning & Evaluation (IITP) grant funded by the Korea government (MSIT) No. 2019-0-00079 (Department of Artificial Intelligence (Korea University)).

References

1. Avants, B.B., Tustison, N.J., Wu, J., Cook, P.A., Gee, J.C.: An open source multivariate framework for N-tissue segmentation with evaluation on public data. Neuroinformatics **9**(4), 381–400 (2011)
2. Bahrami, K., Shi, F., Rekik, I., Gao, Y., Shen, D.: 7T-guided super-resolution of 3T MRI. Med. Phys. **44**(5), 1661–1677 (2017)
3. Balchandani, P., Naidich, T.: Ultra-high-field MR neuroimaging. Am. J. Neuroradiol. **36**(7), 1204–1215 (2015)
4. Buciluǎ, C., Caruana, R., Niculescu-Mizil, A.: Model compression. In: Proceedings of the 12th ACM SIGKDD International Conference on Knowledge Discovery and Data Mining, pp. 535–541 (2006)
5. Chen, H., Dou, Q., Yu, L., Qin, J., Heng, P.A.: VoxResNet: deep voxelwise residual networks for brain segmentation from 3D MR images. NeuroImage **170**, 446–455 (2018)
6. Chen, P., Liu, S., Zhao, H., Jia, J.: Distilling knowledge via knowledge review. In: Proceedings of the IEEE/CVF Conference on Computer Vision and Pattern Recognition, pp. 5008–5017 (2021)
7. Çiçek, Ö., Abdulkadir, A., Lienkamp, S.S., Brox, T., Ronneberger, O.: 3D U-Net: learning dense volumetric segmentation from sparse annotation. In: Ourselin, S., Joskowicz, L., Sabuncu, M.R., Unal, G., Wells, W. (eds.) MICCAI 2016. LNCS, vol. 9901, pp. 424–432. Springer, Cham (2016). https://doi.org/10.1007/978-3-319-46723-8_49
8. Deng, M., et al.: Learning-based 3T brain MRI segmentation with guidance from 7T MRI labeling. Med. Phys. **43**(12), 6588–6597 (2016)
9. He, Y., Carass, A., Zuo, L., Dewey, B.E., Prince, J.L.: Autoencoder based self-supervised test-time adaptation for medical image analysis. Med. Image Anal. **72**, 102136 (2021)
10. Hinton, G., Vinyals, O., Dean, J.: Distilling the knowledge in a neural network. arXiv preprint arXiv:1503.02531 (2015)
11. Isola, P., Zhu, J.Y., Zhou, T., Efros, A.A.: Image-to-image translation with conditional adversarial networks. In: Proceedings of the IEEE Conference on Computer Vision and Pattern Recognition, pp. 1125–1134 (2017)
12. Van der Kolk, A.G., Hendrikse, J., Zwanenburg, J.J., Visser, F., Luijten, P.R.: Clinical applications of 7T MRI in the brain. Eur. J. Radiol. **82**(5), 708–718 (2013)
13. Kumar, P., Nagar, P., Arora, C., Gupta, A.: U-SegNet: fully convolutional neural network based automated brain tissue segmentation tool. In: 2018 25th IEEE International Conference on Image Processing (ICIP), pp. 3503–3507. IEEE (2018)

14. Ma, X., Fu, S.: Position-aware recalibration module: learning from feature semantics and feature position. In: International Joint Conference on Artificial Intelligence, pp. 797–803 (2020)
15. Mao, X., Li, Q., Xie, H., Lau, R.Y., Wang, Z., Paul Smolley, S.: Least squares generative adversarial networks. In: Proceedings of the IEEE International Conference on Computer Vision, pp. 2794–2802 (2017)
16. Nie, D., Wang, L., Adeli, E., Lao, C., Lin, W., Shen, D.: 3-D fully convolutional networks for multimodal isointense infant brain image segmentation. IEEE Trans. Cybern. **49**(3), 1123–1136 (2018)
17. Nöbauer-Huhmann, I.M., et al.: Magnetic resonance imaging contrast enhancement of brain tumors at 3 tesla versus 1.5 tesla. Invest. Radiol. **37**(3), 114–119 (2002)
18. Oak, P.V., Kamathe, R.: Contrast enhancement of brain MRI images using histogram based techniques. Int. J. Innov. Res. Electr. Electron. Instrum. Control Eng. **1**(3), 90–94 (2013)
19. Qu, L., Zhang, Y., Wang, S., Yap, P.T., Shen, D.: Synthesized 7T MRI from 3T MRI via deep learning in spatial and wavelet domains. Med. Image Anal. **62**, 101663 (2020)
20. Wang, H., Li, Y., Wang, Y., Hu, H., Yang, M.H.: Collaborative distillation for ultra-resolution universal style transfer. In: Proceedings of the IEEE/CVF Conference on Computer Vision and Pattern Recognition, pp. 1860–1869 (2020)
21. Wang, L., Xie, C., Zeng, N.: RP-Net: a 3D convolutional neural network for brain segmentation from magnetic resonance imaging. IEEE Access **7**, 39670–39679 (2019)
22. Wang, P., Li, Y., Singh, K.K., Lu, J., Vasconcelos, N.: IMAGINE: image synthesis by image-guided model inversion. In: Proceedings of the IEEE/CVF Conference on Computer Vision and Pattern Recognition, pp. 3681–3690 (2021)
23. Wei, J., et al.: A cascaded nested network for 3T brain MR image segmentation guided by 7T labeling. Pattern Recognit. **124**, 108420 (2021)
24. Xiang, L., et al.: Deep embedding convolutional neural network for synthesizing CT image from T1-weighted MR image. Med. Image Anal. **47**, 31–44 (2018)
25. Zhang, Y., Cheng, J.-Z., Xiang, L., Yap, P.-T., Shen, D.: Dual-domain cascaded regression for synthesizing 7T from 3T MRI. In: Frangi, A.F., Schnabel, J.A., Davatzikos, C., Alberola-López, C., Fichtinger, G. (eds.) MICCAI 2018. LNCS, vol. 11070, pp. 410–417. Springer, Cham (2018). https://doi.org/10.1007/978-3-030-00928-1_47

OnlyCaps-Net, a Capsule only Based Neural Network for 2D and 3D Semantic Segmentation

Savinien Bonheur[1,2]([✉]), Franz Thaler[2,4], Michael Pienn[1], Horst Olschewski[1,3], Horst Bischof[2], and Martin Urschler[5]

[1] Ludwig Boltzmann Institute for Lung Vascular Research, Graz, Austria
bonheur@student.tugraz.at
[2] Institute of Computer Graphics and Vision, Graz University of Technology, Graz, Austria
[3] Department of Internal Medicine, Medical University of Graz, Graz, Austria
[4] Gottfried Schatz Research Center: Biophysics, Medical University of Graz, Graz, Austria
[5] School of Computer Science, University of Auckland, Auckland, New Zealand

Abstract. Since their introduction by Sabour et al., capsule networks have been extended to 2D semantic segmentation with the introduction of convolutional capsules. While extended further to 3D semantic segmentation when mixed with Convolutional Neural Networks (CNNs), no capsule-only network (to the best of our knowledge) has been able to reach CNNs' accuracy on multilabel segmentation tasks. In this work, we propose OnlyCaps-Net, the first competitive capsule-only network for 2D and 3D multi-label semantic segmentation. OnlyCaps-Net improves both capsules' accuracy and inference speed by replacing Sabour et al. squashing with the introduction of two novel squashing functions, i.e. softsquash or unitsquash, and the iterative routing with a new parameter free single pass routing, i.e. unit routing. Additionally, OnlyCaps-Net introduces a new parameter efficient convolutional capsule type, i.e. depthwise separable convolutional capsule.

Keywords: Capsule networks · Convolutional Neural Networks · Multi-label · Semantic segmentation · 2D · 3D

1 Introduction

Allowing the development of new physiological measures, semantic segmentation has seen huge progress thanks to the introduction of CNNs. A more recent architecture introduced in [8], capsule networks, aims at decomposing images into hierarchical features encoded as vectors (i.e. capsules) by learning a specific projection between each lower-level capsule (i.e. child capsule) and higher-level capsule (i.e. parent capsule) pair. Thanks to the feature description of each

© The Author(s), under exclusive license to Springer Nature Switzerland AG 2022
L. Wang et al. (Eds.): MICCAI 2022, LNCS 13435, pp. 340–349, 2022.
https://doi.org/10.1007/978-3-031-16443-9_33

capsule's vector, capsules allow to dynamically weight the contribution of each child capsule to each parent capsule.

Although introduced for classification, capsule networks have been applied to binary semantic segmentation in [4] and extended to multi-label semantic segmentation with Matwo-CapsNet in [1]. Matwo-CapsNet proposed to disentangle feature appearance and pose through two matrices sharing a depthwise filter. Not reaching CNN's accuracy on the Japanese Society of Radiological Technology (JSRT) lungs dataset [9], this approach is limited by its heavy compute and memory footprint as it requires storing and manipulation of two matrices for each feature. Focusing on mixing single vector capsules and CNNs, 3D-UCaps [5] based capsule network outperforms pure CNN's architecture on several 3D multi-label segmentation datasets (i.e. ISeg-2017 dataset [11], Cardiac dataset [10] and Hippocampus dataset [10]). Achieving for the first time (to the best of our knowledge) a competitive capsules-only network for 3D multi-label segmentation, we outperform 3D-UCaps [5] on the Hippocampus dataset [10] by changing the way capsules transform and interact through our contributions:

- We adapt separable depthwise convolution to single feature vector to, (1) reduce the memory requirement for each capsule from $2z^2$ (Matwo-CapsNet) to z, and (2) reduce training time by an order of magnitude.
- We implement two novel squashing functions (i.e. softsquash and unitsquash).
- We introduce unit routing, a parameter-free single pass routing mechanism which projects capsules' predictions onto the n-sphere before adding them.

2   Method

Differing from CNNs, where each feature in each layer is represented by a scalar, capsule networks describe features through a vector of size z. Each vector magnitude is designed to represent the feature existence likelihood while each vector instantiation should encode the feature itself. The instantiation of a capsule is originally created through the weighted association of linearly projected child capsules. The weight associated with each child capsule is calculated dynamically through a routing algorithm to adapt capsule's associations to the input. In the following, we will formalize and motivate the concept of depthwise separable convolutional capsules, introduce two novel squashing functions and investigate routing capsules by projecting them onto a unit n-sphere to form parent features. We first verify that each contribution improves the results through an ablation study on the JSRT dataset [3], then demonstrate the capabilities of our OnlyCaps-Net with a three-fold cross-validation on the computerized tomography (CT) MM-WHS 3D dataset [13], and finally, compare our approach to 3D-UCaps on the Hippocampus dataset [10].

Depthwise Separable Convolutional Capsules. In order to allow our capsules to use spatial relationships for image analysis, we need a mechanism to gather spatial information. In [4], convolutional capsules are used to learn a bias

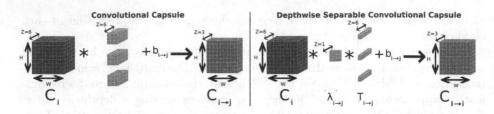

Fig. 1. Convolutional capsule [4] and our proposed depthwise separable convolutional capsule.

and a different pointwise affine projection from domain i to j ($z_i \times z_j + 1$) for each capsule within a $K_x \times K_y$ kernel's window for a total parameter count of $(K_x \times K_y \times z_i \times z_j + 1)$. In this paper, we propose depthwise separable convolutional capsules (Fig. 1), reusing the concept of convolutional capsules but reducing the number of necessary parameters. By setting the child capsule to parent capsule projection to be shared across all possible capsule positions within the kernel's window, we can deconstruct convolutional capsules as a depthwise kernel $\lambda_{\Delta_{x,y}}$ (i.e. spatial filtering) followed by z_j pointwise convolutions (regrouped as a matrix multiplication under $\mathbf{T}_{i \to j}$ of dimension $z_i \times z_j$). This reduces the number of parameters from $(K_x \times K_y \times z_i \times z_j + 1)$ to $(K_x \times K_y + z_i \times z_j + 1)$ and gives convolutional capsules the form of a depthwise separable convolution:

$$\mathbf{C}_{i \to j} = (\sum_{\Delta_{x,y}} \mathbf{C}_{i\Delta_{x,y}} \lambda_{\Delta_{x,y}}) \times \mathbf{T}_{i \to j} + b_{i \to j} \tag{1}$$

where $\mathbf{C}_{i \to j}$ is the prediction of the child capsule $\mathbf{C}_i$ for the parent capsule $\mathbf{C}_j$. Further, $\lambda_{\Delta_{x,y}}$ is a set of learned $K_x \times K_y$ spatial weights, $b_{i \to j}$ is a learned bias, and $\mathbf{T}_{i \to j}$ is a learned matrix multiplication, mapping from the child i capsule's domain to the parent j capsule's domain.

Capsule Unit Routing. Once spatially filtered and projected to the parent space, the predicted parent vectors from all child capsules $\mathbf{C}_{i \to j}$ need to be routed through a weighted sum to form the unsquashed parent capsule $\mathbf{P}_j$

$$\mathbf{P}_j = \sum_i w_{i \to j} \mathbf{C}_{i \to j} \tag{2}$$

In [8] the routing weight $w_{i \to j}$ is calculated through an iterative optimization strategy based on the cross-correlation between vectors of child and parent capsules. Differently, we propose a parameter free single pass routing, unit routing, where $w_{i \to j}$ project each child capsule prediction ($\mathbf{C}_{i \to j}$) onto the unit n-sphere:

$$w_{i \to j} = \frac{1}{\|\mathbf{C}_{i \to j}\|_2} \tag{3}$$

Capsule Squashing. Lastly, the weighted sum of child capsules' prediction is passed through a squashing function to form the output parent capsule. In the following, we propose two new squashing functions to replace the one from [8].

Our first squashing function follows the assumption that each capsule layer should entirely explain its input at every location, i.e. we force the sum of all predicted capsules' lengths (i.e. the predicted capsule's L_2 norm $\|\mathbf{P}_j\|_2$) to represent a probability distribution summing to 1 at each coordinate and across all capsule types through the use of a softmax function:

$$\mathbf{C}_j = softsquash(\mathbf{P}_j) = \frac{\mathbf{P}_j}{\|\mathbf{P}_j\|_2} \times softmax_j(\|\mathbf{P}_j\|_2) \qquad (4)$$

Our second squashing simply projects each capsule onto the unit n-sphere:

$$\mathbf{C}_j = unitsquash(\mathbf{P}_j) = \frac{\mathbf{P}_j}{\|\mathbf{P}_j\|_2} \qquad (5)$$

OnlyCaps-Net Architecture. Similarly as in SegCaps [4], for each pixel (x, y) of either the image or the intermediate layers, a set of capsules is defined, i.e. $\mathbf{C}_i(x, y)$. The predicted segmentation label L at each location (x, y) corresponds to the index of the capsule of the last layer with the highest activation.

3 Experimental Setup and Results

Datasets. <u>JSRT:</u> Consisting of 247 chest X-ray scans manually annotated by van Ginneken et al. [3], the JSRT dataset ground-truth segmentation labels consist of 6 labels, background, left and right lungs, left and right clavicles, and the heart. The JSRT images are resized to a resolution of 128 × 128 pixels and split into a training and testing sets as recommended by [3]. We trained and averaged over 5 full training runs (300 000 iterations) with the augmentation scheme from [1].

<u>MM-WHS:</u> The MM-WHS challenge [13] dataset consists of CT and magnetic resonance imaging (MRI) volumes in which 7 heart substructures were manually segmented. To validate our method, we only used the 60 CT scans (20 for training and 40 for testing). Acquired in several clinics and with different scanners, the scans are of different image quality, resolution, and voxel spacing. Although the maximum size of the CT images is 300 × 300 × 188, using the data pipeline of [6], each scan is resampled to a cropped box of 64 × 64 × 64 voxels with a spacing of 3 × 3 × 3, centered around the heart. Augmented and split into three cross-validation sets as in [6], our networks' results (one training run for each fold and each network configuration) are directly comparable with their results.

<u>Hippocampus:</u> This dataset [10] contains annotated MRI volumes from 90 healthy adults and 105 adults with a non-affective psychotic disorder. It uses 2 labels to segment the hippocampus and parts of the subiculum. Resized to

$32 \times 32 \times 32$ due to memory limitation, our experiments used 3D-UCaps public implementation[1].

Evaluated Networks. <u>JSRT:</u> All our networks are trained with AdaDelta optimizers [12] with default settings and using the pre-processing pipeline from [1].

Training our OnlyCaps-Net on the JSRT dataset with an Nvidia Titan Xp equipped with 12 GB RAM takes $\sim$4 h30 ($\sim$45 h for Matwo-CapsNet). Our architecture follows a U-Net scheme [7] where each layer consists of 7 capsules of length 6 and kernel size 3. U-Net Depth denote a U-Net using CNN's depthwise separable convolution to allow for comparison with our depthwise separable convolutional capsule (used with all OnlyCaps-Net). ConvCaps-Net is a capsule network built from convolutional capsules [4].

<u>MM-WHS:</u> Trained using the AdaDelta optimizer [12] with default settings and pre-processing pipeline from [6], our OnlyCaps-Net uses a U-Net like architecture (kernel size of 5×5) with 8 capsules of size 8 at the first level, 12 capsules of size 11 for each downsampled level and 8 capsules of size 4 for the output layer. Available online[2] and using the 12 GB memory of our Nvidia GeForce RTX 2080 Ti (with the use of memory-efficient gradient propagation [2]), our network requires $\sim$14.5 h to train for 50000 iterations while the U-Net (27 filters by level for 454K parameters) architecture requires $\sim$3.5 h. We do not train any 3D variant of Matwo-CapsNet due to an excessive memory footprint.

<u>Hippocampus:</u> All networks were modified by adding our contributions to 3D-UCaps public code (See footnote 1) (see our github(See footnote 2)). Using an Nvidia TITAN Xp, we observed a 3D-UCaps using convolutional capsules to run $\sim$3.0 iterations/second while using 2.9 GB of memory, while a modified 3D-UCaps, using our depthwise convolutional capsule with unit squashing and unit routing, runs at $\sim$4.3 iterations/second (a $\sim$29% speed up) with a memory footprint of 1.9 GB (a $\sim$35% reduction in memory). Meanwhile our OnlyCaps-Net uses 8.0 GB for 1.2 training iterations/second. We chose not to use any reconstruction loss for any network as we did not observe significant improvement from it. Lastly, the Only-CapsNet implementation is using the SegCaps architecture where the convolutional stem has been removed and all layers have 12 capsules of size 12. We calculate and report statistical significance with a p value of 0.05.

Results. In the JRST experiments (Table 1), the comparison of our network using solely our novel squashing function and depthwise separable convolution (OnlyCaps-Net) show the strength of softsquash squashing and depthwise convolutional capsules by outperforming both a U-Net built on depthwise separable convolution (U-Net Depth) and a capsule network based on convolutional capsule from [4], i.e. ConvCaps-Net. There is further improvement when using

---

[1] https://github.com/VinAIResearch/3D-UCaps.
[2] https://github.com/savinienb/OnlyCaps-Net.

Table 1. Multi-label Dice score for the evaluated networks in % (mean ± standard deviation) on the JSRT dataset. ConvCaps-Net represents a capsules network using convolutional capsule [4] instead of depthwise separable convolutional capsules. Dyn. indicates the use of dynamic routing [8] with 3 iterations. The number of network parameters is shown as multiples of thousands.

Network	#Params	Loss	Squashing	Routing	Lungs		Clavicles		Heart	Mean
					L	R	L	R		
U-Net [1]	42K	Softmax	X	X	**97.36**	97.87	90.87	90.64	94.49	94.25
Matwo-CapsNet [1]	43K	Spread	Psquash/Squash	Dual	97.01	97.45	88.32	87.82	94.37	92.99
U-Net Depth	42K	Softmax	X	X	97.29 (±0.04)	97.83 (±0.10)	90.78 (±0.68)	90.47 (±0.71)	**94.56** (±0.14)	94.18 (±0.14)
ConvCaps-Net	273K	Spread	Softsquash	X	97.16 (±0.07)	97.66 (±0.6)	89.71 (±0.25)	89.9 (±0.25)	94.27 (±0.25)	93.74 (±0.08)
OnlyCaps-Net	41K	Spread	Softsquash	X	97.29 (±0.03)	97.83 (±0.04)	90.78 (±0.26)	90.47 (±0.5)	**94.56** (±0.12)	94.18 (±0.16)
OnlyCaps-Net	41K	Spread	Softsquash	Dyn.	97.17 (±0.06)	97.65 (±0.09)	90.03 (±0.32)	89.06 (±1.02)	94.27 (±0.13)	93.64 (±0.23)
OnlyCaps-Net	41K	Spread	Softsquash	Unit	97.35 (±0.02)	97.88 (±0.04)	91.03 (±0.56)	91.12 (±0.23)	94.42 (±0.08)	94.36 (±0.12)
OnlyCaps-Net	41K	Spread	Squash	Unit	97.28 (±0.05)	97.88 (±0.03)	91.13 (±0.12)	91.01 (±0.29)	94.39 (±0.17)	94.34 (±0.08)
OnlyCaps-Net	41K	Spread	Unitsquash	Unit	97.27 (±0.03)	97.86 (±0.23)	91.16 (±0.12)	**91.33** (±0.19)	94.37 (±0.07)	94.4 (±0.06)
OnlyCaps-Net	41K	Softmax	Unitsquash	Unit	97.32 (±0.05)	**97.89** (±0.03)	**91.28** (±0.28)	**91.33** (±0.17)	94.34 (±0.19)	**94.43** (±0.04)

Table 2. Multi-label Dice scores for the evaluated networks in % (mean ± standard deviation) using three-fold cross validation on the training images of the MM-WHS challenge [6]. Results shown for the individual cardiac structures (LV - left ventricle blood cavity, Myo - myocardium of the left ventricle, RV - right ventricle blood cavity, LA -left atrium blood cavity, RA - right atrium blood cavity, aorta - ascending aorta, PA - pulmonary artery) and the average Dice score over the seven heart substructures. The number of network parameters is shown as multiples of thousands.

Network	#Params	Loss	Squashing	Routing	LV	Myo	RV	LA	RA	aorta	PA	Mean
U-Net [6]	150747K	Softmax	X	X	91.0 (±4.3)	86.1 (±4.2)	88.8 (±3.9)	91.0 (±5.2)	86.5 (±6.0)	94.0 (±6.2)	83.7 (±7.7)	88.7 (±3.3)
Seg-CNN [6]	151899K	Softmax	X	X	92.4 (±3.3)	87.2 (±3.9)	87.9 (±6.5)	**92.4** (±3.6)	87.8 (±6.5)	91.1 (±18.4)	83.3 (±9.1)	88.9 (±4.3)
U-Net small	454K	Softmax	X	X	**92.9** (±2.9)	**88.1** (±4.8)	89.7 (±3.4)	92.0 (±4.1)	88.1 (±5.6)	**94.7** (±3.4)	83.0 (±10.2)	**89.8** (±3.8)
OnlyCaps-Net	443K	Spread	Squash	Unit	92.2 (±3.6)	86.7 (±4.5)	89.0 (±3.6)	91.5 (±4.1)	87.9 (±5.0)	91.2 (±14.8)	81.6 (±9.5)	88.6 (±3.7)
OnlyCaps-Net	443K	Spread	Softsquash	Unit	92.4 (±3.3)	86.7 (±4.8)	88.7 (±4.5)	92.1 (±3.3)	87.8 (±5.8)	93.1 (±6.3)	83.6 (±7.7)	89.2 (±3.5)
OnlyCaps-Net	443K	Spread	Unitsquash	Unit	92.1 (±3.6)	87.4 (±4.4)	**89.9** (±3.3)	92.1 (±3.4)	**88.7** (±4.9)	92.9 (±9.0)	**84.4** (±8.3)	89.6 (±3.1)
OnlyCaps-Net	443K	Softmax	Unitsquash	Unit	92.6 (±3.4)	87.9 (±3.7)	89.4 (±3.6)	91.6 (±4.6)	87.7 (±6.1)	93.6 (±6.7)	84.3 (±8.3)	89.6 (±3.3)

Table 3. Averages of the four cross-fold metrics for the evaluated networks on the 3D Hippocampus dataset. Dice score, Sensitivity and Precision in %. SD represent Standard Distance and Hausdorff represent Hausdorff Distance. Conv. represents the use of convolutional capsules [4] and Depth. the use of depthwise separable convolutional capsules. Dyn. indicates the use of dynamic routing [8]. Softs. and Units. stand for the softsquash and unitsquash function, respectively. Ant. and Post. indicate Anterior and Posterior respectively. The number of network parameters is shown in millions. * indicate statistical significance p<0.05.

Network	#Params	Capsule	Squashing	Routing	Dice		Sensitivity		Precision		SD		Hausdorff	
					Ant.	Post.	Ant.	Post.	Ant.	Post.	Ant.	Post.	Ant.	Post.
3D-UCaps	4.2M	Conv.	Squash	Dyn.	89.13	87.52	88.76	87.92	90.20	88.28	0.43	0.42	**2.24**	2.82
					(±0.23)	(±0.37)	(±0.61)	(±0.61)	(±0.79)	(±0.99)	(±0.01)	(±0.00)	(±0.00)	(±0.26)
3D-UCaps	3.5M	Depth.	Softs.	Unit	89.47	87.40	89.82	87.93	90.19	87.76	0.42	0.41	**2.24**	2.82
					(±0.48)	(±0.35)	(±1.22)	(±0.87)	(±0.67)	(±0.70)	(±0.01)	(±0.02)	(±0.00)	(0.26)
3D-UCaps	3.5M	Depth.	Units.	Unit	89.50	87.68	90.19	87.65*	89.73	88.85	0.42	0.40	**2.24**	**2.64**
					(±0.39)	(±0.36)	(±0.50)	(±0.58)	(±0.90)	(±0.93)	(±0.02)	(±0.01)	(±0.00)	(±0.22)
OnlyCaps-Net	3.1M	Depth.	Softs.	Unit	89.56	88.06	89.34	87.84	**91.00***	**89.3**	**0.40**	**0.39**	2.29	2.91
					(±0.31)	(±0.19)	(±0.63)	(±0.91)	(±0.38)	(±0.66)	(±0.01)	(±0.01)	(±0.11)	(±0.10)
OnlyCaps-Net	3.1M	Depth.	Units.	Unit	**89.78**	**88.23**	**90.61***	**88.75***	89.80	88.17*	0.41	0.40	**2.24**	2.68
					(±0.41)	(±0.75)	(±0.45)	(±0.35)	(±0.75)	(±0.46)	(±0.01)	(±0.02)	(±0.00)	(±0.28)

our novel unit routing algorithm which leads the network to outperform a U-Net with a similar number of parameters. We can furthermore observe that the combination of unitsquash with unit routing leads to our best performing capsule variation. This observation is strengthened by our experiment on the 3D CT MM-WHS (Table 2) where the combination of unit routing and unitsquash achieve better performances than U-Net [6] and Seg-CNN [6] and all other capsules variations while being close to the accuracy of our own U-Net network. Table 3 shows that using depthwise separable convolutional capsules enables a parameter reduction while still performing similarly to a 3D-UCaps using convolutional capsules. Moreover, 3D-UCaps get a slight performance boost when combined with our unit routing and either of our two new squashing functions. To perform a qualitative analysis on the MM-WHS dataset (Fig. 2), we selected the best and worst performing segmentations for each network. Interestingly, there is a rougher appearance of capsules' produced segmentation for Fig. 2(d) and (p). Those artifacts seem to arise from the interaction of the spread loss with the first training set of our cross-validation setup. Indeed, with the two other cross-validation training sets or with a softmax loss, the segmentation output of the capsule network is as smooth as U-Net segmentations. Nevertheless, the spread loss function avoids the right atrium lump (in yellow) and myocardium (in cyan) in (f) and (g). The overall poorer segmentation results on ct-train1018 (e,h) are likely due to acquisition artifacts near the aorta.

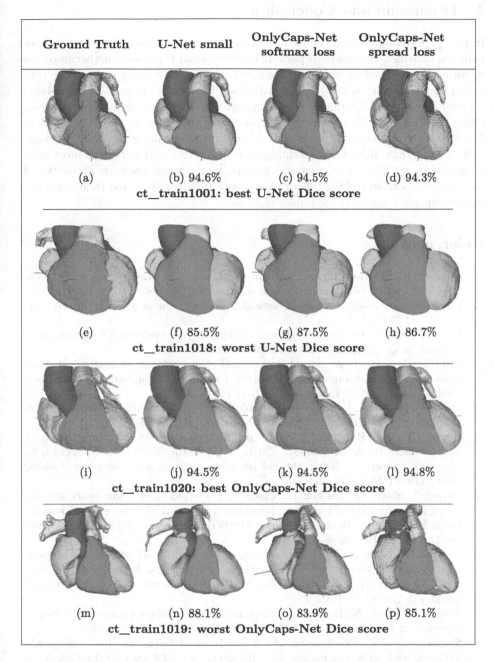

Fig. 2. MM-WHS CT segmentation results with best and worst Dice scores for CNN and capsules. The leftmost column shows the ground truth, the second column shows CNN predictions, the third column shows OnlyCaps-Net trained with a softmax loss and the final column shows OnlyCaps-Net predictions when trained with a spread loss.

4   Discussion and Conclusion

In this paper, we introduced and evaluated a parameter efficient variant of convolution capsules, depthwise separable convolutional capsules. Furthermore, we introduced two novel squashing and a novel capsule routing algorithm. The proposed OnlyCaps-Net is competitive with U-Net for both 2D and 3D multi-label segmentation and outperforms 3D-UCaps while being more parameter efficient. OnlyCaps-Net shows better accuracy, speed, and greater memory efficiency than Matwo-CapsNet. As well, we observed our novel squashing and routing algorithms to increase 3D-UCaps' training speed by $\approx$29% and reduce memory consumption by $\approx$35%. In future work, we would like to investigate the properties of our two proposed squashing functions and routing algorithm, and their influence on the capsule's latent space representation.

References

1. Bonheur, S., Štern, D., Payer, C., Pienn, M., Olschewski, H., Urschler, M.: Matwo-CapsNet: a multi-label semantic segmentation capsules network. In: International Conference on Medical Image Computing and Computer-Assisted Intervention (MICCAI) (2019)
2. Chen, T., Xu, B., Zhang, C., Guestrin, C.: Training deep nets with sublinear memory cost. arXiv preprint arXiv:1604.06174 (2016)
3. van Ginneken, B., Stegmann, M.B., Loog, M.: Segmentation of anatomical structures in chest radiographs using supervised methods: a comparative study on a public database. Med. Image Anal. (MIA) **10**(1) (2006)
4. LaLonde, R., Bagci, U.: Capsules for Object Segmentation. In: International Conference on Medical Imaging with Deep Learning (MIDL) (2018)
5. Nguyen, T., Hua, B.S., Le, N.: 3D-UCaps: 3D capsules UNet for volumetric image segmentation. In: de Bruijne, M., Cattin, P.C., Cotin, S., Padoy, N., Speidel, S., Zheng, Y., Essert, C. (eds.) Medical Image Computing and Computer Assisted Intervention (MICCAI) (2021)
6. Payer, C., Štern, D., Bischof, H., Urschler, M.: Multi-label whole heart segmentation using anatomical label configurations. In: Statistical Atlases and Computational Models of the Heart. ACDC and MMWHS Challenges. STACOM 2017. vol. 10663, pp. 190–198 (2018)
7. c Payer, C., Štern, D., Bischof, H., Urschler, M.: Multi-label whole heart segmentation using anatomical label configurations. In: Statistical Atlases and Computational Models of the Heart. ACDC and MMWHS Challenges. STACOM 2017. vol. 10663, pp. 190–198 (2018)
8. Sabour, S., Frosst, N., Hinton, G.E.: Dynamic routing between capsules. In: Neural Information Processing Systems (NIPS) (2017)
9. Shiraishi, J., et al.: Development of a digital image database for chest radiographs with and without a lung nodule. Am. J. Roentgenol. **174**(1),), 71–74. (2000)
10. Simpson, A.L., et al.: A large annotated medical image dataset for the development and evaluation of segmentation algorithms. ArXiv abs/1902.09063 (2019)
11. Wang, L., et al.: Benchmark on automatic six-month-old infant brain segmentation algorithms: the iseg-2017 challenge. IEEE Trans. Med. Imaging **38**(9) (2019)

12. Zeiler, M.D.: ADADELTA: an adaptive learning rate method. CoRR abs/1212.5 (2012)
13. Zhuang, X., et al.: Evaluation of algorithms for multi-modality whole heart segmentation: an open-access grand challenge. Med. Image Anal. (MIA) **58**, 101537 (2019)

Identifying and Combating Bias in Segmentation Networks by Leveraging Multiple Resolutions

Leonie Henschel[1], David Kügler[1], Derek S. Andrews[2], Christine W. Nordahl[2], and Martin Reuter[1,3,4(✉)]

[1] German Center for Neurodegenerative Diseases (DZNE), Bonn, Germany
`martin.reuter@dzne.de`
[2] MIND Institute and Department of Psychiatry and Behavioral Sciences, UC Davis, Davis, USA
[3] Department of Radiology, Harvard Medical School, Boston, USA
[4] A. A. Martinos Center for Biomedical Imag., Mass. General Hospital, Boston, USA

Abstract. Exploration of bias has significant impact on the transparency and applicability of deep learning pipelines in medical settings, yet is so far woefully understudied. In this paper, we consider two separate groups for which training data is only available at differing image resolutions. For group H, available images and labels are at the preferred high resolution while for group L only deprecated lower resolution data exist. We analyse how this resolution-bias in the data distribution propagates to systematically biased predictions for group L at higher resolutions. Our results demonstrate that single-resolution training settings result in significant loss of volumetric group differences that translate to erroneous segmentations as measured by DSC and subsequent classification failures on the low resolution group. We further explore how training data across resolutions can be used to combat this systematic bias. Specifically, we investigate the effect of image resampling, scale augmentation and resolution independence and demonstrate that biases can effectively be reduced with multi-resolution approaches.

1 Introduction

Over the last years, deep learning networks have been shown to accurately segment brain MRIs to the point that they rival traditional pipelines in reliability, sensitivity and accuracy [1–3]. However, for supervised training all networks rely on reference segmentations with the core assumption of representative training sets. In medical imaging, this assumption is often inherently violated due to limited data availability – specifically for uncommon pathologies, age groups, treatment effects, genetic and ethnic groups as well as resolutions. There are, for example, no manual annotation or disease datasets available at leading sub-millimeter resolutions.

Supplementary Information The online version contains supplementary material available at https://doi.org/10.1007/978-3-031-16443-9_34.

In consequence, networks performing well in the training domain fail to generalize to distributions outside this scope [4]. Unfortunately, most of these effects can only be adequately addressed by acquiring more data.

Datasets with heterogeneous resolutions may, however, be investigated by resizing available ground truth. Using this setup we can address a hitherto unexplored and important question: what effects do (systematic) resolution differences between populations or groups have on segmentation networks? Can we learn segmentation tasks across resolutions? Cross-resolution training is import, because of the trend to acquire images at submillimeter resolutions, where manual labels are not yet available.

The analysis of bias in deep learning (DL) for medical imaging is still in its infancy despite its relevance for fairness [5]. With no established evaluation metrics, recent work employs overlap measures to evaluate racial (cardiac segmentation) [6,7] and motion bias (brain lesion segmentation) [8]. However, overlap measures (e.g. Dice) only detect reduced performance without differentiation between random and systematic errors, as simple simulations reveal. Figure 1 illustrates how similar Dice performance can lead to significantly different volume estimation (shift of the volume distribution in the histogram). Systematic volumetry changes often relate to group differentiations and are a far more robust marker for systematic errors than overlap measures. Therefore, the analysis of bias in traditional neuroimaging pipelines have included signed volume measures to capture consistent over- or undersegmentation biases [9–11]. No work has addressed segmentation bias in DL for neuroanatomical segmentations or multiple resolutions. In a multi-resolution setting, care has to be taken to further differentiate methodological from resolution biases. The latter arise when the spatial representation of a structure can not be captured due to the technical resolution limits, i.e. the voxel size exceeds the size of the structure itself. Tissue borders in particular are prone to get lost at lower resolutions. Our analysis is therefore based on both, accuracy and volume bias to effectively detect systematic segmentation errors.

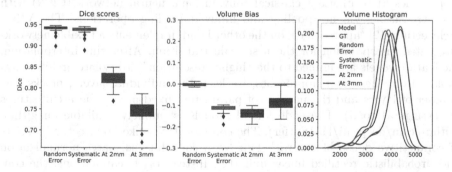

Fig. 1. Simple simulations of Random and Systematic (oversegmentation) Errors illustrate significant distribution shifts despite similar Dice scores. In contrast, Volume Bias highlights this error type. Downsampling (to 2/3 mm) results in undersegmentation.

The multi-resolution question allows two general training strategies - using either a dedicated fixed-resolution network working on (resampled) images or a network that accepts multiple resolutions during training. We evaluate both strategies and show that: i) single resolution networks fail to adequately capture group differences across resolutions and work only on the original distribution they were trained on, ii) training with resampled images alone is not enough to remove this bias, and iii) multi-resolution approaches, i.e. scaling augmentation or resolution independent networks, effectively translate high-resolutional information across the resolution differences and combat the detected bias in segmentation networks.

2   Methodology

We explore four approaches to address unbiased cross-resolution generalization (see Fig. 2): For approach a), group L (at lower resolutions) is omitted from network training. We expect this reduced training dataset diversity will result in low generalization performance for the left-out group L, if the underlying distributions differ significantly (out of distribution effect). Two approaches combine both groups (H and L) for training by image resampling via bilinear interpolation (and lossy label interpolation via majority voting) outside the network: b) upsampling of group L to the higher resolution, or c) scale augmentation during network training. Finally, d) Voxel-size Independent Neural Networks (VINN) [12] achieve resolution independence by internal rescaling shifting the interpolation step into the network itself. This avoids lossy label interpolation while preserving resolution-independence in a multi-resolution training set-up. While approaches (a) and (b) only operate on a single-resolution, (c) and (d) employ a multi-resolution setting.

2.1   Networks

Approaches a) to c) use a classical convolutional neural network (CNN) with scale transitions via maxpooling and index unpooling operations. The VINN architecture [12] (approach d), on the other hand, implements a flexible network-integrated interpolation for the first scale transition. After this interpolation, the feature voxels are unified to the "higher resolution". To isolate architectural choices from these four approaches, we keep the individual layers, blocks, the number of layers and the number of parameters fixed across the architectures in settings a) to d). The code for all models is publicly available on github (github.com/deep-mi/FastSurfer). The common UNet-like architecture consists of competitive dense blocks with four 3×3 convolutions, batch-normalization and probabilistic rectified linear unit [12]. Instead of concatenations, the connecting local skip connections employ feature competition through maxout [13]. For each approach, we train one network on 2D slices per anatomical plane. During inference, we aggregate per-plane probabilities to better exploit the 3D context. This 2.5D approach has recently been shown to outperform state-of-the-art 3D approaches for whole brain segmentation [14].

3   Experiments

To ensure the presence of differences between group H and L, we select two neuroimaging segmentation tasks with known group differences: Cortical segmentation in adults and children (Task 1) and hippocampal segmentation with groups stratified based on their hippocampal volume (Task 2).

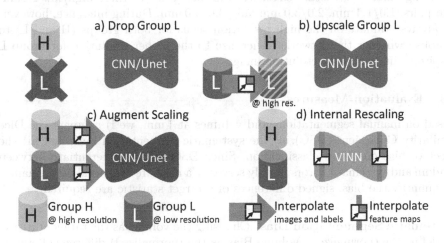

Fig. 2. Four approaches to harmonize the resolution-biased datasets for segmentation network: a) drop the lower resolution data, b) upsample the low-resolution dataset to the higher resolution, c) randomly rescale both datasets as augmentation, and d) incorporate a resampling to the higher resolution into the network.

3.1   Task 1 – Cortical Segmentation in Adults and Children

To answer whether the inclusion of low-resolutional children can overcome the group bias in high resolutional evaluation, we construct a dataset with 90 participants (45 adults, 45 children). The manual cortical segmentations and corresponding T1-weighted intensity images for the adults group ($\geq$20 years) originate from the open-source Mindboggle-101 [15] dataset. From the UC-Davis MIND Institute Autism Phenome Project [16], we obtain manually corrected cortical segmentations and corresponding T1-weighted intensity images from non-autistic children (2.3 to 4.5 years). The total set of 90 scans is split into 50/10/30 for training/validation/testing, each balanced between age groups.

3.2   Task 2 – Stratified Hippocampal Segmentation

Using manual hippocampus segmentations and the corresponding T1-weighted intensity images from the Harmonized Protocol (HarP) [17,18], the MICCAI 2012 Multi-atlas labeling challenge (MALC) [19] and the Internet Brain Segmentation Repository (IBSR), provided by the Center for Morphometric Analysis (CMA) at Massachusetts General Hospital [20], we construct two groups by selecting the 40 smallest (all either AD or MCI) and 40 largest (dominated by controls) hippocampi. Keeping volume groups balanced, we split them into training/validation/testing sets (n = 50/10/20). Akin to Experiment 1, small hippocampi images (group L) are downsampled to a low resolution before training, while images from group H (large hippocampi) reside at a higher resolution.

To analyze the performance of approaches a) to d) in multiple scenarios including different structure sizes, we train networks for three high/low resolution pairs: 1.0/1.4 mm, 2.0/3.0 mm, and 3.0/4.0 mm. During inference, however, we always evaluate on the high resolution scans for both groups (H and L) to establish whether the networks generalize to the higher resolution for group L despite training on low-resolution images.

3.3   Evaluation Measures

Based on manual segmentations and volumes at 1 mm, we i) compare the Dice Similarity Coefficient (DSC), ii) the systematic segmentation bias, and iii) the effect of bias on group classification. Since DSC cannot differentiate between random and systematic errors, it only serves as a measure of general performance. To quantify the bias, signed differences of a target statistic are required.

Systematic Segmentation Bias. Choosing the volume as the target statistic, we define the (normalized) Volume Bias as the (normalized) difference between predicted (Volume($f(x)$)) and ground truth volume (Volume(y)).

$$\text{Volume Bias(Group } i) = E_{(x,y) \sim \text{Group } i} \left[\frac{\text{Volume}(f(x)) - \text{Volume}(y)}{\text{Volume}(y)} \right] \quad (1)$$

A value of zero indicates an unbiased estimate with > 0 implying over- and < 0 under-segmentation. To eliminate the contribution of random errors, we form an expected value over the analyzed group. Since our group-wise test sets contain only 15 (Task 1) or 10 (Task 2) participants, we estimate the expected value by the median and additionally plot its distribution.

Group Classification. We use the area under the Receiver operating characteristics curve (AUC) to investigate the effect of segmentation bias on down-stream analysis. Overall, the task explores, whether "information" of known volumetry effects is obfuscated by (training/dataset) bias which is highly relevant for group analysis. Here, we calculate the AUC for every possible discrimination threshold of the gray matter volumetry with the correct age group assignment (children versus adults) as the predictor variable of interest.

3.4   Training

We train all networks under equal hardware, hyper-parameter and dataset settings unless stated otherwise. Hyper-parameters were not changed from the original publications except for the internal resolution of VINN [12] (set to the higher image resolution for all tasks). Independent models are trained for 70 epochs with mini batch size of 16 using the AdamW optimizer [21], a cosine annealing learning rate schedule [22], an initial learning rate of 0.001 and momentum of 0.95. The number of epochs between two warm restarts is initially set to 10 and subsequently increased by a factor of two. Based on the validation set, we select the best training state per network and assured convergence of all models.

4   Results

First, we analyze how well the different approaches can reproduce the modes of volume distributions for both tasks. Figure 3 combines the distributions from multiple experiments to show three origins for systematic differences: 1. the different groups, 2. the approach used for prediction (method bias), and 3. the resolution of the analysis (resolution bias). Figure 3 clearly illustrates that approaches (b)–(d) are able to separate the different modes of the two groups in most experiments. The prominent exception is "CNN/UNet (Group H only)" (black line). For Task 1 (top row), the Children distribution is basically non exis-tent thus confirming that segmentation networks do not automatically generalize to unseen distributions. Similarly, for Task 2 (bottom row), the distribution for small volumes shifts to the right with decreasing resolutions. A similar trend can be observed for the resampled images "CNN/UNet (resampled)" (red line) - the volume distribution shifts to the left and converges to a uni-modal distribution.

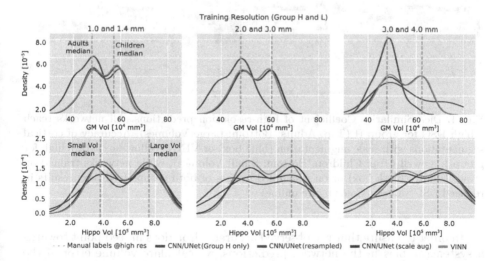

Fig. 3. Segmentation bias in single resolution networks. CNN/UNet (black) trained on group H (Adults) only fails to replicate modes of volume distributions (vertical lines) in out of distribution samples (Children). Multi-resolution networks (CNN + scale augmentation, blue and VINN, yellow) efficiently transfer information during training across resolutions and match underlying distribution of ground truth labels. (Color figure online)

The findings from the histogram distribution are further substantiated by the calculated accuracy measures (Fig. 4). First, all approaches perform equally well on the high-resolution group encountered during training (group H, Adult and Large Volumes). The drop in DSC to 63 for the low resolution pair (3.0 mm versus 4.0 mm) on the gray matter results from lost cortical details in comparison to the 1.0 mm ground truth (i.e. deep sulci are not distinguishable by a complete voxel anymore). This problem is less pronounced on the hippocampus due to its compact shape. Second, cross resolution transfer works best for approaches (c) and (d) (i.e. CNN + scale augmentation (blue box) or VINN (yellow box)). The DSC reaches the highest value for both experiments on the left-out high-resolution group L (children and small volumes) and furthermore stays relatively constant across the different resolution pairs with the highest cortical DSC of 94.84, 82.84 and 76.07 for VINN, respectively (hippocampus: 86.4, 69.67, 61.72). As expected, accuracy on group L is lowest for approach (a) (black box). In accordance with results from Fig. 3, approach (b) (red box) decreases segmentation performance for small hippocampi.

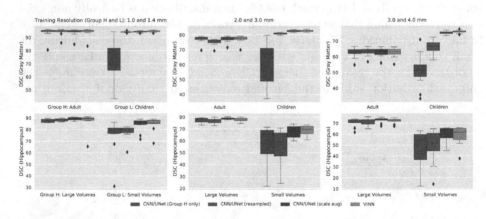

Fig. 4. Dice Similarity Coefficient of high-resolution predictions. All networks reach a high DSC for group H (Top: Adults, Bottom: Large Volumes). Accuracy of cortical segmentations with scale augmentation (blue box) or VINN (yellow box) stays high for group L as well (Top: Children, Bottom: Small Volumes), while networks trained on group H only (black box) or on resampled images of both groups (red box) results in reduced DSC values. (Color figure online)

In order to ensure that the observed accuracy drop does indeed point towards a systematic bias in the network predictions, we calculate volume error of the cortical and hippocampal structures. For the multi-resolution approaches (c) and (d) (scale augmentation and VINN), the cortex segmentation of children at high-resolution MRI is consistent with a median volume error close to 0 (Fig. 5, blue and yellow box). The single resolution approach (a) consistently underestimates the cortical gray matter volume for resolutions 1.0 and 2.0 mm by around 20% (black box). Approach (b) shows a bias with shifts to lower resolutions (red box, 13% at 3.0 versus 4.0 mm). This effect is increasingly apparent on the

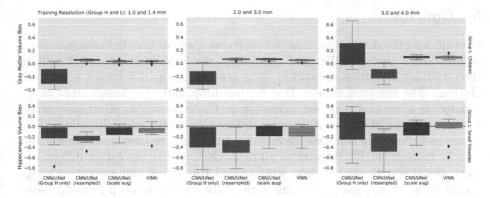

Fig. 5. Volume bias in network predictions. Single resolution approaches show systematic biases in the volume estimations of group L (black and red box). Volume estimates based on VINN (yellow box) and CNN with scale augmentation (blue box) change only marginal indicating increased robustness to resolution variances of group L. (Color figure online)

high-resolution participants with small hippocampi in experiment 2. Here, the resampling leads to a constant underestimation of the hippocampal volume by 23.5 to 39%. Multi-resolution approaches (c) and (d) are again more consistent with a variance in volume measures across resolutions between only 2 and 6%.

Taken together, the results point towards a systematic accuracy bias in single resolution segmentation networks. Next, we evaluate if the detected bias is strong enough to carry over to down-stream analysis. The incompetence of the single resolution approach (a) to translate group differences across resolutions clearly reduces classification performance (Table 1, AUC of 0.36). The method bias hence actively undermines group separation. At 3.0 mm, the AUC recovers probably because the failure to estimate volumes at all rather than the actual volume value becomes the basis for the decision threshold. The accuracy results in Fig. 4 corroborate this hypothesis. Interestingly, the volume bias also translates to a classification bias for approach (b) at the lower resolutions (3.0 mm). Here, the AUC drops by more than 50% to 0.4 when attempting to classify children versus adults. Only scale augmentation and VINN preserve a high AUC of above 0.95 across all resolution pairs.

Table 1. Area under curve (AUC) for classification of "Children" based on volumetry information. Biased segmentations from CNN/UNet significantly decrease classification performance. Scale augmentation and VINN successfully diminish biases across resolutions and achieve AUCs above 0.92.

Training Res Group H & L	CNN/UNet (Group H only)	CNN/UNet (resampled)	CNN/UNet (scale aug)	VINN	Manual labels
1.0 & 1.4 mm	**0.36**	0.93	0.92	0.92	0.90
2.0 & 3.0 mm	**0.35**	0.95	0.94	0.94	0.90
3.0 & 4.0 mm	0.89	**0.4**	0.96	0.96	0.90

5   Conclusion

Overall, we provide the first comprehensive analysis of bias between groups in segmentation networks. We show that single resolution networks work well on their training distribution but fail to generalize group differences across resolutions. The detected volume bias further propagates to classification tasks. Finally, we show that scale augmentation in CNNs as well as alternative architectures with build-in resolution independence like VINN can help reduce these biases and effectively translate information from high- to lower-resolution scans. The findings are highly relevant to all fields dealing with resolution differences. Care should specifically be taken, when considering future high-resolution image acquisitions: subsequent processing with single-resolution deep-learning based analysis pipelines should be avoided.

Acknowledgements. This work was supported by the Federal Ministry of Education and Research of Germany (031L0206), by NIH (R01 LM012719, R01 AG064027, R56 MH121426, and P41 EB030006), and the Helmholtz Foundation (project DeGen). Funding for the APP dataset was provided by NIH and UC Davis MIND Institute. The study team would like to acknowledge Devani Cordero, of the Martinos Center of Biomedical Engineering and thank all of the families and children for their participation. Further, we thank and acknowledge the providers of the datasets listed in Sect. 3.

References

1. Henschel, L., Conjeti, S., Estrada, S., Diers, K., Fischl, B., Reuter, M.: FastSurfer - a fast and accurate deep learning based neuroimaging pipeline. NeuroImage **219**, 117012 (2020). https://doi.org/10.1016/j.neuroimage.2020.117012, https://doi.org/10.1016%2Fj.neuroimage.2020.117012

2. Iglesias, J.E., et al.: Joint super-resolution and synthesis of 1 mm isotropic MP-rage volumes from clinical MRI exams with scans of different orientation, resolution and contrast. Neuroimage **237**, 118206 (2021). https://doi.org/10.1016/j.neuroimage.2021.118206

3. Huo, Y., et al.: 3D whole brain segmentation using spatially localized atlas network tiles. Neuroimage **194**, 105–119 (2019)

4. Gonzalez, C., Gotkowski, K., Bucher, A., Fischbach, R., Kaltenborn, I., Mukhopadhyay, A.: Detecting when pre-trained nnU-Net models fail silently for Covid-19 lung lesion segmentation. In: de Bruijne, M., et al. (eds.) MICCAI 2021. LNCS, vol. 12907, pp. 304–314. Springer, Cham (2021). https://doi.org/10.1007/978-3-030-87234-2_29

5. Chen, R.J., et al.: Algorithm fairness in AI for medicine and healthcare. arXiv preprint arXiv:2110.00603 (2021)

6. Puyol-Antón, E., et al.: Fairness in cardiac MR image analysis: an investigation of bias due to data imbalance in deep learning based segmentation. In: de Bruijne, M., et al. (eds.) MICCAI 2021. LNCS, vol. 12903, pp. 413–423. Springer, Cham (2021). https://doi.org/10.1007/978-3-030-87199-4_39

7. Puyol Anton, E., et al.: Fairness in AI: are deep learning-based CMR segmentation algorithms biased? Eur. Heart J. **42** (2021). https://doi.org/10.1093/eurheartj/ehab724.3055

8. Mathai, T.S., Wang, Y., Cross, N.: Assessing lesion segmentation bias of neural networks on motion corrupted brain MRI. In: Crimi, A., Bakas, S. (eds.) Brainlesion: Glioma, Multiple Sclerosis, Stroke and Traumatic Brain Injuries, pp. 143–156. Springer, Cham (2021). https://doi.org/10.1007/978-3-030-72084-1_14

9. Akudjedu, T.N., et al.: A comparative study of segmentation techniques for the quantification of brain subcortical volume. Brain Imaging Behav. **12**(6), 1678–1695 (2018). https://doi.org/10.1007/s11682-018-9835-y

10. Herten, A., Konrad, K., Krinzinger, H., Seitz, J., von Polier, G.G.: Accuracy and bias of automatic hippocampal segmentation in children and adolescents. Brain Struct. Function **224**(2), 795–810 (2019). https://doi.org/10.1007/s00429-018-1802-2, http://link.springer.com/10.1007/s00429-018-1802-2

11. Popordanoska, T., Bertels, J., Vandermeulen, D., Maes, F., Blaschko, M.B.: On the relationship between calibrated predictors and unbiased volume estimation. In: de Bruijne, M., et al. (eds.) Medical Image Computing and Computer Assisted Intervention - MICCAI 2021, vol. 12901, pp. 678–688. Springer, Cham (2021). https://doi.org/10.1007/978-3-030-87193-2_64, https://link.springer.com/10.1007/978-3-030-87193-2_64

12. Henschel, L., Kügler, D., Reuter, M.: FastSurferVINN: building resolution-independence into deep learning segmentation methods-a solution for High-Res brain MRI. Neuroimage **251**, 118933 (2022). https://doi.org/10.1016/j.neuroimage.2022.118933

13. Goodfellow, I.J., Warde-Farley, D., Mirza, M., Courville, A., Bengio, Y.: Maxout networks. In: Proceedings of the 30th International Conference on International Conference on Machine Learning, vol. 28, pp. III-1319 (2013). JMLR.org

14. Roy, S., Kügler, D., Reuter, M.: Are 2.5D approaches superior to 3D deep networks in whole brain segmentation? In: Medical Imaging with Deep Learning (2022). https://openreview.net/forum?id=Ob62JPB_CDF

15. Klein, A., Tourville, J.: 101 labeled brain images and a consistent human cortical labeling protocol. Front. Neurosci. **6**, 171 (2012)

16. Nordahl, C.W., et al.: The autism phenome project: toward identifying clinically meaningful subgroups of autism. Front. Neurosci. **15**, 786220 (2022). https://doi.org/10.3389/fnins.2021.786220

17. Mueller, S.G., et al.: Ways toward an early diagnosis in Alzheimer's disease: The Alzheimer's Disease Neuroimaging Initiative (ADNI). Alzheimer's Dement. **1**(1), 55–66 (2005)

18. Boccardi, M., et al.: EADC-ADNI Working Group on the Harmonized protocol for manual hippocampal segmentation and for the Alzheimer's disease neuroimaging initiative: training labels for hippocampal segmentation based on the EADC-ADNI harmonized hippocampal protocol. Alzheimer's Dement. **11**(2), 175–183 (2015). https://doi.org/10.1016/j.jalz.2014.12.002

19. Landman, B.A., Warfield, S.K.: MICCAI 2012: Workshop on Multi-atlas Labeling. éditeur non identifié (2012)

20. (2005). http://www.cma.mgh.harvard.edu/ibsr/, Publisher: Centre for Morphometric Analysis Internet Brain Segmentation Repository

21. Loshchilov, I., Hutter, F.: Decoupled weight decay regularization. In: International Conference on Learning Representations (2019). https://openreview.net/forum?id=Bkg6RiCqY7

22. Loshchilov, I., Hutter, F.: SGDR: stochastic gradient descent with warm restarts. In: International Conference on Learning Representations (2017). https://openreview.net/forum?id=Skq89Scxx

Joint Modeling of Image and Label Statistics for Enhancing Model Generalizability of Medical Image Segmentation

Shangqi Gao, Hangqi Zhou, Yibo Gao, and Xiahai Zhuang[✉]

School of Data Science, Fudan University, Shanghai 200433, China
zxh@fudan.edu.cn
https://www.sdspeople.fudan.edu.cn/zhuangxiahai/

Abstract. Although supervised deep-learning has achieved promising performance in medical image segmentation, many methods cannot generalize well on unseen data, limiting their real-world applicability. To address this problem, we propose a deep learning-based Bayesian framework, which jointly models image and label statistics, utilizing the domain-irrelevant contour of a medical image for segmentation. Specifically, we first decompose an image into components of contour and basis. Then, we model the expected label as a variable only related to the contour. Finally, we develop a variational Bayesian framework to infer the posterior distributions of these variables, including the contour, the basis, and the label. The framework is implemented with neural networks, thus is referred to as deep Bayesian segmentation. Results on the task of cross-sequence cardiac MRI segmentation show that our method set a new state of the art for model generalizability. Particularly, the BayeSeg model trained with LGE MRI generalized well on T2 images and outperformed other models with great margins, *i.e.*, over 0.47 in terms of average Dice. Our code is available at https://zmiclab.github.io/projects.html.

Keywords: Bayesian segmentation · Image decomposition · Model generalizability · Deep learning

1 Introduction

Medical image segmentation is a task of assigning specific class for each anatomical structure. Thanks to the advance of deep learning, learning-based methods

This work was funded by the National Natural Science Foundation of China (grant no. 61971142, 62111530195 and 62011540404) and the development fund for Shanghai talents (no. 2020015).

Supplementary Information The online version contains supplementary material available at https://doi.org/10.1007/978-3-031-16443-9_35.

L. Wang et al. (Eds.): MICCAI 2022, LNCS 13435, pp. 360–369, 2022.
https://doi.org/10.1007/978-3-031-16443-9_35

achieve promising performance in medical image segmentation [1–3]. However, many methods require a large number of images with manual labels for supervised learning [4,5], which limits their applications. For cardiac magnetic resonance (CMR) image segmentation, repeatedly labeling multi-sequence CMR image requires more labor of experts, and therefore is expensive [9,10]. Besides, the models trained at one site often cannot perform well at the other site [6]. Therefore, exploring segmentation methods with better generalizability is attractive and challenging.

Much effort has been made to train an end-to-end network by supervised learning. U-Net is one of the widely used networks, since it is more suitable for image segmentation [1]. Training deep neural networks in a supervised way often requires a lot of labeled data [4], but manual labeling of medical image requires professional knowledge and is very expensive. However, small training dataset can result in the problems of over-fitting and overconfidence, which will mislead the clinical diagnosis [6,14]. To solve the problems, Kohl et al. [7] proposed a probabilistic U-Net (PU-Net) for segmentation of ambiguous images by learning the distribution of segmentation. Their results showed that PU-Net could produce the possible segmentation results as well as the frequencies of occurring. Recently, Isensee et al. [3] developed a self-configuring method, i.e., nnU-Net, for learning-based medical image segmentation. This model could automatically configure its preprocessing, network architecture, training, and postprocessing, and achieves state-of-the-art performance on many tasks. Nevertheless, current learning-based methods deliver unsatisfactory performance when applied to unseen tasks [8], and improving generalizability of deep learning models is very challenging.

In this work, we propose a new Bayesian segmentation (BayeSeg) framework to promote model generalizability by joint modeling of image and label statistics. To the best of our knowledge, this is the first attempt of combining image decomposition, image segmentation, and deep learning. Concretely, we first decompose an image into two parts. One is the contour of this image, and the other is the basis approximating its intensity. Both the contour and basis are unknown, and we assign hierarchical Bayesian priors to model their statistics. After that, since the contour of an image is more likely to be sequence-independent, site-independent, and even modality-independent, we try to generate a label from the contour by explicitly modeling of label statistics. Finally, given an image, we build neural networks to infer the posterior distributions of the contour, basis, and label. Being different from many deep learning models that try to learn a deterministic segmentation from a given image, BayeSeg is aimed to learn the distribution of segmentation.

Our contributions are summarized as follows:

- We propose a new Bayesian segmentation framework, i.e., BayeSeg, by joint modeling of image and label statistics. Concretely, we decompose an image into the contour and basis, and assign hierarchical Bayesian priors to model the statistics of the contour, basis, and expected label.

- We solve the model by developing a variational Bayesian approach of computing the posterior distributions of the contour, basis, and label, and build a deep learning architecture of implementing the approach.
- We validate BayeSeg on the tasks of cross-sequence segmentation and cross-site segmentation, and show the superior generalizability of BayeSeg for unseen tasks.

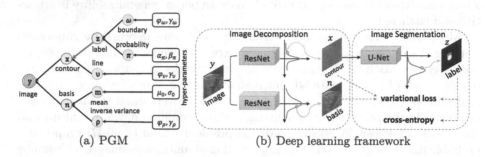

(a) PGM   (b) Deep learning framework

Fig. 1. The framework of Bayesian segmentation (BayeSeg). (a) shows the probabilistic graphical model (PGM) of BayeSeg. Here, the blue circle denotes an observed image, green and orange circles denote unknown variables, and white rectangles denote hyperparameters. (b) presents the deep learning framework of BayeSeg. Given an image, we first use ResNets to infer the posterior distributions of the contour and basis, and obtain their random samples. Then, we use U-Net to infer the posterior distribution of label, and the resulting random sample is a segmentation. Please refer to Sect. 2.3 for the details of network architecture and training strategy. (Color figure online)

2   Methodology

We propose a Bayesian segmentation (BayeSeg) framework to improve the generalizability of deep learning models. Many learning-based methods trained on one sequence MR images cannot generalize well to the other sequence or site data [6]. To solve the challenge, we propose the BayeSeg mainly consisting of two parts, i.e., (1) statistical modeling of image and label as shown in Fig. 1(a), and (2) statistical inference of image and label as shown in Fig. 1(b). At the first stage, we build a probabilistic graphical model (PGM) for the modeling of image and label. That is we decompose an image into its contour and basis, and only the contour is related to an expected label. At the second stage, we first build two residual networks (ResNets) to infer the posterior distributions of the contour and basis, respectively. Then, we build a U-Net to estimate the posterior distribution of the label. An intuitive understanding of "contour" and "basis" is shape and appearance. Since shape is domain-irrelevant, the model predicting a label from the contour will have better generalizability.

Figure 1 shows the framework of statistical modeling and inference of the proposed BayeSeg. For the statistical modeling as shown in Fig. 1(a), we first decompose an image y into its basis n and contour x. The former is a Gaussian variable depending on the mean m and the inverse variance ρ. The latter depends on the expected label z and the line v for detecting the edges of contour. Similarly, the label z depends on the segmentation boundary ω and the segmentation probability π of all classes. Finally, Gamma priors are assigned to ρ, v, and ω, a Beta prior is assigned to π, and a Gaussian prior is assigned to m. Figure 1(b) shows the deep learning framework of inferring related variables. The outputs of ResNets and U-Net will be jointly used to compute a variational loss, which is the key of improving model generalizability.

2.1 Statistical Modeling of Image and Label

This section shows the statistical modeling of image and label. Given an image sampled from the variable $y \in \mathbb{R}^{d_y}$, where d_y denotes the dimension of y, we decompose y into the sum of a contour x and a basis n, i.e., $y = x + n$. Then, the basis n is modeled by a normal distribution with a mean $m \in \mathbb{R}^{d_y}$ and a covariance $diag(\rho)^{-1} \in \mathbb{R}^{d_y \times d_y}$. Moreover, the contour x is modeled by a simultaneous autoregressive model (SAR) [11] depending on the expected label $z \in \mathbb{R}^{d_y \times K}$, the line $v \in \mathbb{R}^{d_y}$ indicating edges of the contour, and the matrix $D_x \in \mathbb{R}^{d_y \times d_y}$ describing a neighboring system of the contour, where K is the number of classes for segmentation. Finally, the observation likelihood of an image y can be expressed as

$$p(y|x, m, \rho) = \mathcal{N}(x + m, diag(\rho)^{-1}). \tag{1}$$

The basis n of an image is determined by a normal distribution, that is, $p(n|m, \rho) = \mathcal{N}(n|m, diag(\rho)^{-1})$. Specifically, we assign a Gaussian prior to m, i.e., $p(m|\mu_0, \sigma_0) = \mathcal{N}(m|\mu_0, \sigma_0^{-1}I)$, and a Gamma prior to ρ, namely, $p(\rho|\phi_\rho, \gamma_\rho) = \prod_{i=1}^{d_y} \mathcal{G}(\rho_i|\phi_{\rho i}, \gamma_{\rho i})$. Here, I denotes an identity matrix, $\mu_0, \sigma_0, \phi_\rho$, and γ_ρ are predefined hyper-parameters, and $\mathcal{G}(\cdot, \cdot)$ represents the Gamma distribution.

The contour x of an image is determined by a SAR mainly depending on the expected label z and the line v,

$$p(x|z, v) = \prod_{k=1}^K \mathcal{N}(x|0, [D_x^T diag(z_k v) D_x]^{-1}), \tag{2}$$

where, z_k denotes the segmentation of the k-th class, and $D_x = I - B_x$ is nonsingular. B_x describes a neighboring system of each pixel. For examples, if the values of B_x for the nearest four pixels equals to 0.25 while others are zeros, then D_x is aimed to compute the average difference of each pixel with its four neighbors. The line v can indicate the edges of the contour, and it is assigned a Gamma prior, i.e., $p(v|\phi_{vi}, \gamma_{vi}) = \prod_{i=1}^{d_y} \mathcal{G}(v_i|\phi_{vi}, \gamma_{vi})$.

The label z is modeled by another SAR depending on the segmentation boundary $\omega \in \mathbb{R}^{d_y \times K}$ and the segmentation probability $\pi \in \mathbb{R}^K$ of all classes, namely,

$$p(z|\pi, \omega) = \prod_{k=1}^K \mathcal{N}(z|0, [-\ln(1 - \pi_k)D_z^T diag(\omega_k)D_z]^{-1}), \tag{3}$$

where, the definition of D_z is the same as D_x in Eq. (2); ω_k can indicate the boundary of the k-th segmentation z_k; and π_k denotes the probability of a pixel belonging to the k-th class. Finally, we assign Gamma prior to ω, i.e., $p(\omega) = \prod_{i=1}^{d_y} \prod_{k=1}^{K} \mathcal{G}(\omega_{ki} | \phi_{\omega ki}, \gamma_{\omega ki})$, and give Beta prior to π, namely, $p(\pi) = \prod_{k=1}^{K} \mathcal{B}(\pi_k | \alpha_{\pi k}, \beta_{\pi k})$. *The details of Gaussian distribution, Gamma distribution, and Beta distribution are provided in our supplementary material.*

2.2 Variational Inference of Image and Label

This section shows a variational method of inferring the contour, basis and label given an image y by maximum a *posteriori* (MAP) estimation. Let $\psi = \{m, \rho, x, v, z, \omega, \pi\}$ denote the set of all variables to infer, then our aim is to infer the posterior distribution $p(\psi | y)$. Since direct computation is intractable, we use variational Bayesian (VB) method [12] to solve the problem. Concretely, we approximate the posterior distribution $p(\psi | y)$ via a variational distribution $q(\psi)$ by assuming the variables in ψ are independent, namely,

$$q(\psi) = q(m)q(\rho)q(x)q(v)q(z)q(\omega)q(\pi). \tag{4}$$

After that, we minimize the KL divergence between $q(\psi)$ and $p(\psi | y)$, and which results in our final variational loss as follows,

$$\min_{q(\psi)} \mathcal{L}_{var} = \text{KL}(q(\psi) \| p(\psi)) - \mathbb{E}[\ln p(y | \psi)] \tag{5}$$

The details of further unfolding the variational loss are provided in our supplementary material.

2.3 Neural Networks and Training Strategy

This section shows the network architecture of achieving the variational inference and the training strategy for image segmentation. As Fig. 1(b) shows, *at the decomposition stage*, we adopt two ResNets [13] to separately infer the variational posteriors of the contour x and basis n, i.e., $q(x)$ and $q(n)$, respectively. The ResNet of inferring the contour consists of 10 residual blocks, and each block has a structure of "Conv + ReLU + Conv". The output of this ResNet has two channels. One is the element-wise mean of the contour, and the other is its element-wise variance. The contour x in the figure denotes a random sample from $q(x)$. The ResNet of inferring the basis consists of 6 residual blocks, and each block has a structure of "Conv + BN + ReLU + Conv + BN". Similarly, this ResNet will output the mean and variance of the basis, and the basis n is randomly sampled from its variational posterior distribution. *At the segmentation stage*, we adopt a U-Net [1] to infer the variational posterior of the label z, i.e., $q(z)$. The output of this U-Net has $2K$ channels. The first K channels denote the element-wise mean of the label, and the left channels represent its element-wise variance. The label z in Fig. 1(b) is a random sample from the resulting posterior distribution, and it will be taken as a stochastic segmentation for training.

BayeSeg is trained in an end-to-end manner by balancing between cross-entropy and the variational loss in (5). For convenience, the cross-entropy between a stochastic segmentation and the provided manual segmentation is notated as $\mathcal{L}_{ce}$. Then, our total loss of training BayeSeg is given by,

$$\min_{q(\psi)} \mathcal{L}_{ce} + \lambda \mathcal{L}_{var}, \tag{6}$$

where, the balancing weight λ is set to 100 in our experiments. Besides, other hyper-parameters in Fig. 1(a) is summarized as follows. Each element of ϕ. of related variables is set to 2; $\alpha_\pi = 2$ and $\beta_\pi = 2$ for the segmentation probability π; $\mu_0 = 0$ and $\sigma_0 = 1$ for the mean of the basis; The elements of γ_ρ, γ_υ, and γ_ω are set to 10^{-6}, 10^{-8}, and 10^{-4}, respectively. Note that BayeSeg could properly decompose an image into the contour and basis due to the priors with respect to υ and ρ. That is the contour and basis are adaptively balanced after selecting proper γ_ρ and γ_υ.

3   Experiments

3.1   Tasks and Datasets

We used the LGE of MSCMRseg [9] to train models. To validate the performance on the task of cross-sequence segmentation, we tested models using LGE and T2 of MSCMRseg. To validate the performance on the task of cross-site segmentation, we tested models using LGE of MSCMRseg and ACDC [14].

MSCMRseg [9,10] was provided by MICCAI'19 Multi-sequence Cardiac MR Segmentation Challenge. This dataset consists of 45 multi-sequence CMR images, including LGE, C0, and T2. Each case comes from the same patient who underwent cardiomyopathy. The manual segmentation results of left ventricle (LV), right ventricle (RV), and myocardium (Myo) for all images are available. In this study, we randomly split the 45 cases into three sets consisting of 25, 5, and 15 cases, respectively. Then, we only used the 25 LGE images for training, and the 5 LGE images for validation. Finally, we tested models on the 15 multi-sequence cases to show the performance of cross-sequence segmentation.

ACDC [14] was provided by MICCAI'17 Automatic Cardiac Diagnosis Challenge. This dataset consists of shot-axis cardiac cine-MRIs of 100 patients for training, and of 50 patients for test. Only the manual segmentation results of training data are provided for LV, RV, and Myo during the end-diastolic (ED) and end-systolic (ES) phases. In our study, we tested models using the 100 training images to show the performance of cross-site segmentation.

BayeSeg was implemented by Pytorch, and trained by Adam optimizer with the initial learning rate to be 10^{-4}. The learning rate was dropped every 500 epochs by a factor of 0.1, and the training was stopped when up to 2000 epochs. At the test stage, we took the mean of $\mathbf{z}$ as the final segmentation label, since the variational posterior $q(\mathbf{z})$ is a Gaussian distribution whose mode is its mean. All experiments were run on a TITAN RTX GPU with 24G memory.

3.2 Cross-Sequence Segmentation

To study the performance of BayeSeg on the task cross-sequence segmentation, we trained four models using the 25 LGE images of MSCMRseg. Concretely, we trained a U-Net, which had the same architecture as the U-Net [1] in Fig. 1(b), by minimizing the cross-entropy. Then, we trained PU-Net [7] on the same dataset using its public code and default settings. Moreover, we trained a baseline, which shares the same architecture as BayeSeg, without using the variational loss in (6). Finally, we trained the BayeSeg by minimizing the total loss in (6). For fair comparisons, all models were trained using consistent data augmentation, including random flip and rotation. In the test stage, we evaluated all models on the 15 multi-sequence images, i.e., LGE and T2, and reported the Dices of LV, Myo, and RV as well as average Dice.

Table 1. Evaluation on the task of cross-sequence cardiac segmentation. *Note that all models were only trained on LGE of MSCMRseg, but tested on LGE and T2.* Here, G denotes the drop of average Dice, and it is used to measure model generalizability.

Method	LGE of MSCMRseg (15 samples)				T2 of MSCMRseg (15 samples)				G
	LV	Myo	RV	Avg	LV	Myo	RV	Avg	
U-Net	.855±.045	.727±.064	.733±.097	.772±.093	.203±.183	.095±.093	.055±.062	.118±.139	.654
PU-Net	**.898±.027**	.768±.056	.729±.089	.798±.096	.279±.162	.166±.122	.195±.130	.213±.147	.585
Baseline	.893±.023	**.783±.045**	.727±.069	.801±.085	.481±.129	.117±.079	.090±.123	.230±.211	.571
BayeSeg	.887±.028	.774±.048	**.763±.060**	**.808±.073**	**.846±.119**	**.731±.117**	**.528±.206**	**.701±.202**	**.107**

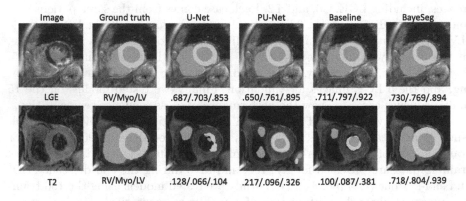

Fig. 2. Visualization of results on the task of cross-sequence segmentation. Here, we choose the median case of BayeSeg according to the Dices of 15 LGE images.

Table 1 reports the results of compared methods. One can see that PU-Net, Baseline, and BayeSeg achieved comparable performance, when the training and test sequences were consistent. If the training sequence and test sequence are different, the performance of BayeSeg dropped weakly, but that of others decreased dramatically. Besides, BayeSeg greatly outperformed Baseline in this case, which demonstrates that the variational loss, induced by joint modeling of image and label statistics, is the key of improving model generalizability. To qualitatively evaluate all models, we chose the median case of BayeSeg according to the average Dices of 15 LGE images, and visualized the segmentation results in Fig. 2. This figure shows BayeSeg delivers the best performance in segmenting the unseen T2 sequence, which again confirms the effectiveness of our framework in improving model generalizability.

3.3 Cross-Site Segmentation

To study the performance of BayeSeg on the task of cross-center segmentation, we tested all models trained in the previous section on ACDC. Table 2 reports the results of these models. This table showed that the performance of all methods dropped when the training and test samples came from two different sites, but BayeSeg delivered the least performance drop. Therefore, BayeSeg generalized well on the unseen samples from the different site, thanks to the joint modeling of image and label statistics.

Table 2. Evaluation on the task of cross-site cardiac segmentation. *Note that all models were only trained on LGE of MSCMRseg, but tested on LGE and ACDC.* Here, G denotes the drop of average Dice, and it is used to measure model generalizability.

Method	LGE of MSCMRseg (15 samples)				ACDC (100 samples)				G
	LV	Myo	RV	Avg	LV	Myo	RV	Avg	
U-Net	.855 ± .045	.727 ± .064	.733 ± .097	.772 ± .093	.721 ± .187	.602 ± .183	.659 ± .202	.660 ± .197	.112
PU-Net	**.898 ± .027**	.768 ± .056	.729 ± .089	.798 ± .096	.743 ± .152	.641 ± .146	.604 ± .215	.663 ± .184	.126
Baseline	.893 ± .023	**.783 ± .045**	.727 ± .069	.801 ± .085	.776 ± .134	.667 ± .150	.585 ± .227	.676 ± .192	.125
BayeSeg	.887 ± .028	.774 ± .048	**.763 ± .060**	**.808 ± .073**	**.792 ± .130**	**.694 ± .123**	**.659 ± .175**	**.715 ± .155**	**.093**

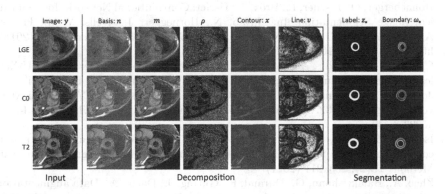

Fig. 3. Visualization of posteriors inferred by BayeSeg. Here, the subscript $*$ denotes the result of myocardium.

3.4   Interpretation of Joint Modeling

In this section we interpreted the joint modeling of image and label statistics. Figure 3 shows the posteriors inferred by our BayeSeg for three different sequences of MSCMRseg. One can see that, at the decomposition stage, an image was mainly decomposed into its basis and contour. The basis n was modeled as a Gaussian distribution with the mean m and the inverse variance ρ. It was an approximation of the image, and therefore x was left as the contour. To avoid the smoothness of this contour, we assigned the line v to detect its edges. The large values of v indeed showed the smooth areas of contour, while the small values indicate the edges. At the segmentation stage, we choose to segment the contour, since it is more likely to be sequence-independent, site-independent, and even modality-independent. To achieve better segmentation around the boundary of some object, such as myocardium, we assigned the ω to detect the segmentation boundary. This variable successfully indicated the inner and outer boundaries of myocardium, as shown in Fig. 3.

4   Conclusion

In this work, we proposed a new Bayesian segmentation framework by joint modeling of image and label statistics. Concretely, we decomposed an image into its basis and contour, and estimated the segmentation of this image from the more stable contour. Our experiments have shown that the proposed framework could address the problem of over-fitting and greatly improve the generalizability of deep learning models.

References

1. Ronneberger, O., Fischer, P., Brox, T.: U-Net: Convolutional Networks for Biomedical Image Segmentation. In: Navab, N., Hornegger, J., Wells, W.M., Frangi, A.F. (eds.) MICCAI 2015. LNCS, vol. 9351, pp. 234–241. Springer, Cham (2015). https://doi.org/10.1007/978-3-319-24574-4_28
2. Çiçek, Ö., Abdulkadir, A., Lienkamp, S.S., Brox, T., Ronneberger, O.: 3D U-Net: learning dense volumetric segmentation from sparse annotation. In: Ourselin, S., Joskowicz, L., Sabuncu, M.R., Unal, G., Wells, W. (eds.) MICCAI 2016. LNCS, vol. 9901, pp. 424–432. Springer, Cham (2016). https://doi.org/10.1007/978-3-319-46723-8_49
3. Isensee, F., Jaeger, P., Kohl, S., Petersen, J., MaierHein, K.: nnU-Net: a self-configuring method for deep learning-based biomedical image segmentation. Nat. Methods **18**(2), 203–211 (2021)
4. Zhao, A., Balakrishnan, G., Durand, F., Guttag, J., Dalca, A.: Data augmentation using learned transformations for one-shot medical image segmentation. In: Proceedings of the IEEE Conference on Computer Vision and Pattern Recognition, pp. 8543–8553. IEEE (2019)
5. Chen, H., Dou, Q., Yu, L., Qin, J., Heng, P.: VoxResNet: deep voxelwise residual networks for brain segmentation from 3D MR images. Neuroimage **170**, 446–455 (2018)

6. Li, Z., Kamnitsas, K., Glocker, B.: Overfitting of neural nets under class imbalance: analysis and improvements for segmentation. In: Shen, C., et al. (eds.) MICCAI 2019. LNCS, vol. 11766, pp. 402–410. Springer, Cham (2019). https://doi.org/10.1007/978-3-030-32248-9_45

7. Kohl, S., et al.: A probabilistic U-net for segmentation of ambiguous images. In: Proceedings of the 32nd International Conference on Neural Information Processing Systems, pp. 6965–6975. ACM (2018)

8. Cheng, O., et al.: Causality-inspired single-source domain generalization for medical image segmentation. arXiv:2111.12525 (2021)

9. Zhuang, X.: Multivariate mixture model for myocardial segmentation combining multi-source images. IEEE Trans. Pattern Anal. Mach. Intell. **41**(12), 2933–2946 (2019)

10. Zhuang, X.: Multivariate mixture model for cardiac segmentation from multi-sequence MRI. In: Ourselin, S., Joskowicz, L., Sabuncu, M.R., Unal, G., Wells, W. (eds.) MICCAI 2016. LNCS, vol. 9901, pp. 581–588. Springer, Cham (2016). https://doi.org/10.1007/978-3-319-46723-8_67

11. Shekhar, S., Xiong, H.: Simultaneous autoregressive model (SAR). In: Shekhar, S., Xiong, H. (eds) Encyclopedia of GIS. Springer, Boston (2008). https://doi.org/10.1007/978-0-387-35973-1_1217

12. Blei, D., Kucukelbir, A., McAuliffe, J.: Variational inference: a review for statisticians. J. Am. Stat. Assoc. **112**(518), 859–877 (2017)

13. He, K., Zhang, X., Sun, J., Ren, S.: Deep residual learning for image recognition. In: Proceedings of the IEEE Conference on Computer Vision and Pattern Recognition, pp. 770–778. IEEE (2016)

14. Berbard, O., et al.: Deep learning techniques for automatic MRI cardiac multi-structures segmentation and diagnosis: is the problem solved? IEEE Trans. Med. Imaging **37**(11), 2514–2525 (2018)

Transformer Based Feature Fusion for Left Ventricle Segmentation in 4D Flow MRI

Xiaowu Sun[1], Li-Hsin Cheng[1], Sven Plein[2], Pankaj Garg[3],
and Rob J. van der Geest[1(✉)]

[1] Division of Image Processing, Department of Radiology, Leiden University Medical Center,
Leiden, The Netherlands
R.J.van_der_Geest@lumc.nl
[2] Leeds Institute of Cardiovascular and Metabolic Medicine, University of Leeds, Leeds, UK
[3] Norwich Medical School, University of East Anglia, Norwich, UK

Abstract. Four-dimensional flow magnetic resonance imaging (4D Flow MRI)
enables visualization of intra-cardiac blood flow and quantification of cardiac
function using time-resolved three directional velocity data. Segmentation of cardiac 4D flow data is a big challenge due to the extremely poor contrast between the
blood pool and myocardium. The magnitude and velocity images from a 4D flow
acquisition provide complementary information, but how to extract and fuse these
features efficiently is unknown. Automated cardiac segmentation methods from
4D flow MRI have not been fully investigated yet. In this paper, we take the velocity and magnitude image as the inputs of two branches separately, then propose
a Transformer based cross- and self-fusion layer to explore the inter-relationship
from two modalities and model the intra-relationship in the same modality. A large
in-house dataset of 104 subjects (91,182 2D images) was used to train and evaluate our model using several metrics including the Dice, Average Surface Distance
(ASD), end-diastolic volume (EDV), end-systolic volume (ESV), Left Ventricle
Ejection Fraction (LVEF) and Kinetic Energy (KE). Our method achieved a mean
Dice of 86.52%, and ASD of 2.51 mm. Evaluation on the clinical parameters
demonstrated competitive results, yielding a Pearson correlation coefficient of
83.26%, 97.4%, 96.97% and 98.92% for LVEF, EDV, ESV and KE respectively.
Code is available at github.com/xsunn/4DFlowLVSeg.

Keywords: LV segmentation · 4D Flow MRI · Feature fusion · Transformer

1 Introduction

Quantitative assessment of left ventricular (LV) function from magnetic resonance imaging (MRI) is typically based on the use of short-axis multi-slice cine MRI due to its
excellent image quality [1, 2]. Recently, four-dimensional (4D) Flow MRI has been introduced, encoding blood flow velocity in all three spatial directions and time dimension.

Supplementary Information The online version contains supplementary material available at
https://doi.org/10.1007/978-3-031-16443-9_36.

4D Flow MRI can be used for detailed analysis of intra-cardiac blood flow hemodynamics, providing additional information over conventional cine MRI. The segmentation of the cardiac cavities is an important step to derive quantitative blood flow results, such as the total LV kinetic energy (KE) [3]. 4D Flow MRI generates four image volumes including a magnitude image and three velocity images, one for each spatial dimension. Figure 1 shows an example of magnitude and velocity images from one slice out of a 4D Flow MRI data set. The example highlights the extremely poor contrast between the heart chambers and the myocardium in the 4D Flow data. Therefore, most authors have used segmentations derived from co-registered short-axis cine MR in order to quantify ventricular blood flow parameters from the 4D Flow data. However, this relies on accurate spatial and temporal registration of the two MR sequences. Inconsistent breath-hold positioning may introduce spatial misalignment while heart rate differences will result in temporal mismatch between the acquisitions. The aim of the current work was therefore to develop an automated method for LV segmentation from 4D Flow MRI data, not requiring additional cine MRI data.

Fig. 1. A sample of cardiac 4D Flow data in short-axis view. The first image is the magnitude image, and the last three images are the velocities in x, y and z dimensions respectively.

Since U-Net [4] was proposed, convolutional neural networks (CNNs) have been predominant in the task of medical image segmentation. Many variants of U-Net have been proposed further improving the performance. For instance, nnU-Net [5] introducing automated self-configuring outperformed most existing approaches on 23 diverse public datasets. Although those CNN based networks have achieved an excellent performance, restricted by the locality of convolutional kernels, they cannot capture long-distance relations [6, 7].

Transformer is considered as an alternative model using its self-attention mechanism to overcome the limitation of CNN. Transformer was designed firstly for natural language processing (NLP) tasks such as machine translation and document classification. More recently, Transformer-based approaches were introduced in medical image processing. TransUnet [8] applied a CNN-Transformer hybrid encoder and pure CNN decoder for segmentation. However, TransUnet still uses convolutional layers as the main building blocks. Inspired by the Swin Transformer [9], Cao proposed a U-Net-like pure Transformer based segmentation model which uses hierarchical Swin Transformer as the encoder and a symmetric Swin Transformer with patch expanding layer as the decoder [10]. Other Transformer-based networks [11–13] also mark the success of Transformer in medical image segmentation and reconstruction.

Although numerous deep learning-based segmentation methods have been proposed in various modalities, the automatic segmentation of the LV directly from 4D Flow data has not been explored yet. A specific challenge is that the magnitude and velocity images of a 4D Flow acquisition have different information content and should be considered as different modalities. Moreover due to velocity noise, a careful fusion method is needed to avoid redundancy or insufficient feature integration [14, 15].

In this paper, we present, to the best of our knowledge, the first study to segment the LV directly from 4D Flow MRI data. Our main contributions are: (1) we propose two self- and cross-attention-based methods to fuse the information from different modalities in 4D Flow data; (2) we evaluate our method in a large 4D Flow dataset using multiple segmentation and clinical evaluation metrics.

2   Method

2.1   Attention Mechanism

Attention mechanism, mapping the queries and a set of keys-value pairs to an output, is the fundamental component in Transformer. In this section, we first introduce how the self-attention module models the intra-relationship of features from the same image modality. Then we explain how cross-attention explores the inter-relationship of features from two different modalities. The two attention modules are illustrated in Fig. 2.

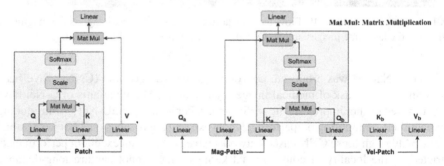

Fig. 2. The structure of self-attention (left) and cross-attention (right) modules.

Self-attention Module. In self-attention module [6], the **Q** (queries), **K** (keys) and **V** (values) are generated from the same modality. **Q** and **K** determine a weight matrix after the scaled dot product which is used to compute the weighted sum of **V** as the output. The computing process can be described as in Eq. (1):

$$Atten(Q_a, K_a, V_a) = softmax(\frac{Q_a K_a^T}{\sqrt{d}})V_a \tag{1}$$

where d is the key dimensionality, and a denotes modality a.

Cross-Attention Module. Although self-attention explores the intra-modality relationship, the inter-modality relationship, such as the relationship between pixels in the magnitude image and velocity image is not explored. The cross-attention module takes two patches as the input to generate the **Q, K,** and **V. V** and **K** are generated from the same modality, while **Q** is derived from another modality. The other operations are kept the same as in self-attention. It can be expressed as Eq. (2). Hence, cross-attention can be adopted to fuse the information from different modalities.

$$Atten(Q_b, K_a, V_a) = softmax(\frac{Q_b K_a^T}{\sqrt{d}})V_a \tag{2}$$

Multi-head Self(Cross)-Attention Module. To consider various attention distributions and multiple aspects of features, the multi-head attention mechanism [6] is introduced. The multi-head attention is the concatenation of h single attentions along the channel dimension followed by a linear projection. Thus, the multi-head attention can be formulated as Eq. (3, 4).

$$MultiHead(Q, K, V) = Concat(H_1, H_2, \cdots, H_h)W^o \tag{3}$$

$$H_i = Atten(Q_i, K_i, V_i) \tag{4}$$

where *Atten* is self-attention or cross-attention, Q_i, K_i, V_i are the i-th vector of Q, K, V. In each single attention head, the channel dimension $d' = d/h$.

2.2 Feature Fusion Layer

To fuse the features generated from the magnitude and velocity images, we proposed a feature fusion layer (FFL). The structure of FFL shown in Fig. 3 contains two branches, each branch has one cross-fusion layer and one self-fusion layer.

Cross-Fusion Layer (CFL). CFL is proposed to fuse the features from different modalities. The structure of CFL is illustrated in the upper dash box in Fig. 3. Given **Q, K** and **V** generated from two modalities, the Multi-head Cross-Attention (MCA) module followed by a linear projection firstly integrate those information. Then the fused features are added to the original input. Subsequently, another two linear projections and one residual connection followed by a normalization layer are used to enhance the fused information.

Self-Fusion Layer (SFL). The lower dash box in Fig. 3 shows the structure of SFL. SFL is a simple stack of Multi-head Self-Attention (MSA), linear projection, residual and normalization layer. Different from CFL, the SFL only uses one input to generate the values for the MSA. CFL aims to fuse the features from different image modalities, SFL further enhances the fused features using self-attention.

Having two feature maps from the magnitude and velocity images respectively, we first transform the feature maps into sequence data using the patch embedding. Specifically, the feature $f \in R^{H \times W \times C}$ is divided into $N = HW/P^2$ patches, where the patch size P is set to 16. The patches are flattened and embedded into a latent D-dimension, obtaining a embedding sequence $e \in R^{N \times D}$. However, dividing feature maps into patches leads to loss of spatial information. Therefore, a learnable positional encoding sequence is added to the embedding sequence to address this issue. Then the sequence data is passed into the FFL. In this work, we used a stack of 4 FFLs as the feature fusion network (FFN).

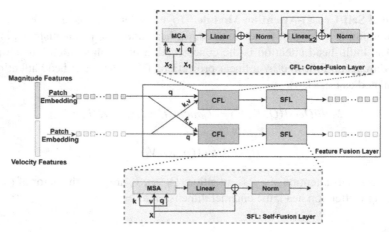

Fig. 3. Structure of feature fusion layer (FFL). The input of the feature fusion layer is two features derived from magnitude and velocity images respectively. The upper box is the structure of cross-fusion layer (CFL) and the lower one is the structure of self-fusion layer (SFL).

2.3   Network Structure

Figure 4 illustrates the proposed segmentation network, which takes the U-Net as the backbone. The encoder uses two parallel branches to extract features from magnitude and velocity image separately. The features at the same level are integrated using the feature fusion network. By doing so, the size of integrated features reduces due to the patch embedding. Hence, the fused features are up-sampled first, then added to the original features as the final aggregated features.

The four-level paired aggregated features derived from the encoder are taken as the inputs to the decoder part. The fused features at the same level generated from the magnitude and velocity branch in the encoder are concatenated followed by a convolutional layer to reduce the number of feature maps. The remaining decoder parts including the up-sampling, convolutional and softmax layers are the same as in U-Net.

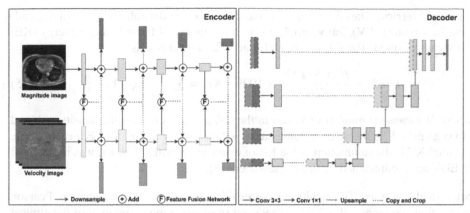

Fig. 4. The architecture of our proposed segmentation network structure. The feature fusion network is a stack of 4 FFLs.

3 Materials

3.1 Dataset

4D flow MRI was performed in 28 healthy volunteers and 76 post-myocardial infarction patients on a 1.5T MR system (Philips Healthcare). The 4D flow acquisition covered the complete LV and was acquired in axial orientation with a voxel size of $3 \times 3 \times 3$ mm^3 and reconstructed into 30 cardiac phases. The other imaging parameters are as follows: flip angle $= 10°$, velocity encoding (VENC) of 150 cm/s, FOV $= 370$–400×370–400 mm^2, echo time (TE) $= 1.88$–3.75 ms, repetition time (TR) $= 4.78$–13.95 ms. In addition, standard cine-MRI was performed in multiple short-axis slices covering the LV from base to apex. More details about the MR acquisition protocol can be found here [16]. The short-axis cine acquisition was used to segment the LV endocardial boundaries in all slices and phases. After rigid registration with the 4D flow acquisition, the defined segmentation served as ground truth segmentation of the 4D flow acquisition. Based on the known short-axis orientation, the 4D flow data was resliced into the short-axis orientation using a slice spacing of 3 mm and a fixed number of 41 slices. The spatial in-plane resolution was defined equal to the available short-axis cine acquisition and varied from 0.83×0.83 mm^2 to 1.19×1.19 mm^2.

Excluding the images without any objects this resulted in 91 182 annotated pairs of 2D images, each pair has one 2D magnitude image and three-directional velocity images. The subjects were randomly split into three parts with 64, 20, 20 (total number of images: 55 825, 17 335 and 18 022) for training, validation and testing respectively. We normalized the magnitude image into [0, 1] using min-max method. The images were cropped into 256×256.

3.2 Evaluation Metrics

Segmentation Metrics. To quantitatively evaluate the segmentation performance, Dice and Average Surface Distance (ASD) were measured.

Clinical Metrics. The clinical metrics, including the end-diastolic volume (EDV), end-systolic volume (ESV), left ventricle ejection fraction (LVEF) and kinetic energy (KE) [3] were measured. The formula of LVEF and KE are defined as:

$$LVEF = \frac{EDV - ESV}{EDV} \times 100\% \quad KE = \sum_{i=1}^{N} \frac{1}{2} \rho_{blood} \cdot V_i \cdot v_i^2 \tag{5}$$

where N means the number of voxels in the LV, ρ_{blood} represents the density of blood (1.06 g/cm^3), V is the voxel volume and v is the velocity magnitude. For each phase, the total KE is the summation of the KE of every voxel within LV. KE was normalized to EDV as recommended by other researchers [3].

Statistical Analysis. The results are expressed as mean $\pm$ standard deviation. Pearson correlation coefficient (PCC) was introduced to measure the correlation of the clinical metrics between the manual and automatic segmentation approaches. Paired evaluation metrics were compared using Wilcoxon-signed-rank test with $P < 0.05$ indicating a significant difference.

The Dice, ASD and KE reported in this work are the mean values as computed over 30 phases per subject.

4 Experiment and Results

All the models were implemented in Pytorch and trained with a NVIDIA Quadro RTX 6000 GPU with 24 GB memory from scratch. We employed Adam as the optimizer with 0.0001 as the learning rate. All of the models were trained for 1000 epochs with a batch size of 15. The sum of Dice loss and cross-entropy loss was used as the loss function. Additionally, due to the complexity of the velocity images, we did not employ any data augmentation methods to enlarge the dataset.

We first evaluated our model against the U-Net, TransUnet [8], and U-NetCon. TransUnet added the self-attention module to the last layer of the encoder. The structure of U-NetCon (shown in the Supplementary) is similar to our proposed network. After removing the feature fusion network, the U-NetCon introduces two U-Net encoders which extract the features from two modalities separately and subsequently, the features from the same level in the encoder are concatenated as the input of the decoder. The

Table 1. Segmentation performance of different methods. Err means the absolute error between the manual and automatic segmentation methods.

Model	Dice (%)	ASD (mm)	EDV-Err (ml)	ESV-Err (ml)	LVEF-Err (%)	KE-Err (µJ/ml)
U-Net	84.62 ± 5.91	2.99 ± 1.66	20.35 ± 31.53	16.01 ± 19.76	7.60 ± 7.10	1.50 ± 1.64
U-NetCon	84.57 ± 6.15	3.19 ± 1.74	22.57 ± 29.46	17.08 ± 24.46	6.11 ± 5.43	0.95 ± 1.94
TransUnet	84.27 ± 5.35	3.09 ± 1.33	18.09 ± 22.91	23.92 ± 16.06	11.79 ± 7.64	0.51 ± 0.48
Ours	**86.52 ± 5.54**	**2.51 ± 1.14**	**9.02 ± 10.03**	**11.86 ± 10.55**	**5.10 ± 4.55**	**0.36 ± 0.34**

input of U-Net and TransUnet is a four-channel stack of one magnitude and three velocity images. Whereas, in our method and U-NetCon, the magnitude and velocity images are taken as two separate input branches.

Table 1 reports the evaluation results of various metrics. It shows our method achieved the best performance for all of the six metrics. In Table 2 the PCC of the clinical metrics derived from different models are presented. Our method performs the best on all clinical metrics demonstrating a high correlation. Comparing the results of U-Net and TransUnet, the Dice and ASD only showed marginal improvement, but the performance decreased in LVEF with a low PCC of 48.7%. In order to evaluate the effectiveness of feature fusion network, we further compared our method to U-NetCon. As compared to U-NetCon, our method improves the Dice by 2% and the PPC by 3%, 9%, 7% and 16% for LVEF, EDV, ESV and KE, respectively, confirming that the proposed feature fusion network efficiently aggregates the features from magnitude and velocity images. More results about the boxplot and correlation comparing the Dice and four clinical parameters derived from our method and U-NetCon can be found in the supplementary.

Table 2. PCC of the clinical metrics derived from manual and automatic segmentation results.

Model	LVEF	EDV	ESV	KE
U-Net	70.65%	84.09%	91.50%	83.76%
U-NetCon	80.61%	88.46%	89.49%	82.46%
TransUnet	48.70%	91.36%	90.33%	97.86%
Ours	**83.26%**	**97.40%**	**96.97%**	**98.92%**

The P-value of Wilcoxon test results between the ground truth and our method in LVEF, EDV, ESV, KE are 0.13, 0.43, 0.35 and 0.43, as shown in Fig. 5. All of those P-values are larger than 0.05, which confirmed that there is no significant different between the clinical parameters derived from the manual and our automatic segmentation.

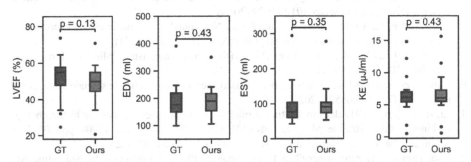

Fig. 5. Box plots comparing four clinical evaluation metrics including EDV, ESV, LVEF and KE derived from the manual segmentation and our prediction. GT represents the ground truth. P-value was computed using Wilcoxon-signed-rank test. P < 0.05 indicate a significant difference between two variables.

5 Conclusion

In this paper, we proposed a Transformer based feature fusion network to aggregate the features from different modalities for LV segmentation in 4D flow MRI data. In the feature fusion network, we introduced a self- and a cross-fusion layer to investigate the inter- and intra-relationship for the features from two different modalities. The proposed method was trained and evaluated in a large in-house dataset and the results of the segmentation accuracy and clinical parameters demonstrate superiority of our method against state-of-arts. We expect that the use of carefully designed data augmentation methods for the velocity images may result in further improvement of the performance of the proposed method.

References

1. Tao, Q., et al.: Deep learning-based method for fully automatic quantification of left ventricle function from cine MR images: a multivendor, multicenter study. Radiology **290**(1), 81–88 (2019)
2. Bai, W., et al.: Automated cardiovascular magnetic resonance image analysis with fully convolutional networks. J. Cardiovasc. Magn. Reson. **20**(1), 1–12 (2018). https://doi.org/10.1186/s12968-018-0471-x
3. Garg, P., et al.: Left ventricular blood flow kinetic energy after myocardial infarction-insights from 4D flow cardiovascular magnetic resonance. J. Cardiovasc. Magn. Reson. **20**(1), 1–15 (2018). https://doi.org/10.1186/s12968-018-0483-6
4. Ronneberger, O., Fischer, P., Brox, T.: U-Net: convolutional networks for biomedical image segmentation. In: Navab, N., Hornegger, J., Wells, W.M., Frangi, A.F. (eds.) MICCAI 2015. LNCS, vol. 9351, pp. 234–241. Springer, Cham (2015). https://doi.org/10.1007/978-3-319-24574-4_28
5. Isensee, F., et al.: nnU-Net: a self-configuring method for deep learning-based biomedical image segmentation. Nat. Methods **18**(2), 203–211 (2021)
6. Vaswani, A., et al.: Attention is all you need. Advances in Neural Information Processing Systems, vol. 30, pp. 5998–6008 (2017)
7. Dosovitskiy, A., et al.: An image is worth 16x16 words: transformers for image recognition at scale. arXiv preprint arXiv:2010.11929 (2020)
8. Chen, J., Lu, Y., Yu, Q., Luo, X., et al.: TransUNet: transformers make strong encoders for medical image segmentation. arXiv preprint arXiv:2102.04306 (2021)
9. Liu, Z., et al.: Swin transformer: hierarchical vision transformer using shifted windows. arXiv preprint arXiv:2103.14030 (2021)
10. Cao, H., et al.: Swin-Unet: Unet-like pure transformer for medical image segmentation. arXiv preprint arXiv:2105.05537 (2021)
11. Li, H., et al.: DT-MIL: deformable transformer for multi-instance learning on histopathological image. In: de Bruijne, M., et al. (eds.) MICCAI 2021. LNCS, vol. 12908, pp. 206–216. Springer, Cham (2021). https://doi.org/10.1007/978-3-030-87237-3_20
12. Luo, Y., et al.: 3D transformer-GAN for high-quality PET reconstruction. In: de Bruijne, M., et al. (eds.) MICCAI 2021. LNCS, vol. 12906, pp. 276–285. Springer, Cham (2021). https://doi.org/10.1007/978-3-030-87231-1_27
13. Ji, Y., et al.: Multi-compound transformer for accurate biomedical image segmentation. In: de Bruijne, M., et al. (eds.) MICCAI 2021. LNCS, vol. 12901, pp. 326–336. Springer, Cham (2021). https://doi.org/10.1007/978-3-030-87193-2_31

14. Berhane, H., et al.: Fully automated 3D aortic segmentation of 4D flow MRI for hemodynamic analysis using deep learning. Magn. Reson. Med. **84**(4), 2204–2218 (2020)
15. Wu, Y., et al.: Automated multi-channel segmentation for the 4D myocardial velocity mapping cardiac MR. In: Medical Imaging 2021: Computer-Aided Diagnosis, vol. 11597. International Society for Optics and Photonics (2021)
16. Garg, P., et al.: Left ventricular thrombus formation in myocardial infarction is associated with altered left ventricular blood flow energetics. Eur. Heart J. Cardiovasc. Imaging **20**(1), 108–117 (2019)

Learning Towards Synchronous Network Memorizability and Generalizability for Continual Segmentation Across Multiple Sites

Jingyang Zhang[1], Peng Xue[1], Ran Gu[2], Yuning Gu[1], Mianxin Liu[1],
Yongsheng Pan[1], Zhiming Cui[1,3], Jiawei Huang[1], Lei Ma[1],
and Dinggang Shen[1,4(✉)]

[1] School of Biomedical Engineering, ShanghaiTech University, Shanghai, China
dgshen@shanghaitech.edu.cn
[2] School of Mechanical and Electrical Engineering, University of Electronic Science
and Technology of China, Chengdu, China
[3] Department of Computer Science, The University of Hong Kong, Hong Kong, China
[4] Shanghai United Imaging Intelligence Co., Ltd., Shanghai, China

Abstract. In clinical practice, a segmentation network is often required
to continually learn on a sequential data stream from multiple sites rather
than a consolidated set, due to the storage cost and privacy restriction.
However, during the continual learning process, existing methods are
usually restricted in either network memorizability on previous sites or
generalizability on unseen sites. This paper aims to tackle the challenging problem of Synchronous Memorizability and Generalizability (SMG)
and to simultaneously improve performance on both previous and unseen
sites, with a novel proposed SMG-learning framework. First, we propose
a Synchronous Gradient Alignment (SGA) objective, which *not only* promotes the network memorizability by enforcing coordinated optimization
for a small exemplar set from previous sites (called replay buffer), *but
also* enhances the generalizability by facilitating site-invariance under
simulated domain shift. Second, to simplify the optimization of SGA
objective, we design a Dual-Meta algorithm that approximates the SGA
objective as dual meta-objectives for optimization without expensive
computation overhead. Third, for efficient rehearsal, we configure the
replay buffer comprehensively considering additional inter-site diversity
to reduce redundancy. Experiments on prostate MRI data sequentially
acquired from six institutes demonstrate that our method can simultaneously achieve higher memorizability and generalizability over state-of-
the-art methods. Code is available at https://github.com/jingyzhang/
SMG-Learning.

J. Zhang and P. Xue—Equal contribution.

Supplementary Information The online version contains supplementary material
available at https://doi.org/10.1007/978-3-031-16443-9_37.

Keywords: Continual segmentation · Generalizability · Memorizability

1   Introduction

Data aggregation of multiple clinical sites [5,15] is desired to train convolutional neural networks for medical image segmentation [20,27,29]. Compared with aggregating multi-site datasets as a large-scaled consolidated set, training network on a sequential data stream is more storage-efficient and privacy-preserving [28], where data of different sites arrives in sequence without consuming storage of most (or even all) data of previous sites [13]. However, consecutively fine-tuning model on only the data from different incoming sites would confront two challenges: 1) weak memorizability [19], causing significant performance drop on previously learned sites with data discrepancy due to their different imaging protocols; 2) poor generalizability [16], decreasing performance on unseen sites with out-of-distribution data [10] and thus impeding direct model deployment in clinical practice [3]. Therefore, a *Synchronous Memorizability and Generalizability (SMG)* is desirable for the network to simultaneously improve performance on both previous and unseen sites during consecutive training on a data stream.

However, prevalent researches have identified network memorizability and generalizability as two isolated tasks with customized solutions, respectively. Specifically, Continual Learning (CL) [4] is proposed to enhance memorizability and mitigate forgetting for old knowledge of previously learned sites [28]. CL usually relies on the storage of a small exemplar set from previous sites (called replay buffer) for rehearsal [2,13,21]. Then a similarity constraint [9,18] is imposed between the gradient directions of losses on the replay buffer and the incoming site, enabling coordinated optimization for previous sites and thus relieving the model forgetting. Besides, Domain Generalization (DG) [30] aims to make a model generalize well to unseen sites, with a key aspect to consider learning invariance [8] across multiple sites with data discrepancy. In DG, the replay buffer collaborated with the data of incoming site serves as a union dataset to provide multi-site distribution for model training. Notably, it is arbitrarily split into virtual-train and virtual-test subsets within each minibatch to simulate real-world domain shift [6,12], boosting feature invariance for network generalizability.

Although above-mentioned methods exhibit advantages on either network memorizability or generalizability, none of them is capable of *Learning with SMG (i.e., SMG-Learning)* that gains both advantages simultaneously [22] due to the following three reasons. First, for network memorizability, CL methods determine coordinated gradient for previous sites by optimizing a unidirectional transfer to the incoming site, which yet biases the learning process and hampers generalization capacity. Second, DG methods enhance feature invariance for a generalized model by using arbitrary data splitting to construct cross-site discrepancy, while without a particular consideration of performance balance between previous and incoming sites to consistently maintain network memorizability. Third, the replay buffer, used in both methods, is commonly populated with representative exemplars [24] from each individual previous site. This ignores data heterogeneity across sites and thus causes rehearsal redundancy, weakening both memorizability and generalizability. These motivate us to integrate the schemes of both CL and DG for SMG-Learning, by exploring gradient with *coordination for previous sites and invariance across various sites* for model updating, and further enforce additional inter-site diversity of representative exemplars in the replay buffer.

In this paper, we present, *to our knowledge for the first*, SMG-Learning framework for continual multi-site segmentation. Our contributions are: 1) We propose a novel Synchronous Gradient Alignment (SGA) objective with two complementary operations, including an orientational alignment that determines coordinated gradient for the previous sites in replay buffer to strengthen memorizability, and an arbitrary alignment that encourages site-invariance across randomly split subsets to promote generalizability; 2) We design a Dual-Meta algorithm which approximates the SGA objective as dual meta-objectives for simplified optimization, avoiding costly computation of the second-order derivatives required in the naive direct optimization; 3) We propose a comprehensive configuration of replay buffer, where each exemplar is sampled considering intra-site representativeness and inter-site diversity, to relieve redundancy and improve rehearsal efficiency. We have evaluated our method on prostate MRI segmentation, using a sequential stream of multi-site datasets for model training. Experimental results show that our method achieves synchronously high memorizability and generalizability, and clearly outperforms the state-of-the-art CL and DG methods.

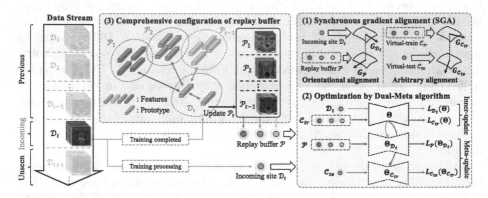

Fig. 1. Overview of our SMG-learning framework for synchronous memorizability and generalizability. In the training process, we use a Synchronous Gradient Alignment (SGA) objective to enforce coordinated optimization for previous sites and enhance invariance across sites (Sect. 2.1). This SGA objective is optimized by Dual-Meta algorithm without costly computation overhead (Sect. 2.2). After completing the training process, we configure the replay buffer comprehensively for efficient rehearsal (Sect. 2.3).

2   Method

For consecutive model training, we use a sequential data stream from T sites. Specifically, in round $t \in [1, T]$ of this procedure, we can obtain only the data $\mathcal{D}_t$ of incoming site, and allow access to a small replay buffer $\mathcal{P} = \{\mathcal{P}_1, \mathcal{P}_2, ..., \mathcal{P}_{t-1}\}$ that stores a small number of exemplars from previous sites for rehearsal. Data $\{\mathcal{D}_{t+1}, ..., \mathcal{D}_T\}$ of unseen sites is not used in the training process. Figure 1 illustrates the proposed SMG-Learning framework, aiming to simultaneously perform well on both previous and unseen sites during the consecutive model training process.

2.1   Synchronous Gradient Alignment (SGA)

Learning only from data $\mathcal{D}_t$ causes overfitting on the incoming site and also cannot enable memorizability and generalizability. Fortunately, replay buffer $\mathcal{P}$ not only allows efficient rehearsal to mitigate model forgetting, but also provides multi-site data distribution for a generalized model. Therefore, it is crucial for SMG-Learning to collaborate the data of $\mathcal{D}_t$ and $\mathcal{P}$ for model training.

Naive Joint Minimization (JM). For improved memorizability, a naive way is to adopt Joint Minimization (JM) for losses on $\mathcal{D}_t$ and $\mathcal{P}$ w.r.t. model parameters Θ for maintaining performance on both incoming and previous sites. Moreover, to strengthen generalizability, we collaborate $\mathcal{D}_t$ and $\mathcal{P}$ as a union set $\mathcal{C}$, and then randomly split it into virtual-train $\mathcal{C}_{tr}$ and virtual-test $\mathcal{C}_{te}$ subsets within each minibatch, with a naive JM defined on both subsets for the training awareness of simulated cross-site discrepancy [6]. These JM objectives are formulated as:

$$L_{\mathrm{JM}}^{\{\mathcal{D}_t, \mathcal{P}\}}(\Theta) = L_{\mathcal{D}_t}(\Theta) + L_{\mathcal{P}}(\Theta), \quad \text{and} \quad L_{\mathrm{JM}}^{\{\mathcal{C}_{tr}, \mathcal{C}_{te}\}}(\Theta) = L_{\mathcal{C}_{tr}}(\Theta) + L_{\mathcal{C}_{te}}(\Theta). \quad (1)$$

However, these naive JM objectives would be optimized asymmetrically, e.g., for $L_{\mathrm{JM}}^{\{\mathcal{D}_t, \mathcal{P}\}}(\Theta)$, a dominated decrease comes from $L_{\mathcal{D}_t}(\Theta)$ which is easier to minimize without domain shift across sites [12], yet with a risk of increasing $L_{\mathcal{P}}(\Theta)$. In this case, the model learns sufficiently on the incoming site while sacrificing performance on previous sites. In addition, the asymmetric optimization for $L_{\mathrm{JM}}^{\{\mathcal{C}_{tr}, \mathcal{C}_{te}\}}(\Theta)$ fails to obtain the consistent minimization on random subsets, and thus cannot extract site-invariant features that generalize well to unseen sites.

Definition of SGA Objective. To solve the problem of naive JM objectives, we need to simultaneously enforce coordinated descent of $L_{\mathcal{D}_t}(\Theta)$ and $L_{\mathcal{P}}(\Theta)$ and extract invariant feature representation from $L_{\mathcal{C}_{tr}}(\Theta)$ and $L_{\mathcal{C}_{te}}(\Theta)$. Let us consider the gradients of losses on $\mathcal{D}_t$, $\mathcal{P}$, $\mathcal{C}_{tr}$ and $\mathcal{C}_{te}$:

$$G_{\mathcal{D}_t} = \frac{\partial L_{\mathcal{D}_t}(\Theta)}{\partial \Theta}, \ G_{\mathcal{P}} = \frac{\partial L_{\mathcal{P}}(\Theta)}{\partial \Theta}, \ G_{\mathcal{C}_{tr}} = \frac{\partial L_{\mathcal{C}_{tr}}(\Theta)}{\partial \Theta}, \ G_{\mathcal{C}_{te}} = \frac{\partial L_{\mathcal{C}_{te}}(\Theta)}{\partial \Theta}. \quad (2)$$

If $G_{\mathcal{D}_t}$ and $G_{\mathcal{P}}$ have a similar direction, i.e., inner product $G_{\mathcal{D}_t} \cdot G_{\mathcal{P}} > 0$, updating model with a step along $G_{\mathcal{D}_t}$ or $G_{\mathcal{P}}$ improves performance on both $\mathcal{D}_t$ and $\mathcal{P}$, achieving coordinated optimization and thus a high memorizability for previous sites. Moreover, if the directions of $G_{\mathcal{C}_{tr}}$ and $G_{\mathcal{C}_{te}}$ are similar, i.e., $G_{\mathcal{C}_{tr}} \cdot G_{\mathcal{C}_{te}} > 0$, the features learned by either step can be invariant to the simulated discrepancy between arbitrarily split $\mathcal{C}_{tr}$ and $\mathcal{C}_{te}$, strengthening potential generalizability.

Based on this observation, we propose to simultaneously maximize $G_{\mathcal{D}_t} \cdot G_{\mathcal{P}}$ for orientational gradient alignment between incoming site and replay buffer, and $G_{\mathcal{C}_{tr}} \cdot G_{\mathcal{C}_{te}}$ for arbitrary gradient alignment across randomly split subsets[1]. They enforce the gradient with coordination for previous sites and invariance across various sites, benefiting to SMG-Learning. Formally, we name our objective as Synchronous Gradient Alignment (SGA), formalized by subtracting $G_{\mathcal{D}_t} \cdot G_{\mathcal{P}}$ and $G_{\mathcal{C}_{tr}} \cdot G_{\mathcal{C}_{te}}$ from the naive JM objectives with weight γ and β:

$$L_{\text{SGA}}(\Theta) = L_{\text{JM}}^{\{\mathcal{D}_t, \mathcal{P}\}}(\Theta) + L_{\text{JM}}^{\{\mathcal{C}_{tr}, \mathcal{C}_{te}\}}(\Theta) - \gamma\, G_{\mathcal{D}_t} \cdot G_{\mathcal{P}} - \beta\, G_{\mathcal{C}_{tr}} \cdot G_{\mathcal{C}_{te}}. \qquad (3)$$

Relationship with CL and DG Methods. If only the orientational alignment is used, the SGA objective can be regarded as a CL variant [9] with a difference: it ensures coordinated descent explicitly without inequality constraints [18]. If we only activate the arbitrary alignment, SGA objective borrows the spirit from DG with random data splitting [12] to simulate domain shift for feature invariance. A unified combination of them in our SGA objective leverages the complementary strengths of CL and DG and thereby makes the SMG-Learning promising.

2.2   Efficient Optimization by Dual-Meta Algorithm

A naive optimization of our SGA objective is costly since it requires to compute second-order derivatives due to the gradient inner product terms. To solve it, we propose a Dual-Meta algorithm that approximates SGA objective as dual meta-objectives for meta-optimization to avoid expensive second-order computation.

Formally, the proposed SGA objective can be approximated as dual meta-objectives by the first-order Taylor expansion with omitted infinitesimal:

$$L_{\text{SGA}}(\Theta) = \underbrace{L_{\mathcal{D}_t}(\Theta) + L_{\mathcal{P}}(\Theta - \gamma G_{\mathcal{D}_t})}_{\text{the first meta-objective}} + \underbrace{L_{\mathcal{C}_{tr}}(\Theta) + L_{\mathcal{C}_{te}}(\Theta - \beta G_{\mathcal{C}_{tr}})}_{\text{the second meta-objective}} \qquad (4)$$

where dual components correspond to the maximization of $G_{\mathcal{D}_t} \cdot G_{\mathcal{P}}$ and $G_{\mathcal{C}_{tr}} \cdot G_{\mathcal{C}_{te}}$ in SGA objective, respectively. Derivation details are given in Appendix B.

Based on this formulation of dual meta-objectives [12], we perform a classical meta-optimization [7] including two stages. First, in inner-update, model parameters Θ are updated on $\mathcal{D}_t$ and $\mathcal{C}_{tr}$ with step size γ and β, respectively:

$$\Theta_{\mathcal{D}_t} = \Theta - \gamma G_{\mathcal{D}_t} \quad \text{and} \quad \Theta_{\mathcal{C}_{tr}} = \Theta - \beta G_{\mathcal{C}_{tr}}. \qquad (5)$$

---

[1] Orientational alignment seems a special case of arbitrary alignment when the subset splitting is $\mathcal{C}_{tr} = \mathcal{D}_t$ and $\mathcal{C}_{te} = \mathcal{P}$, coincidently. However, orientational alignment cannot be omitted with a risk of suffering from potential interference [22], as empirically shown in Appendix A, due to arbitrary alignment for other subset splittings.

Second, in meta-update, we rewrite the dual meta-objectives as $L_{\mathcal{D}_t}(\Theta) + L_{\mathcal{P}}(\Theta_{\mathcal{D}_t}) + L_{\mathcal{C}_{tr}}(\Theta) + L_{\mathcal{C}_{te}}(\Theta_{\mathcal{C}_{tr}})$. Notice that loss $L_{\mathcal{P}}(\Theta_{\mathcal{D}_t})$ and $L_{\mathcal{C}_{te}}(\Theta_{\mathcal{C}_{tr}})$ are computed on $\mathcal{P}$ and $\mathcal{C}_{tr}$ with the updated parameters $\Theta_{\mathcal{D}_t}$ and $\Theta_{\mathcal{C}_{tr}}$, but optimized towards the original parameters Θ where only the first-order derivative is required.

2.3   Comprehensive Configuration of Replay Buffer

After completing the learning round t, we should sample a small exemplar set $\mathcal{P}_t$ from $\mathcal{D}_t$ of incoming site, and then add it into replay buffer $\mathcal{P}$ for rehearsal in the next learning round. For rehearsal efficacy, the selected exemplars are required with intra-site representativeness for the data distribution of $\mathcal{D}_t$. Meanwhile, they are also demanded with inter-site diversity for exemplars of previous sites in $\mathcal{P} = \{\mathcal{P}_1, ..., \mathcal{P}_{t-1}\}$ to reduce redundancy. Therefore, we propose a comprehensive configuration of $\mathcal{P}_t$ that considers both properties for replay buffer updating.

Intra-site Representativeness. Given a subject $s_t^i \in \mathcal{D}_t$, we extract its feature vector $f_\Theta(s_t^i)$ by average-pooling its feature map from the bottleneck layer of the model parameterized with Θ. Its intra-site representativeness is defined as the cosine similarity, i.e., $\text{sim}(a,b) = a^T b/\|a\|\|b\|$, to the site prototype μ_t [24]:

$$R(s_t^i) = \text{sim}(f_\Theta(s_t^i), \mu_t), \qquad \text{with} \qquad \mu_t = \sum\nolimits_{s_t^k \in \mathcal{D}_t} f_\Theta(s_t^k)/N_{\mathcal{D}_t}, \qquad (6)$$

where μ_t is estimated as the mean feature over all $N_{\mathcal{D}_t}$ subjects in $\mathcal{D}_t$. A high similarity $R(s_t^i)$ with respect to μ_t indicates a high representativeness for $\mathcal{D}_t$.

Inter-site Diversity. For a subject $s_t^i \in \mathcal{D}_t$ from a given individual site, inter-site diversity $V(s_t^i)$ measures its diversity w.r.t. all previous sites in replay buffer:

$$V(s_t^i) = \sum\nolimits_{\mathcal{P}_k \in \mathcal{P}} \text{div}(s_t^i, \mathcal{P}_k)/N_{\mathcal{P}}, \quad \text{with} \quad \text{div}(s_t^i, \mathcal{P}_k) = \min_{s_k^j \in \mathcal{P}_k} \text{dsim}(f_\Theta(s_t^i), f_\Theta(s_k^j)),$$
$$(7)$$

where $\text{div}(s_t^i, \mathcal{P}_k)$ denotes the diversity of s_t^i w.r.t. a previous site $\mathcal{P}_k \in \mathcal{P}$ in replay buffer, and the inter-site diversity $V(s_t^i)$ is the average over all $N_{\mathcal{P}}$ previous sites. Notably, $\text{div}(s_t^i, \mathcal{P}_k)$ is defined as the minimal feature dissimilarity, i.e., $\text{dsim}(a,b) = -\text{sim}(a,b)$, of all subjects in $\mathcal{P}_k$, which is insensitive to outliers [26].

Comprehensive Configuration. Based on $R(s_t^i)$ and $V(s_t^i)$, we adopt a hybrid measurement $H(s_t^i)$ with a weight λ and then propose a comprehensive configuration of exemplar set $\mathcal{P}_t$ with size N_e to update replay buffer. The selected s_t^i in $\mathcal{P}_t$ should have the top-N_e hybrid measurement, i.e., satisfying $H(s_t^i) \geq \Gamma(N_e)$:

$$\mathcal{P}_t = \{s_t^i | s_t^i \in \mathcal{D}_t, H(s_t^i) \geq \Gamma(N_e)\}, \quad \text{with} \quad H(s_t^i) = R(s_t^i) + \lambda V(s_t^i). \qquad (8)$$

3   Experiments

Dataset. We employed a well-established multi-site T2-weighted MRI dataset for prostate segmentation [17]: 30 subjects in RUNMC [1] (Site A), 30 subjects in BMC [1] (Site B), 19 subjects in HCRUDB [11] (Site C), 13 subjects in UCL [14] (Site D), 12 subjects in BIDMC [14] (Site E) and 12 subjects in HK [14] (Site F). They were acquired using different protocols with in/through plane resolution ranging from 0.25/2.2-3 to 0.625/3.6 mm. For pre-processing, each image was resized to 384 × 384 in axial plane and normalized to zero mean and unit variance. In each site, we used 60%, 15% and 25% of subjects for training, validation and testing. This dataset is publicly released at https://liuquande. github.io/SAML/.

Experimental Setting. For consecutive model training, we sequentially organized the training sets of site A→B→C→D→E as a training data stream $S_{A \to E}$, while leaving site F as an unseen site without involved in training. Similarly, a reversed stream $S_{F \to B}$ with unseen site A was also validated in experiments.

Table 1. Comparison with state-of-the-arts for network memorizability (Mem.) and generalizability (Gen.), after training on two different data stream $S_{A \to E}$ and $S_{F \to B}$.

Settings		Training data stream $S_{A \to E}$								Training data stream $S_{F \to B}$							
Metrics		DSC(%)↑				ASD(mm)↓				DSC (%)↑				ASD (mm)↓			
		Mem.		Gen.		Mem.		Gen.		Mem.		Gen.		Mem.		Gen.	
		BM	BT	FM	FT	BM	BT	FM	FT	BM	BT	FM	FT	BM	BT	FM	FT
Baseline	FineTuning	62.64	−27.40	82.77	−7.01	5.00	3.84	2.40	1.38	64.20	−24.07	67.87	−24.39	7.57	6.21	11.84	10.72
	JM [24]	73.98	−14.24	83.85	−5.07	3.36	2.01	2.24	1.14	73.81	−13.45	70.39	−20.67	3.18	1.62	5.47	4.16
CL	GEM [18]	79.69	−9.58	84.82	−4.47	2.48	1.23	2.13	1.08	78.15	−9.52	72.31	−19.43	2.41	0.94	3.96	2.70
	C-Meta [9]	81.17	−7.59	85.83	−3.84	2.09	0.81	1.79	0.72	80.88	−6.92	72.33	−19.57	2.14	0.66	3.68	2.47
DG	ISGM [25]	80.49	−8.01	86.22	−2.97	2.25	0.93	1.54	0.45	78.60	−9.03	74.48	−17.04	2.40	0.97	3.46	2.23
	G-Meta [12]	78.98	−10.78	86.31	−2.94	2.33	1.17	1.50	0.43	77.32	−10.15	75.06	−16.70	2.55	1.17	3.43	2.21
SMG-Learning	SGA	81.68	−7.41	86.46	−2.72	1.99	0.76	1.42	0.36	81.72	−6.17	74.82	−16.69	2.06	0.64	3.52	2.30
	SGA(+c)	**83.60**	**−5.25**	**87.18**	**−2.30**	**1.81**	**0.59**	**1.34**	**0.30**	**82.35**	**−5.42**	**77.43**	**−14.05**	2.09	**0.60**	**3.20**	**1.95**

Evaluation Metrics. In each site, we evaluate segmentation accuracy by Dice Score Coefficient (DSC) and Average Surface Distance (ASD). Based on them, we define several specialized metrics [18, 28] to evaluate memorizability and generalizability. First, for memorizability, we define a Backward Measure (BM) as the mean segmentation accuracy over previous and incoming sites for generic evaluation, and a Backward Transfer (BT) as the accuracy degradation on previous sites after learning on an incoming site for quantitative evaluation. Second, to evaluate generalizability, we define a Forward Measure (FM) as the segmentation accuracy on unseen site, and a Forward Transfer (FT) as the gap between the accuracy on unseen site obtained before and after learning on this unseen site. An advanced method should have high BM and FM for DSC while with low values for ASD, and its BT and FT should be as close as possible to zero.

Implementation. We adopted a 2D-UNet [23] as segmentation backbone due to the large variance in slice thickness among different sites. For our SGA objective, both parameter γ and β were empirically set as $5e^{-4}$ [12] for suitable trade-off. Weight λ was set as 1.0 to comprehensively configure replay buffer, where each exemplar set contained only $N_e = 2$ subjects for storage efficiency. The network was trained by Dual-Meta algorithm, using Adam Optimizer with learning rate $5e^{-4}$, batch size 5 and iteration number $20K$ for inner-update and meta-update.

Comparison with State-of-the-Arts. We compare our SMG-Learning framework, including **SGA** (optimized by Dual-Meta algorithm) and **SGA+(c)** (integrated with comprehensive replay buffer), with the state-of-the-art methods: 1) Baselines: **FineTuning** only on the incoming site, and Joint Minimization (**JM**) using only Eq. (1) with representative replay buffer [24]; 2) CL schemes: Gradient Episodic Memory (**GEM**) [18] and Continual Meta-learning (**C-Meta**) [9] for implicitly and explicitly coordinated optimization on previous sites; 3) DG schemes: Inter-Site Gradient Matching (**ISGM**) [25] for feature site-invariance and Generalized Meta-learning (**G-Meta**) [12] on randomly split subsets.

As listed in Table 1, baselines show a poor memorizability with the worst BM and BT values, and a weak generalizability with the worst FM and FT values. Although CL and DG perform better than baselines, their improvements are still limited either on memorizability or generalizability instead of in a synchronous way. The proposed SGA consistently outperforms these methods on almost all metrics. Moreover, SGA+(c) achieves the best results on all two experimental settings, e.g., with an obvious advantage over C-Meta by 2.34% and 1.50% BT for DSC and also over G-Meta by 0.64% and 2.65% FT, indicating both high memorizability and generalizability of our SMG-Learning framework.

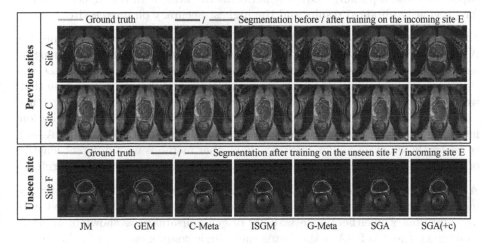

Fig. 2. Segmentation visualization when training on stream $S_{A \to E}$. Upper: results on previous site A and C obtained before (i.e., immediately after training on site A and C, respectively) and after training on incoming site E; Lower: results on unseen site F obtained after training on this unseen site F and after training on the incoming site E.

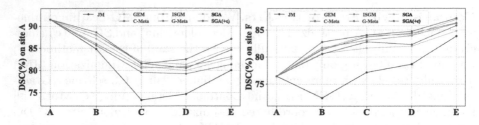

Fig. 3. Curves of DSC on site A (i.e., previous site) and site F (i.e., unseen site) as more sites participate in the consecutive model training ordered by A→B→C→D→E.

Figure 2 visualizes the segmentation results on previous and unseen sites. The red curves on different sites are the results obtained immediately after training on these individual sites. They are close to the ground truth, revealing the high cross-site adaptation capacity of all methods. JM degrades the performance on previous site A and C after learning on incoming site E, and causes inaccuracy on unseen site F if not involved in training. Compared to other methods with only advantage either on previous or unseen site, our SGA achieves a synchronous superiority and especially SGA(+c) performs the best, with the highest overlap for previously obtained results and the highest accuracy on unseen site.

Effect of Training Sequence Length. Figure 3 shows the DSC curves on site A and F, revealing how the network memorizability and generalizability changes, as we gradually increase the training sequence ordered by site A→B→C→D→E. Intuitively, a longer sequence contains more complex multi-site information that would cause more optimization interference, thus aggravating model forgetting for previous site A. Contrarily, its covered more comprehensive data distribution helps the model generalize better to unseen site F. Notably, SGA and SGA(+c) consistently maintain higher accuracy than other methods with different lengths of training data sequence, confirming the stable efficacy of our SMG-Learning.

Ablation Study. C-Meta and G-Meta can be regarded as the variants of SGA using only orientational and arbitrary alignment, respectively. They exhibit limited advantages on either memorizability or generalizability, as shown in Table 1. Combining them in SGA outperforms each single component, and further using comprehensive replay buffer in SGA(+c) offers more superiority owing to its efficient rehearsal. Moreover, SGA optimized by Dual-Meta algorithm achieves a comparable segmentation accuracy with the direct optimization while costs only 3.31% computational time, revealing the efficiency of our Dual-Meta algorithm.

4   Conclusion

This paper presents a novel framework for SMG-Learning. We propose a SGA objective, cooperating with a comprehensive configuration of replay buffer, to enforce coordinated optimization for previous sites and feature invariance across sites. It is optimized by Dual-Meta algorithm without costly second-order computation. In the future, it is of interest to extend our method to federated learning with decentralized data from different sites for better privacy protection.

Acknowledgement. This work was supported in part by National Natural Science Foundation of China (grant number 62131015), and Science and Technology Commission of Shanghai Municipality (STCSM) (grant number 21010502600).

References

1. Bloch, N., et al.: NCI-ISBI 2013 challenge: automated segmentation of prostate structures. Cancer Imaging Arch. (2015)
2. Castro, F.M., Marín-Jiménez, M.J., Guil, N., Schmid, C., Alahari, K.: End-to-end incremental learning. In: Proceedings of the European conference on computer vision (ECCV), pp. 233–248 (2018)
3. Chen, C., Dou, Q., Chen, H., Heng, P.-A.: Semantic-aware generative adversarial nets for unsupervised domain adaptation in chest X-ray segmentation. In: Shi, Y., Suk, H.-I., Liu, M. (eds.) MLMI 2018. LNCS, vol. 11046, pp. 143–151. Springer, Cham (2018). https://doi.org/10.1007/978-3-030-00919-9_17
4. Delange, M., et al.: A continual learning survey: defying forgetting in classification tasks. IEEE Trans. Pattern Analy. Mach. Intell. (2021)
5. Dhruva, S.S., et al.: Aggregating multiple real-world data sources using a patient-centered health-data-sharing platform. NPJ Digit. Med. **3**(1), 1–9 (2020)
6. Dou, Q., Coelho de Castro, D., Kamnitsas, K., Glocker, B.: Domain generalization via model-agnostic learning of semantic features. Adv. Neural Inf. Process. Syst. **32**, 6450–6461 (2019)
7. Finn, C., Abbeel, P., Levine, S.: Model-agnostic meta-learning for fast adaptation of deep networks. In: International Conference on Machine Learning, pp. 1126–1135. PMLR (2017)
8. Ganin, Y.: Domain-adversarial training of neural networks. J. Mach. Learn. Res. **17**(1), 2030–2096 (2016)
9. Gupta, G., Yadav, K., Paull, L.: La-MAML: look-ahead meta learning for continual learning. arXiv preprint arXiv:2007.13904 (2020)
10. Koh, P.W., et al.: Wilds: A benchmark of in-the-wild distribution shifts. In: International Conference on Machine Learning, pp. 5637–5664. PMLR (2021)
11. Lemaître, G., et al.: Computer-aided detection and diagnosis for prostate cancer based on mono and multi-parametric MRI: A review. Comput. Biol. Med. **60**, 8–31 (2015)
12. Li, D., Yang, Y., Song, Y.Z., Hospedales, T.M.: Learning to generalize: meta-learning for domain generalization. In: Thirty-Second AAAI Conference on Artificial Intelligence (2018)
13. Li, Z., Zhong, C., Wang, R., Zheng, W.-S.: Continual learning of new diseases with dual distillation and ensemble strategy. In: Martel, M.A.L., et al. (eds.) MICCAI 2020. LNCS, vol. 12261, pp. 169–178. Springer, Cham (2020). https://doi.org/10.1007/978-3-030-59710-8_17
14. Litjens, G., et al.: Evaluation of prostate segmentation algorithms for MRI: the PROMISE12 challenge. Med. Image Anal. **18**(2), 359–373 (2014)
15. Liu, Q., Dou, Q., Yu, L., Heng, P.A.: MS-Net: Multi-site network for improving prostate segmentation with heterogeneous MRI data. IEEE Trans. Med. Imag. **39**(9), 2713–2724 (2020)

16. Liu, Q., Chen, C., Qin, J., Dou, Q., Heng, P.A.: FedDG: Federated domain generalization on medical image segmentation via episodic learning in continuous frequency space. In: Proceedings of the IEEE/CVF Conference on Computer Vision and Pattern Recognition, pp. 1013–1023 (2021)

17. Liu, Q., Dou, Q., Heng, P.-A.: Shape-aware meta-learning for generalizing prostate MRI segmentation to unseen domains. In: Martel, A.L., et al. (eds.) MICCAI 2020. LNCS, vol. 12262, pp. 475–485. Springer, Cham (2020). https://doi.org/10.1007/978-3-030-59713-9_46

18. Lopez-Paz, D., Ranzato, M.: Gradient episodic memory for continual learning. In: Advances in Neural Information Processing Systems, pp. 6467–6476 (2017)

19. McCloskey, M., Cohen, N.J.: Catastrophic interference in connectionist networks: the sequential learning problem. In: Psychology of Learning and Motivation, vol. 24, pp. 109–165. Elsevier, San Diego (1989)

20. Nie, D., Wang, L., Adeli, E., Lao, C., Lin, W., Shen, D.: 3-d fully convolutional networks for multimodal isointense infant brain image segmentation. IEEE Trans. Cybernet. **49**(3), 1123–1136 (2018)

21. Rebuffi, S.A., Kolesnikov, A., Sperl, G., Lampert, C.H.: ICARL: Incremental classifier and representation learning. In: Proceedings of the IEEE Conference on Computer Vision and Pattern Recognition, pp. 2001–2010 (2017)

22. Riemer, M., et al.: Learning to learn without forgetting by maximizing transfer and minimizing interference. arXiv preprint arXiv:1810.11910 (2018)

23. Ronneberger, O., Fischer, P., Brox, T.: U-Net: convolutional networks for biomedical image segmentation. In: Navab, N., Hornegger, J., Wells, W.M., Frangi, A.F. (eds.) MICCAI 2015. LNCS, vol. 9351, pp. 234–241. Springer, Cham (2015). https://doi.org/10.1007/978-3-319-24574-4_28

24. Sener, O., Savarese, S.: Active learning for convolutional neural networks: a core-set approach. arXiv preprint arXiv:1708.00489 (2017)

25. Shi, Y., et al.: Gradient matching for domain generalization. arXiv preprint arXiv:2104.09937 (2021)

26. Wang, K., Zhang, D., Li, Y., Zhang, R., Lin, L.: Cost-effective active learning for deep image classification. IEEE Trans. Circ. Syst. Video Technol. **27**(12), 2591–2600 (2016)

27. Xiang, L., Wang, Q., Nie, D., Zhang, L., Jin, X., Qiao, Y., Shen, D.: Deep embedding convolutional neural network for synthesizing CT image from T1-weighted MR image. Med. Image Anal. **47**, 31–44 (2018)

28. Zhang, J., Gu, R., Wang, G., Gu, L.: Comprehensive importance-based selective regularization for continual segmentation across multiple sites. In: de Bruijne, M., et al. (eds.) MICCAI 2021. LNCS, vol. 12901, pp. 389–399. Springer, Cham (2021). https://doi.org/10.1007/978-3-030-87193-2_37

29. Zhang, I., et al.: Weakly supervised vessel segmentation in x-ray angiograms by self-paced learning from noisy labels with suggestive annotation. Neurocomputing **417**, 114–127 (2020)

30. Zhou, K., Liu, Z., Qiao, Y., Xiang, T., Loy, C.C.: Domain generalization: a survey. arXiv preprint arXiv:2103.02503 (2021)

Progressive Deep Segmentation of Coronary Artery via Hierarchical Topology Learning

Xiao Zhang[1,2], Jingyang Zhang[2], Lei Ma[2], Peng Xue[2], Yan Hu[2,3], Dijia Wu[2,4],
Yiqiang Zhan[2,4], Jun Feng[1(✉)], and Dinggang Shen[2,4(✉)]

[1] School of Information Science and Technology, Northwest University, Xi'an, China
fengjun@nwu.edu.cn
[2] School of Biomedical Engineering, ShanghaiTech University, Shanghai, China
dgshen@shanghaitech.edu.cn
[3] School of Computer Science and Engineering, The University of New South Wales,
Sydney, Australia
[4] Shanghai United Imaging Intelligence Co., Ltd., Shanghai, China

Abstract. Coronary artery segmentation is a critical yet challenging step in coronary artery stenosis diagnosis. Most existing studies ignore important contextual anatomical information and vascular topologies, leading to limited performance. To this end, this paper proposes a progressive deep-learning based framework for accurate coronary artery segmentation by leveraging contextual anatomical information and vascular topologies. The proposed framework consists of a spatial anatomical dependency (SAD) module and a hierarchical topology learning (HTL) module. Specifically, the SAD module coarsely segments heart chambers and coronary artery for region proposals, and captures spatial relationship between coronary artery and heart chambers. Then, the HTL module adopts a multi-task learning mechanism to improve the coarse coronary artery segmentation by simultaneously predicting the hierarchical vascular topologies i.e., key points, centerlines, and neighboring cube-connectivity. Extensive evaluations, ablation studies, and comparisons with existing methods show that our method achieves state-of-the-art segmentation performance.

Keywords: Coronary artery segmentation · Spatial anatomical dependency · Hierarchical topology representation · Multi-task learning

1 Introduction

Coronary artery stenosis (CAS), one of the most common cardiovascular diseases, is a narrowing of a coronary artery lumen caused by coronary artery atherosclerosis [2,14]. Coronary computed tomography angiography (CCTA) is a non-invasive imaging technique, which is routinely employed to visualize the narrowing artery lumen [7,12]. Precise CAS diagnosis relies on accurate segmentation of coronary artery from CCTA images, since measuring the diameter of

L. Wang et al. (Eds.): MICCAI 2022, LNCS 13435, pp. 391–400, 2022.
https://doi.org/10.1007/978-3-031-16443-9_38

the narrowing artery lumen is the important criteria of CAS grading quantification in clinical practice [1,20].

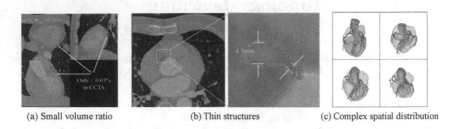

(a) Small volume ratio   (b) Thin structures   (c) Complex spatial distribution

Fig. 1. The challenges of accurate coronary artery segmentation.

However, it is still a challenging task to accurately segment coronary artery in CCTA images due to the following characteristics: 1) Small volume ratio (less than 0.05%) in images (Fig. 1(a)), leading to high-class imbalance between background and coronary artery. 2) Thin structures with lumen diameter ranging from 1mm to 5mm (Fig. 1(b)), leading to numerous hard-to-segment regions and discontinuous segmentation. 3) Patient-specific spatial distribution with various shapes and locations across individuals (Fig. 1(c)), hampering the network to accurately capture the vascular anatomical structures.

Recently, various deep learning-based methods have been proposed for medical image segmentation tasks [9,10,15–17]. The general models such as UNet and its variants [6,18] cannot obtain good performance in coronary artery segmentation, as a result of the extremely complex anatomical structures and patient-specific spatial distribution of coronary artery. Some existing studies attempted to tackle these problems via designing customized networks [3,8,19]. For example, Zhu et al. [19] proposed a modified PSPNet with pyramid pooling module to learn multi-scale global information, which reduces the adverse effect of small volume ratio. In addition, some attention-based methods [3,8] were proposed to learn more fine-grained feature representations in coronary artery segmentation. However, these methods ignored important coronary anatomical and topological properties, resulting in limited segmentation performance. Some other studies employed topology representations [4,13] as the loss constraint. Huang et al. [4] proposed a contour-constrained weighted Hausdorff distance loss, which focuses on the continuity and connectivity of the boundary. More recently, Shit et al. [13] introduced a similarity measurement termed centerline Dice, which calculated on the intersection of the segmentation masks and the skeleton. However, the topology constraints in these methods are incomplete and unable to comprehensively describe the whole topology structures of the coronary artery.

In this paper, we propose a progressive learning-based framework to capture artery-to-heart anatomical dependency and intra-artery hierarchical topology representations. Specifically, a spatial anatomical dependency (SAD) module is developed to coarsely segment the coronary artery and calculate the distance

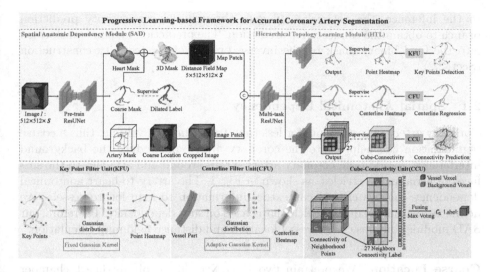

Fig. 2. Overview of the proposed framework. It consists of two components: 1) SAD module including the coarse-grained localization of coronary artery and the distance field map calculation. 2) HTL module using topology structures of key points, centerlines and cube-connectivity. (Color figure online)

field map as spatial relation between artery and heart. It can *not only* reduce the imbalance between vessels and the background, *but also* represent vessel spatial position through the distance map as anatomical dependency consideration. Then a hierarchical topology learning (HTL) module is designed to improve coarse segmentation through learning the coronary structure representations in three levels, i.e. key points, centerlines, and connectivity. It can significantly alleviate the discontinuity of thin branches and patient-specific distribution differences.

Our contributions can be summarized as follows: 1) A novel progressive learning-based framework for accurate coronary artery segmentation; 2) A spatial anatomical dependency module to capture artery-to-heart contextual information; and 3) A hierarchical topology learning module to model coronary structure representations in three levels.

2   Method

As shown in Fig. 2, the proposed framework mainly consists of two modules: 1) spatial anatomical dependency module (SAD) and 2) hierarchical topology learning module (HTL). First, the SAD module is developed to obtain the heart mask and coarse artery mask for the purpose of capturing the anatomic dependency (the distance map) and the coarse-grain spatial localization of vessels, thus removing most background interference. Based on the output of the SAD, a multi-task network is trained to learn the hierarchical topology representations, generating key points, centerlines, and neighboring cube-connectivity.

In the inference, the network only goes through the connectivity prediction branch to obtain the connectivity prediction. The prediction is then transformed into the final binary mask by the inverse process of connectivity construction (Sect. 2.2).

2.1 Spatial Anatomical Dependency

Coronary artery only accounts for less than 0.05% in each volume, thus a coarse segmentation can help locate the coronary artery and reduce the background interference. In addition, coronary artery wraps around and attaches to the surfaces of atria and ventricles, as shown in Fig. 1. The artery-to-heart anatomical dependency conveys critical contextual information to help locate vessels. We define an anatomical distance field map based on such dependency. Therefore, our SAD module contains coarse vessel location and distance field map calculation.

Coarse Location. We pretrain two ResUNets [6] to obtain heart chamber masks and coarse coronary artery mask, respectively. An over-segmented mask of coronary artery ("Dilated Label" in Fig. 2) is generated by dilating the ground truth (with a dilated kernel (5,5,5)) as supervision for coarse coronary artery segmentation.

Anatomical Distance Field Map. Given a coarse vessel segmentation from "Coarse Location", we first extract a point set of coronary artery $P = [p_1, p_2, \cdots, p_N]$ with dimension $N \times 3$, where $n \in [13000, 20000]$. Then, we extract a point set from the surface of each chamber (i.e., left ventricle, right ventricle, left atrium, right atrium, and the whole aorta), denoted as $\{O_1, O_2, \cdots, O_5\}$. To reduce computational cost, each point set of the heart chambers is subsampled to 8000.

The distance field map is obtained by calculating the minimum distance from each point p_n to the five point sets of heart substructures $\{O_1, O_2, \cdots, O_5\}$, defined by Eq. 1. Then, these values are normalized (Eq. 2) to reduce the interference of outliers, and concatenated together to form a 5-channel distance field map F_a, denoted as $F_a = \{(d_{n,1}, d_{n,2}, \cdots, d_{n,5})\}$. The calculation of such a distance field map between P and $O_M (M = 1, 2, \cdots, 5)$ can be formulated as follows:

$$S_{n,M} = \min_{o_m \in O_M} \sqrt{\left(p_n^x - o_m^x\right)^2 + \left(p_n^y - o_m^y\right)^2 + \left(p_n^z - o_m^z\right)^2}, \tag{1}$$

$$d_{n,M} = \begin{cases} \frac{S_{n,M}}{\tau}, & \text{if } \frac{S_{n,M}}{\tau} < 1, \\ 1, & \text{else,} \end{cases} \tag{2}$$

where $S_{n,M}$ denotes the minimum distance from the n-th point p_n on the coronary artery surface to O_M. In addition, $d_{n,M}$ is the normalized representation of $S_{n,M}$, and τ is set as 50.

2.2 Hierarchical Topology Learning

The spatial location and the anatomical distance field map obtained by the SAD module are coarse-grained representations. To learn more refined features of the coronary artery, we propose a hierarchical topology learning (HTL) module containing: 1) key points (bifurcation and endpoints) detection for capturing vascular tree-topology, 2) centerline heatmap regression for capturing skeleton structure, and 3) cube-connectivity prediction for capturing the connected space information.

Key Points Detection. Bifurcation points and endpoints of each coronary branch can provide a strong reference for the vascular tree-like structure. Therefore, we introduce a key points detection branch to guide the segmentation. Compared with the case of regressing coordinates directly, the point heatmap regression focuses on a distribution, which is more suitable for key points detection and produces more robust results. To supervise the learning of key points detection, we propose a key-point filter unit (KFU) to generate a heatmap with a small standard deviation (i.e., $\sigma = 3$ voxels, as shown in Fig. 2, with the light green area) for each point as the ground truth.

Centerline Heatmap Regression. The centerline covers only scattered voxels on each cross-section. A little position bias of those voxels cause a huge error in centerline prediction, so we design a 3D centerline heatmap regression instead of predicting centerline directly.

We use a 3D adaptive Gaussian filter G_t to generate a locally normalized centerline heatmap (i.e., light blue area (centerline filter unit) in Fig. 2) and $G_t \sim N(0, \sigma_t^2)$ (Eq. 4), where σ_t is the standard deviation of the three dimensions (Eq. 3), and is proportional to the width, height, and depth of the individual object. Such a design of G_t ensures that the thick-vessel centerline heatmap has smooth value change, and the thin-vessel heatmap is significantly enhanced. The centerline heatmap generation can be written as:

$$\sigma_t = \sqrt{\frac{1}{L} \sum_{l=1\cdots L} \left(P_t^l - C_t\right)^2} \tag{3}$$

$$G_t = e^{-\frac{\left(P_t^l - C_t\right)^2}{2\sigma_t}} \tag{4}$$

Cube-Connectivity Prediction. The cube-connectivity is defined between the current vessel point and its nearest 26-neighbor points, which provides connections to all vascular voxels. Previous works [5,11] have achieved similar conclusions. Given a voxel of the coronary label P and its all 26-nearest neighbors Q, it results in a $3 \times 3 \times 3$ cube shape.

The center point P and its 26 neighbors Q form a 27-channel vector. Let Q_u represent one of the 26-neighbor points. If both P and Q_u are coronary voxels, then the pair (P, Q_u) is connected and the corresponding position P on the label of the u-th channel is marked as 1, as shown in Fig. 2 (blue area). Otherwise, we mark 0 on the u-th channel to represent disconnected pair. We encode topological relation of all coronary voxels via this cube-connectivity criterion. Therefore, all background voxels are disconnected, which reduces the background interference in the learning process. In the training phase, the loss is computed based on the 27-channel vector.

However, the testing phase requires a fusion process, which follows the inverse process of connectivity construction. For the 27-channel binary mask, two neighbor voxels are connected if and only if the two corresponding labels in the connected cube structure. In this way, 27 predictions are also generated at each voxel, and the max-voting principle is used to select the best result.

2.3   Loss Function

The loss functions of the key points detection branch and the centerline regression branch are both mean squared error (MSE) losses, denoted as $\mathcal{L}_{kp}$ and $\mathcal{L}_{cl}$ respectively. In contrast, we use the average Dice similarity coefficient loss to optimize the cube-connectivity branch, and denote it as $\mathcal{L}_{cc}$.

In the experiment, the three task branches are jointly trained, and the total loss function is defined as:

$$\mathcal{L}_{total} = \mathcal{L}_{cc} + \lambda \left(\mathcal{L}_{kp} + \mathcal{L}_{cl} \right) \tag{5}$$

where λ is the balancing weight empirically set as 0.2 in our experiments.

3   Experiments and Results

3.1   Dataset and Evaluation Metrics

We collected 100 subjects of CCTA volumes with the resolution of $0.3 \times 0.3 \times 0.5\,\mathrm{mm}^3$ and the size of $512 \times 512 \times S_z$, where the slice number S_z varies from 155 to 353. Heart chambers (i.e., left ventricle, right ventricle, left atrium, right atrium, and the whole aorta) and coronary artery were manually labeled by three experienced experts. We randomly cropped sub-volumes with the size of $256 \times 256 \times 64$ as the input for training. In the testing phase, the sliding windows of the same size were utilized and moved in a step of half window size to cover the entire volume images. Finally, the predicted results of different sub-volumes are combined as a final result.

The quantitative results of the proposed framework were reported by Dice Similarity Coefficient (Dice), True Positive Rate (TPR), Mean Intersection over Union (mIoU), Hausdorff Distance (HD), and the Average Surface Distance (ASD).

3.2   Implementation Details

We evaluated the proposed framework by five-fold cross-validation, where 80 subjects were used for training and 20 subjects for testing in each folder. All experiments were implemented using Pytorch on 2 NVIDIA Tesla A100 GPUs. We trained the networks using Adam optimizer with an initial learning rate of 10^{-4}, epoch of 800, and batch size of 4.

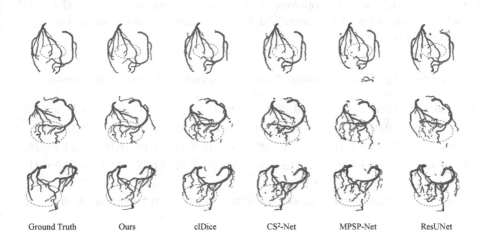

Ground Truth   Ours   clDice   CS²-Net   MPSP-Net   ResUNet

Fig. 3. Comparison of results for three typical subjects by five different methods. The green circles are used to highlight the regions for visual presentation. (Color figure online)

3.3   Comparisons with the State-of-the-Art Methods

To verify the effectiveness of our method, we compared our method with the state-of-the-art deep-learning based vessel segmentation methods, including ResUNet [6], modified PSPNet [19] (denoted as MPSPNet), CS²-Net [8], and clDice [13]. The quantitative results and qualitative results (visual comparison) are presented in Table 1 and Fig. 3, respectively.

Table 1. Quantitative comparison with the state-of-the-art methods.

Method	Dice [%]↑	TPR [%]↑	mIoU [%]↑	ASD [mm]↓	HD [mm]↓
ResUNet [6]	70.25 ± 1.46	76.38 ± 1.50	55.46 ± 1.58	1.53 ± 0.06	9.33 ± 0.22
MPSPNet [19]	73.60 ± 1.17	78.89 ± 1.09	58.55 ± 1.12	0.99 ± 0.06	7.08 ± 0.24
CS²-Net [8]	76.53 ± 1.29	83.47 ± 1.50	62.20 ± 1.47	1.03 ± 0.04	7.33 ± 0.31
clDice [13]	76.72 ± 0.86	84.06 ± 1.26	62.57 ± 1.17	0.82 ± 0.02	$\mathbf{6.29 \pm 0.18}$
Ours	$\mathbf{80.36 \pm 0.93}$	$\mathbf{86.84 \pm 1.14}$	$\mathbf{68.29 \pm 0.95}$	$\mathbf{0.79 \pm 0.04}$	6.50 ± 0.15

Quantitative Results. Coronary artery segmentation results achieved by different methods are summarised in Table 1, which shows that our method significantly outperformed the competing methods. Specifically, compared with the baseline ResUNet, our method achieved significant improvements on all the metrics. In the meantime, our method achieved better performance than clDice (e.g., increasing segmentation accuracy (Dice) from 76.72% to 80.36%). It is worth noting that all these competing methods pay little attention to the artery-to-heart dependency and coronary topology, which usually leads to false positives and vessel discontinuities in segmentation results. These results demonstrate that our proposal is effective.

Table 2. Quantitative results of ablation analysis of different components.

Method	Dice [%]↑	TPR [%]↑	mIoU [%]↑	ASD [mm]↓	HD [mm]↓
bNet	70.25 ± 1.46	76.38 ± 1.50	55.46 ± 1.58	1.53 ± 0.06	9.33 ± 0.22
bNet-s	73.26 ± 1.25	78.67 ± 1.29	58.30 ± 1.22	1.19 ± 0.06	8.54 ± 0.28
bNet-s-kp	75.01 ± 1.30	82.94 ± 1.38	61.46 ± 1.35	1.15 ± 0.04	6.83 ± 0.31
bNet-s-kp-cl	77.89 ± 1.10	83.72 ± 1.33	64.06 ± 1.24	0.93 ± 0.05	7.14 ± 0.27
FullNet	**80.36 ± 0.93**	**86.84 ± 1.14**	**68.29 ± 0.95**	**0.79 ± 0.04**	**6.50 ± 0.15**

Qualitative Visual Comparison. Qualitative results of three typical subjects are shown in Fig. 3. For the simple subject (first row), the five methods could produce satisfactory results in rough appearance structure, but the competing methods are difficult to maintain better details in the local area (bifurcation in the green circle). For hard subjects (last two rows), our method significantly outperforms the competing methods. Because most background is eliminated in the SAD module, the false positive of background rarely appeared in our results. Moreover, through hierarchical topology constraints, our results maintain good topology continuity in each coronary branch, unlike other methods that have a large number of discontinuities. The qualitative results further verify the effectiveness of our method for coronary artery segmentation.

3.4   Ablation Study

We conducted ablation studies to evaluate the effectiveness of the important components in our framework using the following configurations: 1) Using a baseline ResUNet (bNet) to segment coronary artery from raw images directly. 2) Replacing the raw input of bNet by the representations learned by the SAD module (bNet-s). 3) Adding the key points detection (bNet-s-kp), centerline regression (bNet-s-kp-cl), and cube-connectivity prediction (bNet-s-kp-cl-cp, FullNet) on the basis of bNet-s, respectively. Table 2 summarizes the segmentation results of the five experiments.

Effectiveness of Spatial Anatomical Dependency. Compared with the raw input, the SAD module provides richer contextual and anatomical information to guide segmentation. Results of ablation study confirmed the effectiveness of SAD. Especially, compared with the baseline bNet using raw images only, bNet-s (i.e., bNet+SAD) improved the segmentation performance in terms of all metrics (e.g., 3.01% Dice improvement and 0.34 mm ASD improvement, respectively).

Effectiveness of Hierarchical Topology Learning. Results of ablation studies on hierarchical topology learning show that it can improve Dice, TPR, and ASD errors of bNet-s from 73.26%, 78.67%, and 1.19 mm to 80.36%, 86.84%, and 0.79 mm, respectively. This result indicates the effectiveness of the proposed hierarchical topology learning. Moreover, results also show that more hierarchical topology representations yield more accurate segmentation.

4   Conclusion

In this paper, we have proposed a progressive learning-based framework for accurate coronary artery segmentation. This framework first coarsely segments the coronary artery and calculates the contextual distance field maps with respect to heart, and further enhances the coronary artery segmentation by predicting the hierarchical coronary topologies in a multi-task learning manner, including key points, centerlines, and cube-connectivity. According to the evaluation results, our method significantly outperforms the competing methods, demonstrating the effectiveness of our method and its proposed components, as well as the potential applicability in real-world clinical scenarios.

Acknowledgment. This work was supported in part by National Natural Science Foundation of China (grant number 62131015, 62073260), and Science and Technology Commission of Shanghai Municipality (STCSM) (grant number 21010502600).

References

1. Cao, D., et al.: Non-cardiac surgery in patients with coronary artery disease: risk evaluation and periprocedural management. Nat. Rev. Cardiol. **18**(1), 37–57 (2021)
2. Collet, C., et al.: Left main coronary artery disease: pathophysiology, diagnosis, and treatment. Nat. Rev. Cardiol. **15**(6), 321–331 (2018)
3. Hao, D., et al.: Sequential vessel segmentation via deep channel attention network. Neural Netw. **128**, 172–187 (2020)
4. Huang, K., et al.: Coronary wall segmentation in CCTA scans via a hybrid net with contours regularization. In: 2020 IEEE 17th International Symposium on Biomedical Imaging (ISBI), pp. 1743–1747. IEEE (2020)
5. Kampffmeyer, M., Dong, N., Liang, X., Zhang, Y., Xing, E.P.: ConnNet: a long-range relation-aware pixel-connectivity network for salient segmentation. IEEE Trans. Image Process. **28**(5), 2518–2529 (2018)

6. Kerfoot, E., Clough, J., Oksuz, I., Lee, J., King, A.P., Schnabel, J.A.: Left-ventricle quantification using residual U-Net. In: Pop, M., Sermesant, M., Zhao, J., Li, S., McLeod, K., Young, A., Rhode, K., Mansi, T. (eds.) STACOM 2018. LNCS, vol. 11395, pp. 371–380. Springer, Cham (2019). https://doi.org/10.1007/978-3-030-12029-0_40

7. Marano, R., et al.: CCTA in the diagnosis of coronary artery disease. Radiol. Med. (Torino) **125**(11), 1102–1113 (2020). https://doi.org/10.1007/s11547-020-01283-y

8. Mou, L., Zhao, Y., Fu, H., Liu, Y., Cheng, J., Zheng, Y., Su, P., Yang, J., Chen, L., Frangi, A.F., et al.: CS^2-Net: deep learning segmentation of curvilinear structures in medical imaging. Med. Image Anal. **67**, 101874 (2021)

9. Nie, D., Wang, L., Adeli, E., Lao, C., Lin, W., Shen, D.: 3-D fully convolutional networks for multimodal isointense infant brain image segmentation. IEEE Trans. Cybern. **49**(3), 1123–1136 (2018)

10. Qi, Y., et al.: Examinee-Examiner Network: weakly supervised accurate coronary lumen segmentation using centerline constraint. IEEE Trans. Image Process. **30**, 9429–9441 (2021)

11. Qin, Y., et al.: AirwayNet: a voxel-connectivity aware approach for accurate airway segmentation using convolutional neural networks. In: Shen, D., et al. (eds.) MICCAI 2019. LNCS, vol. 11769, pp. 212–220. Springer, Cham (2019). https://doi.org/10.1007/978-3-030-32226-7_24

12. Serruys, P.W., et al.: Coronary computed tomographic angiography for complete assessment of coronary artery disease: JACC state-of-the-art review. J. Am. Coll. Cardiol. **78**(7), 713–736 (2021)

13. Shit, S., et al.: clDice-a novel topology-preserving loss function for tubular structure segmentation. In: Proceedings of the IEEE/CVF Conference on Computer Vision and Pattern Recognition, pp. 16560–16569 (2021)

14. Virani, S.S., Alonso, A., Aparicio, H.J., et al.: Heart disease and stroke statistics-2021 update: a report from the American Heart Association. Circulation **143**(8), e254–e743 (2021)

15. Xiang, L., et al.: Deep embedding convolutional neural network for synthesizing CT image from T1-Weighted MR image. Med. Image Anal. **47**, 31–44 (2018)

16. Zhang, J., et al.: Weakly supervised vessel segmentation in X-ray angiograms by self-paced learning from noisy labels with suggestive annotation. Neurocomputing **417**, 114–127 (2020)

17. Zhang, X., Cui, Z., Feng, J., Song, Y., Wu, D., Shen, D.: CorLab-Net: anatomical dependency-aware point-cloud learning for automatic labeling of coronary arteries. In: Lian, C., Cao, X., Rekik, I., Xu, X., Yan, P. (eds.) MLMI 2021. LNCS, vol. 12966, pp. 576–585. Springer, Cham (2021). https://doi.org/10.1007/978-3-030-87589-3_59

18. Zhou, Z., Rahman Siddiquee, M.M., Tajbakhsh, N., Liang, J.: UNet++: a nested U-Net architecture for medical image segmentation. In: Stoyanov, D., et al. (eds.) DLMIA/ML-CDS -2018. LNCS, vol. 11045, pp. 3–11. Springer, Cham (2018). https://doi.org/10.1007/978-3-030-00889-5_1

19. Zhu, X., Cheng, Z., Wang, S., Chen, X., Lu, G.: Coronary angiography image segmentation based on PSPNet. Comput. Methods Programs Biomed. **200**, 105897 (2021)

20. Zreik, M., et al.: Deep learning analysis of coronary arteries in cardiac CT angiography for detection of patients requiring invasive coronary angiography. IEEE Trans. Med. Imaging **39**(5), 1545–1557 (2020)

Evidence Fusion with Contextual Discounting for Multi-modality Medical Image Segmentation

Ling Huang[1,4(✉)], Thierry Denoeux[1,2], Pierre Vera[3], and Su Ruan[4]

[1] Heudiasyc, CNRS, Université de technologie de Compiègne, Compiègne, France
ling.huang@hds.utc.fr
[2] Institut universitaire de France, Paris, France
[3] Department of Nuclear Medicine, Henri Becquerel Cancer Center, Rouen, France
[4] Quantif, LITIS, University of Rouen Normandy, Rouen, France

Abstract. As information sources are usually imperfect, it is necessary to take into account their reliability in multi-source information fusion tasks. In this paper, we propose a new deep framework allowing us to merge multi-MR image segmentation results using the formalism of Dempster-Shafer theory while taking into account the reliability of different modalities relative to different classes. The framework is composed of an encoder-decoder feature extraction module, an evidential segmentation module that computes a belief function at each voxel for each modality, and a multi-modality evidence fusion module, which assigns a vector of discount rates to each modality evidence and combines the discounted evidence using Dempster's rule. The whole framework is trained by minimizing a new loss function based on a discounted Dice index to increase segmentation accuracy and reliability. The method was evaluated on the BraTs 2021 database of 1251 patients with brain tumors. Quantitative and qualitative results show that our method outperforms the state of the art, and implements an effective new idea for merging multi-information within deep neural networks.

Keywords: Information fusion · Dempster-Shafer theory · Evidence theory · Uncertainty quantification · Contextual discounting · Deep learning · Brain tumor segmentation

1 Introduction

Single-modality medical images often do not contain enough information to reach a reliable diagnosis. This is why physicians generally use multiple sources of information, such as multi-MR images for brain tumor segmentation, or PET-CT images for lymphoma segmentation. The effective fusion of multi-modality

This work was supported by the China Scholarship Council (No. 201808331005). It was carried out in the framework of the Labex MS2T, which was funded by the French Government, through the program "Investments for the future" managed by the National Agency for Research (Reference ANR-11-IDEX-0004-02).

L. Wang et al. (Eds.): MICCAI 2022, LNCS 13435, pp. 401–411, 2022.
https://doi.org/10.1007/978-3-031-16443-9_39

information is of great importance in the medical domain for better disease diagnosis and radiotherapy. Using convolutional neural networks (CNNs), researchers have mainly adopted probabilistic approaches to information fusion, which can be classified into three strategies: image-level fusion such as input data concatenation [19], feature-level fusion such as attention mechanism concatenation [23], and decision-level fusion such as weighted averaging [14]. However, probabilistic fusion is unable to effectively manage conflict that occurs when different labels are assigned to the same voxel based on different modalities. Also, it is important, when combining multi-modality information, to take into account the reliability of the different sources [17].

Dempster-Shafer theory (DST) [4,20], also known as belief function theory or evidence theory, is a formal framework for information modeling, evidence combination, and decision-making with uncertain or imprecise information [6]. Researchers from the medical image community have actively investigated the use of DST for handling uncertain, imprecision sources of information in different medical tasks, such as disease classification [12], tumor segmentation [10,11], multi-modality image fusion [9], etc. In the DST framework, the reliability of a source of information can be taken into account using the discounting operation [20], which transforms each piece of evidence provided by a source into a weaker, less informative one.

In this paper, we propose a multi-modality evidence fusion framework with contextual discounting based on DST and deep learning. To our knowledge, this work is the first attempt to apply evidence theory with contextual discounting to the fusion of deep neural networks. The idea of considering multi-modality images as independent inputs and quantifying their reliability is simple and reasonable. However, modeling the reliability of sources is important and challenging. Our model computes mass functions assigning degrees of belief to each class and an ignorance degree to the whole set of classes. It thus has one more degree of freedom than a probabilistic model, which allows us to model source uncertainty and reliability directly. The contributions of this paper are the following: (1) Four DST-based evidential segmentation modules are used to compute the belief of each voxel belonging to the tumor for four modality MRI images; (2) An evidence-discounting mechanism is applied to each of the single-modality MRI images to take into account its reliability; (3) A multi-modality evidence fusion strategy is then applied to combine the discounted evidence with DST and achieve more reliable results. End-to-end learning is performed by minimizing a new loss function based on the discounting mechanism, allowing us to increase the segmentation performance and reliability.

The overall framework will first be described in Sect. 2 and experimental results will be reported in Sect. 3

2   Methods

Figure 1 summarizes the proposed evidence fusion framework with contextual discounting. It is composed of (1) four encoder-decoder feature extraction (FE) modules (Residual-UNet [15]), (2) four evidential segmentation (ES) modules, and (3) a multi-modality evidence fusion (MMEF) module. Details about the ES and MMEF modules will be given in Sects. 2.1 and 2.2, respectively. The discounted loss function used for optimizing the framework parameters will be described in Sect. 2.3.

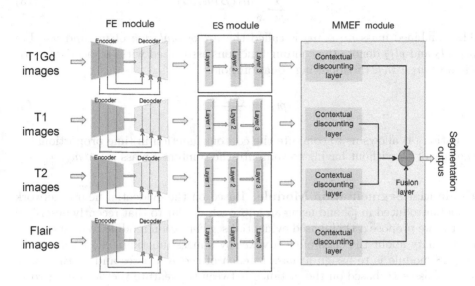

Fig. 1. Multi-modality evidence fusion framework. It is composed of four encoder-decoder feature extraction (FE) modules corresponding to T1Gd, T1, T2 and Flair modality inputs; four evidential segmentation (ES) modules corresponding to each of the inputs; and a multi-modality evidence fusion (MMEF) module.

2.1 Evidential Segmentation

Dempster-Shafer Theory. Before presenting the evidential segmentation module, we first introduce some basic concepts of DST. Let $\Omega = \{\omega_1, \omega_2, \ldots, \omega_K\}$ be a finite set of hypotheses about some question. Evidence about Ω can be represented by a mapping m from 2^Ω to $[0, 1]$ such that $\sum_{A \subseteq \Omega} m(A) = 1$, called a *mass function*. For any hypothesis $A \subseteq \Omega$, the quantity $m(A)$ represents the mass of belief allocated to A and to no more specific proposition. The associated *belief and plausibility functions* are defined, respectively, as

$$Bel(A) = \sum_{\emptyset \neq B \subseteq A} m(B) \quad \text{and} \quad Pl(A) = \sum_{B \cap A \neq \emptyset} m(B), \quad \forall A \subseteq \Omega. \quad (1)$$

The *contour function pl* associated to m is the function that maps each element ω of Ω to its plausibility: $pl(\omega) = Pl(\{\omega\})$.

Two mass functions m_1 and m_2 derived from two independent items of evidence can be combined by *Dempster's rule* [20], denoted as $\oplus$ and formally defined by $(m_1 \oplus m_2)(\emptyset) = 0$ and

$$(m_1 \oplus m_2)(A) = \frac{1}{1-\kappa} \sum_{B \cap C = A} m_1(B)m_2(C), \tag{2}$$

for all $A \subseteq \Omega, A \neq \emptyset$, where κ represents the *degree of conflict* between m_1 and m_2:

$$\kappa = \sum_{B \cap C = \emptyset} m_1(B)m_2(C). \tag{3}$$

The combined mass $m_1 \oplus m_2$ is called the *orthogonal sum* of m_1 and m_2. Let pl_1, pl_2 and pl_{12} denote the contour functions associated with, respectively, m_1, m_2 and $m_1 \oplus m_2$. the following equation holds:

$$pl_{12} = \frac{pl_1 pl_2}{1-\kappa}. \tag{4}$$

Equation (4) allows us to compute the contour function in time proportional to the size of Ω, without having to compute the combined mass $m_1 \oplus m_2$.

Evidential Segmentation Module. Based on the evidential neural network model introduced in [5] and using an approach similar to that recently described in [10], we propose a DST-based evidential segmentation module to quantify the uncertainty about the class of each voxel by a mass function. The basic idea of the ES module is to assign a mass to each of the K classes and to the whole set of classes Ω, based on the distance between the feature vector of each voxel and I prototype centers. The input feature vector can be obtained using any FE module, e.g., Residual-UNet [15], nnUNet [13].

The ES module is composed of an input layer, two hidden layers and an output layer. The first input layer is composed of I units, whose weights vectors are prototypes $p_1, \ldots, p_I$ in input space. The activation of unit i in the prototype layer for input vector x is

$$s_i = \alpha_i \exp(-\gamma_i \|x - p_i\|^2), \tag{5}$$

where $\gamma_i > 0$ and $\alpha_i \in [0,1]$ are two parameters. The second hidden layer computes mass functions m_i representing the evidence of each prototype p_i using the following equations:

$$m_i(\{\omega_k\}) = u_{ik}s_i, \quad k = 1, ..., K \tag{6a}$$
$$m_i(\Omega) = 1 - s_i, \tag{6b}$$

where u_{ik} is the membership degree of prototype i to class ω_k, and $\sum_{k=1}^{K} u_{ik} = 1$. Finally, the third layer combines the I mass functions $m_1, \ldots, m_I$ using Dempster's combination rule (2) to finally obtain a belief function quantifying uncertainty about the class of each voxel.

2.2   Multi-modality Evidence Fusion

In this paper, the problem of quantifying the source reliability is addressed by the discounting operation of DST. Let m be a mass function on Ω and β be a coefficient in $[0,1]$. The *discounting* operation [20] with discount rate $1-\beta$ transforms m into a weaker, less informative mass function $^{\beta}m$ defined as

$$^{\beta}m = \beta\,m + (1-\beta)\,m_?, \tag{7}$$

where $m_?$ is the vacuous mass function defined by $m_?(\Omega) = 1$, and coefficient β is the degree of belief that the source mass function m is reliable [21]. When $\beta = 1$ we accept mass function m provided by the source and take it as a description of our knowledge; when $\beta = 0$, we reject it and we are left with the vacuous mass function $m_?$. In this paper, we focus on the situation when $\beta \in [0,1]$ and combine uncertain evidence with partial reliability using Dempster's rule.

As suggested in [17], the above discounting operation can be extended to *contextual discounting*. This operation can represent richer meta-knowledge regarding the reliability of the source of information in different contexts [7]. It is parameterized by a vector $\boldsymbol{\beta} = (\beta_1, \ldots, \beta_K)$, where β_k is the degree of belief that the source is reliable given that the true class is ω_k. The complete expression of the discounted lass function is given in [17]. Here, we just give the expression of the corresponding contour function, which will be used later:

$$^{\beta}pl(\{\omega_k\}) = 1 - \beta_k + \beta_k pl(\{\omega_k\}), \quad k = 1, \ldots, K, \tag{8}$$

When we have several sources provided by independent evidence, the discounted evidence can be combined by Dempster's rule. Assuming that we have two source of information, let $^{\beta_1}pl_{S_1}$ and $^{\beta_2}pl_{S_2}$ be the discounted contour functions provided, respectively, by sources S_1 and S_2, with discount rate vectors $1-\boldsymbol{\beta}_1$ and $1 - \boldsymbol{\beta}_2$. From (4), the combined contour function is proportional to the product $^{\beta_1}pl_{S_1}{}^{\beta_2}pl_{S_2}$.

2.3   Discounted Dice Loss

We define a new loss function based on discount rates, hereafter referred to as the discounted Dice loss, given as follows:

$$\mathrm{loss}_D = 1 - \frac{2\sum_{n=1}^{N}{}^{\beta}S_n G_n}{\sum_{n=1}^{N}{}^{\beta}S_n + \sum_{n=1}^{N}G_n}, \tag{9a}$$

where $^{\beta}S_n$ is the normalized segmentation output for voxel n by fusing H discounted source information,

$$^{\beta}S_n = \frac{\prod_{h=1}^{H}{}^{\beta^h}pl_{S_h}(\{\omega_k\})}{\sum_{k=1}^{K}\prod_{h=1}^{H}{}^{\beta^h}pl_{S_h}(\{\omega_k\})}, \tag{9b}$$

G_n is the ground truth for voxel n, and N is the number of voxels in the volume.

3   Experimental Results

In this section, we report results of our methods on the brain tumor segmentation dataset. The experimental settings will be described in Sect. 3.1. A comparison with the state of the art and an ablation study will be reported in Sect. 3.2.

3.1   Experiment Settings

BraTS Dataset. We used 1251 MRI scans with size $240 \times 240 \times 155$ voxels coming from the BraTS 2021 Challenge [1,2,16] to evaluate our framework. For each scan, there are four modalities: native (T1), post-contrast T1-weighted (T1Gd), T2-weighted (T2), and T2 Fluid Attenuated Inversion Recovery (FLAIR). Annotations of 1251 scans comprise the GD-enhancing tumor (ET-label 4), the peritumoral edema (ED-label 2), and the necrotic and non-enhancing tumor core (NCR/NET-label1). The task is to segment three semantically meaningful tumor classes: the Enhancing Tumor (ET), the Tumor Core (TC) region (ET+NCR/NET), and the Whole Tumor (WT) region (ET+NCR/NET+ED). Following [19], we divided the 1251 scans into 834, 208, 209 scans for training, validation and testing, respectively. Using the same pre-processing operation as in [19], we performed min-max scaling operation followed by clipping intensity values to standardize all volumes, and cropped/padded the volumes to a fixed size of $128 \times 128 \times 128$ by removing unnecessary background.

Implementation Details. The initial values of parameters α_i and γ_i were set, respectively, to 0.5 and 0.01, and membership degrees u_{ik} were initialized randomly by drawing uniform random numbers and normalizing. The prototypes were initialized by the k-means clustering algorithm. Details about the initialization of ES module parameters can be found in [11]. For the MMEF module, the initial values of parameter β_k was set to 0.5. Each model was trained on the learning set with 100 epochs using the Adam optimization algorithm. The initial learning rate was set to 10^{-3}. The model with the best performance on the validation set was saved as the final model for testing[1].

3.2   Segmentation Results

Segmentation Accuracy. We used the Dice Score and the Hausdorff Distance (HD) as our evaluation metrics. For each patient, we separately computed these two indices for three classes and then averaged indices over the patients, following a similar evaluation strategy as in [19]. We used two baseline models, Residual-UNet [15] and nnUNet [13], as FE modules and applied our proposal based on the two corresponding models to construct our methods named, respectively, MMEF-UNet and MMEF-nnUNet. We compared our methods with two recent transformer-based models (nnFormer [22], VT-UNet [19]), two classical

---

[1] Our code will be available at https://github.com/iWeisskohl.

Table 1. Segmentation results on the BraTS 2021 dataset. The best result is shown in bold, and the second best is underlined.

Methods	Dice score				Hausdorff distance			
	ET	TC	WT	Avg	ET	TC	WT	Avg
UNet [3]	83.39	86.28	89.59	86.42	11.49	6.18	6.15	7.94
VNet [18]	81.04	84.71	90.32	85.36	17.20	7.48	7.53	10.73
nnFormer [22]	82.83	86.48	90.37	86.56	11.66	7.89	8.00	9.18
VT-UNet [19]	85.59	87.41	91.20	88.07	**10.03**	6.29	6.23	<u>7.52</u>
Residual-UNet [15]	85.07	87.61	89.78	87.48	11.76	<u>6.14</u>	6.31	8.07
nnUNet [13]	<u>87.12</u>	**90.31**	<u>91.47</u>	<u>89.68</u>	12.46	11.04	5.97	9.82
MMEF-UNet (Ours)	86.96	87.46	90.68	88.36	<u>10.20</u>	**6.07**	<u>5.29</u>	**7.18**
MMEF-nnUNet (Ours)	**87.26**	<u>90.05</u>	**92.83**	**99.05**	10.09	9.68	**5.10**	8.29

CNN-based methods (UNet [3], V-Net [18]), and the two baseline methods. The quantitative results are reported in Table 1. Our methods outperform the two classical CNN-based models and two recent transformer-based methods according to the Dice score, the best result being obtained by MMEF-nnUNet according to this criterion. In contrast, MMEF-UNet achieves the lowest HD.

Segmentation Reliability. To test the reliability of our methods, we computed the Expected Calibration Error (ECE) [8] for the two baseline methods and our methods. We obtained ECE values of 2.35% and 2.04%, respectively, for Residual-UNet and MMEF-UNet, against ECE values of 4.46% and 4.05%, respectively, for nnUNet and MMEF-nnUNet, respectively. The probabilities computed by our models thus appear to be better calibrated.

Ablation Study. We also investigated the contribution of each module component to the performance of the system. Table 2 highlights the importance of introducing the ES and MMEF modules. Residual-UNet is the baseline model that uses the softmax transformation to map feature vectors to probabilities. Compared to Residual-UNet, Residual-UNet-ES uses the ES module instead of softmax. Residual-UNet-ES-MMEF, our final proposal, fuses the four single modality outputs from Residual-UNet-ES with MMEF module. Compared to the baseline method Residual-UNet, our method, which plugs the ES module after Residual-UNet, improves the segmentation performance based on single T1Gd, T1 and Flair inputs. Furthermore, the use of the MMEF module improves the performance to a large amount compared to any single modality.

Table 2. Segmentation results on BraTS 2021 data (↑ means higher is better).

Methods	Input modality	Dice			
		ET	ED	NRC/NET	Avg
Residual-UNet	T1Gd	84.13 ↑	67.44	72.64	74.73
Residual-UNet-ES	T1Gd	83.92	68.34 ↑	73.29↑	75.18↑
Residual-UNet	T1	53.67	63.51	42.81	53.33
Residual-UNet-ES	T1	57.26↑	67.83↑	54.77↑	59.95↑
Residual-UNet	T2	55.49↑	69.44	51.16	58.70↑
Residual-UNet-ES	T2	54.96	69.84↑	51.27↑	58.69
Residual-UNet	Flair	50.48	75.16	38.06	54.56
Residual-UNet-ES	Flair	53.67↑	77.22↑	49.71↑	60.20↑
Residual-UNet-ES-MMEF	T1Gd,T1,T2,Flair	86.96↑	85.48↑	78.98↑	83.81↑

Table 3. Reliability value β (after training) for classes ET, ED and NRC/NET and the four modalities. Higher values correspond to greater contribution to the segmentation.

β	ET	ED	NRC/NET
T1Gd	0.9996	0.9726	0.9998
T1	0.4900	0.0401	0.2655
T2	0.4814	0.3881	0.4563
Flair	0.0748	0.86207	0.07512

Interpretation of Reliability Coefficients. Table 3 shows the learnt reliability values β for the four modalities with three different classes. The evidence from the T1Gd modality is reliable for ET, ED and NRC classes with the highest β value. In contrast, the evidence from the Flair modality is only reliable for the ED class, and the reliability value β of the T2 modality is only around 0.4 for three classes. The evidence from the T1 modality is less reliable for the three classes compared to the evidence of the other three modalities. These reliability results are consistent with domain knowledge about these modalities reported in [1]. Figure 2 shows the segmentation results of Residual-UNet with the inputs of four concatenated modalities and MMEF-UNet with the inputs of four separate modalities. Our model locates and segments brain tumors precisely, especially the ambiguous voxels located at the tumor boundary.

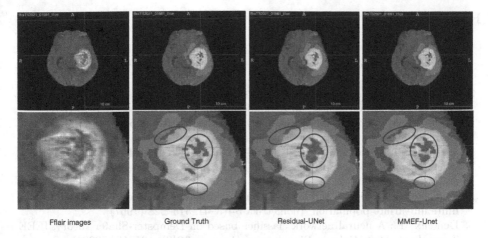

Fflair images   Ground Truth   Residual-UNet   MMEF-Unet

Fig. 2. Visualized segmentation results. The first and the second row are the whole brain with tumor and the detailed tumor region (the main differences are marked in blue circles). The three columns correspond, from left to right, to the Flair image, the ground truth and the segmentation results obtained by Residual-UNet and MMEF-UNet. The green, yellow and red represent the ET, ED and NRC/NET, respectively. (Color figure online)

4   Conclusion

Based on DST, a multi-modality evidence fusion framework considering segmentation uncertainty and source reliability has been proposed for multi-MRI brain tumor segmentation. The ES module performs tumor segmentation and uncertainty quantification, and the MMEF module allows for multi-modality evidence fusion with contextual discounting and Dempster's rule. This work is the first to implement contextual discounting for the fusion of multi-modal information with DST and DNN. The contextual discounting operation allows us to take into account the uncertainty of the different sources of information directly, and it reveals the reliability of different modalities in different contexts. Our method can be used together with any state-of-the-art FE module to get better performance.

Some limitations of this work remain in the computation cost and the segmentation accuracy. We treat single modality images as independent inputs using independent FE and ES modules, which introduces additional computation costs compared to image concatenation methods (e.g., the FLOPs and parameter numbers are equal to 280.07G and 76.85M for UNet-MMEF, against 73.32G, and 19.21M for Residual-UNet). In future research, we will refine our algorithm to improve the accuracy and reliability of our model, and reduce its complexity. We will also explore the possibility of cross-modality evidence fusion for survival prediction tasks.

References

1. Baid, U., et al.: The RSNA-ASNR-MICCAI BraTS 2021 benchmark on brain tumor segmentation and radiogenomic classification. arXiv preprint arXiv:2107.02314 (2021)
2. Bakas, S., et al.: Advancing The Cancer Genome Atlas glioma MRI collections with expert segmentation labels and radiomic features. Sci. Data 4(1), 1–13 (2017)
3. Çiçek, Ö., Abdulkadir, A., Lienkamp, S.S., Brox, T., Ronneberger, O.: 3D U-Net: learning dense volumetric segmentation from sparse annotation. In: Ourselin, S., Joskowicz, L., Sabuncu, M.R., Unal, G., Wells, W. (eds.) MICCAI 2016. LNCS, vol. 9901, pp. 424–432. Springer, Cham (2016). https://doi.org/10.1007/978-3-319-46723-8_49
4. Dempster, A.P.: Upper and lower probability inferences based on a sample from a finite univariate population. Biometrika 54(3–4), 515–528 (1967)
5. Denœux, T.: A neural network classifier based on Dempster-Shafer theory. IEEE Trans. Syst. Man. Cybern. Part A Syst. Humans 30(2), 131–150 (2000)
6. Denœux, T., Dubois, D., Prade, H.: Representations of uncertainty in AI: beyond probability and possibility. In: Marquis, P., Papini, O., Prade, H. (eds.) A Guided Tour of Artificial Intelligence Research, pp. 119–150. Springer, Cham (2020). https://doi.org/10.1007/978-3-030-06164-7_4
7. Denœux, T., Kanjanatarakul, O., Sriboonchitta, S.: A new evidential k-nearest neighbor rule based on contextual discounting with partially supervised learning. Int. J. Approx. Reasoning 113, 287–302 (2019)
8. Guo, C., Pleiss, G., Sun, Y., Weinberger, K.Q.: On calibration of modern neural networks. In: International Conference on Machine Learning, pp. 1321–1330. PMLR (2017)
9. Huang, L., Denœux, T., Tonnelet, D., Decazes, P., Ruan, S.: Deep PET/CT fusion with Dempster-Shafer theory for lymphoma segmentation. In: Lian, C., Cao, X., Rekik, I., Xu, X., Yan, P. (eds.) Machine Learning in Medical Imaging, pp. 30–39. Springer, Cham (2021). https://doi.org/10.1007/978-3-030-87589-3_4
10. Huang, L., Ruan, S., Decazes, P., Denœux, T.: Evidential segmentation of 3D PET/CT images. In: Denœux, T., Lefèvre, E., Liu, Z., Pichon, F. (eds.) Belief Functions: Theory and Applications, pp. 159–167. Springer, Cham (2021). https://doi.org/10.1007/978-3-030-88601-1_16
11. Huang, L., Ruan, S., Decazes, P., Denœux, T.: Lymphoma segmentation from 3D PET-CT images using a deep evidential network. arXiv preprint arXiv:2201.13078 (2022)
12. Huang, L., Ruan, S., Denoeux, T.: Covid-19 classification with deep neural network and belief functions. In: The Fifth International Conference on Biological Information and Biomedical Engineering, pp. 1–4 (2021)
13. Isensee, F., et al.: nnU-Net: self-adapting framework for U-Net-based medical image segmentation. arXiv preprint arXiv:1809.10486 (2018)
14. Kamnitsas, K., et al.: Ensembles of multiple models and architectures for robust brain tumour segmentation. In: Crimi, A., Bakas, S., Kuijf, H., Menze, B., Reyes, M. (eds.) BrainLes 2017. LNCS, vol. 10670, pp. 450–462. Springer, Cham (2018). https://doi.org/10.1007/978-3-319-75238-9_38
15. Kerfoot, E., Clough, J., Oksuz, I., Lee, J., King, A.P., Schnabel, J.A.: Left-ventricle quantification using residual U-Net. In: Pop, M., et al. (eds.) STACOM 2018. LNCS, vol. 11395, pp. 371–380. Springer, Cham (2019). https://doi.org/10.1007/978-3-030-12029-0_40

16. Menze, B.H., et al.: The multimodal brain tumor image segmentation benchmark (brats). IEEE Trans. Med. Imaging **34**(10), 1993–2024 (2014)
17. Mercier, D., Quost, B., Denœux, T.: Refined modeling of sensor reliability in the belief function framework using contextual discounting. Inf. Fusion **9**(2), 246–258 (2008)
18. Milletari, F., Navab, N., Ahmadi, S.A.: V-Net: fully convolutional neural networks for volumetric medical image segmentation. In: 2016 Fourth International Conference on 3D Vision, pp. 565–571. IEEE (2016)
19. Peiris, H., Hayat, M., Chen, Z., Egan, G., Harandi, M.: A volumetric transformer for accurate 3D tumor segmentation. arXiv preprint arXiv:2111.13300 (2021)
20. Shafer, G.: A Mathematical Theory of Evidence, vol. 42. Princeton University Press, Princeton (1976)
21. Smets, P., Kennes, R.: The transferable belief model. Artif. Intell. **66**, 191–243 (1994)
22. Zhou, H.Y., Guo, J., Zhang, Y., Yu, L., Wang, L., Yu, Y.: nnFormer: interleaved transformer for volumetric segmentation. arXiv preprint arXiv:2109.03201 (2021)
23. Zhou, T., Canu, S., Ruan, S.: Fusion based on attention mechanism and context constraint for multi-modal brain tumor segmentation. Comput. Med. Imaging Graph. **86**, 101811 (2020)

Orientation-Guided Graph Convolutional Network for Bone Surface Segmentation

Aimon Rahman[1]([✉]), Wele Gedara Chaminda Bandara[1],
Jeya Maria Jose Valanarasu[1], Ilker Hacihaliloglu[2], and Vishal M. Patel[1]

[1] Johns Hopkins University, Baltimore, USA
arahma30@jhu.edu
[2] University of British Columbia, Vancouver, Canada

Abstract. Due to imaging artifacts and low signal-to-noise ratio in ultrasound images, automatic bone surface segmentation networks often produce fragmented predictions that can hinder the success of ultrasound (US)-guided computer-assisted surgical procedures. Existing pixel-wise predictions often fail to capture the accurate topology of bone tissues due to a lack of supervision to enforce connectivity. In this work, we propose an orientation-guided graph convolutional network to improve connectivity while segmenting the bone surface. We also propose an additional supervision on the orientation of the bone surface to further impose connectivity. We validated our approach on 1042 in vivo US scans of femur, knee, spine, and distal radius. Our approach improves over the state-of-the-art methods by 5.01% in connectivity metric.

Keywords: Graph Convolutional Neural Network · Bone orientation · Ultrasound images

1 Introduction

Computer assisted orthopedic surgery (CAOS) procedures use ultrasound (US) as it offers a cost effective and radiation free alternative to other modalities [8]. Some non-surgical procedures such as [11,16] also use US-based guidance systems. Accurate segmentation of bone surface from US images is essential for improved guidance in these procedures. The overall system accuracy of CAOS procedures is a combination of system errors such as segmentation, registration, tracking, and US probe calibration. Therefore, any improvement in bone segmentation accuracy would have an impact on the overall accuracy of the US-guided CAOS system. Required accuracy limits will depend on the surgical procedures. For example, for spinal fusion surgeries an overall accuracy of <2 mm would be required. Deep convolutional neural networks (CNNs) have been successfully adopted for this task while obtaining a decent performance. An UNet-like method was proposed in [2] to localize vertebra bone surfaces. Methods like [1,12,13] use a filtered feature-guided CNN for robust bone surface segmentation. Recently, a local phase tensor guided CNN is proposed in [14] and validated across US data of various characteristics.

L. Wang et al. (Eds.): MICCAI 2022, LNCS 13435, pp. 412–421, 2022.
https://doi.org/10.1007/978-3-031-16443-9_40

Although the above methods obtain decent segmentation performance with respect to measures like dice score, they do not have any specific measure to ensure bone connectivity. Note that the difference in the number of pixels between the joint and disjoint prediction is usually very less which results only a marginal difference in the dice score. Thus, the segmentation performance metrics do not quantify bone connectivity. However, bone connectivity is essential as a disjoint prediction can result in absurd clinical errors during CAOS procedures. Also, bone surface segmentation maps are used in tasks like bone shadow segmentation. The bone shadow information is essential to guide the orthopedic surgeon to a standardized viewing plane with minimal noise and artifacts. If the bone segmentation map is disjoint, it can adversely affect the shadow predictions as it is very likely for the shadow maps to become discontinuous. Hence, bone connectivity is an import factor to address in bone surface segmentation from US images.

To this end, we propose an **O**rientation-guided **G**raph **C**onvolutional Network (**O-GCN**) for bone surface segmentation. First, we focus on improving connectivity by incorporating graph convolutions in the network architecture. Graph convolutional networks (GCN) are better at capturing relations among arbitrary regions in the input space than CNNs [5] which makes perfect sense for improving bone connectivity. We design a segmentation network in a UNet [10] like fashion but with GCNs instead of CNN blocks. Next, we propose utilizing orientation [4] as an additional supervision to obtain refined segmentation maps with improved connectivity. Learning orientation helps impose a connectivity constraint as learning the bone orientation favours connected bone surfaces. Also, supervising the network for both segmentation and orientation helps use get a more generalized features helping in better segmentation as well as improved connectivity. We validate the proposed O-GCN with 1024 in vivo US scans of femur, knee, spine and distal radius and achieve better performance than recent methods.

In summary, the following are the contributions of this work:

- We propose a new network O-GCN for bone surface segmentation. O-GCN uses GCNs to model a better relationship focusing on improving connectivity and better segmentation.
- We propose an orientation guided supervision for training O-GCN to help impose a connectivity constraint as learning the bone orientation favours connected bone surfaces.
- We conduct extensive experiments using the in vivo US scans of knee, femur, distal radius, spine, and tibia bones collected using two US machines and demonstrate that the proposed method is competitive with recent methods.

2 Method

2.1 Graph Convolution for Bone Segmentation

The main building block of our O-GCN is Graph Convolution module which extracts the connectivity between bone segments by first constructing a graph

$G = (\mathcal{V}, \xi, A)$ in hidden feature domain, and then performing the graph convolution on the constructed graph G as:

$$\widetilde{X} = \sigma(AXW) \tag{1}$$

where $\mathcal{V}$ denotes the nodes, ξ are the edges, A is the similarity/adjacency matrix that describes the similarity between each node, X is the input feature map, W is the learnable weight matrix, $\sigma(\cdot)$ is the non-linear activation function, and $\widetilde{X}$ is the output from the graph convolution module. Different from the standard convolution operation $\widetilde{X} = \sigma(XW)$, the graph convolution incorporates the connectivity between the nodes to the convolution process through the similarity matrix A which aids in improving bone segmentation prediction with less discontinuities.

In particular, given a X of size $c \times h \times w$, where c denotes the number of channels, h and w are the width and height of X, we construct a similarity matrix of size $l \times l$, where $l = h \times w$. Although there are various ways to compute the similarity between two nodes such as Euclidean distance and dot product similarity, we experimentally found that they introduce massive computational overhead and longer inference time making it unsuitable for medical image segmentation. Therefore, to tackle this problem, we propose a learnable similarity matrix $A(X)$, where the parameters of A are adjusted during the learning process which ultimately resulting in less-memory requirement and faster inference time. Concretely, we compute the learnable similarity matrix $A(X)$ as follows:

$$A(X) = \sigma(\phi(X)\Lambda(X)\phi(X)^T), \tag{2}$$

where $\phi(X) \in \mathcal{R}^{l \times m}$ is modeled as 1×1 convolutional layer followed by ReLU, $\Lambda(X) \in \mathcal{R}^{m \times m}$ is a diagonal matrix by Global Average Pooling (GAP) followed by 1×1 convolution, and T denotes the matrix transpose operation. After that, we compute the output from the graph convolution module according to the eqn. (1), where W is modeled as a 1×1 convolutional layer.

2.2   Bone Orientation Learning as an Auxiliary Task

Existing bone segmentation networks only utilize Binary Cross-Entropy (BCE) loss to optimize the network parameters [13,15]. However, we argue that it fails to predict topologically correct and connected bone segments due to the absence of connectivity supervision in the training process. Motivated by this and the human behavior of annotating bones by tracing them in a specific orientation, we formulate an auxiliary bone orientation prediction task to enforce connectivity constraint into the main bone segmentation task [3,4].

To this end, we employ the following procedure to generate the bone orientation maps from the ground-truth bone segmentation maps, that will be further used to calculate auxiliary bone orientation loss during the training process. Given a ground-truth bone segmentation mask (see Fig. 1-(b)), we first obtain bone line strings by skeletonizing the ground-truth bone segmentation map and

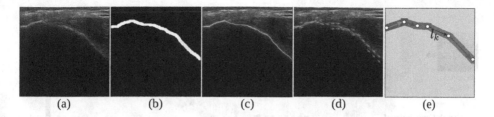

Fig. 1. Generating bone orientation ground-truth from bone segmentation ground-truth. (a) US image. (b) Ground-truth bone segmentation mask. (c) Skeletonized ground-truth bone segmentation map. (d) Ground-truth bone orientation vectors $\vec{v}$. (e) Generated bone orientation ground-truth.

smoothing it using RamerDouglas-Peucker algorithm [6,9] (see Fig. 1-(c)). Let's denote the set of bone line strings as $\{l_1, l_2, \cdots, l_k, \cdots, l_n\}$ and the 2D points connecting a given line string l_k as $\{p_1, p_2, \cdots, p_n\}$. Assuming the surface segmentation is undirected, the points are sorted such that the vectors point from left to right and top to bottom to enforce a neural network to learn the connected representation (see Fig. 1-(c)). Between two consecutive coordinates pairs $(p_1, p_2), \cdots, (p_{n-1}, p_n)$ of s_k we calculate a unit directional vector v' as:

$$\vec{v}_{ij}(x, y) = \frac{p_i(x, y) - p_j(x, y)}{\|p_i(x, y) - p_j(x, y)\|_2^2},\tag{3}$$

where (x, y) are the coordinates of points. Next, we obtain the orientation angle o_r by converting it in to the polar domain as follows:

$$\vec{v}_{ij}(x, y) \equiv \langle 1, \angle o_r \rangle \tag{4}$$

Finally, for each point pair (p_i, p_j), the bone pixels lying within the threshold width of λ_t along the perpendicular direction of l_k, we assign the same orientation value o_r, and for all the other none-bone pixels the non-bone orientation angle o_{nb} is assigned as follows:

$$o_{l_k}(m) = \begin{cases} o_r & \text{if } \left| \vec{v}_\perp \cdot \overrightarrow{(m - p_1)} \right| < \lambda_t \\ o_{nb} & \text{otherwise.} \end{cases} \tag{5}$$

where o denotes the ground truth for orientation learning. In this work, we quantize ground-truth bone orientation angles o into 26 bins; hence formulating the auxiliary bone orientation leaning as pixel-wise multi-class classification problem. Figure 1-(e) shows how the final orientation ground-truth mask looks like for a given bone segmentation ground-truth mask.

2.3   Network Details

As shown in Fig. 2, O-GCN follows a UNet [10] skeleton architecture but with GCNs instead of convolutional blocks. We use 5 GCN blocks in both the encoder

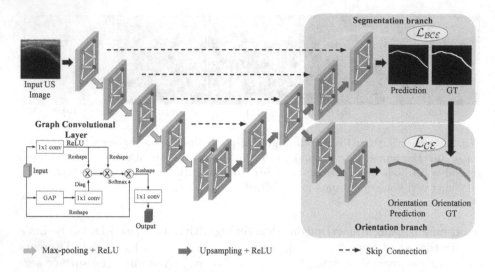

Fig. 2. An overview of the proposed O-GCN.

and decoder. Each GCN block in the encoder is followed by a max-pooling layer and ReLU activation. Each GCN block in the decoder is followed by an upsampling layer and ReLU activation. For upsampling, we use bilinear interpolation. The output of the decoder is passed through a final 1×1 conv layer to get the segmentation map. For orientation learning, we create an auxiliary orientation branch from the 3-rd level of the decoder. Next, we stack two GCN modules with intermediate upsampling and ReLU non-linearity to bring the orientation prediction maps to the same spatial size as input, as shown in Fig. 2. Note that the predicted orientation map $\hat{o}$ has 26 channels as we formulate orientation learning as a pixel-wise multi-class classification problem. The weights and biases of the network are updated by optimizing the following loss,

$$\mathcal{L}_{\text{seg}}(\hat{y}, y) = \text{BCE}(\hat{y}, y) \tag{6}$$

$$\mathcal{L}_{\text{orient}}(\hat{o}, o) = -\sum_{c=0}^{26} o_c \log(\hat{o}_c) \tag{7}$$

$$\mathcal{L}_{total} = \mathcal{L}_{\text{seg}} + 0.5\mathcal{L}_{\text{orient}} \tag{8}$$

where o_c and $\hat{o}_c$ denote the cth class of o and $\hat{o}$, respectively.

3   Experiments and Results

3.1   Dataset

The study is conducted on 1042 ultrasound images of the knee, femur, spine, and radius collected from 25 healthy volunteers with the approval of the institutional review board (IRB). All data are collected using SonixTouch US machine

(Analogic Corporation, Peabody, MA, USA) with 2D C5-2/60 curvilinear and L14-5 linear transducer. A random split of 80:20 is applied for training and testing to the entire dataset based on subjects to avoid data leaking. The dataset is manually segmented by an expert ultrasonographer.

3.2 Implementation Details

The network is trained using a batch size of 32 and trained until the convergence. Adam optimizer with a learning rate of 10^{-4} has been used to optimize the network parameters. All images are normalized between 0 to 1 and resized to 256×256 pixels before feeding to the network. All experiments are performed using a Linux workstation with Intel 3.50 GHz CPU and a 12GB NVidia Titan Xp GPU using the PyTorch framework.

3.3 Evaluation Metrics

Although pixel-wise dice score is a commonly used metric for bone surface segmentation tasks, it is sub-optimal to evaluate the network's ability to produce surface segmentation with accurate connectivity. In the context of bone surface segmentation, manual ground truths are not considered as the absolute gold standard, and detected true positives can be several millimeters away from the ground truth [13]. Dice score will heavily penalize a slight error of bone surface width or its location however a briefly disconnected prediction will be penalized lightly. To this end, we utilize the Average Path Length Similarity (APLS) metric that has been widely used to measure connectivity like in road topography [7]. This measures the difference between ground truth and prediction by summing the differences in optimal path length between two nodes in the ground truth and prediction as follows:

$$
APLS = 1 - \frac{1}{N} \sum_{i}^{N} \min \left\{ 1, \frac{|L(a_i, b_i) - L(a_i', b_i')|}{L(a_i, b_i)} \right\} \tag{9}
$$

where N is the number of unique bone surfaces in the ground truth bone segmentation map, and $L(a_i, b_i) =$ indicates the length between $path(a_i, b_i)$

3.4 Quantitative Comparison

The quantitative results are presented in Table 1. Our proposed method achieves state-of-the-art results in terms of APLS and dice scores. In terms of APLS, the improvement is significant which demonstrates the effectiveness of the proposed network's ability to predict bone surface with accurate connectivity.

3.5 Qualitative Results

For qualitative analysis, we visualize the predictions from Unet [10], MFG-CNN [13], LPT-GCT [15], and our proposed method O-GCN in Fig. 3. Due to low

Table 1. Quantatitive comparison with the current state-of-the-art

Method	SonixTouch	
	APLS (↑)	Dice Score (↑)
Unet [10]	68.38 ± 1.13	80.22 ± 0.13
MFG-CNN [13]	67.65 ± 1.21	82.18 ± 0.21
LPT-GCT [15]	70.13 ± 0.71	82.91 ± 0.16
Ours	**73.65 ± 0.81**	**83.19 ± 0.22**

contrast in input ultrasounds, baseline networks struggle to predict correct bone surfaces. In contrast, our method predicts bone surface without any discontinuity as the network captures long-range dependencies through graph-reasoning and utilizes orientation learning as indirect connectivity supervision.

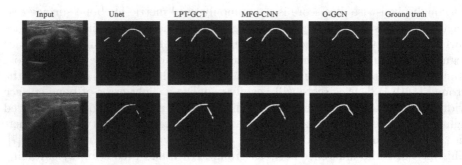

Fig. 3. A qualitative comparison with the state-of-the-art.

3.6   Ablation Study

We perform an ablation study to prove the necessity of GCN-based architecture and orientation loss for bone surface segmentation. Table 2 shows that using GCN-based architecture boosts the performance by 4.40%. Combining both GCN and orientation loss together results in further improvement in both metrics. Additionally, from the qualitative comparison in Fig. 4 it is clear that our method minimizes the fragmentation in bone surface segmentation.

4   Discussion

4.1   Bone Orientation Loss

To study the performance of orientation learning, we choose both convnet and GCN-based architectures. We modify these models by adding additional decoders for dual-task learning. The results in Table 2 show that incorporating

Table 2. Ablation Study

Method	SonixTouch	
	APLS (↑)	Dice Score (↑)
ConvNet Only (Unet) [10]	68.38 ± 1.13	80.22 ± 0.13
ConvNet + Orientation loss	70.11 ± 0.97	81.95 ± 0.31
GCN Only	71.39 ± 0.93	82.77 ± 0.24
GCN + Orientation loss	**73.65 ± 0.81**	**83.19 ± 0.22**

orientation learning as additional supervision improves overall APLS and dice score for both CNN and GCN-based architecture, thus proving that learning two highly related tasks improves the shared encoded feature representation which leads to better performance.

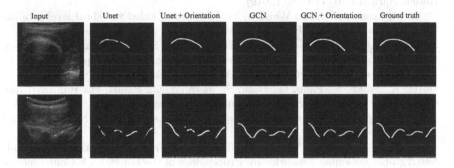

Fig. 4. A qualitative comparison for the ablation study.

4.2   CNN Vs. GCN

Qualitative results from Fig. 4 and quantitative results from Table 2 confirm the effectiveness of graph convolution neural network. CNN has limited capability of learning long-range spatial relations, which is demonstrated in Fig. 4. CNN-based models fail to accurately segment bone surfaces from ultrasound images. In contrast, graph modules accurately segment the entire bone surface even in low contrast images due to their ability to capture long-range dependencies between surface pixels through graph reasoning.

5   Conclusion

In this work, we have introduced a bone-orientation guided graph convolution network (O-GCN) for enhanced bone surface segmentation network. To the best of our knowledge, this is the first study to leverage bone orientation to constraint the model to produce unbroken surface predictions. We also demonstrate the problem with pixel-wise classification for the bone surface segmentation task and utilize APLS metrics to better evaluate the model. We propose using an orientation guided supervision for training O-GCN to help impose a connectivity constraint as learning the bone orientation favours connected bone surfaces. Finally, we demonstrate that our multi-learning framework for related tasks improves the network prediction as the encoder learns better feature representation and obtains better performance than recent methods. Lastly, current work is limited to only 2D US scans, in future the effectiveness of 3D orientation can be explored. Moreover, current work only includes bone images from healthy subjects. Diseased subjects can be included in future works.

References

1. Alsinan, A.Z., Patel, V.M., Hacihaliloglu, I.: Automatic segmentation of bone surfaces from ultrasound using a filter-layer-guided CNN. Int. J. Comput. Assist. Radiol. Surg. **14**(5), 775–783 (2019)
2. Baka, N., Leenstra, S., van Walsum, T.: Ultrasound aided vertebral level localization for lumbar surgery. IEEE Trans. Med. Imaging **36**(10), 2138–2147 (2017)
3. Bandara, W.G.C., Valanarasu, J.M.J., Patel, V.M.: Spin road mapper: extracting roads from aerial images via spatial and interaction space graph reasoning for autonomous driving. arXiv preprint arXiv:2109.07701 (2021)
4. Batra, A., Singh, S., Pang, G., Basu, S., Jawahar, C., Paluri, M.: Improved road connectivity by joint learning of orientation and segmentation. In: Proceedings of the IEEE/CVF Conference on Computer Vision and Pattern Recognition, pp. 10385–10393 (2019)
5. Chen, Y., Rohrbach, M., Yan, Z., Shuicheng, Y., Feng, J., Kalantidis, Y.: Graph-based global reasoning networks. In: Proceedings of the IEEE/CVF Conference on Computer Vision and Pattern Recognition, pp. 433–442 (2019)
6. Douglas, D.H., Peucker, T.K.: Algorithms for the reduction of the number of points required to represent a digitized line or its caricature. Cartograph. Int. J. Geograph. Inf. Geovisualiz. **10**(2), 112–122 (1973)
7. Etten, A.V.: Spacenet road detection and routing challenge part ii - apls implementation (2017). https://medium.com/the-downlinq/spacenet-road-detection-and-routing-challenge-part-ii-apls-implementation-92acd86f4094
8. Hacihaliloglu, I.: Ultrasound imaging and segmentation of bone surfaces: a review. Technology **5**(02), 74–80 (2017)
9. Ramer, U.: An iterative procedure for the polygonal approximation of plane curves. Comput. Graph. Image Process. **1**(3), 244–256 (1972)
10. Ronneberger, O., Fischer, P., Brox, T.: U-Net: convolutional networks for biomedical image segmentation. In: Navab, N., Hornegger, J., Wells, W.M., Frangi, A.F. (eds.) MICCAI 2015. LNCS, vol. 9351, pp. 234–241. Springer, Cham (2015). https://doi.org/10.1007/978-3-319-24574-4_28
11. Seitel, A., et al.: Ultrasound-guided spine anesthesia: feasibility study of a guidance system. Ultras. Med. Biol.**42**(12), 3043–3049 (2016)
12. Villa, M., Dardenne, G., Nasan, M., Letissier, H., Hamitouche, C., Stindel, E.: FCN-based approach for the automatic segmentation of bone surfaces in ultrasound images. Int. J. Comput. Assist. Radiol. Surg. **13**(11), 1707–1716 (2018)

13. Wang, P., Patel, V.M., Hacihaliloglu, I.: Simultaneous segmentation and classification of bone surfaces from ultrasound using a multi-feature guided CNN. In: Frangi, A.F., Schnabel, J.A., Davatzikos, C., Alberola-López, C., Fichtinger, G. (eds.) MICCAI 2018. LNCS, vol. 11073, pp. 134–142. Springer, Cham (2018). https://doi.org/10.1007/978-3-030-00937-3_16
14. Wang, P., Vives, M., Patel, V.M., Hacihaliloglu, I.: Robust real-time bone surfaces segmentation from ultrasound using a local phase tensor-guided CNN. Int. J. Comput. Assist. Radiol. Surg. 15(7), 1127–1135 (2020)
15. Wang, P., Vives, M., Patel, V.M., Hacihaliloglu, I.: Robust real-time bone surfaces segmentation from ultrasound using a local phase tensor-guided CNN. Int. J. Comput. Assist. Radiol. Surg. 15, 1127–1135 (2020)
16. Yamauchi, M., Kawaguchi, R., Sugino, S., Yamakage, M., Honma, E., Namiki, A.: Ultrasound-aided unilateral epidural block for single lower-extremity pain. J. Anesthesia 23(4), 605 608 (2009)

Weakly Supervised Volumetric Image Segmentation with Deformed Templates

Udaranga Wickramasinghe[1(✉)], Patrick Jensen[1,2], Mian Shah[1],
Jiancheng Yang[1,3], and Pascal Fua[1]

[1] EPFL, Lausanne, Switzerland
udaranga.wickramasinghe@epfl.ch
[2] DTU, Lyngby, Denmark
[3] Shanghai Jiao Tong University, Shanghai, China

Abstract. There are many approaches to weakly-supervised training of networks to segment 2D images. By contrast, existing approaches to segmenting volumetric images rely on full-supervision of a subset of 2D slices of the 3D volume. We propose an approach to volume segmentation that is truly weakly-supervised in the sense that we only need to provide a sparse set of 3D points on the surface of target objects instead of detailed 2D masks. We use the 3D points to deform a 3D template so that it roughly matches the target object outlines and we introduce an architecture that exploits the supervision it provides to train a network to find accurate boundaries. We evaluate our approach on Computed Tomography (CT), Magnetic Resonance Imagery (MRI) and Electron Microscopy (EM) image datasets and show that it substantially reduces the required amount of effort.

1 Introduction

State-of-the-Art volumetric segmentation techniques rely on Convolutional Neural Networks (CNNs) operating on image volumes [3,23]. However, their performance depends critically on obtaining enough annotated data, which itself requires expert knowledge and is both tedious and expensive.

Weakly-supervised image segmentation techniques can be used to mitigate this problem. They typically rely on tag annotations [8,10] or coarse object annotations in the form of point annotations [20,33], bounding box annotations [9], scribbles [33] or approximate target shapes [14]. However, these techniques have been mostly demonstrated in 2D and do not provide enough information when segmenting complex shapes, such as the liver or the hippocampus. For 3D volume segmentation, the dominant approach is to fully label a subset of 2D slices [3,5]. This is often referred to as *weak supervision*, even though it requires full supervision within individual slices.

Supplementary Information The online version contains supplementary material available at https://doi.org/10.1007/978-3-031-16443-9_41.

By contrast, we propose a truly weakly-supervised approach that only requires a sparse set of 3D points on the surface of the target objects instead of the usual 2D masks in selected slices. Given an appropriate user-interface, this is much faster and easier because it eliminates the need to painstakingly outline fine details, as shown in Fig. 1. To this end, we introduce the **Weak-Net** architecture depicted by Fig. 2. It comprises two U-Net-like networks. The first one produces a segmentation map that matches a rough model of the target object obtained from the 3D point annotations. The second one takes the map as input and uses it to reconstruct the original image, which forces the segmentation boundaries to be accurate even though those of the template are not.

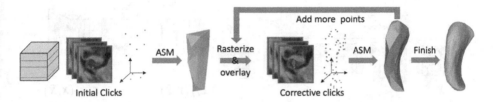

Fig. 1. Iterative annotation strategy. The annotator provides a few 3D points, which we fit to the template using an Active Surface Model (ASM) [28]. The result is rasterized and overlaid on the images. The annotator can then add more and deform the template again as needed.

We evaluate the performance of **Weak-Net** on Computed Tomography (CT), Magnetic Resonance Imagery (MRI) and Electron Microscopy (EM) datasets. We show that it outperforms the standard approach to weak-supervision in 3D at a reduced supervision cost. More specifically, we can deliver the same accuracy as when fully annotating 2D slices for less than a third of the annotation effort. This matters because annotators typically are experts whose time is both scarce and valuable. Furthermore, it creates the basis for interactive annotating strategies that deliver the full accuracy at a lower cost than full supervision.

2   Related Work

We review current approaches to weak-supervision for 2D and 3D image segmentation. We then discuss using atlases to segment biomedical image volumes.

2.1   Weakly-Supervised Image Segmentation

Segmenting 2D Images. Tag and box annotations are among the weakest forms of annotations used to segment natural and medical images [6,9,10]. However, they rarely provide enough supervision for accurate results. By contrast, point annotations [20], scribbles [33], and approximate shape annotation [14] can be used to provide useful shape information.

The annotation process can be sped-up using dynamic programming [19] or deformable contours [12]. This makes it possible to mark only a subset of points along a contour and have the system refine it to match the target object boundary. Unfortunately, these algorithms are hard to deploy effectively in medical imagery because there are many contours besides those of interest and they can easily confuse these algorithms. This is addressed in [21] by introducing *deep deformable contours* that only need a simple approximate contour for initialization purposes. In [2,15], the annotator is brought into the loop by giving corrective clicks when necessary. However, these deep contours require fully labelled data for their own training.

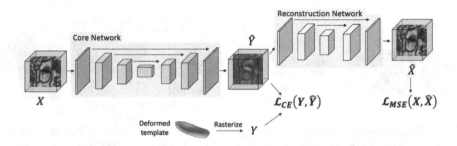

Fig. 2. Weak-Net architecture. A first U-Net takes the image $\mathbf{X}$ as input and outputs a segmentation $\hat{\mathbf{Y}}$, which is in turn fed to a second U-Net that outputs a reconstructed image $\hat{\mathbf{X}}$. Training is achieved by jointly minimizing L_{mse} and L_{ce}. This encourages $\hat{\mathbf{X}}$ to resemble the original image and $\hat{\mathbf{Y}}$ to be similar to $\mathbf{Y}$, a roughly aligned version of a template.

In the result section, we show that supplying a few 3D points, as we do, is faster than using these techniques to delineate whole 2D contours in slices, as is often done.

Segmenting 3D Volumes. By far the most prevalent approach to weak-supervision for 3D segmentation, is to *fully* annotate subsets of slices from the training volumes [3]. Even though this reduces the segmentation effort, the annotator still has to carefully trace the object boundary in those slices. The effort can be reduced by using scribbles in individual slices [5] or any of the semi-automated techniques described above. Unfortunately, the effectiveness of those that do not require training is limited [12,19] while the others [2,15,21] require full supervision and are not applicable in our scenario.

Part of the problem is that these annotation techniques operate purely in 2D without exploiting the 3D nature of the data. Because we do so, we only need a comparatively small number of point annotations for effective training.

2.2   Template Based Approaches

Shape priors have long been used for image segmentation [7] and much recent work use shape priors in conjunction with deep nets [18,29] For medical imaging, these priors are usually supplied in the form of sophisticated templates known as Probabilistic Atlases (PAs) that assign to each pixel or voxel a probability of belonging to a specific class. They are typically built by fusing multiple manually annotated images and used as auxiliary CNN inputs to provide localization priors that help the network find structures of interest. The PAs can be either very detailed, as in [25], to model structures that are known in detail or very rough, as in [27], to deal with 3D structures whose shape can vary significantly.

PAs built by annotating points have been used in medical imaging [11,22] as a source of prior information. They are often referred to as seed layers and come in two main flavors, Gaussian priors [22] or binary seed layers [11]. These seed layers either indicate points inside the object [11,32] or points on the boundary of the object [17,26]. In the context of deep learning, PAs have been mostly used in fully-supervised approaches [11,22,25]. An exception is the work of [25]. However, they are only used for pre-training purposes. Another are the one-shot and few-shot learning-based segmentation algorithms of [4]. However, they require a few fully annotated target objects as the atlases, which we do not.

2.3   Image Reconstruction

Image reconstruction is used for semi-supervised [16] and unsupervised [31] image segmentation as an auxiliary task to improve the results. In this work, we demonstrate that this idea also applies in a weakly-supervised setting to improve the segmentations produced by the core-network trained with rough annotations. In our framework, the image reconstruction network helps refine the rough initial shapes we obtain from the annotations. There are alternative approaches to boundary refinement [1] but they are designed for 2D segmentation and extending them to 3D would be non-trivial.

3   Method

Our approach to training a network involves an annotator providing only a sparse set of 3D points on the surface of target objects for each image volume, as opposed to carefully annotating several individual slices in each. These points are used to deform a template, such as those shown in Fig. 1, using active surface models [28]. This provides a rough indication of where the target object boundary is. **Weak-Net** uses it to learn weights that yield accurate object boundaries. In short, we provide minimal human input at training time so that, at inference time, the trained network can be used without human intervention.

3.1   Network Architecture and Losses

Weak-Net is depicted by Fig. 2 and comprises two separate U-Net networks [3]. The first takes as input an image volume $\mathbf{X}$ of $D \times H \times W$, where D, H, W stand for depth, height and width. It outputs a tensor $\hat{\mathbf{Y}}$ of dimension $D \times H \times W$ that stores the probabilities of each voxel belonging to the foreground. The second takes $\hat{\mathbf{Y}}$ as input and yields $\hat{\mathbf{X}}$, which is of the same dimension as $\mathbf{X}$. Ideally, $\hat{\mathbf{Y}}$ should be the desired segmentation and $\hat{\mathbf{X}}$ should be equal to $\mathbf{X}$.

To train **Weak-Net**, we minimize a weighted sum of two losses

$$\mathcal{L} = \mathcal{L}_{ce} + \lambda \mathcal{L}_{mse}, \tag{1}$$

$$\mathcal{L}_{ce} = -\sum_{i,j,k} \mathbf{Y}_{i,j,k} \log(\hat{\mathbf{Y}}_{i,j,k}), \quad \mathcal{L}_{mse} = \sum_{i,j,k} \mathbf{W}^{mse}_{i,j,k}(\mathbf{X}_{i,j,k} - \hat{\mathbf{X}}_{i,j,k})^2,$$

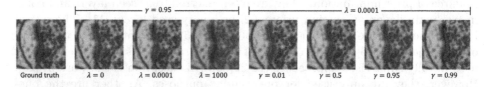

Fig. 3. Impact of the λ parameter in the loss of Eq. 1 and the thresholding parameter γ. (Col. 2–4) When $\lambda = 0$, the segmentation is very similar to the template. When λ is large, the segmentation follows boundaries that exist in the image but are not necessarily the right ones. In between, the boundaries are correct. (Col. 5–8) We set λ to 10^{-4} and vary γ.

where $\mathbf{Y}$ is the rasterized template fitted to the target object and λ is a scalar that controls the influence of the second loss. $\mathcal{L}_{ce}$ is a standard cross entropy loss whose minimization promotes similarity between the rasterized template $\mathbf{Y}$ and the segmentation $\hat{\mathbf{Y}}$. $\mathcal{L}_{mse}$ is a voxel-wise mean squared error in which the individual voxels are given weights $W^{mse}_{i,j,k}$ whose value is high within a distance d from boundaries in $\mathbf{Y}$ and low elsewhere. At inference time, we only use the first U-Net, which we will refer to as our *core network*, and obtain the final segmentation by thresholding its output $\hat{\mathbf{Y}}$ using a threshold γ.

Minimizing $\mathcal{L}_{ce}$ during training ensures that the segmentations will be roughly correct. However, because the template can only be expected to provide a coarse depiction of the object, this is not enough. We therefore also minimize $\mathcal{L}_{mse}$ to force the network to yield accurate boundaries. Figure 3 illustrates the influence of the γ and λ parameters on a real image. In practice, the results are insensitive to how they are chosen over a wide range. However, minimizing $\mathcal{L}_{ce}$ alone ($\lambda = 0$) produces boundaries in $\hat{\mathbf{Y}}$ that are exactly those of the template while minimizing $\mathcal{L}_{mse}$ alone ($\lambda = 1000$) yields boundaries that exist in the image but are not necessarily those we are looking for. Minimizing a properly weighted sum of the two yields segmentations that conform to the template while matching actual image boundaries.

3.2 Template Deformation

The template $\mathbf{Y}$ of Eq. 1 should approximately match the target structure. Hence, the annotator should supply points that are distributed across the object surface. These points are then used to deform the template. In practice, For structures of genus 0, we start from a simple spherical template but more complex ones are possible. As we increase the number of points, we get increasingly refined templates, as shown in Fig. 4.

To perform this deformation interactively, we developed a GUI that relies on Active Surface Models (ASMs) [28] implemented as a MITK [30] plugin. It lets the annotator supply a few points by clicking on 2D cross sections of the input image volume. The ASM then deforms the template in real-time and overlays it on the image data, both as 2D cross sections and 3D surface renderings. The annotator can then add more points wherever the deformed template is too far from the target organ's boundary and iterate as often as necessary. This effectively puts the human in the loop in a painless and practical way. We illustrate this in a video that can be found in the supp. material.

(a)   (b)   (c)   (d)   (e)

Fig. 4. Deformed templates. (a, b, c, d, e) Deformed template using $N = 25, 50, 125, 250$, and 3661 user-supplied 3D points when segmenting a liver. $N = 3661$ corresponds to full annotations in all slices.

4   Experiments

4.1   Datasets, Metrics and Baseline

We use three datasets acquired using MRI, FIB-SEM, and CT. MRI dataset consists of 260 labeled MRI image cubes of hippocampuses from the Medical Segmentation Decathlon [24]. CT image dataset consists of 20 labeled CT image cubes of the liver from the CHAOS challenge [13]. EM Dataset consist of 26 image cubes of Synaptic Junctions from a $500 \times 500 \times 200$ FIB-SEM image stack of a mouse cortex [29].

Our aim is to produce segmentations of these image values such that they are above a given quality threshold using as few annotations as possible. To quantify this goal, we use two metrics, one for quality and the other for annotation effort. We use the standard IoU metric [3] to quantify segmentation quality. We use the number of points provided by the annotator as a proxy for amount of effort. When the annotator provides individual points, this is clearly proportional to the time spent. When the annotator outlines contours on 2D slices, this becomes the sum of the contour lengths in each slice. It could be argued that this is an overestimate because the task is easier. However, in our experience it is not because precisely outlining a contour requires deliberation.

As discussed in Sect. 2.1, the generally accepted way to provide weak-supervision for 3D image segmentation is to fully annotate a few 2D slices [3]. To provide a baseline, we therefore use this approach to train a 3D U-Net, as described in [3]. For a fair comparison, we use the same one as in **Weak-Net**.

4.2   Comparative Results

We exploit the real-time performance of active surface models to enable the annotator to provide a few points, deform the template accordingly, and then add more points where the deformed shape is not satisfactory. We first benchmark this scenario using human annotations and then provide results using simulated annotation on the object surfaces to eliminate the subjective element it contains.

Table 1. Performance given human annotations. Values are given as mean±std. **Weak-Net** consistently outperforms the baseline in terms of both IoU and annotation effort required to achieve it.

		Hippocampus	Liver	Syn. junction
Baseline	IoU (%)	71.2 ± 0.9	81.9 ± 0.7	68.2 ± 0.9
	Annotation effort (%)	12.4%	6.9%	10.8%
Weak-Net	IoU (%)	74.2 ± 0.7	84.2 ± 1.0	71.5 ± 1.4
	Annotation effort (%)	10.1%	6.8%	9.2%
Full annotation		79.3 ± 0.4	87.3 ± 0.3	73.3 ± 0.6

Table 2. Annotation time. Values are given as mean ± std. Times are in minutes.

	Hippocampus	Liver
Manual Contouring	6.7 ± 0.7 min	125 ± 19.3 min
3D Region Growing	5.4 ± 1.2 min	75.4 ± 30.5 min
3D Fast Marching	5.7 ± 1.4 min	86.1 ± 25.3 min
Ours	4.6 ± 0.9 min	12.3 ± 1.6 min

4.3   Human Annotations

Weak vs Full Supervision. We compare **Weak-Net** and the baseline against providing full supervision and report the results in Table 1. **Weak-Net** delivers higher IoU numbers than the baseline on all three datasets. It delivers 92 to 97% of the accuracy that can be achieved with full supervision for only 7 to 10% of the annotation effort, as defined above.

Annotation Time. In the results of Table 1, we use the number of points supplied by the annotators to gauge the annotation effort. To complement this, we asked the annotators to manually annotate some images using other MITK tools [30], slice-by-slice *manual contouring* of the borders, *3D Region Growing*, and *3D Fast Marching*. *3D Region Growing* and *3D Fast Marching* produce many false positive and negative regions, as shown in the supp. material. Therefore, we had to perform slice-by-slice corrections to obtain final segmentation using these two tools. In Table 2, we report the average time it took to fully annotate a single sample using each tool. As the Hippocampus volumes are small, point annotation is only ~1.5x faster than annotating the full volume. For the large Liver volumes however, point annotation is ~10x faster, which is significant.

Simulated Annotations. To eliminate subjectivity from our experiments, we use the fact that our datasets are fully annotated to simulate the annotation process using the algorithm described in the supp. material. It is driven by two numbers, N the number of points per sample and P the number of samples we annotate. To keep the number of experiments within a manageable range, we vary both N and P when experimenting on the Hippocampus dataset and only N for the other two. To provide a baseline, we randomly pick a number of slices from three image planes to be annotated and use the ground-truth annotations for these slices. When selecting them, we check that they contain the target object. When evaluating the baseline, we vary the number of slices we use and of samples we annotate.

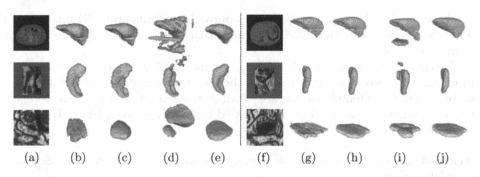

(a)   (b)   (c)   (d)   (e)   (f)   (g)   (h)   (i)   (j)

Fig. 5. Segmentation results. (a + f) A slice from the input volume (b + g) Ground truth (c + h) U-Net with full-supervision (d + i) Corresponding Baseline (e) **Weak-Net** trained with 25 points per sample. (f) **Weak-Net** trained with 125 points per sample.

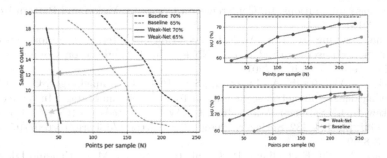

Fig. 6. Performance given synthetic annotations (left). Annotation effort required to achieve 65% and 70% IoU on the hippocampus. The blue and red arrows indicate the effort reduction our approach delivers. (right) IoU as function of the number of points used (N) for the synaptic junction and liver dataset sets. The dashed line denotes fully supervised results. (Color figure online)

We present qualitative results of the experiments in Fig. 5. On the left of Fig. 6, we plot the total annotation effort, that is, the product of the number of points per sample and the number of samples, required to attain a target IoU—either 65% or 70%—in the Hippocampus dataset. As there are many ways to achieve a given IoU by increasing one number while decreasing the other, we draw *iso-IoU* curves. The baseline ones are dashed and ours are full and clearly to the left of the dashed ones. In other words, we need significantly less effort to achieve a similar result. On the right of Fig. 6, we report our results on the other two datasets that contain fewer samples. Hence we used all training samples and plot the IoU as a function of the number of points per sample. For the same number of points, our approach consistently outperforms the baseline. When using it, we cannot annotate less than one slice and, hence, reduce the annotation effort below a certain level, which is why the blue curves extend further to the left than the yellow ones.

5   Conclusion

We have presented a weakly supervised approach to segmenting 3D image volumes that outperforms more traditional approaches that rely on fully annotating individual 2D slices.

It relies on deforming simple spherical templates that incorporate no shape prior. In future work, to further reduce the annotation burden, we will develop more sophisticated templates that are parameterized in terms of low-dimensional latent vectors and can therefore be deformed by specifying even fewer 3D points than we do now.

Acknowledgment. This work was supported in part by a Swiss National Science Foundation grant.

References

1. Acuna, D., Kar, A., Fidler, S.: Devil is in the edges: learning semantic boundaries from noisy annotations. In: Conference on Computer Vision and Pattern Recognition (2019)
2. Akuna, D., Ling, H., Kar, A., Fidler, S.: Efficient interactive annotation of segmentation datasets with Polygon-RNN++. In: Conference on Computer Vision and Pattern Recognition (2018)
3. Çiçek, Ö., Abdulkadir, A., Lienkamp, S.S., Brox, T., Ronneberger, O.: 3D U-Net: learning dense volumetric segmentation from sparse annotation. In: Ourselin, S., Joskowicz, L., Sabuncu, M.R., Unal, G., Wells, W. (eds.) MICCAI 2016. LNCS, vol. 9901, pp. 424–432. Springer, Cham (2016). https://doi.org/10.1007/978-3-319-46723-8_49
4. Dalca, A.V., Yu, E., Golland, P., Fischl, B., Sabuncu, M.R., Eugenio Iglesias, J.: Unsupervised deep learning for Bayesian brain MRI segmentation. In: Shen, D., Liu, T., Peters, T.M., Staib, L.H., Essert, C., Zhou, S., Yap, P.-T., Khan, A. (eds.) MICCAI 2019. LNCS, vol. 11766, pp. 356–365. Springer, Cham (2019). https://doi.org/10.1007/978-3-030-32248-9_40

5. Dorent, R., et al.: Scribble-based domain adaptation via co-segmentation. In: Martel, A.L., et al. (eds.) MICCAI 2020. LNCS, vol. 12261, pp. 479–489. Springer, Cham (2020). https://doi.org/10.1007/978-3-030-59710-8_47

6. Feng, X., Yang, J., Laine, A.F., Angelini, E.D.: Discriminative localization in CNNs for weakly-supervised segmentation of pulmonary nodules. In: Descoteaux, M., Maier-Hein, L., Franz, A., Jannin, P., Collins, D.L., Duchesne, S. (eds.) MICCAI 2017. LNCS, vol. 10435, pp. 568–576. Springer, Cham (2017). https://doi.org/10.1007/978-3-319-66179-7_65

7. Freedman, D., Zhang, T.: Interactive graph-cut based segmentation with shape priors. In: Conference on Computer Vision and Pattern Recognition, pp. 755–62 (2005)

8. Ge, W., Yanga, S., Yu, Y.: Multi-evidence filtering and fusion for multi-label classification, object detection and semantic segmentation based on weakly supervised learning. In: Conference on Computer Vision and Pattern Recognition (2018)

9. Hsu, C., Hsu, K., Tsai, C., Lin, Y., Chuang, Y.: Weakly supervised instance segmentation using the bounding box tightness prior. In: Advances in Neural Information Processing Systems (2019)

10. Huang, Z., Wang, X., Wang, J., Liu, W., Wang, J.: Weakly-supervised semantic segmentation network with deep seeded region growing. In: Conference on Computer Vision and Pattern Recognition (2018)

11. Januszewski, M., Jain, V.: High-precision automated reconstruction of neurons with flood-filling networks. Nat. Methods 15, 605–610 (2018)

12. Kass, M., Witkin, A., Terzopoulos, D.: Snakes: active contour models. Int. J. Comput. Vis. 1(4), 321–331 (1988)

13. Kavur, A., Selver, M.: CHAOS challenge - combined (CT-MR) healthy abdominal organ segmentation. arXiv Preprint (2020)

14. Khoreva, A., Benenson, R., Hosang, J., Hein, M., Schiele, B.: Simple does it: weakly supervised instance and semantic segmentation. In: Conference on Computer Vision and Pattern Recognition, pp. 1665–1674 (2017)

15. Ling, H., Gao, J., Kar, A., Chen, W., Fidler, S.: Fast interactive object annotation with curve-GCN. In: Conference on Computer Vision and Pattern Recognition, pp. 5257–5266 (2019)

16. Liu, X., Thermos, S., O'Neil, A., Tsaftaris, S.A.: Semi-supervised meta-learning with disentanglement for domain-generalised medical image segmentation. In: de Bruijne, M., Cattin, P.C., Cotin, S., Padoy, N., Speidel, S., Zheng, Y., Essert, C. (eds.) MICCAI 2021. LNCS, vol. 12902, pp. 307–317. Springer, Cham (2021). https://doi.org/10.1007/978-3-030-87196-3_29

17. Maninis, K., Caelles, S., Pont-Tuset, J., Gool, L.: Deep extreme cut: from extreme points to object segmentation. In: Conference on Computer Vision and Pattern Recognition (2018)

18. Mirikharaji, Z., Hamarneh, G.: Star shape prior in fully convolutional networks for skin lesion segmentation. In: Frangi, A.F., Schnabel, J.A., Davatzikos, C., Alberola-López, C., Fichtinger, G. (eds.) MICCAI 2018. LNCS, vol. 11073, pp. 737–745. Springer, Cham (2018). https://doi.org/10.1007/978-3-030-00937-3_84

19. Mortensen, E., Barrett, W.: Intelligent scissors for image composition. In: ACM SIGGRAPH, pp. 191–198, August 1995

20. Bearman, A., Russakovsky, O., Ferrari, V., Fei-Fei, L.: What's the point: semantic segmentation with point supervision. In: Leibe, B., Matas, J., Sebe, N., Welling, M. (eds.) ECCV 2016. LNCS, vol. 9911, pp. 549–565. Springer, Cham (2016). https://doi.org/10.1007/978-3-319-46478-7_34

21. Peng, S., Jiang, W., Pi, H., Li, X., Bao, H., Zhou, X.: Deep snake for real-time instance segmentation. In: Conference on Computer Vision and Pattern Recognition (2020)

22. Roth, H., et al.: Weakly supervised segmentation from extreme points. In: Zhou, L., et al. (eds.) LABELS/HAL-MICCAI/CuRIOUS 2019. LNCS, vol. 11851, pp. 42–50. Springer, Cham (2019). https://doi.org/10.1007/978-3-030-33642-4_5

23. Shvets, A., Iglovikov, V.: Automatic instrument segmentation in robot-assisted surgery using deep learning. arXiv Preprint (2018)

24. Simpson, A., Menze, B.: A large annotated medical image dataset for the development and evaluation of segmentation algorithms. arXiv Preprint (2019)

25. Spitzer, H., Kiwitz, K., Amunts, K., Harmeling, S., Dickscheid, T.: Improving cytoarchitectonic segmentation of human brain areas with self-supervised Siamese networks. In: Frangi, A.F., Schnabel, J.A., Davatzikos, C., Alberola-López, C., Fichtinger, G. (eds.) MICCAI 2018. LNCS, vol. 11072, pp. 663–671. Springer, Cham (2018). https://doi.org/10.1007/978-3-030-00931-1_76

26. Wang, Z., Acuna, D., Ling, H., Kar, A., Fidler, S.: Object instance annotation with deep extreme level set evolution. In: European Conference on Computer Vision (2020)

27. Wickramasinghe, U., Knott, G., Fua, P.: Probabilistic atlases to enforce topological constraints. In: Shen, D., Liu, T., Peters, T.M., Staib, L.H., Essert, C., Zhou, S., Yap, P.-T., Khan, A. (eds.) MICCAI 2019. LNCS, vol. 11764, pp. 218–226. Springer, Cham (2019). https://doi.org/10.1007/978-3-030-32239-7_25

28. Wickramasinghe, U., Knott, G., Fua, P.: Deep active surface models. In: Conference on Computer Vision and Pattern Recognition (2021)

29. Wickramasinghe, U., Remelli, E., Knott, G., Fua, P.: Voxel2Mesh: 3D mesh model generation from volumetric data. In: Martel, A.L., Abolmaesumi, P., Stoyanov, D., Mateus, D., Zuluaga, M.A., Zhou, S.K., Racoceanu, D., Joskowicz, L. (eds.) MICCAI 2020. LNCS, vol. 12264, pp. 299–308. Springer, Cham (2020). https://doi.org/10.1007/978-3-030-59719-1_30

30. Wolf, I., et al.: The medical imaging interaction toolkit (MITK): a toolkit facilitating the creation of interactive software by extending VTK and ITK. In: Medical Imaging 2004: Visualization, Image-Guided Procedures, and Display (2004)

31. Xia, X., Kulis, B.: W-Net: a deep model for fully unsupervised image segmentation. arXiv Preprint (2017)

32. Yang, L., Wang, Y., Xiong, X., Yang, J., Katsaggelos, A.: Efficient video object segmentation via network modulation. In: Conference on Computer Vision and Pattern Recognition (2018)

33. Zhao, T., Yin, Z.: Weakly supervised cell segmentation by point annotation. IEEE Trans. Med. Imaging 40, 2736–2747 (2020)

Implicit Neural Representations for Medical Imaging Segmentation

Muhammad Osama Khan[1(✉)] and Yi Fang[1,2]

[1] New York University, New York, USA
{osama.khan,yfang}@nyu.edu
[2] New York University Abu Dhabi, Abu Dhabi, UAE

Abstract. 3D signals in medical imaging, such as CT scans, are usually parameterized as a *discrete* grid of voxels. For instance, existing state-of-the-art organ segmentation methods learn *discrete* segmentation maps. Unfortunately, the memory requirements of such methods grow *cubically* with increasing spatial resolution, which makes them unsuitable for processing high resolution scans. To overcome this, we design an Implicit Organ Segmentation Network (IOSNet) that utilizes *continuous* Implicit Neural Representations and has several useful properties. Firstly, the IOSNet decoder memory is roughly *constant* and *independent* of the spatial resolution since it parameterizes the segmentation map as a *continuous* function. Secondly, IOSNet converges much faster than *discrete* voxel based methods due to its ability to accurately segment organs irrespective of organ sizes, thereby alleviating size imbalance issues without requiring any auxiliary tricks. Thirdly, IOSNet naturally supports superresolution (i.e. sampling at arbitrary resolutions during inference) due to its *continuous* learnt representations. Moreover, despite using a simple lightweight decoder, IOSNet consistently outperforms the *discrete* specialized segmentation architecture UNet. Hence, our approach demonstrates that Implicit Neural Representations are well-suited for medical imaging applications, especially for processing high-resolution 3D medical scans.

Keywords: Implicit Neural Representations · Segmentation

1 Introduction

Head and neck cancer has one of the highest incidence globally [26]. During radiotherapy treatment of head and neck cancer, it is important to accurately delineate organs at risk (OARs) (i.e. healthy organs) in order to avoid exposing them to unnecessary radiation and limit risk of downstream complications [10].

Supplementary Information The online version contains supplementary material available at https://doi.org/10.1007/978-3-031-16443-9_42.

3D UNet [5] and its variants have achieved remarkable performances on several medical imaging segmentation tasks, including head and neck OARs segmentation. Initially, 3D convolutional methods that required specific pre- and postprocessing were developed [7,12,21,25]. Later, however, Zhu et al. [28] developed AnatomyNet, which predicted segmentations for all organs in one pass with limited additional processing. Owing to the substantial size variations between large and small organs in head and neck (H&N), it can be difficult to train networks to jointly optimize both small and large organ segmentation. Hence, segmentation-by-detection strategies were utilized by FocusNet [8] and UaNet [24], where the locations of organs were first predicted and then the corresponding smaller regions were processed to yield organ segmentations. Building on this, SOARS [9] used stratified learning (using large organs as guides for small organ segmentation) and neural architecture search (automatically learning network architectures for different organ sizes) to alleviate the size imbalance issue. Despite using different techniques, all of the aforementioned methods use convolutional architectures (e.g. variants of 3D UNet [5]) and learn *discrete* segmentation maps.

In contrast to these UNet based methods that utilize *discrete* grid-based output representations, we propose IOSNet that learns via *continuous* Implicit Neural Representations (INRs). For an organ in a CT scan, the *continuous* segmentation function is a mapping from a point (x, y, z) in the CT scan to whether that point is inside the organ (1) or not (0). Since INRs encode the underlying signal as a *continuous* function, the memory required to encode them is *independent* of the spatial resolution. This is in stark contrast to grid-based *discrete* representations where the memory required increases *cubically* with increasing spatial resolution. Hence, INRs are a viable and memory efficient alternative to encode large 3D medical scans. Moreover, INRs learn shape boundaries instead of voxel distributions [3], which facilitates model convergence. Additionally, due to their *continuous* nature, INRs naturally allow for super-resolution.

Prior works in 3D shape representation initially demonstrated the superiority of INRs for representing shape geometries (IM-Net [3], OccNet [15], DeepSDF [18]). Since then, INRs have been used for a wide range of other tasks, including representing appearances [16,17,22,23] and performing semantic segmentation [2,14]. Whereas initial models represented relatively simple objects, later methods [1,4,13,19] designed techniques for preserving fine-grained details of large 3D scenes. For example, IF-Nets [4] was able to retain fine details by conditioning the decoder on the deep features at the (x, y, z) coordinates instead of conditioning it on the raw point coordinates. We build on top of these works and show how INRs provide a strong framework that is well-suited to medical imaging, and can thereby help alleviate some of the key challenges (large memory footprints and organ size imbalance). To summarize, our main contributions are as follows:

- In contrast to existing medical imaging segmentation methods that learn *discrete* grid-based representations, we design an Implicit Organ Segmentation Network (IOSNet), which learns *continuous* segmentation functions.
- IOSNet outperforms the specialized segmentation architecture UNet at a wide range of spatial resolutions, despite using a simple lightweight decoder.

– Due to the *continuous* Implicit Neural Representations, IOSNet is considerably more memory efficient, accurately segments organs regardless of organ size, converges faster, and naturally allows super-resolution.

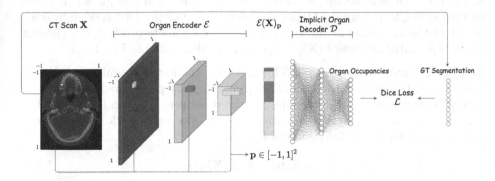

Fig. 1. IOSNet: In contrast to existing segmentation methods that use *discrete* grid-based output representations, IOSNet learns a *continuous* segmentation function s_i. s_i maps a point $\mathbf{p} \in \mathbb{R}^3$ in the scan to an occupancy value, indicating whether the point lies inside the organ (1) or not (0). See Sect. 2.1 for details. 2D for illustration only.

2   Proposed Approach

Firstly, we provide an overview of our proposed Implicit Organ Segmentation Network (IOSNet). Next, we detail the data preparation steps to aid learning via INRs. Lastly, we mention the implementation details used to train IOSNet.

2.1   Method Overview

In contrast to existing algorithms for head and neck OARs segmentation which use *discrete* output representations, we propose to learn a *continuous* function for segmenting organs. Concretely, we learn a *continuous* segmentation (occupancy) function s_i for each organ $i \in \{0, 1, \ldots, R\}$ where R is the number of OARs. Each of these segmentation masks s_i learns the segmentation (occupancy) for a *continuous* point $\mathbf{p} \in \mathbb{R}^3$ in the scan:

$$s_i : \mathbb{R}^3 \rightarrow [0, 1] \tag{1}$$

In order to preserve fine details, we follow the recently proposed IF-Nets [4] designed for 3D shape reconstruction and completion. As illustrated in Fig. 1, IOSNet consists of two main components, Organ Encoder and Implicit Organ Decoder, which are described next.

Organ Encoder ($\mathcal{E}$): This is a neural network (e.g. CNN) encoder that processes the input CT scan, yielding several learned feature maps. Feature maps corresponding to earlier layers learn local features whereas those at later layers learn global characteristics about the organs. These local and global features enable the network to accurately segment organs of varied sizes.

Point Features: Since the Implicit Organ Decoder $\mathcal{D}$ segments each point $\mathbf{p}$ in the CT scan individually, we need to extract the (local and global) features corresponding to $\mathbf{p}$ from the Organ Encoder feature maps. This is done by evaluating the feature maps at the normalized scan coordinate $\mathbf{p} \in [-1, 1]^3$, yielding the concatenated features $\mathcal{E}(\mathbf{X})_\mathbf{p}$ at point $\mathbf{p}$, as illustrated in Fig. 1.

Implicit Organ Decoder ($\mathcal{D}$): This is a fully connected neural network that takes the global and local features $\mathcal{E}(\mathbf{X})_\mathbf{p}$ at the point $\mathbf{p}$ and predicts the occupancy for each organ, indicating whether the point $\mathbf{p}$ in the scan lies inside that organ (1) or not (0). Thereby, $\mathcal{E}$ learns a *continuous* segmentation (occupancy) function for each of the OARs, given by:

$$s(\mathbf{X}, \mathbf{p}) = \mathcal{D}(\mathcal{E}(\mathbf{X})_\mathbf{p}) \tag{2}$$

Training: In order to train our network, we need a dataset of (scan coordinate, label) $(\mathbf{p}_i, \mathbf{s}_i)$ pairs. Section 2.2 describes how this dataset is created. Given this dataset with N sampled points per scan, we use the following loss function to train the network:

$$\mathcal{L} = L(\{s(\mathbf{X}, \mathbf{p}_i), \mathbf{s}_i\}_{i=1}^N) \tag{3}$$

where $L(\cdot)$ is the Dice loss, and $s(\mathbf{X}, \mathbf{p}_i)$ and $\mathbf{s}_i$ are the predicted organ occupancies and ground-truth segmentations for the ith point respectively.

Inference: During inference, we evaluate the Implicit Organ Decoder $\mathcal{D}$ at each point $\mathbf{p}$ on the grid, yielding the desired segmentation mask. Owing to the *continuous* segmentation function s_i, we can predict a segmentation mask of any arbitrary resolution during inference by evaluating on grid points of the desired resolution. Section 3.3 shows super-resolution experiments to illustrate this.

2.2   Implementation

Data Preparation: In this work, we use the head and neck OARs segmentation dataset provided by Zhu et al. [28], which consists of 261 train and 10 test CT scans, with 9 annotated OARs. Note that the dataset has been compiled from several sources [6,20,27] and we assume that the shapes are roughly aligned. We need to sample this dataset to generate (scan coordinate, label) $(\mathbf{p}_i, \mathbf{s}_i)$ pairs for training IOSNet. To avoid excessive computation, our sampling strategy samples more points close to organs of interest. Concretely, for each organ, we sample K points within the organ bounding box (with a padding of P). Additionally, we sample M points throughout the scan. For both of the above samplings, we employ Latin hypercube sampling to ensure that the sampled points are spread out. Hence, our sampling strategy is computationally efficient (does not sample too many background points) and ensures a good coverage of the entire scan, with emphasis on regions near the organs of interest. The data sampling parameters are $K = 4000$, $P = 4$, and $M = 5000$.

Training Details: To ensure a fair comparison with other segmentation methods, we use UNet encoder for learning the feature maps in Sect. 2.1. Implicit Organ Decoder, on the other hand, is a relatively lightweight network with 4 MLP layers $(1024, 1024, 1024, 10)$. We use PyTorch[1] (1.8.1) and Monai[2] (0.8.0) to implement IOSNet, which is trained for 200 epochs with the Adam optimizer and a learning rate of 0.0005 on a NVIDIA RTX8000 GPU.

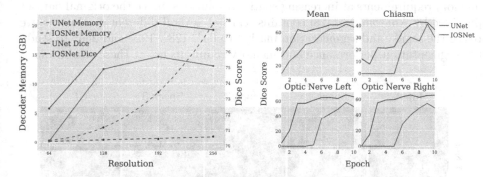

Fig. 2. Left (Sect. 3.1): IOSNet performs better and is significantly more memory efficient with increasing spatial resolution than UNet. **Right** (Sect. 3.2): Dice score (y-axis) vs epoch (x-axis) plots for mean (across all 9 OARs) and small organs (Chiasm, Optic Nerve Left & Right). IOSNet converges much faster, especially on small organs.

3   Experiments

In this section, we compare our IOSNet, which learns *continuous* segmentation functions, with UNet, which outputs *discrete* segmentation maps. Although improved methods [8,9,24,28] have been developed for H&N OARs segmentation in recent years, these methods build on top of UNet by either using specialized architectures (e.g. Squeeze and Excitation blocks [11]) or specialized techniques such as segmentation-by-detection, stratified learning or neural architecture search. These strategies themselves are rather generic and can yield improvements with IOSNet as well. More importantly, all of these methods also learn *discrete* segmentation maps similar to UNet. Hence, to illustrate the unique properties of IOSNet, we use UNet for comparison, which provides a competitive baseline with SOTA without using any extra bells and whistles. For completeness, however, we also compare against these methods in the supplementary.

   Section 3.1 compares IOSNet and UNet in terms of segmentation performance and memory requirements with increasing spatial resolution. This is followed by an analysis of small organs segmentation and convergence speed in Sect. 3.2. Section 3.3 illustrates the efficacy of the IOSNet continuous representations for super-resolution whereas Sect. 3.4 studies its inference efficiency and robustness

---

[1] https://pytorch.org/.
[2] https://monai.io/.

to different dataset splits. Lastly, Sect. 3.5 presents ablation studies on different network architectures and sampling schemes. Unless indicated otherwise, all the experiments in this section report results in terms of the Dice coefficient.

3.1   Segmentation Performance and Memory Requirements

Here, we compare IOSNet and UNet in terms of segmentation performance and memory requirements at increasing spatial resolutions. Since the original dataset provided by AnatomyNet [28] contains scans at several different resolutions, we resize the scans (and segmentation masks) to 64^3, 128^3, 192^3 and 256^3 to study the segmentation performance and memory requirements at these resolutions.

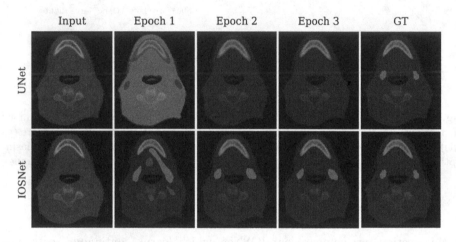

Fig. 3. UNet learns voxel distributions and hence struggles to segment even medium sized organs (Submandibular Gland Left & Right) initially whereas IOSNet segments them accurately within the first few epochs. GT: Ground-truth.

IOSNet consistently outperforms UNet on the Dice score across all spatial resolutions, as illustrated by Fig. 2 (left). Additionally, IOSNet is much more memory efficient during training, especially at higher resolutions. Since both IOSNet and UNet use identical encoders, we study the memory consumption of the respective decoders with increasing resolution. In contrast to the UNet decoder, where memory grows *cubically* with increasing resolution, the memory consumption of the IOSNet decoder stays roughly *constant* (modulo the number of sampled points) and is *independent* of the spatial resolution. This is a direct consequence of the *continuous* nature of the representations and is especially beneficial for large 3D medical scans, where the memory consumption of *discrete* voxel grids grows intractably fast. Please refer to the supplementary for overall (encoder + decoder) memory consumption.

3.2 Small Organs and Model Convergence

Motivated by the observation that INRs based decoders learn shape boundaries whereas CNN decoders learn voxel distributions [3], we compare IOSNet and UNet via the visual quality of the segmentation results. The first epoch visualizations in Fig. 3 illustrate that UNet tries to learn the entire voxel distribution whereas IOSNet focuses specifically around the organs of interest and hence is more efficient. Moreover, since UNet learns voxel distributions, it favors the larger organs and struggles to generate accurate predictions for the smaller ones. In contrast, IOSNet is able to accurately segment even the small organs within a few epochs of training due to the point-wise continuous occupancy objective, which directly learns shape boundaries. Learning accurate segmentations for all organs irrespective of organ size naturally allows IOSNet training to converge much quickly than UNet as indicated by Fig. 2 (right). This is especially evident for the small organs (Chiasm, Optic Nerve Left, and Optic Nerve Right).

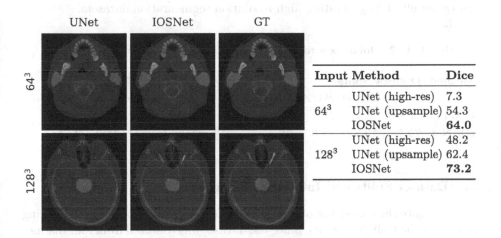

Input	Method	Dice
	UNet (high-res)	7.3
64^3	UNet (upsample)	54.3
	IOSNet	**64.0**
	UNet (high-res)	48.2
128^3	UNet (upsample)	62.4
	IOSNet	**73.2**

Fig. 4. Left: UNet and IOSNet segmentation super-resolution results for models trained on resized 64^3 (top row) and 128^3 (bottom row) scans. Super-resolution segmentation maps are compared against the ground-truth (original) segmentation maps. **Right:** Dice scores for UNet vs IOSNet super-resolution.

3.3 Super-Resolution

Owing to the *continuous* nature of the representations learnt, we can sample at any arbitrary resolution during inference regardless of the training scans resolution. To illustrate this, we train two IOSNet models on 64^3 and 128^3 scans respectively. Without any retraining, we then compute the segmentation masks for the *original* dataset by adjusting the sampling grid dimensions according to the desired output resolution. Observe that the *original* dataset contains scans with several different resolutions and hence super-resolution on this dataset is particularly challenging. Similarly, we train two UNet models on 64^3 and 128^3 scans respectively. Here, we try two types of super-resolution. Firstly, keeping the network weights frozen, we use the high resolution scans as input, which guarantees that the output has the desired spatial resolution (UNet high-res). Secondly, we use the low resolution scans (64^3 or 128^3) as input and instead upsample the output segmentation maps (UNet upsample). Figure 4 (left) and Fig. 4 (right) compares IOSNet and UNet in terms of visual results and Dice scores respectively. For UNet, the naive upsampling strategy performs better, illustrating that UNet is unable to deal with the input resolution change between training and inference. In contrast, IOSNet significantly outperforms both variants of UNet owing to its *continuous* representation that naturally allows for super-resolution. Observe from the visual results in Fig. 4 (left) that IOSNet tends to produce sharp boundaries (top row) and retains fine details such as segmenting small organs (bottom row) due to its *continuous* representation. This is in contrast to UNet which produces blurry boundaries and completely misses small organs. Via this super-resolution of segmentation maps, IOSNet can help reduce annotation effort in generating high resolution segmentation datasets.

Table 1. Left: Robustness to train/test splits. **Right:** Inference efficiency (s).

Method	Dataset Splits				Input	UNet	IOSNet		
	231/40	241/30	251/20	261/10			Naive	AMP	MISE [15]
UNet	71.1	73.3	74.4	77.1	128^3	**0.08**	2.75	1.57	**0.45**
IOSNet	72.7	73.8	75.6	**78.3**	256^3	**0.38**	29.1	15.0	**2.51**

3.4 Dataset Splits and Inference Efficiency

We investigate the robustness of IOSNet to different dataset splits by rebalancing the sizes of the train/test splits. Since the dataset was collected from four distinct sources, not all of the scans are annotated for each of the 9 OARs (see [28] for details). Hence, we choose 40 as the maximum size of the test set since only about 40 samples have annotations for all 9 OARs. Across all the dataset splits, IOSNet consistently outperforms UNet as illustrated by Table 1 (left).

Next, we study the inference efficiency of IOSNet against UNet (Table 1 (right)). The naive strategy of computing segmentations for each voxel of a high-resolution scan can be quite time consuming. Hence, we explore two techniques for improving the inference efficiency. Firstly, we use automatic mixed precision (AMP), which increases inference speed without degrading model performance by intelligently choosing to use 16-bit instead of 32-bit floating point computations for some steps. Secondly, we leverage the optimized Multiresolution IsoSurface Extraction (MISE) [15] algorithm, which does not need to densely evaluate at all points of a high-resolution grid since it incrementally builds an octree (i.e. starting from a lower resolution, it densely evaluates only blocks that are at organ boundaries). MISE [15] significantly increases inference speed and hence, overall, the IOSNet benefits (continuous representations, memory efficiency, faster convergence, super-resolution) far outweigh the minor reduction in inference efficiency.

3.5 Network Architectures and Sampling Strategies

Table 2 (left) presents an ablation study on different network architectures. We compare IOSNet with UNet, which learns *discrete* output representations, and with BAENet [2] and IF-Nets [4] that learn *continuous* Implicit Neural Representations. IF-Nets (128) and IF-Nets (256) refer to the 128^3 and 256^3 voxel models respectively. Overall, IOSNet learns *continuous* segmentation functions and outperforms the other models.

We try two different sampling schemes – 1) uniform (sampling randomly throughout scan), and 2) organ (described in Sect. 2.2). With sparse sampling ($\sim$30k), organ sampling performs best since it captures both the interior and exterior of organs without sampling too many points from background. However, with dense sampling ($\sim$150k), the naive uniform sampling yields the best results.

Table 2. Left: Network architectures ablation. **Right:** Sampling strategies ablation.

Method	Continuous	Dice
UNet [5]	✗	77.1
BAENet [2]	✓	24.2
IF-Nets (128) [4]	✓	71.8
IF-Nets (256) [4]	✓	73.5
IOSNet	✓	**78.3**

Sampling	Points	Dice
Uniform	30k	56.9
Organ	30k	**77.2**
Uniform	150k	**78.3**
Organ	150k	77.6

4   Conclusion

We introduce a novel Implicit Organ Segmentation Network (IOSNet) which learns *continuous* segmentation functions in contrast to existing methods that learn *discrete* representations. In addition to outperforming the discrete specialized segmentation architecture UNet, IOSNet yields considerable memory savings, accurately segments organs regardless of organ size, converges faster, and naturally allows super-resolution due to its *continuous* representations. Hence, our work highlights the efficacy of continuous Implicit Neural Representations for medical imaging, which we hope will be further explored in the future.

References

1. Chabra, R., et al.: Deep local shapes: learning local SDF priors for detailed 3D reconstruction. In: Vedaldi, A., Bischof, H., Brox, T., Frahm, J.-M. (eds.) ECCV 2020. LNCS, vol. 12374, pp. 608–625. Springer, Cham (2020). https://doi.org/10.1007/978-3-030-58526-6_36

2. Chen, Z., Yin, K., Fisher, M., Chaudhuri, S., Zhang, H.: BAE-NET: branched autoencoder for shape co-segmentation. In: Proceedings of the IEEE/CVF International Conference on Computer Vision, pp. 8490–8499 (2019)
3. Chen, Z., Zhang, H.: Learning implicit fields for generative shape modeling. In: Proceedings of the IEEE/CVF Conference on Computer Vision and Pattern Recognition, pp. 5939–5948 (2019)
4. Chibane, J., Alldieck, T., Pons-Moll, G.: Implicit functions in feature space for 3D shape reconstruction and completion. In: Proceedings of the IEEE/CVF Conference on Computer Vision and Pattern Recognition, pp. 6970–6981 (2020)
5. Çiçek, Ö., Abdulkadir, A., Lienkamp, S.S., Brox, T., Ronneberger, O.: 3D U-Net: learning dense volumetric segmentation from sparse annotation. In: Ourselin, S., Joskowicz, L., Sabuncu, M.R., Unal, G., Wells, W. (eds.) MICCAI 2016. LNCS, vol. 9901, pp. 424–432. Springer, Cham (2016). https://doi.org/10.1007/978-3-319-46723-8_49
6. Clark, K., et al.: The cancer imaging archive (TCIA): maintaining and operating a public information repository. J. Digit. Imaging **26**(6), 1045–1057 (2013)
7. Fritscher, K., Raudaschl, P., Zaffino, P., Spadea, M.F., Sharp, G.C., Schubert, R.: Deep neural networks for fast segmentation of 3D medical images. In: Ourselin, S., Joskowicz, L., Sabuncu, M.R., Unal, G., Wells, W. (eds.) MICCAI 2016. LNCS, vol. 9901, pp. 158–165. Springer, Cham (2016). https://doi.org/10.1007/978-3-319-46723-8_19
8. Gao, Y., et al.: FocusNet: imbalanced large and small organ segmentation with an end-to-end deep neural network for head and neck CT images. In: Shen, D., et al. (eds.) MICCAI 2019. LNCS, vol. 11766, pp. 829–838. Springer, Cham (2019). https://doi.org/10.1007/978-3-030-32248-9_92
9. Guo, D., et al.: Organ at risk segmentation for head and neck cancer using stratified learning and neural architecture search. In: Proceedings of the IEEE/CVF Conference on Computer Vision and Pattern Recognition, pp. 4223–4232 (2020)
10. Harari, P.M., Song, S., Tomé, W.A.: Emphasizing conformal avoidance versus target definition for IMRT planning in head-and-neck cancer. Int. J. Radiat. Oncol. Biol. Phys. **77**(3), 950–958 (2010)
11. Hu, J., Shen, L., Sun, G.: Squeeze-and-excitation networks. In: Proceedings of the IEEE Conference on Computer Vision and Pattern Recognition, pp. 7132–7141 (2018)
12. Ibragimov, B., Xing, L.: Segmentation of organs-at-risks in head and neck CT images using convolutional neural networks. Med. phys. **44**(2), 547–557 (2017)
13. Jiang, C., et al.: Local implicit grid representations for 3D scenes. In: Proceedings of the IEEE/CVF Conference on Computer Vision and Pattern Recognition, pp. 6001–6010 (2020)
14. Kohli, A.P.S., Sitzmann, V., Wetzstein, G.: Semantic implicit neural scene representations with semi-supervised training. In: 2020 International Conference on 3D Vision (3DV), pp. 423–433. IEEE (2020)
15. Mescheder, L., Oechsle, M., Niemeyer, M., Nowozin, S., Geiger, A.: Occupancy networks: learning 3D reconstruction in function space. In: Proceedings of the IEEE/CVF Conference on Computer Vision and Pattern Recognition, pp. 4460–4470 (2019)
16. Mildenhall, B., Srinivasan, P.P., Tancik, M., Barron, J.T., Ramamoorthi, R., Ng, R.: NeRF: representing scenes as neural radiance fields for view synthesis. In: Vedaldi, A., Bischof, H., Brox, T., Frahm, J.-M. (eds.) ECCV 2020. LNCS, vol. 12346, pp. 405–421. Springer, Cham (2020). https://doi.org/10.1007/978-3-030-58452-8_24

17. Oechsle, M., Mescheder, L., Niemeyer, M., Strauss, T., Geiger, A.: Texture fields: learning texture representations in function space. In: Proceedings of the IEEE/CVF International Conference on Computer Vision, pp. 4531–4540 (2019)
18. Park, J.J., Florence, P., Straub, J., Newcombe, R., Lovegrove, S.: DeepSDF: learning continuous signed distance functions for shape representation. In: Proceedings of the IEEE/CVF Conference on Computer Vision and Pattern Recognition, pp. 165–174 (2019)
19. Peng, S., Niemeyer, M., Mescheder, L., Pollefeys, M., Geiger, A.: Convolutional occupancy networks. In: Vedaldi, A., Bischof, H., Brox, T., Frahm, J.-M. (eds.) ECCV 2020. LNCS, vol. 12348, pp. 523–540. Springer, Cham (2020). https://doi.org/10.1007/978-3-030-58580-8_31
20. Raudaschl, P.F., et al.: Evaluation of segmentation methods on head and neck CT: auto-segmentation challenge 2015. Med. Phys. **44**(5), 2020–2036 (2017)
21. Ren, X., et al.: Interleaved 3D-CNN s for joint segmentation of small-volume structures in head and neck CT images. Med. Phys. **45**(5), 2063–2075 (2018)
22. Saito, S., Huang, Z., Natsume, R., Morishima, S., Kanazawa, A., Li, H.: PIFu: pixel-aligned implicit function for high-resolution clothed human digitization. In: Proceedings of the IEEE/CVF International Conference on Computer Vision, pp. 2304–2314 (2019)
23. Sitzmann, V., Zollhöfer, M., Wetzstein, G.: Scene representation networks: continuous 3D-structure-aware neural scene representations. arXiv preprint arXiv:1906.01618 (2019)
24. Tang, H., et al.: Clinically applicable deep learning framework for organs at risk delineation in CT images. Nat. Mach. Intell. **1**(10), 480–491 (2019)
25. Tong, N., Gou, S., Yang, S., Ruan, D., Sheng, K.: Fully automatic multi-organ segmentation for head and neck cancer radiotherapy using shape representation model constrained fully convolutional neural networks. Med. Phys. **45**(10), 4558–4567 (2018)
26. Torre, L.A., Bray, F., Siegel, R.L., Ferlay, J., Lortet-Tieulent, J., Jemal, A.: Global cancer statistics, 2012. CA Cancer J. Clin. **65**(2), 87–108 (2015)
27. Vallieres, M., et al.: Radiomics strategies for risk assessment of tumour failure in head-and-neck cancer. Sci. Rep. **7**(1), 1–14 (2017)
28. Zhu, W., et al.: AnatomyNet: deep learning for fast and fully automated whole-volume segmentation of head and neck anatomy. Med. Phys. **46**(2), 576–589 (2019)

ModDrop++: A Dynamic Filter Network with Intra-subject Co-training for Multiple Sclerosis Lesion Segmentation with Missing Modalities

Han Liu[1]([✉]), Yubo Fan[1], Hao Li[2], Jiacheng Wang[1], Dewei Hu[2], Can Cui[1], Ho Hin Lee[1], Huahong Zhang[1], and Ipek Oguz[1,2]

[1] Department of Computer Science, Vanderbilt University, Nashville, USA
han.liu@vanderbilt.edu
[2] Department of Electrical and Computer Engineering, Vanderbilt University, Nashville, USA

Abstract. Multiple Sclerosis (MS) is a chronic neuroinflammatory disease and multi-modality MRIs are routinely used to monitor MS lesions. Many automatic MS lesion segmentation models have been developed and have reached human-level performance. However, most established methods assume the MRI modalities used during training are also available during testing, which is not guaranteed in clinical practice. Previously, a training strategy termed Modality Dropout (ModDrop) has been applied to MS lesion segmentation to achieve the state-of-the-art performance with missing modality. In this paper, we present a novel method dubbed ModDrop++ to train a unified network adaptive to an arbitrary number of input MRI sequences. ModDrop++ upgrades the main idea of ModDrop in two key ways. First, we devise a plug-and-play dynamic head and adopt a filter scaling strategy to improve the expressiveness of the network. Second, we design a co-training strategy to leverage the intra-subject relation between full modality and missing modality. Specifically, the intra-subject co-training strategy aims to guide the dynamic head to generate similar feature representations between the full- and missing-modality data from the same subject. We use two public MS datasets to show the superiority of ModDrop++. Source code and trained models are available at https://github.com/han-liu/ModDropPlusPlus.

Keywords: Multiple Sclerosis · Magnetic Resonance Imaging · Missing modality · Modality Dropout · Dynamic filters · Co-training

1 Introduction

Multiple Sclerosis (MS) is a chronic inflammatory neurological disease characterized by focal lesions and diffuse neurodegeneration [15]. Magnetic Resonance

Supplementary Information The online version contains supplementary material available at https://doi.org/10.1007/978-3-031-16443-9_43.

Imaging (MRI) is crucial for the diagnosis and monitoring of MS lesions [13]. To characterize different types of MS lesions, multiple MR imaging modalities such as T1w, T2w, Proton Density (PD), Fluid-attenuated inversion recovery (FLAIR), and T1-contrast enhanced (T1CE) are routinely collected in clinical practice. There have been many automatic MS lesion segmentation algorithms in the medical imaging community and some even achieved human-level performances [5,23,26]. Most established methods have been developed based on the assumption that the imaging modalities used during training are available during deployment. However, in clinical practice, it is not possible to always meet this assumption because (1) the number and categories of MRI sequences acquired at different sites can vary and (2) the acquired MR sequences for a specific patient may be unusable due to poor image quality. The modality mismatch between the training and testing data, i.e., the missing modality problem, significantly limits the use of the existing methods during deployment.

A straightforward way to tackle this problem is to train independent models for each missing-modality condition but this requires a long training time. To overcome this issue, a training strategy termed Modality Dropout (*ModDrop*) [20] has been developed and widely used in various fields such as computer vision [4], dialogue systems [21] and medical imaging [16,18,22]. Particularly, for MS lesion segmentation, Feng *et al.* [7] adopted ModDrop and achieved the state-of-the-art performance to handle MRIs with missing sequences. Similar to the regular dropout for neurons, in ModDrop, some modalities are randomly dropped-out during training and a unified network is trained to create feature representations that are robust to all missing conditions [6]. ModDrop is easily applicable to any existing models without modifying the network architectures. However, it may suffer from two limitations: (1) regardless of different missing conditions, ModDrop always forces the network to learn a single set of parameters, which may limit the expressiveness of the network, and (2) ModDrop does not leverage the intra-subject relation between full- and missing-modality data.

Our work is heavily inspired by dynamic filter networks [14] and co-training strategies. In dynamic filter networks, the filter parameters can be dynamically generated conditioned on either an input image [14], latent features [11], or a pre-defined task code [27]. In medical image analysis, dynamic filter networks have been successfully used to improve the model accuracy and efficiency in a variety of tasks such as image translation [25], partial label training [27], and domain generalization [11]. However, the effectiveness of dynamic filter networks on missing modality issue has not been explored. On the other hand, co-training strategies have been widely used for knowledge distillation to minimize the gap between multi- and uni-modalities. For instance, KD-net [10] can distill knowledge from multi-modal data via co-training to improve the performance of a uni-modal model. However, KD-net suffers from inefficiency because each uni-modal model has to be distilled from the multi-modal model separately.

We propose a novel training method, dubbed *ModDrop++*, to address missing modalities for MS lesion segmentation. Our novel contributions are as follows:

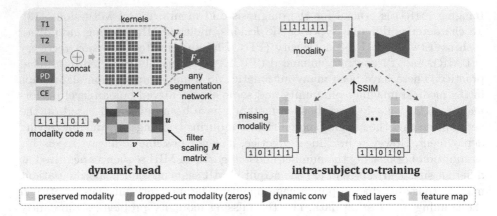

Fig. 1. The two key upgrades in ModDrop++: a plug-and-play dynamic head (left) and an intra-subject co-training strategy (right). The dynamic head aims to improve the network expressiveness by learning a set of filter scaling matrices to adaptively adjust the first convolution layer for each missing condition. The intra-subject co-training aims to transfer the knowledge between the full-modality data and the missing-modality data of the same subject, which can guide the dynamic head to produce similar feature representations even when multiple modalities are absent.

– We devise a plug-and-play dynamic head to improve the network expressiveness. The dynamic head can adaptively generate the model parameters conditioned on the missing condition. To the best of our knowledge, this is the first study to apply the dynamic filters for the missing modality problem.
– We introduce a co-training strategy to leverage the intra-subject relation between the full- and missing-modality data.
– Built upon the leading method [26] of the ISBI challenge [5], our experiments on two publicly available MS datasets show the superiority of ModDrop++.

2   Methods

2.1   Preliminaries

ModDrop++ aims to train a unified MS lesion segmentation model that can be adaptable to an arbitrary number of available MRI sequences (modalities) during deployment. In particular, it inherits the training scheme of the classic ModDrop. Let x and y be the input data of K modalities and the label. During training, a shared network F is trained using the input data with image modalities randomly dropped-out (replaced by zeros). There are $2^K - 1$ unique missing modality configurations, represented by:

$$\tilde{x}^k = \delta^k x^k, k \in \{1, ..., K\}, \tag{1}$$

where δ^k is a Bernoulli selector variable that can take on values in $\{0, 1\}$ for each modality k. The frequency of each possible configuration $i \in \{1, ..., 2^K - 1\}$ being activated during training is determined by the probability p^k for each δ^k. Let θ_s be the model parameters of the shared network and L_t be the loss function for the downstream task. The learning objective of the traditional ModDrop can be expressed as

$$\operatorname*{argmin}_{\theta_s} L_t(y, F(\tilde{x}_i|\theta_s)) \tag{2}$$

In this work, we build upon ModDrop and upgrade it to ModDrop++ by introducing (1) a plug-and-play dynamic head and (2) an intra-subject co-training strategy, as illustrated in Fig. 1. The details of these upgrades are as follows.

2.2 Dynamic Head with Filter Scaling

In ModDrop, a single set of model parameters θ is learned to handle all possible missing conditions. This may limit the expressiveness of the network and lead to suboptimal performance for competing missing conditions. This can get severe with an increasing number of total modalities, since K modalities leads to $2^K - 1$ missing conditions. As shown in Fig. 1 (left), to improve the network expressiveness, we devise a dynamic head D to adaptively generate model parameters conditioned on the availability of input modalities. We use a binary modality code $m \in \mathbb{R}^K$ where 0/1 represent the absence/presence of each modality.

To mitigate the large input variation caused by artificially zero-ed channels, we use the dynamic head to generate the parameters for the first convolutional layer F_d. Suppose the number of input channels and output channels are u and v, and the kernel size is $p \times q$. Typically, all parameters in the dynamic convolution layer are generated individually based on a given prior. In our scenario, the dynamic head is asked to learn to generate a total number of $uvpq + b$ parameters from the modality code m (b parameters for bias). However, this mapping might be too difficult to learn for the dynamic head, and our networks fail to converge in preliminary experiments.

To address this issue, we propose to update our dynamic filters with a filter scaling strategy, which was originally designed for unsupervised image-to-image translation [2]. Our goal is that kernels should contribute differently for inputs with different dropped-out modalities. The task of our dynamic head then becomes to learn to generate a filter scaling matrix $M \in \mathbb{R}^{u \times v}$ for each missing condition, where each element in M represents the contribution (scale) of the corresponding kernel at each missing condition. The kernel weights in F_d are updated with the corresponding scale factor by a scalar multiplication. The kernel biases are updated in the same manner. This filter scaling strategy reduces the number of learnable parameters from $uvpq + b$ to $uv + b$, which preserves the dynamic nature of our network but with a much simpler learning task.

2.3 Intra-subject Co-training

With our dynamic head, our network can be considered to have two components: the first convolutional layer F_d with dynamic filters and the subsequent fixed layers F_s with static filters. The feature map obtained by F_d can thus be expressed as $f_i = F_d(\tilde{x}_i|m_i)$, which are further passed to F_s to obtain the final prediction. One assumption of our co-training strategy is that networks can perform the best when fed with full-modality data x. Based on this assumption, we can improve the segmentation performance for the missing-modality data $\tilde{x}_i$ when f_i becomes similar to $f = F_d(x|\mathbb{1})$, because f and f_i are passed to the same network F_s with static filters. Motivated by this, we develop an intra-subject co-training strategy to guide the dynamic head to learn to maximize the similarity between f and f_i. Specifically, we forward the full-modality data x and missing-modality data $\tilde{x}_i$ from the same subject to F_d in parallel to obtain the associated feature maps f and f_i, which are further passed to F_s to obtain the final outputs. Besides the downstream task loss L_t for both outputs, we also optimize a similarity loss L_{sim} to maximize the similarity between f and f_i (Fig. 1, right panel). Therefore, the overall training objective of ModDrop++ can be expressed as follows:

$$\underset{\theta_s, \theta_d}{\mathrm{argmin}}\, \alpha L_t(y, F_s(f)) + \beta L_t(y, F_s(f_i)) - \gamma L_{sim}(f, f_i) \tag{3}$$

where α, β and γ are the coefficients to weigh the importance of each loss term.

Specifically, ModDrop++ uses the Structural Similarity Index (SSIM) loss as L_{sim}. The SSIM has been widely used as a quality metric for the similarity between two images [9]. Recently, Wang *et al.* [24] showed that the SSIM can also be an effective loss to measure the similarity in feature space. The features that we aim to maximize the similarity of are from F_d and thus contain mostly low-level structural features, which can be nicely captured by the SSIM loss.

2.4   Implementation Details

Without loss of generality, we develop ModDrop++ based on the public implementation of the method leading the ISBI challenge leaderboard [5]. This model uses a 2D fully convolutional densely connected network with concatenated 2.5D stacked slices from each modality as the input [26]. Our dynamic head is a 2D convolutional layer with kernel size of 1×1. During training, we use the focal loss [19] for lesion segmentation, which has been shown to achieve higher Dice coefficient than other losses [26]. The focusing parameter and the weighting factor of class imbalance are set as 0.25 and 2 respectively as found in [26]. For intra-subject co-training, we use the SSIM loss with a window size of 11×11 as the similarity loss.

In our overall loss function, the weighting factors α, β and γ are empirically set as 1, 1 and 0.05. As in [26], we set the initial learning rate as 0.25 (linearly decayed after 100 epochs) with a batch size of 16. All the networks are totally optimized to 300 epochs with an Adam optimizer with a momentum of 0.5.

2.5   Datasets and Evaluation Metrics

In our experiments, two public 3T MS datasets are used. **UMCL Multi-rater Consensus Dataset** [17]. This dataset, referred to as **UMCL** (University Medical Center Ljubljana), is a cohort of 30 patients with 3D FLAIR, 2D T1w,

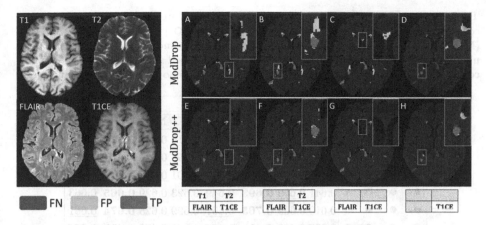

Fig. 2. Qualitative comparisons of ModDrop and ModDrop++ (ours). The left panel displays the four available modalities of a testing sample from the UMCL dataset. The right panel shows the segmentation results by ModDrop (row 1) and ModDrop++ (row 2). The major differences between ModDrop and ModDrop++ are zoomed in on the upper-right panels. The number of available modalities from left to right column decreases from 4 to 1. (Color figure online)

2D T1CE and 2D T2w scans. The manual segmentation is delineated by three experts and a consensus was reached via discussion. **Longitudinal MS Lesion Segmentation Challenge (ISBI)** [5]. This dataset contains 21 timepoints from 5 subjects with T1w, T2w, FLAIR, and PD, and two separate manual delineations. For both datasets, we follow the same pre-processing procedures as in [26] by performing intensity normalization for each image with kernel density estimation. We randomly split the dataset subject-wise with a ratio of 4:1 for training and testing.

We use the evaluation metrics in the ISBI challenge [5] including Dice similarity coefficient (DSC), positive predictive value (PPV), true positive rate (TPR), lesion-wise false positive rate (LFPR), lesion-wise true positive rate (LTPR), volume difference (VD), and volume correlation (Corr). To evaluate the overall performance, we follow [5] and compute the overall score (SC) as

$$SC = \frac{DSC}{8} + \frac{PPV}{8} + \frac{1-LFPR}{4} + \frac{LTPR}{4} + \frac{Corr}{4}.$$

3 Experimental Results

We compare ModDrop++ and ModDrop, which has been demonstrated to be very effective for our application [7], in all possible configurations of available MRI sequences. We assess the impact of the dynamic head and the co-training strategy by evaluating the performance of ModDrop with dynamic head but without intra-subject co-training. We also compare the performances of the unified models against independent models trained for each specific missing condition. The independent models are trained with consistent input modalities and are thus

Table 1. UMCL results. Dice and overall score (SC) are presented for all possible configurations of MRI modalities being either absent (○) or present (●). ModDrop (MD), ModDrop with only the dynamic head (MD++) and ModDrop++ (MD++) are unified models, while independent models (IM) are trained specifically for each missing condition. The best and second best performances are denoted by **bold** and underline.

Modalities				DSC(↑)				SC(↑)			
T1	Flair	T2	CE	MD	MD+	MD++	IM	MD	MD+	MD++	IM
●	●	●	●	0.633	0.663	**0.704**	0.690	0.610	0.633	**0.676**	0.652
○	●	●	●	0.611	0.652	**0.706**	0.698	0.596	0.624	0.673	**0.674**
●	○	●	●	0.400	0.453	0.474	**0.512**	0.401	0.443	0.442	**0.472**
●	●	○	●	0.641	0.659	**0.693**	0.649	0.623	0.629	**0.665**	0.663
●	●	●	○	0.610	0.639	**0.705**	0.697	0.599	0.625	**0.674**	0.661
○	○	●	●	0.377	0.439	0.458	**0.482**	0.379	0.430	0.431	**0.462**
○	●	○	●	0.629	0.653	**0.696**	0.692	0.600	0.621	0.671	**0.676**
○	●	●	○	0.613	0.651	0.704	**0.711**	0.589	0.618	**0.676**	0.655
●	○	○	●	0.325	0.329	0.331	**0.382**	0.303	0.338	0.343	**0.400**
●	○	●	○	0.382	0.427	0.451	**0.471**	0.399	0.424	0.427	**0.459**
●	●	○	○	0.631	0.650	**0.696**	0.683	0.600	0.622	**0.670**	0.658
●	○	○	○	0.275	0.304	0.324	**0.400**	0.288	0.326	0.318	**0.398**
○	●	○	○	0.612	0.642	0.693	**0.702**	0.576	0.607	0.665	**0.670**
○	○	●	○	0.338	0.412	0.420	**0.459**	0.341	0.411	0.403	**0.456**
○	○	○	●	0.270	0.304	0.304	**0.362**	0.292	0.333	0.330	**0.367**

more likely to outperform the unified models. These models can be considered as the upper bounds for each missing condition.

Quantitative Results. In Tables 1 and 2, we show the DSC and SC of the compared methods for all possible combinations of MRI modalities, on the UMCL and ISBI 2015 datasets respectively. Remaining evaluation metrics can be found in the supplemental materials. First, we observe that the lesion segmentation performance for all compared methods drops considerably when FLAIR is absent. This is expected because FLAIR is indeed important for lesion identification tasks in real clinical settings and the most commonly used modality for manual lesion delineation. On both datasets, by adding the dynamic head alone (Mod-Drop++), the performances of all missing conditions are improved *consistently* (except when only T1w is available on the ISBI dataset). Moreover, when combining the intra-subject co-training with the dynamic head (ModDrop++), the performance can be further improved, demonstrating the effectiveness of the co-training strategy. Lastly, we observe that the performance gap between the unified model and independent models are substantially minimized by upgrading ModDrop to ModDrop++.

Qualitative Results. In Fig. 2, we qualitatively compare ModDrop++ (bottom row) against ModDrop (top row) with different number of available modalities.

Table 2. Quantitative results on ISBI 2015 dataset. The best and second best performances are denoted by **bold** and <u>underline</u>.

Modalities				DSC(↑)				SC(↑)			
T1	Flair	T2	PD	MD	MD+	MD++	IM	MD	MD+	MD++	IM
●	●	●	●	0.733	0.756	<u>0.774</u>	**0.793**	0.717	0.747	<u>0.771</u>	**0.800**
○	●	●	●	0.719	0.752	<u>0.768</u>	**0.789**	0.707	0.737	<u>0.766</u>	**0.779**
●	○	●	●	0.691	0.694	<u>0.695</u>	**0.735**	0.717	0.723	<u>0.730</u>	**0.750**
●	●	○	●	0.725	0.743	<u>0.768</u>	**0.775**	0.713	0.738	<u>0.760</u>	**0.775**
●	●	●	○	0.725	0.750	<u>0.774</u>	**0.778**	0.701	0.742	<u>0.767</u>	**0.787**
○	○	●	●	0.684	0.693	<u>0.695</u>	**0.727**	0.708	0.727	<u>0.734</u>	**0.748**
○	●	○	●	0.710	0.735	<u>0.766</u>	**0.772**	0.708	0.727	<u>0.772</u>	**0.776**
○	●	●	○	0.710	0.742	<u>0.772</u>	**0.782**	0.693	0.726	<u>0.772</u>	**0.773**
●	○	○	●	0.668	<u>0.679</u>	0.677	**0.720**	0.697	<u>0.733</u>	0.729	**0.739**
●	○	●	○	0.684	0.688	<u>0.696</u>	**0.731**	0.692	0.715	<u>0.727</u>	**0.748**
●	●	○	○	0.717	0.747	<u>0.777</u>	**0.779**	0.704	<u>0.715</u>	**0.793**	**0.793**
●	○	○	○	<u>0.656</u>	0.633	0.616	**0.717**	0.675	<u>0.700</u>	0.685	**0.749**
○	●	○	○	0.680	0.709	<u>0.742</u>	**0.773**	0.684	0.715	<u>0.770</u>	**0.782**
○	○	●	○	0.680	0.686	<u>0.694</u>	**0.715**	0.680	0.712	<u>0.721</u>	**0.736**
○	○	○	●	0.647	0.649	<u>0.655</u>	**0.682**	0.674	<u>0.724</u>	0.716	**0.725**

The major differences between ModDrop and ModDrop++ are zoomed in on the upper-right panels. With multiple available modalities (columns 1–3), we find that there are fewer false positives (FP, green) and false negatives (FN, red) for ModDrop++ than for ModDrop. When there is only a single modality available (column 4), more FNs can be observed in both ModDrop and ModDrop++, but ModDrop++ still produces better results than ModDrop with fewer FPs and FNs.

4   Discussion and Conclusion

The main advantages of ModDrop++ lie in the following two aspects: (1) it allows us to train one unified model for all missing conditions, which is applicable and efficient in clinical practice. Moreover, a unified model that can take an arbitrary combination of modalities is also a pre-requisite to study transfer learning and domain generalization problems for multi-modal data: otherwise only common modalities from source and target domains can be considered as in [1,3,12], rather than leveraging the rich information from the distinct modalities. (2) with the plug-and-play dynamic head and the intra-subject co-training strategy, ModDrop++ can be easily applied to any existing CNN models.

Other methods for missing modalities exist, such as HeMIS [8]. However, these methods typically require changing model architectures and the impact on performances caused by such changes may not be trivial, especially when a specific model architecture has been proved very effective on certain applications. Therefore, the focus of our study is to investigate the feasibility of improving a widely used training strategy for missing modality problems (ModDrop) without modifications on model architecture, and minimizing the performance gap between the unified model and the modality-specific models. Though our experiments demonstrate the effectiveness of ModDrop++, we can still observe performance gaps between ModDrop++ and modality-specific models from the results on the ISBI dataset, suggesting the potential for further improvement. One limitation of our study is the dataset size, especially the ISBI dataset, where, even though each subject has 4–5 time points, only 5 unique subjects are available. Validation in a larger cohort thus remains as future work.

Acknowledgements. This work was supported in part by the NIH grant R01-NS094456, in part by the NIH and NIBIB grant T32EB021937, and by the Advanced Computing Center for Research and Education (ACCRE) of Vanderbilt University.

References

1. Ackaouy, A., Courty, N., Vallée, E., Commowick, O., Barillot, C., Galassi, F.: Unsupervised domain adaptation with optimal transport in multi-site segmentation of multiple sclerosis lesions from MRI data. Front. Comput. Neurosci. **14**, 19 (2020)
2. Alharbi, Y., Smith, N., Wonka, P.: Latent filter scaling for multimodal unsupervised image-to-image translation. In: Proceedings of the IEEE/CVF Conference on Computer Vision and Pattern Recognition, pp. 1458–1466 (2019)
3. Aslani, S., Murino, V., Dayan, M., Tam, R., Sona, D., Hamarneh, G.: Scanner invariant multiple sclerosis lesion segmentation from MRI. In: 2020 IEEE 17th International Symposium on Biomedical Imaging (ISBI), pp. 781–785. IEEE (2020)
4. de Blois, S., Garon, M., Gagné, C., Lalonde, J.F.: Input dropout for spatially aligned modalities. In: 2020 IEEE International Conference on Image Processing (ICIP), pp. 733–737. IEEE (2020)
5. Carass, A., et al.: Longitudinal multiple sclerosis lesion segmentation: resource and challenge. NeuroImage **148**, 77–102 (2017)
6. Cheerla, A., Gevaert, O.: Deep learning with multimodal representation for pancancer prognosis prediction. Bioinformatics **35**(14), i446–i454 (2019)
7. Feng, Y., Pan, H., Meyer, C., Feng, X.: A self-adaptive network for multiple sclerosis lesion segmentation from multi-contrast MRI with various imaging sequences. In: 2019 IEEE 16th International Symposium on Biomedical Imaging (ISBI 2019), pp. 472–475. IEEE (2019)
8. Havaei, M., Guizard, N., Chapados, N., Bengio, Y.: HeMIS: hetero-modal image segmentation. In: Ourselin, S., Joskowicz, L., Sabuncu, M.R., Unal, G., Wells, W. (eds.) MICCAI 2016. LNCS, vol. 9901, pp. 469–477. Springer, Cham (2016). https://doi.org/10.1007/978-3-319-46723-8_54
9. Hore, A., Ziou, D.: Image quality metrics: PSNR vs. SSIM. In: 2010 20th international Conference on Pattern Recognition, pp. 2366–2369. IEEE (2010)
10. Hu, M., et al.: Knowledge distillation from multi-modal to mono-modal segmentation networks. In: Martel, A.L., et al. (eds.) MICCAI 2020. LNCS, vol. 12261, pp. 772–781. Springer, Cham (2020). https://doi.org/10.1007/978-3-030-59710-8_75

11. Hu, S., Liao, Z., Zhang, J., Xia, Y.: Domain and content adaptive convolution for domain generalization in medical image segmentation. arXiv preprint arXiv:2109.05676 (2021)
12. Kamraoui, R.A., Ta, V.T., Tourdias, T., Mansencal, B., Manjon, J.V., Coupé, P.: DeepLesionBrain: towards a broader deep-learning generalization for multiple sclerosis lesion segmentation. Med. Image Anal. **76**, 102312 (2022)
13. Kaunzner, U.W., Gauthier, S.A.: MRI in the assessment and monitoring of multiple sclerosis: an update on best practice. Ther. Adv. Neurol. Disord. **10**(6), 247–261 (2017)
14. Klein, B., Wolf, L., Afek, Y.: A dynamic convolutional layer for short range weather prediction. In: Proceedings of the IEEE Conference on Computer Vision and Pattern Recognition, pp. 4840–4848 (2015)
15. Kolasinski, J., et al.: A combined post-mortem magnetic resonance imaging and quantitative histological study of multiple sclerosis pathology. Brain **135**(10), 2938–2951 (2012)
16. La Rosa, F., et al.: Automated detection of cortical lesions in multiple sclerosis patients with 7T MRI. In: Martel, A.L., et al. (eds.) MICCAI 2020. LNCS, vol. 12264, pp. 584–593. Springer, Cham (2020). https://doi.org/10.1007/978-3-030-59719-1_57
17. Lesjak, Ž, Galimzianova, A., Koren, A., Lukin, M., Pernuš, F., Likar, B., Špiclin, Ž: A novel public MR image dataset of multiple sclerosis patients with lesion segmentations based on multi-rater consensus. Neuroinformatics **16**(1), 51–63 (2018)
18. Li, X., et al.: 3D multi-scale FCN with random modality voxel dropout learning for intervertebral disc localization and segmentation from multi-modality MR images. Med. Image Anal. **45**, 41–54 (2018)
19. Lin, T.Y., Goyal, P., Girshick, R., He, K., Dollár, P.: Focal loss for dense object detection. In: Proceedings of the IEEE International Conference on Computer Vision, pp. 2980–2988 (2017)
20. Neverova, N., Wolf, C., Taylor, G., Nebout, F.: ModDrop: adaptive multi-modal gesture recognition. IEEE Trans. Pattern Anal. Mach. Intell. **38**(8), 1692–1706 (2015)
21. Sun, R., Chen, B., Zhou, Q., Li, Y., Cao, Y., Zheng, H.T.: A non-hierarchical attention network with modality dropout for textual response generation in multimodal dialogue systems. arXiv preprint arXiv:2110.09702 (2021)
22. van Tulder, G., de Bruijne, M.: Learning cross-modality representations from multimodal images. IEEE Trans. Med. Imaging **38**(2), 638–648 (2018)
23. Valverde, S., et al.: Improving automated multiple sclerosis lesion segmentation with a cascaded 3d convolutional neural network approach. NeuroImage **155**, 159–168 (2017)
24. Wang, Z., et al.: Model pruning based on quantified similarity of feature maps. arXiv preprint arXiv:2105.06052 (2021)
25. Yang, H., Sun, J., Yang, L., Xu, Z.: A unified hyper-GAN model for unpaired multi-contrast MR image translation. In: de Bruijne, M., Cattin, P.C., Cotin, S., Padoy, N., Speidel, S., Zheng, Y., Essert, C. (eds.) MICCAI 2021. LNCS, vol. 12903, pp. 127–137. Springer, Cham (2021). https://doi.org/10.1007/978-3-030-87199-4_12
26. Zhang, H., et al.: Multiple sclerosis lesion segmentation with Tiramisu and 2.5D stacked slices. In: Shen, D., et al. (eds.) MICCAI 2019. LNCS, vol. 11766, pp. 338–346. Springer, Cham (2019). https://doi.org/10.1007/978-3-030-32248-9_38
27. Zhang, J., Xie, Y., Xia, Y., Shen, C.: DoDnet: learning to segment multi-organ and tumors from multiple partially labeled datasets. In: Proceedings of the IEEE/CVF Conference on Computer Vision and Pattern Recognition, pp. 1195–1204 (2021)

DOMINO: Domain-Aware Model Calibration in Medical Image Segmentation

Skylar E. Stolte[1], Kyle Volle[2], Aprinda Indahlastari[3,4], Alejandro Albizu[3,5], Adam J. Woods[3,4,5], Kevin Brink[6], Matthew Hale[2], and Ruogu Fang[1,3,7(✉)]

[1] J. Crayton Pruitt Family Department of Biomedical Engineering, Herbert Wertheim College of Engineering, University of Florida (UF), Gainesville, USA
[2] Department of Mechanical and Aerospace Engineering, Herbert Wertheim College of Engineering, UF, Gainesville, USA
[3] Center for Cognitive Aging and Memory, McKnight Brain Institute, UF, Gainesville, USA
[4] Department of Clinical and Health Psychology, College of Public Health and Health Professions, UF, Gainesville, USA
[5] Department of Neuroscience, College of Medicine, UF, Gainesville, USA
[6] United States Air Force Research Laboratory, Eglin Air Force Base, FL, USA
[7] Department of Electrical and Computer Engineering, Herbert Wertheim College of Engineering, UF, Gainesville, USA
ruogu.fang@ufl.edu

Abstract. Model calibration measures the agreement between the predicted probability estimates and the true correctness likelihood. Proper model calibration is vital for high-risk applications. Unfortunately, modern deep neural networks are poorly calibrated, compromising trustworthiness and reliability. Medical image segmentation particularly suffers from this due to the natural uncertainty of tissue boundaries. This is exasperated by their loss functions, which favor overconfidence in the majority classes. We address these challenges with DOMINO, a domain-aware model calibration method that leverages the semantic confusability and hierarchical similarity between class labels. Our experiments demonstrate that our DOMINO-calibrated deep neural networks outperform non-calibrated models and state-of-the-art morphometric methods in head image segmentation. Our results show that our method can consistently achieve better calibration, higher accuracy, and faster inference times than these methods, especially on rarer classes. This performance is attributed to our domain-aware regularization to inform semantic model calibration. These findings show the importance of semantic ties between class labels in building confidence in deep learning models. The framework has the potential to improve the trustworthiness and reliability of generic medical image segmentation models. The code for this article is available at: https://github.com/lab-smile/DOMINO.

Supplementary Information The online version contains supplementary material available at https://doi.org/10.1007/978-3-031-16443-9_44.

Keywords: Image segmentation · Machine learning uncertainty ·
Model calibration · Model generalizability · Whole head MRI

1   Introduction

Machine learning calibration measures the agreement between the predicted probability estimates and the true correctness likelihood [8]. Proper calibration is vital for high-risk applications. Modern deep neural networks (DNNs) achieve impressive accuracy at poor calibration [8]. Incorrectly calibrated DNNs are unreliable on out-of-distribution data and don't know when they are likely to be incorrect. This discrepancy leaves them vulnerable in critical decision-making such as self-driving cars, surgical robots, and disease subtyping On the other hand, well-calibrated models are less certain when incorrect and comparably certain when correct. Their reliable confidence establishes trustworthiness.

We hypothesize that domain-aware model calibration that leverages the *semantic confusability* and *hierarchical similarity* among class labels can yield well-calibrated and higher-performing models. To test this hypothesis, we have chosen medical image segmentation because it is fundamental in medical image analysis. Overly-confident tissue boundaries can introduce significant errors in brain volume estimations [4]. Head image segmentation is prone to errors due to fine tissue boundaries, tissue imbalance, and low contrast. These challenges can make open-source software fall short on patient sub-populations [3,12,17]. Errors in head segmentation can lead to downstream errors in clinical pipelines, like in estimating parameters for non-invasive brain stimulation [2,11].

Hence, we address uncertainty in medical image segmentation by introducing DOMINO, a framework that leverages domain information among class labels to calibrate DNNs. Unlike prior works that push class means to be orthogonal [15], we assume some class labels are naturally similar. The choice of the loss function is important to calibration because loss drives how a model learns [16]. Medical image segmentation still largely relies on standard losses [1]. We extend these approaches with domain-aware loss regularization to improve model calibration. We study two regularization schemes that are based on confusion matrices (CM) and hierarchical classes (HC). The former imposes a penalty based on class confusability when using a standard network on a held-out data subset. The latter groups labels into hierarchical classes based on common tissue properties.

2   Domain-Aware Model Calibration

2.1   U-Net Transformers (UNETR) Model

We employ UNETR [9] as our base model due to its superior segmentation performance. UNETR utilizes a U-Net architecture with a transformer encoder. This approach combats the relative locality of convolutional layers in fully convolutional networks (FCNs). Transformers have revolutionized Natural Language Processing due to superior long-range learning [18]. Transformers encode images as sequences of one-dimensional patch embeddings. Self-attention modules learn weighted sums from hidden layers. Hence, UNETR reformulates 3D image segmentation as sequence-to-sequence predictions. Skip connections pass the transformer's global context to a traditional FCN decoder. The decoder concatenates local information with the global multi-scale information from the encoder. This paper refers to un-regularized UNETR as UNETR-Base.

2.2 Domain-Aware Loss Regularization

Concept. Our penalty addresses a deficit with cross-entropy (CE) loss in uncertainty. CE loss maximizes the output of the ground truth label. Due to this, the network increases the true label logit more than the incorrect label logits. The resulting networks are overly confident in their predictions. Meanwhile, the non-selected classes' softmax outputs do not represent the true likelihood. Our work introduces more meaningful uncertainty by penalizing incorrect classes. Specifically, we assume that some classes are more similar to others. Network presentation often pushes class means to all be orthogonal to one another [15]. Such networks assume that all classes are equally separable. This assumption fights the natural similarities between certain classes. Thus, we hypothesize that a network can learn better class representation by taking advantage of *class similarities*, rather than fighting them. Our methods apply to classification and segmentation. This treats segmentation as pixel-wise classification [13].

Derivation. Our regularization term adds to any loss function as follows:

$$\mathcal{L}(y, \hat{y}) + \beta(y')(W)(\hat{y}) \tag{1}$$

where $\mathcal{L}$ is a suitable loss function (we use $DiceCE$ which is a combination of Dice score and cross-entropy), y is the one-hot encoded true label, and $\hat{y}$ is the softmax output. β can take on any value between zero and one. W represents a generic regularization term of size $N \times N$, where N is the number of classes. The diagonals are zero, whereas the off-diagonals represent the penalties for confusing classes. We propose two domain-aware approaches to design W as below.

Confusion Matrix (UNETR-CM). Confusion matrix-based calibration utilizes the natural confusability among class labels using a non-calibrated DNN. First, we train UNETR-base without regularization on the training set. Then, we evaluate the trained model on a held-out validation set to generate a confusion matrix for all classes. The loss regularization is computed as below:

$$W_{ij} = S \cdot \frac{I_i - C_{ij}}{I_i} \tag{2}$$

Here, i and j represent the row and column indices, respectively. C is the confusion matrix generated when UNETR-Base is applied on a held-out validation set and normalized by class prevalence. W_{ij} represents any given matrix entity. I_i is i^{th} row of the identity matrix. Thus $W_{ii} = 0$ so there is no penalty for the correct class. Finally, S is a scaling factor to make the regularization weights more significant. We set $S = 3$ based on empirical experiments; however, jointly varying β and S can change the balance of the loss function. Low values for both result in no regularization; too high and it begins to affect model accuracy. The correct values for these hyperparameters will depend on the model and dataset.

Hierarchical Class (UNETR-HC). Here, we regularize using hierarchical relationships between semantic labels. Hierarchical groups are more likely to have similar properties than inter-group classes. Hence, confusion within groups can facilitate more informed and safer mistakes when wrong. Table 1 shows the hierarchy for our head segmentation. We define the matrix penalty in Fig. 1b by considering which classes are subsets of the same super-class. In Fig. 1b, each row represents the penalties for confusing the given class with any other class. The maximum penalty is 3, and penalties are manually lowered within the

Table 1. Hierarchical class groupings. *Eyes are considered to fall within CSF and soft tissue due to have aqueous and fibrous components.

Hierarchical groupings	Tissues
Background (BG)	BG
White matter (WM)	WM
Grey matter (GM)	GM
Cerebrospinal fluid (CSF)	CSF, Eyes*
Bone	Cancellous bone, cortical bone
Soft tissue	Skin, fat, muscle, eyes*
Air	Air
Major artery (Blood)	Blood

groups of Table 1. The eye class is considered close to two groupings. This matrix penalty is more subjective than UNETR-CM, but it incorporates domain knowledge.

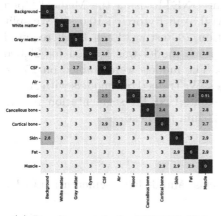

(a) Penalty matrix for UNETR-CM

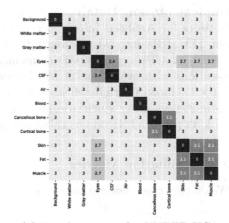

(b) Penalty matrix for UNETR-HC

Fig. 1. Computed matrix penalties (W terms) for both experiments

3   Experiments and Results

3.1   Dataset

This study uses data from a Phase III clinical trial on cognitive training and non-invasive brain stimulation for cognitive improvements. The study recruited participants between 65–89 years old and with age-related cognitive decline. The trial was approved by all relevant Institutional Review Boards. Structural T1-weighted magnetic resonance images (MRIs) were obtained using a 32-channel, receive-only head coil from a 3-T Siemens MAGNETOM Prisma MRI scanner. MPRAGE sequence parameters: repetition time = 1800 ms; echo time = 2.26 ms; flip angle = 8°; field of view = 256 × 256 × 256 mm; voxel size = 1 mm^3.

Ground Truth. Trained staff segmented the T1 MRIs into 11 tissues using semi-automated segmentation. These 11 tissues included muscle, fat, skin, cortical bone, cancellous bone, major artery (blood), air, cerebrospinal fluid (CSF), eyes, grey matter (GM), and white matter (WM). Semi-automated segmentation consists of automated segmentation followed by manual correction. First, base segmentations for WM, GM, and bone were obtained using Headreco, while air was generated in the Statistical Parametric Mapping toolbox (SPM12). Next, these automatic outputs were manually corrected using ScanIP Simpleware^{TM} (version 2018.12, Synopsys, Inc., Mountain View, USA). Bone was separated into cancellous and cortical tissue using thresholding and morphology. Blood, skin, fat, muscle, and eyes (sclera and lens) were manually segmented in Simpleware. CSF was generated by subtracting the other ten tissues from the entire head. The resulting 11 tissue masks served as the ground truths for learned segmentation.

Implementation Details. We implement UNETR using the Medical Open Network for Artificial Intelligence (MONAI-0.8) in Pytorch 1.10.0 [6]. We split our 113 MRIs into 93 training/10 validation/10 testing. Each DNN required 1 GPU, 4 CPUs, and 30 GB of memory. Each model was trained for 25,000 iterations with evaluation at 500 intervals. The models were trained on 256 × 256 × 256 images with batch sizes of 2 images. We trained our models with Adam optimization using stochastic gradient descent. UNETR segmentation results took 3 s per head. Headreco takes roughly 20 min per head.

3.2   Evaluation Metrics

We employ the following metrics on the 11-class and 6-class segmentation tasks.

Dice. represent the overlap of two binary masks [5]: $Dice = \frac{2|Y \cap \hat{Y}|}{|Y| + |\hat{Y}|}$ where Y and $\hat{Y}$ represent the ground truth mask and generated mask for a given tissue, respectively. A perfect overlap between these two generates a Dice score of 1, whereas a 0 represents no mask overlap.

Hausdorff Distance (Hausdorff). calculates the average distances between the closest points in two data subsets [7,10]. Hausdorff distances are generally more robust than Dice in respect to the precise boundaries.

$$H(Y, \hat{Y}) = max(h(Y, \hat{Y}), h(\hat{Y}, Y)) \tag{3}$$

$$h(Y, \hat{Y}) = \max_{y \in Y}(\min_{\hat{y} \in \hat{Y}}(d(y, \hat{y}))), \quad h(\hat{Y}, Y) = \max_{\hat{y} \in \hat{Y}}(\min_{y \in Y}(d(\hat{y}, y))) \tag{4}$$

where y represents a point in Y and $\hat{y}$ represents a point in $\hat{Y}$. $H(Y, \hat{Y})$ is the overall modified Hausdorff distance, whereas $h(Y, \hat{Y})$ and $h(\hat{Y}, Y)$ are directed Hausdorff distances. $d(y, \hat{y})$ and $d(\hat{y}, y)$ are Euclidean distances. Smaller the Hausdorff distance indicates better segmentation.

Top-N Accuracy. Top-N accuracy measures how often your true class falls within your top N highest softmax outputs. This metric reflects meaning in the outputs that were not the selected class. For instance, higher Top-2 and Top-3 predictions can show that a well-calibrated makes reasonable mistakes that are supported by the data, rather than random misclassifications.

Calibration Curves. show the relationship between the predicted probability estimates and the true correctness likelihood. These plots are meant for binary classification, so for segmentation one class "positive" is compared to the rest "negative". The prevalence of positive classes is compared to predicted certainty for that class. Perfect calibration is a straight line from the origin to (1,1).

3.3 Calibrated Models Outperform UNETR-Base on 11-Classes

Qualitative Analysis. Figure 2 shows that UNETR-HC best captures the fine detail of the boundary between GM and CSF. This observation is noticeable in the upper left and upper right "grooves" in the light blue (CSF) color. UNETR-HC attempts to tract out these regions and label them as CSF, whereas the UNETR-Base and UNETR-CM assign more of these pixels as GM. This boundary is a major challenge in automatic segmentation due to partial volume effects.

Quantitative Comparison. Figure 3 and Table 2 show the Dice, Hausdorff, and Top-N. UNETR-CM performs best in Dice and Top-N accuracy, whereas UNETR-CM and UNETR-HC outperform UNETR-Base in Hausdorff. Hence, UNETR-CM classifies the most pixels correctly, whereas both models capture tissue boundaries.

Table 2. Top-N accuracy on 11 classes

Method	Top-1	Top-2	Top-3
UNETR-Base	0.876	0.979	0.990
UNETR-HC	0.891	0.984	0.993
UNETR-CM	**0.895**	**0.986**	**0.996**

3.4   Calibrated UNETR Outperforms or Performs Comparably to Headreco in 6-Class Segmentation

Qualitative Analysis. We compare 6-classes because the current field standard in head segmentation (e.g., Headreco) provides different tissues than our method. For example, Headreco [14] uses 8 tissues and SPM uses 6 tissues. Thus, we had to combine tissues into groups for a fair comparison. We combine DOMINO classes that are subsets of Headreco classes; for

Table 3. Top-N accuracy on 6 classes

Method	Top-1	Top-2	Top-3
Headreco	0.905	0.977	0.983
UNETR-Base	0.913	0.993	0.998
UNETR-HC	0.924	0.995	0.998
UNETR-CM	**0.928**	**0.996**	**0.999**

example, cancellous and cortical bone are both labeled as bone. Figure 4 shows the results for our models and Headreco. Differences are highlighted with white rectangles. Our methods show comparable or superior performance to Headreco across all tissue types.

Quantitative Comparison. Figure 5 and Table 3 show the Dice, Hausdorff, and top-1/2/3 accuracy on 6-classes. Calibrated UNETR is comparable to Headreco in WM, GM, and CSF; our models outperform Headreco in Air, Bone, and Soft tissue. UNETR-HC's Hausdorff shows that the regularization can improve 6-class

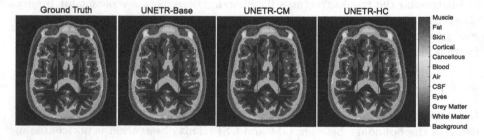

Fig. 2. Sample image slice for 11-tissue segmentation. The red squares show that UNETR-HC captures the GM - CSF boundary better than other methods (Color figure online)

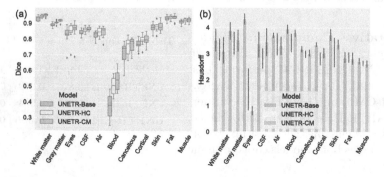

Fig. 3. (a) Dice scores and (b) Hausdorff distances in 11-class segmentation.

segmentation without retraining. UNETR-CM performs the best in Top-1/2/3 accuracy. Figure 6 shows that DOMINO achieves better calibration than UNETR-Base. All algorithms are approximately evenly calibrated on GM and air. Our methods are better calibrated than Headreco on WM, CSF, bone, and soft tissue.

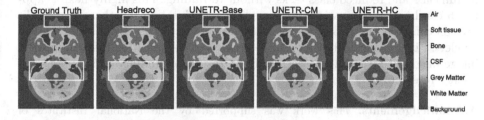

Fig. 4. Sample image slice for 6-tissue segmentation. The white squares highlight important regions where our methods outperformed Headreco

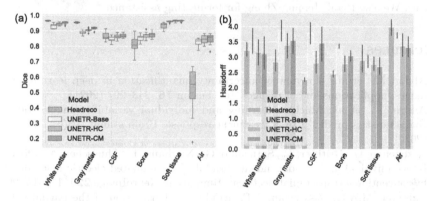

Fig. 5. (a) Dice scores and (b) Hausdorff distances in 6-class segmentation.

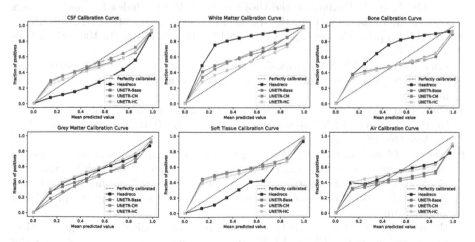

Fig. 6. Calibration curves for 6-class problem.

4   Conclusions

There is often a trade-off between performance and calibration. This work proposes a novel domain-aware calibration method that improves model calibration, top-N accuracy, and segmentation metrics. The calibrated models perform well on full class and reduced class tasks without retraining. This highly-flexible approach can be applied to widespread medical segmentation. Further, model calibration can help improve cross-talk between automated algorithms and manual labelers. Finally, our calibration can be applied to classification tasks in medical image diagnosis. We will release DOMINO to the community to support open science research.

Acknowledgements. This work was supported by the National Institutes of Health/National Institute on Aging (NIA RF1AG071469, NIA R01AG054077), the National Science Foundation (1908299), and the NSF-AFRL INTERN Supplement (2130885). We acknowledge NVIDIA AI Technology Center (NVAITC) for their suggestions. We also thank Jiaqing Zhang for formatting assistance.

References

1. Abdar, M., et al.: A review of uncertainty quantification in deep learning: techniques, applications and challenges. Inf. Fusion **76**, 243–297 (2021)
2. Albizu, A., et al.: Machine learning and individual variability in electric field characteristics predict TDCS treatment response. Brain stimul. **13**(6), 1753–1764 (2020)
3. Antonenko, D., Grittner, U., Saturnino, G., Nierhaus, T., Thielscher, A., Flöel, A.: Inter-individual and age-dependent variability in simulated electric fields induced by conventional transcranial electrical stimulation. NeuroImage **224**, 117413 (2021)
4. Ballester, M.A.G., Zisserman, A.P., Brady, M.: Estimation of the partial volume effect in MRI. Med. Image Anal. **6**(4), 389–405 (2002)
5. Bertels, J., et al.: Optimizing the Dice score and Jaccard index for medical image segmentation: theory and practice. In: Shen, D., et al. (eds.) MICCAI 2019. LNCS, vol. 11765, pp. 92–100. Springer, Cham (2019). https://doi.org/10.1007/978-3-030-32245-8_11
6. Consortium, M.: MONAI: medical open network for AI, March 2020. https://doi.org/10.5281/zenodo.6114127. If you use this software, please cite it using these metadata
7. Dubuisson, M.P., Jain, A.K.: A modified Hausdorff distance for object matching. In: Proceedings of 12th International Conference on Pattern Recognition, vol. 1, pp. 566–568. IEEE (1994)
8. Guo, C., Pleiss, G., Sun, Y., Weinberger, K.Q.: On calibration of modern neural networks. In: Precup, D., Teh, Y.W. (eds.) Proceedings of the 34th International Conference on Machine Learning. Proceedings of Machine Learning Research, vol. 70, pp. 1321–1330. PMLR, 06–11 August 2017. https://proceedings.mlr.press/v70/guo17a.html
9. Hatamizadeh, A., et al.: UNETR: transformers for 3D medical image segmentation. In: Proceedings of the IEEE/CVF Winter Conference on Applications of Computer Vision, pp. 574–584 (2022)

10. Huttenlocher, D.P., Klanderman, G.A., Rucklidge, W.J.: Comparing images using the Hausdorff distance. IEEE Trans. Pattern Anal. Mach. Intell. **15**(9), 850–863 (1993)
11. Indahlastari, A., et al.: Individualized tDCS modeling predicts functional connectivity changes within the working memory network in older adults. Brain Stimulation **14**(5), 1205–1215 (2021)
12. Indahlastari, A., et al.: Modeling transcranial electrical stimulation in the aging brain. Brain stimul. **13**(3), 664–674 (2020)
13. Jadon, S.: A survey of loss functions for semantic segmentation. In: 2020 IEEE Conference on Computational Intelligence in Bioinformatics and Computational Biology (CIBCB), pp. 1–7. IEEE (2020)
14. Nielsen, J.D., et al.: Automatic skull segmentation from MR images for realistic volume conductor models of the head: assessment of the state-of-the-art. Neuroimage **174**, 587–598 (2018)
15. Papyan, V., Han, X., Donoho, D.L.: Prevalence of neural collapse during the terminal phase of deep learning training. Proc. Natl. Acad. Sci. **117**(40), 24652–24663 (2020)
16. Taghanaki, S.A., et al.: Combo loss: handling input and output imbalance in multiorgan segmentation. Comput. Med. Imaging Graph. **75**, 24–33 (2019)
17. Wilke, M., Schmithorst, V., Holland, S.: Normative pediatric brain data for spatial normalization and segmentation differs from standard adult data. Magn. Reson. Med. **50**(4), 749–757 (2003)
18. Wolf, T., et al.: Huggingface's transformers: state-of-the-art natural language processing (2020)

iSegFormer: Interactive Segmentation via Transformers with Application to 3D Knee MR Images

Qin Liu, Zhenlin Xu, Yining Jiao, and Marc Niethammer[✉]

Department of Computer Science, University of North Carolina at Chapel Hill,
Chapel Hill, USA
{qinliu19,zhenlinx,jyn,mn}@cs.unc.edu

Abstract. Interactive image segmentation has been widely applied to obtain high-quality voxel-level labels for medical images. The recent success of Transformers on various vision tasks has paved the road for developing Transformer-based interactive image segmentation approaches. However, these approaches remain unexplored and, in particular, have not been developed for 3D medical image segmentation. To fill this research gap, we investigate Transformer-based interactive image segmentation and its application to 3D medical images. This is a nontrivial task due to two main challenges: 1) limited memory for computationally inefficient Transformers and 2) limited labels for 3D medical images. To tackle the first challenge, we propose iSegFormer, a memory-efficient Transformer that combines a Swin Transformer with a lightweight multilayer perceptron (MLP) decoder. To address the second challenge, we pretrain iSegFormer on large amount of unlabeled datasets and then finetune it with only a limited number of segmented 2D slices. We further propagate the 2D segmentations obtained by iSegFormer to unsegmented slices in 3D images using a pre-existing segmentation propagation model pretrained on videos. We evaluate iSegFormer on the public OAI-ZIB dataset for interactive knee cartilage segmentation. Evaluation results show that iSegFormer outperforms its convolutional neural network (CNN) counterparts on interactive 2D knee cartilage segmentation, with competitive computational efficiency. When propagating the 2D interactive segmentations of 5 slices to other unprocessed slices within the same 3D volume, we achieve 82.2% Dice score for 3D knee cartilage segmentation. Code is available at https://github.com/uncbiag/iSegFormer.

1 Introduction

Deep learning-based approaches for semantic and instance segmentation have achieved tremendous success for medical images [1,2]. However, these approaches are data hungry and heavily rely on the availability of large-scale voxel-level segmentations, which to obtain requires significant labor, time, and expertise [3]. Interactive image segmentation, which aims to extract image objects using limited human interactions, is an

Supplementary Information The online version contains supplementary material available at https://doi.org/10.1007/978-3-031-16443-9_45.

efficient way to obtain these segmentations [4]. Hence, significant work is ongoing to explore interactive segmentation approaches [4–6].

Existing state-of-the-art interactive segmentation methods are all CNN-based, leveraging the good representation ability of CNNs [6,7]. Although these CNN-based methods have achieved excellent performance, they suffer from limited receptive fields and cannot learn global and long-range semantic information well due to the inductive bias of locality and weight sharing [8]. Several techniques have been proposed to address these problems, such as atrous convolutional layers [9] and non-local blocks [10]. A recent research direction is to replace CNN with vision Transformer (ViT), which can naturally capture long-range dependencies through the self-attention mechanism [11, 12]. Following this direction, various vision Transformers have been proposed for medical image segmentation [13], paving the road for developing Transformer-based interactive image segmentation approaches. However, these approaches remain unexplored and, in particular, have not been developed for 3D medical images.

In this work, we aim to fill this research gap by investigating Transformer-based interactive image segmentation and its application to 3D medical images. This is a challenging task due to: 1) limited memory for computationally inefficient Transformers and 2) limited labels for 3D medical images. To tackle the first challenge, we propose iSegFormer, a memory-efficient Transformer that combines a Swin Transformer with a lightweight multilayer perceptron (MLP) decoder. With the efficient Swin Transformer blocks for hierarchical self-attention and the simple MLP decoder for aggregating both local and global attention, iSegFormer learns powerful representations while achieving high computational efficiencies. To address the second challenge, we pretrain iSegFormer on large amount of unlabeled datasets and then finetune it with only a limited number of segmented 2D slices. To extend iSegFormer to 3D interactive segmentation, we further combine it with a segmentation propagation module that propagates segmented 2D slices to unlabeled ones in the same image volume. When the propagated segmentations are not as desired, the user can refine them and start a new round of propagation if necessary. Specifically, we combine iSegFormer with STCN [14], which achieves state-of-the-art results on interactive video object segmentation. We use a pretrained STCN model without finetuning on medical images.

We evaluate iSegFormer on the public OAI-ZIB dataset for interactive knee cartilage segmentation. Evaluation results show that iSegFormer outperforms its convolutional neural network (CNN) counterparts on interactive 2D knee cartilage segmentation, with competitive computational efficiency. When propagating 2D interactive segmentation of 5 slices to other unprocessed slices within the same 3D volume, the propagation model achieves a Dice score of 82.2% for 3D knee cartilage segmentation. Finally, we show that iSegFormer combined with the segmentation propagation model results in an efficient framework for interactive 3D medical image segmentation.

Our contributions are as follows:

1) We propose iSegFormer, a memory-efficient Transformer that combines a Swin Transformer with a lightweight MLP decoder, for interactive image segmentation.
2) iSegFormer outperforms its CNN counterparts for interactive 2D knee cartilage segmentation on the OAI-ZIB dataset with comparable computational efficiency

with CNNs. To the best of our knowledge, iSegFormer is the first Transformer-based approach for interactive medical image segmentation.
3) We further show that iSegFormer can be easily extended to interactive 3D medical image segmentation by combining it with a pre-existing segmentation propagation model trained on videos.

2   Related Work

Interactive Medical Image Segmentation. Existing interactive segmentation methods for medical images are all CNN-based [15–18], partially inspired by the seminal work [4]. MIDeepSeg [18] proposes a click-based approach that encodes foreground and background clicks through Gaussian-smoothed maps, which serve as the input to the CNN encoder-decoder. Recently, MONAI Label [15] proposes an open-source framework for CNN-based interactive segmentation of 3D medical images, which consists of both click-based and scribbles-based interactive segmentation algorithms. In this work, we are interested in Transformer-based interactive segmentation, which has not been well-explored, especially for medical images. Given the recent success of vision Transformers for automatic medical image segmentation [19], it is a natural extension for applying Transformers to interactive medical image segmentation. Specifically, we apply to the challenging knee cartilage segmentation from MR images [20].

Vision Transformers. The vision Transformer (ViT) [21] first shows that a pure Transformer can achieve state-of-the-art performance for image classification. Pyramid vision Transformer [22] further shows that ViT can also achieve comparable performance with CNNs in dense prediction tasks. SegFormer [12] proposes an efficient segmentation approach that combines a hierarchically structured Transformer encoder with a light-weight decoder using MLPs, demonstrating the state-of-the-art segmentation performance compared with CNNs. Swin Transformer [23] is a breakthrough that shows the superiority of hierarchical vision Transformer over CNN as a general vision backbone. Meanwhile, different ViTs have also been proposed in automatic medical image segmentation [13,24,25]. Among these methods, Swin-Unet [13] and UTNet [25] are "U-shaped" networks inspired by Unets [26]. Comparing with these methods, iSegFormer is more efficient considering its efficient Swin Transformer encoder and light-wighted MLP decoder.

Interactive Video Object Segmentation (iVOS). iVOS aims at extracting high-quality segmentation masks of a target video object through two modules: a 2D interactive segmentation module and a segmentation propagation module [27,28]. MiVOS [28] decouples the two modules and train them independently. During inference, MiVOS first interactively segments one or several frames in a video, followed by propagating the segmented frames to the unsegmented ones. STCN [14] further improves the segmentation propagation module in MiVOS by directly encoding the query and memory frames without re-encoding the mask features for every object. Inspired by the observation that iVOS pipeline can be directly applied to 3D medical images, we extend iSegFormer to interactive 3D medical image segmentation by combining it with an existing STCN model trained on videos, leading to a very promising results even without fine-tuning.

3   Method

The proposed iSegFormer is a Transformer-based interactive 2D image segmentation approach that combines a Swin Transformer with a lightweight MLP decoder. As shown in Fig. 1, it can be easily extended to a 3D interactive segmentation approach by combining it with a segmentation propagation module (i.e., STCN [14]). This 3D interactive segmentation approach consists of an iSegFormer for obtaining 2D segmentation from user interactions and a segmentation propagation module that propagates segmented slices to unsegmented ones, resulting in a 3D segmentation. If the propagated segmentation results are not desired, the user can refine them with further interactions and start a new round of propagation if necessary.

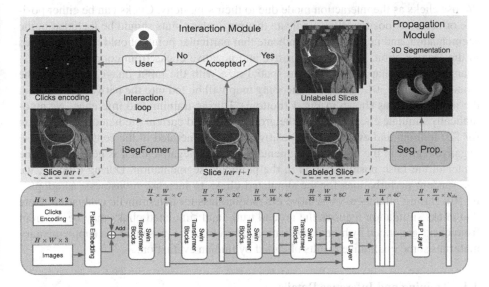

Fig. 1. Illustration of iSegFormer for interactive 3D medical image segmentation. Initially, the user selects and interactively annotates one slice (i.e., the user clicks on points) to produce a segmentation, followed by propagating the segmented slice to the unsegmented ones. The iSegFormer architecture is shown at the bottom. The output of iSegFormer will be upsampled to the orignal image size.

3.1   Network Architecture of iSegFormer

The network architecture of iSegFormer is shown in Fig. 1 (bottom). It uses a Swin Transformer as the segmentation backbone and two light-weight MLP layers as the decoder to produce segmentation. Specifically, there are four Swin Transformer blocks for hierarchical self-attention and a simple MLP decoder that first aggregates both local and global attention and then produces the segmentation as the output. The input of iSegFormer is the concatenation of image and clicks encoding map (introduced in Sect. 3.2). Since we want to make use of existing pretrained Swin Transformer models on ImageNet-21 k [29], we do not change the number of input channels of the Swin

Transformer blocks. To achieve this, we use element-wise addition instead of concatenation for merging image features and clicks encoding features after the patch embedding layers, which are linear projection layers that produce patch embeddings for self-attention. Note that there are two separate patch embedding layers in iSegFormer (one for the input image and the other for the clicks encoding map), though Fig. 1 only shows one for brevity. The clicks embedding is essential for extending a segmentation model to an interactive segmentation model as it transforms user's interactions from clicks to feature maps that can be fed into the network. For medical images which typically only have one gray channel, we simply replicate the gray channel to RGB format.

3.2 Clicks Encoding and Simulation

We use clicks as the interaction mode due to their simplicity. Clicks can be either positive or negative: positive clicks indicate that particular points should be included in the segmentation, and negative clicks indicate that particular points should not be included in the segmentation. We encode positive and negative clicks from coordinates to a 2-channel feature map with the same spatial size with the input image, following the strategy used in [30]. The clicks encoding map will be fed into the network along with the input image, as shown in Fig. 1 (bottom). During training and inference, we automatically simulate clicks based on the ground truth and current predicted segmentation for fast training and evaluation. A positive click is generated in the center of the false negative region in the predicted segmentation, and a negative click is generated in the center of the false positive region in the predicted segmentation. During training, we add random perturbations for the simulated clicks to increase robustness, as adopted in [6]. During inference, we remove the randomness for determinstic evaluation. Note that clicks simulation requires the ground truth, and simulated clicks may be different from clicks generated by human evaluation. Therefore, we present in the supplementary materials some qualitative results obtained by human evaluation.

3.3 Training and Inference Details

For fair comparison with RITM [6], we adopt most of the hyper-parameters used in RITM for training and inference. The iSegFormer models are trained in a class-agnostic binary segmentation task with the normalized focal loss function (NFL) [31]. We randomly crop the image to the size of 320×480 for training. We adopt the same data augmentation techniques with RITM [6] including random scaling and resizing. We implement iSegFormer using Pytorch with Adam optimizer. All experiments are conducted on a NVIDIA A6000 GPU. All models are trained 55 epochs with batch size as 32 (except the SegFormer and HRFormer models in Fig. 3). More details please refer to our codebase.

3.4 Extending to Interactive 3D Image Segmentation

iSegFormer can be easily extended to a 3D interactive segmentation approach by combining it with a segmentation propagation module (i.e., STCN [14]). Since this is not our main contribution, we introduce the details in the appendix.

4   Experiments

Datasets. The OAI-ZIB [32] dataset consists of 507 3D MR images with segmentations for femur, tibia, tibial cartilage, and femoral cartilage. In this work, we only consider cartilage segmentation. Each 3D image contains 160 slices of size of 384×384. We split the dataset randomly into 407 images for training, 50 images for validation, and 50 images for testing. Since we are interested in the problem setting where the segmentations for the 3D images are limited, we only use three segmented slices of each image in the training and validation sets for developing iSegFormer, resulting in 1521 training slices, 150 validation slices, and 150 testing slices. The three slices are selected at a fixed interval (i.e., slice 40, 80, and 120). We also use 9 other public datasets in our cross-domain evaluation experiments. Please refer to Sect. 4.1 for details.

Evaluation Metrics. We use Number of Clicks (NoC) to measure the number of clicks required to achieve a predefined Intersection over Union (IoU) between predicted and ground truth segmentations. For example, NoC@85% measures the number of clicks required to obtain 85% IoU. We use an automatic evaluation procedure to simulate clicks during inference and report the quantitative results, following the practices used in [6]. We also perform a human evaluation for a qualitative study. For measuring the 3D segmentation results, we use the Dice Similarity Coefficient (DSC), sensitivity (SEN), and the positive predictive value (PPV).

4.1   Results of Interactive 2D Image Segmentation

We compare iSegFormer with RITM [6], the state-of-the-art CNN-based approach for interactive 2D femoral and tibial cartilage segmentation. Both RITM and iSegFormer are implemented on two segmentation backbones. For RITM, the backbones are UNet [26] and HRNet32 [33]. For iSegFormer, the backbones are Swin Transformer's base and large models. For fair comparison, all the models are trained on the OAI-ZIB training set under the same training settings.

Table 1. Evaluation on the OAI-ZIB test set for femoral and tibial cartilage segmentation. "Mem" denotes the GPU memory consumption for inference. "SPC" represents second per click. "≥20@85" measures the number of difficult cases that require more than 20 clicks to achieve 85% IoU. We report mean and standard deviation for the NoC metrics.

Model	Mem (M)	SPC (ms)	NoC@80%	NoC@85%	NoC@90%	≥20@85	≥20@90
RITM-UNet	2680	56	9.74 (8.76)	15.28 (7.39)	17.79 (3.54)	102	144
RITM-HR32	2763	82	8.47 (8.12)	14.48 (7.82)	18.85 (2.47)	94	138
Ours-swin-B	2797	64	7.25 (7.74)	**11.67** (8.19)	**17.03** (6.05)	**68**	**115**
Ours-swin-L	3755	89	**6.91** (7.59)	11.77 (8.26)	17.57 (5.45)	70	123

Table 1 reports the comparison results for tibial and femoral cartilage segmentation on the 150 slices of the OAI-ZIB testing set. The results show that iSegFormer outperforms its CNN counterparts with very competitive speed and GPU memory consumption, demonstrating the effectiveness and efficiency of iSegFormer for interactive segmentation.

Comparison with Other Transformer Backbones. To further demonstrate the efficiency of iSegFormer, we also implemented iSegFormer using two recently proposed Transformer backbones for segmentation: HRFormer [11] and SegFormer [12]. As shown in Table 3, our proposed Swin Transformer-based segmentation backbone is much more memory-efficient than the other Transformer-based backbones.

Cross-Domain Evaluation. We have shown that iSegFormer outperforms CNNs when trained with only 1,221 labeled 2D slices (labeling such a dataset amounts to labeling 8 3D images with 160 slices). However, in many applications no segmented slices are available, for example, when studying new medical image datasets. Therefore, it is important to generalize the trained interactive segmentation models to unseen objects or objects in different domains. In this cross-domain evaluation, we train iSegFormer and RITM models on the COCO+LVIS [34] dataset, which contains millions of high-quality labels for natural images. Then we test the model on 5 natural image datasets (GrabCut, Berkeley, DAVIS, PascalVOC, and SBD) and 3 medical image datasets (ssTEM, BraTS, and OAI-ZIB).

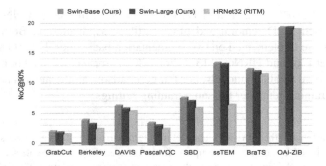

Fig. 2. Cross-domain evaluation results. Models are trained on the COCO+LVIS dataset and tested on 5 natural datasets (GrabCut, Berkeley, DAVIS, PascalVOC, and SBD) and 3 medical image datasets (ssTEM, BraTS, and OAI-ZIB).

The results are shown in Fig. 2. Although there is still a significant performance gap between in-domain and out-of-domain evaluations, both CNN and Transformer models generalize reasonably well to medical image datasets. Note that our models do not outperform the CNN counterpart in this experiment. We argue that HRNet is the best performing model in RITM with well-tuned hyper parameters, while we adopt most of their hyper parameters for Transformers and spend little effort in tuning them.

4.2 Results on Segmentation Propagation

Given the interactively segmented 2D slices obtained by iSegFormer, we now interested in 3D segmentation via a segmentation propagation model released by STCN [14]. The results in Table 2 show that with more segmented slices, the propagation results get better. With only 5 segmented slices, it achieves a Dice score of 82.2% for femoral cartilage segmentation. This is a very promising result considering that the segmentation propagation model was not trained on the medical images. We hope this preliminary experiment would attract more research effort in transferring knowledge from video domain to the medical imaging domain.

Table 2. Femoral cartilage segmentation using 1, 3, 5, or 10 segmented slices for propagation. The propagation model is trained on video.

#Slices	DSC (%)	SEN (%)	PPV (%)	IoU (%)
1	55.1	61.7	55.3	38.0
3	78.7	85.8	73.1	64.9
5	82.2	87.1	77.9	70.6
10	85.1	88.1	88.3	74.1

Table 3. Memory and speed comparison between different segmentation backbones.

Backbone	Params (M)	Mem	SPC
HRNet32	41	2763 M	82 ms
SegFormer-B5	28	>5 G	>0.2 s
HRFormer-B	50	>5 G	>0.2 s
Swin-B	88	2797 M	64 ms
Swin-L	197	3755 M	89 ms

4.3 Ablation Study

We demonstrated in Sect. 4.1 that iSegFormer performed better than its CNN counterparts. Other than the architecture difference, the biggest difference comes from the pretraining settings. In Sect. 4.1, our iSegFormer models are pretrained on ImageNet-21 k, while the CNN models have two pretraining steps: first pretrained on ImageNet-21 k and then finetuned on the COCO+LVIS dataset. In this study, we adopt different pretrain settings for a more fair comparison between Transformer and CNN models. Note that the pre-training task can be either classification (Cls) or interactive segmentation (iSeg). As shown in Table 4, pretraining on Image21k is essential for the success of iSegFormer. More details are included in the supplementary materials.

Table 4. Ablation study on pretraining strategies. The pretraining task can be either classification (Cls) or interactive segmentation (iSeg).

Pretrain dataset	Pretrain task	Fine tune	Swin-B		Swin-L		HRNet32	
			NoC@80	NoC@85	NoC@80	NoC@85	NoC@80	NoC@85
N/A	N/A	✓	19.69	19.99	18.85	19.91	15.47	18.99
ImageNet-21K	Cls	✓	7.25	11.67	6.91	11.77	17.19	19.61
COCO+LVIS	iSeg	✗	15.48	17.19	15.09	17.45	14.58	16.93
COCO+LVIS	iSeg	✓	12.11	15.73	9.00	13.29	8.47	14.48
OAI (w/o GT)	iSeg	✗	18.49	19.65	18.89	19.72	16.35	18.48
OAI (w/o GT)	iSeg	✓	12.67	15.87	13.01	16.41	7.93	12.81

5   Conclusion

We proposed iSegFormer, a memory-efficient Transformer that combined a Swin Transformer with a lightweight multilayer perceptron (MLP) decoder. iSegFormer outperformed its CNN counterparts on the interactive 2D knee cartilage segmentation while achieving comparable computational efficiency with CNNs. We further extended iSegFormer for interactive 3D knee cartilage segmentation by combining it with a pre-existing segmentation propagation model trained on videos, achieving promising results even without finetuning the segmentation propagation model.

Acknowledgement. Research reported in this publication was supported by the National Institutes of Health (NIH) under award number NIH 1R01AR072013. The content is solely the responsibility of the authors and does not necessarily represent the official views of the NIH. Data related to knee osteoarthritis used in the preparation of this manuscript were obtained from the controlled access datasets distributed from the Osteoarthritis Initiative (OAI), a data repository housed within the NIMH Data Archive (NDA). OAI is a collaborative informatics system created by the National Institute of Mental Health and the National Institute of Arthritis, Musculoskeletal and Skin Diseases (NIAMS) to provide a worldwide resource to quicken the pace of biomarker identification, scientific investigation and OA drug development. Dataset identifier(s): NIMH Data Archive Collection ID: 2343.

References

1. Wang, R., Lei, T., Cui, R., Zhang, B., Meng, H., Nandi, A.K.: Medical image segmentation using deep learning: a survey. In: IET Image Processing, Jan 2022
2. Shen, D., Wu, G., Suk, H.-I.: Deep learning in medical image analysis. Annu. Rev. Biomed. Eng. **19**, 221–248 (2017)
3. Tajbakhsh, N., Jeyaseelan, L., Li, Q., Chiang, J.N., Wu, Z., Ding, X.: Embracing imperfect datasets: A review of deep learning solutions for medical image segmentation. Med. Image Anal. **63**, 101693 (2020)
4. Xu, N., Price, B., Cohen, S., Yang, J., Huang, T.S.: Deep interactive object selection. In: CVPR, pp. 373–381 (2016)
5. Xu, N., Price, B., Cohen, S., Yang, J., Huang, T.: Deep grabcut for object selection. arXiv preprint arXiv:1707.00243 (2017)
6. Sofiiuk, K., Petrov, I.A., Konushin, A.: Reviving iterative training with mask guidance for interactive segmentation. arXiv preprint arXiv:2102.06583 (2021)
7. Zhang, S., Liew, J.H., Wei, Y., Wei, S., Zhao, Y.: Interactive object segmentation with inside-outside guidance. In: CVPR, pp. 12234–12244 (2020)
8. Cohen, N., Shashua, A.: Inductive bias of deep convolutional networks through pooling geometry. arXiv preprint arXiv:1605.06743 (2016)
9. Chen, L.-C., Papandreou, G., Kokkinos, I., Murphy, K., Yuille, A.L.: Deeplab: Semantic image segmentation with deep convolutional nets, atrous convolution, and fully connected crfs. IEEE Trans. Pattern Anal. Mach. Intell. **40**(4), 834–848 (2017)
10. Wang, X., Girshick, R., Gupta, A., He, K.: Non-local neural networks. In: CVPR, pp. 7794–7803 (2018)
11. Yuan, Y., et al.: Hrformer: high-resolution vision transformer for dense predict. Adv. Neural. Inf. Process. Syst. **34**, 7281–7293 (2021)

12. Xie, E., Wang, W., Yu, Z., Anandkumar, A., Alvarez, J.M., Luo, P.: Segformer: simple and efficient design for semantic segmentation with transformers. Adv. Neural. Inf. Process. Syst. **34**, 12077–12090 (2021)

13. Cao, H.: Swin-unet: Unet-like pure transformer for medical image segmentation. arXiv preprint arXiv:2105.05537 (2021)

14. Cheng, H.K., Tai, Y.-W., Tang, C.-K.: Rethinking space-time networks with improved memory coverage for efficient video object segmentation. In: Advances in Neural Information Processing Systems, vol. 34 (2021)

15. Diaz-Pinto, A., et al.: Monai label: a framework for ai-assisted interactive labeling of 3d medical images. arXiv preprint arXiv:2203.12362 (2022)

16. Chao, C.-H., Cheng, H.-T., Ho, T.-Y., Lu, L., Sun, M.: Interactive radiotherapy target delineation with 3d-fused context propagation. arXiv preprint arXiv:2012.06873 (2020)

17. Sakinis, T.: Interactive segmentation of medical images through fully convolutional neural networks. arXiv preprint arXiv:1903.08205 (2019)

18. Luo, X., et al.: Mideepseg: Minimally interactive segmentation of unseen objects from medical images using deep learning. Med. Image Anal. **72**, 102102 (2021)

19. Chen, J.: Transunet: transformers make strong encoders for medical image segmentation. arXiv preprint arXiv:2102.04306 (2021)

20. Liu, Q., et al.: Multi-class gradient harmonized dice loss with application to knee MR image segmentation,. In: Shen, D., et al. (eds.) MICCAI 2019. LNCS, vol. 11769, pp. 86–94. Springer, Cham (2019). https://doi.org/10.1007/978-3-030-32226-7_10

21. Dosovitskiy, A., et al.: An image is worth 16×16 words: Transformers for image recognition at scale. arXiv preprint arXiv:2010.11929 (2020)

22. Wang, W., et al.: Pyramid vision transformer: a versatile backbone for dense prediction without convolutions. In: ICCV, pp. 568–578 (2021)

23. Liu, Z.: Swin transformer: hierarchical vision transformer using shifted windows. In: ICCV, pp. 10012–10022 (2021)

24. Zhang, Y., Liu, H., Hu, Q.: TransFuse: fusing transformers and CNNs for medical image segmentation. In: de Bruijne, M., et al. (eds.) MICCAI 2021. LNCS, vol. 12901, pp. 14–24. Springer, Cham (2021). https://doi.org/10.1007/978-3-030-87193-2_2

25. Gao, Y., Zhou, M., Metaxas, D.N.: UTNet: a hybrid transformer architecture for medical image segmentation. In: de Bruijne, M., et al. (eds.) MICCAI 2021. LNCS, vol. 12903, pp. 61–71. Springer, Cham (2021). https://doi.org/10.1007/978-3-030-87199-4_6

26. Ronneberger, O., Fischer, P., Brox, T.: U-Net: convolutional networks for biomedical image segmentation. In: Navab, N., Hornegger, J., Wells, W.M., Frangi, A.F. (eds.) MICCAI 2015. LNCS, vol. 9351, pp. 234–241. Springer, Cham (2015). https://doi.org/10.1007/978-3-319-24574-4_28

27. Oh, S.W., Lee, J.-Y., Xu, N., Kim, S.J.: Fast user-guided video object segmentation by interaction-and-propagation networks. In: CVPR, pp. 5247–5256 (2019)

28. Cheng, H.K., Tai, Y.-W., Tang, C.-K.: Modular interactive video object segmentation: Interaction-to-mask, propagation and difference-aware fusion. In: CVPR, pp. 5559–5568 (2021)

29. Deng, J., Dong, W., Socher, R., Li, L.-J., Li, K., Fei-Fei, L.: Imagenet: a large-scale hierarchical image database. In: CVPR, pp. 248–255. IEEE (2009)

30. Benenson, R., Popov, S., Ferrari, V.: Large-scale interactive object segmentation with human annotators. In: CVPR, pp. 11700–11709 (2019)

31. Sofiiuk, K., Barinova, O., Konushin, A.: Adaptis: adaptive instance selection network. In: ICCV, pp. 7355–7363 (2019)

32. Ambellan, F., Tack, A., Ehlke, M., Zachow, S.: Automated segmentation of knee bone and cartilage combining statistical shape knowledge and convolutional neural networks: Data from the osteoarthritis initiative. Med. Image Anal. **52**, 109–118 (2019)

33. Wang, J., et al.: Deep high-resolution representation learning for visual recognition. IEEE Trans. Pattern Anal. Mach. Intell. (2020)
34. Lin, T.-Y., et al.: Microsoft COCO: common objects in context. In: Fleet, D., Pajdla, T., Schiele, B., Tuytelaars, T. (eds.) ECCV 2014. LNCS, vol. 8693, pp. 740–755. Springer, Cham (2014). https://doi.org/10.1007/978-3-319-10602-1_48

Patcher: Patch Transformers with Mixture of Experts for Precise Medical Image Segmentation

Yanglan Ou[1]([✉]), Ye Yuan[2], Xiaolei Huang[1], Stephen T. C. Wong[3],
John Volpi[4], James Z. Wang[1], and Kelvin Wong[3]

[1] The Pennsylvania State University, University Park, State College, PA, USA
yxo43@psu.edu
[2] Carnegie Mellon University, Pittsburgh, PA, USA
[3] TT and WF Chao Center for BRAIN and Houston Methodist Cancer Center,
Houston Methodist Hospital, Houston, TX, USA
[4] Eddy Scurlock Comprehensive Stroke Center, Department of Neurology,
Houston Methodist Hospital, Houston, TX, USA

Abstract. We present a new encoder-decoder Vision Transformer architecture, Patcher, for medical image segmentation. Unlike standard Vision Transformers, it employs Patcher blocks that segment an image into large patches, each of which is further divided into small patches. Transformers are applied to the small patches within a large patch, which constrains the receptive field of each pixel. We intentionally make the large patches overlap to enhance intra-patch communication. The encoder employs a cascade of Patcher blocks with increasing receptive fields to extract features from local to global levels. This design allows Patcher to benefit from both the coarse-to-fine feature extraction common in CNNs and the superior spatial relationship modeling of Transformers. We also propose a new mixture-of-experts (MoE) based decoder, which treats the feature maps from the encoder as experts and selects a suitable set of expert features to predict the label for each pixel. The use of MoE enables better specializations of the expert features and reduces interference between them during inference. Extensive experiments demonstrate that Patcher outperforms state-of-the-art Transformer- and CNN-based approaches significantly on stroke lesion segmentation and polyp segmentation. Code for Patcher is released to facilitate related research. (Code: https://github.com/YanglanOu/patcher.git.).

Keywords: Medical image segmentation · Vision transformers · Mixture of experts

Supplementary Information The online version contains supplementary material available at https://doi.org/10.1007/978-3-031-16443-9_46.

1   Introduction

Deep learning-based medical image segmentation has many important applications in computer-aided diagnosis and treatment. Until recently, the field of medical image segmentation has mainly been dominated by convolutional neural networks (CNNs). U-Net and its variants [2,3,8,12,15,21] are a representative class of CNN-based models, which are often the preferred networks for image segmentation. These models mainly adopt an encoder-decoder architecture, where an encoder uses a cascade of CNN layers with increasing receptive fields to capture both local and global features while a decoder leverages skip-connections and deconvolutional layers to effectively combine the local and global features into the final prediction. Despite the tremendous success of CNN-based models, their drawbacks start to become apparent as their performance saturates. First, CNNs are suboptimal at modeling global context. While increasing the depth of CNN-based models can enlarge the receptive fields, it also leads to problems such as diminishing feature reuse [16], where low-level features are diluted by consecutive multiplications. Second, the translation invariance of CNNs is a double-edged sword – it allows CNNs to generalize better, but also severely constrains their ability to reason about the spatial relationships between pixels.

Originally designed for natural language processing tasks, Transformers [17] have recently become popular in computer vision and image analysis domains since the invention of ViT [4]. By design, Vision Transformers have the ability to address the two aforementioned drawbacks of CNNs: (1) They can effectively model the global context by segmenting an image into patches and applying self-attention to them; (2) The use of positional encodings makes Transformers be able to model spatial relationships between patches. Due to these advantages, many Vision Transformer models have been proposed for image segmentation. For instance, SETR [20] employs a ViT-based encoder in an encoder-decoder architecture to attain superior performance over CNNs. Recently, Swin Transformer [10] adopts a hierarchical design to improve the efficiency of Vision Transformers and achieves SOTA results on various vision tasks including image segmentation. The success of Vision Transformers has attracted significant attention in the domain of medical image segmentation. For example, Swin-Unet [1], LambdaUNet [13], and U-NetR [5] replace convolutional layers with Transformers in a U-Net-like architecture. Other models like TransUNet [2], U-Net Transformer [14], and TransFuse [19] adopt a hybrid approach where they use Transformers to capture global context and convolutional layers to extract local features. So far, most prior works utilize Transformers mainly to extract patch-level features instead of fine pixel-level features. Given Transformers' strong abilities to model spatial relationships, we believe there is an opportunity to fully leverage Transformers for extracting fine-grained pixel-level features without delegating the extraction task to convolutional layers.

To this end, we propose a new encoder-decoder Vision Transformer architecture, Patcher, that uses Transformers for extracting fine-grained local features in addition to global features. Its key component is the Patcher block, which segments an image into large patches (e.g., 32×32), each of which is further

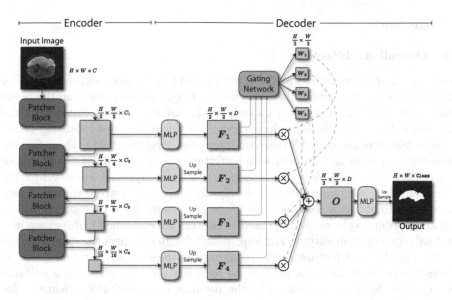

Fig. 1. Model overview of Patcher. The encoder uses a cascade of Patcher blocks to extract expert features from local to global levels. The MoE-based decoder uses a gating network to select a suitable set of expert features for the prediction of each pixel.

divided into small patches (*e.g.*, 2 × 2). Transformers are applied to the small patches within each large patch to extract pixel-level features. Each large patch constrains the receptive fields of the pixels inside, and we intentionally make the large patches overlap to enhance intra-patch communication. The encoder employs a cascade of Patcher blocks with increasing receptive fields to output a sequence of feature maps extracted from local to global levels. This design allows Patcher to combine the best of both worlds: it shares the classic coarse-to-fine feature extraction common in CNNs and also enjoys Transformers' superior spatial relationship modeling power to capture low-level details. Moreover, we observe that image segmentation models mainly require local features for some pixels (*e.g.*, edge pixels) while relying more on global features for other pixels (*e.g.*, pixels inside a global shape). This motivates us to further propose a new mixture-of-experts (MoE) based decoder. It treats the feature maps from the encoder as experts and learns a gating network to select a suitable set of expert features to predict the label for each pixel. By using MoE, the model can learn more specialized and disentangled expert feature maps and reduce interference between them during inference.

The *contributions* of this paper are as follows. First, we propose a new Vision Transformer architecture that can effectively make use of large and small patches to focus on global and local context. Second, we propose a new MoE-based decoder that enables better specialization and disentanglement of feature maps, which substantially improves performance. Lastly, extensive experiments demonstrate that our method outperforms SOTA Transformer- and CNN-based approaches significantly on stroke lesion segmentation and polyp segmentation.

2   Method

2.1   Overall Architecture

The network architecture of Patcher is outlined in Fig. 1. Given an input image of size $H \times W \times C$, Patcher first uses an encoder to extract features from the input image. The encoder contains a cascade of Transformer-based Patcher blocks (detailed in Sec. 2.2), which produce a sequence of feature maps capturing visual features from local to global levels with increasing receptive fields. These feature maps are then input to a decoder with a mixture-of-experts (MoE) design, where each of the feature maps from the encoder serves as an expert. A four-layer gating network in the decoder outputs weight maps for the expert feature maps and uses the weights to obtain a combined feature map. A multi-layer perceptron (MLP) and an up-sampling layer are then used to process the combined feature map into the final segmentation output. The MoE-based design increases the specialization of different levels of features while reducing the interference between them. It allows the network to make predictions for each pixel by choosing a suitable set of expert features. For example, the network may need global features for pixels inside a particular global shape, while it may require local features to capture fine details at segmentation boundaries. Finally, we use the standard binary cross-entropy (BCE) loss for image segmentation to train Patcher.

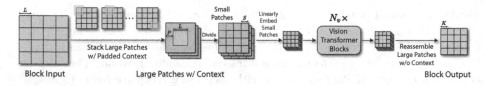

Fig. 2. Overview of the Patcher block. The input is segmented into large patches with overlapping context, each of which is further divided into small patches. The small patches are processed by a sequence of Vision Transformer blocks to extract fine-grained features. The final output is produced by reassembling the large patches.

2.2   Patcher Block

The key component inside the Patcher encoder is the Patcher block, which is a versatile Transformer-based module that can extract visual features from the block input at different spatial levels. The overview of the Patcher block is outlined in Fig. 2. We first divide the block input of spatial dimensions $H \times W$ into an $N_h \times N_w$ grid of large patches. Each of the large patches is of size $L \times L$, where $N_h = H/L$ and $N_w = W/L$. We further pad each large patch with P pixels from neighboring patches on each side, forming a patch of size $(L+2P) \times (L+2P)$. We stack the large patches with padded context along the batch dimension (batch size $B = B_0 N_h N_w$ where B_0 is the original batch size), so different large patches will not interfere with each other in the subsequent operations. Each large patch defines a receptive field similar to the kernel in CNNs, with the difference that all pixels inside the large patch share the same receptive field. Therefore, it is crucial to have the padded context since it enlarges the receptive field of pixels, which is especially important for pixels close to the patch boundaries. Next, we partition the stacked large patches further into an $M_h \times M_w$ grid of small patches, each of size $S \times S$, which have a limited receptive field as defined by the large patch, therefore focusing on local context modeling. Similar to prior work, we linearly embed all pixels inside each small patch into a token, and the tokens of all small patches form a sequence. We then use N_v Vision Transformer blocks [4] to process the sequence to model the relationship between patches and extract useful visual features. Inspired by SegFormer [18], we do not use positional encodings but mix convolutional layers inside the MLPs of the Transformer to capture spatial relationships. We also use the efficient self-attention in SegFormer to further reduce the computational cost. We provide more details of the Vision Transformer blocks in the supplementary materials. The output feature maps of the Transformer blocks have a spatial dimension of $M_h \times M_w$ with batch size B. We take the center $K \times K$ area from the feature maps where $K = L/S$, which excludes the padded context and corresponds to the actual large patches. We then reassemble the large patches based on their locations in the original image to form the final output with spatial dimensions $\frac{H}{S} \times \frac{W}{S}$. There are two important hyperparameters of the Patcher block: (1) large patch size L, which defines the receptive field and allows feature extraction at local or global levels; (2) padded context size P, which controls how much information from neighboring large patches are used.

2.3   Patcher Encoder

As shown in Fig. 1, the Patcher encoder employs a cascade of four Patcher blocks to produce four feature maps with decreasing spatial dimensions and increasing receptive fields. The small patch size S is set to 2 for all the blocks, which means the spatial dimensions are halved after each block. The large patch size L and padded context size P are set to 32 and 8, respectively. By setting L and P the same for all the blocks, we allow deeper Patcher blocks to have a larger receptive field, therefore gradually shifting the focus of the blocks from capturing local features to global features. This encoder design mirrors the behavior of its CNN counterparts such as U-Net [15], which has been proven effective. Therefore, the Patcher encoder combines the best of both worlds – it not only benefits from the superior spatial relationship modeling of Transformers but also enjoys the effective coarse-to-fine feature extraction of CNNs.

2.4   Mixture of Experts Decoder

The decoder follows an MoE design, where it treats the four feature maps from the encoder as experts. As illustrated in Fig. 1, the decoder first uses pixel-wise MLPs to process each of the feature maps and then upsamples them to the size of the first feature maps, i.e., $\frac{H}{2} \times \frac{W}{2} \times D$, where D is the number of channels after the MLPs. We use $[\boldsymbol{F}_1, \boldsymbol{F}_2, \boldsymbol{F}_3, \boldsymbol{F}_4]$ to denote the upsampled features, which are also called expert features. Next, a gating network takes the expert features as input and produces the weight maps $[W_1, W_2, W_3, W_4]$ for the expert feature maps, where each weight map is of size $\frac{H}{2} \times \frac{W}{2}$. The weight maps sum to 1 per pixel, $i.e.$, $W_1 + W_2 + W_3 + W_4 = 1$. The gating network first concatenates all the expert feature maps along channels and uses several convolutional layers and a final softmax layer to process the concatenated features into the weight maps. We then use the weight maps to produce the combined feature map O:

$$O = \sum_{i=1}^{4} W_i \odot F_i, \tag{1}$$

where $\odot$ denotes pixel-wise multiplication. The combined feature map O then passes through another MLP to predict the segmentation logits before being upsampled to the original image size. The MoE design of the decoder allows the network to learn more specialized feature maps and reduce the interference between them. For the prediction of each pixel, the gating function chooses a suitable set of features by weighing the importance of global $vs.$ local features.

3   Experiments

Datasets. We perform experiments on two important medical image segmentation tasks. (1) **Stroke lesion segmentation**: We collect 99 acute ischemic stroke cases for this research with help from Houston Methodist hospital. The dataset contains 99 cases (2,451 images) in total. Each image has two channels, the eADC and DWI, from ischemic stroke patients. We use 67 cases (1,652 images) for training, 20 cases (499 images) for validation, and 12 cases (300 images) for testing. The test data has a high diversity of lesion sizes, locations, and stroke types. (2) **Polyp segmentation**: We further conduct experiments on a public dataset of gastrointestinal polyp images, Kvasir-SEG [7]. The dataset contains 1,000 RGB images and corresponding segmentation masks. We randomly split the dataset with a ratio of 8:1:1 for training, validation, and testing.

Implementation Details. We train our models on two NVIDIA RTX 6000 GPUs using a batch size of 8, which takes around 12 h. For both stroke lesion and polyp segmentation, the input image is scaled to 256×256. During training, we randomly scale the image with a ratio from 0.7 to 2.0 and crop the image to 256×256 again. For the encoder's four Patcher blocks, we use $[64, 128, 320, 512]$ for the Transformer embedding dimensions and $[3, 6, 40, 3]$ for the number of Transformer blocks N_v. For the decoder, all the MLPs have hidden dimensions $[256, 256]$ with ReLU activations. The expert feature maps $[F_1, F_2, F_3, F_4]$ have $D = 256$ channels. The gating network uses four 3×3 convolutional layers with channels $[256, 256, 256, 4]$ and ReLU activations. For stroke lesion segmentation, we use the AdamW [11] optimizer with an initial learning rate of $6e - 5$. For polyp segmentation, we use the Adam [9] optimizer with an initial learning rate of $1e - 4$. We adopt a polynomial-decay learning rate schedule for both datasets. We also use the Intersection-over-Union (IoU) loss in addition to the BCE loss for polyp segmentation, which improves model training.

3.1   Comparison with State-of-the-Art Methods

We compare our model, Patcher, with SOTA Transformer- and CNN-based segmentation models – SETR [20], SegFormer [18], Swin Transformer [10], U-Net [15], AttnUNet [12], and TransUNet [2] – using their released code. We use two common metrics for image segmentation – dice score coefficient (DSC) and IoU.

Table 1. Quantitative comparison for stroke lesion segmentation.

Method	DSC	IoU
UNet [15]	84.54	82.07
TransUNet [2]	87.37	83.14
AttnUNet [12]	85.30	83.28
SETR [20]	82.88	77.40
SegFormer [18]	81.45	79.56
Swin transformer [10]	84.74	80.73
Patcher (ours)	**88.32**	**83.88**

Table 2. Quantitative comparison for polyp segmentation on Kvasir-SEG [7].

Method	DSC	IoU
UNet [15]	78.89	67.81
TransUNet [2]	82.80	70.86
AttnUNet [12]	77.08	61.48
SETR [20]	75.30	61.60
SegFormer [18]	87.39	78.81
Swin transformer [10]	85.58	77.64
Patcher (ours)	**90.67**	**84.31**

Quantitative Results. The results for stroke lesion segmentation and polyp segmentation are shown in Tables 1 and 2, respectively. We can observe that Patcher outperforms the Transformer- and CNN-based baselines significantly. It is worth noting that the two segmentation tasks evaluate different aspects of the models. Stroke lesion segmentation requires the model to capture local details, and CNN-based models such as TransUnet [2] and AttnUNet [12] perform better than Transformer-based baselines. However, these CNN-based models perform much worse than Transformer-based models for polyp segmentation, *e.g.,* 82.80 (TransUNet) *vs.* 90.26 (Patcher), which is because polyp segmentation relies on better modeling of the global context where Transformers-based models excel. We notice that some papers report higher scores on Kvasir-SEG [6,19]. However, [6,19] use different train/test splits and do not have a validation split. In contrast, we use a validation split to select the best model to avoid overfitting to test data. On both tasks, Patcher consistently outperforms the baselines, which shows its superior ability to model both local details and global context.

Visual Results. In Fig. 3, we provide a visual comparison of the segmentation maps from various models, where we paint correct predictions in green, false positives (FPs) in red, and false negatives (FNs) in white. We can observe that the baselines often have FPs and FNs, do not predict the segmentation boundaries well, and often miss small lesions. In contrast, Patcher has much fewer FPs and FNs, and can predict the segmentation boundaries accurately, even for very small lesions (Fig. 3, rows 2 and 4). This is due to Patcher's effective use of Transformers for capturing fine-grained pixel-level details. We also provide a visual comparison for polyp segmentation in the supplementary materials.

Visualization of MoE Weight Maps. To better understand the MoE decoder, in Fig. 4, we visualize the MoE weight maps $[W_1, W_2, W_3, W_4]$ for the expert features for stroke lesion segmentation. We can see that MoE allows different weight maps to focus on different areas: W_1 focuses on local details, W_2 focuses on the global context inside the brain, W_3 focuses on the boundaries of the brain, and W_4 focuses on areas of the brain that are complementary to the lesions.

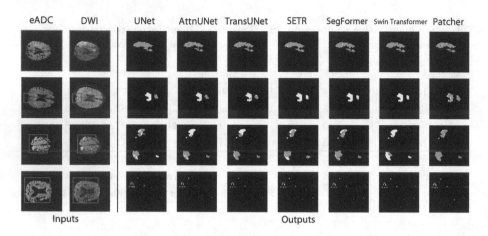

Fig. 3. Visualization of stroke lesion segmentation. The outputs correspond to the red box. We highlight correct predictions (green), false positives (red), and false negatives (white). Patcher outputs more accurate segmentation especially for small lesions. (Color figure online)

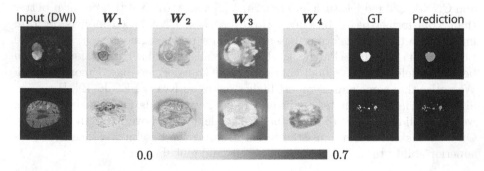

Fig. 4. Visualization of four MoE weight maps $[W_1, W_2, W_3, W_4]$. Different weight maps focus on different areas (local details, global context, brain boundaries, *etc.*)

3.2   Ablation Study

We first perform ablation studies to evaluate the importance of the Patcher encoder and MoE decoder by replacing them with popular encoder or decoder designs. As shown in Table 3, when replacing the Patcher encoder with SETR [20] or Swing Transformer [10], the performance decreases significantly. Similarly, when replacing the MoE decoder with SETR-PUP [20] and U-Net [15], the performance also drops considerably. This validates the importance of both designs.

We also conduct experiments to study the effect of varying the large patch size L and padded context size P. As shown in Table 4, both L and P need to be carefully selected as they need to have enough context (not too small) while also focusing mainly on local features (not too large) to attain better performance.

Table 3. Ablation studies of the Patcher encoder and MoE decoder.

Encoder	Decoder	DSC	IoU
SETR	MoE (ours)	77.23	72.69
Swin Transformer	MoE (ours)	85.29	81.47
Patcher (ours)	SETR-PUP	85.84	79.79
Patcher (ours)	U-Net	87.27	83.06
Patcher (ours)	MoE (ours)	**88.32**	**83.88**

Table 4. Effect of padded context size P and large patch size L for the 4 Patcher blocks.

$L = [32,32,32,32]$			$P = 8$		
Context P	DSC	IoU	Large Patch L	DSC	IoU
0	86.85	83.40	[64,64,64,32]	85.50	83.55
4	85.12	83.36	[64,64,32,32]	87.70	83.78
8	**88.32**	**83.88**	[32,32,32,32]	**88.32**	**83.88**
16	86.88	83.31	[32,16,16,16]	87.15	83.53

4 Conclusions

In this paper, we proposed Patcher, a new Vision Transformer architecture for medical image segmentation that can extract fine-grained pixel-level features with Transformers only. By stacking a cascade of encoder blocks with increasing receptive fields, Patcher can extract both local and global features effectively. We also proposed a new MoE-based decoder that uses a gating network to select a suitable set of expert features from the encoder to output the prediction for each pixel. The use of MoE enables better specializations of the expert features and reduces interference between them. Extensive experiments on stroke lesion and polyp segmentation indicate that our method can be applied to various medical image data. We hope the use of MoE in our work can provide a new perspective on Transformer-based architecture for medical imaging. We also look forward to extending Patcher to other medical diagnosis tasks, such as classification and landmark detection.

References

1. Cao, H., et al.: Swin-Unet: Unet-like pure transformer for medical image segmentation. arXiv preprint arXiv:2105.05537 (2021)
2. Chen, J., et al.: TransUNet: Transformers make strong encoders for medical image segmentation. arXiv preprint arXiv:2102.04306 (2021)
3. Çiçek, Ö., Abdulkadir, A., Lienkamp, S.S., Brox, T., Ronneberger, O.: 3D U-Net: learning dense volumetric segmentation from sparse annotation. In: Ourselin, S., Joskowicz, L., Sabuncu, M.R., Unal, G., Wells, W. (eds.) MICCAI 2016. LNCS, vol. 9901, pp. 424–432. Springer, Cham (2016). https://doi.org/10.1007/978-3-319-46723-8_49
4. Dosovitskiy, A., et al.: An image is worth 16 × 16 words: Transformers for image recognition at scale. arXiv preprint arXiv:2010.11929 (2020)

5. Hatamizadeh, A., et al.: UNETR: Transformers for 3D medical image segmentation. In: Proceedings of the IEEE/CVF Winter Conference on Applications of Computer Vision, pp. 574–584 (2022)

6. Huang, C.H., Wu, H.Y., Lin, Y.L.: HarDNet-MSEG: a simple encoder-decoder polyp segmentation neural network that achieves over 0.9 mean dice and 86 FPS. arXiv preprint arXiv:2101.07172 (2021)

7. Jha, D., et al.: Kvasir-SEG: a segmented polyp dataset. In: Ro, Y.M., et al. (eds.) MMM 2020. LNCS, vol. 11962, pp. 451–462. Springer, Cham (2020). https://doi.org/10.1007/978-3-030-37734-2_37

8. Jha, D., et al.: ResUNet++: An advanced architecture for medical image segmentation. In: Proceedings of the IEEE International Symposium on Multimedia (ISM), pp. 225–2255. IEEE (2019)

9. Kingma, D.P., Ba, J.: Adam: A method for stochastic optimization. arXiv preprint arXiv:1412.6980 (2014)

10. Liu, Z., et al.: Swin Transformer: Hierarchical vision transformer using shifted windows. Proceedings of the International Conference on Computer Vision (ICCV) (2021)

11. Loshchilov, I., Hutter, F.: Decoupled weight decay regularization. arXiv preprint arXiv:1711.05101 (2017)

12. Oktay, O., et al.: Attention U-Net: Learning where to look for the pancreas. arXiv preprint arXiv:1804.03999 (2018)

13. Ou, Y., et al.: LambdaUNet: 2.5D Stroke lesion segmentation of diffusion-weighted MR images. In: de Bruijne, M., et al. (eds.) MICCAI 2021. LNCS, vol. 12901, pp. 731–741. Springer, Cham (2021). https://doi.org/10.1007/978-3-030-87193-2_69

14. Petit, O., Thome, N., Rambour, C., Themyr, L., Collins, T., Soler, L.: U-Net transformer: self and cross attention for medical image segmentation. In: Lian, C., Cao, X., Rekik, I., Xu, X., Yan, P. (eds.) MLMI 2021. LNCS, vol. 12966, pp. 267–276. Springer, Cham (2021). https://doi.org/10.1007/978-3-030-87589-3_28

15. Ronneberger, O., Fischer, P., Brox, T.: U-Net: Convolutional networks for biomedical image segmentation. In: Navab, N., Hornegger, J., Wells, W.M., Frangi, A.F. (eds.) MICCAI 2015. LNCS, vol. 9351, pp. 234–241. Springer, Cham (2015). https://doi.org/10.1007/978-3-319-24574-4_28

16. Srivastava, R.K., Greff, K., Schmidhuber, J.: Highway networks. arXiv preprint arXiv:1505.00387 (2015)

17. Vaswani, A., et al.: Attention is all you need. In: Advances in Neural Information Processing Systems, vol. 30 (2017)

18. Xie, E., Wang, W., Yu, Z., Anandkumar, A., Alvarez, J.M., Luo, P.: SegFormer: Simple and efficient design for semantic segmentation with transformers. arXiv preprint arXiv:2105.15203 (2021)

19. Zhang, Y., Liu, H., Hu, Q.: TransFuse: Fusing transformers and cnns for medical image segmentation. In: de Bruijne, M., et al. (eds.) MICCAI 2021. LNCS, vol. 12901, pp. 14–24. Springer, Cham (2021). https://doi.org/10.1007/978-3-030-87193-2_2

20. Zheng, S., et al.: Rethinking semantic segmentation from a sequence-to-sequence perspective with transformers. In: Proceedings of the IEEE/CVF Conference on Computer Vision and Pattern Recognition, pp. 6881–6890 (2021)

21. Zhou, Z., Rahman Siddiquee, M.M., Tajbakhsh, N., Liang, J.: UNet++: A nested U-Net architecture for medical image segmentation. In: Stoyanov, D., et al. (eds.) DLMIA/ML-CDS -2018. LNCS, vol. 11045, pp. 3–11. Springer, Cham (2018). https://doi.org/10.1007/978-3-030-00889-5_1

TransFusion: Multi-view Divergent Fusion for Medical Image Segmentation with Transformers

Di Liu[1], Yunhe Gao[1], Qilong Zhangli[1], Ligong Han[1], Xiaoxiao He[1],
Zhaoyang Xia[1], Song Wen[1], Qi Chang[1], Zhennan Yan[2], Mu Zhou[2],
and Dimitris Metaxas[1(✉)]

[1] Department of Computer Science, Rutgers University, New Jeresy, USA
dnm@cs.rutgers.edu
[2] SenseBrain Research, California, USA

Abstract. Combining information from multi-view images is crucial to improve the performance and robustness of automated methods for disease diagnosis. However, due to the non-alignment characteristics of multi-view images, building correlation and data fusion across views largely remain an open problem. In this study, we present TransFusion, a Transformer-based architecture to merge divergent multi-view imaging information using convolutional layers and powerful attention mechanisms. In particular, the Divergent Fusion Attention (DiFA) module is proposed for rich cross-view context modeling and semantic dependency mining, addressing the critical issue of capturing long-range correlations between unaligned data from different image views. We further propose the Multi-Scale Attention (MSA) to collect global correspondence of multi-scale feature representations. We evaluate TransFusion on the Multi-Disease, Multi-View & Multi-Center Right Ventricular Segmentation in Cardiac MRI (M&Ms-2) challenge cohort. TransFusion demonstrates leading performance against the state-of-the-art methods and opens up new perspectives for multi-view imaging integration towards robust medical image segmentation.

Keywords: Cardiac MRI · Segmentation · Deep learning

1 Introduction

Multi-view medical image analysis allows us to combine the strengths from different views towards fully understanding of cardiac abnormalities, wall motion, and outcome diagnosis in clinical workflows [14,20,26,28–30,33]. For instance, cardiac MRI (cMRI) given appropriate MR acquisition settings can produce images with high intra-slice resolution and low inter-slice resolution [13,21,23,24]. Short-axis cMRI analysis is prone to fail in the base and the apex [2,5,15] due to low inter-slice resolution and possible misalignment. Many of these issues and missing information can be resolved with long-axis cMRI scans [3,6,11,19]. Nevertheless, finding corresponding structures and fusing multi-view information is

L. Wang et al. (Eds.): MICCAI 2022, LNCS 13435, pp. 485–495, 2022.
https://doi.org/10.1007/978-3-031-16443-9_47

challenging, because the locations of the regions of interest in different views can be relatively far apart (see Fig. 1). Due to this distinct structure unalignment among views, long-range correspondence modeling is highly desired. Convolution on simply concatenated unaligned multi-view images lacks the ability to find such correspondences [18, 29].

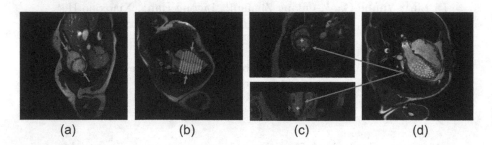

Fig. 1. (a) and (b): The correspondence of ventricular annotation between the short-axis and long-axis views. (c) and (d) show an example of how the cross markers in (c) aggregate features from corresponding dots in (d) via learned attention map using DiFA module. The bottom image in (c) is the sagittal view of short-axis image stack with low resolution.

Transformers with the self-attention mechanism are gaining momentum in medical image analysis [4, 8–10, 12, 32]. The transformer self-attention mechanism implements pairwise interactions to dynamically aggregate long-range dependencies of feature representations according to the input content from various data modalities [27]. TransUNet [7] adopts Transformer blocks to collect global correlations on top of ResNet backbone. UTNet [10] instead incorporates interleaved transformer blocks and convolution blocks for small medical dataset. MCTrans [16] employs a Transformer Cross-Attention (TCA) module to collect context information from feature maps of different scales. However, these approaches are designed for single-view and can be sub-optimal for complex segmentation tasks due to the absence of considering semantic dependencies of different scales and views, which are critical for enhancing clinical lesion assessment.

In this paper, we propose TransFusion to merge divergent information from multiple views and scales, using powerful attention mechanisms for medical image segmentation. We employ convolutional layers for local feature extraction and develop strong attentive mechanisms to aggregate long-range correlation information from cross-scale and cross-view feature representations. To gather long-range dependencies among image views, we propose the Divergent Fusion Attention (DiFA) module to achieve cross-view context modeling and semantic dependency mining. Further, the Multi-Scale Attention (MSA) module is designed to collect global correspondence of multi-scale feature representations to ensure feature consistency at different levels of the pyramidal architecture. We evaluate TransFusion on the Multi-Disease, Multi-View & Multi-Center Right Ventricular Segmentation (M&Ms-2) challenge cohort. TransFusion demonstrates leading performance against the state-of-the-art methods and holds the promise for a wide range of medical image segmentation tasks. To the best of our knowledge, our method is the first work to apply Transformer to multi-view medical image segmentation tasks, which also shows potential in wide applications for unaligned or multi-modality data.

2   Methods

Figure 2 highlights the proposed TransFusion framework. The inputs to the model are images from M views. As the size, modality, and even dimension can be different among views, TransFusion has a sub-network for each view, where each sub-network can be an arbitrary network structure designed for any input image modality. Further, TransFusion has the ability to find correspondences and align features from heterogeneous inputs. In our experiments, 3D short-axis cMRI and 2D long-axis cMRI are used as multi-view inputs. For the purpose of efficiency, 3D UTNet [10] and 2D UTNet are used as the backbone of sub-networks for the two inputs, which consist of interleaved Residual blocks and efficient Transformer blocks. As seen in Fig. 2, for joint modeling of multi-view data and fusing multi-scale features, we apply the Divergent Fusion Attention (DiFA) block and multi-scale attention block (MSA) on the middle- and high-level token maps (Fig. 2 grey modules). The three outputs of DiFA at different scales are further folded back to their corresponding scales of decoders as indicated by the blue, green and orange arrows. The convolution stem consists of multiple convolution layers and down-samplings to reduce the spatial resolution by 4×. The Res+Trans Block is the stack of a Residual Block and a Transformer Block (see Fig. 3).

2.1   Revisiting the Self-attention Mechanism

Transformers use the Multi-Head Self-Attention (MHSA) module for the modeling of data dependencies without considering their distance in different representation sub-spaces [1,17,27]. The outputs of multiple heads are concatenated through the Feed-forward Network (FFN). Given an input feature map $X \in \mathbb{R}^{C \times H \times W}$, where C, H, W are the number of input channels, the spatial height and width, respectively (e.g. short and long axis cardiac MR images). The input X is first projected to query Q, key K and value V, through three linear projections. Here $Q, K, V \in \mathbb{R}^{d \times H \times W}$, where d represents the embedding dimensions of each head. Q, K, V are flattened and transposed into three token sequences with the same size $n \times d$ where $n = HW$. The output of self-attention layer can be denoted by a scaled dot-product as

$$\text{Attn}(Q,K,V) = \text{softmax}(\frac{QK^T}{\sqrt{d}})V, \tag{1}$$

where $\text{softmax}(\frac{QK^T}{\sqrt{d}}) \in \mathbb{R}^{n \times n}$ is the attention matrix, which computes the pairwise similarity between each token of queries and keys, and then is used as the weights to collect context information from the values. This mechanism allows the modeling of long-range data dependencies for global feature aggregation.

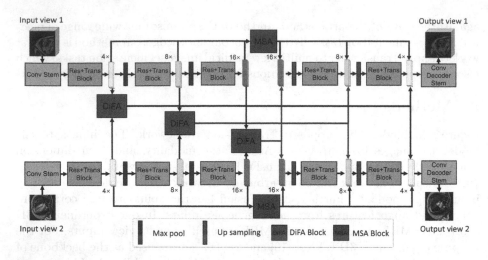

Fig. 2. Overview of the Transfusion architecture. The Divergent Fusion Attention (DiFA) module enables rich cross-view context modeling and semantic dependency mining, capturing long-range dependencies between unaligned data from different image views. For each view, the proposed Multi-Scale Attention (MSA) module collects global correspondence of multi-scale feature representations.

2.2 Divergent Fusion Attention Module (DiFA)

The self-attention mechanism offers global feature aggregation that is built upon single input features. For multi-view tasks, we seek to jointly model inputs from different views. We extend the self-attention to be applicable for multi-view fusion and thus introduce the Divergent Fusion Attention (DiFA) mechanism. The proposed DiFA aims to find correspondences in unaligned data and complement the missing information among multi-view images such as long and short axis cardiac images. For instance, the high-resolution information of the long-axis cardiac image can be used to complement the base and the apex the ventricles on the short-axis image. We take 3D short-axis and 2D long-axis cMRI inputs as an example shown in Fig. 3.

Given token maps $f_0, f_1, ..., f_M \in \mathbb{R}^{C \times N_m}$ from M views, where N_m represents the token number in a single view (e.g. $N_m = H_m \times W_m$ for 2D input or $N_m = D_m \times H_m \times W_m$ for 3D input, $m \in \{0, ..., M\}$). All token maps f_m are linearly projected into $Q_m, K_m, V_m \in \mathbb{R}^{d \times N_m}$. The main idea is to enhance a specific view representation with context information from all non-target views. We first concatenate all non-target keys and values as $\tilde{K}_m, \tilde{V}_m \in \mathbb{R}^{d \times \sum_{i=0, i \neq m}^{M} N_i}$. The attention matrix is further computed by the target query Q_m and the concatenated non-target embedding $\tilde{K}_m$. Then the updated $\bar{f}_m$ is:

$$\bar{f}_m = \text{DiFA}(Q_m, \tilde{K}_m, \tilde{V}_m) = \text{softmax}(\frac{Q_m \tilde{K}_m^T}{\sqrt{d}})\tilde{V}_m \tag{2}$$

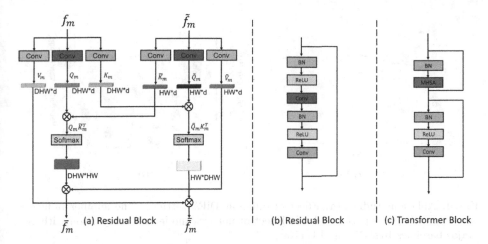

Fig. 3. Illustration of the (a) Divergent Fusion Attention Module (DiFA) (b) Residual block with pre-activation and (c) Transformer block.

We then repeat DiFA for all other views. In this way, the divergent context information from different views are fused in high-dimensional spaces to refine the initial representation from the target view only. Notably, as the inputs from different views are not aligned, the absolute or relative positional encoding are not used in our DiFA module. Further, the design of DiFA is not limited to multi-view images, but applied all data that is not well aligned, such as different modalities from 2D and 3D data.

2.3   Multi-Scale Attention (MSA) Module

To fuse the multi-scale representations in the hierarchical backbone, we introduce the Multi-Scale Attention (MSA) module to learn this contextual dependencies of inter-scale features. In practice, given the m-th view input, the encoder hierarchically extract features and obtain multiple-level feature f_m^l, where l denotes the level of down-sampling. Then the output features $f_m^l \in \mathbb{R}^{C^l \times N_m^l}$ for every level are flattened and concatenated as the input of a Transformer block, which consists of a Multi-Head Self-Attention module and a Feed-forward Network. Following the feature extraction using CNN and Self-Attention, the MSA block is applied to fuse scale divergence. A single MSA layer is formulated as:

$$\bar{f}_m^0, \cdots, \bar{f}_m^L = \text{FFN}(\text{Attn}(\text{Cat}(f_m^0, \cdots, f_m^L)))   \tag{3}$$

where $f_m^l = \text{Attn}(Q_m^l, K_m^l, V_m^l)$ denotes the flattened output of the Self-Attention Encoder layer at level l. The feature map of MSA is then folded back to the decoder at corresponding scales for further interaction and prediction.

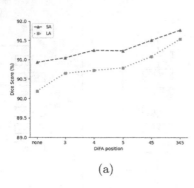 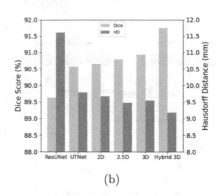

(a)   (b)

Fig. 4. Ablation studies. (a) Effect of different DiFA settings. The number indicates which layer DiFA is performed. (b) Effect of network mode and comparison with two major baselines ResUNet and UTNet.

3   Experiments

3.1   Datasets and Settings

The proposed TransFusion is systematically evaluated on the Multi-Disease, Multi-View & Multi-Center Right Ventricular Segmentation in Cardiac MRI (M&Ms-2) challenge cohort [3]. The M&Ms-2 challenge cohort contains 160 scans, which were collected from subjects with different RV and LV pathologies in three clinical centers from Spain using three different magnetic resonance scanner vendors (Siemens, General Electric, and Philips). The acquired cine MR images were delineated by experienced clinical doctors, including the left ventricle (LV), right ventricle (RV) and the left ventricular myocardium (Myo). The whole cine MR sequences with varied number of frames are provided, but only ED and ES frames are labeled in both short-axis and long-axis views. Note that we treat the stack of short-axis images as 3D data, while the long-axis four-chamber view as 2D data. The training set includes subjects of normal, dilated left ventricle (DLV), hypertrophic cardiomyopathy (HC), congenital arrhythmogenesis (CA), tetralogy of fallot (TF) and interatrial comunication (IC).

We randomly shuffle the 160 samples and evaluate all models with 5-fold cross validation. All models are trained with Dice loss and focal loss, with batch size 32 and Adam optimizer for 300 epochs. The learning rate is 0.001 and decayed linearly by 99% per epoch. Models are trained with eight Quadro RTX 8000 GPUs in PyTorch framework. For data preprocessing, all images are resampled to 1.25 mm in spacing. Data augmentation strategies are utilized, including random histogram matching, rotation, shifting, scaling, elastic deformation and mirroring. Dice score and Hausdorff distance are used as evaluation metrics.

3.2   Results

Table 1 compares the performance of TransFusion with several state-of-the-art approaches. ResUNet applies residual blocks as UNet building blocks [25].

Table 1. Comparison of methods in terms of Dice and Hausdorff Distance. (T) indicates the target view of segmentation. The bold texts mark the best performance. The methods marked with stars indicate they use multi-view inputs.

Methods	Input type		Dice(%)				HD(mm)			
	SA(T)	LA	LV	RV	Myo	Avg	LV	RV	Myo	Avg
UNet	✓	–	87.02	88.85	79.07	84.98	13.78	12.10	12.23	12.70
ResUNet	✓	–	87.98	89.63	79.28	85.63	13.80	11.61	12.09	12.50
DLA	✓	–	88.27	89.88	80.23	86.13	13.25	10.84	12.31	12.13
InfoTrans*	✓	✓	88.24	90.41	80.25	86.30	12.41	10.98	12.83	12.07
rDLA*	✓	✓	88.64	90.28	80.78	86.57	12.74	10.31	12.49	11.85
TransUNet	✓	–	87.91	88.69	78.67	85.09	13.80	10.29	13.45	12.51
MCTrans	✓	–	88.52	89.90	80.08	86.17	12.29	9.92	13.28	11.83
MCTrans*	✓	✓	87.79	89.22	79.37	85.46	**11.28**	9.36	13.84	11.49
UTNet	✓	–	87.52	90.57	80.20	86.10	12.03	9.78	13.72	11.84
UTNet*	✓	✓	87.74	90.82	80.71	86.42	11.79	9.11	13.41	11.44
Proposed*	✓	✓	**89.52**	**91.75**	**81.46**	**87.58**	11.31	**9.18**	**11.96**	**10.82**
Methods	Input type		Dice(%)				HD(mm)			
	SA	LA(T)	LV	RV	Myo	Avg	LV	RV	Myo	Avg
UNet	–	✓	87.26	88.20	79.96	85.14	13.04	8.76	12.24	11.35
ResUNet	–	✓	87.61	88.41	80.12	85.38	12.72	8.39	11.28	10.80
DLA	–	✓	88.37	89.38	80.35	86.03	11.74	7.04	10.79	9.86
InfoTrans*	✓	✓	88.21	89.11	80.55	85.96	12.47	7.23	10.21	9.97
rDLA*	✓	✓	88.71	89.71	81.05	86.49	11.12	6.83	10.42	9.46
TransUNet	–	✓	87.91	88.23	79.05	85.06	12.02	8.14	11.21	10.46
MCTrans	–	✓	88.42	88.19	79.47	85.36	11.78	7.65	10.76	10.06
MCTrans*	✓	✓	88.81	88.61	79.94	85.79	11.52	7.02	10.07	9.54
UTNet	–	✓	86.93	89.07	80.48	85.49	11.47	6.35	10.02	9.28
UTNet*	✓	✓	87.36	90.42	81.02	86.27	11.13	5.91	9.81	8.95
Proposed*	✓	✓	**89.78**	**91.52**	**81.79**	**87.70**	**10.25**	**5.12**	**8.69**	**8.02**

MCTrans [16] introduces a cross-attention block between the encoder and decoder to gather cross-scale dependencies of the feature maps. Refined DLA (rDLA) [22] bases its backbone on a leading CNN architecture, Deep Layer Aggregation (DLA) [31], and aggregates context information from cross-view through a refinement stage. InfoTrans* [18] similarly aggregates information from cross-views using information transition. The result shows our TransFusion outperforms the other methods in terms of both short- and long-axis cardiac MR segmentation.

We provide a further comparison by applying the proposed DiFA module to these existing baselines for multi-view segmentation tasks as indicated as starred methods in Table 1. The starred MCTrans and UTNet perform better than the single-view networks due to the mutual feature aggregation of the proposed DiFA. Figure 5 further shows that Transfusion displays leading performance as compared to other competitive approaches in multi-view segmentation results.

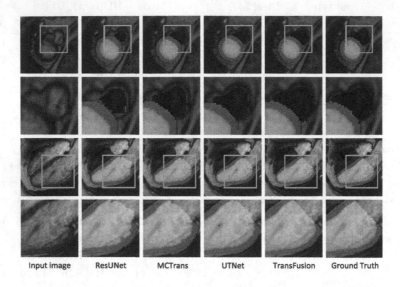

| Input image | ResUNet | MCTrans | UTNet | TransFusion | Ground Truth |

Fig. 5. Segmentation results of short-(top) and long-axis (bottom) images.

3.3   Ablation Studies

Sensitivity to the DiFA Setting. We evaluate the performance of the Divergent Fusion Attention (DiFA) modules of TransFusion by the segmentation accuracy. As seen in Fig. 4(a), the x-axis number indicates the level where DiFAs are located, e.g., '45' means two DiFAs are performed respectively in the fourth and fifth level of the encoder. As the level goes up, the DiFA module is able to collect more detailed and divergent information from the other views. Note that the curve saturates when adding to the fourth level. However, when adding DiFA to multiple levels, we recognize that the learned cross-view prior helps construct identified context dependencies and improve the performance constantly with limited additional computational cost.

Analysis of the Network Mode. Figure 4(b) shows the performance using different network modes. The 2.5D mode combines two neighboring slices of short-axis sample and forms a three-channel input, allowing a local spatial information fusion for segmentation. The proposed TransFusion applies a hybrid 3D + 2D mode and outperforms other baselines, which is attributed to the alignment of spatial information between short- and long-axis data. For each short-axis sample, as more through-plane textures are included to construct the cross-view dependencies with the corresponding long-axis image, our TransFusion can better build up global context dependencies through the DiFA modules.

4 Conclusion

We have proposed TransFusion, a powerful Transformer architecture, to merge critical cross-view information towards enhanced segmentation performance using convolutional layers and powerful attentive mechanisms for medical images. The proposed Multi-Scale Attention (MSA) and Divergent Fusion Attention (DiFA) modules allow rich cross-scale and cross-view context modeling and semantic dependency mining, effectively addressing the issues of capturing long-range dependencies within as well as between different scales and views. This strong ability further opens up new perspectives for applying TransFusion on more downstream view-fusion tasks in medical imaging and computer vision.

References

1. Bahdanau, D., Cho, K., Bengio, Y.: Neural machine translation by jointly learning to align and translate. arXiv preprint arXiv:1409.0473 (2014)
2. Bernard, O., et al.: Deep learning techniques for automatic mri cardiac multi-structures segmentation and diagnosis: is the problem solved? IEEE Trans. Med. Imaging **37**(11), 2514–2525 (2018)
3. Campello, V.M., et al.: Multi-centre, multi-vendor and multi-disease cardiac segmentation: the m&ms challenge. IEEE Trans. Med. Imaging **40**(12), 3543–3554 (2021)
4. Cao, H., et al.: Swin-unet: Unet-like pure transformer for medical image segmentation. arXiv preprint arXiv:2105.05537 (2021)
5. Chang, Q., Yan, Z., Lou, Y., Axel, L., Metaxas, D.N.: Soft-label guided semi-supervised learning for bi-ventricle segmentation in cardiac cine mri. In: 2020 IEEE 17th International Symposium on Biomedical Imaging (ISBI), pp. 1752–1755. IEEE (2020)
6. Chang, Q., et al.: Deeprecon: Joint 2d cardiac segmentation and 3d volume reconstruction via a structure-specific generative method. arXiv preprint arXiv:2206.07163 (2022)
7. Chen, J., et al.: Transunet: Transformers make strong encoders for medical image segmentation. arXiv preprint arXiv:2102.04306 (2021)
8. Dosovitskiy, A., et al.: An image is worth 16×16 words: Transformers for image recognition at scale. arXiv preprint arXiv:2010.11929 (2020)
9. Gao, Y., Zhou, M., Liu, D., Metaxas, D.: A multi-scale transformer for medical image segmentation: Architectures, model efficiency, and benchmarks. arXiv preprint arXiv:2203.00131 (2022)
10. Gao, Y., Zhou, M., Metaxas, D.N.: UTNet: a hybrid transformer architecture for medical image segmentation. In: de Bruijne, M., et al. (eds.) MICCAI 2021. LNCS, vol. 12903, pp. 61–71. Springer, Cham (2021). https://doi.org/10.1007/978-3-030-87199-4_6
11. Ge, C., Liu, D., Liu, J., Liu, B., Xin, Y.: Automated recognition of arrhythmia using deep neural networks for 12-lead electrocardiograms with fractional time-frequency domain extension. J. Med. Imaging Health Inf. **10**(11), 2764–2767 (2020)
12. Hatamizadeh, A., et al.: Unetr: Transformers for 3d medical image segmentation. In: Proceedings of the IEEE/CVF Winter Conference on Applications of Computer Vision, pp. 574–584 (2022)

13. He, X., Tan, C., Qiao, Y., Tan, V., Metaxas, D., Li, K.: Effective 3d humerus and scapula extraction using low-contrast and high-shape-variability mr data. In: Medical Imaging 2019: Biomedical Applications in Molecular, Structural, and Functional Imaging, vol. 10953, p. 109530O. International Society for Optics and Photonics (2019)

14. Hu, J.B., Guan, A., Zhangli, Q., Sayadi, L.R., Hamdan, U.S., Vyas, R.M.: Harnessing machine-learning to personalize cleft lip markings. Plastic Reconstr. Surg. Glob. Open **8**(9S), 150–151 (2020)

15. Hu, Z., Metaxas, D., Axel, L.: In vivo strain and stress estimation of the heart left and right ventricles from mri images. Med. Image Anal. **7**(4), 435–444 (2003)

16. Ji, Y., Zhang, R., Wang, H., Li, Z., Wu, L., Zhang, S., Luo, P.: Multi-compound transformer for accurate biomedical image segmentation. In: de Bruijne, M., et al. (eds.) MICCAI 2021. LNCS, vol. 12901, pp. 326–336. Springer, Cham (2021). https://doi.org/10.1007/978-3-030-87193-2_31

17. Kim, Y., Denton, C., Hoang, L., Rush, A.M.: Structured attention networks. arXiv preprint arXiv:1702.00887 (2017)

18. Li, L., Ding, W., Huang, L., Zhuang, X.: Right ventricular segmentation from short- and long-axis mris via information transition. arXiv preprint arXiv:2109.02171 (2021)

19. Liu, D., Ge, C., Xin, Y., Li, Q., Tao, R.: Dispersion correction for optical coherence tomography by the stepped detection algorithm in the fractional fourier domain. Opt. Express **28**(5), 5919–5935 (2020)

20. Liu, D., Liu, J., Liu, Y., Tao, R., Prince, J.L., Carass, A.: Label super resolution for 3d magnetic resonance images using deformable u-net. In: Medical Imaging 2021: Image Processing, vol. 11596, p. 1159628. International Society for Optics and Photonics (2021)

21. Liu, D., Xin, Y., Li, Q., Tao, R.: Dispersion correction for optical coherence tomography by parameter estimation in fractional fourier domain. In: 2019 IEEE International Conference on Mechatronics and Automation (ICMA), pp. 674–678. IEEE (2019)

22. Liu, D., Yan, Z., Chang, Q., Axel, L., Metaxas, D.N.: Refined deep layer aggregation for multi-disease, multi-view & multi-center cardiac mr segmentation. In: STACOM 2021. LNCS, vol. 13131, pp. 315–322. Springer, Cham (2022). https://doi.org/10.1007/978-3-030-93722-5_34

23. Petitjean, C., Dacher, J.N.: A review of segmentation methods in short axis cardiac mr images. Med. Image Anal. **15**(2), 169–184 (2011)

24. Remedios, S.W., Han, S., Dewey, B.E., Pham, D.L., Prince, J.L., Carass, A.: Joint image and label self-super-resolution. In: Svoboda, D., Burgos, N., Wolterink, J.M., Zhao, C. (eds.) SASHIMI 2021. LNCS, vol. 12965, pp. 14–23. Springer, Cham (2021). https://doi.org/10.1007/978-3-030-87592-3_2

25. Ronneberger, O., Fischer, P., Brox, T.: U-Net: Convolutional networks for biomedical image segmentation. In: Navab, N., Hornegger, J., Wells, W.M., Frangi, A.F. (eds.) MICCAI 2015. LNCS, vol. 9351, pp. 234–241. Springer, Cham (2015). https://doi.org/10.1007/978-3-319-24574-4_28

26. Tian, Y., Peng, X., Zhao, L., Zhang, S., Metaxas, D.N.: Cr-gan: learning complete representations for multi-view generation. arXiv preprint arXiv:1806.11191 (2018)

27. Vaswani, A., et al.: Attention is all you need. In: Advances in Neural Information Processing Systems, pp. 5998–6008 (2017)

28. Vigneault, D.M., Xie, W., Ho, C.Y., Bluemke, D.A., Noble, J.A.: ω-net (omega-net): fully automatic, multi-view cardiac mr detection, orientation, and segmentation with deep neural networks. Med. Image Anal. **48**, 95–106 (2018)

29. Wang, S., et al.: A multi-view deep convolutional neural networks for lung nodule segmentation. In: 2017 39th Annual International Conference of the IEEE Engineering in Medicine and Biology Society (EMBC), pp. 1752–1755. IEEE (2017)

30. Xia, Y., et al.: Uncertainty-aware multi-view co-training for semi-supervised medical image segmentation and domain adaptation. Med. Image Anal. **65**, 101766 (2020)

31. Yu, F., Wang, D., Shelhamer, E., Darrell, T.: Deep layer aggregation. In: Proceedings of the IEEE Conference on Computer Vision and Pattern Recognition, pp. 2403–2412 (2018)

32. Zhangli, Q., et al.: Region proposal rectification towards robust instance segmentation of biological images. arXiv preprint arXiv:2203.02846 (2022)

33. Zhao, C., et al.: Applications of a deep learning method for anti-aliasing and super-resolution in mri. Magn. Reson. Imaging **64**, 132–141 (2019)

CorticalFlow++: Boosting Cortical Surface Reconstruction Accuracy, Regularity, and Interoperability

Rodrigo Santa Cruz[1,2]([✉])(iD), Léo Lebrat[1,2], Darren Fu[3], Pierrick Bourgeat[1], Jurgen Fripp[1], Clinton Fookes[2], and Olivier Salvado[1,2]

[1] CSIRO, Brisbane, Australia
[2] Queensland University of Technology, Brisbane, Australia
rodrigo.santacruz@csiro.au
[3] University of Queensland, Brisbane, Australia

Abstract. The problem of Cortical Surface Reconstruction from magnetic resonance imaging has been traditionally addressed using lengthy pipelines of image processing techniques like FreeSurfer, CAT, or CIVET. These frameworks require very long runtimes deemed unfeasible for real-time applications and unpractical for large-scale studies. Recently, supervised deep learning approaches have been introduced to speed up this task cutting down the reconstruction time from hours to seconds. Using the state-of-the-art CorticalFlow model as a blueprint, this paper proposes three modifications to improve its accuracy and interoperability with existing surface analysis tools, while not sacrificing its fast inference time and low GPU memory consumption. First, we employ a more accurate ODE solver to reduce the diffeomorphic mapping approximation error. Second, we devise a routine to produce smoother template meshes avoiding mesh artifacts caused by sharp edges in CorticalFlow's convex-hull based template. Last, we recast pial surface prediction as the deformation of the predicted white surface leading to a one-to-one mapping between white and pial surface vertices. This mapping is essential to many existing surface analysis tools for cortical morphometry. We name the resulting method CorticalFlow++. Using large-scale datasets, we demonstrate the proposed changes provide more geometric accuracy and surface regularity while keeping the reconstruction time and GPU memory requirements almost unchanged.

Keywords: Cortical surface reconstruction · CorticalFlow · 3D deep learning

---

R. Santa Cruz and L. Lebrat—Equal contribution.
Our code is made available at: https://bitbucket.csiro.au/projects/CRCPMAX/repos/corticalflow/browse.

L. Wang et al. (Eds.): MICCAI 2022, LNCS 13435, pp. 496–505, 2022.
https://doi.org/10.1007/978-3-031-16443-9_48

1 Introduction

The problem of cortical surface reconstruction (CSR) consists of estimating triangular meshes for the inner and outer cortical surfaces from a magnetic resonance image (MRI). It is a pivotal problem in Neuroimaging and a fundamental task in clinical studies of neurodegenerative diseases [4] and psychological disorders [21]. Traditionally, this problem is tackled by extensive pipelines of handcrafted image processing algorithms [6,13,18,25] which are subject to careful hyperparameter tuning (e.g., thresholds, iteration numbers, and convergence criterion) and very long runtimes.

To overcome these issues, deep learning (DL) based approaches have been proposed recently [9,15,17,23]. These methods can directly predict cortical surfaces geometrically close to those produced by traditional methods while reducing their processing time from hours to seconds. More specifically, the current state-of-the-art DL model for CSR, named CorticalFlow [15], consists of a sequence of diffeomorphic deformation modules that learn to deform a template mesh towards the target cortical surface from an input MRI. This method takes only a few seconds to produce cortical surfaces with sub-voxel accuracy.

In this paper, we propose three modifications to improve the accuracy of CorticalFlow and its interoperability to other surface processing tools without increasing its reconstruction time and GPU memory consumption. First, we upgrade the Euler method used by CorticalFlow to solve the flow ODE responsible for computing the diffeomorphic mapping to the fourth-order Runge-Kutta method [20]. This tool provides more accurate ODE solutions leading the model to better approximate the target surfaces and reducing the number of self-intersecting faces on the reconstructed meshes. Second, instead of using the convex hull of the training surfaces as a template, we propose a simple routine to generate smooth templates that tightly wrap the training surfaces. This new template eases the approximation problem leading to more accurate surfaces while suppressing mesh artifacts in highly curved regions. Finally, inspired by Ma et al. [17], we leverage the estimated white surfaces as the initial mesh template for learning and predicting the pial surfaces. This approach provides a better "starting point" to the approximation problem as well as a one-to-one mapping between white and pial surface vertices which facilitates the use of the generated surfaces in existing surface-based analysis tools [8,24]. In acknowledgment of the CorticalFlow framework, we name our resulting method CorticalFlow++.

Using a large dataset of MRI images and pseudo-ground-truth surfaces, we compare CorticalFlow and CorticalFlow++ performance in the reconstruction of cortical surfaces from MRI on three perspectives: geometric accuracy, mesh regularity, and time and space complexity for the surface reconstruction. We conclude that the proposed CorticalFlow++ improves upon CorticalFlow, on average, by 19.11% in terms of Chamfer distance and 56.77% in terms of the percentage of self-intersecting faces. Additionally, it adds only half a second to the final surface reconstruction time while keeping the same GPU memory budget.

2   Related Work

Traditional cortical surface reconstruction frameworks like FreeSurfer [6], Brain-Suite [25], and CIVET [18] involve two major steps usually accomplished by lengthy sequences of image processing techniques. They first voxel-wise segment the input MRI, then fit surfaces enclosing the gray matter tissue delimiting the brain cortex. More specifically, the widely used FreeSurfer V6 framework for cortical surface analysis from MRI uses an atlas-based segmentation [7] and a deformable model [3] for surface fitting on these segmented volumes. Recently, Henschel et al. [9] accelerated this framework with a modern DL brain segmentation model and a fast spectral mesh processing algorithm for spherical mapping, cutting down FreeSurfer's processing time to one hour.

In contrast, supervised deep learning approaches leverage large datasets of MRIs and pseudo-ground-truth produced with traditional methods to train high-capacity neural networks to predict surfaces directly from the MRI. DeepCSR [23] trains an implicit surface predictor and CorticalFlow [15] learns a diffeomorphic deformable model. Similarly, PialNN [17] also learns a deformable model but focuses only on pial surface reconstruction, receiving as input the white matter surface generated with traditional methods.

Building upon the success and the powerful framework of CorticalFlow, this paper proposes to address its main limitations, aiming to improve its accuracy and interoperability with existent surface analysis tools, but without severely degrading its inference time and GPU memory consumption.

3   Method

In this Section, we present the proposed method CorticalFlow^{++}. We start by reviewing the original CorticalFlow framework and its main components, then we introduce our proposed changes.

3.1   CorticalFlow Framework

Originally proposed by Lebrat et al. [15], CorticalFlow consists of a chain of deformation blocks that receives as input a 3-dimensional MRI $\mathbf{I} \in \mathbb{R}^{H \times W \times D}$ and deforms an initial template mesh $\mathcal{T}$ producing the desired cortical surface mesh. As shown in Fig. 1, a CorticalFlow's deformation block comprises a U-Net [22] and a Diffeomorphic Mesh Deformation module (DMD) [15]. The U-Net in the i-th deformation block, denoted as $\text{UNet}_{\theta_i}^i$ and parametrized by θ_i, outputs a stationary flow field $\mathbf{U}_i \in \mathbb{R}^{H \times W \times D \times 3}$, while it receives as input the channel-wise concatenation of the input MRI and all the previous blocks' U-Nets outputs $\{\mathbf{U}_1, \ldots, \mathbf{U}_{i-1}\}$. The predicted flow field encodes how the template mesh should be deformed to approximate the target cortical surface. The DMD module receives as input the block's UNet predicted flow field $\mathbf{U}_i$ and computes a diffeomorphic mapping $\Phi : [0,1] \times \mathbb{R}^3 \rightarrow \mathbb{R}^3$ for each vertex position $\mathbf{x} \in \mathbb{R}^3$ in

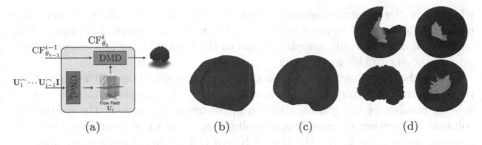

Fig. 1. (a) CorticalFlow's Deformation block. (b) Convex-hull based CorticalFlow's mesh template. (c) CorticalFlow^{++}'s proposed template. (d) Examples of mesh artifacts caused by sharp edges in CorticalFlow's template.

the resulting mesh computed by the previous deformation block. Formally, the DMD module solves the flow ODE,

$$\frac{d\Phi(s; \mathbf{x})}{ds} = \mathbf{U}(\Phi(s; \mathbf{x})), \text{ with } \Phi(0; \mathbf{x}) = \mathbf{x}, \tag{1}$$

using the forward Euler method [5]. This flow ODE formulation preserves the topology of the initial template mesh without producing self-intersecting faces. Hence, CorticalFlow with k deformations (CF^k) can be written using the following recurrence,

$$\mathrm{CF}^1_{\theta_1}(\mathbf{I}, \mathcal{T}) = \mathrm{DMD}(\mathrm{UNet}^1_{\theta_1}(\mathbf{I}), \mathcal{T}))$$

$$\mathrm{CF}^{i+1}_{\theta_{i+1}}(\mathbf{I}, \mathcal{T}) = \mathrm{DMD}(\mathrm{UNet}^{i+1}_{\theta_{i+1}}(\mathbf{U}_1^\frown \cdots \mathbf{U}_i^\frown \mathbf{I}), \mathrm{CF}^i_{\theta_i}(\mathbf{I}, \mathcal{T})) \text{ for } i \geq 1, \tag{2}$$

with $\mathbf{A}^\frown\mathbf{B}$ denoting the channel-wise concatenation of the tensors $\mathbf{A}$ and $\mathbf{B}$. As in [15], we focus on CorticalFlow with three deformation blocks (CF^3).

In order to train CF^3, the authors propose a multi-stage approach where the deformation blocks are trained successively keeping previous blocks' weights frozen and using template meshes of increasing resolution $\mathcal{T}_i$. Mathematically, at each stage i, they optimize the following objective,

$$\arg\min_{\theta_i} \sum_{(\mathbf{I}, \mathbf{S}) \in \mathcal{D}} \mathcal{L}(\mathrm{CF}^i_{\theta_i}(\mathbf{I}, \mathcal{T}_i), S), \tag{3}$$

where $\mathcal{D}$ is a dataset composed of pairs of MRIs and their respective triangle meshes S representing some cortical surface. $\mathcal{L}(\cdot, \cdot)$ is the training loss composed of mesh edge loss [27] and Chamfer Distance loss [27]. Note that for each cortical surface and brain hemisphere, a separate CorticalFlow is trained independently.

3.2 Higher Order ODE Solver

DMD modules in CorticalFlow solve the flow ODE (1) using the forward Euler method which consists of an iterative method defined by the following integration step:

$$\hat{\Phi}(h, \mathbf{x}) = \mathbf{x} + h\mathbf{U}(\mathbf{x}), \tag{4}$$

where h is the algorithm step-size and $\mathbf{U}(\mathbf{x})$ is the linear interpolation of a predicted flow field $\mathbf{U}$ at mesh vertex position $\mathbf{x}$. $\hat{\Phi}$ is a numerical approximation of the true diffeomorphic mapping Φ due to the interpolation and discretization errors inherent to this application.

Since the Euler method only provides an approximation of the continuous problem with an error that decreases linearly as h decreases, it may require a large number of integration steps to approximate accurately the continuous solution. Otherwise, the resulting mapping may cease to be invertible. For these reasons, we propose to use the Runge-Kutta [20] explicit fourth-order approximation method also known as $RK4$. The integration step of this method consists of the weighted average of four slopes estimated at the beginning, two different midpoints, and end of the step size interval. For our stationary vector field, the $RK4$ integration step is defined as,

$$\hat{\Phi}(h, \mathbf{x}) = \mathbf{x} + \frac{1}{6}\left[k_1 + 2k_2 + 2k_3 + k_4\right], \tag{5}$$

where $k_1 = \mathbf{U}(\mathbf{x})$, $k_2 = \mathbf{U}\left(\mathbf{x} + h\frac{k_1}{2}\right)$, $k_3 = \mathbf{U}\left(\mathbf{x} + h\frac{k_2}{2}\right)$, and $k_4 = \mathbf{U}(\mathbf{x} + hk_3)$ are the averaged slopes.

3.3   Smooth Templates

CorticalFlow's template mesh consists of the convex-hull of all cortical surface meshes in the training set. This approach has two main shortcomings:

1. Even target surfaces with small differences between them can lead to a "loose" template. Consequently, the model has to learn "large" deformations making the smooth approximation problem harder. For pial surfaces, this problem is even worse because some template regions may lay outside the image bounds where the predicted flow field is undefined.
2. The convex-hull is defined by a set of intersecting planes which leads to sharp edges as shown in Fig. 1b. These edges are very hard to be unfolded by a smooth deformable model like CorticalFlow, because it requires non-smooth deformations with drastic local changes of direction in its flow field representation. Hence, these edges may remain in the predicted mesh as undesirable artifacts (see Fig. 1d).

To overcome these issues, we develop genus zero smooth mesh templates that tightly wrap all training meshes. We first compute a signed distance map for every target surface mesh in the training set by computing the largest 3D bounding box that contains these meshes, create 512^3 voxel-grids into this bounding box, and populate these voxel-grids with the signed distance to each target mesh. These signed distance maps are implicit representations of the target meshes where voxels with positives values are outside of the mesh and voxels with negative values are inside of the mesh. Then, by thresholding the binary union of these maps and running the standard marching cubes algorithm [16], we obtain a template mesh that is very tight around all training surfaces. However, this template mesh looks "blocky" with many small sharp edges and undesired topological defects. The template mesh is thus smoothed using the Laplacian smoothing algorithm [10, 26] and re-meshed with Delaunay triangulation [1]. The result is a smooth template mesh with spherical topology tightly wrapping all training set surfaces (see Fig. 1c). Finally, we apply a topology preserving mesh subdivision algorithm [2] to generate template meshes at different resolutions which are required to train CorticalFlow. Implementations of the used algorithms are available in the Libigl [12] and MeshLab [2] toolboxes.

3.4   White to Pial Surface Morphing

In Lebrat et al. [15], a separate CorticalFlow model is trained for each cortical surface (i.e., pial and white) and each brain hemisphere (i.e., left and right). This approach leads to reconstructed surface meshes without a one-to-one mapping between the vertices in the white and pial surfaces on the same brain hemisphere. In the absence of this mapping, many existent surface analysis tools can not process the generated surfaces. Additionally, as also observed in Ma et al. [17], the pial surface may only differ from the white surface by a "small" deformation thanks to the natural anatomical agreement between these surfaces. Therefore, we propose to predict pial surfaces by learning to deform the predicted white surfaces instead of using a pial surface template mesh.

Formally, the resulting model for pial surfaces with k deformation blocks (CFP^k) can be restated by the following recurrence,

$$\text{CFP}^1_{\theta_1}(\mathbf{I}, \mathcal{T}^w) = \text{DMD}(\text{UNet}^1_{\theta_1}(\mathbf{I}), \text{CFW}^{k'}(\mathbf{I}, \mathcal{T}^w))$$
$$\text{CFP}^{i+1}_{\theta_{i+1}}(\mathbf{I}, \mathcal{T}^w) = \text{DMD}(\text{UNet}^{i+1}_{\theta_{i+1}}(\mathbf{U}_1^\frown \cdots \mathbf{U}_i^\frown \mathbf{I}), \text{CFP}_i(\mathbf{I}, \mathcal{T}^w)) \quad \text{for } i \geq 1, \quad (6)$$

where CFW$^{k'}$ is a CorticalFlow model with k' deformation blocks pretrained to reconstruct the same hemisphere white surface as described in Sect. 3.1 and $\mathcal{T}^w$ is its respective template mesh for white surfaces generated as described in Sect. 3.3. Note that the formulation of CFW$^{k'}$ remains the one stated in Eq. 2 and we only use template meshes for the white surfaces since the pial surfaces are obtained by deforming the predicted white surface.

4   Experiments

We now evaluate CorticalFlow^{++} in the cortical surface reconstruction problem. First, using the CSR benchmark introduced by Santa Cruz et al. [23], we quantify the performance impact of each proposed modification separately. This benchmark consists of 3,876 MRI images extracted from the Alzheimer's Disease Neuroimaging Initiative (ADNI)[11] and their respective pseudo-ground-truth surfaces generated with the FreeSurfer V6.0 cross-sectional pipeline. We strictly follow the benchmark's data splits and evaluation protocol. We refer the reader to [15, 23] for full details on this dataset. As evaluation metrics for measuring

Table 1. Cortical Surface Reconstruction Benchmark [23]. ↓ indicates smaller metric value is better, while ↑ indicates greater metric value is better.

	Left Pial Surface				Right Pial Surface			
	CH(mm) ↓	HD(mm) ↓	CHN ↑	% SIF ↓	CH(mm) ↓	HD(mm) ↓	CHN ↑	% SIF ↓
CorticalFlow	0.681	0.802	0.932	0.686	0.693	0.815	0.929	1.239
3.22 sec / 2.82 GB	(±0.098)	(±0.049)	(±0.006)	(±0.469)	(±0.091)	(±0.046)	(±0.006)	(±0.629)
CorticalFlow + RK4	0.629	0.761	0.937	0.502	0.580	0.751	0.943	0.280
3.55 sec / 2.82 GB	(±0.100)	(±0.042)	(±0.006)	(±0.196)	(±0.082)	(±0.038)	(±0.006)	(±0.133)
CorticalFlow + W2P	0.545	0.730	0.943	0.188	0.540	0.729	0.945	0.176
3.63 sec / 2.82 GB	(±0.082)	(±0.037)	(±0.006)	(±0.116)	(±0.075)	(±0.033)	(±0.006)	(±0.134)
PialNN†	5.500	2.793	0.792	4.730	5.948	2.898	0.789	4.537
white gen. + 0.880 secs / 1.92 GB	(±0.786)	(±0.220)	(±0.009)	(±0.841)	(±0.811)	(±0.212)	(±0.011)	(±0.815)
PialNN*	1.388	1.251	0.864	10.507	1.374	1.236	0.863	11.159
white gen. + 0.880 secs / 1.92 GB	(±0.223)	(±0.120)	(±0.011)	(±1.908)	(±0.217)	(±0.115)	(±0.011)	(±1.930)
CorticalFlow++	0.529	0.721	0.946	0.069	0.528	0.723	0.946	0.099
3.76 sec / 2.82 GB	(±0.088)	(±0.036)	(±0.006)	(±0.060)	(±0.074)	(±0.031)	(±0.005)	(±0.093)
	Left White Surface				Right White Surface			
	CH(mm) ↓	HD(mm) ↓	CHN ↑	% SIF ↓	CH(mm) ↓	HD(mm) ↓	CHN ↑	% SIF ↓
CorticalFlow	0.608	0.785	0.941	0.033	0.599	0.783	0.942	0.030
3.22 sec / 2.82 GB	(±0.098)	(±0.060)	(±0.007)	(±0.030)	(±0.093)	(±0.059)	(±0.007)	(±0.029)
CorticalFlow + RK4	0.540	0.733	0.948	0.042	0.517	0.716	0.951	0.010
3.55 sec / 2.82 GB	(±0.107)	(±0.055)	(±0.006)	(±0.039)	(±0.089)	(±0.044)	(±0.006)	(±0.023)
CorticalFlow + NEWTPL	0.598	0.780	0.942	0.030	0.558	0.747	0.945	0.104
3.04 sec / 2.82 GB	(±0.101)	(±0.062)	(±0.007)	(±0.027)	(±0.091)	(±0.053)	(±0.006)	(±0.079)
CorticalFlow++	0.514	0.712	0.952	0.017	0.510	0.711	0.952	0.031
3.35 sec / 2.82 GB	(±0.090)	(±0.044)	(±0.006)	(±0.023)	(±0.083)	(±0.040)	(±0.006)	(±0.040)

Fig. 2. Predicted cortical surfaces color-coded with the distance to the pseudo-ground-truth surfaces. See our supplementary materials for more examples.

geometric accuracy, we use Chamfer distance (CH), Hausdorff distance (HD), and Chamfer normals (CHN). These distances are computed for point clouds of 200k points uniformly sampled from the predicted and target surfaces. As a measure of surface regularity, we compute the percentage of self-intersecting faces (%SIF) using PyMeshLab [19]. Finally, we also report the average time (in seconds) and the maximum GPU memory (in GB) required by the evaluated methods to reconstruct a cortical surface. Table 1 presents these results.

Table 2. Out-of-train-distribution evaluation on OASIS3 [14]. ↓ indicates smaller metric value is better, while ↑ indicates greater metric value is better.

	Left Pial Surface				Right Pial Surface			
	CH(*mm*) ↓	HD(*mm*) ↓	CHN ↑	% SIF ↓	CH(*mm*) ↓	HD(*mm*) ↓	CHN ↑	% SIF ↓
CorticalFlow	0.677	0.803	0.923	0.594	0.724	0.845	0.918	1.467
3.22 sec / 2.82 GB	(±0.099)	(±0.056)	(±0.007)	(±0.319)	(±0.106)	(±0.063)	(±0.007)	(±0.519)
PialNN†	5.426	2.763	0.781	4.281	5.944	2.873	0.777	4.033
white gen. + 0.880 secs / 1.92 GB	(±0.486)	(±0.144)	(±0.009)	(±0.709)	(±0.503)	(±0.139)	(±0.009)	(±0.671)
PialNN*	1.307	1.243	0.860	9.661	1.264	1.211	0.857	10.482
white gen. + 0.880 secs / 1.92 GB	(±0.202)	(±0.119)	(±0.013)	(±1.604)	(±0.194)	(±0.117)	(±0.013)	(±1.678)
CorticalFlow++	0.520	0.711	0.935	0.136	0.528	0.727	0.935	0.245
3.76 sec / 2.82 GB	(±0.082)	(±0.044)	(±0.006)	(±0.096)	(±0.079)	(±0.047)	(±0.006)	(±0.167)

We observe that the adoption of the $RK4$ ODE solver (CorticalFlow + RK4) and the white to pial surface morphing formulation (CorticalFlow + W2P) significantly improves the geometric accuracy and surface regularity of the CorticalFlow baseline. For instance, we noticed an average decrease of 12.2% and 21.02% in chamfer distance, respectively. Likewise, the percentage of self-intersecting faces reduces on average by 35.90% and 79.19%, respectively. On the other hand, the proposed new templates (CorticalFlow + NEWTPL) present a modest improvement in those criteria, but greatly succeeds in suppressing mesh artifacts as qualitatively shown in Fig. 2. Additionally, none of these changes incur a significant increase in reconstruction time or memory consumption due to their GPU friendly nature. All together, these modifications in CorticalFlow++ establish a new state-of-the-art method for cortical surface reconstruction.

Since the white to pial morphing approach has been previously introduced in [17], we compare CorticalFlow++ and PialNN [17] on the pial surface reconstruction. For this comparison, we report the performance of the publicly available pretrained PialNN model[1] (PialNN†) as well as training it by ourselves in the CSR benchmark training dataset (PialNN*). CorticalFlow++ compares favourably to both PialNN variants and importantly, it does not need traditional methods to generate white surfaces.

Finally, we extend our evaluation to an out-of-training-distribution dataset. More specifically, we use the trained models from the previous experiment to reconstruct cortical surfaces for a subset of MRIs extracted from the OASIS3 dataset [14]. These generated surfaces are also compared to FreeSurfer V6.0 pseudo-ground-truth surfaces using the same evaluation metrics described above. As shown in Table 2, CorticalFlow++ significantly outperforms CorticalFlow and PialNN, while presenting comparable surface reconstruction runtime and GPU memory consumption.

---

[1] https://github.com/m-qiang/PialNN.

5   Conclusion

This paper tackles some limitations of CorticalFlow, the current state-of-the-art model for Cortical surface reconstruction from MRI, in order to improve its accuracy, regularity, and interoperability without sacrificing its computational requirements for inference (reconstruction time and maximum GPU memory consumption). The resulting method, CorticalFlow++, achieves state-of-the-art performance on geometric accuracy and surface regularity while keeping the GPU memory consumption constant and adding less than a second to the entire surface reconstruction process.

Compliance with Ethical Standards

This research was approved by CSIRO ethics 2020 068 LR.

Acknowledgement. This work was funded in part through an Australian Department of Industry, Energy and Resources CRC-P project between CSIRO, Maxwell Plus and I-Med Radiology Network.

References

1. Bobenko, A.I., Springborn, B.A.: A discrete Laplace-beltrami operator for simplicial surfaces. Discrete Comput. Geom, **38**(4), 740–756 (2007)
2. Cignoni, P., Callieri, M., Corsini, M., Dellepiane, M., Ganovelli, F., Ranzuglia, G.: MeshLab: an open-source mesh processing tool. In: Scarano, V., Chiara, R.D., Erra, U. (eds.) Eurographics Italian Chapter Conference. The Eurographics Association (2008). https://doi.org/10.2312/LocalChapterEvents/ItalChap/ItalianChapConf2008/129-136
3. Dale, A.M., Fischl, B., Sereno, M.I.: Cortical surface-based analysis: I. segmentation and surface reconstruction. Neuroimage **9**(2), 179–194 (1999)
4. Du, A.T., et al.: Different regional patterns of cortical thinning in Alzheimer's disease and frontotemporal dementia. Brain **130**(4), 1159–1166 (2007)
5. Euler, L.: Institutiones Calculi Integralis, vol. 4. Academia Imperialis Scientiarum (1794)
6. Fischl, B.: Freesurfer. Neuroimage **62**(2), 774–781 (2012)
7. Fischl, B., et al.: Whole brain segmentation: automated labeling of neuroanatomical structures in the human brain. Neuron **33**(3), 341–355 (2002)
8. Fischl, B., Sereno, M.I., Dale, A.M.: Cortical surface-based analysis: II: inflation, flattening, and a surface-based coordinate system. Neuroimage **9**(2), 195–207 (1999)
9. Henschel, L., Conjeti, S., Estrada, S., Diers, K., Fischl, B., Reuter, M.: FastSurfer - a fast and accurate deep learning based neuroimaging pipeline. NeuroImage **219**, 117012 (2020)
10. Herrmann, L.R.: Laplacian-isoparametric grid generation scheme. J. Eng. Mech. Div. **102**(5), 749–756 (1976)
11. Jack Jr., C.R., et al.: The Alzheimer's disease neuroimaging initiative (ADNI): MRI methods. J. Magn. Reson. Imaging **27**(4), 685–691 (2008)

12. Jacobson, A., Panozzo, D., et al.: libigl: a simple C++ geometry processing library (2018). https://libigl.github.io/
13. Kim, J.S., et al.: Automated 3-D extraction and evaluation of the inner and outer cortical surfaces using a Laplacian map and partial volume effect classification. Neuroimage **27**(1), 210–221 (2005)
14. LaMontagne, P.J., et al.: Oasis-3: longitudinal neuroimaging, clinical, and cognitive dataset for normal aging and Alzheimer disease. MedRxiv (2019)
15. Lebrat, L., et al.: CorticalFlow: a diffeomorphic mesh transformer network for cortical surface reconstruction. In: Advances in Neural Information Processing Systems 34 (2021)
16. Lorensen, W.E., Cline, H.E.: Marching cubes: a high resolution 3D surface construction algorithm. ACM SIGGRAPH Comput. Graph. **21**(4), 163–169 (1987)
17. Ma, Q., Robinson, E.C., Kainz, B., Rueckert, D., Alansary, A.: PialNN: a fast deep learning framework for cortical pial surface reconstruction. In: Abdulkadir, A., et al. (eds.) MLCN 2021. LNCS, vol. 13001, pp. 73–81. Springer, Cham (2021). https://doi.org/10.1007/978-3-030-87586-2_8
18. MacDonald, D., Kabani, N., Avis, D., Evans, A.C.: Automated 3-D extraction of inner and outer surfaces of cerebral cortex from MRI. NeuroImage **12**(3), 340–356 (2000)
19. Muntoni, A., Cignoni, P.: PyMeshLab, January 2021. https://doi.org/10.5281/zenodo.4438750
20. Press, W., Flannery, B., Teukolsky, S.A., Vetterling, W.: Runge-Kutta Method. Numerical Recipes in FORTRAN: The Art of Scientific Computing, pp. 704–716 (1992)
21. Rimol, L.M., et al.: Cortical volume, surface area, and thickness in schizophrenia and bipolar disorder. Biol. Psychiatry **71**(6), 552–560 (2012)
22. Ronneberger, O., Fischer, P., Brox, T.: U-Net: convolutional networks for biomedical image segmentation. In: Navab, N., Hornegger, J., Wells, W.M., Frangi, A.F. (eds.) MICCAI 2015. LNCS, vol. 9351, pp. 234–241. Springer, Cham (2015). https://doi.org/10.1007/978-3-319-24574-4_28
23. Santa Cruz, R., Lebrat, L., Bourgeat, P., Fookes, C., Fripp, J., Salvado, O.: DeepCSR: A 3D deep learning approach for cortical surface reconstruction. In: Proceedings of the IEEE/CVF Winter Conference on Applications of Computer Vision, pp. 806–815 (2021)
24. Schaer, M., Cuadra, M.B., Tamarit, L., Lazeyras, F., Eliez, S., Thiran, J.P.: A surface-based approach to quantify local cortical gyrification. IEEE Trans. Med. imaging **27**(2), 161–170 (2008)
25. Shattuck, D.W., Leahy, R.M.: BrainSuite: an automated cortical surface identification tool. Med. Image Anal. **6**(2), 129–142 (2002)
26. Sorkine, O., Cohen-Or, D., Lipman, Y., Alexa, M., Rössl, C., Seidel, H.P.: Laplacian surface editing. In: Proceedings of the 2004 Eurographics/ACM SIGGRAPH Symposium on Geometry Processing, pp. 175–184 (2004)
27. Wang, N., Zhang, Y., Li, Z., Fu, Y., Liu, W., Jiang, Y.-G.: Pixel2Mesh: generating 3D mesh models from single RGB images. In: Ferrari, V., Hebert, M., Sminchisescu, C., Weiss, Y. (eds.) ECCV 2018. LNCS, vol. 11215, pp. 55–71. Springer, Cham (2018). https://doi.org/10.1007/978-3-030-01252-6_4

Carbon Footprint of Selecting and Training Deep Learning Models for Medical Image Analysis

Raghavendra Selvan[1,2]([✉]), Nikhil Bhagwat[3], Lasse F. Wolff Anthony[1], Benjamin Kanding[1], and Erik B. Dam[1]

[1] Department of Computer Science, University of Copenhagen, Copenhagen, Denmark
raghav@di.ku.dk
[2] Department of Neuroscience, University of Copenhagen, Copenhagen, Denmark
[3] McGill University, Montreal, Canada

Abstract. The increasing energy consumption and carbon footprint of deep learning (DL) due to growing compute requirements has become a cause of concern. In this work, we focus on the carbon footprint of developing DL models for medical image analysis (MIA), where volumetric images of high spatial resolution are handled. We present and compare the features of four tools from literature to quantify the carbon footprint of DL. Using one of these tools we estimate the carbon footprint of medical image segmentation pipelines. We choose nnU-net as the proxy for a medical image segmentation pipeline and experiment on three common datasets. With our work we hope to inform on the increasing energy costs incurred by MIA. We discuss simple strategies to cut-down the environmental impact that can make model selection and training processes more efficient.

Keywords: Energy consumption · Carbon emissions · Image segmentation · Deep learning

1 Introduction

"Global warming of 1.5 °C *and* 2 °C *will be exceeded during the 21st century unless* deep reductions[1] *in carbon dioxide (CO2) and other greenhouse gas emissions occur in the coming decades"*, reads one of the key points in the most recent assessment report of the Intergovernmental Panel on Climate Change (IPCC-2021) [20]. Without sounding as alarmists, the purpose of this work is to shine light on the increasing energy- and carbon- costs of developing advanced

---

[1] Underlining by us for emphasis.

Supplementary Information The online version contains supplementary material available at https://doi.org/10.1007/978-3-031-16443-9_49.

machine learning (ML) models within the, relatively small but rapidly growing, domain of medical image analysis (MIA).

The massive progress in computer vision tasks, as witnessed in MIA applications, in the last decade has been enabled by the class of ML methods that are now under the umbrella of Deep Learning (DL) [19,26]. The progress of DL methods have been accelerated by the access to big data sources and big compute resources. According to some estimates, the compute required for DL methods has been doubling every 3.4–6 months since 2010 [2,30]. However, this increasing compute also results in a proportional increase in demand for energy production. In 2010, energy production was responsible for about 35% of the global anthropogenic green house gas emissions [8]. The broad adoption and success of DL, while exciting, can also evolve into becoming a significant contributor to climate change due to its growing energy consumption.

Fig. 1. Word cloud showing the trends of papers in MICCAI'21. *Deep Learning* and *image segmentation* continue to be the dominant trend.

In MIA, the ingested data are not usually *big* in "sample size" compared to other natural image datasets (e.g. Imagenet [9]) but they are *big data*, in terms of their feature set as they can be volumetric and/or of higher spatial resolution (e.g. $256 \times 256 \times 256$ T1-weighted magnetic resonance imaging (MRI) scans). This necessitates novel model customization and selection strategies for model training. Nevertheless, several successes (e.g. U-net [24]), have generated huge interest in application of DL models in MIA. Particularly, the overall trend of increasing DL for tasks such as segmentation is also captured in venues such as MICCAI [7], where *deep learning* and *image segmentation* continue to be the most prominent trends, as captured in the keyword cloud in Fig. 1.[2]

In this work, we focus on studying the carbon footprint of DL for MIA using image segmentation as a case study. Segmentation could be a useful representative study, as in MICCAI-2021 there were a total of 143 papers out of the 531 accepted ones with *segment* in their title i.e., around 27% of all papers tackle medical image segmentation in some form or another [7]. We investigate the energy costs and carbon footprint of performing model selection and training

---

[2] https://www.jasondavies.com/wordcloud/.

the nnU-net segmentation model [15] on three popular datasets. In line with the MICCAI-2022 recommendations on reporting energy costs of training DL models, we present several methods from literature and compare their features. Finally, we present five simple steps to ensure good practices when developing and training DL models for MIA.

We present our work as a meta-analytic study to 1) benchmark carbon footprint of DL methods for MIA and 2) present tools and practices to reduce the compute costs of DL models during future pipeline development in MIA.

2   Related Work

In natural language processing (NLP), where some of the largest DL models are in use, the research community has taken notice of the growing energy consumption and its adverse impact on environmental sustainability and equitable research [4,32]. To draw attention to these issues, initiatives such as dedicated conference tracks focused on sustainable NLP methods are gaining traction.[3] There is also growing concern of the environmental impact within bioinformatics where massive data processing is carried out, and their sustainability is being discussed in recent publications focusing on the carbon costs of storage in data banks [25] and high performance computing [12,18]. Closer to the MIA community is the initiative by The Organization for Human Brain Mapping (OHBM) which has instituted a dedicated special interest group (SIG) to focus on the environmental sustainability of brain imaging. Members of this SIG have published an action plan for neuroscientists with suggestions ranging from optimizing computational pipelines to researchers flying less for academic purposes, including to conferences [23].[4]

3   Methods for Carbon Footprinting

Any DL model development undergoes three phases until deployment: Model Selection, Training, Inference. Widely reported performance measures of DL models such as the computation time primarily focus on the inference time, and in some cases on the training time of the final chosen model. While these two measures are informative they do not paint the complete picture. Any DL practitioner would agree that the primary resource intensive process in DL is during the model selection phase. This is due to the massive hyperparameter space for searching network architectures, and the resulting iterative model selection process. Moreover, the model selection and training procedures are highly iterative in practice, and many large projects implement them as continual learning. This further underscores the need for better estimation of the recurring costs associated with these procedures.

---

[3] https://2021.eacl.org/news/green-and-sustainable-nlp.
[4] https://ohbm-environment.org/.

The accounting of these costs is contingent on several factors. Reporting the computation time of model development, training and inference can improve transparency in DL to some extent. However, as the computation time is dependent on the infrastructure at the disposal of the researchers, only reporting the computation time provides a skewed measure of performance even when the hardware resources are described.

On the other hand, reporting the energy consumption (instead of computation time) of the model development provides a more holistic view that is largely agnostic of the specific hardware used.

Table 1. Feature comparison of four popular carbon tracking tools for ML. The table shows if the tools can be used for tracking (Track.), predicting (Pred.) and visualizing (Vis.) the carbon emissions. Further, if there are options to obtain some form of carbon impact statements (Report), if the tool can be installed as a python package (Pip) and the ease of integrating APIs for extending functionalities are also shown.

Tool	Track.	Pred.	Report	Pip	Vis.	API	CPU
MLEC [17]	✗	✗	✓	✗	✗	✗	✗
EIT [13]	✓	✗	✓	✓	✗	✗	✓
CarbonTracker [3]	✓	✓	✓	✓	✗	✗	✓
CodeCarbon [27]	✓	✗	✓	✓	✓	✓	✓

In the past couple of years, several tools have been developed to help researchers estimate the energy costs and carbon emissions of their model development; four of these with different features are summarized in Table 1 and described below:

1. **Machine Learning Emissions Calculator (MLEC)** [17]: This is a self-reporting interface where users input training time, hardware and geographic location post-hoc to estimate the carbon emissions. It currently only estimates based on the energy consumption of only the graphics processing unit (GPU).
2. **Experiment-Impact-Tracker (EIT)** [13]: This tool can be used to track the energy costs of CPU and GPU during the model training by embedding few lines of code into the training scripts. The tool can be tweaked to fetch real-time carbon intensity; currently this feature is supported for California (US).
3. **Carbontracker** [3]: This tool is similar to EIT in several aspects. In addition to the tracking capabilities of EIT, Carbontracker can also predict the energy consumption based on even a single run of the model configuration. This can be useful for performing model selection based on carbon emissions without having to train all the way. This tool currently provides real-time carbon emissions for Denmark and the United Kingdom, and can be extended to other locations.

4. **CodeCarbon** [27]: This is the most recent addition to the carbon tracking tools and is the best maintained. It provides tracking capabilities like EIT and Carbontracker, along with comprehensive visualizations. The application programming interface (API) integration capabilities are best supported in CodeCarbon which can be useful for extending additional functionalities.

All the four tools are open-source (MIT License), provide easy report generation that summarize the energy consumption and present interpretable forms of carbon emission statements f.x, as distance travelled by car to equal the carbon emissions.[5]

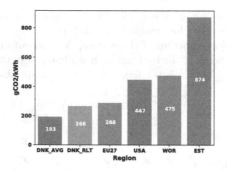

Fig. 2. Average carbon intensity over an year, which is the ratio of CO2 emissions from public electricity production and gross electricity production, for different regions of the world [11]. DNK_RLT is based on the real-time monitoring of the carbon intensity in Denmark over a period of one week in 02/2022. The remaining values are based on the average intensities for Denmark (DNK_AVG), Estonia (EST) and EU countries (EU-27) as reported by government agencies in the European Union (EU) [11] and USA [33]. The global average (WOR) is obtained from [10]. In general, higher ratio of coal powered energy production increases the average carbon intensity, as is evident when comparing DNK_AVG to EST.

4   Data and Experiments

We use medical image segmentation as a case study to present some observations on the energy costs and carbon emissions of performing model selection and training them. We use the recent nnU-net [15] as the segmentation method as it integrates a comprehensive and automatic framework for segmentation model development, including preprocessing and post-processing steps.

Data: We present experiments on three diverse MIA datasets for segmentation:

1. **MONuSeg** [16], which is a multi-organ nuclei segmentation dataset in 2D. The dataset consists of 30 images for training/validation and 14 for testing purposes. Each image is of 1000×1000 px resolution and comprising several thousand nuclei annotations per image.

---

[5] When the trackers cannot fetch real-time carbon intensity of energy for the specific geographic location, most resort to using some average estimate from a look-up table.

2. **Heart** segmentation dataset from the medical decathlon [31] and consists of 30 mono-modal MRI scans of the entire heart, with the segmentation target of the left atrium. The dataset is split into 20 for training and 10 for testing.
3. **Brain** segmentation dataset, which is also from the medical decathlon [31] consisting of 4D volumes (multiple-modalities per image) with different types of tumours as the target for segmentation. The dataset consists of 484 volumes for training and 266 for testing purposes.

Experimental Set-Up: We use the official version of nnU-net implemented in Pytorch [22] and integrate Carbontracker [3] into the network training script to track/predict the energy consumption and carbon emissions, as shown in Appendix B in the online supplementary material [29]. All experiments were performed on a desktop workstation with GeForce RTX 3090 GPU with 24 GB memory, Intel-i7 processor and 32 GB memory. The default configuration of nnU-net uses five-fold cross validation on the training set and a maximum of 1000 epochs per training fold. We follow the K-fold cross validation set-up, with $K = 5$ for all datasets and train MONuSeg for 1000 epochs, Heart dataset for 100 epochs and Brain dataset for 50 epochs to reduce the computational cost. We use the predictive model in Carbontracker to report the energy consumption for the full 1000 epochs based on shorter runs [3]. Carbontracker provides real-time carbon emissions for only few regions. To demonstrate the difference in carbon emissions if the experiments were carried out in different geographical regions, we use the average carbon intensity (gCO2/kWh) queried from several sources and are shown along with the sources in Fig. 2.

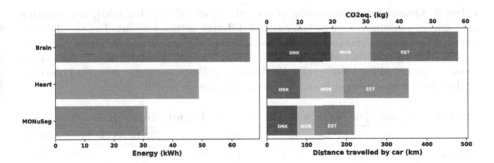

Fig. 3. (left) Total predicted energy consumed (kWh) over the five-fold cross validation for the three datasets using nnU-net[15]. For MONuSeg the predicted (orange) and the actual energy consumptions are shown, which are almost the same. (right) Carbon cost due to the training on the three datasets reported in CO2eq.(kg) and equivalent distance travelled by car (km). The carbon intensity and distance are also reported for three geographic regions (Denmark:DNK, Estonia: EST, Global: WOR) based on the regional average carbon intensities. All measurements were tracked/predicted using Carbontracker[3].

Results: The total predicted energy cost of running nnU-net on the three datasets over the five-fold cross validation is shown in Fig. 3 (left). For MONuSeg we have also plotted the predicted energy consumption after the first epoch of training (with transparency and green error bars) and we see there is no substantial difference between the actual- and predicted- energy consumption as estimated from Carbontracker. As the Brain dataset is 4D and has many more data points, the aggregate energy cost is highest compared to the Heart- and MONuSeg datasets.

In Fig. 3 (right) and Table 2, the carbon emissions for performing these experiments in different regions (Denmark: DNK, Global average: WOR, Estonia: EST) are reported, along with the corresponding distance travelled by car as reported from Carbontracker. Due to the high average carbon intensity in EST, the carbon emissions are easily three times higher than in DNK or in other words, the carbon emissions in EST are equivalent to driving about 480 km compared to 146 km for a single nnU-net run on the Brain dataset.

The default configuration of nnU-net uses automatic mixed precision (AMP) scheme which is known to be efficient by reducing computational overhead by adaptively reducing the quantization (float32 to float16) without significant loss in performance [21]. To investigate the reduction in energy costs due to AMP, we re-run all the experiments for 10 epochs using full precision. We then use Carbontracker to predict the energy consumption and carbon emissions for 1000 epochs. These results are reported in Table 2 where a clear reduction in total energy consumed and the carbon emissions (reported as distance travelled) for all three datasets is noticed, with larger gains in the 3D experiments.

Table 2. Quantitative comparison of the *total* (over all training folds) computation time, energy consumed and carbon costs of different settings reported as the distance travelled by car, for different datasets and different regions. To demonstrate the influence of training precision we compare the full and mixed precision modes. All experiments are reported for 1000 epochs. For Heart and Brain datasets, they were based on predicted estimates using 100 and 50 epoch runs, respectively.

Dataset	Comp. time	Energy (kWh)		DNK-RLT (km)		Region (km)		
		mixed	*full*	*mixed*	*full*	DNK_AVG	EST	WOR
MONuSeg	61 h 55 m	30.2	35.1	75.5	80.2	66.8	219.6	119.3
Heart	116 h 40 m	48.9	65.8	83.2	128.5	108.1	355.1	192.9
Brain	368 h 20 m	66.1	84.6	160.8	170.4	146.1	479.8	260.4

5   Discussions

In Table 2, we reported the carbon emissions (as distance travelled by car) for Denmark under two settings: DNK_RLT and DNK_AVG, differing based on their average carbon intensity. DNK_RLT is based on real-time carbon emissions which during the course of the experiments was about 266 gCo2/kWh, whereas the

DNK_AVG retrieved from [11] for 2018 is $193\,\mathrm{gCO2/kWh}$. The difference could be due to the increased loads during winters in Denmark, caused by additional coal powered energy. This highlights the need for *tracking* the carbon emissions in real-time instead of self-reporting and also to integrate APIs for live carbon intensity fetching for all the regions into the carbon tracking tools presented in Sect. 3.

We used nnU-net as a proxy for extensive hyperparameter tuning, as it also integrates hyperparameter selection for medical image segmentation. Whenever possible, DL practitioners should use improved *hyperparameter optimization* strategies other than grid search. For instance, simple strategies like random search [5] to more complex methods such as hypernetworks [6,14] can significantly reduce the computation time needed during model selection.

Using *energy-efficient hardware* and settings can also help reduce carbon emissions. For instance, power management techniques like dynamic voltage and frequency scaling (DVFS) have been shown to conserve up to 23.1% energy consumption by operating at the optimal core frequency instead of the default one [35]. In instances where the user does not have full control over the hardware configuration, accessing cloud-based hardware can be more efficient as some of the datacenters might have better power usage effectiveness (PUE) than locally maintained hardware.

Figure 3 (right) points to an interesting observation. Depending on the region where the training is performed, the carbon emissions can vary drastically. This is due to the regional carbon intensity as shown in Fig. 2 which reflects the extent of clean and coal-powered energy in specific geographic locations. In the past couple of years, major datacenters are offering options to select datacenter locations for computations. Whenever users are able to control this, the preferred choice should be to use infrastructure that uses cleaner energy. Further, even within a given geographic location the carbon intensity can vary depending on the time of the day [3,13]. Using carbon tracking tools can provide insight into choosing the optimal times to schedule jobs in order to reduce the carbon emissions. Job schedulers like slurm can use this information to minimize the energy consumption.

As reported in Table 2, the carbon emissions of using nnU-net on the two datasets with *AMP* is consistently lower than operating in full precision mode. Frameworks such as Pytorch [22] and Tensorflow offer easy settings to operate in AMP without loss in performance[21].

There are several small steps users can take to reduce the carbon emissions of their model development pipeline. We summarize the discussion from above into five key points, or our THETA-guidelines for reducing carbon emissions during model development:

1. **T**rack-log-report for control and transparency
2. **H**yperparameter optimization frameworks instead of grid search
3. **E**nergy-efficient hardware and settings are useful
4. **T**raining location and time of day/year are important
5. **A**utomatic mixed precision training always

In addition to these steps, open-science practices such as making pretrained models and pre-processed data available can reduce the repeated computational cost and improve reproducibility in MIA.

6   Conclusions

As a final note, consider MICCAI-2021 where 143/531 accepted papers had *segment* in their title. Many reviewers look for strong baseline comparison, and nnU-net is a thorough baseline for performing model comparison. If we assume, all these papers reported experiments on at least one 2D dataset, the total carbon emissions due to these papers would be $143 \times 14.7 \approx 2102$ kgCO2eq, where we use the global average carbon emissions from Fig. 3. However, MICCAI-2021 had an acceptance rate of about 30%. This implies the underlying energy costs of training a baseline model on one 2D dataset is about three-fold. i.e. $2102 \times 3 = 6306$ kgCO2eq. This is the annual carbon footprint of about 27 people from a low income country, where the annual per-capita carbon footprint 236.7 kgCO2eq [34]! These estimates are based on very conservative estimates as most papers report more than one baseline and on multiple datasets.

Most of the suggestions in this work, including the THETA-guidelines, encourage more efficient methods or use of cleaner energy. Using more efficient methods or clean energy alone will not reduce the environmental impact of DL in MIA, as more efficient methods could result in growing demands and hence increased energy consumption. This effect, known as Jevons paradox, could nullify the gains made by improving efficiency [1]. Some suggestions to counteract this paradox is to use rationing or quota of resources. For instance, model selection could be constrained to the carbon budget, which was briefly addressed in [28].

In conclusion, we tried to quantify the energy cost of developing medical image segmentation models on three datasets using nnU-net, which performs an exhaustive hyperparameter tuning. We presented four carbon tracking tools from literature and made a comparative assessment of their features. Based on our experiments we recommended the THETA-guidelines for reducing the carbon footprint when using DL in MIA. We hope that the community will find our suggestions on tools and practices useful; not only in reporting the energy costs but also to act on reducing them.

In line with the recommendations in this work, we report the carbon footprint of all the experiments in this work using the report generated from Carbontracker[3]:

The training of models in this work is estimated to use 39.9 kWh of electricity contributing to 11.4 kg of CO2eq. This is equivalent to 94.9 km travelled by car.

Acknowledgments. The authors would like to thank members of OHBM SEA-SIG community for insightful and thought-provoking discussions on environmental sustainability and MIA.

References

1. Alcott, B.: Jevons' paradox. Ecol. Econ. **54**(1), 9–21 (2005)
2. Amodei, D., Hernandez, D., Sastry, G., Clark, J., Brockman, G., Sutskever, I.: AI and compute (2018). https://blog.openai.com/aiand-compute
3. Anthony, L.F.W., Kanding, B., Selvan, R.: Carbontracker: tracking and predicting the carbon footprint of training deep learning models. In: ICML Workshop on Challenges in Deploying and monitoring Machine Learning Systems, July 2020. arXiv:2007.03051
4. Bender, E.M., Gebru, T., McMillan-Major, A., Shmitchell, S.: On the dangers of stochastic parrots: can language models be too big? In: Proceedings of the 2021 ACM Conference on Fairness, Accountability, and Transparency, pp. 610–623 (2021)
5. Bergstra, J., Bengio, Y.: Random search for hyper-parameter optimization. J. Mach. Learn. Res. **13**, 281–305 (2012)
6. Brock, A., Lim, T., Ritchie, J., Weston, N.: SMASH: one-shot model architecture search through hypernetworks. In: International Conference on Learning Representations (2018). https://openreview.net/forum?id=rydeCEhs-
7. de Bruijne, M., et al. (eds.): MICCAI 2021, Part III. LNCS, vol. 12903. Springer, Cham (2021). https://doi.org/10.1007/978-3-030-87199-4
8. Change, I.C., et al.: Mitigation of climate change. Contribution of Working Group III to the Fifth Assessment Report of the Intergovernmental Panel on Climate Change, vol. 1454, p. 147 (2014)
9. Fei-Fei, L., Deng, J., Li, K.: ImageNet: Constructing a large-scale image database. J. Vis. **9**(8), 1037 (2009)
10. Energy, G.: CO2 Status Report—The Latest Trends in Energy and Emissions in 2018. International Energy Agency (2019). https://www.iea.org/reports/global-energy-co2-status-report-2019/emissions
11. European-Environment-Agency: Greenhouse gas emission intensity of electricity generation by country (2022). https://www.eea.europa.eu/data-and-maps/daviz/co2-emission-intensity-9/
12. Grealey, J., et al.: The carbon footprint of bioinformatics. BioRxiv (2021)
13. Henderson, P., Hu, J., Romoff, J., Brunskill, E., Jurafsky, D., Pineau, J.: Towards the systematic reporting of the energy and carbon footprints of machine learning. J. Mach. Learn. Res. **21**(248), 1–43 (2020)
14. Hoopes, A., Hoffmann, M., Fischl, B., Guttag, J., Dalca, A.V.: HyperMorph: amortized hyperparameter learning for image registration. In: Feragen, A., Sommer, S., Schnabel, J., Nielsen, M. (eds.) IPMI 2021. LNCS, vol. 12729, pp. 3–17. Springer, Cham (2021). https://doi.org/10.1007/978-3-030-78191-0_1
15. Isensee, F., Jaeger, P.F., Kohl, S.A., Petersen, J., Maier-Hein, K.H.: nnU-Net: a self-configuring method for deep learning-based biomedical image segmentation. Nat. Methods **18**(2), 203–211 (2021)
16. Kumar, N., et al.: A multi-organ nucleus segmentation challenge. IEEE Trans. Med. Imaging **39**(5), 1380–1391 (2019)
17. Lacoste, A., Luccioni, A., Schmidt, V., Dandres, T.: Quantifying the carbon emissions of machine learning. Technical report (2019)
18. Lannelongue, L.: Carbon footprint: the (not so) hidden cost of high performance computing. ITNOW **63**(4), 12–13 (2021)
19. LeCun, Y., Bengio, Y., Hinton, G.: Deep learning. Nature **521**(7553), 436–444 (2015)

20. Masson-Delmotte, V., Zhai, P., Pirani, A., Connors, S., et al. (eds.): Summary for Policymakers. Climate Change 2021: The Physical Science Basis. Contribution of Working Group I to the Sixth Assessment Report of the Intergovernmental Panel on Climate Change (2021)

21. Micikevicius, P., et al.: Mixed precision training. In: International Conference on Learning Representations (2018). https://openreview.net/forum?id=r1gs9JgRZ

22. Paszke, A., et al.: PyTorch: an imperative style, high-performance deep learning library. In: Advances in Neural Information Processing Systems, pp. 8024–8035 (2019)

23. Rae, C., Farley, M., Jeffery, K., Urai, A.E.: Climate crisis and ecological emergency: why they concern (neuro) scientists, and what we can do (2021)

24. Ronneberger, O., Fischer, P., Brox, T.: U-Net: convolutional networks for biomedical image segmentation. In: Navab, N., Hornegger, J., Wells, W.M., Frangi, A.F. (eds.) MICCAI 2015. LNCS, vol. 9351, pp. 234–241. Springer, Cham (2015). https://doi.org/10.1007/978-3-319-24574-4_28 http://arxiv.org/abs/1505.04597

25. Samuel, G., Lucivero, F., Lucassen, A.: Sustainable biobanks: a case study for a green global bioethics. Glob. Bioeth. **33**(1), 50–64 (2022)

26. Schmidhuber, J.: Deep learning in neural networks: an overview. Neural Netw. **61**, 85–117 (2015)

27. Schmidt, V., et al.: CodeCarbon: estimate and track carbon emissions from machine learning computing (2021). https://doi.org/10.5281/zenodo.4658424

28. Selvan, R.: Carbon footprint driven deep learning model selection for medical imaging. In: Medical Imaging with Deep Learning (Short Paper Track) (2021)

29. Selvan, R., Bhagwat, N., Anthony, L.F.W., Kanding, B., Dam, E.B.: Carbon footprint of selecting and training deep learning models for medical image analysis (supplementary material). arXiv preprint arXiv:2203.02202 (2022)

30. Sevilla, J., Heim, L., Ho, A., Besiroglu, T., Hobbhahn, M., Villalobos, P.: Compute trends across three eras of machine learning. arXiv preprint arXiv:2202.05924 (2022)

31. Simpson, A.L., et al.: A large annotated medical image dataset for the development and evaluation of segmentation algorithms. arXiv preprint arXiv:1902.09063 (2019)

32. Strubell, E., Ganesh, A., McCallum, A.: Energy and policy considerations for deep learning in NLP. In: Proceedings of the 57th Annual Meeting of the Association for Computational Linguistics, pp. 3645–3650 (2019)

33. U.S. Energy-Information-Administration: United states electricity profile 2020 (2022). https://www.eia.gov/electricity/state/unitedstates/

34. Worlbank-DataBank: Sustainable Development Goals (SDGs): CO2 emissions per capita (2022). Data retrieved from World Development Indicators on 28/02/2022. https://databank.worldbank.org/source/sustainable-development-goals-(sdgs)

35. Yang, T.J., Chen, Y.H., Sze, V.: Designing energy-efficient convolutional neural networks using energy-aware pruning. In: Proceedings - 30th IEEE Conference on Computer Vision and Pattern Recognition, CVPR 2017. vol. 2017, pp. 6071–6079, November 2017. https://doi.org/10.1109/CVPR.2017.643. https://arxiv.org/abs/1611.05128

SMESwin Unet: Merging CNN and Transformer for Medical Image Segmentation

Ziheng Wang[1], Xiongkuo Min[2], Fangyu Shi[2], Ruinian Jin[1], Saida S. Nawrin[1], Ichen Yu[3], and Ryoichi Nagatomi[1,3(✉)]

[1] Division of Biomedical Engineering for Health and Welfare,
Graduate School of Biomedical Engineering, Tohoku University, Sendai, Japan
{wang.ziheng.a3,ruinian.jin.d2,saida.salima.nawrin.s2,
ryoichi.nagatomi.c4}@tohoku.ac.jp
[2] The Institute of Image Communication and Network Engineering,
Shanghai Jiao Tong University, Shanghai, China
{minxiongkuo,fangyu.shi}@sjtu.edu.cn
[3] Department of Medicine and Science in Sports and Exercise,
Graduate School of Medicine, Tohoku University, Sendai, Japan
yuichen@med.tohoku.ac.jp

Abstract. Vision transformer is the new favorite paradigm in medical image segmentation since last year, which surpassed the traditional CNN counterparts in quantitative metrics. The significant advantage of ViTs is to utilize the attention layers to model global relations between tokens. However, the increased representation capacity of ViTs comes with corresponding shortcomings: short of CNN's inductive biases (locality), translation invariance, and hierarchical structure of visual information. Consequently, well-trained ViTs require more data than CNNs. As high quality data in medical imaging area is always limited, we propose SMESwin UNet. Firstly, based on Channel-wise Cross fusion Transformer (CCT) we fuse multi-scale semantic features and attention maps by designing a compound structure with CNN and ViTs (named MCCT). Secondly, we introduce superpixel by dividing the pixel-level feature into district-level to avoid the interference of meaningless parts of the image. Finally, we used External Attention to consider the correlations among all data samples, which may further reduce the limitation of small datasets. According to our experiments, the proposed superpixel and MCCT-based Swin Unet (SMESwin Unet) achieves better performance than CNNs and other Transformer-based architectures on three medical image segmentation datasets (nucleus, cells, and glands).

1 Introduction

Medical image segmentation aims to improve the accuracy and efficiency of medical examinations by identifying and highlighting anatomical or pathological changes in images. Due to the rapid emergence of deep learning techniques,

L. Wang et al. (Eds.): MICCAI 2022, LNCS 13435, pp. 517–526, 2022.
https://doi.org/10.1007/978-3-031-16443-9_50

it has become widely used in the field of medical imaging [1]. Since 2015, UNet structures began to dominate this field where ViT-based UNets began to take in the position in early 2021.

U-Net [2], a CNNs based model which consists of a symmetric encoder-decoder network with skip-connections to improve detail preservation, has had considerable success in a range of medical imaging applications. 3D U-Net [3], Res-UNet [4], UNet++ [5], and UNet3+ [6] continue such practices and become representative descendants. However, as the explicit global and long-range semantic information interaction [2] may help in constructing a robust model, the intrinsic locality of convolution operation makes CNN fail to model this relationship. Researchers naturally turn to ViT [7] for more possibilities: TransFuse [8] and MedT [9] combine CNNs and Transformers in one architecture as two branches, where global dependency and low-level spatial details can be captured more effectively. Swin UNet [10] is the first pure Transformer-based U-shaped network and outperforms models with full CNNs or the hybrid of Transformers and CNNs.

Although the aforementioned ViT-based networks achieved good performance (around 90% in GlaS dataset and 80% in MoNuSeg dataset), there are always rooms for improvement. It was observed that the transformer-based models work well only when they were trained on large-scale datasets [11]. Specifically, ViT is trained with JFT-300M [11] which has 303 million labeled high-resolution images but performs worse than ResNets [12] with similar capacity when trained on ImageNet-1K [13] (1.3 million samples).

As the scales of high-quality samples in medical datasets, on average, are not very large. A burning issue arises: how to train the ViT-based model more sufficiently for small datasets. To overcome this limitation, MedT [9] uses a parallel-training strategy consist of Transformer and CNN, which may focus on finer details presented in the local patches and shows a good performance. UCtransnet [14] replaces skip connection with a multi-scale Channel Cross fusion with Transformer (CCT), which may effectively connect features from encoder and decoder to eliminate the ambiguity. However, UCTransnet [14] still utilizes CNN for both shallow and deep information. Therefore, incorporating CNN for local information extraction and ViT for semantic information modeling in skip connection is possible to enhance the model performance in datasets of small volumes.

Inspired by the above works, we combine CNN and ViT, and uptake the superpixel conception into a novel model, which is dubbed as superpixel, mixed skip connection and External attention-based Swin Unet (named SMESwin UNet). To be concrete, we first incorporate superpixel which is an image segmentation technique. According to the similarity of features between neighboring pixels, superpixel divide the pixel-level feature into district-level to generate the low-level encoding feature map where remove redundant information. Then we propose a mixed skip connection (named MCCT) by combining CNN for local details description and ViTs to encode multi-scale features through global self-attention. Finally, we add External Attention (EA) [15] in MCCT, which implicitly considers the correlations among all data samples. The new skip connection module can be easily embedded in and applied for any UNet-based architectures. To the best of our knowledge, this is the first attempt to merge superpixel and EA in skip connection.

Extensive experiments show that our SMESwin UNet improves segmentation pipelines by absolute gains of 2.05% Dice, 0.3% Dice and 0.5% Dice over Swin UNet on GlaS [16], MoNuSeg [17] and WBCs [18] datasets, respectively. Moreover, we made a ablation study to investigate how the components of SMESwin Unet work. The results demonstrate that the superpixel, MCCT and External attention scheme generally leads to a better performance. We found that SMESwin UNet can serve as a strong skip connection scheme for medical image segmentation.

2   SMESwin Unet for Medical Image Segmentation

The overall architecture of the proposed SMESwin Unet is presented in Fig. 1. Current ViT-based segmentation architectures, such as MedT [9] and Swin Unet [10], blend the ViT with Unet by plugging the ViT module into the encoder and decoder. In this study, we use the SOTA architecture Swin Unet [10] as the basic encoder-decoder structure. As we mentioned before, the information from ViTs and CNNs are different, medical image segmentation datasets may not be enough for training ViTs. Therefore, we reconstruct Swin Unet [10] by optimizing the skip connection (named MCCT) based on CCT [14]. MCCT allows combining deep, global, multi-scale feature maps from ViT layers and shallow, low-level, fine-grained feature maps from the CNN layer through the skip connection. Further, we utilize superpixel segmentation on the shallow CNN-feature maps to remove meaningless information. Besides, to capture the correlation among all data samples, we add EA [15] in MCCT.

2.1   Encoder-Decoder Architecture

We follow the Swin Unet [10] encoder-decoder architecture. The input images $\mathbf{x} \in \mathbb{R}^{H \times W \times 3}$ are split into non-overlapping patches d with a resolution of $\frac{H}{4} \times \frac{W}{4}$. The patches are then flattened and passed into a linear embedding layer with the output dimension C_i, and the raw embedding sequence $\mathbf{e} \in \mathbb{R}^{d \times C_i}$ is obtained. The transformed patch tokens pass through several Swin Transformer blocks [19] to extract global information and patch merging layers which can down-sample and increase dimension. The decoder is consisted of Swin Transformer blocks [19] and patch expanding layers. A patch expanding layer is to perform up-sampling with 2× up-sampling of resolution. Specially, the last patch expanding layer performs 4× up-sampling to restore the feature maps to the input resolution. Finally, these up-sampled features pass through a linear projection layer to output the pixel-level segmentation predictions.

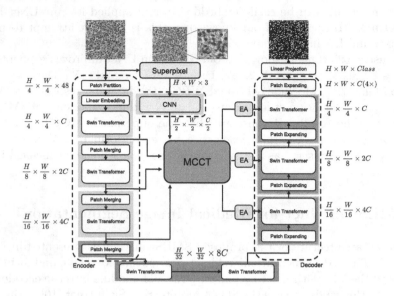

Fig. 1. The main architecture of SMESwin Unet which optimizes skip connection for training.

2.2   Superpixel

We reduce the invalid information from the original input image by a superpixel segmentation branch. To minimize the complexity of the model, we use The simple linear iterative clustering (SLIC) [20] which does not need any training for superpixel segmentation. The key steps are as follow:

1) **Clustering center** Assume the original image includes P pixels and N clusters for superpixels. The distance between adjacent clustering centers approximates to $S = \sqrt{P/N}$. In the process of clustering center initialization, within the scope of window n × n, replace the original clustering center with the minimum gradient position.

2) **Clustering threshold value** With two pixels from images i and j, the color dissimilarity of pixels $\mathbf{d}_{lab}$ could be defined as:

$$\mathbf{d}_{lab} = \sqrt{\left(l_i - l_j\right)^2 + \left(a_i - a_j\right)^2 + \left(b_i - b_j\right)^2}, \tag{1}$$

and $\mathbf{d}_{xy}$ is the Euclidean distance of pixels i and j, as:

$$\mathbf{d}_{xy} = \sqrt{\left(x_i - x_j\right)^2 + \left(y_i - y_j\right)^2}, \tag{2}$$

continuously, the similarity between them through LAB color space values and XY coordinates, could be defined as:

$$D_s = \mathbf{d}_{lab} + \frac{m}{S}\mathbf{d}_{xy}, \tag{3}$$

where D_s is the clustering threshold value in which bigger value indicates higher pixel similarity; m is a balance parameter which controls color differentiation and space distance proportion of the clustering threshold. SLIC utilizes the clustering center to search the $2S \times 2S$ neighborhood: (1) if the similarity of the pixels is above the clustering threshold, the pixel should cluster to the corresponding center, and (2) assign the clustering center's sign mark to the corresponding pixels. Repeat this iterative process until convergence. Then SLIC generates the feature map with superpixel segmentation, and pass the feature map to a simple CNN in the next skip connection module.

2.3   MCCT

Following CCT in UCTransnet [14], we propose a novel module (MCCT) which mixes CNN and ViT. MCCT has better detail capture ability and needs fewer calculations than CCT. As shown in Fig. 2, the two major differences between MCCT and CCT are: 1) one new connection with CNN, and 2) the fourth connection is deleted. In order to keep the detail information, we use a CNN layer (d_0) for the superpixel feature map. For the features from CNN layer, tokenization occurs from CCT [14] by flattening the encoder features into 2D patches with the size of 4×4, which has the same size with the patch tokens from transformer blocks. We utilize tokens $\mathbf{T}_i \in \mathbb{R}^{\frac{HW}{i^2} \times C_i}$ as the queries and $\mathbf{T}_\Sigma = \text{Concat}(\mathbf{T}_0, \mathbf{T}_1, \mathbf{T}_2, \mathbf{T}_3)$ as key and value in Multi-head Cross-Attention [14]. Then, following by a Multi-Layer Perceptron to encode channel and dependencies for refining features, as:

$$\mathbf{Q}_i = \mathbf{T}_i W_{\mathbf{Q}_i}, \mathbf{K} = \mathbf{T}_\Sigma W_{\mathbf{K}}, \mathbf{V} = \mathbf{T}_\Sigma W_{\mathbf{V}}, \tag{4}$$

where $W_{\mathbf{Q}_i} \in \mathbb{R}^{C_i \times d}$, $W_{\mathbf{K}} \in \mathbb{R}^{C_\Sigma \times d}$, $W_{\mathbf{V}} \in \mathbb{R}^{C_\Sigma \times d}$ are weights of different inputs, C_i ($i = 0, 1, 2, 3$) are the channel dimensions of the encoding layers (one CNN layer and three Swin transformer blocks), our implementation consists of: $C_0 = 48$, $C_1 = 96$, $C_2 = 192$, $C_3 = 384$, and d is the sequence length. Then according to MCCT, with $\mathbf{Q}_i \in \mathbb{R}^{C_i \times d}$, $\mathbf{K} \in \mathbb{R}^{C_\Sigma \times d}$, $\mathbf{V} \in \mathbb{R}^{C_\Sigma \times d}$, produce the similarity matrices $\mathbf{SM}_i$ which weight value $\mathbf{V}$ through a cross-attention (CA) [14] mechanism, as:

$$\text{CA}_i = \mathbf{SM}_i \mathbf{V}^\top = \sigma \left[\psi \left(\frac{\mathbf{Q}_i^\top \mathbf{K}}{\sqrt{C_\Sigma}} \right) \right] \mathbf{V}^\top, \tag{5}$$

where $\psi(\cdot)$ denote the instance normalization [21], and $\sigma(\cdot)$ is the softmax function. In a N-head attention situation, the feature map after multi-head cross-attention [14] is calculated as:

$$\begin{aligned} \mathbf{F}_i = &\left(\text{CA}_i^1 + \text{CA}_i^2 +, \ldots, +\text{CA}_i^N\right)/N + \\ &\text{MLP}\left(\mathbf{Q}_i + \left(\text{CA}_i^1 + \text{CA}_i^2 +, \ldots, +\text{CA}_i^N\right)/N\right), \end{aligned} \tag{6}$$

where, to build a 4-layer transformer [14], the process in Eq. (6) is repeated four times.

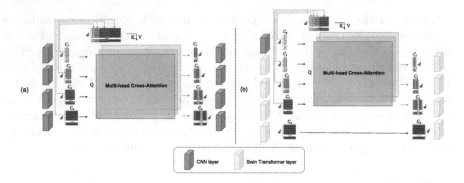

Fig. 2. The comparison of CCT (a) and MCCT (b).

2.4   External Attention

Hereinafter, for full application of correlations to small sample data sets, we employ the EA [15] which uses a cascade of two linear layers and two normalization layers, it can be easily implemented as a pair of small, external and shared memories. EA facilitates better learning of the features in the same scale in the whole dataset and better grasp of the differences between different data. Here, using different memory units M_k and M_v as the key and value, then we can get output as:

$$A_i = \text{Norm}\left(\mathbf{F}_i M_{k_i}^T\right)$$
$$O_i = A_i M_{vi}, \tag{7}$$

where an input feature map $\mathbf{F}_i \in \mathbb{R}^{d \times C_i}$ and d is the number of elements and C is the number of feature dimensions, A_i is the attention matrix of the i-th feature. Finally, the three outputs of the layer O_1, O_2, and O_3 are reconstructed and concatenated with the decoder features D_1, D_2, and D_3, respectively.

3   Experiments and Results

3.1   Dataset

We use GlaS (Gland) [16], MoNuSeg (Nuclei) [17] and WBCs (cell) [18] for evaluating our method. GlaS contains microscopic images of Hematoxylin and Eosin (H&E) stained slides and the corresponding ground truth annotations by expert pathologists. GlaS has 85 images for training and 80 for testing. Images at 40x magnification of H&E stained tissue were used to create MoNuSeg which contains images across multiple organs and patients. MoNuSeg dataset has 30 images with around 22000 nuclear boundary annotations for training and 14 images with over 7000 nuclear boundary annotations for testing. WBCs dataset selected cells are from one of the main five categories of WBCs: Neutrophils, Lymphocytes, Eosinophils, Monocytes, or Basophils. WBCs has 1214 images for training and 250 for testing.

3.2 Implementation Details

For our experiments, we use a workstation equipped with one GeForce GTX 3080 GPU with 10 G memory. All experiments are done in PyTorch. Specially, because of the high-resolution of original images of MoNuSeg, we split 1 image into 25 patches with a resolution of 224×224 for training models, and merge patches into the original image to calculate the loss or evaluation metrics. For GlaS and MoNuSeg, we resize the input resolution of 224×224. For all applications, models are trained for 1000 epochs. Adam optimizer with a learning rate of 1×10^{-3} was employed and One-cycle is used as the learning rate policy. The batch size is set to 12. We use multiple augmentations as follows: random scaling, translation, rotation, random blur and random brightness contrast. For all datasets, we use dice coefficient (Dice) and mean Intersection over Union (mIoU) as evaluation metrics.

3.3 Experiments

Comparison with State-of-the-Art Methods. To demonstrate the effectiveness of SMESwin Unet, we compare the segmentation performance with three UNet-based methods: UNet++ [5], Attention U-Net [22] and MultiResUNet [23], and two transformer-based segmentation methods: UCTransNet [14] and Swin Unet [10]. The same settings as SMESwin Unet are used in the experiment.

Quantitative results are showed in Table 1, where the best results are bold-faced. Table 1 shows that even for datasets with relatively fewer images like GlaS [16] and WBCs [18], our method performs better than CNN and Transformer baselines, and shows good performance in MoNuSeg [17]. Overall, these observations suggest that the three proposed modules can be incorporated into UNet-like structures for improving segmentation performance. However, the inconsistent performance may arise from the different characteristics of the datasets, the segment objects in MoNuSeg are smaller than other 2 datasets, this may imply that our method might be more applicable for big objects.

Table 1. The results on GlaS, MoNuSeg and WBCs datasets

Methods	GlaS		MoNuSeg		WBCs	
	Dice (%)	mIoU (%)	Dice (%)	mIoU (%)	Dice (%)	mIoU (%)
UNet++	88.76	80.70	**81.17**	**69.29**	87.05	77.64
Attention U-Net	89.94	82.54	79.74	67.19	85.84	75.70
MultiResUNet	88.31	80.26	78.96	66.45	80.74	68.45
UCTransNet	89.58	82.02	80.73	68.42	92.59	86.35
Swin Unet	90.11	82.79	80.85	67.97	92.71	86.53
SMESwin Unet	**92.20**	**86.13**	81.13	68.35	**93.21**	**87.43**

The segmentation results of different methods are visualized in Fig. 3. It shows that our SMESwin Unet generates more accurate segmentation results, which are more consistent with the ground truth than the results of the baseline models, and limit the confusing false-positive lesions. As shown in these observations, SMESwin Unet has improved segmentation capabilities while preserving details of the shape.

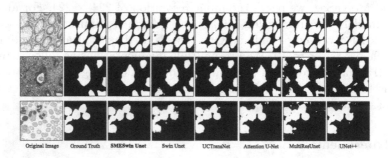

Original Image   Ground Truth   SMESwin Unet   Swin Unet   UCTransNet   Attention U-Net   MultiResUnet   UNet++

Fig. 3. Qualitative results on sample test images from GlaS and WBCs datasets. SMESwin Unet which makes better use of optimized skip connection for training performs better than the other compared methods.

Ablation Studies. For the ablation study, we use the GlaS dataset. As shown in Table 2, we first start with EA in skip connection. Then we replace skip connection with MCCT by the fusion of CNN and Transformer. This shows that compound structure with CNN and ViT may get decent performance. Then, we add the EA to MCCT making it a EAMCCT module. Finally, we add the superpixel to the EAMCCT, which replaces the original input feature with superpixel segmentation. The ablation study shows that each individual component of SMESwin Unet provides a useful contribution to improve performance.

Table 2. Ablation study on the impact of different modules on GlaS dataset

Modules	Dice (%)	mIoU (%)
Baseline	90.11	82.79
Baseline + EA (d1)	90.81	83.81
Baseline + EA (d2)	90.40	83.13
Baseline + EA (d3)	89.72	82.05
Baseline + EA (d1 + d2 + d3)	91.01	84.26
Baseline + CCT	91.05	84.25
Baseline + MCCT	91.06	84.45
Baseline + EA + MCCT	91.62	85.17
Baseline + Superpixel + EA + MCCT	**92.20**	**86.20**

'Baseline' denotes Swin Unet; 'EA' denotes the External attention; 'd1, d2, d3' denotes the first, second and third encoding layer; 'MCCT' denotes the mixed Channel-wise Cross Fusion Transformer. The best results are boldfaced.

4 Conclusion

In this work, we introduced a superpixel and EA-based Segmentation network (SMESwin Unet) from the CNN and ViT mixed-wise perspective to provide the precise and reliable automatic segmentation of medical images. By taking advantage of Superpixel, EA, CNN and Transformer-mixed skip connection in Swin Unet, SMESwin Unet improves the baseline results in medical image segmentation on three datasets. Extensive experiments show the advantages of the SMESwin Unet which successfully narrows the limitation of data size. With regard to this study, limitations need to be noted, one is the usage of the superpixel may hurt the feasibility, and the other one is the lack of the evaluation in 3D datasets and the evaluation metric PQ for nuclei segmentation. Further research regarding the efficiency of mixing CNN and Transformer in skip connection would be worthwhile.

References

1. Lei, T., Wang, R.S., Wan, Y., Du, X.G., Meng, H.Y., Asoke, N.: Medical image segmentation using deep learning: a survey. arXiv:2009.13120 (2020)
2. Ronneberger, O., Fischer, P., Brox, T.: U-Net: convolutional Networks for Biomedical Image Segmentation. arXiv:1505.04597 (2015)
3. Çiçek, Ö., Abdulkadir, A., Lienkamp, S.S., Brox, T., Ronneberger, O.: 3D U-Net: learning dense volumetric segmentation from sparse annotation. In: Ourselin, S., Joskowicz, L., Sabuncu, M.R., Unal, G., Wells, W. (eds.) MICCAI 2016. LNCS, vol. 9901, pp. 424–432. Springer, Cham (2016). https://doi.org/10.1007/978-3-319-46723-8_49
4. Xiao, X., S. Lian., Z. Luo., S. Li.: Weighted Res-UNet for high-quality retina vessel segmentation. In: 2018 9th International Conference on Information Technology in Medicine and Education (ITME), pp. 327–331 (2018)
5. Zhou, Z., Rahman Siddiquee, M.M., Tajbakhsh, N., Liang, J.: UNet++: a nested U-Net architecture for medical image segmentation. In: Stoyanov, D., Taylor, Z., Carneiro, G., Syeda-Mahmood, T., Martel, A., Maier-Hein, L., Tavares, J.M.R.S., Bradley, A., Papa, J.P., Belagiannis, V., Nascimento, J.C., Lu, Z., Conjeti, S., Moradi, M., Greenspan, H., Madabhushi, A. (eds.) DLMIA/ML-CDS -2018. LNCS, vol. 11045, pp. 3–11. Springer, Cham (2018). https://doi.org/10.1007/978-3-030-00889-5_1
6. Huang, H., Lin, L., Tong, R., Hu, H., Wu, J.: UNet 3+: a full-scale connected UNet for medical image segmentation. arXiv:2004.08790v1 (2020)
7. Vaswani., A., et al.: Attention is all you need. In: Advances in Neural Information Processing Systems, vol. 30 (2017)
8. Zhang, Y., Liu, H., Hu, Q.: TransFuse: fusing transformers and CNNs for medical image segmentation. In: de Bruijne, M., Cattin, P.C., Cotin, S., Padoy, N., Speidel, S., Zheng, Y., Essert, C. (eds.) MICCAI 2021. LNCS, vol. 12901, pp. 14–24. Springer, Cham (2021). https://doi.org/10.1007/978-3-030-87193-2_2
9. Valanarasu, J.M.J., Oza, P., Hacihaliloglu, I., Patel, V.M.: Medical transformer: gated axial-attention for medical image segmentation. In: de Bruijne, M., et al. (eds.) MICCAI 2021. LNCS, vol. 12901, pp. 36–46. Springer, Cham (2021). https://doi.org/10.1007/978-3-030-87193-2_4

10. Cao, H., et al.: Swin-Unet: Unet-like pure transformer for medical image segmentation. arXiv:2105.05537 (2021)
11. Dosovitskiy, A., et al.: An image is worth 16x16 words: transformers for image recognition at scale. arXiv:2010.11929 (2020)
12. He, K., Zhang, X., Ren, S., Sun, J.: Deep residual learning for image recognition. In: 2016 IEEE Conference on Computer Vision and Pattern Recognition (CVPR), USA , pp. 770–778 (2016)
13. Russakovsky, O., et al.: ImageNet large scale visual recognition challenge. Int. J. Comput. Vis. **115**(3), 211–252 (2015). https://doi.org/10.1007/s11263-015-0816-y
14. Wang H, Cao P, Wang J, Zaiane O.R.: UCTransNet: rethinking the skip connections in U-Net from a channel-wise perspective with transformer. arXiv:2109.04335 (2021)
15. Guo, M.-H., Liu, Z.-N., Mu, T.-J., Hu, S.-M.: Beyond self-attention: external attention using two linear layers for visual tasks. arXiv:2105.02358 (2021)
16. Sirinukunwattana, K., et al.: Gland segmentation in colon histology images: the GlaS challenge contest. arXiv:1603.00275 (2016)
17. Kumar, N., Verma, R., Sharma, S., Bhargava, S., Vahadane, A., Sethi, A.: A dataset and a technique for generalized nuclear segmentation for computational pathology. IEEE Trans. Med. Imaging **6**(7), 1550–1560 (2017)
18. Koohbanani, N.A., Jahanifar, M., Tajadin, N.Z., Rajpoot, N.: NuClick: a deep learning framework for interactive segmentation of microscopic images. Med. Image Anal. **65**, 101771 (2020)
19. Liu Z., et al.: Swin transformer: hierarchical vision transformer using shifted windows. In: Proceedings of the IEEE/CVF International Conference on Computer Vision, pp. 10012–10022 (2021)
20. Radhakrishna, A., Appu, S., Kevin, S., Aur'elien, L., Pascal, F., Sabine, S.: SLIC superpixels compared to state-of-the-art superpixel methods. IEEE Trans. Pattern Anal. Mach. Intell. (TPAMI) **34**(11), 2274–2282 (2012)
21. Ulyanov, D., Vedaldi, A., Lempitsky, V. Instance normalization: the missing ingredient for fast stylization. arXiv:1607.08022 (2017)
22. Oktay, O., et al.: Attention U-Net: learning where to look for the pancreas. arXiv:1804.03999 (2018)
23. Ibtehaz, N., Sohel Rahman, M.: MultiResUNet: rethinking the U-Net architecture for multimodal biomedical image segmentation. Neural Netw, **121**, 74–87 (2020)

The Dice Loss in the Context of Missing or Empty Labels: Introducing Φ and ϵ

Sofie Tilborghs[1,2]([✉]), Jeroen Bertels[1,2], David Robben[1,2,3],
Dirk Vandermeulen[1,2], and Frederik Maes[1,2]

[1] Department of Electrical Engineering, ESAT/PSI, KU Leuven, Leuven, Belgium
{sofie.tilborghs,jeroen.bertels}@kuleuven.be
[2] Medical Imaging Research Center, UZ Leuven, Leuven, Belgium
[3] icometrix, Kolonel Begaultlaan 1b/12, Leuven, Belgium

Abstract. Albeit the Dice loss is one of the dominant loss functions in medical image segmentation, most research omits a closer look at its derivative, i.e. the real motor of the optimization when using gradient descent. In this paper, we highlight the peculiar action of the Dice loss in the presence of missing or empty labels. First, we formulate a theoretical basis that gives a general description of the Dice loss and its derivative. It turns out that the choice of the reduction dimensions Φ and the smoothing term ϵ is non-trivial and greatly influences its behavior. We find and propose heuristic combinations of Φ and ϵ that work in a segmentation setting with either missing or empty labels. Second, we empirically validate these findings in a binary and multiclass segmentation setting using two publicly available datasets. We confirm that the choice of Φ and ϵ is indeed pivotal. With Φ chosen such that the reductions happen over a single batch (and class) element and with a negligible ϵ, the Dice loss deals with missing labels naturally and performs similarly compared to recent adaptations specific for missing labels. With Φ chosen such that the reductions happen over multiple batch elements or with a heuristic value for ϵ, the Dice loss handles empty labels correctly. We believe that this work highlights some essential perspectives and hope that it encourages researchers to better describe their exact implementation of the Dice loss in future work.

1 Introduction

The *Dice loss* was introduced in [5] and [13] as a loss function for binary image segmentation taking care of the class imbalance between foreground and background often present in medical applications. The *generalized Dice loss* [16] extended this idea to multiclass segmentation tasks, thereby taking into account the class imbalance that is present across different classes. In parallel, the Jaccard loss was introduced in the wider computer vision field for the same purpose [14,17]. More recently, it has been shown that one can use either Dice or Jaccard loss during training to effectively optimize both metrics at test time [6].

S. Tilborghs and J. Bertels—Contributed equally to this work.

L. Wang et al. (Eds.): MICCAI 2022, LNCS 13435, pp. 527–537, 2022.
https://doi.org/10.1007/978-3-031-16443-9_51

The use of the Dice loss in popular and state-of-the-art methods such as No New-Net [9] has only fueled its dominant usage across the entire field of medical image segmentation. Despite its fast and wide adoption, research that explores the underlying mechanisms is remarkably limited and mostly focuses on the loss value itself building further on the concept of *risk minimization* [8]. Regarding model calibration and inherent uncertainty, for example, some intuitions behind the typical hard and poorly calibrated predictions were exposed in [4], thereby focusing on the potential volume bias as a result of using the Dice loss. Regarding semi-supervised learning, adaptations to the original formulations were proposed to deal with "missing" labels [7,15], i.e. a label that is missing in the ground truth even though it is present in the image.

In this work, we further contribute to a deeper understanding of the specific implementation of the Dice loss, especially in the context of missing and empty labels. In contrast to missing labels, "empty" labels are labels that are not present in the image (and hence also not in the ground truth). We will first take a closer look at the derivative, i.e. the real motor of the underlying optimization when using gradient descent, in Sect. 2. Although [13] and [16] report the derivative, it is not being discussed in detail, nor is any reasoning behind the choice of the reduction dimensions Φ given (Sect. 2.1). When the smoothing term ϵ is mentioned, no details are given and its effect is underestimated by merely linking it with numerical stability [16] and convergence issues [9]. In fact, we find that both Φ and ϵ are intertwined, and that their choice is non-trivial and pivotal in the presence of missing or empty labels. To confirm and validate these findings, we set up two empirical settings with missing or empty labels in Sects. 3 and 4. Indeed, we can make or break the segmentation task depending on the exact implementation of the Dice loss.

2   Bells and Whistles of the Dice Loss: Φ and ϵ

In a CNN-based setting, the weights $\theta \in \Theta$ are often updated using gradient descent. For this purpose, the loss function ℓ computes a real valued cost $\ell(Y, \tilde{Y})$ based on the comparison between the ground truth Y and its prediction $\tilde{Y}$ in each iteration. Y and $\tilde{Y}$ contain the values $y_{b,c,i}$ and $\tilde{y}_{b,c,i}$, respectively, pointing to the value for a semantic class $c \in \mathcal{C} = [C]$ at an index $i \in \mathcal{I} = [I]$ (e.g. a voxel) of a batch element $b \in \mathcal{B} = [B]$ (Fig. 1). The exact update of each θ depends on $d\ell(Y, \tilde{Y})/d\theta$, which can be computed via the generalized chain rule. With $\omega = (b, c, i) \in \Omega = \mathcal{B} \times \mathcal{C} \times \mathcal{I}$, we can write:

$$\frac{d\ell(Y, \tilde{Y})}{d\theta} = \sum_{b \in \mathcal{B}} \sum_{c \in \mathcal{C}} \sum_{i \in \mathcal{I}} \frac{\partial \ell(Y, \tilde{Y})}{\partial \tilde{y}_{b,c,i}} \frac{\partial \tilde{y}_{b,c,i}}{\partial \theta} = \sum_{\omega \in \Omega} \frac{\partial \ell(Y, \tilde{Y})}{\partial \tilde{y}_\omega} \frac{\partial \tilde{y}_\omega}{\partial \theta}. \tag{1}$$

The Dice similarity coefficient (DSC) over a subset $\phi \subset \Omega$ is defined as:

$$\mathrm{DSC}(Y_\phi, \tilde{Y}_\phi) = \frac{2|Y_\phi \cap \tilde{Y}_\phi|}{|Y_\phi| + |\tilde{Y}_\phi|}. \tag{2}$$

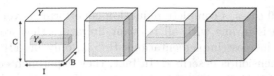

Fig. 1. Schematic representation of Y, having a batch, class and image dimension, respectively with $|\mathcal{B}| = $ B, $|\mathcal{C}| = $ C and $|\mathcal{I}| = $ I (similarly for $\tilde{Y}$). The choice of Φ, i.e. a family of subsets ϕ over Ω defines the extent of the reductions in sDSC($Y_\phi, \tilde{Y}_\phi$). From left to right, we see how the choice of Φ, and thus an example subset ϕ in blue, is different between the image-wise (DL$_\text{I}$), class-wise (DL$_\text{CI}$), batch-wise (DL$_\text{BI}$) and all-wise (DL$_\text{BCI}$) implementation of DL.

This formulation of DSC($Y_\phi, \tilde{Y}_\phi$) requires Y and $\tilde{Y}$ to contain values in $\{0, 1\}$. In order to be differentiable and handle values in $[0, 1]$, relaxations such as the *soft* DSC (sDSC) are used [5,13]. Furthermore, in order to allow both Y and $\hat{Y}$ to be empty, a smoothing term ϵ is added to the nominator and denominator such that DSC($Y_\phi, \tilde{Y}_\phi$) = 1 in case both Y and $\tilde{Y}$ are empty. This results in the more general formulation of the Dice loss (DL) computed over a number of subsets $\Phi = \{\phi\}$:

$$\text{DL}(Y, \tilde{Y}) = 1 - \frac{1}{|\Phi|} \sum_{\phi \in \Phi} \text{sDSC}(Y_\phi, \tilde{Y}_\phi) = 1 - \frac{1}{|\Phi|} \sum_{\phi \in \Phi} \frac{2\sum_{\varphi \in \phi} y_\varphi \tilde{y}_\varphi + \epsilon}{\sum_{\varphi \in \phi}(y_\varphi + \tilde{y}_\varphi) + \epsilon}. \quad (3)$$

Note that typically all ϕ are equal in size and define a partition over the domain Ω, such that $\bigcup_{\phi \in \Phi} \phi = \Omega$ and $\bigcap_{\phi \in \Phi} \phi = 0$. In $d\text{DL}(Y, \tilde{Y})/d\theta$ from Eq. 1, the derivative $\partial \text{DL}(Y, \tilde{Y})/\partial \tilde{y}_\omega$ acts as a scaling factor. In order to understand the underlying optimization mechanisms we can thus analyze $\partial \text{DL}(Y, \tilde{Y})/\partial \tilde{y}_\omega$. Given that all ϕ are disjoint, this can be written as:

$$\frac{\partial \text{DL}(Y, \tilde{Y})}{\partial \tilde{y}_\omega} = -\frac{1}{|\Phi|} \left(\frac{2y_\omega}{\sum_{\varphi \in \phi^\omega}(y_\varphi + \tilde{y}_\varphi) + \epsilon} - \frac{2\sum_{\varphi \in \phi^\omega} y_\varphi \tilde{y}_\varphi + \epsilon}{\left(\sum_{\varphi \in \phi^\omega}(y_\varphi + \tilde{y}_\varphi) + \epsilon\right)^2} \right), \quad (4)$$

with ϕ^ω the subset that contains ω. As such, it becomes clear that the specific action of DL depends on the exact configuration of the partition Φ of Ω and the choice of ϵ. Next, we describe the most common choices of Φ and ϵ in practice. Then, we investigate their effects in the context of missing or empty labels. Finally, we present a simple heuristic to tune both.

2.1   Configuration of Φ and ϵ in Practice

In Fig. 1, we depict four straightforward choices for Φ. We define these as the *image-wise*, *class-wise*, *batch-wise* or *all-wise* DL implementation, respectively DL$_\text{I}$, DL$_\text{CI}$, DL$_\text{BI}$ and DL$_\text{BCI}$, thus referring to the dimensions over which a complete reduction (i.e. the summations $\sum_{\varphi \in \phi}$ in Eq. 3 and Eq. 4) is performed.

We see that in all cases, a complete reduction is performed over the set of image indices $\mathcal{I}$, which is in line with all relevant literature that we consulted. Furthermore, while in most implementations $B > 1$, only in [11] the exact usage of the batch dimension is described. In fact, they experimented with both $DL_\mathbb{I}$ and $DL_{\mathbb{BI}}$, and found the latter to be superior for head and neck organs at risk segmentation in radiotherapy. Based on the context, we assume that most other contributions [5,6,9,10,13,18] used $DL_\mathbb{I}$, although we cannot rule out the use of $DL_{\mathbb{BI}}$. Similarly, we assume that in [16] $DL_{\mathbb{CI}}$ was used (with additionally weighting the contribution of each class inversely proportional to the object size), although we cannot rule out the use of $DL_{\mathbb{BCI}}$.

Note that in Eq. 3 and Eq. 4 we have assumed the choice for Φ and ϵ to be fixed. As such, the loss value or gradients only vary across different iterations due to a different sampling of Y and $\tilde{Y}$. Relaxing this assumption allows us to view the *leaf Dice loss* from [7] as a special case of choosing Φ. Being developed in the context of missing labels, the partition Φ of Ω is altered each iteration by substituting each ϕ with $\emptyset$ if $\sum_\varphi^\phi y_\varphi = 0$. Similarly, the *marginal Dice loss* from [15] adapts Φ every iteration by treating the missing labels as background and summing the predicted probabilities of unlabeled classes to the background prediction before calculating the loss.

Based on our own experience, ϵ is generally chosen to be small (e.g. 10^{-7}). However, most research does not include ϵ in their loss formulation, nor do they mention its exact value. We do find brief mentions related to convergence issues [9] (without further information) or numerical stability in the case of empty labels [10,16] (to avoid division by zero in Eq. 3 and Eq. 4).

2.2   Effect of Φ and ϵ on Missing or Empty Labels

When inspecting the derivative given in Eq. 4, we notice that in a way $\partial DL/\partial \tilde{y}_\omega$ does not depend on $\tilde{y}_\omega$ itself. Instead, the contributions of $\tilde{y}_\varphi$ are aggregated over the reduction dimensions, resulting in a global effect of prediction $\tilde{Y}_\phi$. Consequently, the derivative in a subset ϕ takes only two distinct values corresponding to $y_\omega = 0$ or $y_\omega = 1$. This is in contrast to the derivative shown in [13] who used a L^2 norm-based relaxation, which causes the gradients to be different for every ω if $\tilde{y}_\omega$ is different. If we work further with the L^1 norm-based relaxation (following the vast majority of implementations) and assuming that $\sum_{\varphi \in \phi^\omega} \tilde{y}_\varphi \gg \epsilon$, we see that $\partial DL/\partial \tilde{y}_\omega$ will be negligible for missing or empty ground truth labels. Exploiting this property, we can either avoid having to implement specific losses for missing labels, or we can learn to predict empty maps with a good configuration of Φ. Regarding the former, we simply need to make sure $\sum_{\varphi \in \phi^\omega} y_\varphi = 0$ for each map that contains missing labels which can be achieved by using the image-wise implementation $DL_\mathbb{I}$. Regarding the latter, non-zero gradients are required for empty maps. Hence, we want to choose ϕ large enough to avoid $\sum_{\varphi \in \phi^\omega} y_\varphi = 0$ for which a batch-wise implementation $DL_{\mathbb{BI}}$ is suitable.

2.3   A Simple Heuristic for Tuning ϵ to Learn from Empty Maps

We hypothesized that we can learn to predict empty maps by using the batch-wise implementation DL_{BI}. However, due to memory constraints and trade-off with receptive field, it is often not possible to go for large batch sizes. In the limits when $B = 1$ we find that $DL_I = DL_{BI}$, and thus the gradients of empty maps will be negligible. Hence, we want to mimic the behavior of DL_{BI} with $B \gg 1$, but using DL_I. This can be achieved by tuning ϵ to increase the derivative for empty labels $y_\omega = 0$. A very simple strategy would be to let $\partial DL(Y, \tilde{Y})/\partial \tilde{y}_\omega$ for $y_\omega = 0$ be equal in case of (i) DL_{BI} with infinite batch size such that $\sum_{\varphi \in \phi^\omega} y_\varphi \neq 0$ and negligible ϵ and (ii) DL_I with non-negligible epsilon and $\sum_{\varphi \in \phi^\omega} y_\varphi = 0$. If we set $\sum_{\varphi \in \phi^\omega} \tilde{y}_\varphi = \hat{v}$ we get:

$$\frac{2 \sum_{\varphi \in \phi^\omega} y_\varphi \tilde{y}_\varphi}{\left(\sum_{\varphi \in \phi^\omega} (y_\varphi + \tilde{y}_\varphi)\right)^2} = \frac{\epsilon}{\left(\sum_{\varphi \in \phi^\omega} \tilde{y}_\varphi + \epsilon\right)^2} \Rightarrow \frac{2a\hat{v}}{(b\hat{v})^2} = \frac{\epsilon}{(\hat{v} + \epsilon)^2}, \quad (5)$$

with a and b variables to express the intersection and union as a function of $\hat{v}$. We can easily see that when we assume the overlap to be around 50%, thus $a \approx 1/2$, and $\sum_{\varphi \in \phi^\omega} y_\varphi \approx \sum_{\varphi \in \phi^\omega} \tilde{y}_\varphi = \hat{v}$, thus $b \approx 2$, we can find $\epsilon \approx \hat{v}$. It is further reasonable to assume that after some iterations $\hat{v} \approx \mathbb{E} \sum_{\varphi \in \phi^\omega} y_\varphi$, thus setting $\epsilon = \hat{v}$ will allow DL to learn empty maps.

3   Experimental Setup

To confirm empirically the observed effects of Φ and ϵ on missing or empty labels (Sect. 2.2), and to test our simple heuristic choice of ϵ (Sect. 2.3), we perform experiments using three implementations of DL on two different public datasets.

Setups I, BI and I_ϵ: In I and BI, respectively DL_I and DL_{BI} are used to calculate the Dice loss (Sect. 2.1). The difference between I and I_ϵ is that we use a negligible value for epsilon $\epsilon = 10^{-7}$ in I and use the heuristic from Sect. 2.3 to set $\epsilon = \mathbb{E} \sum_{\varphi \in \phi^\omega} y_\varphi$ in I_ϵ. From Sect. 2.2, we expect I (any B) and BI ($B = 1$) to successfully ignore missing labels during training, still segmenting these at test time. Vice versa, we expect BI ($B > 1$) and I_ϵ (any B) to successfully learn what maps should be empty and thus output empty maps at test time.

BRATS: For our purpose, we resort to the binary segmentation of whole brain tumors on pre-operative MRI in BRATS 2018 [1,2,12]. The BRATS 2018 training dataset consists of 75 subjects with a lower grade glioma (LGG) and 210 subjects with a glioblastoma (HGG). To construct a partially labeled dataset for the missing and empty label tasks, we substitute the ground truth segmentations of the LGGs with empty maps during training. In light of missing labels, we would like the CNN to successfully segment LGGs at test time. In light of empty maps, we would like the CNN to output empty maps for LGGs at test time.

Based on the ground truths of the entire dataset, in $\mathbb{I}_\epsilon$ we need to set $\epsilon = 8,789$ or $\epsilon = 12,412$ when we use the partially or fully labeled dataset for training, respectively.

ACDC: The ACDC dataset [3] consists of cardiac MRI of 100 subjects. Labels for left ventricular (LV) cavity, LV myocardium and right ventricle (RV) are available in end-diastole (ED) and end-systole (ES). To create a structured partially labeled dataset, we remove the myocardium labels in ES. This is a realistic scenario since segmenting the myocardium only in ED is common in clinical practice. More specifically, ED and ES were sampled in the ratio 3/1 for $\mathbb{I}_\epsilon$, resulting in ϵ being equal to 13,741 and 19,893 on average for the myocardium class during partially or fully labeled training, respectively. For LV and RV, ϵ was 21,339 and 18,993, respectively. We ignored the background map when calculating DL. Since we hypothesize that $DL_{\mathbb{I}}$ is able to ignore missing labels, we compare $\mathbb{I}$ to the marginal Dice loss [15] and the leaf Dice loss [7], two loss functions designed in particular to deal with missing labels.

Implementation Details: We start from the exact same preprocessing, CNN architecture and training parameters as in No New-Net [9]. The images of the BRATS dataset were first resampled to an isotropic voxel size of $2 \times 2 \times 2\,\text{mm}^3$, such that we could work with a smaller output segment size of $80 \times 80 \times 48$ voxels as to be able to vary B in $\{1, 2, 4, 8\}$. Since we are working with a binary segmentation task we have $C = 1$ and use a single sigmoid activation in the final layer. For ACDC, the images were first resampled to $192 \times 192 \times 48$ with a voxel size of $1.56 \times 1.56 \times 2.5\,\text{mm}^3$. The aforementioned CNN architecture was modified to use batch normalization and pReLU activations. To compensate the anisotropic voxel size, we used a combination of $3 \times 3 \times 3$ and $3 \times 3 \times 1$ convolutions and omitted the first max-pooling for the third dimension. These experiments were only performed for $B = 2$. In this multiclass segmentation task, we use a softmax activation in the final layer to obtain four output maps.

Statistical Performance: All experiments were performed under a five-fold cross-validation scheme, making sure each subject was only present in one of the five partitions. Significant differences were assessed with non-parametric bootstrapping, making no assumptions on the distribution of the results [2]. Results were considered statistically significant if the p-value was below 5%.

4   Results

Table 1 reports the mean DSC and mean volume difference (ΔV) between the fully labeled validation set and the predictions for tumor (BRATS) and myocardium (ACDC). For both the label that was always available (HGG or MYO_{ED}) and the label that was not present in the partially labeled training dataset (LGG or MYO_{ES}), we can make two observations. First, configurations

Table 1. Mean DSC and mean ΔV. HGG and MYO_{ED} are always present during training while LGG and MYO_{ES} are replaced by empty maps under partial labeling. Configurations that we expect to learn to predict empty maps are highlighted (since we used a fully labeled validation set, we expect lower DSC and ΔV). Comparing partial with full labeling, inferior (p $<$ 0.05) results are indicated in italic.

	Labeling	B	DSC						ΔV [ml]					
			HGG/MYO_{ED}			LGG/MYO_{ES}			HGG/MYO_{ED}			LGG/MYO_{ES}		
			$\mathbb{I}$	$\mathbb{BI}$	$\mathbb{I}_\epsilon$	$\mathbb{I}$	$\mathbb{BI}$	$\mathbb{I}_\epsilon$	$\mathbb{I}$	$\mathbb{BI}$	$\mathbb{I}_\epsilon$	$\mathbb{I}$	$\mathbb{BI}$	$\mathbb{I}_\epsilon$
BRATS	Full	1	0.89	0.89	0.89	0.89	0.89	0.89	−4	−4	−7	−8	−7	−9
		2	0.89	0.89	0.89	0.89	0.88	0.88	−5	−5	−7	−9	−10	−12
		4	0.89	0.89	0.89	0.88	0.90	0.89	−6	−5	−7	−11	−7	−11
		8	0.89	0.89	0.89	0.89	0.88	0.89	−6	−4	−6	−12	−10	−9
	Partial	1	0.89	0.89	*0.83*	*0.88*	0.88	*0.23*	−5	−5	*−12*	*−11*	*−12*	*−88*
		2	0.89	*0.83*	*0.82*	0.88	*0.24*	*0.16*	−6	*−12*	*−13*	*−12*	*−89*	*−96*
		4	0.89	*0.82*	*0.83*	0.88	*0.20*	*0.20*	−6	*−12*	*−12*	*−15*	*−93*	*−94*
		8	0.89	*0.82*	*0.83*	0.88	*0.20*	*0.23*	−6	*−12*	*−12*	*−14*	*−94*	*−90*
ACDC	Full	2	0.88	0.88	0.87	0.89	0.89	0.89	−1	0	−2	−3	0	−3
	Partial	2	*0.88*	*0.80*	*0.80*	*0.88*	*0.08*	*0.06*	0	*−11*	*−14*	*−5*	*−129*	*−131*

$\mathbb{I}$ and $\mathbb{BI}$ (B = 1) delivered a comparable segmentation performance (in terms of both DSC and ΔV) compared to using a fully labeled training dataset. Second, using configurations $\mathbb{BI}$ (B $>$ 1) and $\mathbb{I}_\epsilon$ the performance was consistently inferior. In this case, the CNN starts to learn when it needs to output empty maps. As a result, when calculating the DSC and ΔV with respect to a fully labeled validation dataset, we expect both metrics to remain similar for HGG and MYO_{ES}. On the other hand, we expect a mean DSC of 0 and a $|\Delta V|$ close to the mean volume of LGG or MYO_{ES}. Note that this is not the case due to the incorrect classification of LGG or MYO_{ES} as HGG or MYO_{ED}, respectively. Figure 2 shows the Receiver Operating Characteristic (ROC) curves when using a partially labeled training dataset with the goal to detect HGG or MYO_{ED} based on a threshold on the predicted volume at test time. For both tasks, we achieved an Area Under the Curve (AUC) of around 0.9. Figure 3 shows an example segmentation.

When comparing $\mathbb{I}$ with the marginal Dice loss [15] and the leaf Dice loss [7], no significant differences between any method for myocardium ($MYO_{ED} = 0.88$, $MYO_{ES} = 0.88$), LV ($LV_{ED} = 0.96$, $LV_{ES} = 0.92$) and RV ($RV_{ED} = 0.93$, $RV_{ES} = 0.86 - 0.87$) were found in both ED and ES.

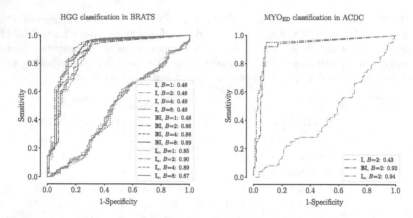

Fig. 2. ROC analysis if we want to detect the label that was always present during training by using different thresholds on the predicted volume. In the legend we also report the AUC for each setting.

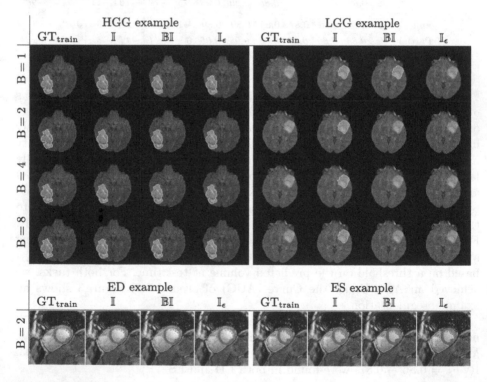

Fig. 3. Segmentation examples for BRATS (top) and ACDC (bottom). The ground truths for LGG and MYO_{ES} were replaced with empty maps during training (GT_{train}).

5   Discussion

The experiments confirmed the analysis from Sect. 2.2 that $DL_{\mathbb{I}}$ (equal to $DL_{\mathbb{BI}}$ when $B = 1$) ignores missing labels during training and that it can be used in the context of missing labels naively. On the other hand, we confirmed that $DL_{\mathbb{BI}}$ (with $B > 1$) and $DL_{\mathbb{I}}$ (with a heuristic choice of ϵ) can effectively learn to predict empty labels, e.g. for classification purposes or to be used with small patch sizes.

When heuristically determining ϵ for configuring $\mathbb{I}_\epsilon$ (Eq. 5), we only focused on the derivative for $y_\omega = 0$. Of course, by adapting ϵ, the derivative for $y_\omega = 1$ will also change. Nonetheless, our experiments showed that $\mathbb{I}_\epsilon$ can achieve the expected behavior, indicating that the effect on the derivative for $y_\omega = 1$ is only minor compared to $y_\omega = 0$. We wish to derive a more exact formulation of the optimal value of ϵ in future work. We expect this optimal ϵ to depend on the distribution between the classes, object size and other labels that might be present. Furthermore, it would be interesting to study the transition between the near-perfect prediction for the missing class ($DL_{\mathbb{I}}$ with small ϵ) and the prediction of empty labels for the missing class ($DL_{\mathbb{I}}$ with large ϵ).

All the code necessary for exact replication of the results including preprocessing, training scripts, statistical analysis, etc. was released to encourage further analysis on this topic (https://github.com/JeroenBertels/dicegrad).

6   Conclusion

We showed that the choice of the reduction dimensions Φ and the smoothing term ϵ for the Dice loss is non-trivial and greatly influences its behavior in the context of missing or empty labels. We believe that this work highlights some essential perspectives and hope that it encourages researchers to better describe their exact implementation of the Dice loss in the future.

Acknowledgement. This research received funding from the Flemish Government under the "Onderzoeksprogramma Artificiële intelligentie (AI) Vlaanderen" programme and is also partially funded by KU Leuven Internal Funds C24/18/047 (F. Maes).

References

1. Bakas, S., et al.: Advancing The Cancer Genome Atlas glioma MRI collections with expert segmentation labels and radiomic features. Sci. Data **4**(1), 170117 (2017). https://doi.org/10.1038/sdata.2017.117, http://www.nature.com/articles/sdata2017117

2. Bakas, S., Reyes, M., Jakab, A., Bauer, S., Rempfler, M., et al.: Identifying the best machine learning algorithms for brain tumor segmentation, progression assessment, and overall survival prediction in the BRATS challenge. arXiv, November 2018. http://arxiv.org/abs/1811.02629

3. Bernard, O., et al.: Deep learning techniques for automatic MRI cardiac multi-structures segmentation and diagnosis: is the problem solved? IEEE Trans. Med. Imaging **37**(11), 2514–2525 (2018). https://doi.org/10.1109/TMI.2018.2837502

4. Bertels, J., Robben, D., Vandermeulen, D., Suetens, P.: Theoretical analysis and experimental validation of volume bias of soft Dice optimized segmentation maps in the context of inherent uncertainty. Med. Image Anal. **67**, 101833 (2021). https://doi.org/10.1016/j.media.2020.101833, https://linkinghub.elsevier.com/retrieve/pii/S1361841520301973

5. Drozdzal, M., Vorontsov, E., Chartrand, G., Kadoury, S., Pal, C.: The importance of skip connections in biomedical image segmentation. In: Carneiro, G., et al. (eds.) LABELS/DLMIA -2016. LNCS, vol. 10008, pp. 179–187. Springer, Cham (2016). https://doi.org/10.1007/978-3-319-46976-8_19

6. Eelbode, T., et al.: Optimization for medical image segmentation: theory and practice when evaluating with dice score or Jaccard index. IEEE Trans. Med. Imaging **39**(11), 3679–3690 (2020). https://doi.org/10.1109/TMI.2020.3002417, https://ieeexplore.ieee.org/document/9116807/

7. Fidon, L., et al.: Label-set loss functions for partial supervision: application to fetal brain 3D MRI parcellation. In: de Bruijne, M., et al. (eds.) MICCAI 2021. LNCS, vol. 12902, pp. 647–657. Springer, Cham (2021). https://doi.org/10.1007/978-3-030-87196-3_60

8. Goodfellow, I., Bengio, Y., Courville, A.: Deep Learning. MIT Press, Cambridge (2016). http://www.deeplearningbook.org

9. Isensee, F., Kickingereder, P., Wick, W., Bendszus, M., Maier-Hein, K.H.: No New-Net. In: Crimi, A., Bakas, S., Kuijf, H., Keyvan, F., Reyes, M., van Walsum, T. (eds.) BrainLes 2018. LNCS, vol. 11384, pp. 234–244. Springer, Cham (2019). https://doi.org/10.1007/978-3-030-11726-9_21

10. Jadon, S.: A survey of loss functions for semantic segmentation. In: 2020 IEEE Conference on Computational Intelligence in Bioinformatics and Computational Biology, CIBCB 2020 (2020). https://doi.org/10.1109/CIBCB48159.2020.9277638

11. Kodym, O., Španěl, M., Herout, A.: Segmentation of head and neck organs at risk using CNN with batch dice loss. In: Brox, T., Bruhn, A., Fritz, M. (eds.) GCPR 2018. LNCS, vol. 11269, pp. 105–114. Springer, Cham (2019). https://doi.org/10.1007/978-3-030-12939-2_8

12. Menze, B.H., et al.: The multimodal brain tumor image segmentation benchmark (BRATS). IEEE Trans. Med. Imaging **34**(10), 1993–2024 (2015). https://doi.org/10.1109/TMI.2014.2377694, http://ieeexplore.ieee.org/document/6975210/

13. Milletari, F., Navab, N., Ahmadi, S.A.: V-Net: fully convolutional neural networks for volumetric medical image segmentation. In: Proceedings - 2016 4th International Conference on 3D Vision, 3DV 2016, pp. 565–571 (2016). https://doi.org/10.1109/3DV.2016.79

14. Nowozin, S.: Optimal decisions from probabilistic models: the intersection-over-union case. In: 2014 IEEE Conference on Computer Vision and Pattern Recognition, pp. 548–555. IEEE, June 2014. https://doi.org/10.1109/CVPR.2014.77, http://ieeexplore.ieee.org/document/6909471/

15. Shi, G., Xiao, L., Chen, Y., Zhou, S.K.: Marginal loss and exclusion loss for partially supervised multi-organ segmentation. Med. Image Anal. **70**, 101979 (2021). https://doi.org/10.1016/j.media.2021.101979

16. Sudre, C.H., Li, W., Vercauteren, T., Ourselin, S., Jorge Cardoso, M.: Generalised dice overlap as a deep learning loss function for highly unbalanced segmentations. In: Cardoso, M.J., et al. (eds.) DLMIA/ML-CDS -2017. LNCS, vol. 10553, pp. 240–248. Springer, Cham (2017). https://doi.org/10.1007/978-3-319-67558-9_28

17. Tarlow, D., Adams, R.P.: Revisiting uncertainty in graph cut solutions. In: 2012 IEEE Conference on Computer Vision and Pattern Recognition, pp. 2440–2447. IEEE, June 2012. https://doi.org/10.1109/CVPR.2012.6247958, http://ieeexplore. ieee.org/document/6247958/
18. Yeung, M., Sala, E., Schönlieb, C.B., Rundo, L.: Unified focal loss: generalising dice and cross entropy-based losses to handle class imbalanced medical image segmentation. Computerized Med. Imaging Graph. **95**, 102026 (2021, 2022). https:// doi.org/10.1016/j.compmedimag.2021.102026

Robust Segmentation of Brain MRI in the Wild with Hierarchical CNNs and No Retraining

Benjamin Billot[1]([✉]), Colin Magdamo[2], Steven E. Arnold[2], Sudeshna Das[2], and Juan Eugenio Iglesias[1,3,4]

[1] Centre for Medical Image Computing, University College London, London, UK
`benjamin.billot.18@ucl.ac.uk`
[2] Department of Neurology, Massachusetts General Hospital, Boston, USA
[3] Martinos Center for Biomedical Imaging, Massachusetts General Hospital and Harvard Medical School, Boston, USA
[4] Computer Science and Artificial Intelligence Laboratory, Massachusetts Institute of Technology, Cambridge, USA

Abstract. Retrospective analysis of brain MRI scans acquired in the clinic has the potential to enable neuroimaging studies with sample sizes much larger than those found in research datasets. However, analysing such clinical images "in the wild" is challenging, since subjects are scanned with highly variable protocols (MR contrast, resolution, orientation, etc.). Nevertheless, recent advances in convolutional neural networks (CNNs) and domain randomisation for image segmentation, best represented by the publicly available method *SynthSeg*, may enable morphometry of clinical MRI at scale. In this work, we first evaluate *SynthSeg* on an uncurated, heterogeneous dataset of more than 10,000 scans acquired at Massachusetts General Hospital. We show that *SynthSeg* is generally robust, but frequently falters on scans with low signal-to-noise ratio or poor tissue contrast. Next, we propose *SynthSeg*$^+$, a novel method that greatly mitigates these problems using a hierarchy of conditional segmentation and denoising CNNs. We show that this method is considerably more robust than *SynthSeg*, while also outperforming cascaded networks and state-of-the-art segmentation denoising methods. Finally, we apply our approach to a proof-of-concept volumetric study of ageing, where it closely replicates atrophy patterns observed in research studies conducted on high-quality, 1 mm, T1-weighted scans. The code and trained model are publicly available at https://github.com/BBillot/SynthSeg.

Keywords: Clinical MRI · Brain · Segmentation

1 Introduction

Neuroimaging with MRI is of paramount importance in the understanding of the morphology and connectivity of the human brain. *In vivo* MR imaging has

Supplementary Information The online version contains supplementary material available at https://doi.org/10.1007/978-3-031-16443-9_52.

L. Wang et al. (Eds.): MICCAI 2022, LNCS 13435, pp. 538–548, 2022.
https://doi.org/10.1007/978-3-031-16443-9_52

been widely adopted in research, where studies rely most often on prospective datasets of high-quality brain scans. Meanwhile, clinical MRI datasets remain largely unexplored in neuroimaging studies, despite their much higher abundance (e.g., 10 million brain scans were acquired in the US in 2019 [32]). Analysing such datasets is highly desirable, since it would enable sample sizes in the millions, which is much higher than the current largest research studies, which include tens of thousands subjects (e.g., ENIGMA [18] or UK BioBank [3]).

The use of clinical data in neuroimaging studies has been mainly hindered by the high variability in acquisition protocols. As opposed to high resolution (HR) scans used in research, physicians usually prefer low resolution (LR) acquisitions with fewer slices (for faster inspection), which span a large range of orientations, slice spacings and slice thicknesses. Moreover, clinical scans employ numerous MRI contrasts to highlight different tissue properties.

Overall, no segmentation method can robustly adapt to such variability. Manual labelling is the gold standard in segmentation techniques, but it remains too tedious for large-scale clinical applications. An alternative would be to only consider subjects with high-quality acquisitions (e.g., 1 mm T1 scans), as these can be easily analysed with neuroimaging softwares [5,15]. However, this would enormously decrease the effective sample size of clinical datasets, where such scans are seldom available. Hence, there is a clear need for a robust automated segmentation tool that can adapt to clinical scans of any MRI contrast and resolution.

Contrast-invariance has traditionally been achieved via Bayesian segmentation strategies with unsupervised likelihood model [33]. Unfortunately, these methods are highly sensitive to partial volume effects (PV) caused by changes in resolution [10]. This problem can partly be mitigated by directly modelling PV within the Bayesian framework [38]. However, this strategy quickly becomes intractable for scans with decreasing resolutions and increasing number of labels, thus limiting its application to large clinical datasets.

Recent automated segmentation methods rely on supervised convolutional neural networks (CNNs) [22,28,34]. While CNNs obtain fast and accurate results on their training domain, they are fragile to changes in resolution [17,31] and MRI contrast [2,21], even within the same MRI modality [23,26]. Although data augmentation can improve robustness in intra-modality scenarios [39], CNNs still need to be retrained for each new MRI contrast and resolution. This issue has sparked a vivid interest in domain adaptation schemes, where CNNs are trained to generalise to a specific target domain [9,23]. However, these methods still need to be retrained for every new target resolution or MRI contrast, which makes them impractical to apply on highly heterogeneous clinical data.

Very recently, a publicly available method named *SynthSeg* [6] has been proposed for out-of-the-box segmentation of brain scans of any contrast and resolution. *SynthSeg* relies on a 3D UNet trained on synthetic scans generated with a domain randomisation approach [36]. While *SynthSeg* yields excellent generalisation compared with previous techniques, it still lacks robustness when applied to clinical scans with low signal-to-noise (SNR) ratio or poor tissue contrast.

Improvements in robustness have previously been tackled with hierarchical models, where a first CNN performs a simpler preliminary task (e.g., predicting

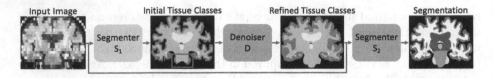

Fig. 1. Overview of the proposed architecture. A first network S_1 outputs initial segmentations of four tissue classes. Robustness is then improved by refining these with a denoiser D (e.g., red box). Final segmentations are obtained with a second segmenter S_2, which takes as inputs the image and the robust estimates of the four tissue classes. (Color figure online)

an initial mask [35], or pre-segmenting at low resolution [19]), and the results are refined by a second network trained for the target task. However, these methods often remain insufficient to capture high-order topological relations, which is a well-known problem for CNNs [29]. A possible solution is to use conditional random fields for postprocessing [22], but these often struggle to model relations between multiple labels at different scales. Recent methods now seek to improve semantic correctness either by aligning predictions and ground truths in latent space during training [30], or by using denoising CNNs [27]. Although these methods have shown promising results in relatively simple cases (i.e., 2D images with few labels), they are yet to be demonstrated in more complex setups.

In this work, we present *SynthSeg+*, a novel architecture for robust segmentation of clinical MRI scans of any contrast and resolution without retraining. Specifically, we build on the domain randomisation strategy introduced by *Synth-Seg*, and propose a hierarchy of conditional segmentation and denoising CNNs for improved robustness and semantic correctness. We evaluate this method on more than 10,000 highly heterogeneous clinical scans, directly taken from the picture archiving communication system (PACS) of Massachusetts General Hospital (MGH). *SynthSeg+* yields considerably enhanced robustness compared to *Synth-Seg*, while also outperforming cascaded networks and state-of-the-art denoising methods.

2   Methods

2.1   Hierarchical Conditional Architecture

We propose an architecture that relies on three hierarchical CNN modules (Fig. 1). This design aims at efficiently subdividing the target segmentation task into intermediate steps that are easier to perform, and thus less prone to errors. For this purpose, a first network S_1 is trained to produce coarse initial segmentations of the input images. More precisely, these initial segmentations only contain four labels that group brain regions into classes of similar tissue types and intensities (cerebral white matter, cerebral grey matter, cerebrospinal fluid, and cerebellum). These classes are easier to discriminate than individual regions.

The output of S_1 is then fed to a denoising network D [27] in order to increase the robustness of the initial segmentations. By modelling high-level

(a) Deformed Labels   (b) HR image   (c) LR image   (d) Training data for S_1 (e,f) Training data for S_2

Fig. 2. Generative model to train S_1 and S_2. A label map is deformed (a) and used to generate an HR scan (b), from which we simulate a LR scan of random resolution (c). S_1 is trained to produce HR segmentations of the four coarse tissue classes from the LR image, which is upsampled to HR space for convenience (d). S_2 is trained to estimate all target labels, using both the image and soft tissue segmentations (e,f). During training, we corrupt these tissue maps to model the errors made by the denoiser D.

relations between tissue types, D seeks to correct potential semantic inconsistencies introduced by S_1 (e.g., cerebral grey matter in the cerebellum). Moreover, it also enables recovery from large mistakes in the initial tissue classes, which sometimes occur for scans with low SNR, poor contrast, or very low resolution.

Final segmentations are obtained with a second segmenter S_2, which takes as input the image and the corrected tissue classes given by D, thus combining the robustness of D with the accuracy of S_2. In practice, S_2 learns to subdivide the initial tissue predictions into the target labels, as well as to refine the boundaries given by D, which are often excessively smooth (e.g., the cortex in Fig. 1).

2.2 Training Scheme for the Segmentation Modules

S_1 and S_2 are trained separately with a domain randomisation strategy [36]. Specifically, we use synthetic data sampled on the fly from a generative model that only needs a set of label maps as input, and whose parameters (contrast, resolution, artefacts, etc.) are drawn from uninformative priors (Supplementary Table S1). As a result, S_1 and S_2 are exposed to vastly varying examples, which forces them to learn contrast- and resolution-agnostic features. Training image-target pairs for S_1 are generated with a procedure similar to *SynthSeg* [6]:

(a) We draw a label map from a set of 3D segmentations with N labels. We assume that these are defined on a grid of J voxels at high resolution r_{HR} (here 1 mm isotropic). We then spatially augment the segmentation with a nonlinear transform (a smooth stationary velocity field [4]) as well as three rotations, scalings, shearings, and translations. We call this augmented map L (Fig. 2a).

(b) We obtain an image $G = \{G_j\}_{j=1}^{J}$ by sampling a Gaussian Mixture Model (GMM) conditioned on L. All means and variances $\{\mu_n, \sigma_n^2\}_{n=1}^{N}$ are sampled from uniform priors to obtain a different random contrast at each minibatch [7]. We also corrupt G by a random bias field B, sampled in logarithmic domain:

$$p(G|L, \{\mu_n, \sigma_n^2\}_{n=1}^{N}) = \prod_{j=1}^{J} \frac{B_j}{\sqrt{2\pi\sigma_{L_j}^2}} \exp[-\frac{1}{2\sigma_{L_j}^2}(G_j B_j - \mu_{L_j})^2] \qquad (1)$$

An image I_{HR} is then formed by normalising the intensities of G in $[0, 1]$, and nonlinearly augmenting them with a random voxel-wise exponentiation (Fig. 2b).

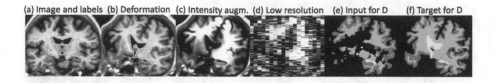

Fig. 3. Degradation model to train D. (a) A real image and its labels are spatially deformed (b, f). The image is degraded (c,d) and fed to S_1 to obtain the input for D (e).

(c) We then simulate LR scans with PV. This is achieved by blurring I_{HR} with a Gaussian kernel K of random standard deviation (to model slice thickness), and subsampling it to a random low resolution r_{sp} (to simulate slice spacing). Finally, we form an image I_{LR} by modelling the scanner noise with an additive field $\mathcal{E}$, sampled from a zero-mean Gaussian of random variance (Fig. 2c):

$$I_{LR} = Resample(I_{HR} * K, r_{sp}) + \mathcal{E}. \tag{2}$$

(d) The final training image I is obtained by resampling I_{LR} back to r_{HR}, while the target segmentation for S_1 is built by grouping the cerebral regions of L into the four tissue classes (Fig. 2d). We emphasise that test scans will also be resampled to r_{HR}, since this enables us to segment on the target HR grid.

Training data for S_2 are sampled from the same model, but with two differences. First, the ground truths now contain all target labels (Fig. 2e). Second, we still build tissue segmentations as in *(d)* (since these are needed as inputs to S_2), but we now represent them as soft probability maps, which we randomly dilate/erode and spatially deform to model D imperfections at test-time (Fig. 2f).

2.3   Training Scheme for the Denoising Module

Recent denoising methods are mostly based on supervised CNNs trained to recover ground truth segmentations from artificially corrupted versions of the same maps [23,24,27]. However, the employed corruption strategies are often handcrafted (random erosion and dilation, swapping of labels, etc.), and thus do not accurately capture errors made by the segmentation method to correct.

Instead, we propose to employ examples representative of S_1 errors, obtained by degrading real images, and feeding them to the trained S_1. D is then trained to map the outputs of S_1 back to their ground truths. Images are degraded on the fly with the same steps as Sect. 2.2 (except for the GMM, since we now use real images): spatial deformation, bias field, voxel-wise exponentiation, simulation of low resolution, and noise injection (Fig. 3). In practice, these corruptions use considerably wider prior parameter distributions than in Sect. 2.2 (Table S1), in order to ensure a high probability of erroneous segmentations from S_1.

2.4 Implementation Details

All modules are trained separately using the average soft Dice as loss function [28]. The segmentation modules S_1 and S_2 use the same 3D UNet architecture as in [6]. Briefly, it comprises 5 levels of 2 convolution layers. All layers use $3 \times 3 \times 3$ kernels and ELU activation [12], except for the last layer, which employs a softmax. The first layer has 24 feature maps; this number is doubled after each max-pooling, and halved after each upsampling. Meanwhile, the denoiser D uses a lighter structure chosen with a validation set: one convolution per layer, with a constant number of 16 features. Importantly, we delete the skip connections between the top two levels, to reach a compromise between UNets, where high-level skip connections propagate potential errors in the input segmentations to correct; and auto-encoders, with excessive bottleneck-induced smoothness.

3 Experiments and Results

3.1 Brain MRI Datasets

The training dataset consists of 1020 1 mm T1 brain scans (500 from ADNI [20], 500 from HCP [37], 20 from a private dataset [16]), which are available with a combination of manual and automated labels for 44 regions (Table S2). We note that S_1 and S_2 are only trained with the label maps, and that using subjects from varied populations enables us to increase robustness to morphological variability.

We evaluate $SynthSeg^+$ on 10,520 uncurated scans from the PACS of MGH. These are obtained from 1,047 MRI sessions of distinct subjects (Fig. S3), using a huge range of resolutions and contrasts (T1, T2, FLAIR, B0 channels from diffusion, among others). Among all sessions, 62 include T1 scans at maximum 1.3 mm resolution, for which we obtain label maps by running FreeSurfer [15]. Next, we propagate the resulting labels with rigid registration to all the scans of the corresponding sessions, which provides us with silver standard segmentations for 520 scans. Finally, these are split between validation (20), and testing (500), while the other 10,000 scans are used for indirect evaluation.

3.2 Competing Methods

We evaluate $SynthSeg^+$ (i.e., $S_1 + D + S_2$) against four approaches.

SynthSeg [6]: We use the publicly available model for testing.

Cascaded Networks [35] ($S_1 + S_2$): we ablate the denoiser D to obtain an architecture that is representative of classical cascaded networks.

Denoiser [27] (**SynthSeg** $+D$): A state-of-the-art method for denoising by postprocessing, where a denoiser (D) is appended to the method to correct (*SynthSeg*). Here, D is trained as in Sect. 2.3 to correct *all* target labels.

Cascaded Networks with Appended Denoiser ($S_1 + S_2 + D$): A combination of the cascaded architecture with the denoising network D.

All networks are trained for 300,000 steps (7 days on a Nvidia RTX6000) with the Adam optimiser [25]. Based on [6], we use Keras [11] and Tensorflow [1]. Inference takes between 8 and 12 s for all methods on the same GPU.

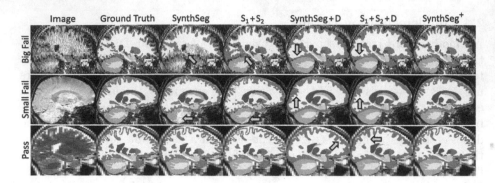

Fig. 4. Segmentations obtained by all methods for scans where *SynthSeg* shows large ("Big Fail"), mild ("Small Fail"), or no errors ("Pass"). Arrows indicate major mistakes. *SynthSeg*$^+$ yields outstanding results given the low SNR, poor tissue contrast, or low resolution of the inputs. Note that appending D considerably smooths segmentations.

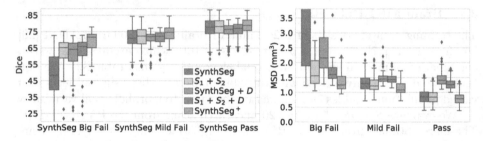

Fig. 5. Dice scores and mean surface distances (MSD) for 500 heterogeneous clinical scans. Results are presented based on the outcome of a QC analysis performed on the segmentations of *SynthSeg*: Big Fails (82 cases), Mild Fails (103), and Passes (315).

3.3   Quantitative Analysis

First, we evaluate all methods on the 500 scans with ground truth. For visualisation purposes, we subdivide these scans into three classes based on the outcome of a visual quality control (QC) performed on the segmentations of *SynthSeg* (Fig. 4, third column): large errors ("Big fails", 82 cases), mild errors ("Small Fails", 103 cases) and good segmentations ("Passes", 315 cases).

Figure 5 reveals that decomposing the target segmentation task into easier steps considerably improves robustness [35]. Indeed, the cascaded networks $S_1 + S_2$ outperforms *SynthSeg* by 16.5 Dice points on the worst scans (i.e., Big Fails), while presenting much less outliers. Moreover, this strategy slightly increases accuracy, and leads to improvements by up to 2 points for small fails.

In comparison, using a denoiser for postprocessing also improves robustness, but leads to a non-negligible loss in accuracy for Mild Fails and Passes, leading to an increase of at least 0.25 mm in mean surface distance (MSD) compared with *SynthSeg*. This is due to the fact that D cannot accurately model convoluted

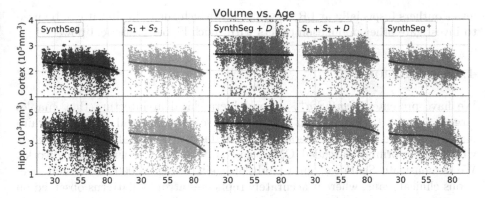

Fig. 6. Trajectories of cortical and hippocampal volumes with age (10,000 scans).

boundaries and returns very smoothed segmentations (e.g., the cortex in Fig. 4). Moreover, since D does not have access to the input scan, its segmentations may deviate from the original anatomy, which may introduce biases when analysing populations with morphologies different from the training set.

Remarkably, $SynthSeg^+$ yields an outstanding robustness (see Fig. S4 for examples of failures) and obtains the best scores in all three categories, with an outstanding improvement of 23.5 Dice points over $SynthSeg$ for Big Fails. In comparison with cascaded networks, inserting a denoiser D between S_1 and S_2 enables us to obtain robust tissue segmentations, which consistently improves scores by 2 to 5 Dice points. Finally, integrating D within our framework (rather than using it for postprocessing) enables $SynthSeg^+$ to exploit both images and prior information when predicting final segmentations. As a result, our approach is more accurate than the two methods using D for postprocessing (Fig. 4), and outperforms them by at least 0.35 mm in MSD and 3 Dice points.

3.4 Volumetric Study

We now conduct a proof-of-concept volumetric study on the held-out 10,000 scans. Specifically, we analyse age-related atrophy using the volumes estimated by all methods. Our ageing model includes: B-splines with 10 equally spaced knots and soft constraints for monotonicity, linear terms for slice spacing in each acquisition direction, and a bias for gender. We then fit this model numerically by minimising the sum of squares of the residuals with the L-BFGS-B method [8].

Figure 6 shows the scatter plot and B-spline fits for the cerebral cortex and hippocampus (see Fig. S5 for the other regions). It reveals that the two methods using D for postprocessing detect no or small atrophy, and over-estimate the volumes relatively to the other methods (due to the smoothing effect). Remarkably, $SynthSeg^+$ yields atrophy curves very close to those obtained in recent studies conducted on scans of much higher quality (i.e., 1 mm T1 scans) [13,14]. While $SynthSeg$ and $S_1 + S_2$ yield similar average trajectories, $SynthSeg^+$ produces far fewer outliers (especially at LR, see Fig. S6), which suggests that it can be used to investigate other population effects with much higher statistical power.

4   Conclusion

We have presented *SynthSeg*$^+$, a novel hierarchical architecture that enables large-scale robust segmentation of brain MRI scans in the wild, without retraining. Our method shows considerably improved robustness relatively to *SynthSeg*, while outperforming cascaded CNNs and state-of-the-art denoising networks. We demonstrate *SynthSeg*$^+$ in a study of ageing using 10,000 highly heterogeneous clinical scans, where it accurately replicates atrophy patterns observed on research data of much higher quality. By releasing the trained model, we aim at greatly facilitating the adoption of neuroimaging studies in the clinic, which has the potential to highly improve our understanding of neurological disorders.

Acknowledgement. This work is supported by the European Research Council (ERC Starting Grant 677697), the EPSRC-funded UCL Centre for Doctoral Training in Medical Imaging (EP/L016478/1), the Department of Health's NIHR-funded Biomedical Research Centre at UCLH, Alzheimer's Research UK (ARUK-IRG2019A-003), and the NIH (1R01AG070988, 1RF1MH123195).

References

1. Abadi, M., Barham, P., Chen, J., Chen, Z., Davis, A.: TensorFlow: a system for large-scale machine learning. In: Symposium on Operating Systems Design and Implementation, pp. 265–283 (2016)
2. Akkus, Z., Galimzianova, A., Hoogi, A., Rubin, D.L., Erickson, B.J.: Deep learning for brain MRI segmentation: state of the art and future directions. J. Digit. Imaging **30**(4), 449–459 (2017). https://doi.org/10.1007/s10278-017-9983-4
3. Alfaro-Almagro, F., Jenkinson, M., Bangerter, N., Andersson, J., Griffanti, L., Douaud, G., et al.: Image processing and quality control for the first 10,000 brain imaging datasets from UK biobank. Neuroimage **166**, 400–424 (2018)
4. Arsigny, V., Commowick, O., Pennec, X., Ayache, N.: A log-Euclidean framework for statistics on diffeomorphisms. In: Medical Image Computing and Computer Assisted Intervention, pp. 924–931 (2006)
5. Ashburner, J., Friston, K.: Unified segmentation. Neuroimage **26**(3), 39–51 (2005)
6. Billot, B., et al.: SynthSeg: domain randomisation for segmentation of brain scans of any contrast and resolution. arXiv:2107.09559 [cs] (2021)
7. Billot, B., Greve, D., Van Leemput, K., Fischl, B., Iglesias, J.E., Dalca, A.: A learning strategy for contrast-agnostic MRI segmentation. In: Medical Imaging with Deep Learning, pp. 75–93 (2020)
8. Byrd, R., Lu, P., Nocedal, J., Zhu, C.: A limited memory algorithm for bound constrained optimization. J. Sci. Comput. **16**, 1190–1208 (1995)
9. Chen, C., Dou, Q., Chen, H., Qin, J., Heng, P.A.: Synergistic image and feature adaptation: towards cross-modality domain adaptation for medical image segmentation. AAAI Conf. Artif. Intell. **33**, 65–72 (2019)

10. Choi, H., Haynor, D., Kim, Y.: Partial volume tissue classification of multichannel magnetic resonance images-a mixel model. IEEE Trans. Med. Imaging **10**, 395–407 (1991)
11. Chollet, F.: Keras (2015). https://keras.io
12. Clevert, D.A., Unterthiner, T., Hochreiter, S.: Fast and accurate deep network learning by exponential linear units (ELUs). arXiv:1511.07289 [cs] (2016)
13. Coupé, P., Catheline, G., Lanuza, E., Manjón, J.: Towards a unified analysis of brain maturation and aging across the entire lifespan: a MRI analysis. Hum. Brain Mapp. **38**(11), 5501–5518 (2017)
14. Dima, D., Modabbernia, A., Papachristou, E., Doucet, G.E., et al.: Subcortical volumes across the lifespan: Data from 18,605 healthy individuals aged 3–90 years. Hum. Brain Mapp. **43**(1), 452–469 (2022)
15. Fischl, B.: FreeSurfer. NeuroImage **62**, 774–781 (2012)
16. Fischl, B., Salat, D., Busa, E., Albert, M., et al.: Whole brain segmentation: automated labeling of neuroanatomical structures in the human brain. Neuron **33**, 41–55 (2002)
17. Ghafoorian, M., et al.: Transfer learning for domain adaptation in MRI: application in brain lesion segmentation. In: Descoteaux, M., Maier-Hein, L., Franz, A., Jannin, P., Collins, D.L., Duchesne, S. (eds.) MICCAI 2017. LNCS, vol. 10435, pp. 516–524. Springer, Cham (2017). https://doi.org/10.1007/978-3-319-66179-7_59
18. Hibar, D., Stein, J., Renteria, M., Arias-Vasquez, A., et al.: Common genetic variants influence human subcortical brain structures. Nature **520**, 224–229 (2015)
19. Isensee, F., Jaeger, P., Kohl, S., Petersen, J., Maier-Hein, K.: nnU-Net: a self-configuring method for deep learning-based biomedical image segmentation. Nat. Methods **18**(2), 203–211 (2021)
20. Jack, C.R., Bernstein, M., Fox, N., Thompson, P., Alexander, G., Harvey, D., et al.: The Alzheimer's disease neuroimaging initiative (ADNI): MRI methods. J. Magn. Reson. Imaging JMRI **27**(4), 685–691 (2008)
21. Jog, A., Hoopes, A., Greve, D., Van Leemput, K., Fischl, B.: PSACNN: pulse sequence adaptive fast whole brain segmentation. Neuroimage **199**, 553–569 (2019)
22. Kamnitsas, K., et al.: Efficient multi-scale 3D CNN with fully connected CRF for accurate brain lesion segmentation. Med. Image Anal. **36**, 61–78 (2017)
23. Karani, N., Erdil, E., Chaitanya, K., Konukoglu, E.: Test-time adaptable neural networks for robust medical image segmentation. Med. Image Anal. **68**, 101907 (2021)
24. Khan, M., Gajendran, M., Lee, Y., Khan, M.: Deep neural architectures for medical image semantic segmentation: review. IEEE Access **9**, 83002–83024 (2021)
25. Kingma, D., Ba, J.: Adam: a method for stochastic optimization. arXiv:1412.6980 [cs] (2017)
26. Kushibar, K., et al.: Supervised domain adaptation for automatic sub-cortical brain structure segmentation with minimal user interaction. Sci. Rep. **9**, 6742 (2019)
27. Larrazabal, A., Martínez, C., Glocker, B., Ferrante, E.: Post-DAE: anatomically plausible segmentation via post-processing with denoising autoencoders. IEEE Trans. Med. Imaging **39**(12), 3813–3820 (2020)
28. Milletari, F., Navab, N., Ahmadi, S.: V-Net: fully convolutional neural networks for volumetric medical image segmentation. In: International Conference on 3D Vision, pp. 565–571 (2016)
29. Nosrati, M., Hamarneh, G.: Incorporating prior knowledge in medical image segmentation: a survey. arXiv:1607.01092 [cs] (2016)

30. Oktay, O., Ferrante, E., Kamnitsas, K., Heinrich, M., Bai, W., Caballero, J., et al.: Anatomically constrained neural networks (ACNNs): application to cardiac image enhancement and segmentation. IEEE Trans. Med. Imaging **37**(2), 384–395 (2018)

31. Orbes-Arteaga, M., et al.: Multi-domain adaptation in brain MRI through paired consistency and adversarial learning. In: Wang, Q., et al. (eds.) DART/MIL3ID -2019. LNCS, vol. 11795, pp. 54–62. Springer, Cham (2019). https://doi.org/10.1007/978-3-030-33391-1_7

32. Oren, O., Kebebew, E., Ioannidis, J.: Curbing unnecessary and wasted diagnostic imaging. JAMA **321**(3), 245–246 (2019)

33. Puonti, O., Iglesias, J.E., Van Leemput, K.: Fast and sequence-adaptive whole-brain segmentation using parametric Bayesian modeling. Neuroimage **143**, 235–249 (2016)

34. Ronneberger, O., Fischer, P., Brox, T.: U-Net: convolutional networks for biomedical image segmentation. In: Navab, N., Hornegger, J., Wells, W.M., Frangi, A.F. (eds.) MICCAI 2015. LNCS, vol. 9351, pp. 234–241. Springer, Cham (2015). https://doi.org/10.1007/978-3-319-24574-4_28

35. Roth, H.R., Oda, H., Zhou, X., Shimizu, N., Yang, Y., Hayashi, Y., et al.: An application of cascaded 3D fully convolutional networks for medical image segmentation. Comput. Med. Imaging Graph. **66**, 90–99 (2018)

36. Tobin, J., Fong, R., Ray, A., Schneider, J., Zaremba, W., Abbeel, P.: Domain randomization for transferring deep neural networks from simulation to the real world. In: International Conference on Intelligent Robots, pp. 23–30 (2017)

37. Van Essen, D., Ugurbil, K., Auerbach, E., et al.: The human connectome project: a data acquisition perspective. Neuroimage **62**(4), 2222–2231 (2012)

38. Van Leemput, K., Maes, F., Vandermeulen, D., Suetens, P.: A unifying framework for partial volume segmentation of brain MR images. IEEE Trans. Med. Imaging **22**, 105–119 (2003)

39. Zhang, L., Wang, X., Yang, D., Roth, H., Xu, D., Xu, Z., et al.: Generalizing deep learning for medical image segmentation to unseen domains via deep stacked transformation. IEEE Trans. Med. Imaging **39**(7), 2531–2540 (2020)

Deep Reinforcement Learning for Small Bowel Path Tracking Using Different Types of Annotations

Seung Yeon Shin[✉] and Ronald M. Summers

Imaging Biomarkers and Computer-Aided Diagnosis Laboratory,
Radiology and Imaging Sciences, Clinical Center, National Institutes of Health,
Bethesda, MD, USA
{seungyeon.shin,rms}@nih.gov

Abstract. Small bowel path tracking is a challenging problem considering its many folds and contact along its course. For the same reason, it is very costly to achieve the ground-truth (GT) path of the small bowel in 3D. In this work, we propose to train a deep reinforcement learning tracker using datasets with different types of annotations. Specifically, we utilize CT scans that have only GT small bowel segmentation as well as ones with the GT path. It is enabled by designing a unique environment that is compatible for both, including a reward definable even without the GT path. The performed experiments proved the validity of the proposed method. The proposed method holds a high degree of usability in this problem by being able to utilize the scans with weak annotations, and thus by possibly reducing the required annotation cost.

Keywords: Small bowel path tracking · Reinforcement learning · Annotation type · Abdominal computed tomography

1 Introduction

The small bowel is the longest section of the digestive tract (about 6 m). Despite its length, it is pliable and has many folds so that it can fit into the abdominal cavity [1]. As a consequence, it has abundant contact between parts of itself as well as with adjacent organs such as the large bowel. Its appearance also varies dynamically along its course since bowel contents are locally highly variable.

Identifying the small bowel from a computed tomography (CT) scan is a prerequisite for inspection of its normalcy, including detection of diseases that can exist in the small bowel. To replace the manual interpretation, which is laborious and time-consuming, there have been researches on automatic small bowel segmentation [15,16,22]. Considering the difficulty of labeling the small bowel,

Supplementary Information The online version contains supplementary material available at https://doi.org/10.1007/978-3-031-16443-9_53.

these works have focused on data-efficient methods by incorporating a shape prior [15] or by developing an unsupervised domain adaptation method [16].

Although the segmentation could provide fundamental information for subsequent clinical tasks by separating the small bowel from the other tissues and organs, it is insufficient to identify its course precisely due to the aforementioned contact issue. To achieve this complementary information, there have been attempts to develop path tracking methods for the small bowel in recent years [6,11,17,18]. While a deep orientation classifier is trained to predict the direction of the small bowel in cine-MR scans [6], the 3D U-Net [3] is used to estimate the distance from the small bowel centerlines for each voxel [11]. CT scans that contain air-inflated bowels, where the walls are more distinguishable from the lumen than with other internal material, were used in [11]. On account of the labeling difficulty again, their networks were trained and evaluated using sparse annotation (several axial slices) [11], or with a set of bowel segments (88 mm long on average) [6]. In [17], a graph-based method that does not entail training with GT path is presented. Although its predicted path showed a high coverage on the entire course of the small bowel in routine CT scans, it is affected by the quality of the accompanying segmentation prediction.

Fig. 1. Main idea of this paper. We utilize scans that have only ground-truth (GT) segmentation, *SegmSet*, as well as ones with GT path, *PathSet*, to train a deep reinforcement learning (DRL) tracker for the small bowel. The GT path and segmentation are shown in red and green, respectively.

Over the past several years, deep reinforcement learning (DRL) based approaches have been proposed to trace elongated structures in biomedical images, including axons [4], the aorta [21], and coronary arteries [9]. The RL formulation for each application was validated by comparing with a supervised learning counterpart, which predicts moving actions as output labels from independent inputs [9,21]. Despite the benefit, their reward is based on a distance to ground-truth (GT) path, thus precluding the training when the GT path is unavailable.

In this paper, we propose to train a DRL tracker for the small bowel using different types of annotations. Specifically, we make use of scans that have only GT segmentation as well as ones with GT path (Fig. 1). To facilitate this, the environment our agent interacts with and the accompanying reward should not necessarily require the GT path for their definition, and should be compatible regardless of whether the GT path is available or not. We note that the annotation cost of the small bowel path is much higher than that of segmentation. Our experimental dataset will be introduced first to explain the necessity of using different types of annotations in the following section.

2   Method

2.1   Dataset

Our dataset consists of 30 intravenous and oral contrast-enhanced abdominal CT scans. Before the scan, a positive oral contrast medium, Gastrografin, was taken. All scans were done during the portal venous phase. They were cropped manually along the z-axis to include from the diaphragm through the pelvis. Then, they were resampled to have isotropic voxels of 1.5^3 mm^3.

Two types of GT labels, which are path and segmentation of the small bowel, are included in our dataset. It is separated into two subsets depending on whether or not the GT path is included. The first subset includes 10 scans that have both labels. The remaining 20 scans have only the GT segmentation. We call the subsets *PathSet* and *SegmSet*, respectively. An experienced radiologist used "Segment Editor" module in 3DSlicer [5] to acquire the GTs. Our GTs include the entire small bowel from the pylorus to the ileocecal junction.

For *SegmSet*, the small bowel was first drawn roughly using a large brush, and thresholded and manually fixed. This segmentation annotation took a couple of hours for each scan. The start and end coordinates of the small bowel were also recorded. For *PathSet*, the GT path is drawn as an interpolated curve connecting a series of manually placed points inside the bowel. This is exceptionally time-consuming and took one full day for each scan. For these scans that have the GT path, segmentation can be acquired more easily by growing the path. A thresholding and manual correction were done too. Our work is originated from the difficulty of achieving the GT path in a larger scale. It would be nice if we can utilize the scans that are without the GT path for training a better tracker. *PathSet* is used for cross validation while *SegmSet* is used for training. We note that our GT path is all connected for the entire small bowel. Compared with the sparse [11] and segmented [6] ones from the previous works, ours are more appropriate to see the tracking capability for the entire course.

2.2   Environment

In RL, an agent learns a policy π from episodes generated by interacting with an environment. We use one image per episode. An agent (tracker) is initialized at a position within the small bowel, and moves within the image until certain conditions are met. The conditions are: 1) finding one end of the small bowel, 2) leaving the image, or 3) being at the maximum time step T. One more termination condition of zero movement is used in test time. At every time step t, the agent performs an action a_t (movement), which is predicted by our actor network, and receives a reward r_t and a new state s_{t+1} as result. An episode, which is a sequence of (s_t, a_t, r_t), is generated by iterating this. Each of the state space, action space, and reward will be explained in the following sections.

State. At each time step t, local image patches, of size 60^3 mm^3, centered at the current position p_t is used to represent the agent's state. They are the input to

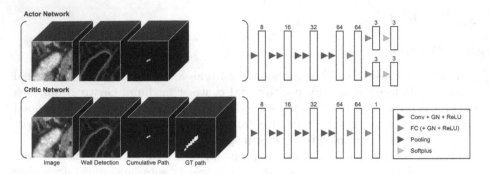

Fig. 2. Architectures of our networks and their input. Boxes on the right represent feature maps. The number of channels (dimensions) is denoted on top of each feature map. GN represents the group normalization [19]. All convolution layers have $3 \times 3 \times 3$ kernels. All fully connected (FC) layers are followed by a GN and a ReLU excepting the very last ones in each network.

our networks. Figure 2 shows example input patches. Due to the contact issue, preventing the tracker from penetrating the bowel walls is critical in this problem. To better provide this awareness to the agent, we use the wall detection as input in addition to an 'ordinary' image patch. In our dataset, the lumen appears brighter than the walls due to the oral contrast. We detect the walls by finding valleys in an input volume using the Meijering filter [10] as in [17]. We also provide a local patch of the agent's cumulative path in the current episode as in [4]. It denotes the previously visited voxels around the current position, which is binary.

In this work, we use an actor-critic algorithm [8]. The actor decides which action to take (policy), and the critic judges how good the actor's action was (value function). They are learned simultaneously, and the critic helps reduce the variance of the policy updates. The critic is used only during training. In DRL, the policy and value function are approximated by neural networks. To this end, we use separate networks, namely, actor and critic networks. Especially, we use the asymmetric actor-critic algorithm [12]. An additional information of the GT path is provided only to the critic network. It helps training the critic network and thus allows for better updates for the actor [4,12]. All the explained input patches are concatenated and fed into the networks. For *SegmSet*, the GT path is not available, and a zero tensor is used instead.

Action. The action is chosen based on the output of the actor network, which are three pairs of the parameters (α, β) of the beta distribution. The beta distribution has a finite support $[0, 1]$ so that can constrain the action space more easily than using the Gaussian [2]. Each pair is responsible for movement along each axis. A softplus operation is applied to the output of the last fully connected layers, and then 1 is added to ensure every α and β is greater than 1. When this condition is met, the beta distributions are unimodal. In training, a

---

Algorithm 1. Reward for small bowel path tracking

Input: Current agent position p_t, current action a_t, ground-truth segmentation image I^{segm}, wall detection image I^{wall}, cumulative path image I^{cum}, geodesic distance transform (GDT) image I^{GDT}, maximum GDT value achieved so far v_{max}^{GDT}, maximum allowed increase of GD per step θ, predefined reward scale r^{val1}, r^{val2}

Output: Reward r_t

1: **if** $a_t == \vec{0}$ **then**
2:　　$r_t = -r^{val1}$　　　　　　　　　　　　　　　　　　▷ zero movement penalty
3: **else**
4:　　$r_t = 0$
5:　　Next agent position $p_{t+1} \leftarrow p_t + a_t$
6:　　Set of voxels S on the line between p_t and p_{t+1}
7:　　Next GDT value $v_{t+1}^{GDT} \leftarrow I^{GDT}[p_{t+1}]$
8:　　**if** $v_{t+1}^{GDT} > v_{max}^{GDT}$ **then**
9:　　　　$\Delta v^{GDT} = v_{t+1}^{GDT} - v_{max}^{GDT}$
10:　　　**if** $\Delta v^{GDT} > \theta$ **then**
11:　　　　　$r_t = -r^{val2}$　　　　　　　　　▷ penalty on an abrupt increase of GD
12:　　　**else**
13:　　　　　$r_t = \frac{\Delta v^{GDT}}{\theta} \times r^{val2}$　　　　　　　　　　▷ GDT-based reward
14:　　　**end if**
15:　　　$v_{max}^{GDT} \leftarrow v_{t+1}^{GDT}$
16:　　**end if**
17:　　$r_t -= (\frac{1}{|S|}\sum_S I^{wall}[s]) \times r^{val2}$　　　　　　　▷ wall-based penalty
18:　　**if** $\sum_S I^{cum}[s] > 0$ **then**
19:　　　$r_t -= r^{val1}$　　　　　　　　　　　　　　　　　　▷ revisiting penalty
20:　　**end if**
21:　　**if** $I^{segm}[p_{t+1}] == False$ **then**
22:　　　$r_t = -r^{val1}$　　　　　　　　　　　　　▷ out-of-segmentation penalty
23:　　**end if**
24: **end if**
25: **return** r_t

---

value is sampled from each beta distribution. This probabilistic sampling allows for exploration of the agent. The sampled value within $[0, 1]$ is mapped to an actual displacement within $[-d_{step}, d_{step}]$, and rounded off to move on the image volume. In test, the mode of the distribution, $\frac{\alpha-1}{\alpha+\beta-2}$, is deterministically used.

Reward. Our reward calculation for each step is summarized in Algorithm 1. A key point is to use the geodesic distance transform (GDT). We hypothesized the reward function based on the Euclidean distance to the closest point on the GT path, which is used in [9,21], would be inappropriate for our problem since the small bowel has local paths that are spatially close each other. It may encourage crossing the walls. The corresponding experiment will be discussed in Sect. 3. Figure 3 shows toy examples of the GDT. Despite the contact issue, when the GT path is available (Fig. 3 (b)), it reflects the bowel path well. We incentivize increasing the distance from the start on the GDT. One more benefit of using the GDT is its applicability to the scans without the GT path (Fig. 3 (c)), namely *SegmSet*. Being of

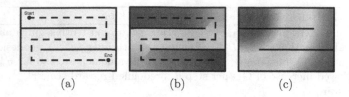

Fig. 3. Toy examples of the geodesic distance transform (GDT), which is computed by the fast marching method [20]. (a) Toy small bowel with folds and contact. The solid and dashed lines denote the bowel wall and GT path, respectively. (b) GDT from the start when the GT path is given. The distance gradually increases from blue to red. (c) GDT when only segmentation is given. (Color figure online)

lower quality, it still encodes the distance from the start roughly. Our objective is to utilize *SegmSet* as well as *PathSet* for training. To make our reward compatible for both, we do not penalize a decrease of GD. We only incentivize achieving a higher value than the maximum achieved so far. This encourages the agent to travel more to achieve a higher maximum, even though it looks to become closer to the start temporarily on the lumpy small bowel.

Another important factor is to use the bowel wall detection. We penalize passing the detected walls. Revisiting the previous path and leaving the small bowel segmentation are also penalized. For calculation of the revisiting penalty, the cumulative path is stored also as cylinders of radius 6 mm. When an episode terminates, we incentivize/penalize finally according to the coverage of the cylinderized cumulative path on the segmentation, c. We provide the final reward of $c \times r^{final}$ if the agent arrives at the end, otherwise $(c-1) \times r^{final}$. In summary, our reward encourages the agent to travel to get farther from the start while not penetrating the walls, not revisiting the previous path, and not leaving the small bowel. This can be pursued even without the GT path.

2.3   Training and Testing

We now can make our environment using both *PathSet* and *SegmSet*. We run four episodes in parallel to collect training data, where two of them are from *PathSet* and the other two are from *SegmSet*. In training, a tracking can start from either of two ends of the small bowel, or from a middle point on the GT path with a probability of 0.3 to diversify the episodes. Since *SegmSet* has no GT path, this middle start point is chosen more carefully as a non-wall small bowel voxel. It always starts at the pylorus in test time. Each episode continues until the conditions mentioned in Sect. 2.2 are met. We used proximal policy optimization (PPO) [14] to train our actor/critic networks. The detailed training algorithm can be found in supplementary material.

2.4   Evaluation Details

The maximum displacement along each axis per step d_{step} and the maximum time step per episode T were set to 10 mm and 800, respectively. The cell length in GDT was set to 1.5, which is the same with the voxel size of the input volume. The maximum allowed increase of GD per step $\theta = \sqrt{3} * 10^2$ was used accordingly. The reward scale $r^{val1} = 4$, $r^{val2} = 6$, and $r^{final} = 100$ were chosen by experiments. The discount factor $\gamma = 0.99$, generalized advantage estimation (GAE) [13] parameter $\lambda = 0.95$, clipping parameter $\epsilon = 0.2$, and entropy coefficient of 0.001 were used for PPO. A learning rate of 10^{-5}, minibatch size of 32, and the number of epochs 5 were used for both networks.

We implemented our networks, including ones for the comparable methods, using PyTorch 1.8.2. We used Adam optimizers [7] for training. We used a NVIDIA Tesla V100 32GB GPU to conduct experiments.

We provide the maximum length of the GT path that is tracked without making an error, which is used in [17], as an evaluation metric. When comparing the predicted path with the GT path, a static distance tolerance of 10 mm was used. We note that this metric is more important than the others used in [17], including the precision, recall, and curve-to-curve distance, since they are computed in disregard of the order of the tracking, and the error of crossing the walls. Instead, we provide more information for the maximum length metric.

3   Results

3.1   Quantitative Evaluation

Table 1 shows quantitative results of the proposed method and comparable methods. We first provide the performance of the methods that use *PathSet* for training. Since the GT path is available for the entire small bowel in this set, we used the full supervision to train the network proposed in [11], which was originally trained using sparse annotation. Nevertheless, it presented a difficulty on predicting the precise centerline DT for our dataset, which is the key of their method. Using the reward function based on the Euclidean distance to the GT path, 'Zhang et al. [21]', which was originally proposed for the aorta, had trouble with learning a successful policy for the small bowel as well. Our method using the same training set showed a better performance.

The next group of methods are ones that use *SegmSet*. The 'TSP' method of [17] entails small bowel segmentation, thus the set can be used to train a segmentation network. The proposed method performed reasonably when trained using only *SegmSet*, implying our formulation is operational even without the GT path. By expanding the set a little, 'Ours (s) (0/25)' showed a slightly better performance than 'Ours (p)'. Considering the respective annotation costs for *PathSet* and *SegmSet*, which were a full day versus a couple of hours per scan, it could be a possible replacement.

Finally, the full method 'Ours (p+s)', which uses the both sets for training, performed the best. We can see that including a small number of scans from *PathSet* for training helps increase the performance more by comparing with 'Ours (s) (0/25)' again. The use of the wall-related components, namely, the wall input patch and wall-based penalty, are important in the proposed method. It is highlighted when those were eliminated, 'Ours (p+s) w/o wall'.

Table 1. Quantitative comparison of different methods. Statistics on the maximum length of the GT path that is tracked without making an error are presented in *mm*. N^{th} denotes the Nth percentile. Each method is categorized into one of the supervised learning (SL), RL, and graph based methods. The first three and the next three are ones that use *PathSet* or *SegmSet* for training, respectively. The last two are trained using both sets. The numbers of scans used for training are shown. All ten scans in *PathSet* were evaluated using a 2-fold cross validation except for 'TSP [17]' and 'Ours (s) (0/20)', which are trained using only *SegmSet*, thus do not need the cross validation. '0/25' denotes using additional five scans from *PathSet* as *SegmSet* to test the remaining five. Refer to the text for the explanation on each method.

Method	CAT	# Tr (*PathSet*/*SegmSet*)	Mean	Std	Median	20th	80th	Max
DT [11]	SL	5/0	342.4	236.6	308.0	156.0	518.0	910.0
Zhang et al. [21]	RL	5/0	496.9	250.6	441.8	340.3	607.5	1153.7
Ours (p)	RL	5/0	1333.0	515.6	1143.3	929.9	1873.1	2456.7
TSP [17]	Graph	0/20	810.0	193.6	837.0	582.0	1060.0	1162.0
Ours (s)	RL	0/20	1267.3	447.5	1243.3	1011.9	1604.5	2370.1
Ours (s)	RL	0/25	1393.0	478.4	1302.9	1119.4	1925.4	**2532.8**
Ours (p+s)	RL	5/20	**1519.0**	500.0	**1317.2**	**1168.7**	**2183.6**	2394.0
Ours (p+s) w/o wall	RL	5/20	1050.0	369.4	1007.5	668.7	1523.9	1567.2

3.2 Qualitative Evaluation

Figure 4 shows example path tracking results. The result of our full method, which utilizes both of *PathSet* and *SegmSet*, is compared to that of using only *PathSet*. The use of an easier-to-acquire dataset, *SegmSet*, helped achieve a better

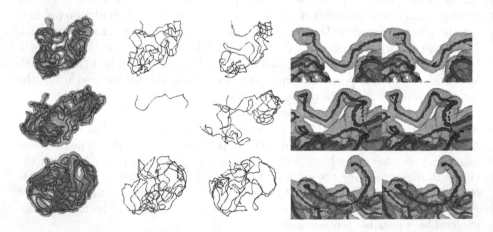

Fig. 4. Example path tracking results. Each row represents different cases. The columns, from left, represent GT path (red) and segmentation (green), result corresponding to 'Ours (p)' in Table 1, result corresponding to 'Ours (p+s)' in Table 1, a part of the tracked path corresponding to the second and third columns, respectively. In the last two columns, only a selected local path is shown to highlight the difference, and each step movement is drawn with different colors. (Color figure online)

tracker in a situation where acquiring the GT path in a larger scale is infeasible due to the exceptionally high annotation cost. Note that in the first row, the result of 'Ours (p)' has more tracked path overall, but it is achieved by crossing the wall at the first contact area it visits. The full method showed less crossing, resulting in an increase in the maximum length metric.

4 Conclusion

We have presented a novel DRL method for small bowel path tracking, which can learn even from scans that are without the GT path. The experimental results showed that it is possible to train a reasonable tracker without using the GT path, and that utilizing those weakly annotated scans together with ones having the GT path can render a better tracker than using either of them alone. Considering the annotation difficulty of the GT path, the proposed method could reduce the annotation cost required for training. The tracked path can provide better information on the small bowel structure than segmentation. Especially, it could be useful for image-guided intervention where a device approaches according to the identified structure.

Acknowledgments. We thank Dr. James Gulley for patient referral and for providing access to CT scans. This research was supported by the Intramural Research Program of the National Institutes of Health, Clinical Center. The research used the high performance computing facilities of the NIH Biowulf cluster.

References

1. Small bowel obstruction (2019). https://my.clevelandclinic.org/health/diseases/15850-small-bowel-obstruction
2. Chou, P.W., Maturana, D., Scherer, S.: Improving stochastic policy gradients in continuous control with deep reinforcement learning using the beta distribution. In: Precup, D., Teh, Y.W. (eds.) Proceedings of the 34th International Conference on Machine Learning. Proceedings of Machine Learning Research, vol. 70, pp. 834–843. PMLR, 06–11 August 2017. https://proceedings.mlr.press/v70/chou17a.html
3. Çiçek, Ö., Abdulkadir, A., Lienkamp, S.S., Brox, T., Ronneberger, O.: 3D U-Net: learning dense volumetric segmentation from sparse annotation. In: Ourselin, S., Joskowicz, L., Sabuncu, M.R., Unal, G., Wells, W. (eds.) MICCAI 2016. LNCS, vol. 9901, pp. 424–432. Springer, Cham (2016). https://doi.org/10.1007/978-3-319-46723-8_49
4. Dai, T., et al.: Deep reinforcement learning for subpixel neural tracking. In: Cardoso, M.J., et al. (eds.) Proceedings of the 2nd International Conference on Medical Imaging with Deep Learning. Proceedings of Machine Learning Research, vol. 102, pp. 130–150. PMLR, 08–10 July 2019. https://proceedings.mlr.press/v102/dai19a.html
5. Fedorov, A., et al.: 3D slicer as an image computing platform for the quantitative imaging network. Magn. Reson. Imaging **30**(9), 1323–1341 (2012). https://doi.org/10.1016/j.mri.2012.05.001

6. van Harten, L., de Jonge, C., Stoker, J., Isgum, I.: Untangling the small intestine in 3D cine-MRI using deep stochastic tracking. In: Medical Imaging with Deep Learning (2021). https://openreview.net/forum?id=cfYAFR6s6iJ
7. Kingma, D.P., Ba, J.: Adam: a method for stochastic optimization. In: Bengio, Y., LeCun, Y. (eds.) 3rd International Conference on Learning Representations, ICLR 2015, San Diego, CA, USA, 7–9 May 2015, Conference Track Proceedings (2015). http://arxiv.org/abs/1412.6980
8. Konda, V., Tsitsiklis, J.: Actor-critic algorithms. In: Solla, S., Leen, T., Müller, K. (eds.) Advances in Neural Information Processing Systems, vol. 12. MIT Press (1999)
9. Li, Z., Xia, Q., Hu, Z., Wang, W., Xu, L., Zhang, S.: A deep reinforced tree-traversal agent for coronary artery centerline extraction. In: de Bruijne, M., et al. (eds.) MICCAI 2021. LNCS, vol. 12905, pp. 418–428. Springer, Cham (2021). https://doi.org/10.1007/978-3-030-87240-3_40
10. Meijering, E., Jacob, M., Sarria, J.C., Steiner, P., Hirling, H., Unser, M.: Design and validation of a tool for neurite tracing and analysis in fluorescence microscopy images. Cytometry Part A **58A**(2), 167–176 (2004). https://doi.org/10.1002/cyto.a.20022
11. Oda, H., et al.: Intestinal region reconstruction of ileus cases from 3D CT images based on graphical representation and its visualization. In: Mazurowski, M.A., Drukker, K. (eds.) Medical Imaging 2021: Computer-Aided Diagnosis. vol. 11597, pp. 388–395. International Society for Optics and Photonics, SPIE (2021). https://doi.org/10.1117/12.2581261
12. Pinto, L., Andrychowicz, M., Welinder, P., Zaremba, W., Abbeel, P.: Asymmetric actor critic for image-based robot learning. arXiv preprint arXiv:1710.06542 (2017)
13. Schulman, J., Moritz, P., Levine, S., Jordan, M.I., Abbeel, P.: High-dimensional continuous control using generalized advantage estimation. In: Bengio, Y., LeCun, Y. (eds.) 4th International Conference on Learning Representations, ICLR 2016, San Juan, Puerto Rico, 2–4 May 2016, Conference Track Proceedings (2016)
14. Schulman, J., Wolski, F., Dhariwal, P., Radford, A., Klimov, O.: Proximal policy optimization algorithms. arXiv preprint arXiv:1707.06347 (2017)
15. Shin, S.Y., Lee, S., Elton, D., Gulley, J.L., Summers, R.M.: Deep small bowel segmentation with cylindrical topological constraints. In: Martel, A.L., et al. (eds.) MICCAI 2020. LNCS, vol. 12264, pp. 207–215. Springer, Cham (2020). https://doi.org/10.1007/978-3-030-59719-1_21
16. Shin, S.Y., Lee, S., Summers, R.M.: Unsupervised domain adaptation for small bowel segmentation using disentangled representation. In: de Bruijne, M., et al. (eds.) Medical Image Computing and Computer Assisted Intervention - MICCAI 2021, pp. 282–292. Springer, Cham (2021). https://doi.org/10.1007/978-3-030-87199-4_27
17. Shin, S.Y., Lee, S., Summers, R.M.: A graph-theoretic algorithm for small bowel path tracking in CT scans. In: Drukker, K., Iftekharuddin, K.M., Lu, H., Mazurowski, M.A., Muramatsu, C., Samala, R.K. (eds.) Medical Imaging 2022: Computer-Aided Diagnosis, vol. 12033, pp. 863–868. International Society for Optics and Photonics, SPIE (2022). https://doi.org/10.1117/12.2611878
18. Shin, S.Y., Lee, S., Summers, R.M.: Graph-based small bowel path tracking with cylindrical constraints. In: 2022 IEEE 19th International Symposium on Biomedical Imaging (ISBI), pp. 1–5 (2022). https://doi.org/10.1109/ISBI52829.2022.9761423
19. Wu, Y., He, K.: Group normalization. In: Proceedings of the European Conference on Computer Vision (ECCV), September 2018

20. Yatziv, L., Bartesaghi, A., Sapiro, G.: O(n) implementation of the fast marching algorithm. J. Comput. Phys. **212**(2), 393–399 (2006). https://doi.org/10.1016/j.jcp.2005.08.005
21. Zhang, P., Wang, F., Zheng, Y.: Deep reinforcement learning for vessel centerline tracing in multi-modality 3D volumes. In: Frangi, A.F., Schnabel, J.A., Davatzikos, C., Alberola-López, C., Fichtinger, G. (eds.) MICCAI 2018. LNCS, vol. 11073, pp. 755–763. Springer, Cham (2018). https://doi.org/10.1007/978-3-030-00937-3_86
22. Zhang, W., et al.: Mesenteric vasculature-guided small bowel segmentation on 3-D CT. IEEE Trans. Med. Imaging **32**(11), 2006–2021 (2013). https://doi.org/10.1109/TMI.2013.2271487

Efficient Population Based Hyperparameter Scheduling for Medical Image Segmentation

Yufan He$^{(\boxtimes)}$, Dong Yang, Andriy Myronenko, and Daguang Xu

NVidia, Santa Clara, USA
yufanh@nvidia.com

Abstract. The training hyperparameters (learning rate, augmentation policies, e.t.c) are key factors affecting the performance of deep networks for medical image segmentation. Manual or automatic hyperparameter optimization (HPO) is used to improve the performance. However, manual tuning is infeasible for a large number of parameters, and existing automatic HPO methods like Bayesian optimization are extremely time consuming. Moreover, they can only find a fixed set of hyperparameters. Population based training (PBT) has shown its ability to find dynamic hyperparameters and has fast search speed by using parallel training processes. However, it is still expensive for large 3D medical image datasets with limited GPUs, and the performance lower bound is unknown. In this paper, we focus on improving the network performance using hyperparameter scheduling via PBT with limited computation cost. The core idea is to train the network with a default setting from prior knowledge, and finetune using PBT based hyperparameter scheduling. Our method can achieve 1%–3% performance improvements over default setting while only taking 3%–10% computation cost of training from scratch using PBT.

Keywords: Hyperparameter optimization · Population based training · Bayesian optimization · Medical image segmentation

1 Introduction

The success of deep networks relies heavily on the correct setting of training hyperparameters. Hyperparameters like learning rate, choice of optimizers, and augmentation policies can greatly affect the performance of 3D medical image segmentation networks. This has led to the automatic hyperparameter optimization (HPO) algorithms. The basic algorithms include random search [3], grid search, and Bayesian optimization [8,18,19]. Random search trains the deep network with the randomly sampled hyperparameters, which is inefficient. Bayesian optimization samples the hyperparameters based on a Bayesian model, which is

Supplementary Information The online version contains supplementary material available at https://doi.org/10.1007/978-3-031-16443-9_54.

updated during searching. For medical imaging, Tran et al. [20] applied bayesian optimization (Gaussian Processes) on 2D echocardiography segmentation and trained 100 jobs. Yang et al. [22] and Nath et al. [14] used reinforcement learning and searched on 50 GPUs for 2 days. Hoopes et al. [7] proposed a specific hyper-network to generate hyperparameters for registration tasks. The core of those methods are the same: explore and interpolate the manifold of hyperparameter θ and deep network performance p, as shown in Fig. 1 (a). The major drawback is that the evaluation for each sampled hyperparameter requires expensive deep network retraining. Many works focus on reducing the evaluation cost by early stopping the network retraining [4,11,12]. For example, Bayesian optimization hyperband (BOHB) [4] samples hyperparameters using a Tree Parzen estimators (TPE) [2], and the hyperparameters with good performances are allocated more training budget for more accurate evaluations (Hyperband [12]). However, if the network requires long training epochs to converge, aggressive early stopping may cause inaccurate evaluation. Moreover, the searched hyperparameters are fixed for the network training, which is sub-optimal.

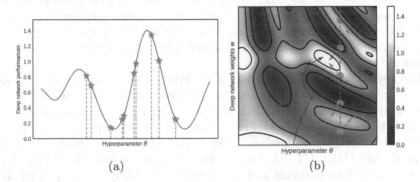

(a)   (b)

Fig. 1. (a) The hyperparameters θ and best validation performances p of the network trained with θ. Each dashed line represents a sample of hyperparameters. (b) The manifold when considering the deep network parameters (weights, w). The green arrow represents the PBT from scratch, and the blue dashed line represents training with a default setting. The blue dots are the validation checkpoints. The blue arrows represent PBT searching starting from validation checkpoints. (Color figure online)

Population based training (PBT) [10,15] solves the HPO problems by introducing parallel workers and evolutionary strategies. Each parallel train-ing process (worker) trains the network with different hyperparameters for a step (e.g. 1 epoch). The workers with top performances keep the hyperparame-ters unchanged. The rest workers shall load the hyperparameters and network weights from the top performing ones (exploit), then continue training with mutated hyperparameters (explore). All the workers will continue and repeat the process. If a deep network needs to be trained for N epochs with certain hyperparameters, the total training epochs for PBT are also N if ideally without computation resource limit, since the workers are running in parallel. Meanwhile, the searched hyperparameters are dynamic across training steps. This method is suitable for training large scale deep networks and shows superior performance in reinforcement learning. However, the computation resource is limited in real practice (e.g. a station with 8 GPUs), and data parallel is usually used to increase the batch size for 3D medical images (training one network using all 8 GPUs, and no GPUs left for parallel workers). The training cost becomes $N \times W$, where W is the number of PBT workers and needs to be large enough to reach certain performance. Can we reduce N and adapt the early stopping methods for evaluation mentioned above? PBT evaluates different hyperparameters every single step (e.g. 1 epoch) and already uses "early stopping". To understand the problem, we plot the network performance manifold related to both the hyperparameters θ and deep network weights w (simulated in 2D for intuitive explanation), as shown in Fig. 1 (b). Figure 1 (a) is a sub-space of Fig. 1 (b), where the deep network performance (the y axis) is the maximum performance of all w given a θ. The blue dashed line represents the training of w using certain θ and each blue dot means a validation checkpoint. The green arrow represents the best worker in each step for the PBT algorithm. The large training cost comes from the large number of steps needed to converge from scratch. Meanwhile, the green arrow explores the manifold from random initialled states without any convergence guarantee [23], thus the performance lower bound cannot be estimated beforehand.

The key to reduce training costs is to reduce the steps the network needs to converge from scratch. However, the "No Free Lunch Theorem" [17,21] suggests that no optimization method offers a "shortcut" unless using prior knowledge about the problem. We observed that a major difference between medical image and natural image analysis tasks is that the targets (brain, lung, e.t.c) and modalities (MRI, CT, e.t.c) for medical images are limited, and the expert knowledge might be obtained and transferred. For example, nn-UNet [9] shows that hyperparameters generated by an expert understanding of the medical data can achieve outstanding performance on most medical segmentation challenges. In many scenarios, we already have a good set of hyperparameters for a task, and what we really need is an efficient and robust method to further improve the performance (e.g. compete in challenges). If we train the network using the hyperparameters from expert knowledge (denoted as the default setting) and w is updated along the dotted blue line in Fig. 1 (b), we can start from the validation checkpoints and use PBT for finetuning (blue arrows). This greatly accelerates the training and the results is at least as good as the default setting. Based on this observation, we propose a fast performance improvement method using PBT and finetuning. The proposed method can be applied to any given task and achieve significant performance improvement with limited computation cost. As far as we know, our method is (1) the first work applying PBT for large scale 3D medical image segmentation, and can (2) reduce training cost to 3%–10% of the original PBT and makes it computationally feasible, (3) the performance lower bound is bounded by the default setting performance, thus making the whole process controllable. (4) The mutation step (explore the hyperparameter space) in original PBT relies on random heuristics and lacks theoretical guarantees [15], and we implemented the TPE based Bayesian sampler which is theoretically sound and does not need a heuristic exploring algorithm as in [6,10]. TPE model has shown its efficacy and simplicity as used by BOHB [4], so we use TPE instead of the Gaussian process as used in [15]. Besides, we implemented the PBT with sequential workers, where each worker uses all GPUs for data parallel and is more suitable for 3D medical imaging tasks. We performed experiments on four datasets from the MSD challenge [1] using two network structures. The results show the efficacy and simplicity of our proposed method.

2   Method

Our method can be added to any training pipeline. Conventionally, given default hyperparameters, the network is trained for N epochs and validated for V times (network weights are saved as checkpoints). The checkpoint with the best validation accuracy is deployed. Our method tries to improve the performance of the network from this training pipeline, and we have two steps: 1) select checkpoints to finetune 2) apply PBT to the selected checkpoints.

Checkpoint Selection
If only finetune from the checkpoint with the highest validation accuracy, the final network can only reach the local maximum around that point. We can start from multiple checkpoints to explore more local maximums, as shown in Fig. 1 (b). If the checkpoint has low validation accuracy, the checkpoint may be located at a bad position, and explore around is a waste of computation. If two checkpoints are too close, they may be located around the same local maximum, and the exploration is redundant. Given checkpoints $c_{e_1}, c_{e_2}, \cdots, c_{e_V}$ validated at epoch $e_1, e_2, \cdots, e_V$ with performances $p_{e_1}, p_{e_2}, \cdots, p_{e_V}$, we select a list of checkpoints $S = \{c_{e_i} | p_{e_i} > 0.95 * p_{e_m}\}$ with good enough performances, where $p_{e_m} = max(p_{e_1}, p_{e_2}, \cdots, p_{e_V})$. We chose three checkpoints $c_{e_i}, c_{e_j}, c_{e_k}, e_i < e_j < e_k$ from S, which satisfies

$$\max_{i,j,k}(\min(e_j - e_i, e_k - e_j)) \quad \forall i, j, k \in S, m \in \{i, j, k\}. \tag{1}$$

The epoch differences $e_j - e_i$ and $e_k - e_j$ is used as distances between the checkpoints, and we want to keep checkpoints as far as possible, so we maximize the minimum value of these two distances. Meanwhile, we make sure the best checkpoint c_{e_m} is among these three selected checkpoints ($m \in \{i, j, k\}$). If several sets of $\{i, j, k\}$ have the same value in Eq. 1, the set with $\max_{i,j,k}(\max(e_j - e_i, e_k - e_j))$ is selected.

Population Based Training with TPE
We run PBT starting from $c_{e_i}, c_{e_j}, c_{e_k}$. The algorithm is shown in Alg. 1. We use $W = 27$ workers, and each worker runs sequentially. In the first step, the first worker loads the checkpoint c_{e_i} (the same procedure for c_{e_j} and c_{e_k}) and

randomly samples hyperparameters h from configuration space H. The network Φ is trained for B = 1 epoch with h. The validation result and h will be saved into a set R, and a TPE model will be fitted if the elements in R is large enough (the details of TPE fitting and sampling can be found in BOHB[1] [4]). All the following workers in this step will load checkpoint c_{e_i}. If the TPE model is not fitted or a random number $p \sim \text{Uniform}(0,1)$ is smaller than $\sigma = 0.3$, h is randomly sampled; otherwise the TPE model is used. After all workers finish training and validation in the step, the top $\eta = 3$ workers ranked by validation accuracies will not change their hyperparameters or load weights from other workers. The rest W $- \eta$ workers will randomly load checkpoints from these top workers and sample new h. Then all workers will continue to the next step for in total S = 50 steps.

Algorithm 1 Population Based Training for Finutuning

Input: checkpoints $c_{e_i}, c_{e_j}, c_{e_k}$, network Φ, hyperparameter search space H, number of search steps S and epochs B in each step, number of workers W, number of top workers η, random rate σ

1: **for** $c \in \{c_{e_i}, c_{e_j}, c_{e_k}\}$ **do**
2:     Initialize TPE model M $= \emptyset$, performance record R $= \emptyset$
3:     Initialize top workers set $T_w = \emptyset$, checkpoints set $C_w = \emptyset$
4:     **for** s in $\{1, 2, \cdots, \text{S}\}$ **do**
5:         Initialize performance record SR $= \emptyset$
6:         **for** w in $\{1, 2, \cdots, \text{W}\}$ **do**
7:             **if** $w \notin T_w$ **then**
8:                 **if** M $= \emptyset$ or $p \sim \text{Uniform}(0,1) < \sigma$ **then**
9:                     Sample random hyperparameters h from H
10:                    **else**
11:                        Sample h using M
12:                    **if** $C_w \neq \emptyset$ **then**
13:                        Load a random checkpoint from C_w into Φ
14:                    **else**
15:                        Load checkpoint c into Φ
16:                **else**
17:                    Use the hyperparameters h of worker w
18:                    Load checkpoint that corresponds to w from C_w
19:                Train Φ for B epochs with hyperparameters h
20:                Get validation performance p_w of Φ and save checkpoints
21:                SR $=$ SR $\cup \{p_w\}$, R $=$ R $\cup \{(p_w, h)\}$
22:                **if** $|R|_0 >$ Minimum samples needed for fitting TPE **then**
23:                    Fit new TPE model M using R
24:            Find η top workers, $T_w = \{w | w \in \text{top } \eta \text{ workers in SR}\}$
25:            Define set $C_w =$ {most recent checkpoint of $w | w \in T_w$}

---

3   Experiments

Datasets and Default Setting. Four datasets from the MSD challenge [1] are used: Task01 Brain Tumour (484 multi-modal MR images), Task05 Prostate (32

---

[1] https://github.com/automl/HpBandSter.

Table 1. Hyperparamter search space H. The values in "Range" are the choices for "Categorical", and are the min and max values for others. {Aug} contains nine random augmentation probabilities. Details about augmentations and losses can be found at https://monai.io/ and DiNTS (See footnote 2).

Parameter	Learning rate	Optimizer	Weight decay	Loss function	Background crop ratio	{Aug}
Distribution	LogUniform	Categorical	LogUniform	Categorical	Uniform	Uniform
Range	[1e−4, 0.2]	[Sgd, Adam]	[1e−5, 1e−1]	[DiceLoss, DiceCELoss, DiceFocalLoss]	[0.1, 0.9]	[0, 1]

MR/ADC images), Task06 Lung Tumour (64 CT images), and Task07 Pancreas Tumour (282 CT images). We train the standard U-Net [16] and the SegResNet [13] (1st in Brats18 challenge) with a sophisticated training pipeline with hyperparameters from DiNTS [5][2] (2st place in MSD live challenge). Each dataset is equally split into five folds, and the network is trained on the first four (5000 epochs for Task05, Task06, and 1500 epochs for Task01, Task07) and validated (every 100 epochs) on the last fold. Our method tries to find the optimal hyperparameter schedule to fit data, and we report validation results.

Search Setting. The hyperparameter search space H is shown in Table 1. H defines how h is sampled using random sample and TPE model. We search 14 parameters: learning rate, optimizer, weight decay, loss function, background crop ratio (background patches sampling ratio), and 9 random augmentations probabilities. PBT starts from three checkpoints, and run with 27 workers (W = 27) for S = 50 steps (B = 1, one epoch per step) using 8 NVIDIA V100 GPU.

PBT Finetuning Results. The validation curve (average Dice score for all segmentation classes) of the U-Net and SegResNet on the four datasets are shown in Fig. 3. We show the best PBT validation score among W = 27 workers in each step. The training using the default setting validates every 100 epochs, while PBT validates every 1 epoch for 50 epochs. We resume the default setting training (including optimizer states) from these three checkpoints and validates every epoch for 50 epochs ("default-continue"). The best Dice score of PBT finetuning in all the steps and the Dice score of the default setting are also listed in Fig. 3. We can see that PBT finetuning can achieve 1%–3% Dice improvement over the default setting rapidly for most experiments, while the "default-continue" follows the trend of the default setting validation curve. Intuitively, the "default-continue" explores along the blue dashed line (y axis), while PBT can explore both x and y axes thus can reach better performance, as shown in Fig. 1 (b).

---

[2] https://github.com/Project-MONAI/research-contributions/tree/master/DiNTS.

Comparison with PBT from Scratch. We also show a comparison with PBT from scratch. The validation curve of PBT from scratch for Task05 Prostate and Task06 Lung Tumour using U-Net are shown in Fig. 2 (total 5000 epochs, S = 100, B = 50, W = 9 and 27 for Task05, W = 9 only for Task06 due to the training cost). We can see that the PBT performance is influenced by the number of workers. PBT from scratch shows faster convergence and achieves better results with more workers (W = 27) than the default setting. However, it is worse than the default setting if the worker number is small (W = 9). This is also observed on the Task01 Brain Tumour (PBT W = 9 Dice:0.68, default Dice:0.73) and Task07 Pancreas Tumour (PBT W = 9 Dice:0.49 default Dice:0.53, details are in the supplementary material). Our PBT finetuning (W = 27) greatly reduces the performance gap between default setting and PBT from scratch (W = 27). However, our method is also influenced by the worker number. A smaller worker number (W = 9) will provide less improvement, but is still better than PBT from scratch (W = 9) and the default setting. Compared with our PBT finetuning, PBT training from scratch can potentially find better local performance maximums than searching around default settings, however, the major obstacle is the computation cost. For large datasets like Task01 and Task07, training with default setting takes 9 h on an 8 GPU station (1500 epochs), and a 27 worker PBT process will cost ten days (27 × 1500 epochs), while PBT finetuning only takes 20 h (3 × 27 × 50 epochs) and achieves 10 times speedup. For Task05 and Task06, the default setting takes 0.67 and 9 h respectively for 5000 epochs, PBT finetuning achieves 33 times speedup (3 × 27 × 50 epochs) compared to PBT (W = 27) from scratch (27×5000 epochs).

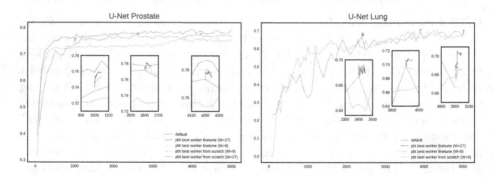

Fig. 2. Validation curve of PBT training from scratch and our PBT finetuning on prostate and lung datasets, with varying worker number W = 9 and W = 27.

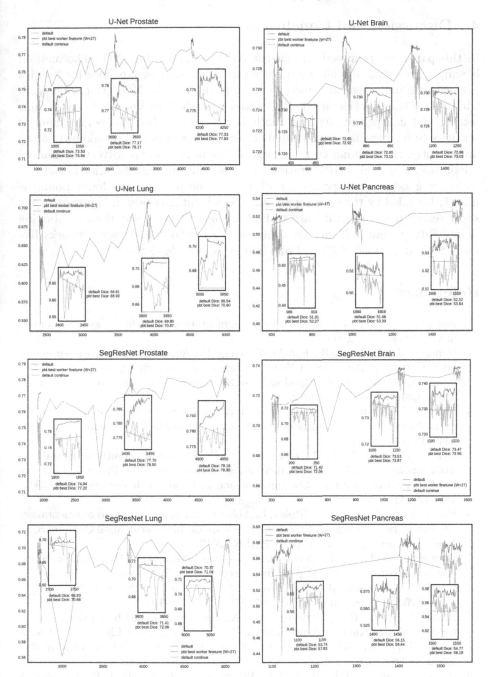

Fig. 3. Validation curve (Dice score, y-axis) of default hyperparameters ("default"), PBT finetuning ("pbt best worker finetune", results of the best worker in each step), and continue training using default hyperparameters ("default-continue"). x-axis is the training epoch number. The best Dice score of PBT finetuning in all the steps and the Dice score of the default training are listed besides the zoomed in windows for each checkpoint.

4   Discussion and Conclusion

In this paper, we proposed an efficient HPO method using PBT. The method is simple, robust, and widely applicable to most training pipelines for a quick performance improvement. Our experiments show a 1%–3% performance gain with 10–33 times acceleration. PBT from scratch does not need any default setting and can potentially find much better results, but the performance is sensitive to settings like worker number, and the lower bound is unknown. More importantly, the training cost for large datasets is too expensive with limited GPUs (e.g. a typical 8 GPU station). Our method utilizes the prior knowledge to locate good starting points for PBT, and the performance lower bound is usually bounded by the starting points. The advantage is also its limitation, the short search time makes it hard to find a much better result than its starting point. Future work will include applying the methods to larger networks like DiNTS [5] and nn-UNet [9] and obtaining challenge leaderboard results.

References

1. Antonelli, M., et al.: The medical segmentation decathlon. arXiv preprint arXiv:2106.05735 (2021)
2. Bergstra, J., Bardenet, R., Bengio, Y., Kégl, B.: Algorithms for hyper-parameter optimization. In: Advances in Neural Information Processing Systems 24 (2011)
3. Bergstra, J., Bengio, Y.: Random search for hyper-parameter optimization. J. Mach. Learn. Res. **13**(2) (2012)
4. Falkner, S., Klein, A., Hutter, F.: BOHB: robust and efficient hyperparameter optimization at scale. In: International Conference on Machine Learning, pp. 1437–1446. PMLR (2018)
5. He, Y., Yang, D., Roth, H., Zhao, C., Xu, D.: DiNTS: differentiable neural network topology search for 3D medical image segmentation. In: Proceedings of the IEEE/CVF Conference on Computer Vision and Pattern Recognition, pp. 5841–5850 (2021)
6. Ho, D., Liang, E., Chen, X., Stoica, I., Abbeel, P.: Population based augmentation: efficient learning of augmentation policy schedules. In: International Conference on Machine Learning, pp. 2731–2741. PMLR (2019)
7. Hoopes, A., Hoffmann, M., Fischl, B., Guttag, J., Dalca, A.V.: HyperMorph: amortized hyperparameter learning for image registration. In: Feragen, A., Sommer, S., Schnabel, J., Nielsen, M. (eds.) IPMI 2021. LNCS, vol. 12729, pp. 3–17. Springer, Cham (2021). https://doi.org/10.1007/978-3-030-78191-0_1
8. Hutter, F., Hoos, H.H., Leyton-Brown, K.: Sequential model-based optimization for general algorithm configuration. In: Coello, C.A.C. (ed.) LION 2011. LNCS, vol. 6683, pp. 507–523. Springer, Heidelberg (2011). https://doi.org/10.1007/978-3-642-25566-3_40
9. Isensee, F., Jaeger, P.F., Kohl, S.A., Petersen, J., Maier-Hein, K.H.: nnU-Net: a self-configuring method for deep learning-based biomedical image segmentation. Nat. Methods **18**(2), 203–211 (2021)
10. Jaderberg, M., et al.: Population based training of neural networks. arXiv preprint arXiv:1711.09846 (2017)

11. Jamieson, K., Talwalkar, A.: Non-stochastic best arm identification and hyperparameter optimization. In: Artificial Intelligence and Statistics, pp. 240–248. PMLR (2016)
12. Li, L., Jamieson, K., DeSalvo, G., Rostamizadeh, A., Talwalkar, A.: Hyperband: a novel bandit-based approach to hyperparameter optimization. J. Mach. Learn. Res. **18**(1), 6765–6816 (2017)
13. Myronenko, A.: 3D MRI brain tumor segmentation using autoencoder regularization. In: Crimi, A., Bakas, S., Kuijf, H., Keyvan, F., Reyes, M., van Walsum, T. (eds.) BrainLes 2018. LNCS, vol. 11384, pp. 311–320. Springer, Cham (2019). https://doi.org/10.1007/978-3-030-11726-9_28
14. Nath, V., et al.: The power of proxy data and proxy networks for hyper-parameter optimization in medical image segmentation. In: de Bruijne, M., et al. (eds.) MICCAI 2021. LNCS, vol. 12903, pp. 456 465. Springer, Cham (2021). https://doi.org/10.1007/978-3-030-87199-4_43
15. Parker-Holder, J., Nguyen, V., Roberts, S.J.: Provably efficient online hyperparameter optimization with population-based bandits. Adv. Neural Inf. Process. Syst. **33**, 17200–17211 (2020)
16. Ronneberger, O., Fischer, P., Brox, T.: U-Net: convolutional networks for biomedical image segmentation. In: Navab, N., Hornegger, J., Wells, W.M., Frangi, A.F. (eds.) MICCAI 2015. LNCS, vol. 9351, pp. 234–241. Springer, Cham (2015). https://doi.org/10.1007/978-3-319-24574-4_28
17. Schaffer, C.: A conservation law for generalization performance. In: Machine Learning Proceedings, pp. 259–265. Elsevier (1994)
18. Snoek, J., et al.: Scalable Bayesian optimization using deep neural networks. In: International Conference on Machine Learning, pp. 2171–2180. PMLR (2015)
19. Srinivas, N., Krause, A., Kakade, S., Seeger, M.: Gaussian process optimization in the bandit detting: no tegret and experimental design. In: Proceedings of the 27th International Conference on Machine Learning. No. CONF, Omnipress (2010)
20. Tran, T., Stough, J.V., Zhang, X., Haggerty, C.M.: Bayesian optimization of 2D echocardiography Segmentation. In: International Symposium on Biomedical Imaging (ISBI), pp. 1007–1011. IEEE (2021)
21. Wolpert, D.H., Macready, W.G.: No free lunch theorems for optimization. IEEE Trans. Evolut. Comput. **1**(1), 67–82 (1997)
22. Yang, D., Roth, H., Xu, Z., Milletari, F., Zhang, L., Xu, D.: Searching learning strategy with reinforcement learning for 3D medical image segmentation. In: Shen, D., et al. (eds.) MICCAI 2019. LNCS, vol. 11765, pp. 3–11. Springer, Cham (2019). https://doi.org/10.1007/978-3-030-32245-8_1
23. Yu, T., Zhu, H.: Hyper-parameter optimization: a review of algorithms and applications. arXiv preprint arXiv:2003.05689 (2020)

Atlas-Based Semantic Segmentation
of Prostate Zones

Jiazhen Zhang[1], Rajesh Venkataraman[5], Lawrence H. Staib[1,3,4],
and John A. Onofrey[1,2,4(✉)]

[1] Department of Radiology and Biomedical Imaging,
Yale University, New Haven, CT, USA
{jiazhen.zhang,lawrence.staib,john.onofrey}@yale.edu
[2] Department of Urology, Yale University, New Haven, CT, USA
[3] Department of Electrical Engineering, Yale University, New Haven, CT, USA
[4] Department of Biomedical Engineering, Yale University, New Haven, CT, USA
[5] Eigen Health, Grass Valley, CA, USA
rajesh.venkataraman@eigen.com

Abstract. Segmentation of the prostate into specific anatomical zones is important for radiological assessment of prostate cancer in magnetic resonance imaging (MRI). Of particular interest is segmenting the prostate into two regions of interest: the central gland (CG) and peripheral zone (PZ). In this paper, we propose to integrate an anatomical atlas of prostate zone shape into a deep learning semantic segmentation framework to segment the CG and PZ in T2-weighted MRI. Our approach incorporates anatomical information in the form of a probabilistic prostate zone atlas and utilizes a dynamically controlled hyperparameter to combine the atlas with the semantic segmentation result. In addition to providing significantly improved segmentation performance, this hyperparameter is capable of being dynamically adjusted during the inference stage to provide users with a mechanism to refine the segmentation. We validate our approach using an external test dataset and demonstrate Dice similarity coefficient values (mean±SD) of 0.91 ± 0.05 for the CG and 0.77 ± 0.16 for the PZ that significantly improves upon the baseline segmentation results without the atlas. All code is publicly available on GitHub: https://github.com/OnofreyLab/prostate_atlas_segm_miccai2022.

Keywords: Deep learning · Image segmentation · Probabilistic atlas · Prostate · MRI

1 Introduction

Prostate cancer (PCa) is one of the most commonly occurring forms of cancer, accounting for 27% of all cancer in men, and one of the major causes

---

Supplementary Information The online version contains supplementary material available at https://doi.org/10.1007/978-3-031-16443-9_55.

L. Wang et al. (Eds.): MICCAI 2022, LNCS 13435, pp. 570–579, 2022.
https://doi.org/10.1007/978-3-031-16443-9_55

of cancer-related death in the U.S [16]. Multi-parametric magnetic resonance imaging (mpMRI) that combines T2-weighted (T2W) imaging with functional sequences, such as diffusion-weighted imaging (DWI), and dynamic contrast enhanced (DCE) imaging has led to improved capabilities for detecting, localizing, and staging PCa [4]. The Prostate Imaging - Reporting and Data System version 2 (PI-RADS) was designed to standardize the reporting of PCa [17] with recommended imaging acquisition parameters for mpMRI (a minimum of T2W MRI, DWI and its derived apparent diffusion coefficient (ADC) image, and DCE imaging), simplified radiology reporting, and rules for lesion scoring assessment categories to stratify levels of PCa. PI-RADS divides the prostate into four clinically relevant zones: the peripheral zone (PZ), the transition zone (TZ), central zone (CZ), and anterior fibromuscular stroma (AFS). Here, we consider the TZ, CZ, AFS, and urethra as a single combined central gland (CG) area. PI-RADS scores are dependent on localization of lesions with respect to specific prostatic zones, and accurate scoring is necessary to identify needle biopsy targets for histopathological diagnosis of disease. Therefore, a roubust method for reproducible, automatic segmentation of prostate zones may enable the consistent assignment of mpMRI lesion location since manual segmentation of prostate zones is a time-consuming process, dependent on reader experience and expertise, and may also help relieve clinician cognitive workload [13].

A large number of segmentation approaches have been proposed for prostate whole gland (WG) segmentation in T2W MRI [11]. Semantic segmentation techniques based on the U-net deep learning architecture [15] have demonstrated excellent performance in prostate segmentation challenges [6]. These deep learning methods have also been applied to segmenting prostatic zonal anatomy [2,3,6,13]. However, semantic information captured by U-net may not be sufficient to describe the heterogeneous anatomic structures of the prostate and indiscernible borders between CG and PZ, resulting in inconsistent and suboptimal segmentation performance. U-net-based semantic segmentation algorithms are trained using supervised learning, and learn the segmentation on a voxelwise basis but do not incorporate global shape directly into the segmentation process without the use of large receptive fields [12]. An atlas of anatomical structure has the potential to provide global shape information.

In this study, we proposed a new deep learning approach for automated semantic segmentation of prostate zones by including a prior shape information from an anatomical atlas. Our method is designed to fit within the existing commercial ProFuseCAD image-guided biopsy planning software (Eigen Health, Grass Valley, CA). The novel contributions of this method are the integration of a 3D spatial probabilistic atlas prior into a semantic segmentation framework. We construct this probabilistic spatial atlas using ground-truth CG segmentations from a large set of publicly available prostate segmentations. In addition, motivated by HyperMorph [5], rather than inferring one segmentation, which could be sub-optimal due to the heterogeneity among different subjects, we would like to have the freedom to dynamically explore additional segmentation predictions by controlling the relative weight given to the atlas prior during inference.

Our experimental results using an external, publicly available dataset demonstrate that incorporating the atlas during training and setting different weights of the atlas during testing improves the accuracy of the segmentation results and yields more accurate quantitative results compared with the conventional training strategy.

2   Methods

Our method includes two parts: (1) construction of a probabilistic prostate zone atlas (Sect. 2.1); and (2) an atlas-based deep learning semantic segmentation framework (Sect. 2.2).

2.1   Prostate Zone Atlas Construction and Image Pre-processing

We utilize $N_{\mathrm{Atlas}} = 98$ subjects from the publicly available PROSTATEx dataset [1,10] to construct a probabilistic atlas of prostate zones. For each subject i, this dataset contains anatomical T2W MRI, denoted I_i, and corresponding 3D segmentations for four classes: PZ, TZ, AFS, and urethra [13]. We combine all four labels to create a WG mask M_i and combine the TZ, AFS and urethra classes into a single class to compose the CG label image L_i. We denote the set of images $\mathcal{A} = \{I_i, M_i, L_i | i = 1, \ldots, N_{\mathrm{Atlas}}\}$. Using the WG mask, we spatially normalize the segmentation data by performing the cropping and resizing. All images are first resampled to have standard voxel spacing $1.0 \times 1.0 \times 3.0$ mm and then cropped around the WG mask with a margin of 8 voxels on all sides. The cropped images are resized to $96 \times 96 \times 32$ voxels using nearest neighbor interpolation, and then we perform one-hot encoding to split the label image L_i into two channels (background and CG). These steps effectively perform an affine transformation from each subject's *native space* to the *atlas space*, where we spatially normalize each subject's data by gland size/shape. With all data in atlas space, we then create a voxelwise probabilistic atlas of CG shape L_{Atlas}, where $L_{\mathrm{Atlas}} = \bar{L} = \frac{1}{N_{\mathrm{Atlas}}} \sum_{i=1}^{N_{\mathrm{Atlas}}} L_i$ is the mean probability of observing the background/CG labels across the population at each voxel (see Atlas in Fig. 1). To increase the number of samples in this atlas and take advantage of anatomical symmetry, we perform left-right flipping of all label images L_i, effectively doubling the number of samples in L_{Atlas}. Both the atlas L_{Atlas} and the pre-processing routine are then used as inputs to our proposed segmentation methodology.

2.2   Atlas-Based Semantic Segmentation Architecture

Our approach segments the prostate into PZ and CG in T2W MRI. Our framework assumes that a segmentation of the WG is available, which is part of the ProFuseCAD system where WG segmentation is performed in a semi-automated manner by a trained radiologist [8]. Using this formulation allows us to formulate the three class (background, PZ, CG) segmentation problem as a two class (background, CG) segmentation problem where the PZ segmentation result is

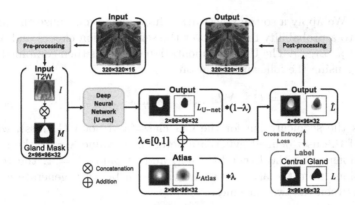

Fig. 1. Atlas-based semantic segmentation framework. Our proposed framework consists of four modules: (1) pre-processing to transform imaging data from the *native space* to the *atlas space*; (2) a deep neural network to segment the CG with the aid of a probabilistic CG shape atlas L_{Atlas}; (3) a hyperparameter λ to dynamically adjust the atlas weight; and (4) post-processing to transform segmentation results from the *atlas space* back to the *native space*.

calculated by subtracting the CG segmentation result from the WG mask. The input to the network consists of the following set of data $\{I, M, L_{Atlas}, \lambda\}$, where I is the T2W MRI, M is the WG segmentation mask, L_{Atlas} is the segmentation atlas (Sect. 2.1), and $\lambda \in [0, 1]$ is an atlas-weighting hyperparameter. The network performs semantic segmentation within the atlas space defined by L_{Atlas}, and we detail the pre- and post-processing routines necessary to convert data from the native space to the atlas space and back again. Figure 1 illustrates the segmentation framework.

Pre-processing. We first pre-process the T2W images and binary WG masks $\{I, M\}$ to spatially normalize the imaging data to the atlas space defined by L_{Atlas}. This spatial normalization process follows the same cropping and resizing process detailed in Sect. 2.1 and results in the data being transformed to 3D volumes of $96 \times 96 \times 32$ voxels. We then normalize the T2W image intensities using a percentile normalization strategy, which scales and shifts the intensities to have unit interquartile range [14] with values between -0.5 and 0.5.

Segmentation Network Design. Our proposed segmentation architecture utilizes a modified version of the fully-convolutional U-net architecture [7] to segment the CG. The normalized T2W image and WG mask are concatenated along the channel dimension as a form of early-fusion and passed to the segmentation network. The entire $96 \times 96 \times 32$ voxel image volume is passed as the input and the U-net architecture outputs a 3D segmentation estimate of the same size. The network operates on 3D images using $3 \times 3 \times 3$ convolution kernels with batch normalization applied after each convolutional layer and before the ReLU. We apply a softmax activation function to normalize the outputs of the U-net to a probability function over the segmentation classes and we denote this output as $L_{\text{U-net}}$. The U-net segmentation output is then combined with the atlas L_{Atlas} using the following equation:

$$\hat{L}(\lambda) = (1 - \lambda)L_{\text{U-net}} + \lambda L_{\text{Atlas}}, \quad \lambda \in [0, 1]. \tag{1}$$

Here, $\hat{L}$ is the segmentation for the CG class (and not CG class), which is a function of the user-defined hyperparameter λ. At value $\lambda = 0$, the results are segmentations using the U-net by itself, and at value $\lambda = 1$, the results are segmentations from the atlas by itself. Values in between generate a weighted average between the two extremes.

Post-processing. The segmentation $\hat{L}$ exists in atlas space defined by L_{Atlas}. We post-process the images to both calculate the PZ segmentation and return the segmentation to native T2W space. First, we calculate the PZ segmentation by subtracting $\hat{L}$ thresholded at 0.5 from the WG segmentation mask M. We now have one-hot segmentation results for three classes: background, PZ and CG. The three class segmentation is then transformed back to native T2W space by inverting the affine resizing operation, undoing the cropping, and inverting the voxel spacing standardization step. Linear interpolation is used to transform the one-hot $\hat{L}$ results to the native space to avoid interpolation artifacts going from low resolution atlas space to the high resolution native space. Finally, an argmax function is used to perform the final semantic segmentation.

2.3   Model Training

We train the network in a supervised manner using the PROSTATEx data and probabilistic atlas L_{Atlas} (Sect. 2.1) by minimizing the cross entropy loss between the ground-truth segmentation labels L and predicted results $\hat{L}$. The training data consists of 98 unique subjects, and we augment our training set by randomly flipping images left-right to take advantage of anatomical symmetries. We train the network using a mini-batch size of 16. For each sample within each mini-batch of each epoch, the value of λ is randomly set according to a uniform distribution between 0 and 1. Dropout is applied with a rate of 0.3 during training to prevent over-fitting, however, is removed in the testing phase. The total number of trainable parameters are 1,188,785. The network is trained with the Adam optimizer with initial learning rate of 5×10^{-3} and decay factor of 0.98 after each 10 epoch. Training requires 3,000 epochs at which time the network training loss converged. The training requires $\sim$12 h on a server with Intel Xeon Gold 5218 processors, 256 GB RAM, and an NVIDIA Quadro RTX 8000 GPU (48 GB RAM). We implement our framework in Python (version 3.8.5) using PyTorch (version 1.10.0) and MONAI (version 0.6.0).

2.4   Model Inference

In the inference phase, users define the λ value for the segmentation of image I (and WG mask M). Given input image pairs $\{I, M\}$, we apply the same pre-processing transformations as described in Sect. 2.2 and perform segmentation using the specified λ value. The segmentation estimate is then post-processed as described in Sect. 2.2 and the final PZ and CG segmentations are displayed in native T2W imaging space. Because the U-net segmentation results are independent of λ, the U-net segmentation only needs to be performed once, and users have the ability to dynamically adjust the value of λ to update the segmentation estimate in near real-time.

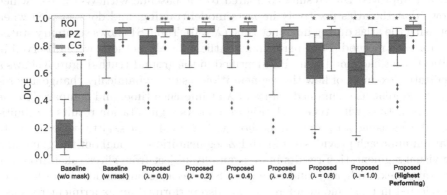

Fig. 2. Quantitative results. Dice overlap segmentation results comparing the ground truth, the baseline results, and atlas-based results. Boxplots show median and inter-quartile range and outliers. * and ** indicate statistically significant results ($p <$ 0.001) for PZ and CG, respectively, compared to baseline with WG mask results.

3   Experiments and Results

We utilize the publicly available NCI-ISBI 2013 Challenge: Automated Segmentation of Prostate-3T Structures (Prostate-3T) dataset [9] as an external testing resource to validate our approach. Prostate-3T contains T2W images from 30 subjects acquired on a Siemens 3T TrioTim scanner with pelvic phased-array coil. Each subject has 3D ground-truth segmentations for two classes: CG and PZ. We combine the CG and PZ labels to create a binary WG mask and use this along with the T2W images as input to our proposed segmentation method.

We compare our proposed approach with two baseline segmentation models: (i) a U-net segmentation model without WG mask (only using the T2W image as input) to perform three class segmentation (background, PZ, CG) to achieve the same segmentation results of the models with WG mask; and (ii) a U-net segmentation model with WG mask that does not use the probabilistic atlas. All models used the same U-net architecture, training data, and were trained with the same optimization parameters.

We assessed segmentation results both quantitatively and qualitatively. To quantitatively assess segmentation performance, we computed the Dice similarity coefficient between the ground-truth and predicted segmentations for both the PZ and CG. Figure 2 shows a boxplot of PZ and CG Dice values from our proposed method with different λ settings compared to the two baseline methods. The baseline results (not surprisingly) demonstrate that use of the WG mask as an additional input image dramatically improved PZ and CG segmentation performance. We evaluated statistical significance using a Wilcoxon signed-rank test with $p < 0.001$ by comparing our proposed method's segmentation performance and the baseline method without WG mask to the baseline method with WG mask. The proposed approach using $\lambda \leq 0.4$ demonstrated statistically significant improved Dice results compared to the baseline with WG mask, which shows that the atlas positively impacts model training, and does so even when not using the atlas during inference time ($\lambda = 0$). Table S1 shows summary statistics for each method. Poor performing outliers for all methods were observed in subjects that had small or no PZ identified in the ground-truth. Figure 3 shows a qualitative example of how the segmentation results dynamically change according to different user-supplied values of λ at inference time, and compares to the baseline method with WG mask. Figure 4 shows segmentation results for multiple subjects using our proposed method with $\lambda = 0.4$. The sagittal views indicate that our approach provides a smooth PZ segmentation throughout the gland and we visual segmentation results in an example subject from slices taken from the prostate apex, midgland, and base (Fig. S1). Because our model allows for λ to be selected in real-time at inference, we also performed an experiment to select the maximum Dice from all λ values for each subject, which simulates best-case human-in-the-loop performance and results in small improvements for the CG. 14 of 30 subjects utilized the atlas to different degrees, which indicates that our method's real-time refinement affects segmentation performance (Fig. S2).

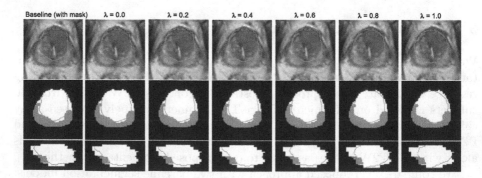

Fig. 3. Dynamic segmentation with λ for one subject. CG (orange) and PZ (blue) segmentation results comparing the proposed method using various λ values for a single subject to the baseline method with WG mask. The top row shows the axial views of segmentation results superimposed on T2W images. The middle and bottom rows show segmentation results superimposed on ground-truth labels (gray: PZ, white: CG) in axial and sagittal views, respectively. Images are displayed in atlas space. (Color figure online)

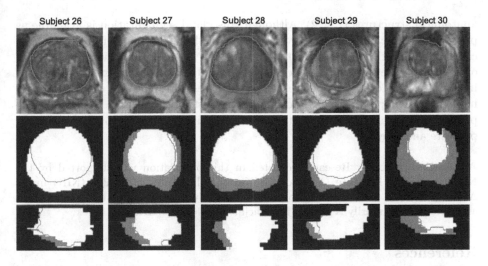

Fig. 4. Segmentation results for five subjects. CG (orange) and PZ (blue) segmentation results using our proposed method ($\lambda = 0.4$) for 5 subjects overlayed on T2W images. Results show slices from the middle of prostate gland in axial views (top and middle rows) and sagittal view (bottom row). Images are displayed in atlas space. (Color figure online)

4   Discussion and Conclusion

In this work, we proposed a novel framework for the automated semantic segmentation that incorporates prior spatial knowledge in the form of a probabilistic segmentation mask into a standard U-net framework with a dynamically controlled weighting hyperparameter. Our results demonstrate that the use of this atlas improves segmentation performance compared to baseline results without the atlas. Furthermore, our method provides users with the ability to dynamically control the atlas weighting parameter (λ) in real-time during inference. This feature can be integrated into GUI software to perform real-time inspection by clinicians for robust and accurate segmentation results. In some subjects (Subject 26 and 29 in Fig. 4), our results appear more realistic than the ground-truth, but these require further assessment by clinical radiologists. Additionally, we report segmentation results using an external, publicly-available test dataset (Prostate-3T) that is completely independent of our model training dataset (PROSTATEx). While it is challenging to directly compare results from others due to different testing datasets being used, our results (mean±SD Dice) (CG: 0.91 ± 0.05, PZ: 0.77 ± 0.16) using a single T2W modality compare favorably to state-of-the-art segmentation results [3] (CG: 0.84 ± 0.18, PZ: 0.72 ± 0.20) and to nnU-net [6] (CG: 0.85 ± 0.05, PZ: 0.75 ± 0.05) using bi-parametric MRI (both T2W and ADC), which requires an additional registration step.

One of the limitations of the current work is that we make the assumption of having a gland segmentation available. Our baseline results highlight the importance of a two-stage segmentation approach when delineating ROIs within larger anatomical structures (Fig. 2). Although this gland segmentation is available as part of the clinical routine of ProFuseCAD, in the future, we aim to test a cascaded segmentation approach utilizing existing WG segmentation methods [6] or to incorporate the initial gland segmentation task within an end-to-end training approach. Additional future work will investigate the use of higher-order moments of shape statistics into our probabilistic atlas rather than using the population mean, which may not be representative of the distribution of diverse prostate anatomy.

Acknowledgements. Research reported in this publication was supported by the National Cancer Institute (NCI) of the National Institutes of Health (NIH) under Award Number R42 CA224888. The content is solely the responsibility of the authors and does not necessarily represent the official views of the NIH.

References

1. Armato, S.G., et al.: PROSTATEx challenges for computerized classification of prostate lesions from multiparametric magnetic resonance images. J. Med. Imaging **5**(4), 044501 (2018)
2. Bohlender, S., Oksuz, I., Mukhopadhyay, A.: A survey on shape-constraint deep learning for medical image segmentation. arXiv preprint arXiv:2101.07721 (2021)
3. Cuocolo, R., et al.: Deep learning whole-gland and zonal prostate segmentation on a public MRI dataset. J. Magn. Reson. Imaging, **54**, 452–459 (2021). jmri.27585
4. Hoeks, C.M.A., et al.: Prostate cancer: multiparametric MR imaging for detection, localization, and staging. Radiology **261**(1), 46–66 (2011)
5. Hoopes, A., Hoffmann, M., Fischl, B., Guttag, J., Dalca, A.V.: HyperMorph: amortized hyperparameter learning for image registration. In: Feragen, A., Sommer, S., Schnabel, J., Nielsen, M. (eds.) IPMI 2021. LNCS, vol. 12729, pp. 3–17. Springer, Cham (2021). https://doi.org/10.1007/978-3-030-78191-0_1
6. Isensee, F., Jaeger, P.F., Kohl, S.A., Petersen, J., Maier-Hein, K.H.: nnU-Net: a self-configuring method for deep learning-based biomedical image segmentation. Nat. Methods **18**(2), 203–211 (2021)
7. Kerfoot, E., Clough, J., Oksuz, I., Lee, J., King, A.P., Schnabel, J.A.: Left-ventricle quantification using residual U-Net. In: Pop, M., Sermesant, M., Zhao, J., Li, S., McLeod, K., Young, A., Rhode, K., Mansi, T. (eds.) STACOM 2018. LNCS, vol. 11395, pp. 371–380. Springer, Cham (2019). https://doi.org/10.1007/978-3-030-12029-0_40
8. Ladak, H.M., Mao, F., Wang, Y., Downey, D.B., Steinman, D.A., Fenster, A.: Prostate boundary segmentation from 2D ultrasound images. Med. Phys. **27**(8), 1777–1788 (2000)
9. Litjens, G., Futterer, J., Huisman, H.: Data from prostate-3t: the cancer imaging archive (2015)
10. Litjens, G., Debats, O., Barentsz, J., Karssemeijer, N., Huisman, H.: Computeraided detection of prostate cancer in MRI. IEEE Trans. Med. Imaging **33**(5), 1083–1092 (2014)
11. Litjens, G., et al.: Evaluation of prostate segmentation algorithms for MRI: the promise12 challenge. Med. Image Anal. **18**(2), 359–373 (2014)

12. Luo, W., Li, Y., Urtasun, R., Zemel, R.: Understanding the effective receptive field in deep convolutional neural networks. In: Lee, D.D., Sugiyama, M., Luxburg, U.V., Guyon, I., Garnett, R. (eds.) Advances in Neural Information Processing Systems 29, pp. 4898–4906. Curran Associates, Inc. (2016)
13. Meyer, A., et al.: Towards patient-individual PI-RADS v2 sector map: CNN for automatic segmentation of prostatic zones from T2-weighted MRI. In: 2019 IEEE 16th International Symposium on Biomedical Imaging (ISBI 2019), pp. 696–700. IEEE (2019)
14. Onofrey, J.A., et al.: Generalizable multi-site training and testing of deep neural networks using image normalization. In: 2019 IEEE 16th International Symposium on Biomedical Imaging (ISBI 2019), pp. 348–351. IEEE (2019)
15. Ronneberger, O., Fischer, P., Brox, T.: U-Net: convolutional networks for biomedical image segmentation. In: Navab, N., Hornegger, J., Wells, W.M., Frangi, A.F. (eds.) MICCAI 2015. LNCS, vol. 9351, pp. 234–241. Springer, Cham (2015). https://doi.org/10.1007/978-3-319-24574-4_28
16. Siegel, R.L., Miller, K.D., Fuchs, H.E., Jemal, A.: Cancer statistics, 2022. CA Cancer J. Clin. **72**(1), 7–33 (2022)
17. Weinreb, J.C., et al.: PI-RADS prostate imaging - reporting and data system: 2015, version 2. Eur. Urol. **69**(1), 16–40 (2016)

Bayesian Pseudo Labels: Expectation Maximization for Robust and Efficient Semi-supervised Segmentation

Mou-Cheng Xu[1]([✉]), Yukun Zhou[1], Chen Jin[1], Marius de Groot[2],
Daniel C. Alexander[1], Neil P. Oxtoby[1], Yipeng Hu[1], and Joseph Jacob[1]

[1] Centre for Medical Image Computing, University College London, London, UK
xumoucheng28@gmail.com
[2] GlaxoSmithKline Research and Development, Stevenage, UK
https://github.com/moucheng2017/EMSSL/

Abstract. This paper concerns pseudo labelling in segmentation. Our contribution is fourfold. Firstly, we present a new formulation of pseudo-labelling as an Expectation-Maximization (EM) algorithm for clear statistical interpretation. Secondly, we propose a semi-supervised medical image segmentation method purely based on the original pseudo labelling, namely SegPL. We demonstrate SegPL is a competitive approach against state-of-the-art consistency regularisation based methods on semi-supervised segmentation on a 2D multi-class MRI brain tumour segmentation task and a 3D binary CT lung vessel segmentation task. The simplicity of SegPL allows less computational cost comparing to prior methods. Thirdly, we demonstrate that the effectiveness of SegPL may originate from its robustness against out-of-distribution noises and adversarial attacks. Lastly, under the EM framework, we introduce a probabilistic generalisation of SegPL via variational inference, which learns a dynamic threshold for pseudo labelling during the training. We show that SegPL with variational inference can perform uncertainty estimation on par with the gold-standard method Deep Ensemble.

Keywords: Semi-supervised segmentation · Pseudo labels ·
Expectation-maximization · Variational inference · Uncertainty ·
Probabilistic modelling · Out-of-distribution · Adversarial robustness

1 Introduction

Image segmentation is a fundamental component of medical image analysis, essential for subsequent clinical tasks such as computer-aided-diagnosis and disease progression modelling. In contrast to natural images, the acquisition of pixel-wise labels for medical images requires input from clinical specialists making such labels expensive to obtain. Semi-supervised learning is a promising approach to address label scarcity by leveraging information contained within

---

N. P. Oxtoby, Y. Hu and J. Jacob—Joint Senior Authorships.

L. Wang et al. (Eds.): MICCAI 2022, LNCS 13435, pp. 580–590, 2022.
https://doi.org/10.1007/978-3-031-16443-9_56

the additional unlabelled datasets. Consistency regularisation methods are the current state-of-the-art strategy for semi-supervised learning.

Consistency regularisation methods [2,3,23,24,27] make the networks invariant to perturbations at the input level, the feature level [8,18] or the network architectural level [6]. These methods heavily rely on specifically designed perturbations at the input, feature or the network level, which lacks transferability across different tasks [21]. In contrast, pseudo-labelling is a simple and general approach which was proposed for semi-supervised image classification. Pseudo-labelling [14] generates pseudo-labels using a fixed threshold and uses pseudo-labels to supervise the training of unlabelled images. A recent work [21] argued that pseudo labelling on its own can achieve results comparable with consistency regularisation methods in image classification.

Recent works in semi-supervised segmentation with consistency regularisation are complex in their model design and training. For example, a recent work [18] proposed a model with multiple decoders, where the features in each decoder are also augmented differently, before the consistency regularisation is applied on predictions of different decoders. Another recent method [6] proposed an ensemble approach and applied the consistency regularisation between the predictions of different models. Other approaches include applying different data augmentation on the input in order to create two different predictions. One of the predictions is then used as a pseudo-label to supervise the other prediction [23]. The aforementioned related works are all tested as baselines in our experiments presented in Sect. 4.

In this paper, we propose a straightforward pseudo label-based algorithm called SegPL, and report its improved or comparable efficacy on two medical image segmentation applications following comparison with all tested alternatives. We highlight the simplicity of the proposed method, compared to existing approaches, which allows efficient model development and training. We also explore the benefits of pseudo labels at improving model generalisation with respect to out-of-distribution noise and model robustness against adversarial attacks in segmentation. Theoretically, we as well provide a new perspective of pseudo-labelling as an EM algorithm. Lastly, under the EM framework, we show that SegPL with variational inference can learn the threshold for pseudo-labelling leading to uncertainty quantification as well as Deep Ensemble [13]. A high level introduction of SegPL can be found in Fig. 1.

2   Pseudo Labelling as Expectation-Maximization

We show that training with pseudo labels is an Expectation-Maximization (EM) algorithm using a binary segmentation case. Multi-class segmentation can be turned into a combination of multiple binary segmentation via one-hot encoding and applying Sigmoid on each class individually (multi-channel Sigmoid).

Let $X = \{(x_1, y_1), ...(x_L, y_L), x_{L+1}, ...x_N\}$ be the available training images, where $X_L = \{(x_l, y_l) : l \in (1, ..., L)\}$ is a batch of L labelled images; $X_U = \{x_u : u \in (L + 1, ..., N)\}$ is an additional batch of unlabelled images. Our goal is to optimise a segmentation network (θ) so that we maximise the marginal

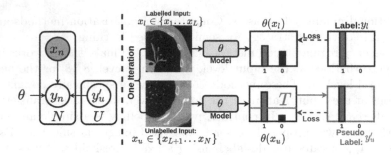

Fig. 1. Left: Graphical model of pseudo-labelling. Pseudo-label y'_n is generated using data x_u and model θ, therefore, pseudo-labelling can be seen as the E-step in Expectation-Maximization. The M-step updates θ using y'_n and data X. Right: Pseudo-labelling for binary segmentation. In our 1st implementation, namely SegPL, the threshold T is fixed for selecting the pseudo labels. In our 2nd implementation, namely SegPL-VI, the threshold T is dynamic and learnt via variational inference.

likelihood of $P(X, \theta)$ w.r.t θ. Given that pseudo-labelling is a generative task, we can cast pseudo-labels ($Y'_U = \{y'_u \in (L+1, ..N)\}$) as latent variables. Now our optimisation goal is to maximise the likelihood of $P(X, Y'_U, \theta)$. We can solve θ and Y'_U iteratively like the Expectation-Maximization (EM) [4] algorithm.

At the nth iteration, the E-step generates the latent variable by estimating its posterior with the old model (θ^{n-1}). In the original pseudo-labelling [14], pseudo-labels are generated according to its maximum predicted probability. For example in the binary setting, the pseudo-labels are generated using a fixed threshold value (T) which is normally set up as 0.5. Pseudo-labelling is equivalent to applying the plug-in principle [9] to estimate the posterior of Y'_U using its empirical estimation. Hence, the pseudo-labelling is the E-step:

$$y'_u = \mathbb{1}(\theta^{n-1}(x_u) > T = 0.5) \tag{1}$$

At the M-step of iteration n, Y'_U is fixed and we update the old model θ^{n-1} using the estimated latent variables (pseudo-labels Y'_U) and the data X including both the labelled and unlabelled data:

$$\theta^n := \underset{\theta}{\operatorname{argmax}}\, p(\theta^{n_1}|X, \theta^{n-1}, Y'_n) \tag{2}$$

Equation 2 can be solved by stochastic gradient descent given an objective function. We use Dice loss function ($f_{dice}(.)$) [17] as the objective function and we weight the unsupervised learning with a hyper-parameter α, the Eq. 2 and Eq. 1 can thereby be extended to a loss function over the whole dataset (supervised learning part L_L and unsupervised learning part L_U):

$$\mathcal{L}_{SegPL} = \alpha \underbrace{\frac{1}{N-L} \sum_{u=L+1}^{N} f_{dice}(\theta^{n-1}(x_u), y'_u)}_{\mathcal{L}_U} + \underbrace{\frac{1}{L} \sum_{l=1}^{L} f_{dice}(\theta^{n-1}(x_l), y_l)}_{\mathcal{L}_L} \tag{3}$$

3   Generalisation of Pseudo Labels via Variational Inference for Segmentation

We now derive a generalisation of SegPL under the same EM framework. As shown in the E-step in Eq. 1, the pseudo-labelling depends on the threshold value T, so finding the pseudo-label y'_u is the same with finding the threshold value T. We therefore can recast the learning objective of EM as estimating the likelihood $P(X, T, \theta)$ w.r.t θ. The benefit of flexible T is that different stages of training might need different optimal T to avoid wrong pseudo labels which cause noisy training [21]. The E-step is now estimating the posterior of T. According to the Bayes' rule, the posterior of T is $p(T|X, \theta) \propto p(X|\theta, T)p(T)$. The E-step at iteration n is:

$$p(T_n = i|X, \theta^{n-1}) = \frac{\prod_{u=L+1}^{N} p(x_u|\theta^{n-1}, T_n = i)p(T_n = i)}{\sum_{j \in [0,1]} \prod_{u=L+1}^{N} p(x_u|\theta^{n-1}, T_n = j)p(T_n = j)} \quad (4)$$

As shown in Eq. 4, the denominator is computationally intractable as there are infinite possible values between 0 and 1. This also confirms that the empirical estimation in Eq. 1 is actually necessary although not optimal. We use variational inference to approximate the true posterior of $p(T|X)$. We use new parameters ϕ^1 to parametrize $p(T|X)$ given the image features, assuming that T is drawn from a Normal distribution between 0 and 1:

$$(\mu, Log(\sigma)) = \phi(\theta(X)) \quad (5)$$

$$p(T|X, \theta, \phi) \approx \mathcal{N}(\mu, \sigma) \quad (6)$$

Following [11], we introduce another surrogate prior distribution over T and denote this as $q(\beta), \beta \sim \mathcal{N}(\mu_\beta, \sigma_\beta)$. The E-step now minimises $KL(p(T)||q(\beta))$. The M-step is the same with Eq. 3 but with a learnt threshold T. The learning objective $P(X, T, \theta)$ has supervised learning $P(X_L, T, \theta)$ and unsupervised learning part $P(X_U, T, \theta)$. Although $P(X_L, T, \theta)$ is not changed, $P(X_U, T, \theta)$ now estimates the Evidence Lower Bound (ELBO):

$$Log(P(X_U, T, \theta, \phi)) \geq \sum_{u=L+1}^{N} \mathbb{E}_{T \sim P(T)}[Log(P(y_u|x_n, T, \theta, \phi))] - KL(p(T)||q(\beta)) \quad (7)$$

where Log is taken for expression simplicity in Eq. 7, y_n is the prediction of unlabelled image x_u. The final loss function is:

$$\mathcal{L}_{SegPL}^{VI} = \mathcal{L}_{SegPL} + \underbrace{Log(\sigma_\beta) - Log(\sigma) + \frac{\sigma^2 + (\mu - \mu_\beta)^2}{2 * (\sigma_\beta)^2} - 0.5}_{KL(p(T)||q(\beta)), \beta \sim \mathcal{N}(\mu_\beta, \sigma_\beta)} \quad (8)$$

where $\mathcal{L}_{SegPL}$ can be found in Eq. 3 but with a learnt threshold T. Different data sets need different priors. Priors are chosen empirically for different data sets. $\beta \sim \mathcal{N}(0.5, 0.1)$ for BRATS and $\beta \sim \mathcal{N}(0.4, 0.1)$ for CARVE.

---

[1] ϕ: Average pooling layer followed by one 3×3 convolutional block including ReLU and normalisation layer, then two 1×1 convolutional layers for μ and $Log(\sigma)$.

4   Experiments and Results

Segmentation Baselines.[2] We compare SegPL to both supervised and semi-supervised learning methods. We use U-net [22] in SegPL as an example of segmentation network. Partly due to computational constraints, for 3D experiments we used a 3D U-net with 8 channels in the first encoder such that unlabelled data can be included in the same batch. For 2D experiments, we used a 2D U-net with 16 channels in the first encoder. The first baseline utilises supervised training on the backbone and is trained with labelled data denoted as "Sup". We compared SegPL with state-of-the-art consistency based methods: 1) "cross pseudo supervision" or CPS [6], which is considered the current state-of-the-art for semi-supervised segmentation; 2) another recent state-of-the-art model "cross consistency training" [18], denoted as "CCT", due to hardware restriction, our implementation shares most of the decoders apart from the last convolutional block; 3) a classic model called "FixMatch" (FM) [23]. To adapt FixMatch for a segmentation task, we added Gaussian noise as weak augmentation and "RandomAug" [7] for strong augmentation; 4) "self-loop [15]", which solves a self-supervised jigsaw problem as pre-training and combines with pseudo-labelling.

CARVE 2014. The classification of pulmonary arteries and veins (CARVE) dataset [5] comprises 10 fully annotated non-contrast low-dose thoracic CT scans. Each case has between 399 and 498 images, acquired at various spatial resolutions ranging from (282×426) to (302×474). We randomly select 1 case for labelled training, 2 cases for unlabelled training, 1 case for validation and the remaining 5 cases for testing. All image and label volumes were cropped to $176 \times 176 \times 3$. To test the influence of the number of labelled training data, we prepared four sets of labelled training volumes with differing numbers of labelled volumes at: 2, 5, 10, 20. Normalisation was performed at case wise. Data curation resulted in 479 volumes for testing, which is equivalent to 1437 images.

BRATS 2018. BRATS 2018 [16] comprises 210 high-grade glioma and 76 low-grade glioma MRI cases. Each case contains 155 slices. We focus on multi-class segmentation of sub-regions of tumours in high grade gliomas (HGG). All slices were centre-cropped to 176×176. We prepared three different sets of 2D slices for labelled training data: 50 slices from one case, 150 slices from one case and 300 slices from two cases. We use another 2 cases for unlabelled training data and 1 case for validation. 50 HGG cases were randomly sampled for testing. Case-wise normalisation was performed and all modalities were concatenated. A total of 3433 images were included for testing.

4.1   SegPL Outperforms Baselines with Less Computational Resource and Less Training Time

As shown in Table 2, Table 1 and Fig. 3, SegPL consistently outperformed the state-of-the-art semi-supervised segmentation methods and the supervised

---

[2] Training: Adam optimiser [10]. Our code is implemented using Pytorch 1.0 [20], trained with a TITAN V GPU. See Appendix for details of hyper-parameters.

Table 1. Our model vs Baselines on a binary vessel segmentation task on 3D CT images of the CARVE dataset. Metric is Intersection over Union (IoU (↑) in %). All 3D networks have 8 channels in the first encoder. Avg performance of 5 training.

Data	Supervised	Semi-supervised			
Labelled volumes	3D U-net [22] (2015)	FixMatch [23] (2020)	CCT [18] (2020)	CPS [6] (2021)	SegPL (Ours, 2022)	SegPL+VI (Ours, 2022)
2	56.79 ± 6.44	62.35 ± 7.87	51.71 ± 7.31	66.67 ± 8.16	**69.44 ± 6.38**	70.65 ± 6.33
5	58.28 ± 8.85	60.80 ± 5.74	55.32 ± 9.05	70.61 ± 7.09	76.52 ± 9.20	**73.33 ± 8.61**
10	67.93 ± 6.19	72.10 ± 8.45	66.94 ± 12.22	75.19 ± 7.72	**79.51 ± 8.14**	79.73 ± 7.24
20	81.40 ± 7.45	80.68 ± 7.36	80.58 ± 7.31	81.65 ± 7.51	**83.08 ± 7.57**	83.41 ± 7.14
Computational need						
Train (s)	1014	2674	4129	2730	1601	1715
Flops	6.22	12.44	8.3	12.44	6.22	6.23
Para (K)	626.74	626.74	646.74	1253.48	626.74	630.0

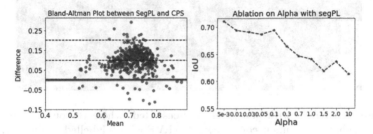

Fig. 2. Left: SegPL statistically outperforms the best performing baseline CPS when trained on 2 labelled volumes from the CARVE dataset. Each data point represents a single testing image. Right: Ablation study on the α which is the weight on the unsupervised learning part. Different data might need different α.

counterpart for different labelled data regimes, on both 2D multi-class and 3D binary segmentation tasks. The Bland-Altman plot in Fig. 2. Left shows SegPL performs significantly better than the best performed baseline CPS in an extremely low data regime (2 labelled volume on CARVE). We also applied Mann Whitney test on results when trained with 2 labelled volume and CARVE and the p-value was less than 1e−4. As shown in bottom three rows in Table 1, SegPL also is the most close one to the supervised learning counterpart in terms of computational cost, among all of the tested methods. As shown in Table 2,[3] SegPL-VI further improves segmentation performance on the MRI based brain tumour segmentation task. It also appears that the variational inference is sensitive to the data because we didn't observe obvious performance gain with SegPL-VI on CARVE. The performance of SegPL-VI on CARVE could be further improved with more hyper-parameter searching.

---

[3] Blue: 2nd best. Red: best.

Table 2. Our model vs Baselines on multi-class tumour segmentation on 2D MRI images of BRATS. Metric is Intersection over Union (IoU (↑) in %). All 2D networks have 16 channels in the first encoder. Avg performance of 5 runs.

Data	Supervised	Semi-supervised				
Labelled slices	2D U-net [22] (2015)	Self-loop [15] (2020)	FixMatch [23] (2020)	CPS [6] (2021)	SegPL (Ours, 2022)	SegPL+VI (Ours, 2022)
50	54.08 ± 10.65	65.91 ± 10.17	67.35 ± 9.68	63.89 ± 11.54	**70.60 ± 12.57**	71.20 ± 12.77
150	64.24 ± 8.31	68.45 ± 11.82	69.54 ± 12.89	69.69 ± 6.22	**71.35 ± 9.38**	72.93 ± 12.97
300	67.49 ± 11.40	70.80 ± 11.97	70.84 ± 9.37	71.24 ± 10.80	**72.60 ± 10.78**	75.12 ± 13.31

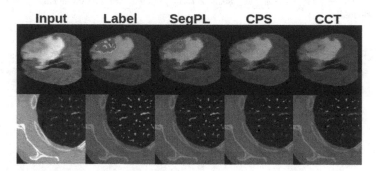

Fig. 3. Row1: BRATS with 300 labelled slices. Red: whole tumour. Green: tumour core. Blue: enhancing tumour core. Row2: CARVE with 5 labelled volumes. Red: false positive. Green: true positive. Blue: false negative. Yellow: ground truth. (Color figure online)

4.2   Ablation Studies

Ablation studies are performed using BRATS with 150 labelled slices. The suitable range for α is $5e-3$ to 0.1 as shown in Fig. 2. Right. The other ablation studies can be found in the Appendix. We found that SegPL needs large learning rate from at least 0.01. We noticed that SegPL is not sensitive to the warm-up schedule of α. The appropriate range of the ratio between unlabelled images to labelled images in each batch is between 2 to 10. Different data sets might need different ranges of α for optimal performances.

4.3   Better Generalisation on Out-of-Distribution (OOD) Samples and Better Robustness Against Adversarial Attack

We mimic OOD image noise resulting from variations in scan acquisition parameters and patient characteristics that are characteristic of medical imaging. We simulated OOD samples with unseen random contrast and Gaussian noise on CARVE and mix-up [28] to create new testing samples. Specifically, for a given original testing image x_t, we applied random contrast and noise augmentation on x_t to derive OOD samples x_t'. We arrived at the testing sample ($\hat{x}_t$) via $\gamma x_t' + (1 - \gamma)x_t$. As shown in Fig. 4, as testing difficulty increases, the performances across all baselines drop exponentially. SegPL outperformed all of the

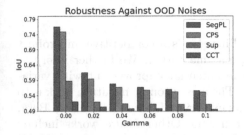

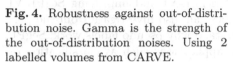

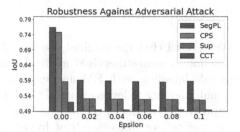

Fig. 4. Robustness against out-of-distribution noise. Gamma is the strength of the out-of-distribution noises. Using 2 labelled volumes from CARVE.

Fig. 5. Robustness against adversarial attack. Epsilon is the strength of the FGSM [12] attack. Using 2 labelled volumes from CARVE.

baselines across all of the tested experimental settings. The findings suggest that SegPL is more robust when testing on OOD samples and achieves better generalisation performance against that from the baselines. As robustness has become important in privacy-preserving collaborative learning among hospitals, we also examined SegPL's robustness against adversarial attack using fast gradient sign method (FGSM) [12]. With increasing strength of adversarial attack (Epsilon), all the networks suffered performance drop. Yet, as shown in Fig. 5, SegPL suffered much less than the baselines.

4.4 Uncertainty Estimation with SegPL-VI

SegPL-VI is trained with stochastic threshold for unlabelled data therefore not suffering from posterior collapse. Consequently, SegPL can generate plausible segmentation during inference using stochastic thresholds. We tested the uncertainty estimation of SegPL-VI with the common metric Brier score on models trained with 5 labelled volumes of CARVE dataset. We compared SegPL-VI with Deep Ensemble, which is regarded as a gold-standard baseline for uncertainty estimation [13,19]. Both SegPL-VI and Deep Ensemble achieved the same Brier score at 0.97, using 5 Monte Carlo samples.

5 Related Work

Although we are the first to solely use the original pseudo labelling in semi-supervised segmentation of medical images, a few prior works explored the use of pseudo labels in other forms. Bai [1] post-processed pseudo-labels via conditional random field. Wu [26] proposed sharpened cycled pseudo labels along with a two-headed network. Wang [25] post-processed pseudo-labels with uncertainty-aware refinements for their attention based models.

6   Conclusion

We verified that the original pseudo-labelling [14] is a competitive and robust baseline in semi-supervised medical image segmentation. We further interpret pseudo-labelling with EM and explore the potential improvement with variational inference following generalised EM. The most prominent future work is to improve the quality of the pseudo labels by exploring different priors (e.g. Beta, Categorical) or auto-search of hyper parameters. Other future works include investigating the convergence property of SegPL-VI and improving SegPL-VI at uncertainty quantification. Future studies can also investigate from the perspective of model calibration [27] to understand why pseudo labels work. SegPL and SegPL-VI can also be used in other applications such as classification or registration.

Acknowledgements. We thank the anonymous reviewer 3 for the suggestion of using Beta prior and an interesting discussion about posterior collapse issue in uncertainty quantification. MCX is supported by GlaxoSmithKline (BIDS3000034123) and UCL Engineering Dean's Prize. NPO is supported by a UKRI Future Leaders Fellowship (MR/S03546X/1). DCA is supported by UK EPSRC grants M020533, R006032, R014019, V034537, Wellcome Trust UNS113739. JJ is supported by Wellcome Trust Clinical Research Career Development Fellowship 209,553/Z/17/Z. NPO, DCA, and JJ are supported by the NIHR UCLH Biomedical Research Centre, UK.

References

1. Bai, W., et al.: Semi-supervised learning for network-based cardiac MR image segmentation. In: Descoteaux, M., Maier-Hein, L., Franz, A., Jannin, P., Collins, D.L., Duchesne, S. (eds.) MICCAI 2017. LNCS, vol. 10434, pp. 253–260. Springer, Cham (2017). https://doi.org/10.1007/978-3-319-66185-8_29
2. Berthelot, D., et al.: ReMixMatch: semi-supervised learning with distribution alignment and augmentation anchoring. In: International Conference on Learning Representation (ICLR) (2020)
3. Berthelot, D., Carlini, N., Goodfellow, I., Oliver, A., Papernot, N., Raffel, C.: MixMatch: a holistic approach to semi-supervised learning. In: Neural Information Processing Systems (NeurIPS) (2019)
4. Bishop, C.M.: Pattern Recognition and Machine Learning (Information Science and Statistics). Springer, Heidelberg (2006)
5. Charbonnier, J.P., Brink, M., Ciompi, F., Scholten, E.T., Schaefer-Prokop, C.M., van Rikxoort, E.M.: Automatic pulmonary artery-vein separation and classification in computed tomography using tree partitioning and peripheral vessel matching. IEEE Trans. Med. Imaging **35**, 882–892 (2015)
6. Chen, X., Yuan, Y., Zeng, G., Wang, J.: Semi-supervised semantic segmentation with cross pseudo supervision. In: Computer Vision and Pattern Recognition (CVPR) (2021)
7. Cubuk, E.D., Zoph, B., Shlens, J., Le, Q.V.: RandAugment: practical automated data augmentation with a reduced search space. In: Neural Information Processing Systems (NeurIPS) (2020)

8. French, G., Laine, S., Aila, T., Mackiewicz, M., Finalyson, G.: Semi-supervised semantic segmentation needs strong, varied perturbations. In: British Machine Vision Conference (BMVC) (2020)

9. Grandvalet, Y., Bengio, Y.: Semi-supervised learning by entropy minimization. In: Neural Information Processing Systems (NeurIPS) (2004)

10. Kingma, D.P., Ba, J.: Adam: a method for stochastic optimization. In: International Conference on Learning Representation (ICLR) (2015)

11. Kingma, D.P., Welling, M.: Auto-encoding variational Bayes. In: ICLR (2014)

12. Kurakin, A., Goodfellow, I.J., Bengio, S.: Adversarial examples in the physical world. In: International Conference on Learning Representation Workshop (2017)

13. Lakshminarayanan, B., Pritzel, A., Blundell, C.: Simple and scalable predictive uncertainty estimation using deep ensembles. In: Neural Information Processing System (NeurIPS) (2017)

14. Lee, D.H.: Pseudo-label: the simple and efficient semi-supervised learning method for deep neural networks. In: ICML Workshop on Challenges in Representation Learning (2013)

15. Li, Y., Chen, J., Xie, X., Ma, K., Zheng, Y.: Self-loop uncertainty: a novel pseudo-label for semi-supervised medical image segmentation. In: Martel, A.L., et al. (eds.) MICCAI 2020. LNCS, vol. 12261, pp. 614–623. Springer, Cham (2020). https://doi.org/10.1007/978-3-030-59710-8_60

16. Menze, B.H., et al.: The multimodal brain tumor image segmentation benchmark (BRATS). IEEE Trans. Med. Imaging **34**, 1993–2024 (2015)

17. Milletari, F., Navab, N., Ahmadi, S.A.: V-Net: fully convolutional neural networks for volumetric medical image segmentation. In: International Conference on 3D Vision (3DV) (2016)

18. Ouali, Y., Hudelot, C., Tami, M.: Semi-supervised semantic segmentation with cross-consistency training. In: Computer Vision and Pattern Recognition (CVPR) (2020)

19. Ovadia, Y., et al.: Can you trust your model's uncertainty? Evaluating predictive uncertainty under dataset shift. In: Advances in Neural Information Processing Systems (NeurIPS) (2019)

20. Paszke, A., et al.: PyTorch: an imperative style, high-performance deep learning library. In: Neural Information Processing System (NeurIPS) (2019)

21. Rizve, M.N., Duarte, K., Rawat, Y.S., Shah, M.: In defense of pseudo-labelling: an uncertainty-aware pseudo-label selective framework for semi-supervised learning. In: International Conference on Learning Representation (ICLR) (2021)

22. Ronneberger, O., Fischer, P., Brox, T.: U-Net: convolutional networks for biomedical image segmentation. In: Navab, N., Hornegger, J., Wells, W.M., Frangi, A.F. (eds.) MICCAI 2015. LNCS, vol. 9351, pp. 234–241. Springer, Cham (2015). https://doi.org/10.1007/978-3-319-24574-4_28

23. Sohn, K., et al.: FixMatch: simplifying semi-supervised learning with consistency and confidence. In: Neural Information Processing Systems (NeurIPS) (2020)

24. Tarvainen, A., Valpola, H.: Mean teachers are better role models: weight-averaged consistency targets improve semi-supervised deep learning results. In: Neural Information Processing Systems (NeurIPS) (2017)

25. Wang, G., et al.: Semi-supervised segmentation of radiation induced pulmonary fibrosis from lung CT scans with multi-scale guided dense attention. IEEE Trans. Med. Imaging **41**, 531–542 (2014)

26. Wu, Y., Xu, M., Ge, Z., Cai, J., Zhang, L.: Semi-supervised left atrium segmentation with mutual consistency training. In: de Bruijne, M., et al. (eds.) MICCAI 2021. LNCS, vol. 12902, pp. 297–306. Springer, Cham (2021). https://doi.org/10.1007/978-3-030-87196-3_28
27. Xu, M.C., et al.: Learning morphological feature perturbations for calibrated semi supervised segmentation. In: International Conference on Medical Imaging with Deep Learning (MIDL) (2022)
28. Zhang, H., Cisse, M., Dauphin, Y.N., Lopez-Paz, D.: mixup: beyond empirical risk minimization. In: International Conference on Learning Representation (ICLR) (2018)

Enforcing Connectivity of 3D Linear Structures Using Their 2D Projections

Doruk Oner[1]([⊠])[ID], Hussein Osman[1][ID], Mateusz Koziński[2][ID], and Pascal Fua[1][ID]

[1] Computer Vision Laboratory, École Polytechnique Fédérale de Lausanne, Lausanne, Switzerland
doruk.oner@epfl.ch
[2] Institute of Computer Graphics and Vision, Graz University of Technology, Graz, Austria

Abstract. Many biological and medical tasks require the delineation of 3D curvilinear structures such as blood vessels and neurites from image volumes. This is typically done using neural networks trained by minimizing voxel-wise loss functions that do not capture the topological properties of these structures. As a result, the connectivity of the recovered structures is often wrong, which lessens their usefulness. In this paper, we propose to improve the 3D connectivity of our results by minimizing a sum of topology-aware losses on their 2D projections. This suffices to increase the accuracy and to reduce the annotation effort required to provide the required annotated training data.

Keywords: Delineation · Neurons · Microscopy scans · Topology

1 Introduction

Delineating 3D curvilinear structures, such as veins and arteries visible in computed tomography (CT) scans, or dendrites and axons revealed by light microscopy (LM) scans, is central to many applications. State-of-the-art algorithms typically rely on deep networks trained to classify each voxel as either foreground or background by minimizing a voxel-wise loss. Networks trained this way are good at voxel classification but nevertheless prone to topological errors, such as unwarranted gaps in the linear structures and false interconnections between them. This mostly occurs when vessels and neuronal projections appear as thin but densely woven structures and misclassifying a few voxels can disrupt their connectivity without much influence on voxel-wise accuracy. These errors greatly reduce the usefulness of the resulting arborization models. Correcting them requires manual interventions, which is very time consuming when

---

D. Onerand and H. Osman—The two authors contributed equally to this paper.

Supplementary Information The online version contains supplementary material available at https://doi.org/10.1007/978-3-031-16443-9_57.

performed on whole-brain microscopy scans or whole-organ CT scans, especially at scales sufficiently large to produce statistically significant results.

In other words, networks trained by minimizing losses such as the Cross Entropy and the Mean Squared Error, which are sums of per-voxel terms *independent* of all other voxels, struggle to learn patterns formed *jointly* by groups of voxels [6,14,24]. A promising approach to addressing this issue is to develop *topology-aware* loss functions that evaluate patterns emerging from predictions for multiple voxels. This includes perceptual losses [23], loss functions based on persistent homology [6,14], and a loss that enforces continuity of linear structures by penalizing interconnections between background regions on their opposite sides [24]. The latter has proved to be more effective for 2D images than the others but does not naturally generalize to 3D volumes.

In this paper, we start from the observation that continuity of a 3D linear structure implies continuity of its 2D projections. Hence, we can use a topology-aware loss, such as the one of [24], to penalize connectivity errors in 2D projections of the 3D predictions, thereby indirectly penalizing the errors in the 3D originals. This also means that we can use 2D annotations, which are much easier to obtain than full 3D ones, to train a 3D network. This is close in spirit to the approach of [18] in which delineation networks are trained by minimizing a loss function in maximum intensity projections of the predictions and the annotations.

We demonstrate the effectiveness of our approach for delineating neurons in light microscopy scans and tracing blood vessels in Magnetic Resonance Angiography scans. Not only do we produce topologically correct delineations, but we also reduce the annotation effort required to train the networks.

2   Related Work

Over the years, many approaches to delineating 3D linear structures have been proposed. They range from hand-designing filters that are sensitive to tubular structures [9,19,29] to learning such filters [1,34] using support vector machines [15], gradient boost [28], or decision trees [30].

Neural networks have now become the dominant technique [11,12,20,21,25, 33]. They are often trained by minimizing pixel-wise loss functions, such as the cross-entropy or the mean square error. As a result, the delineations they produce often feature topological mistakes, such as unwarranted gaps or false connections. This occurs because it often takes very few mislabeled pixels to significantly alter the topology with little impact on the pixel-wise accuracy.

Specialized solutions to this problem have been proposed in the form of loss functions comparing the topology of the predictions to that of the annotations. For example, the perceptual loss [22] has been shown to be sensitive to topological differences between the prediction and the ground truth, but cannot be guaranteed to penalize all of them. Persistent Homology [7] is an elegant approach to describing and comparing topological structures. It has been used to define topology-oriented loss functions [4,6,13]. Unfortunately, computing this loss is computationally intensive and error-prone because it does not account for the location of topological structures. clDice [27] is a loss function that employs a soft skeletonization algorithm to compare the topology of the prediction and the annotation, but it is designed for volumetric segmentation, whereas we focus on tracing linear structures given their centerlines.

For delineation of 2D road networks, existing approaches are outperformed by the method of [24] that repurposes the MALIS loss initially proposed to help better segment electron microscopy scans [2,10] to improve the topology of reconstructed loopy curvilinear networks. Unfortunately, the algorithm of [24] can only operate in 2D. In this paper, we show how it can nevertheless be exploited for 3D delineation in volumetric images.

3 Approach

We train a deep network to regress the distance from each voxel of the input 3D image $\mathbf{x}$ to the center of the nearest linear structure. We denote the predicted 3D distance map by $\mathbf{y}$. The annotations are given in the form of a graph with nodes in 3D space. We denote the set of edges of this graph by $\mathcal{E}$. From the annotation graph, we compute the truncated ground truth distance map $\hat{\mathbf{y}}$. For a voxel $\mathbf{p}$, $\hat{\mathbf{y}}[p] = \min((\min_{\epsilon \in \mathcal{E}} d_{p\epsilon}), d_{\max})$, where $d_{p\epsilon}$ is the distance from $\mathbf{p}$ to the annotation edge ϵ, and $d_{\max}$ is the truncation distance set to 15 pixels.

A simple way to train our deep net is to minimize a Mean Squared Error loss $L_{\mathrm{MSE}}(\mathbf{y}, \hat{\mathbf{y}})$ for all training images. As discussed in Sect. 2, minimizing such a voxel-wise loss does not guarantee that connectivity is preserved because mislabeling only a few voxels is enough to disrupt it. For 2D images, this problem has been addressed with a loss term L_{TOPO} that effectively enforces continuity of 2D linear structures [24]. Unfortunately, it is limited to 2D data by design and cannot be extended to 3D. To bypass this limitation, we leverage the observation that continuity of 3D structures implies continuity of their 2D projections and evaluate L_{TOPO} on 2D projections of the 3D predicted and ground truth distance maps. We introduce the 2D connectivity-oriented loss term and the technique to train 3D deep networks on 2D projections in the following subsections.

3.1 Connectivity Loss

In this section, we recall the intuition behind the connectivity-oriented loss term L_{TOPO} of [24]. We refer the reader to the original publication for a more detailed explanation. As illustrated by Fig. 1(a), a path connecting pixels on opposite sides of a linear structure must cross that structure and should therefore contain at least one pixel $\mathbf{p}$ such that the predicted distance map $\mathbf{y}[\mathbf{p}] = 0$. If $\mathbf{y}$ contains erroneous disconnections, then it is possible to construct a path that connects pixels on opposite sides of the structure, but only crosses pixels with predicted distance values larger than zero, as depicted by Fig. 1(b). In particular, a *maximin* path, that is, the path with largest smallest pixel among all possible paths between the same end points, is guaranteed to pass through an interruption of the linear structure, if it exists. L_{TOPO} minimizes the smallest pixel on the maximin path between each pair of end points that belong to background regions on the opposite sides of annotated linear structures, shown in Fig. 1(c). It has proven effective in enforcing connectivity of 2D linear structures, but cannot be extended to 3D, because 3D linear structures do not subdivide 3D volumes into disjoint background regions.

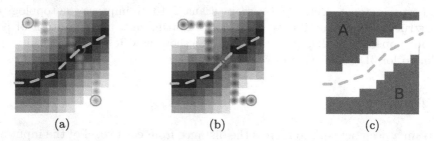

(a)   (b)   (c)

Fig. 1. The intuition behind L_{TOPO}. *(a)* In a perfect distance map, any path connecting pixels on the opposite sides of an annotation line (dashed, magenta) crosses a zero-valued pixel (red circle). *(b)* If a distance map has erroneously high-valued pixels along the annotation line, the maximin path (violet) between the same pixels crosses one of them (red circle). *(c)* The connectivity-oriented loss L_{TOPO} is a sum of the smallest values crossed by maximin paths connecting pixels from different background regions. The background regions are computed by first dilating the annotation (dilated annotation shown in white), to accommodate possible annotation inaccuracy. (Color figure online)

3.2   Projected Connectively Loss

The key observation underlying our approach is that 3D continuity of three-dimensional curvilinear structure, represented as a depth-map, implies its continuity in 2D minimum-intensity projections of the depth map. The reverse is not true: a projection of a discontinuous 3D depth-map might appear continuous if it is taken along the direction tangent to the linear structure at discontinuity. However, even in such case, the discontinuity appears in other projections, taken along directions orthogonal to the direction of the first projection, as shown in Fig. 2. In general, given three orthogonal projections of a 3D volume, each discontinuity appears in at least two of them, unless it is occluded by other linear structures. Hence, we evaluate the topology-enforcing loss L_{TOPO} on projections of the predicted and ground truth distance maps along

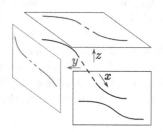

Fig. 2.   Disconnections in 3D linear structures appear in at least two out of three orthogonal projections, unless the structure is occluded.

the principal directions. Let $\mathbf{y}^i$ be the min-intensity projection of $\mathbf{y}$ along direction i, where i can be one of the axes x, y, or z and the corresponding projection of $\hat{\mathbf{y}}$ be $\hat{\mathbf{y}}^i$. We take our connectivity-enforcing loss to be

$$L_{\mathrm{conn}}(\mathbf{y}, \hat{\mathbf{y}}) = \sum_{i \in \{x,y,z\}} L_{\mathrm{TOPO}}(\mathbf{y}^i, \hat{\mathbf{y}}^i), \tag{1}$$

where L_{TOPO} is the 2D connectivity loss of [24] discussed above. This loss can easily be differentiated with respect to the values of $\mathbf{y}$, as the minimum-intensity projection is just a column-wise *min* operation.

3.3 Total Loss

The total loss that we minimize can therefore be written as

$$L_{3D}(\mathbf{y}, \hat{\mathbf{y}}) = L_{\text{MSE}}(\mathbf{y}, \hat{\mathbf{y}}) + \alpha L_{\text{conn}}(\mathbf{y}, \hat{\mathbf{y}}), \tag{2}$$

where α is a scalar that weighs the influence of the two terms. As discussed above, L_{MSE} can be simply computed as the mean squared difference between the predicted and ground truth 3D distance maps.

This is a perfectly valid choice when 3D annotations are available, but such annotations are typically hard to obtain. Fortunately, it has been shown in [18] that one can train a network to perform 3D volumetric delineation given *only* 2D annotations in Maximum Intensity Projections. This saves time because manually delineating in 2D is much easier than in 3D. Since we impose our connectivity constraints on projections along the axes x, y, and z, it makes sense to also provide annotations only for the corresponding projections of the input volume $\mathbf{x}$, and generate from them ground truth distance maps $\hat{\mathbf{y}}_x$, $\hat{\mathbf{y}}_y$, and $\hat{\mathbf{y}}_z$. To replace the 3D ground truth $\hat{\mathbf{y}}$, that L_{3D} requires, we can rewrite our total loss as

$$L_{2D}(\mathbf{y}, \hat{\mathbf{y}}_x, \hat{\mathbf{y}}_y, \hat{\mathbf{y}}_z) = \sum_{i \in \{x,y,z\}} L_{\text{MSE}}(\mathbf{y}^i, \hat{\mathbf{y}}_i) + \alpha \sum_{i \in \{x,y,z\}} L_{\text{TOPO}}(\mathbf{y}^i, \hat{\mathbf{y}}_i), \tag{3}$$

where the Mean Squared Error is evaluated on the minimum-intensity projections of the predicted distance map and the distance map produced for the 2D annotation of data projection.

4 Experiments

4.1 Datasets

We tested our approach on three data sets. The **Brain** comprises 14 light microscopy scans of mouse brain, sized $250 \times 250 \times 250$. We use 10 of them for training and 4 as a validation test. **Neurons** contains 13 light microscopy scans of mouse neurons, sized $216 \times 238 \times 151$. We use 10 for training and 3 for validation. **MRA** is a publicly available set of Magnetic Resonance Angiography brain scans [3]. We crop them to size $416 \times 320 \times 28$ by removing their empty margins, and use 31 annotated scans for training and 11 for validation. A sample image from each data set can be found in Fig. 4.

4.2 Metrics

We use the following performance metrics. **CCQ** [32], *correctness*, *completeness*, and *quality* are similar to precision, recall, and the F1 score, but predicted foreground voxels are counted as true positives if they are closer than 3 voxels away from the ground truth ones. **APLS** [8] is defined as the mean of relative length differences between shortest paths connecting corresponding pairs of randomly selected end points in the ground truth and predicted graphs. **TLTS** [31] is the fraction of shortest paths in the prediction that are less than 15% longer or shorter than the corresponding ground truth paths.

Table 1. Comparative results. A *U-Net* trained with our loss function outperforms existing methods by a considerable margin in terms of the topology-aware metrics. The improvement in terms of the pixel-wise metrics is smaller but still there on average.

Dataset	Methods	Pixel-wise			Topology-aware	
		Corr	Comp	Qual	APLS	TLTS
Brain	**MSE-3D**	96.6	93.5	90.5	77.4	81.7
	MSE-2D	**98.3**	93.6	92.1	76.2	80.1
	CE	97.3	96.7	94.2	71.0	81.2
	Perc	97.6	96.7	94.5	76.6	84.1
	PHomo	97.5	**96.9**	94.7	81.5	83.9
	OURS-3D	**98.3**	96.7	**95.1**	87.1	**89.6**
	OURS-2D	97.8	96.3	94.3	**91.6**	87.4
Neurons	**MSE-3D**	80.6	83.5	69.5	62.9	69.1
	MSE-2D	78.4	83.5	67.9	65.6	71.8
	CE	79.5	82.6	68.1	61.2	68.6
	Perc	80.1	85.0	70.2	68.9	74.5
	PHomo	**81.3**	84.8	**71.0**	69.4	75.2
	OURS-3D	79.9	**86.4**	70.9	75.1	80.2
	OURS-2D	80.3	85.5	70.7	**76.3**	**81.2**
MRA	**MSE-3D**	84.9	81.2	70.8	58.5	60.4
	MSE-2D	83.0	82.3	70.3	58.7	59.6
	CE	**85.7**	81.1	71.3	58.8	60.0
	Perc	83.4	83.9	71.9	60.9	64.5
	PHomo	85.3	83.5	72.8	62.1	65.2
	OURS-3D	81.5	**89.5**	**74.3**	**70.7**	**72.0**
	OURS-2D	80.3	87.3	71.8	70.5	**71.9**

4.3   Architectures and Baselines

In all experiments, we use a 3D U-Net [26] with three max-pooling layers and two convolutional blocks. The first layer has 32 filters. Each convolution is followed by a batch-norm and dropout with a probability of 0.1. We used a batch size of 4. For data augmentation, we randomly crop volumes of size $96 \times 96 \times 96$ and flip them over the three axes. The networks were trained for $50k$ iterations with Adam [16], with the learning rate of $1e - 3$ and weight decay of $1e - 3$. At test time, the predicted distance map is thresholded at 2 and skeletonized to obtain centerlines. To compute the *TLTS* and *APLS*, we extract graphs from the prediction and the ground-truth, based on voxel connectivity.

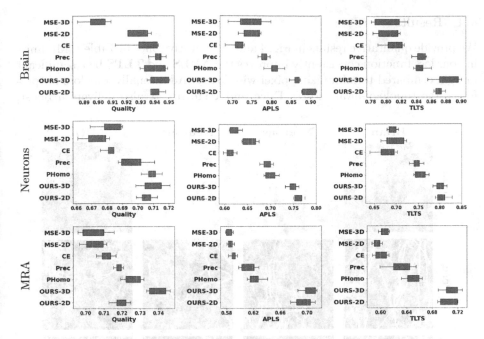

Fig. 3. Median and quartiles over five training runs of scores attained by networks trained with different loss functions. Minimizing our topology-aware loss results in significantly higher score values than minimizing hte baselines.

As discussed in Sect. 3.3, we can train our network by minimizing either L_{3D} (2) or L_{2D} (3). Recall that computing L_{3D} requires 3D annotations, while 2D annotations suffice to compute L_{2D}. We will refer to these approaches as **OURS-3D** and **OURS-2D**. We compare the results we obtain in this way to:

- **MSE-3D.** L_{MSE} between 3D predictions and ground truths.
- **MSE-2D.** L_{MSE} between 2D ground truth and projected predictions [18].
- **CE.** 3D binary segmentation trained with Cross-Entropy (CE).
- **Perc.** A weighted sum of CE and a perceptual loss function that compares feature maps computed for the ground truth and predicted distance maps [23]. To extract the feature maps, we use a ResNet50 architecture pre-trained with 23 different biomedical datasets [5].
- **PHomo.** A weighted sum of CE and a loss based on Persistent Homology [13].

For **Perc** and **PHomo**, we use the weighing coefficients recommended in the original publications. For **OURS-3D** and **OURS-2D**, α is set to $1e-3$ and β to 0.1. These values are selected empirically, based on the ablation study we provide in the supplementary material. We used windows of size 48 pixel to calculate L_{TOPO}.

4.4   Results

We provide qualitative results in Fig. 4 and quantitative ones in Table 1. Minimizing our loss function consistently improves the **APLS** and **TLTS** by a significant margin compared to minimizing pixel-wise losses. Additionally, our loss outperforms the topology-aware losses **Perc** and **PHomo**. It also delivers a boost,

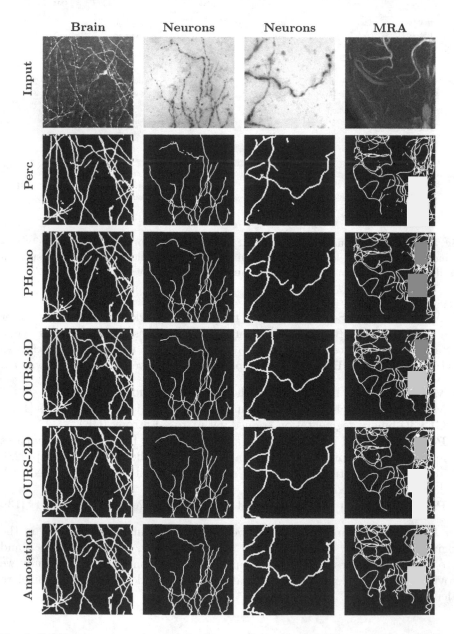

Fig. 4. Qualitative comparison of the test results in three different datasets. The connectivity improves significantly when our approach is used.

albeit only on average, in terms of the **CCQ**. Box plots in Fig. 3, show that these conclusions hold when variance of the scores is taken into account.

On average, **OURS-2D** achieves performance that are slightly lower than those of comparable performance to **OURS-3D**. However, annotating 2D slices instead of 3D stacks significantly reduces the time required to annotate, as shown in the user study conducted in [17]. Thus, when annotation effort is a concern, **OURS-2D** is an excellent alternative to **OURS-3D**.

5   Conclusion and Future Work

We proposed a loss function that enforces topological consistency in 2D projections. Training a deep net with our loss greatly improves the 3D connectivity of its outputs and reduces the annotation effort required to obtain training data. In our current implementation, we use projection direction independently of the shape of the delineated structures. However, some projections are more informative than others. To further improve delineation accuracy while reducing the required annotation effort, we will develop algorithms for automatic selection of the optimal projection direction for different parts of the volume, so that we can use less than three projections.

Acknowledgements. DO received support from the Swiss National Science Foundation under Sinergia grant number 177237. MK was supported by the FWF Austrian Science Fund Lise Meitner grant no. M3374. The author's version of the manuscript is available from arxiv.org.

References

1. Breitenreicher, D., Sofka, M., Britzen, S., Zhou, S.K.: Hierarchical discriminative framework for detecting tubular structures in 3D images. In: Gee, J.C., Joshi, S., Pohl, K.M., Wells, W.M., Zöllei, L. (eds.) IPMI 2013. LNCS, vol. 7917, pp. 328–339. Springer, Heidelberg (2013). https://doi.org/10.1007/978-3-642-38868-2_28
2. Briggman, K., Denk, W., Seung, S., Helmstaedter, M., Turaga, S.: Maximin affinity learning of image segmentation. In: Advances in Neural Information Processing Systems, pp. 1865–1873 (2009)
3. Bullitt, E., et al.: Vessel tortuosity and brain tumor malignancy: a blinded study. Acad. Radiol. **12**(10), 1232–1240 (2005)
4. Byrne, N., Clough, J.R., Montana, G., King, A.P.: A persistent homology-based topological loss function for multi-class CNN segmentation of cardiac MRI. In: Puyol Anton, E., et al. (eds.) STACOM 2020. LNCS, vol. 12592, pp. 3–13. Springer, Cham (2021). https://doi.org/10.1007/978-3-030-68107-4_1
5. Chen, S., Ma, K., Zheng, Y.: Med3D: transfer learning for 3D medical image analysis. arXiv preprint (2019)
6. Clough, J., Byrne, N., Oksuz, I., Zimmer, V.A., Schnabel, J.A., King, A.: A topological loss function for deep-learning based image segmentation using persistent homology. IEEE Trans. Pattern Anal. Mach. Intell. (2020)
7. Edelsbrunner, H., Harer, J.: Persistent homology - a survey. Contemp. Math. **453**, 257–282 (2008)

8. Van Etten, A., Lindenbaum, D., Bacastow, T.: Spacenet: A Remote Sensing Dataset and Challenge Series. arXiv preprint (2018)
9. Frangi, A.F., Niessen, W.J., Vincken, K.L., Viergever, M.A.: Multiscale vessel enhancement filtering. In: Wells, W.M., Colchester, A., Delp, S. (eds.) MICCAI 1998. LNCS, vol. 1496, pp. 130–137. Springer, Heidelberg (1998). https://doi.org/10.1007/BFb0056195
10. Funke, J., et al.: Large scale image segmentation with structured loss based deep learning for connectome reconstruction. IEEE Trans. Pattern Anal. Mach. Intell. **41**(7), 1669–1680 (2018)
11. Ganin, Y., Lempitsky, V.: N4-fields: neural network nearest neighbor fields for image transforms. In: Asian Conference on Computer Vision, pp. 536–551 (2014)
12. Guo, Z., et al.: Deepcenterline: a multi-task fully convolutional network for centerline extraction. In: IPMI, pp. 441–453 (2019)
13. Hu, X., Li, F., Samaras, D., Chen, C.: Topology-preserving deep image segmentation. In: Advances in Neural Information Processing Systems, pp. 5658–5669 (2019)
14. Hu, X., Wang, Y., Fuxin, L., Samaras, D., Chen, C.: Topology-aware segmentation using discrete Morse theory. In: International Conference on Learning Representations (2021)
15. Huang, X., Zhang, L.: Road centreline extraction from high-resolution imagery based on multiscale structural features and support vector machines. Int. J. Rem. Sens. **30**, 1977–1987 (2009)
16. Kingma, D.P., Ba, J.: Adam: a method for stochastic optimisation. In: International Conference on Learning Representations (2015)
17. Koziński, M., Mosińska, A., Salzmann, M., Fua, P.: Learning to segment 3D linear structures using only 2D annotations. In: Conference on Medical Image Computing and Computer Assisted Intervention, pp. 283–291 (2018) Learning to Segment 3D Linear Structures Using Only 2D Annotations. In *Conference on Medical Image Computing and Computer Assisted Intervention*, pages 283–291, 2018
18. Koziński, M., Mosinska, A., Salzmann, M., Fua, P.: Tracing in 2D to reduce the annotation effort for 3D deep delineation of linear structures. Med. Image Anal. **60** (2020)
19. Law, M.W.K., Chung, A.C.S.: Three dimensional curvilinear structure detection using optimally oriented flux. In: Forsyth, D., Torr, P., Zisserman, A. (eds.) ECCV 2008. LNCS, vol. 5305, pp. 368–382. Springer, Heidelberg (2008). https://doi.org/10.1007/978-3-540-88693-8_27
20. Maninis, K.K., Pont-Tuset, J., Arbeláez, P., Van Gool, L.: Deep retinal image understanding. In: Conference on Medical Image Computing and Computer Assisted Intervention, pp. 140–148 (2016)
21. Mnih, V., Hinton, G.E.: Learning to detect roads in high-resolution aerial images. In: Daniilidis, K., Maragos, P., Paragios, N. (eds.) ECCV 2010. LNCS, vol. 6316, pp. 210–223. Springer, Heidelberg (2010). https://doi.org/10.1007/978-3-642-15567-3_16
22. Mosińska, A., Kozinski, M., Fua, P.: Joint segmentation and path classification of curvilinear structures. IEEE Trans. Pattern Anal. Mach. Intell. **42**(6), 1515–1521 (2020)
23. Mosińska, A., Marquez-Neila, P., Kozinski, M., Fua, P.: Beyond the pixel-wise loss for topology-aware delineation. In: Conference on Computer Vision and Pattern Recognition, pp. 3136–3145 (2018)

24. Oner, D., Koziński, M., Citraro, L., Dadap, N.C., Konings, A.G., Fua, P.: Promoting connectivity of network-like structures by enforcing region separation. IEEE Trans. Pattern Anal. Mach. Intell. (2021)

25. Peng, H., Zhou, Z., Meijering, E., Zhao, T., Ascoli, G.A., Hawrylycz, M.: Automatic tracing of ultra-volumes of neuronal images. Nat. Methods **14**, 332–333 (2017)

26. Ronneberger, O., Fischer, P., Brox, T.: U-Net: convolutional networks for biomedical image segmentation. In: Conference on Medical Image Computing and Computer Assisted Intervention, pp. 234–241 (2015)

27. Shit, S., et al.: cldice - a novel topology-preserving loss function for tubular structure segmentation. In: CVPR, pp. 16560–16569. Computer Vision Foundation/IEEE (2021)

28. Sironi, A., Lepetit, V., Fua, P.: Multiscale centerline detection by learning a scale-space distance transform. In: Conference on Computer Vision and Pattern Recognition (2014)

29. Turetken, E., Becker, C., Glowacki, P., Benmansour, F., Fua, P.: Detecting irregular curvilinear structures in gray scale and color imagery using multi-directional oriented flux. In: International Conference on Computer Vision, pp. 1553–1560 (2013)

30. Turetken, E., Benmansour, F., Andres, B., Glowacki, P., Pfister, H., Fua, P.: Reconstructing curvilinear networks using path classifiers and integer programming. IEEE Trans. Pattern Anal. Mach. Intell. **38**(12), 2515–2530 (2016)

31. Wegner, J.D., Montoya-Zegarra, J.A., Schindler, K.: A higher-order CRF model for road network extraction. In: Conference on Computer Vision and Pattern Recognition, pp. 1698–1705 (2013)

32. Wiedemann, C., Heipke, C., Mayer, H., Jamet, O.: Empirical evaluation of automatically extracted road axes. In: Empirical Evaluation Techniques in Computer Vision, pp. 172–187 (1998)

33. Wolterink, J., van Hamersvelt, R., Viergever, M., Leiner, T., Isgum, I.: Coronary artery centerline extraction in cardiac CT angiography using a CNN-based orientation classifier. Med.Image Anal. **51**, 46–60 (2019)

34. Wu, D., et al.: A learning based deformable template matching method for automatic rib centerline extraction and labeling in CT images. In: Conference on Computer Vision and Pattern Recognition (2012)

RT-DNAS: Real-Time Constrained Differentiable Neural Architecture Search for 3D Cardiac Cine MRI Segmentation

Qing Lu[1]($\boxtimes$), Xiaowei Xu[2], Shunjie Dong[3], Cong Hao[4], Lei Yang[5], Cheng Zhuo[3], and Yiyu Shi[1]

[1] University of Notre Dame, Notre Dame, IN, USA
{qlu2,yshi4}@nd.edu
[2] Guangdong Provincial People's Hospital, Guangzhou, China
xxu8@nd.edu
[3] Zhejiang University, Hangzhou, China
{sj_dong,czhuo}@zju.edu.cn
[4] Georgia Institute of Technolog, Atlanta, GA, USA
callie.hao@gatech.edu
[5] University of New Mexico, Albuquerque, NM, USA
leiyang@unm.edu

Abstract. Accurately segmenting temporal frames of cine magnetic resonance imaging (MRI) is a crucial step in various real-time MRI guided cardiac interventions. To achieve fast and accurate visual assistance, there are strict requirements on the maximum latency and minimum throughput of the segmentation framework. State-of-the-art neural networks on this task are mostly hand-crafted to satisfy these constraints while achieving high accuracy. On the other hand, existing literature has demonstrated the power of neural architecture search (NAS) in automatically identifying the best neural architectures for various medical applications, within which differentiable NAS is a prevailing and efficient approach. However, they are mostly guided by accuracy, sometimes with computation complexity, but the importance of real-time constraints are overlooked. A major challenge is that such constraints are non-differentiable and thus are not compatible with the widely used differentiable NAS frameworks. In this paper, we present a strategy that can directly handle real-time constraints in differentiable NAS frameworks, named **RT-DNAS**. Experiments on extended 2017 MICCAI ACDC dataset show that compared with state-of-the-art manually and automatically designed architectures, RT-DNAS is able to identify neural architectures that can achieve better accuracy while satisfying the real-time constraints.

Keywords: Neural architecture search · Real-time segmentation · Cardiac cine MRI

Supplementary Information The online version contains supplementary material available at https://doi.org/10.1007/978-3-031-16443-9_58.

L. Wang et al. (Eds.): MICCAI 2022, LNCS 13435, pp. 602–612, 2022.
https://doi.org/10.1007/978-3-031-16443-9_58

1 Introduction

The rapid advance in real-time cine Magnetic Resonance Imaging (MRI) has been deployed for visual assistance in various cardiac interventions, such as aortic valve replacement [13], cardiac electroanatomic mapping and ablation [15], electrophysiology for atrial arrhythmias [19], intracardiac catheter navigation [6], and myocardial chemoablation [17]. In these applications, the cine MRI needs to be segmented on-the-fly, for which deep neural networks (DNNs) have been a prevailing choice. The cine MRI requires a reconstruction rate of 22 frames per second (FPS) [9,18], setting the minimum throughput for segmentation using DNNs. In addition, the latency of the network should be less than 50 ms to avoid any lags that can be perceived by human cyes [1]. ICA-UNet [21] is the state-of-the-art DNN that can maintain an acceptable accuracy while meeting these timing constraints, with a latency of 39 ms and a throughput of 28.3 FPS.

However, it is observed that the latency and throughput of ICA-UNet are well above the requirements. Since the MRI frames can only be reconstructed at 22 FPS, achieving a throughput higher than that would not yield any additional benefits but only introduce computation waste. Similarly, as suggested in [1], human vision cannot differentiate any latency below 50 ms. Therefore, there is still room to further trade off throughput and latency for higher accuracy. Unfortunately, ICA-UNet is manually designed, making the exploration of such tradeoff very difficult if not impossible.

Differentiable neural architecture search (NAS) has been recently applied to automatically and efficiently identify the best neural architecture for various biomedical applications [12,14,27]. A majority of these works focus on improving the accuracy as the only metric of interest [2,7,16]. Most recently, researchers have realized that it is crucial to also consider network latency/throughput for biomedical image segmentation applications [5,8,22,23,26]. They use network complexity, in terms of either the number of parameters or the number of floating point operations (FLOPs), as a proxy for latency/throughput estimation. This is primarily because latency/throughput are non-differentiable and can hardly be incorporated into any differentiable NAS frameworks. However, as demonstrated in [10], the latency/throughput can vary significantly even for networks of the same size and FLOPs, due to the difference in computation dependencies that are decided by the network topology. As such, there is no guarantee that the network architectures identified by these frameworks can satisfy the real-time constraints. As such, they cannot be extended to address the 3D cardiac cine MRI segmentation needed for surgical visual assistance.

To address these challenges and to directly incorporate real-time constraint into neural architecture search for 3D cardiac cine MRI segmentation, in this work, we propose a novel NAS approach, termed **RT-DNAS**, where the latency constraint can be incorporated into the differentiable search paradigm. Different from existing NAS approaches where both accuracy and network complexity are considered as optimization objectives, our goal is to *achieve as high accuracy as possible while satisfying the latency and throughput constraints*. It is achieved by using a genetic algorithm (GA) to decide the final network architecture.

Experiments on extended 2017 MICCAI ACDC dataset show that RT-DNAS can achieve the highest accuracy while satisfying the real-time constraints compared with both manual designed ICA-UNet and state-of-the-art NAS frameworks.

2   Method

We build our framework based on MS-NAS [25], the state-of-the-art NAS approach for medical image segmentation with competitive performance and implementation flexibility. An overview of the framework is illustrated in Fig. 1.

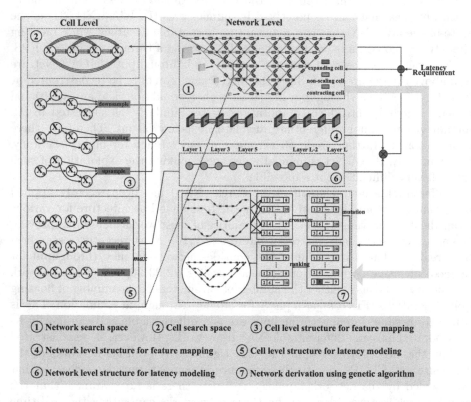

Fig. 1. Overview of the RT-DNAS workflow. We adopt the MS-NAS two-level search space. To incorporate real-time performance, the model finds favorable cell structures within latency constraint. A modified decoding method using genetic algorithm is proposed to derive the network structure.

2.1   Search Space and Method

Inspired by MS-NAS, we define a *network architecture search space* with various types of network design choices, including operation type, feature scale, feature fusion mechanism, and the connections between different operations and cells.

We define the entire NAS search space in a hierarchical manner: assume the network has L layers; each layer can be composed of multiple *cells*; each *cell* is composed of multiple *operations*.

Cell Level. A cell can be represented by a directed acyclic graph (DAG) where the vertices are tensors, and the edges represent the dependence between them. For an input tensor X_0, a cell produces N new tensors X_1, X_2, ..., X_N, and then process all of them using a scaling operation which is the only difference among cell types. The computation of each tensor depends on all its previous tensors (including X_0) and an operation o. We relax the search of the connections and operations by associating an coefficient to each combination of input tensor, output tensor, and operation. Denote the coefficient corresponding to the edge from X_j to X_i by operation o as $v_{i,j}^o$, and the output with performing o on X_j is $o(X_j)$, we can then compute X_i as

$$X_i = \sum_{j<i} \sum_{o \in \mathcal{O}} \frac{exp(v_{i,j}^o)}{\sum_{j'<i} \sum_{o' \in \mathcal{O}} exp(v_{i,j'}^{o'})} o(X_j) \tag{1}$$

where $\mathcal{O}$ is the operation set including, depth-wise separable convolution, dilated convolution, etc. It is worth mentioning that relaxing the connection and operation search separately as in MS-NAS can easily cause the gradient of relaxation coefficients vanish, so we directly search input-operation pairs instead.

Network Level. The network level search builds a network consisting of multiple paths from input to output; each path is formed by L cells in series. This is completed by two phases. In the first phase, all the paths are combined to generate the feature maps at all scales. Following the original MS-NAS, we relax the selection of different types of cells, i.e. expanding cell, non-scaling cell, and contracting cell, where their output can merge in the supernet topology (Fig. 1). As a result, the performance of each path can be evaluated by the proximal accumulated weight along the cells it covers. In the second phase, final networks are derived from the best paths based on such evaluation method. To take real-time efficiency into account, we propose a second search procedure using genetic algorithm to decode the network level structure as described in Sect. 2.3.

2.2 Latency Constraint Incorporation

Searching architectures for excellent real-time performance requires considering not only accuracy but also latency and throughput. However, since throughput is naturally coupled with latency (the higher the latency, the lower the throughput), to reduce search burden, we will incorporate latency only in the search process. If the identified architectures cannot satisfy the throughput constraint, we can simply lower the maximum latency and run the search again. As such, the goal of RT-DNAS is to discover the Pareto frontier in the trade-offs between the two metrics. Inspired by an early work targeting a specific hardware accelerator [11], we define the loss function as follows:

$$\mathcal{L} = \mathcal{L}_a \cdot \mathcal{L}_t + \lambda e^{L_t - L_{ub}} \tag{2}$$

where $\mathcal{L}_a$ is the cross-entropy loss of the network evaluated using training data, $\mathcal{L}_t$ is the loss for latency to ensure real-time performance; L_t is the latency achieved by the network candidate; L_{ub} is a constant representing the latency upper bound, which is set to be 50 ms as advised by [1]; λ is a large penalty factor, indicating that if the network architecture latency excels the upper bound, the loss will increase exponentially, so that infeasible solutions can be eliminated as early as possible. Since it is difficult to measure the real latency on the target device during the search, we estimate the latency $\mathcal{L}_t$ in two steps, including cell level and network level, similar to the search space construction.

Cell Level. For each tensor-to-tensor connection, there are $|\mathcal{O}|$ operations in parallel and the latency of the connection equals to the selected operation. Therefore, a reasonable estimation of *latency expectation* is the weighted sum of all the latency values from tensor i to tensor j, as follows:

$$Lat_{i,j} = \sum_{j<i} \sum_{o\in\mathcal{O}} \frac{exp(v_{i,j}^o)}{\sum_{j'<i}\sum_{o'\in\mathcal{O}} exp(v_{i,j'}^{o'})} Lat_{i,j}^o(X_j) \qquad (3)$$

where $Lat_{i,j}^o$ is the latency of operation o with input X_j, a constant for a specific operation on a fixed-size tensor. With the latency between any pair of tensors available, the latency of the cell is computed as:

$$Lat_{cell} = max_{\mathbf{x}\in\mathcal{X}} Path(\mathbf{x}) + S_{cell} \qquad (4)$$

where S_{cell} is a constant representing the latency of scaling operation; $\mathbf{x}$ is a set of the tensors that form a path from X_0 to X_N; $\mathcal{X}$ is the set of all possible paths. Note that Eq. 4 can be solved by finding the longest path within the cell.

Network Level. The overall latency of a network architecture can be estimated using the summation of the latency of each layer. Similar to cell level latency computation, we take a weighted sum of all the cell latency values to compute layer latency expectation based on the probability of each cell occurring in a network path as follows:

$$L_t = \sum_{l\leq L} \sum_{c\in\mathcal{C}^l} k_c^l Lat_{cell} \qquad (5)$$

where k_c^l is the probability of cell c being sampled from l-th layer. The details about the computation of k_c^l are presented in the supplementary materials.

2.3 Architecture Derivation

After the search, the architecture parameters $\mathcal{W}$ are decided. The final network can be derived by combining a group of paths formed by a sequence of cells from input to output. Note that different paths usually have overlapping segments in the network topology, which offers opportunities to further reduce the latency. Existing work MS-NAS constructs the network by finding a fixed number of paths with highest accumulated weights to maximize the accuracy but does not consider network latency. In fact, the weight-latency ratio of different paths varies significantly, so that carefully selecting the paths is important to reducing the network latency.

In RT-DNAS, we optimize the network path selection using genetic algorithm (GA) to build faster networks without significance in accuracy degradation. We define the following concepts in GA. *Gene:* a legal end-to-end path sampled from the backbone graph; *Individual:* a network formed by N_l paths; *Fitness:* the overall network latency of an individual. To keep the high accuracy, a gene pool is firstly built by top paths with high accumulated weights, and only the selected genes can exist during the evolution. We apply two operations to generate new graph, crossover and mutation. The crossover allows two individuals to exchange a portion of their genes and mutation replaces a random gene of an individual with another gene from the pool. To prevent duplication, the paths are sorted by their weights and the order of genes is maintained throughout the process.

3 Experiments

3.1 Experiment Setup

For performance evaluation, following existing works, we employ the data of the extended 2017 MICCAI ACDC challenge dataset made available by MSU-Net [20] with labels on all the frames in the training data. The dataset covers right ventricle (LV), myocardium (MYO), and left ventricle (LV) labeled by experienced radiologists from segmentation frames. There are totally 150 exams in this dataset each from a different patient, where the training set of 100 exams are employed to search the architecture. Further, 5-fold cross validation is performed to evaluate the resultant network using Dice score.

To implement RT-DNAS, we follow the guidelines presented in [25]. The skeleton is constructed by 10 layers of modules, three cells in each module, and three blocks in each cell. max-pooling and bilinear upsampling are used for contracting and expanding the feature size respectively, both by a factor of 2. We also adopt partial channel connection technique [24] to reduce memory cost by forcing 3/4 channels not going through the selected operations. We spend 40 epochs tuning the architectural parameters using an SGD optimizer with momentum of 0.9, learning rate from 0.01 to 0.001, and weight decay of 0.0003.

After all the parameters are fixed, networks consisting of different numbers of paths (N_l) are investigated and derived using GA method presented in Sect. 2.3. The initial population is set to 20, genes are randomly selected from a pool consisting of the top 10 paths, and the search loop runs for 100 generations. The evolution of the cell structures with highest probabilities as well as the final architecture obtained by RT-DNAS are visualized in the supplementary material.

We compare RT-DNAS with five baseline NAS frameworks. ICA-UNet [21] is a manually designed network that attains state-of-the-art performance on this dataset while satisfying real-time constraints. Hardware-aware NAS [26]

Table 1. Comparison of Dice, latency (LT, in ms) and throughput (TP, in FPS) of sub-networks formed by RT-DNAS with network level search carried out using the proposed GA algorithm (which considers real-time constraints) and the *Dijkstra* algorithm used in the original MS-NAS (which cannot consider real-time constraints). N_l is the number of consecutive paths involved.

Methods	RV	MYO	LV	Average	LT	TP
Dijkstra ($N_l = 3$)	.865 ± .024	.843 ± .021	.877 ± .022	.849 ± .032	27	39.8
Dijkstra ($N_l = 5$)	.911 ± .033	.881 ± .028	.925 ± .019	.905 ± .020	40	32.0
Dijkstra ($N_l = 7$)	.928 ± .026	.882 ± .017	.943 ± .020	.918 ± .032	48	22.1
Genetic ($N_l = 3$)	.857 ± .034	.843 ± .020	.878 ± .015	.859 ± .027	26	40.4
Genetic ($N_l = 5$)	.910 ± .026	.889 ± .039	.932 ± .022	.910 ± .031	35	34.1
Genetic ($N_l = 7$)	.925 ± .023	.891 ± .019	.950 ± .016	.922 ± .028	41	28.9

Table 2. Comparison of Dice score, latency (LT, in ms) and throughput (TP, in FPS) of the state-of-the-art real-time caridac cine MRI segmentation method ICA-UNet (with two settings used in the original paper), and the best networks discovered by Hardware-aware NAS, BiX-NAS, and the proposed RT-DNAS. The latency cannot exceed 50 ms while throughput should be no less than 22 FPS. Violations of latency and throughput constraints are marked in red.

Methods	RV	MYO	LV	Average	LT	TP
ICA-UNet (n = 3) [21]	.900 ± .023	.869 ± .027	.934 ± .013	.901 ± .017	35	31.6
ICA-UNet (n = 4) [21]	.921 ± .017	.888 ± .034	.952 ± .011	.920 ± .019	39	28.3
HW-Aware NAS [26]	.922 ± .015	.910 ± .042	.958 ± .019	.930 ± .020	52	19.5
BiX-NAS [22]	.916 ± .021	.908 ± .012	.946 ± .033	.923 ± .021	49	20.1
MS-NAS [25]	.928 ± .011	.911 ± .024	.961 ± .025	.933 ± .023	55	18.8
DeU-NAS [4]	.961 ± .015	.927 ± .025	.971 ± .018	.953 ± .019	121	10.2
RT-DNAS ($N_l = 6$)	.924±.010	.890±.017	.939±.030	.918±.018	37	31.8
RT-DNAS ($N_l = 7$)	.925±.023	.891±.019	.950±.016	.922±.028	41	28.9
RT-DNAS ($N_l = 8$)	**.933±.018**	**.902±.029**	**.959±.011**	**.931±.019**	46	**23.4**

and BiX-NAS [22] are NAS frameworks for MRI segmentation with hardware performance objectives, where the former includes both accuracy and FLOPs, and the latter includes architectural optimization. They aim at reducing computation complexity and parameter number, but do not explicitly take latency/throughput as hard constraints. MS-NAS [25] and DeU-NAS [4] are two NAS frameworks driven by only accuracy. For reference, we calibrate their timing performance of the reported networks based on reproduced implementations.

All the networks are executed using CUDA/CuDNN [3]. All experiments, including latency and throughput profiling, run on a machine with 16 cores of Intel Xeon E5-2620 v4 CPU, 256G memory, and an NVIDIA Tesla P100 GPU.

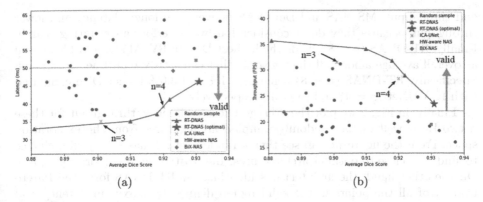

Fig. 2. Accuracy v.s. (a) latency and (b) throughput for randomly sampled architectures, ICA-UNet (n = 3, 4), the best architectures identified by hardware-aware NAS, BiX-NAS, and RT-DNAS. We can see that the Pareto frontier of the points in the valid region is formed by the architectures from RT-DNAS.

3.2   Results

We start with an ablation study to show the benefits of deploying the GA method instead of the *Dijkstra* algorithm used in the original MS-NAS in determining the network level structure. Based on the identified cell structures by RT-DNAS, we construct sub-networks using both *Dijkstra* and GA methods, and the performance of the resulting networks are summarized in Table 1. Comparing the two algorithms with the same number of consecutive paths N_l, it can be clearly observed that GA method gives higher throughput and lower latency, with slightly higher accuracy. This convincingly demonstrates the necessity and effectiveness of deploying GA in the network level search to handle the real-time constraints. It is also concluded that N_l is an effective tuning knob in balancing between the accuracy and timing performance: the more paths involved in the sub-network, the higher Dice score can be achieved, and the slower the network runs.

We collect the best networks obtained by RT-DNAS in terms of the tradeoff between accuracy and efficiency and compare with the baseline NAS approaches. There are two reported versions of ICA-UNet in [21] (with a hpyerparameter n equal to 3 and 4 respectively). As shown in Table 2, RT-DNAS ($N_l = 6$) has slightly longer latency and similar throughput compared with ICA-UNet (n = 3), but higher average Dice. Similar comparison can be observed between RT-DNAS ($N_l = 7$) and ICA-UNet (n = 4). Moreover, all these networks have a latency well below 50 ms and a throughput well above 22 FPS, suggesting additional room to enhance Dice. For hardware-aware NAS and BiX-NAS, although achieving higher average Dice, they violate the real-time constraint in terms of either latency or throughput. This echos our discussion earlier that simply considering network computation complexity or parameter number cannot guarantee that the resulting networks will meet the real-time constraints on latency and throughput. MS-NAS and DeU-NAS deliver even longer latency and lower throughput because they don't consider hardware performance during search. Finally, RT-DNAS (n = 8) yields the highest Dice in RV, MYO, and LV as well as overall average among all the networks that can satisfy real-time constraints. Specifically, RT-DNAS (n = 8) achieves 1.1% and 3.0% higher Dice compared with ICA-UNet (n = 4) and (n = 3), respectively.

Finally, in Fig. 2 we plot the average Dice v.s. latency/throughput for these networks as well as the randomly sampled architectures from the exploration space. From the figure we can see that a random search has a very slim chance to find any close-to-optimal architectures, due to the large exploration space. On the other hand, the architectures identified by RT-DNAS form the Pareto frontier of all the points in the valid region defined by maximum latency and minimum throughput, suggesting that they can indeed offer feasible designs with best tradeoff. Visualization of segmentation results by RT-DNAS (n = 8) along with the corresponding ground truth is included in the supplementary material.

4   Conclusion

In this work we proposed a latency-aware differentiable NAS framework for cine magnetic resonance image segmentation, named RT-DNAS. By incorporating hardware latency into the search objective, RT-DNAS is able to achieve the highest accuracy while perfectly meeting the real-time constraint, comparing with state-of-the-art manually and automatically designed architectures. It is demonstrated that using neural architecture search is a promising direction to real-time medical intervention using deep learning.

References

1. Annett, M., Ng, A., Dietz, P., Bischof, W.F., Gupta, A.: How low should we go? Understanding the perception of latency while inking. In: Graphics Interface 2014, pp. 167–174. AK Peters/CRC Press (2020)
2. Bosma, M., Dushatskiy, A., Grewal, M., Alderliesten, T., Bosman, P.A.: Mixed-block neural architecture search for medical image segmentation. In: SPIE Medical Imaging (2022)
3. Chetlur, S., et al.: cuDNN: Efficient primitives for deep learning. arXiv preprint arXiv:1410.0759 (2014)
4. Dong, S., et al.: DeU-Net 2.0: Enhanced deformable U-Net for 3D cardiac cine MRI segmentation. Medical Image Analysis 78, 102389 (2022)
5. Fernandes, F.E., Yen, G.G.: Automatic searching and pruning of deep neural networks for medical imaging diagnostic. IEEE Trans. Neural Netw. Learn. Syst. 32(12), 5664–5674 (2021)
6. Gaspar, T., Piorkowski, C., Gutberlet, M., Hindricks, G.: Three-dimensional real-time MRI-guided intracardiac catheter navigation. Eur. Heart J. 35(9), 589–589 (2014)

7. He, Y., Yang, D., Roth, H., Zhao, C., Xu, D.: Dints: differentiable neural network topology search for 3d medical image segmentation. In: Proceedings of the IEEE/CVF Conference on Computer Vision and Pattern Recognition, pp. 5841–5850 (2021)
8. Huang, Z., Wang, Z., Gu, L., et al.: Adwu-Net: adaptive depth and width U-Net for medical image segmentation by differentiable neural architecture search. In: Medical Imaging with Deep Learning (2021)
9. Iltis, P.W., Frahm, J., Voit, D., Joseph, A.A., Schoonderwaldt, E., Altenmüller, E.: High-speed real-time magnetic resonance imaging of fast tongue movements in elite horn players. Quant. Imaging Med. Surg. **5**(3), 374 (2015)
10. Jiang, W., et al.: Hardware/software co-exploration of neural architectures. IEEE Trans. Comput.-Aided Des. Integr. Circ. Syst. **39**(12), 4805–4815 (2020)
11. Li, Y., et al.: EDD: efficient differentiable DNN architecture and implementation co-search for embedded AI solutions. In: 2020 57th ACM/IEEE Design Automation Conference, DAC 2020. Proceedings - Design Automation Conference, Institute of Electrical and Electronics Engineers Inc., USA, July 2020
12. Lu, Z., et al.: Multi-scale and multi-level neural architecture search for low-dose ct denoising. arXiv preprint arXiv:2103.12995 (2021)
13. McVeigh, E.R., Guttman, M.A., Lederman, R.J., Li, M., Kocaturk, O., Hunt, T., Kozlov, S., Horvath, K.A.: Real-time interactive Mri-guided cardiac surgery: Aortic valve replacement using a direct apical approach. Mag. Resona. Med. **56**(5), 958–964 (2006)
14. Peng, C., et al.: Hypersegnas: bridging one-shot neural architecture search with 3d medical image segmentation using hypernet. arXiv preprint arXiv:2112.10652 (2021)
15. Radau, P.E., et al.: VURTIGO: visualization platform for real-time, MRI-guided cardiac electroanatomic mapping. In: Camara, O., Konukoglu, E., Pop, M., Rhode, K., Sermesant, M., Young, A. (eds.) STACOM 2011. LNCS, vol. 7085, pp. 244–253. Springer, Heidelberg (2012). https://doi.org/10.1007/978-3-642-28326-0_25
16. Radau, Z., et al.: VURTIGO: visualization platform for real-time, MRI-guided cardiac electroanatomic mapping. In: Camara, O., Konukoglu, E., Pop, M., Rhode, K., Sermesant, M., Young, A. (eds.) STACOM 2011. LNCS, vol. 7085, pp. 244–253. Springer, Heidelberg (2012). https://doi.org/10.1007/978-3-642-28326-0_25
17. Rogers, T., et al.: Transcatheter myocardial needle chemoablation during real-time magnetic resonance imaging: a new approach to ablation therapy for rhythm disorders. Circulat. Arrhyth. Electrophysiol. **9**(4), e003926 (2016)
18. Schaetz, S., Voit, D., Frahm, J., Uecker, M.: Accelerated computing in magnetic resonance imaging: real-time imaging using nonlinear inverse reconstruction. In: Computational and Mathematical Methods in Medicine 2017 (2017)
19. Vergara, C.R., et al.: Real-time magnetic resonance imaging-guided radiofrequency atrial ablation and visualization of lesion formation at 3 tesla. Heart Rhythm **8**(2), 295–303 (2011)
20. Wang, T., et al.: MSU-Net: multiscale statistical u-net for real-time 3d cardiac MRI video segmentation. In: Shen, S., et al. (eds.) MICCAI 2019. LNCS, vol. 11765, pp. 614–622. Springer, Cham (2019). https://doi.org/10.1007/978-3-030-32245-8_68
21. Wang, T., et al.: ICA-UNet: ICA inspired statistical UNet for real-time 3D cardiac cine MRI segmentation. In: Martel, M.L., et al. (eds.) MICCAI 2020. LNCS, vol. 12266, pp. 447–457. Springer, Cham (2020). https://doi.org/10.1007/978-3-030-59725-2_43

22. Wang, X., et al.: BiX-NAS: searching efficient bi-directional architecture for medical image segmentation. In: de Bruijne, M., et al. (eds.) MICCAI 2021. LNCS, vol. 12901, pp. 229–238. Springer, Cham (2021). https://doi.org/10.1007/978-3-030-87193-2_22

23. Xu, S., Quan, H.: ECT-NAS: Searching efficient CNN-transformers architecture for medical image segmentation. In: 2021 IEEE International Conference on Bioinformatics and Biomedicine (BIBM), pp. 1601–1604. IEEE (2021)

24. Xu, Y., et al.: PC-darts: partial channel connections for memory-efficient architecture search. arXiv preprint arXiv:1907.05737 (2019)

25. Yan, X., Jiang, W., Shi, Y., Zhuo, C.: MS-NAS: multi-scale neural architecture search for medical image segmentation. In: Martel, A.L., et al. (eds.) MICCAI 2020. LNCS, vol. 12261, pp. 388–397. Springer, Cham (2020). https://doi.org/10.1007/978-3-030-59710-8_38

26. Zeng, D., et al.: Towards cardiac intervention assistance: hardware-aware neural architecture exploration for real-time 3D cardiac cine MRI segmentation. In: Proceedings of the 39th International Conference on Computer-Aided Design, pp. 1–8 (2020)

27. Zhu, Y., Meijering, E.: Automatic improvement of deep learning-based cell segmentation in time-lapse microscopy by neural architecture search. Bioinformatics **37**(24), 4844–4850 (2021)

Integration of Imaging
with Non-imaging Biomarkers

Lesion Guided Explainable Few Weak-Shot Medical Report Generation

Jinghan Sun[1,2], Dong Wei[2], Liansheng Wang[1(✉)], and Yefeng Zheng[2]

[1] Xiamen University, Xiamen, China
jhsun@stu.xmu.edu.cn, lswang@xmu.edu.cn
[2] Tencent Healthcare (Shenzhen) Co., Ltd., Tencent Jarvis Lab, Shenzhen, China
{donwei,yefengzheng}@tencent.com

Abstract. Medical images are widely used in clinical practice for diagnosis. Automatically generating interpretable medical reports can reduce radiologists' burden and facilitate timely care. However, most existing approaches to automatic report generation require sufficient labeled data for training. In addition, the learned model can only generate reports for the training classes, lacking the ability to adapt to previously unseen novel diseases. To this end, we propose a lesion guided explainable few weak-shot medical report generation framework that learns correlation between seen and novel classes through visual and semantic feature alignment, aiming to generate medical reports for diseases not observed in training. It integrates a lesion-centric feature extractor and a Transformer-based report generation module. Concretely, the lesion-centric feature extractor detects the abnormal regions and learns correlations between seen and novel classes with multi-view (visual and lexical) embeddings. Then, features of the detected regions and corresponding embeddings are concatenated as multi-view input to the report generation module for explainable report generation, including text descriptions and corresponding abnormal regions detected in the images. We conduct experiments on FFA-IR, a dataset providing explainable annotations, showing that our framework outperforms others on report generation for novel diseases.

Keywords: Medical report generation · Few weak-shot learning · Multi-view learning

1 Introduction

Medical images are widely used in clinics for diagnosis, prognosis, and therapy planning. Radiologists usually write imaging reports based on their expertise

---

J. Sun and D. Wei—Equal contribution; J. Sun contributed to this work during an internship at Tencent.

Supplementary Information The online version contains supplementary material available at https://doi.org/10.1007/978-3-031-16443-9_59.

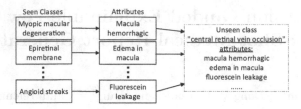

Fig. 1. Illustration of our motivation: the seen classes are used to learn a mapping between visual and lexical features in the training stage, which is then transferred to generate reports for previously unseen novel classes.

and experience while observing abnormal areas in the images. However, due to the demand for expert knowledge and large number of images, this process is laborious, time-consuming, and error-prone. To relieve this burden, a large body of literature [4,11,12,17,26] has investigated automatic radiology report generation. However, there are still some practical limitations of these methods. A notable one is that the trained generator can only generate reports for the training classes and the performance will degrade on novel diseases. Several few-shot learning (FSL) report generation methods [8,9] have been proposed to tackle this problem, which require the ground truth reports of the novel diseases to fine-tune the network. However, in clinical practice, ground truth reports for the novel diseases can be unavailable, too.

To enable the model to recognize previously unseen classes, zero-shot learning (ZSL) [13] is proposed for natural image tasks. In ZSL [6,10,25], a mapping function is learned to project visual input to a semantic space, where the relationship between seen and novel classes is extracted from an auxiliary source, e.g., word embeddings. Being inspired, we propose to build a connection between seen training diseases and novel testing diseases through learning common attributes. Taking Fig. 1 for example, although in the training stage the model does not see a report of the novel class—central retinal vein occlusion, we can make it learn a mapping from visual features to semantic features, with which the correlation between seen and novel diseases can be established. To ensure a well alignment of both seen and novel classes' semantic features, we propose a multi-view embedding ensemble strategy where lexical embeddings [24] of the diseases are employed for feature calibration. Then, when a novel disease sample is presented, the model can make use of the calibrated correlation for report generation by assembling descriptions of the common attributes. However, ZSL approaches often cannot produce optimal results, as the learned relationship between seen and novel classes may be weak and ambiguous.

Alternatively, some researchers propose to provide weak annotations for the novel classes, which are more accessible than full annotations and meanwhile can provide more definite information about the novel classes than ZSL. This sort of task is known as weak-shot learning [3,27]. However, for the task of quickly adapting to generate reports for novel diseases, collecting enough weakly annotated samples can be prohibitive, too. Thus, it is desirable to allow *few* weak-shot learning in this scenario.

In this work, we propose multi-view lesion guided few weak-shot learning for explainable medical report generation. Concretely, we first extract regional lesion features based on Faster R-CNN [21] to guide the network to capture the noteworthy features and output detection results to provide explainable prediction for reports. In addition, we adopt the weight imprinting strategy [20] based on a few novel samples that are weakly annotated with labeled bounding boxes to capture novel class features in the testing stage. Then, we project the visual features into semantic space and construct a soft target of all classes. The soft target indicates the correlation between the lexical embeddings of seen classes and novel classes, which force the network to learn about the novel classes although only seen classes are available for training. Finally, the visual features and lexical embeddings are concatenated as multi-view features and fed into the generation model to generate reports. In summary, the contributions of this paper reside in: (1) This is the first study that proposes few weak-shot medical report generation, which can generate explainable reports of novel classes without any training report. (2) We extract lesion proposal features instead of global visual features to guide the model for more focused report generation. (3) We propose a soft target of lexical embeddings, through which features of previously unseen classes can be effectively learned. Furthermore, visual features and lexical embeddings are concatenated as the input to provide multi-view knowledge for the generation model. (4) We conduct experiments on the Fundus Fluorescein Angiography Images and Reports dataset (FFA-IR) [15] with six experimental settings. The results demonstrate the superiority of the proposed method.

2   Methods

Problem Setting. We denote the set of disease classes as $C = C^{\text{seen}} \cup C^{\text{nov}}$, where C^{seen} and C^{nov} denote the sets of seen and novel diseases, respectively, satisfying $C^{\text{seen}} \cap C^{\text{nov}} = \emptyset$. Each case in the dataset contains images of a patient at different periods: $\{I_1, \ldots, I_N\}$, where $N \geq 1$ is the total number of images in one case. Notably, N is not a parameter but a property of the data; the proposed method can also generalize for non-periodical data, e.g., N views in a chest X-ray. For seen classes, explainable annotations including a set of labeled bounding boxes $B = \{(X, y)\}$ for the images and a report $R = \{r_1, \ldots, r_T\}, r_t \in \mathbb{V}$ are provided for each case, where X specifies the location and size of each box, y is the disease class label and $y \in \{1, \ldots, |C^{\text{seen}}|\}$, T is the length of reports, $\mathbb{V}$ is a vocabulary of all possible tokens. For novel diseases, we sample K images for each class to compose the support set S, with all remaining images composing the query set Q. Different from existing weak-shot methods that provide weak annotations for all samples of the novel classes, we only provide weak annotations (i.e., labeled bounding boxes B in this work) for the few samples in the support set S. It is worth noting that our architecture also has the potential to use other weak annotations, which we leave for future work. Given S, the target is to generate explainable medical reports for the query set Q by transferring the explainable knowledge of seen diseases. To this end, we propose a novel few weak-shot task for medical report generation.

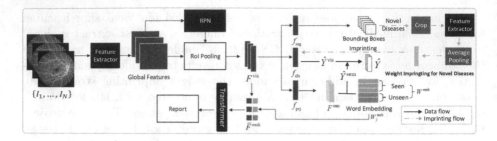

Fig. 2. Overview of the proposed approach.

Method Overview. Figure 2 shows an overview of our method. Given the images of a patient, we first extract global features from each image I. Then the region proposal network (RPN) generates proposals of lesion regions and uses region of interest (ROI) pooling to extract regional visual features F^{vis}. F^{vis} are sent into three branches: 1. box regression branch (f_{reg}) to detect the location and size of lesion regions; 2. classification branch (f_{cls}) to predict category of each region; 3. semantic prediction branch (f_{prj}) to project visual features to semantic features (F^{sem}) and learn the relationship between seen and novel diseases. Finally, we concatenate visual features and their corresponding word embeddings, and send the concatenated features $\bar{F}^{mult}$ into Transformer [22] to generate the report. In the inference stage, we employ the weight imprinting technique (top right of Fig. 2) to enable f_{cls} to recognize novel diseases given the few weak-shot samples.

Lesion-Centric Feature Extractor. The key ingredient of report generation is the performance of extracted visual features. However, existing report generation frameworks [4,11] usually use global features for the generation model, which impedes the quality of generated reports. We propose to extract features from bounded lesion regions to provide the report generation model with more representative knowledge. Our visual extractor is built on Faster R-CNN [21], a popular object detection algorithm. In the first stage, Faster R-CNN extracts a global feature for an image. In the second stage, the region proposal network generates lesion region proposals. Then the network uses ROI pooling to extract fixed-size visual feature F^{vis} for each region proposal. These features are fed into two fully connection layers—a box regression layer f_{reg} and a box classification layer f_{cls} parameterized by θ^{reg} and θ^{cls}, respectively. For f_{cls}, we adopt the weight imprinting mechanism [20] to enable few weak-shot report generation (described later), and enforce $\left\| F^{vis} \right\|_2 = 1$ via L2 normalization. The weight matrix of θ^{cls} can be viewed as a learnable visual embedding dictionary of the diseases. Then, the detection loss is defined as:

$$\mathcal{L}_{det} = \mathcal{L}_{cls}\left((\hat{Y}^{vis} + \hat{Y}^{sem})/2, y \right) + \mathcal{L}_{reg}\left(f_{reg}(F^{vis}), X \right), \tag{1}$$

where $\hat{Y}^{\mathrm{vis}} = [\hat{p}_1^{\mathrm{vis}}, \ldots, \hat{p}_{|C^*|}^{\mathrm{vis}}]$ is the visual prediction by f_{cls} after softmax, C^* must be C^{seen} for training with seen diseases but can be any of C^{seen}, C^{nov}, or C in different test settings, $\hat{Y}^{\mathrm{sem}}$ is the semantic prediction (described below), y is the class label, and X is the location and size of each box. The classification loss $\mathcal{L}_{\mathrm{cls}}$ is the cross-entropy loss. For regression, we adopt the smooth L1 loss: $\mathcal{L}_{\mathrm{reg}} = 0.5x^2$ if $|x| < 1$ and $|x| - 0.5$ otherwise.

Multi-view Embedding Ensemble. Report generation is an image-to-text task, so visual and semantic features both contain valuable information to improve report generation. The Faster R-CNN is trained on seen diseases but neglects the contents of novel diseases. As the reports of novel diseases are not available, we can learn from the lexical embeddings of novel diseases to guide the visual feature learning. To make the model transferable over novel diseases through learning relationship between seen and novel diseases, we first create a semantic space and project visual feature F^{vis} into this semantic space via linear projection: $F^{\mathrm{sem}} = f_{\mathrm{prj}}(F^{\mathrm{vis}}) = \theta^{\mathrm{prj}} F^{\mathrm{vis}}$, where f_{prj} is a projection function parameterized by θ^{prj}. Then we obtain lexical embeddings $W^{\mathrm{emb}} \in \mathbb{R}^{|C| \times d}$ of both seen and novel diseases from a pretrained word embedding model such as BioBert [14], where d is the dimension of each embedding. Instead of directly maximizing the similarity between visual features and their corresponding lexical embeddings, we build a soft label for exploiting the semantic relationship between seen and novel diseases. Specifically, we define the soft label for class y as:

$$L_y^{\mathrm{s}} = [l_1^{\mathrm{s}}, \ldots, l_{|C|}^{\mathrm{s}}]^T, \text{ where } l_c^{\mathrm{s}} = \mathrm{Sim}(W_y^{\mathrm{emb}}, W_c^{\mathrm{emb}}), \qquad (2)$$

and Sim is the cosine similarity function to measure the similarity of class y to class c in the semantic space. Thus, L_y^{s} collects class y's similarity to all classes, both seen and novel, establishing correlations between them. The loss function in the semantic space is $\mathcal{L}_{\mathrm{kl}}(\mathrm{Sim}(F^{\mathrm{sem}}, W^{\mathrm{emb}}), L^{\mathrm{s}})$, where $\mathcal{L}_{\mathrm{kl}}$ is the Kullback-Leibler divergence loss. In addition, we define $\hat{Y}^{\mathrm{sem}} = \mathrm{softmax}(\mathrm{Sim}(F^{\mathrm{sem}}, W^{\mathrm{emb}}))$ as the semantic prediction to assist the detection (Eq. (1)). Thus, we not only force the network to learn the relationship between seen and novel diseases through $\mathcal{L}_{\mathrm{kl}}$, but also boost object detection with the semantic information.

Lesion Guided Report Generation. Conventional report generation approaches usually treated the lesions and non-lesion regions equally. In contrast, we propose to merge lesion guided visual and semantic features to improve report generation. Concretely, given a set of lesion proposals $\hat{B}$ predicted in all images of a patient, and the corresponding visual features $\{F^{\mathrm{vis}}\}$ and labels $\{\hat{Y}\}$, where $\hat{Y} = (\hat{Y}^{\mathrm{vis}} + \hat{Y}^{\mathrm{sem}})/2 = [\hat{p}_1, \ldots, \hat{p}_{|C^*|}]$ and $\sum_{c=1}^{|C^*|} \hat{p}_c = 1$, the corresponding set of class predictions can be obtained by $\{\hat{y} = \mathrm{argmax}_c \, \hat{p}_c\}$. Then, we obtain the corresponding lexical embeddings $\{W_{\hat{y}}^{\mathrm{emb}}\}$. Next, a multi-view feature is obtained by concatenating the visual feature and lexical embedding $F^{\mathrm{mult}} = [F^{\mathrm{vis}}, W_{\hat{y}}^{\mathrm{emb}}]$. Finally, the input feature to the report generator is obtained as the average multi-view feature over all lesion proposals, weighted by each proposal's prediction probability: $\bar{F}^{\text{mult}} = \frac{1}{|\hat{B}|}\sum_{b=1}^{|\hat{B}|} \max(\hat{Y}_b) F_b^{\text{mult}}$.

For report generation, we employ the commonly used Transformer [1, 15, 22] architecture. Denoting the report generator by f_{trans}, then we have $\{\hat{r}_1, \ldots, \hat{r}_T\} = f_{\text{trans}}(\bar{F}^{\text{mult}})$. The loss function is defined as:

$$\mathcal{L}_{\text{gen}} = -\log p\left(\hat{r}_T \mid \hat{r}_1, \ldots, \hat{r}_{T-1}; \theta^{\text{trans}}\right), \tag{3}$$

where θ^{trans} are the parameters of the Transformer. For more details of the Transformer, we refer readers to [22].

The overall optimization objective of our method is defined as:

$$\mathcal{L} = \mathcal{L}_{\text{det}}/|\hat{B}| + \mathcal{L}_{\text{kl}}/|\hat{B}| + \mathcal{L}_{\text{gen}}. \tag{4}$$

Few Weak-Shot Report Generation for Novel Diseases. In the inference stage, to ensure that the model can accurately recognize previously unseen novel diseases with only few weakly labeled examples in the support set S, we employ the weight imprinting scheme [20]. Concretely, given a few examples for a novel disease c, we first make use of the box annotations B to crop the lesion regions for each class. Then, we obtain the visual features $\{F_b^{\text{vis}}\}_{b=1}^{|B|}$ by feeding the cropped regions to the global visual extractor (without using the RPN or ROI pooling). We compute new weights for the classification layer (f_{cls}) by averaging the normalized visual features: $\theta_c^{\text{cls}} = (\frac{1}{|B|}\sum F_b^{\text{vis}})/\|\frac{1}{|B|}\sum F_b^{\text{vis}}\|_2$, where θ_c^{cls} is the classification weights for the novel class c. As discussed in [20], the embeddings of novel classes may not have a unimodal distribution and influence the detection accuracy. Therefore, we conduct fine-tuning to diminish the bias in the learned embedding space after the weight embedding. To generate explainable reports for novel classes in the query set Q, we first predict the bounding boxes and classification results (both visual and semantic) through forward propagation and then apply non-maximum suppression with a threshold of 0.5 to obtain the final detections. Then the visual features and corresponding lexical embeddings are concatenated and input to the Transformer to generate reports.

3   Experiments

Dataset and Evaluation Metrics. FFA-IR [15] is a new benchmark based on fundus fluorescein angiography images and reports. FFA-IR includes annotations of 46 categories of lesions including 315 cases with 12,166 lesion regions. For each case, we use the English reports, lesion category, lesion regions and the images. We conduct experiments with two different splits of train (seen), validation and test (novel) classes of 32/5/9 and 34/5/7 classes, respectively, to verify our method's robustness to different task scenarios. For the training set, we first select all images containing lesions of seen diseases. Then, we exclude those images with any lesion of the novel diseases to make sure the novel classes are not observed during training stage. For testing, K is set to five (i.e., five

Table 1. The performance of our framework and exiting SOTA methods. B * N denotes N-gram score of BLEU. Best results are shown in bold.

	Split	Model	B1	B2	B3	B4	METEOR	ROUGE	CIDER
Test-Seen	32/5/9	Grounded [28]	0.477	0.346	0.253	0.190	**0.233**	**0.407**	0.511
		R2Gen [4]	0.472	0.347	0.247	0.184	0.215	0.402	0.460
		Ours	**0.491**	**0.357**	**0.263**	**0.200**	0.223	0.398	**0.519**
	34/5/7	Grounded [28]	0.494	0.356	**0.262**	0.197	0.220	**0.410**	0.558
		R2Gen [4]	0.459	0.311	0.222	0.171	0.222	0.383	0.541
		Ours	**0.511**	**0.357**	0.262	0.199	**0.223**	0.401	**0.571**
Test-Novel	32/5/9	Grounded [28]	0.466	0.341	0.239	0.183	**0.220**	0.375	0.472
		R2Gen [4]	0.434	0.324	0.232	0.175	0.208	0.374	0.331
		Ours	**0.481**	**0.349**	**0.256**	**0.193**	0.217	**0.409**	**0.496**
	34/5/7	Grounded [28]	0.435	0.348	0.257	0.212	**0.230**	**0.417**	0.551
		R2Gen [4]	0.402	0.320	0.211	0.209	0.209	0.403	0.492
		Ours	**0.510**	**0.371**	**0.276**	**0.217**	0.228	0.404	**0.564**
Test-Mix	32/5/9	Grounded [28]	0.453	0.343	**0.267**	**0.207**	0.174	0.392	0.507
		R2Gen [4]	0.464	0.321	0.239	0.183	0.175	0.375	0.412
		Ours	**0.490**	**0.356**	0.262	0.199	**0.222**	**0.398**	**0.511**
	34/5/7	Grounded [28]	0.426	0.297	0.218	0.170	0.210	**0.386**	0.535
		R2Gen [4]	0.456	0.309	**0.224**	0.163	0.208	0.384	0.445
		Ours	**0.459**	**0.311**	0.222	**0.171**	**0.222**	0.383	**0.541**

examples for each novel disease class) for the support set S. We use three different settings following [29]: Test-Seen, Test-Novel and Test-Mix. Test-Seen data contain images of only seen diseases, Test-Novel contain images of only novel diseases, and Test-Mix contain images of both seen and novel classes—which is a more challenging task known as generalized weak shot. We employ seven commonly used metrics to evaluate the quality of generated reports, including: BLEU (1- to 4-gram) [18], CIDER [23], METEOR [2], and ROUGE [16].

Implementation. PyTorch [19] (1.4.0) is used for experiments. We use ResNet-101 [7] pretrained on ImageNet [5] as the visual extractor of Faster R-CNN. We train the network for 100 epochs with a mini-batch of four cases on four Tesla V100 GPUs. The SGD optimizer is used with a momentum 0.9 and a decay of 0.0005. We set initial learning rate to 10^{-4} and decay at 60 and 80 epochs by multiplying by 0.1, respectively. All images are resized to 224×224 pixels. For the rest, we follow the optimal training strategies suggested in [15]. The source code is available at: https://github.com/jinghanSunn/Few-weak-shot-RG.

Comparison to State-of-the-Art (SOTA) Methods. We compare our method with Grounded [28] and R2Gen [4], for their excellent performance in medical report generation. In Table 1, we can see that our method outperforms existing SOTA methods in most metrics (5 to 6 of 7 metrics) on both split configurations and three test settings. The results demonstrate the effectiveness of our method for the few weak-shot report generation task. The performance of the 32/5/9 split is slightly worse than that of 34/5/7, because the 32/5/9 split task is more difficult as the number of training data and diversity are reduced while more

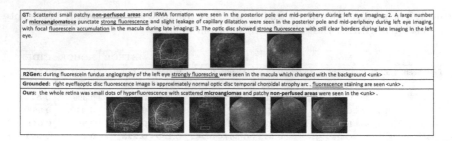

Fig. 3. The visualization of ground truth report (GT), and reports generated by R2Gen [4], Grounded [28] and ours. The green and blue boxes are lesion regions of two novel classes. The white boxes in our results are predicted abnormalities that do not belong to these two classes. (Color figure online)

novel classes need to be discriminated. Figure 3 shows example reports generated by our method, R2Gen [4] and Grounded [28], from which we make the following observations. First, in the ground truth report, only three images have labeled bounding boxes which may cause a great loss of critical diagnostic information. Second, R2Gen [4] and Grounded [28] cannot generate satisfactory description of novel classes. They only perceives non-critical knowledge, such as "strongly fluorescing". Last, our method generates a report including descriptions of the novel diseases, with corresponding lesion regions detected in images of various periods for explainable reporting.

To further validate the efficacy of our method with varying K values, we conduct experiments with the Test-Novel setting on the test set of 34/5/9 split (supplement Table 1). Results show that larger K values generally yield better results, as expected, and that our method overall outperforms Grounded and R2Gen with different K values.

Ablation Study. We conduct ablation studies to validate the effects of our proposed components on the validation set of the 34/5/7 split. From the results in Table 2, we make the following observations. First, we replace the Faster R-CNN based visual extractor with ResNet-101 [7] (Ablation-1). Results show the Faster R-CNN based visual extractor (Ablation-2) achieves better performance to vanilla CNN (Ablation-1) as Faster R-CNN can capture lesion-centric features. Second, equipping with imprinted weights (Ablation-3) assists the model to capture novel diseases' features, improving the generation performance. Third, combining semantic embeddings with visual features (Ablation-4) outperforms single-view method in four of the seven metrics. However, it is worse than our complete model as Ablation-4 lacks the ability to learn the relationship via soft labels, with which the correlation between seen and novel diseases can be established. With all these components, our complete model yields the best overall performance, achieving the best results for six metrics and the second best for the remaining one metric.

Table 2. Ablation study on validation set with the 34/5/7 split. B*N denotes N-gram score of BLEU [18]. Best results are shown in bold.

	Faster R-CNN	Imprinted	Multi-view	Soft label	B1	B2	B3	B4	METEOR	ROUGE	CIDER
Ablation-1					0.459	0.306	0.261	0.231	0.211	0.410	0.507
Ablation-2	✓				0.488	0.323	0.262	0.232	0.197	0.387	0.532
Ablation-3	✓	✓			0.508	0.339	0.274	0.242	0.250	**0.412**	0.560
Ablation-4	✓	✓	✓		0.513	0.377	0.279	0.223	0.229	0.389	0.565
Ours	✓	✓	✓	✓	**0.524**	**0.389**	**0.280**	**0.251**	**0.254**	0.411	**0.568**

4 Conclusion

In this work, we introduced a novel few weak-shot report generation task, where no ground truth report but only a few box-annotated image samples of novel diseases were available. To tackle this new task, we proposed a multi-view lesion guided report generation framework. We first detected lesion regions to extract lesion-centric visual features and output abnormal regions. Then we projected visual features into semantic space where a word embedding model was utilized to learn the relationship between seen and novel classes. Finally, we combined the multi-view features for report generation. The reports were accompanied by detected abnormal regions in the input images to provide enhanced explainability. Our method generally outperformed other state-of-art approaches to medical report generation in most metrics (5 to 6 of 7 metrics) and six experimental settings. In the future, we plan to employ more advanced few-shot detection methods to improve the absolute scores.

Acknowledgement. This work was supported by the Scientific and Technical Innovation 2030 - "New Generation Artificial Intelligence" Project (No. 2020AAA0104100) and the National Key Research and Development Program of China (2019YFE0113900).

References

1. Amjoud, A.B., Amrouch, M.: Automatic generation of chest X-ray reports using a transformer-based deep learning model. In: 2021 Fifth International Conference on Intelligent Computing in Data Sciences (ICDS), pp. 1–5. IEEE (2021)
2. Banerjee, S., Lavie, A.: METEOR: an automatic metric for MT evaluation with improved correlation with human judgments. In: Proceedings of the ACL Workshop on Intrinsic and Extrinsic Evaluation Measures for Machine Translation and/or Summarization, pp. 65–72 (2005)
3. Chen, J., Niu, L., Liu, L., Zhang, L.: Weak-shot fine-grained classification via similarity transfer. In: Advances in Neural Information Processing Systems, vol. 34 (2021)
4. Chen, Z., Song, Y., Chang, T.H., Wan, X.: Generating radiology reports via memory-driven transformer. In: Proceedings of the 2020 Conference on Empirical Methods in Natural Language Processing (EMNLP), pp. 1439–1449 (2020)

5. Deng, J., Dong, W., Socher, R., Li, L.J., Li, K., Fei-Fei, L.: ImageNet: a large-scale hierarchical image database. In: 2009 IEEE Conference on Computer Vision and Pattern Recognition, pp. 248–255. IEEE (2009)
6. Han, Z., Fu, Z., Chen, S., Yang, J.: Contrastive embedding for generalized zero-shot learning. In: Proceedings of the IEEE/CVF Conference on Computer Vision and Pattern Recognition, pp. 2371–2381 (2021)
7. He, K., Zhang, X., Ren, S., Sun, J.: Deep residual learning for image recognition. In: Proceedings of the IEEE Conference on Computer Vision and Pattern Recognition, pp. 770–778 (2016)
8. Jia, X., et al.: Radiology report generation for rare diseases via few-shot transformer. In: 2021 IEEE International Conference on Bioinformatics and Biomedicine (BIBM), pp. 1347–1352. IEEE (2021)
9. Jia, X., Xiong, Y., Zhang, J., Zhang, Y., Zhu, Y.: Few-shot radiology report generation for rare diseases. In: 2020 IEEE International Conference on Bioinformatics and Biomedicine (BIBM), pp. 601–608. IEEE (2020)
10. Jiang, H., Wang, R., Shan, S., Chen, X.: Transferable contrastive network for generalized zero-shot learning. In: Proceedings of the IEEE/CVF International Conference on Computer Vision, pp. 9765–9774 (2019)
11. Jing, B., Wang, Z., Xing, E.: Show, describe and conclude: on exploiting the structure information of chest X-ray reports. In: Proceedings of the 57th Annual Meeting of the Association for Computational Linguistics, pp. 6570–6580 (2019)
12. Johnson, A.E., et al.: MIMIC-CXR-JPG, a large publicly available database of labeled chest radiographs. arXiv preprint arXiv:1901.07042 (2019)
13. Larochelle, H., Erhan, D., Bengio, Y.: Zero-data learning of new tasks. In: AAAI, vol. 1, p. 3 (2008)
14. Lee, J., et al.: BioBERT: a pre-trained biomedical language representation model for biomedical text mining. Bioinformatics 36(4), 1234–1240 (2020)
15. Li, M., et al.: FFA-IR: towards an explainable and reliable medical report generation benchmark. In: Thirty-fifth Conference on Neural Information Processing Systems Datasets and Benchmarks Track (Round 2) (2021)
16. Lin, C.Y.: ROUGE: a package for automatic evaluation of summaries. In: Text Summarization Branches Out, pp. 74–81 (2004)
17. Liu, G., et al.: Clinically accurate chest X-ray report generation. In: Machine Learning for Healthcare Conference, pp. 249–269. PMLR (2019)
18. Papineni, K., Roukos, S., Ward, T., Zhu, W.J.: BLEU: a method for automatic evaluation of machine translation. In: Proceedings of the 40th Annual Meeting of the Association for Computational Linguistics, pp. 311–318 (2002)
19. Paszke, A., et al.: PyTorch: an imperative style, high-performance deep learning library. In: Advances in Neural Information Processing Systems, vol. 32 (2019)
20. Qi, H., Brown, M., Lowe, D.G.: Low-shot learning with imprinted weights. In: Proceedings of the IEEE Conference on Computer Vision and Pattern Recognition, pp. 5822–5830 (2018)
21. Ren, S., He, K., Girshick, R., Sun, J.: Faster R-CNN: towards real-time object detection with region proposal networks. In: Advances in Neural Information Processing Systems, vol. 28 (2015)
22. Vaswani, A., et al.: Attention is all you need. In: Advances in Neural Information Processing Systems, vol. 30 (2017)
23. Vedantam, R., Lawrence Zitnick, C., Parikh, D.: CIDEr: consensus-based image description evaluation. In: Proceedings of the IEEE Conference on Computer Vision and Pattern Recognition, pp. 4566–4575 (2015)

24. Wang, W., Zheng, V.W., Yu, H., Miao, C.: A survey of zero-shot learning: settings, methods, and applications. ACM Trans. Intell. Syst. Technol. (TIST) **10**(2), 1–37 (2019)

25. Xian, Y., Schiele, B., Akata, Z.: Zero-shot learning-the good, the bad and the ugly. In: Proceedings of the IEEE Conference on Computer Vision and Pattern Recognition, pp. 4582–4591 (2017)

26. Xue, Y., et al.: Multimodal recurrent model with attention for automated radiology report generation. In: Frangi, A.F., Schnabel, J.A., Davatzikos, C., Alberola-López, C., Fichtinger, G. (eds.) MICCAI 2018. LNCS, vol. 11070, pp. 457–466. Springer, Cham (2018). https://doi.org/10.1007/978-3-030-00928-1_52

27. Zhou, S., Niu, L., Si, J., Qian, C., Zhang, L.: Weak-shot semantic segmentation by transferring semantic affinity and boundary. arXiv preprint arXiv:2110.01519 (2021)

28. Zhou, Y., Wang, M., Liu, D., Hu, Z., Zhang, H.: More grounded image captioning by distilling image-text matching model. In: Proceedings of the IEEE/CVF Conference on Computer Vision and Pattern Recognition, pp. 4777–4786 (2020)

29. Zhu, P., Wang, H., Saligrama, V.: Zero shot detection. IEEE Trans. Circuits Syst. Video Technol. **30**(4), 998–1010 (2019)

Survival Prediction of Brain Cancer with Incomplete Radiology, Pathology, Genomic, and Demographic Data

Can Cui[1], Han Liu[1], Quan Liu[1], Ruining Deng[1], Zuhayr Asad[1], Yaohong Wang[2],

Shilin Zhao[2], Haichun Yang[2], Bennett A. Landman[1], and Yuankai Huo[1(✉)]

[1] Vanderbilt University, Nashville, TN 37235, USA
yuankai.huo@vanderbilt.edu
[2] Vanderbilt University Medical Center, Nashville, TN 37215, USA

Abstract. Integrating cross-department multi-modal data (e.g., radiology, pathology, genomic, and demographic data) is ubiquitous in brain cancer diagnosis and survival prediction. To date, such an integration is typically conducted by human physicians (and panels of experts), which can be subjective and semi-quantitative. Recent advances in multi-modal deep learning, however, have opened a door to leverage such a process in a more objective and quantitative manner. Unfortunately, the prior arts of using four modalities on brain cancer survival prediction are limited by a "complete modalities" setting (i.e., with all modalities available). Thus, there are still open questions on how to effectively predict brain cancer survival from incomplete radiology, pathology, genomic, and demographic data (e.g., one or more modalities might not be collected for a patient). For instance, should we use both complete and incomplete data, and more importantly, how do we use such data? To answer the preceding questions, we generalize the multi-modal learning on cross-department multi-modal data to a missing data setting. Our contribution is three-fold: 1) We introduce a multi-modal learning with missing data (MMD) pipeline with competitive performance and less hardware consumption; 2) We extend multi-modal learning on radiology, pathology, genomic, and demographic data into missing data scenarios; 3) A large-scale public dataset (with 962 patients) is collected to systematically evaluate glioma tumor survival prediction using four modalities. The proposed method improved the C-index of survival prediction from 0.7624 to 0.8053.

Keywords: Multi-modal learning · Survival prediction · Missing modalities

1 Introduction

Multi-modal learning tends to improve model performance by extracting and aggregating the information of the same subject from different modalities, as compared to learning with a single modality [7]. Medical domains are rich with multiple modalities, including image data such as pathology images, radiology images, and non-image

---

Supplementary Information The online version contains supplementary material available at https://doi.org/10.1007/978-3-031-16443-9_60.

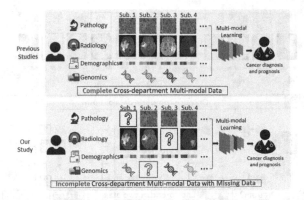

Fig. 1. Multi-modal learning with missing data for cancer survival prediction. This study generalizes multi-modal learning on radiology, pathology, genomic and demographic data from a "complete data" setting (upper panel) to a "missing data" scenario (lower panel), which is ubiquitous in routine practice.

data such as genomic data and demographic information, etc. An increasing number of works have shown that multi-modal learning helps to improve the performance of computer-aided diagnosis and prognosis in many medical applications [15, 24].

Gliomas are the most common primary malignant brain tumors. Diagnosis and prognosis of patients with this kind of tumor are beneficial for treatment decisions and patient management [23]. Some prior studies have already shown the prognostic advances of utilizing pathology images [19], MRI images [4], genomic [8] and clinical features [9] for the survival time prediction of glioma tumors. More recent studies have shown the promising results of integrating multi-modal data for glioma tumor prognosis [10,12,21,27]. Mobadersany et al. [21] concatenated genomic biomarkers and learned pathology image features. Chen et al. [12] proposed to aggregate features from genomic data, pathology images and cell graphs with tensor fusion [27]. More recently, Braman et al. [10] extended tensor fusion with orthogonal regularization to four modalities for more precise survival prediction.

However, the above studies only utilized the comprehensive radiology, pathology, genomic, and demographic data as a "complete data" setting (all modalities are available for every patient). However, complete data are not always guaranteed in routine practice patients (Fig. 1). Discarding patient data with missing modalities (1) extremely aggravates the data scarcity problem of medical applications and increases the risk of overfitting [3], and (2) limits the usage of the model trained by complete modalities in the inference phase because the prediction for data with complete modalities cannot be guaranteed. Cheerla et al. [11] used modality dropout to improve pan-cancer classification when the dataset had missing modalities. Ghosal et al. [14] reconstructed uni-modal features from the multi-modal representation to force the fusion model to keep all the information of uni-modalities. However, these methods did not leverage the available modalities for the modality missing problem, and they did not thoroughly compare the model performance between the training and testing data with and without modality missing, respectively.

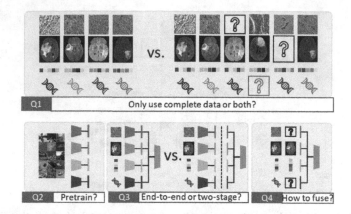

Fig. 2. Four questions to answer for this paper. The experimental answers of the four questions are presented in Sect. 5.

Therefore, there are still open questions on how to effectively learn from incomplete radiology, pathology, genomic, and demographic data (e.g., one or more modalities might not be collected for a patient) (Fig. 2). Specifically, 1) *Would more training data with incomplete modalities enhance the accuracy of prognosis prediction?* 2) *How can we effectively utilize the pre-trained feature extractors?* 3) *Which training strategy should we use, end-to-end or two-stage?* 4) *How should we fuse the multi-modal information under the missing data scenarios?*

In this paper, we propose an effective multi-modal learning with missing data (MMD) pipeline while addressing the above four key questions with comprehensive experimental analyses. The contribution of this paper is three-fold:

- The proposed MMD method is designed for using any combinations of available data as a two-stage plug-and-play method with computational efficiency.
- This study generalizes multi-modal learning on radiology, pathology, genomic, and demographic data into more clinically applicable missing data scenarios.
- The methodological development and data analyses are performed on a large-scale dataset by combining the glioma tumor data in TCGA [1,2], TCIA [13], and BraTs dataset [5,6,20].

2   Methods

The multi-modal fusion learning approach of the proposed MMD method, as well as the methods in prior arts, are presented in Fig. 3. The MMD pipeline is a two-stage framework that is designed for missing data scenarios, including the (1) uni-modal feature embedding, and (2) multi-modal information fusion.

2.1   Uni-Modal Feature Embedding

Low-dimensional features are obtained for each modality by training independent deep survival models using different neural networks.

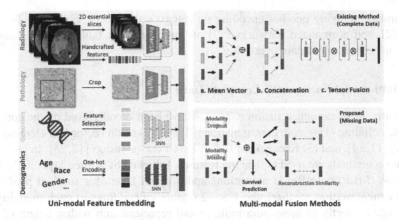

Fig. 3. The pipeline of proposed multi-modal learning model with missing modalities. Uni-modal deep neural networks are used to generate features of different modalities (left panel). Different fusion methods are shown in the right panel.

Radiology Modal. For registered 3D radiology MRI scans, the tumor and edema regions are automatically obtained by a trained nnUNet [16] segmentation model. The 2D essential slices (resized to 120×120) with the largest tumor and edema region from the four modalities are stacked as a four-channel input. To bridge the inconsistency between a pre-trained model and the new input dimensionality, the weights of the 1st channel of a pre-trained ResNet-18 are copied to the 4th channel as the weight initialization. In addition, 318 2D and 3D handcrafted features are extracted by PyRadiomics from tumor volumes to provide extra shape, location, and texture information. The handcrafted features are concatenated to the image features learned by the ResNet-18. Based on our experiments, adding handcrafted features improve the model performance as compared to using CNN features along.

Pathology Modal. Following the prior work [12], ImageNet pre-trained Vgg-16 is used to extract features from pathology images. Same as the radiology modality, only the last layers of the pre-trained network (ResNet and Vgg) is finetuned, as opposed to retraining the entire network. Each pathology image with the size of 1024×1024 is extracted from a tumor region [12,21]. An image patch with the size of 512×512 is extracted from each image in each training iteration for learning. In the testing phase, a sliding window with the size of 512×512 and a stride of 256 is used to crop each image into 9 patches. The average aggregation of the prediction of these 9 patches is the prediction for each pathology image sample.

Genomics and Demographics Modals. The genomic modality consists of 80 DNA features including 79 of the most expressive features from copy number variations (CNV) and one feature from mutation status (binary indication of mutation status for IDH1 gene, 0/1), same as the previous work [10]. For the demographic modality, we use four features including age, gender, race and ethnicity. They are extended to 9 features through applying one-hot encoding on categorical features. Following the prior work [12], self-normalized networks [17] are employed for both genomic data and demographic data to mitigate overfitting.

2.2 Multi-modal Fusion with Complete Data

In prior arts, three canonical fusion strategies have been broadly used on the complete datasets, including (1) feature concatenation [21,26], (2) mean vector (i.e. element-wise average) [11,14], and (3) Kronecker product based tensor fusion [12,25]. In our study, such fusion methods are used to fuse the features, followed by a three-layer multi-layer perceptron (MLP) with ReLU activations and dropout layers for survival prediction. For the **concatenation method**, the 1×32 feature embeddings of each modality are concatenated directly to generate a multi-modal representation with a length of 128. For the **mean vector fusion**, the uni-modal embeddings are firstly extended to 128 dimensional vectors with a modal-specific two-layer perceptron, and a reverse network structure is used to reconstruct the uni-modal embeddings from the mean vector. For the **pathomic fusion method**, we directly extend the tensor fusion with gated attention method proposed by Chen et al. [12] from three to four uni-modal 1×32 embeddings.

2.3 Multi-modal Fusion for Missing Data

In the missing data setting, we extend multi-modal fusion with the MMD method, which fuses the feature embeddings from all available modalities. Among the above three canonical fusion strategies, we employ the mean vector (element-wise average of available embeddings) as the backbone fusion method since its "order-less" nature leads to larger flexibility in using missing data. To further boost fusion capability on missing data, we introduce modality dropout [22] by randomly discarding modalities for training. Only the rest of uni-modal embeddings are calculated for the mean vector.

To encourage the model to learn cross-modality information, modality reconstruction is used in the MMD method to fit a complete multi-modal representation from all available modalities. Moreover, different from previous work [14], for the modalities that are intentionally dropped out, we ask the MMD model to reconstruct such modalities as well, since the original modalities are actually known.

The loss function to supervise the proposed method consists of two terms. The first term is a cox loss that is defined as:

$$\mathcal{L}_{cox} = -\sum_{i:E_i=1} (F_\theta(h_i) - \log \sum_{j:t_i \geq t_j} e^{F_\theta(h_j)}), \tag{1}$$

where the values t_i, E_i, h_i are survival time, censor status and multi-modal representations of sample i. If the death is observed, censor status $E_i = 1$. F_θ denotes the networks which predict the survival hazard (risk of death) from h_i. The second term is the reconstruction loss:

$$\mathcal{L}_{recon} = \frac{1}{\sum_i^N |\alpha_i = 1|} \sum_i^N (\sum_v^V \alpha_i^v \parallel x_i^v, \widetilde{x}_i^v \parallel), \tag{2}$$

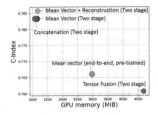

Data type	Number of patients
Pathology	769
Radiology	376
Genomics	697
Demographics	803
Patient with complete modalities	170
Patient with survival label	962

Fig. 4. The left panel shows the GPU memory consumption and c-index of different methods (with the best performance). The right panel shows the number of patients in different modalities.

where x_i^v is the embedding of modality $v \in V$ of sample $i \in N$. $\widetilde{x}_i^v$ is the reconstructed embedding of modality v and α_i^v indicates the original availability of the modality v of subject i. $\alpha_i^v = 0$ when this modality is missing and $|\alpha_i > 0|$ represents the number of available modalities for sample i. $L_2 \ norm$ is used to calculate the difference between the reconstructed uni-modal embeddings and the corresponding real uni-modal embeddings. The overall loss becomes:

$$\mathcal{L}_{total} = \mathcal{L}_{cox} + \lambda \mathcal{L}_{recon} \tag{3}$$

where a weight $\lambda = 1$ is empirically selected to balance two losses.

3   Data and Experimental Setting

We collected a large-scale dataset with both complete and incomplete cross-department glioma tumor data (1698 samples from 962 patients in total) by combining TCGA [1,2], TCIA [13], and the BraTs dataset [5,6,20]. Each patient had at least one modality available. All patients had survival time and censor status for the survival prediction task. The detailed data splits are presented in Fig. 4. Following the train-test splits of previous works [12,21], we enlarged the training set with more patients from the BraTs dataset and used the subset of original testing sets with four modalities available. In the routine clinical situation, the acquisition of pathology and genomic modalities was more invasive to achieve than the non-invasive radiology and demographic data. Thus, we also evaluated scenarios in which pathology and genomic data were missing (Table 3). Each modality was randomly dropped out (set to zero) by an empirically selected rate 0.5 until only one modality was available for a sample. For the end-to-end fusion strategy, the network weights pre-trained by ImageNet or the uni-modal models were loaded, and the original modality data were used as inputs. The uni-modal models and the fusion network were trained together. As for the two-stage strategy, 32-dimensional uni-modal embeddings prepared by trained uni-modal models were used as inputs to train the fusion network only. Specifically, a 512×512 patch was randomly cropped from each pathology image for an embedding in training. To evaluate the performance of the survival prediction, the concordance index (c-index) value was used. The average aggregation of multiple patches from multiple samples of each patient is calculated as the final prediction. Intensity normalized images were augmented by rotation, flipping,

Table 1. Survivals prediction performance with benchmarks

Methods	C-Index
Pathomic fusion [12]	0.7697 ± 0.047
Deep orthogonal fusion [10]	0.7624 ± 0.042
MMD (ours)	**0.8053 ± 0.038**

Table 2. Survival prediction results of different uni-modal model training strategies and fusion strategies using all data or complete data. "C" means training using complete data while "M" means data with missing modalities. Same layers were froze in the training of multi-modal fusion as the uni-modal modals. * Only the weights pre-trained by ImageNet were loaded. ** Weights of uni-modal networks were loaded for finetuning.

Modality	Training strategy	Training data	C-index
Uni-modal (radiology)	Finetune entire network	C+M	0.6957 ± 0.043
Uni-modal (pathology)	Finetune entire network	C+M	0.6803 ± 0.008
Uni-modal (radiology)	Finetune last layers	C+M	**0.7062 ± 0.039**
Uni-modal (pathology)	Finetune last layers	C+M	**0.7319 ± 0.026**
Multi-modal (mean vector)	End-to-end (from scratch*)	C	0.7263 ± 0.027
Multi-modal (mean vector)	End-to-end (from scratch*)	C+M	0.7609 ± 0.016
Multi-modal (mean vector)	End-to-end (finetune**)	C	0.7580 ± 0.030
Multi-modal (mean vector)	End-to-end (finetune**)	C+M	0.7607 ± 0.021
Multi-modal (mean vector)	Two-stage	C	0.7571 ± 0.034
Multi-modal (mean vector)	Two-stage	C+M	**0.7717 ± 0.034**

and color jitter methods in training. The batch size was 64 for uni-modal training, with a learning rate of 0.0005 for images and 0.002 for other data. The batch size of the fusion was set to 8 with a learning rate of 0.0002. All the experiments were run on a 16 GB NVIDIA GPU. The code is available at: https://github.com/cuicathy/MMD_SurvivalPrediction.git

4   Results

Table 1 compared our proposed pipeline with prior benchmarks [10, 12]. Our proposed MMD method achieved superior performance compared with benchmarks. The performance on joint testing sets of pathomic fusion [12] and our work are used for a fair comparison of different methods. The best performance of their released model (the "pathgraphomic" network with attention gates trained by mRNA, DNA and pathology images) was compared with our best model (all data + modality dropout + reconstruction + mean vector) on this subset. We rebuilt the Deep orthogonal fusion following the papers [10, 18] (e.g., uni-modal embeddings were compressed to 8 dimension by a linear layer for tensor fusion), but the four uni-modal structures and features mentioned above trained by all available uni-modal data were used for a fair comparison.

Table 3. Survival prediction results (C-index) of different training strategies

Training					Testing		
Fusion method	Data for uni-modal embedding	Data for multi-modal fusion	Drop-out	Recon	Complete modalities	Pathology missing	Gene and pathology missing
Concatenation	C+M	C			0.7659 ± 0.032	0.7563 ± 0.035	0.7169 ± 0.035
	C+M	C+M			0.7743 ± 0.034	0.7647 ± 0.030	0.7278 ± 0.037
	C+M	C	✓		0.7833 ± 0.030	0.7725 ± 0.031	0.7275 ± 0.037
	C+M	C+M	✓		0.7817 ± 0.029	0.7668 ± 0.032	0.7352 ± 0.034
Tensor fusion	C+M	C			0.7503 ± 0.032	0.7274 ± 0.030	0.6926 ± 0.045
	C+M	C+M			0.7483 ± 0.026	0.7328 ± 0.036	0.6926 ± 0.044
	C+M	C	✓		0.7513 ± 0.035	0.7386 ± 0.035	0.7187 ± 0.040
	C+M	C+M	✓		0.7660 ± 0.031	0.7314 ± 0.034	0.7117 ± 0.035
Mean vector	C	C			0.7597 ± 0.029	0.7464 ± 0.032	0.7013 ± 0.038
	C+M	C			0.7571 ± 0.034	0.7495 ± 0.036	0.7118 ± 0.037
	C+M	C+M			0.7717 ± 0.034	0.7622 ± 0.034	0.7321 ± 0.036
	C	C	✓		0.7740 ± 0.026	0.7644 ± 0.029	0.7081 ± 0.032
	C+M	C	✓		0.7821 ± 0.029	0.7683 ± 0.032	0.7223 ± 0.037
	C+M	C+M	✓		0.7833 ± 0.030	0.7702 ± 0.032	0.7373 ± 0.033
	C+M	C		✓	0.7660 ± 0.029	0.7576 ± 0.032	0.7131 ± 0.034
	C+M	C+M		✓	0.7779 ± 0.032	0.7708 ± 0.033	0.7255 ± 0.037
	C+M	C	✓	✓	0.7812 ± 0.027	0.7609 ± 0.030	0.7164 ± 0.033
	C+M	C+M	✓	✓	**0.7857 ± 0.026**	**0.7808 ± 0.032**	**0.7451 ± 0.034**

* "C" means training using complete data while "M" means missing data. The best performances among all settings are highlighted in **bold**, while the best performances of a certain fusion method (without reconstruction loss) are highlighted by underline.

Results of different training strategies for uni-modal learning and multi-modal fusion are in Table 2. For uni-modal training, only finetuning the last layer of the ImageNet pre-trained network achieved better uni-modal survival prediction than finetuning the whole pre-trained network. When using multi-modal data, two-stage training with fixed uni-modal embeddings is competitive compared with the end-to-end training. Table 3 shows the results of three different multi-modal fusion methods trained by complete data ("C" in the table) or all data ("C+M" in the table), with or without modality dropout. For the mean vector method, the modality dropout, reconstruction, and training with all data consistently improved the performance on testing sets.

5 Ablation Studies for Answering Four Questions

To answer the four questions in Fig. 2, we conducted comprehensive studies as shown in Table 2 and 3.

Q1: Would more training data with incomplete modalities enhance the accuracy of prognosis prediction? Table 2 and 3 results indicated that using both complete and incomplete modalities can improve the c-index for survival prediction.

Q2: How do we effectively utilize the pre-trained feature extractors? Fig. 4 and Table 2 indicated that only finetuning the last layer yields better performance compared with finetuning the entire network in our task.

Q3: Which training strategy should we use, end-to-end or two-stage? Table 2 indicated that the two-stage learning strategy yielded a competitive performance compared with the end-to-end counterparts, and it is memory efficient for more flexibility.

Q4: How should we fuse multi-modal information under the missing data scenarios? The comprehensive ablation studies of different fusion strategies have been provided in Table 3. The proposed MMD approach achieved superior performance among different strategies.

6 Conclusion

In this paper, we generalize the cross-department multi-modal data (radiology, pathology, genomic, and demographic data) deep survival learning from a complete data setting to the more clinically applicable missing data scenarios. First, we presented the MMD method, a multi-modal prognosis pipeline with an effective and efficient design of using both modality complete and incomplete data. Second, with a relatively large-scale cohort, we investigated four key questions of utilizing modality missing data for the brain cancer survival analysis. Future work might include applying this method on external datasets and exploring other uni-modal embedding methods. We hope this study could also be used as a reference for future studies in predicting brain cancer prognosis.

Acknowledgements. This work is supported by the Leona M. and Harry B. Helmsley Charitable Trust grant G-1903-03793, NSF CAREER 1452485. This work is in part based upon data generated by the TCGA Research Network: https://www-cancer-gov.proxy.library.vanderbilt.edu/tcga.

References

1. Pedano, N., et al: Radiology data from the cancer genome atlas low grade glioma [tcga-lgg] collection. Cancer Imaging Arch. (2016). https://doi.org/10.7937/K9/TCIA.2016. L4LTD3TK
2. Scarpace, L., et al: Radiology data from the cancer genome atlas glioblastoma multiforme [tcga-gbm] collection. Cancer Imaging Arch. (2016). https://doi.org/10.7937/K9/TCIA.2016.RNYFUYE9
3. Bach, F.: Breaking the curse of dimensionality with convex neural networks. J. Mach. Learn. Res. **18**(1), 629–681 (2017)
4. Bae, S., et al.: Radiomic MRI phenotyping of glioblastoma: improving survival prediction. Radiology **289**(3), 797–806 (2018)
5. Bakas, S., et al.: Advancing The Cancer Genome atlas glioma MRI collections with expert segmentation labels and radiomic features. Sci. Data **4**(1), 1–13 (2017)
6. Bakas, S., et al.: Identifying the best machine learning algorithms for brain tumor segmentation, progression assessment, and overall survival prediction in the brats challenge. arXiv preprint arXiv:1811.02629 (2018)
7. Baltrušaitis, T., Ahuja, C., Morency, L.P.: Multimodal machine learning: a survey and taxonomy. IEEE Trans. Pattern Anal. Mach. Intell. **41**(2), 423–443 (2018)

8. Beig, N., et al.: Radiogenomic-based survival risk stratification of tumor habitat on Gd-T1w MRI is associated with biological processes in glioblastoma. Clin. Cancer Res. **26**(8), 1866–1876 (2020)

9. Beig, N., et al.: Sexually dimorphic radiogenomic models identify distinct imaging and biological pathways that are prognostic of overall survival in glioblastoma. Neuro Oncol. **23**(2), 251–263 (2021)

10. Braman, N., Gordon, J.W.H., Goossens, E.T., Willis, C., Stumpe, M.C., Venkataraman, J.: Deep orthogonal fusion: multimodal prognostic biomarker discovery integrating radiology, pathology, genomic, and clinical data. In: de Bruijne, M., et al. (eds.) MICCAI 2021. LNCS, vol. 12905, pp. 667–677. Springer, Cham (2021). https://doi.org/10.1007/978-3-030-87240-3_64

11. Cheerla, A., Gevaert, O.: Deep learning with multimodal representation for pancancer prognosis prediction. Bioinformatics **35**(14), i446–i454 (2019)

12. Chen, R.J., et al.: Pathomic fusion: an integrated framework for fusing histopathology and genomic features for cancer diagnosis and prognosis. IEEE Trans. Med. Imaging, 757–770 (2020)

13. Clark, K., et al.: The Cancer Imaging Archive (TCIA): maintaining and operating a public information repository. J. Digit. Imaging **26**(6), 1045–1057 (2013). https://doi.org/10.1007/s10278-013-9622-7

14. Ghosal, S., et al.: G-MIND: an end-to-end multimodal imaging-genetics framework for biomarker identification and disease classification. In: Medical Imaging 2021: Image Processing, vol. 11596, p. 115960C. International Society for Optics and Photonics (2021)

15. Huang, S.C., Pareek, A., Seyyedi, S., Banerjee, I., Lungren, M.P.: Fusion of medical imaging and electronic health records using deep learning: a systematic review and implementation guidelines. NPJ Digit. Med. **3**(1), 1–9 (2020)

16. Isensee, F., Petersen, J., Kohl, S.A., Jäger, P.F., Maier-Hein, K.H.: nnU-Net: breaking the spell on successful medical image segmentation, vol. 1, pp. 1–8. arXiv preprint arXiv:1904.08128 (2019)

17. Klambauer, G., Unterthiner, T., Mayr, A., Hochreiter, S.: Self-normalizing neural networks. In: Advances in Neural Information Processing Systems, vol. 30 (2017)

18. Lezama, J., Qiu, Q., Musé, P., Sapiro, G.: OLE: orthogonal low-rank embedding-a plug and play geometric loss for deep learning. In: Proceedings of the IEEE Conference on Computer Vision and Pattern Recognition, pp. 8109–8118 (2018)

19. Louis, D.N., et al.: The 2016 world health organization classification of tumors of the central nervous system: a summary. Acta Neuropathol. **131**(6), 803–820 (2016)

20. Menze, B.H., et al.: The multimodal brain tumor image segmentation benchmark (BRATS). IEEE Trans. Med. Imaging **34**(10), 1993–2024 (2014)

21. Mobadersany, P., et al.: Predicting cancer outcomes from histology and genomics using convolutional networks. Proc. Natl. Acad. Sci. **115**(13), E2970–E2979 (2018)

22. Neverova, N., Wolf, C., Taylor, G., Nebout, F.: ModDrop: adaptive multi-modal gesture recognition. IEEE Trans. Pattern Anal. Mach. Intell. **38**(8), 1692–1706 (2015)

23. Pereira, S., Pinto, A., Alves, V., Silva, C.A.: Brain tumor segmentation using convolutional neural networks in MRI images. IEEE Trans. Med. Imaging **35**(5), 1240–1251 (2016)

24. Schneider, L., et al.: Integration of deep learning-based image analysis and genomic data in cancer pathology: a systematic review. Eur. J. Cancer **160**, 80–91 (2022)

25. Wang, Z., Li, R., Wang, M., Li, A.: GPDBN: deep bilinear network integrating both genomic data and pathological images for breast cancer prognosis prediction. Bioinformatics **37**(18), 2963–2970 (2021)

26. Yap, J., Yolland, W., Tschandl, P.: Multimodal skin lesion classification using deep learning. Exp. Dermatol. **27**(11), 1261–1267 (2018)

27. Zadeh, A., Chen, M., Poria, S., Cambria, E., Morency, L.P.: Tensor fusion network for multimodal sentiment analysis. arXiv preprint arXiv:1707.07250 (2017)

Discrepancy and Gradient-Guided Multi-modal Knowledge Distillation for Pathological Glioma Grading

Xiaohan Xing[1], Zhen Chen[1], Meilu Zhu[1], Yuenan Hou[2], Zhifan Gao[3],
and Yixuan Yuan[1(✉)]

[1] Department of Electrical Engineering, City University of Hong Kong,
Kowloon, Hong Kong, China
yxyuan.ee@cityu.edu.hk

[2] Shanghai Artificial Intelligence Laboratory, Shanghai, China

[3] School of Biomedical Engineering, Sun Yat-sen University, Guangdong, China

Abstract. The fusion of multi-modal data, e.g., pathology slides and genomic profiles, can provide complementary information and benefit glioma grading. However, genomic profiles are difficult to obtain due to the high costs and technical challenges, thus limiting the clinical applications of multi-modal diagnosis. In this work, we address the clinically relevant problem where paired pathology-genomic data are available during training, while only pathology slides are accessible for inference. To improve the performance of pathological grading models, we present a discrepancy and gradient-guided distillation framework to transfer the privileged knowledge from the multi-modal teacher to the pathology student. For the teacher side, to prepare useful knowledge, we propose a *Discrepancy-induced Contrastive Distillation* (DC-Distill) module that explores reliable contrastive samples with teacher-student discrepancy to regulate the feature distribution of the student. For the student side, as the teacher may include incorrect information, we propose a *Gradient-guided Knowledge Refinement* (GK-Refine) module that builds a knowledge bank and adaptively absorbs the reliable knowledge according to their agreement in the gradient space. Experiments on the TCGA GBM-LGG dataset show that our proposed distillation framework improves the pathological glioma grading significantly and outperforms other KD methods. Notably, with the sole pathology slides, our method achieves comparable performance with existing multi-modal methods. The code is available at https://github.com/CityU-AIM-Group/MultiModal-learning.

Keywords: Knowledge distillation · Missing modality · Glioma grading

1 Introduction

As the most common brain tumors, gliomas can be classified into Grade II-IV by the World Health Organization (WHO) according to their malignancy

[1]. Gliomas across different grades exhibit huge heterogeneity in cell proliferation, mutation of genomic biomarkers, and morphological characteristics of the pathology slides, resulting in various treatment resistance and survival rate [2,3]. Therefore, genomic profiles and pathology slides are widely studied for glioma grading and survival prediction [2,4,5]. Recent studies reveal that multi-modal approaches fusing pathology and genomic data can provide more comprehensive information for glioma grading and prognosis, significantly outperforming the unimodal counterparts [6–9].

Although fusing multi-modal information improves the grading performance, its clinical application is limited since paired pathology-genomic data is not always available in real clinical scenarios. Compared with the easily accessible pathology slides, genomic profiles are difficult to obtain due to the high costs and technical challenges [10]. Since multi-modal training data can be collected from public datasets, this work addresses the clinically relevant problem where paired pathology-genomic data are available during training, while only pathology slides are accessible for inference. To improve the test performance of pathological grading models, it is desirable to explore and imitate the privileged pathology-genomic knowledge during training.

Among solutions to the missing modality problems in other research fields, knowledge distillation (KD) [11] significantly improves the unimodal performance by distilling knowledge from the multi-modal teacher [12–14]. Despite these achievements, there are two crucial issues that have been ignored by existing cross-modal KD methods. (1) To ensure efficient teaching, the teacher should prepare the knowledge according to the needs of the student. The reliable knowledge with discrepancy between the teacher and student is more likely to carry crucial information and improve the student performance [15,16]. (2) Despite the superior performance of the multi-modal teacher, it may include incorrect information (e.g., glioma grading accuracy of current multi-modal methods is less than 80% [6,7]), thus mimicking the teacher's knowledge without distinction might mislead the student [17,18]. In this regard, the student should hold a dialectic perspective and adaptively absorb the teacher's beneficial knowledge while shielding the misleading information.

To tackle the aforementioned issues, we propose a discrepancy and gradient-guided distillation framework to transfer the pathology-genomic knowledge towards more accurate glioma grading with pathology slides. (1) For the teacher side, we propose a *Discrepancy-induced Contrastive Distillation* (DC-Distill) module to prepare knowledge according to the needs of the student. Specifically, built upon the contrastive distillation paradigm, DC-Distill selects the reliable contrastive samples with teacher-student discrepancy to regulate the feature distribution in the student model. (2) For the student side, we propose a *Gradient-guided Knowledge Refinement* (GK-Refine) scheme to selectively absorb the beneficial knowledge from the multi-modal teacher. Specifically, GK-Refine constructs a knowledge bank to provide complementary guidance and adaptively refines multiple knowledge according to their agreement in the

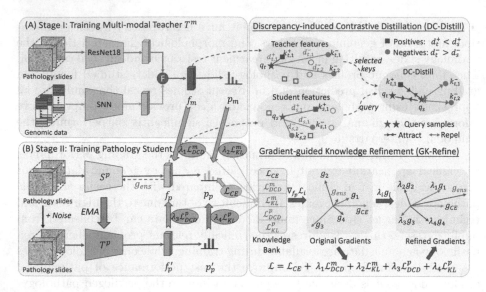

Fig. 1. The framework of our method. (A) In stage I, a multi-modal teacher T^m is trained with pathology slides and genomic data as inputs. (B) In stage II, a pathology student S^p distills knowledge from the multi-modal teacher T^m and the pathology mean-teacher T^p via our proposed *Discrepancy-induced Contrastive Distillation* (DC-Distill) and *Gradient-guided Knowledge Refinement* (GK-Refine) modules.

gradient space. These two modules enable more effective distillation, leading to a more accurate pathology student. The main contributions are summarized as:

- We present the first work that studies the clinically relevant problem of distilling pathology-genomic knowledge to benefit the glioma grading with pathology slides.
- We propose a novel discrepancy and gradient-guided distillation framework, including a DC-Distill module to ensure reliable knowledge preparation from the multi-modal teacher, and a GK-Refine scheme to enable adaptive knowledge absorption for the pathology student model.
- Extensive experiments show that our proposed distillation framework improves the pathological grading remarkably and outperforms state-of-the-art KD methods. With the sole pathology slides, our method achieves comparable glioma grading performance with existing multi-modal methods.

2   Method

An overview of our proposed framework is shown in Fig. 1. In stage I, features extracted from the pathology slides and genomic data are fused via Kronecker product [19] to train a multi-modal teacher T^m. The predictions of each branch (i.e., pathology, genomic, and their combination) are supervised by the cross-entropy loss and mean-teacher supervision [20]. In stage II, a pathology student

S^p is trained by distilling knowledge via the commonly used KL divergence (i.e., $\mathcal{L}_{KL}^m$ and $\mathcal{L}_{KL}^p$) and our proposed *Discrepancy-induced Contrastive Distillation* (DC-Distill with $\mathcal{L}_{DCD}^m$ and $\mathcal{L}_{DCD}^p$) from the fixed multi-modal teacher T^m and the pathology mean-teacher T^p. To maximally absorb the teachers' beneficial information while shielding misleading knowledge, we propose a *Gradient-guided Knowledge Refinement* (GK-Refine) scheme that constructs a knowledge bank with $[\mathcal{L}_{CE}, \mathcal{L}_{DCD}^m, \mathcal{L}_{KL}^m, \mathcal{L}_{DCD}^p, \mathcal{L}_{KL}^p]$ and refines the knowledge according to their agreement in the gradient space. During inference, the pathology model S^p performs glioma grading with the solely available pathology slides.

2.1   Discrepancy-Induced Contrastive Distillation

Previous studies [15,16] indicate that informative samples with teacher-student discrepancy carry important information on why the student lags behind the teacher and enable effective knowledge transfer. Besides, the reliability of the teacher's knowledge is another important issue that affects the KD performance [21]. For effective knowledge distillation from the multi-modal teacher to the pathology student, we propose a *Discrepancy-induced Contrastive Distillation* (DC-Distill) module that allows the teacher to adaptively transfer the informative and reliable knowledge according to the needs of the student.

Considering the promising performance of contrastive KD [22], our DC-Distill utilizes contrastive samples from the teacher to regulate the feature distribution of the student model. Different from the existing contrastive KD method [22] with a single positive key and randomly sampled negative keys, our DC-Distill selects k_P positive keys and k_N negative keys for each query sample according to the aforementioned informative and reliable criteria. With the guidance of class label y, we define the positive candidates for each query q as $\{k^+ : \forall y_{k^+} = y_q\}$ and negative candidates as $\{k^- : \forall y_{k^-} \neq y_q\}$. As illustrated in Fig. 1, in the teacher model, the distance between the positive k_t^+ and the query feature q_t is denoted as d_t^+. For the student model, the distance between the positive k_s^+ and the query feature q_s is denoted as d_s^+. Then, we define the effectiveness of all positive candidates k_t^+ for the query q_s as:

$$E_{q_s,k_t^+} = d_s^+ - d_t^+ = (1 - \frac{q_s \cdot k_s^+}{||q_s|| \times ||k_s^+||}) - (1 - \frac{q_t \cdot k_t^+}{||q_t|| \times ||k_t^+||}), \quad \forall y_{k^+} = y_q, \quad (1)$$

where $\cdot$ is inner product and $||\cdot||$ denotes l_2 norm. Positive E_{q_s,k_t^+} (i.e., $d_t^+ < d_s^+$) indicates the (q, k^+) is more similar in the teacher model than the student model. Since samples from the same class should ideally be similar, thus the sample k_t^+ is regarded as a reliable positive key for q_s and is more likely to convey beneficial intra-class regulation to the student. Therefore, for each query sample q_s in the student model, k_P positives (i.e., k_t^+) with $E_{q_s,k_t^+} > 0$ are selected from the multi-modal teacher to compute the contrastive loss. Note that q_t is naturally regarded as a positive key for q_s since the features of the same sample should be consistent between the teacher and student. Similarly, the effectiveness of all negative candidates k_t^- for the query q_s is defined as:

$$E_{q_s,k_t^-} = d_t^- - d_s^- = (1 - \frac{q_t \cdot k_t^-}{||q_t|| \times ||k_t^-||}) - (1 - \frac{q_s \cdot k_s^-}{||q_s|| \times ||k_s^-||}), \quad \forall y_{k^-} \neq y_q, \quad (2)$$

where positive E_{q_s,k_t^-} (i.e., $d_t^- > d_s^-$) indicates the (q, k^-) is more separable in the teacher model than the student, thus the sample k_t^- is reliable to convey beneficial inter-class regulation to the student. Therefore, we select top k_N negatives (i.e., k_t^-) with the highest E_{q_s,k_t^-} for each query q_s for contrastive loss computation.

For the computation of contrastive KD loss, we follow the common practice in contrastive learning to transform the features to an embedding space via a two-layer MLP followed by L2 normalization [23]. Inspired by the mutual information estimator in [22,24], we define the DC-Distill loss for each query sample q_s as:

$$\mathcal{L}_{DCD}^m = -\frac{1}{k_P} \sum_{k_t^+} (\log \frac{e^{(q_s \cdot k_t^+/\tau)}}{e^{(q_s \cdot k_t^+/\tau)} + \frac{k_N}{M}} + \sum_{k_t^-} \log(1 - \frac{e^{(q_s \cdot k_t^-/\tau)}}{e^{(q_s \cdot k_t^-/\tau)} + \frac{k_N}{M}})), \quad (3)$$

where τ is the temperature that controls the concentration level, and M is the cardinality of the dataset. k_t^+ and k_t^- are the selected positives and negatives according to Eq. 1 and Eq. 2. The DC-Distill loss for the q_t is defined in a similar way to train the feature transformation layers in the teacher model. With the regularization of $\mathcal{L}_{DCD}^m$, the query feature q_s from the pathology student is pulled closer to the k_P positive keys k_t^+ and pushed apart from the k_N negative keys k_t^- in the multi-modal teacher, thus the feature distribution of the student will be better aligned with the teacher.

Compared with the existing contrastive KD method [22] with randomly selected contrastive samples, the proposed DC-Distill explores the informative and reliable contrastive samples based on the teacher-student discrepancy, allowing the multi-modal teacher to adaptively transfer the most useful knowledge to the pathology student.

2.2   Gradient-Guided Knowledge Refinement

Although the multi-modal teacher achieves superior performance than the pathology student, it may still include incorrect information (e.g., glioma grading accuracy is less than 80% [6,7]) and mislead the student [17,18]. Preceding studies employ the ensemble of multiple teachers to provide complementary knowledge and alleviate the misleading guidance [18,25]. However, the knowledge ensemble might be dominated by unreliable knowledge and result in suboptimal distillation [18]. To tackle this dilemma, we propose a *Gradient-guided Knowledge Refinement* (GK-Refine) module that constructs a knowledge bank to provide complementary guidance and allows the student to selectively absorb beneficial knowledge while shielding the misleading information.

To provide complementary knowledge, the GK-Refine introduces another pathology mean-teacher T^p [20], which allows the retrospection of previous lessons and provides robust intra-modal knowledge. Similar to the multi-modal teacher, the pathology mean-teacher guides the student via KL divergence $\mathcal{L}_{KL}^p$ and DC-distill loss $\mathcal{L}_{DCD}^p$ defined in a similar way as in Eq. 3. Then, we construct a knowledge bank for the pathology student, including the knowledge supervised by ground-truth labels (i.e., $\mathcal{L}_{CE}$), the privileged knowledge from the multi-modal teacher (i.e., $\mathcal{L}_{DCD}^m$, $\mathcal{L}_{KL}^m$), and the retrospective knowledge from the pathology mean-teacher (i.e., $\mathcal{L}_{DCD}^p$, $\mathcal{L}_{KL}^p$). Each knowledge would impose a gradient on the pathology student, resulting in the gradient space as $G : [g_{CE} = \nabla_{f_p}\mathcal{L}_{CE}, \ g_1 = \nabla_{f_p}\mathcal{L}_{DCD}^m, \ g_2 = \nabla_{f_p}\mathcal{L}_{KL}^m, \ g_3 = \nabla_{f_p}\mathcal{L}_{DCD}^p, \ g_4 = \nabla_{f_p}\mathcal{L}_{KL}^p]$, where f_p denotes the feature of the pathology student model. As shown in Fig. 1, the ensemble gradient g_{ens} is decided by the gradients of all knowledge. To prevent the g_{ens} from being dominated by unreliable or misleading knowledge, our proposed GK-Refine module refines the knowledge ensemble process by modulating the contributions of different knowledge according to their reliability in the gradient space. Specifically, for the i-th knowledge, we compute the cosine similarity between its gradient g_i and the gradients of all knowledge. Then, we assign the weight λ_i for the i-th knowledge as:

$$\lambda_i = \sum_{g_j \in G} \frac{g_i \cdot g_j}{||g_i|| \times ||g_j||}. \tag{4}$$

Higher λ_i indicates the i-th knowledge shows higher agreement with other knowledge, thus is regarded as more reliable and should make more contribution to the student training [18]. On the contrary, the unreliable knowledge will be assigned with a smaller λ_i due to its contradiction with other knowledge. The overall training loss for the pathology student model is defined as:

$$\mathcal{L} = \mathcal{L}_{CE} + \lambda_1 \cdot \mathcal{L}_{DCD}^m + \lambda_2 \cdot \mathcal{L}_{KL}^m + \lambda_3 \cdot \mathcal{L}_{DCD}^p + \lambda_4 \cdot \mathcal{L}_{KL}^p. \tag{5}$$

By modulating the weights of multiple knowledge, their contributions to the student training are recalibrated based on their reliability. As shown in Fig. 1, after the refinement, the gradients from reliable knowledge are strengthened while the gradients of unreliable knowledge are reduced, thus leading to a better gradient descending direction g_{ens} which benefits the student learning and convergence.

3 Experiments

3.1 Dataset and Implementation Details

Dataset: We evaluated our method on the data obtained from the TCGA-GBM and TCGA-LGG project [26], which contains paired samples of pathology slides and genomic profiles. There are a total of 736 patients with WHO grading labels, including 182 grade II, 205 grade III, and 350 grade IV. Following [6,7], we utilized the curated region-of-interests (ROIs) with size 1024×1024 as inputs of the pathology modality. Since each patient has 1–3 ROIs curated from the pathology

Table 1. The glioma grading performance of our method, **pathological** grading with state-of-the-art knowledge distillation methods, and existing **multi-modal** approaches. Average and standard deviation over 5-fold cross validation are reported. The highest pathological and multi-modal performance is highlighted in **bold**.

Modality	Method	AP (%)	Accuracy (%)	AUC (%)
Pathology	Baseline (w/o KD)	82.28 ± 2.38	73.83 ± 1.88	90.16 ± 1.19
	Hinton et al. [11]	84.17 ± 0.91	74.48 ± 1.06	91.25 ± 0.49
	Passalis et al. [30]	84.32 ± 0.57	74.91 ± 1.08	91.40 ± 0.26
	Hu et al. [13]	84.83 ± 0.83	74.77 ± 1.41	91.61 ± 0.41
	Tung et al. [31]	84.77 ± 0.74	75.01 ± 0.69	91.62 ± 0.38
	Park et al. [32]	84.88 ± 0.93	74.58 ± 1.64	91.64 ± 0.57
	Tian et al. [22]	84.94 ± 1.05	75.24 ± 1.54	91.69 ± 0.59
	Zhu et al. [33]	84.51 ± 0.66	74.44 ± 1.15	91.41 ± 0.39
	DC-Distill	85.25 ± 1.55	75.81 ± 0.86	91.82 ± 0.79
	GK-Refine	85.27 ± 1.27	75.80 ± 0.86	91.89 ± 0.62
	GK-Refine$^-$	84.65 ± 0.99	74.81 ± 1.31	91.53 ± 0.49
	Our method	**86.12 ± 0.99**	**76.78 ± 0.70**	**92.35 ± 0.45**
Multi-modal	Mobadersany et al. [6]	84.40 ± 1.83	74.40 ± 2.09	91.18 ± 0.85
	Chen et al. [7]	85.51 ± 2.53	75.24 ± 2.07	92.10 ± 1.30
	Upper bound	**86.54 ± 1.39**	**77.60 ± 2.04**	**92.80 ± 0.82**

slides, there were 1325 ROI images utilized in our experiment. To maintain the stability of the training process, the ROI images were augmented through random cropping, color jittering, and flipping. For each patient, 80 genomic features were utilized as inputs of the genomic modality, including gene CNV (79) and mutation status (1) obtained from the cBioPortal [27].

We performed 5-fold cross-validation and reported the average test performance over these five folds. The model performance is measured by Average Precision (AP), Accuracy, and Area Under the Curve (AUC).

Implementation: Our method was implemented with the PyTorch library [28]. We employed the ResNet-18 [29] and self-normalizing network (SNN) [7] as the backbone of the pathology and genomic modality, respectively. The network was trained on a NVIDIA RTX 2080ti GPU with the batch size as 8. Adam with $\beta_1 = 0.9$ and $\beta_2 = 0.999$ was used for network optimization. Both the multi-modal teacher and pathology student were trained for 30 epochs. The initial learning rate was set to 0.0005 and linearly decayed to 0 during training. The temperature τ was set as 0.07. For each query sample, the number of positive pairs k_P and negative pairs k_N was empirically set as 20 and 512, respectively.

3.2   Experimental Results

Comparison with KD Methods: Under the framework of distilling multi-modal knowledge for pathological glioma grading, we compared the proposed

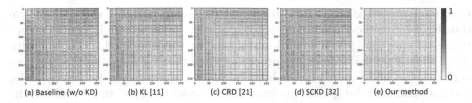

(a) Baseline (w/o KD)   (b) KL [11]   (c) CRD [21]   (d) SCKD [32]   (e) Our method

Fig. 2. Visualization of the distance between the student's and teacher's correlation matrices. Smaller distance (i.e., lighter color) indicates that (e) our method enables more effective knowledge transfer and better alignment of the teacher and student.

method with current state-of-the-art KD methods, including logits-based KD [11,30], feature-based KD [13], relational KD [31,32], contrastive KD [22], and student customized KD [33]. For a fair comparison, all KD methods distilled knowledge from the same multi-modal teacher (i.e., upper bound, the last row in Table 1) to the same pathology student (i.e., ResNet-18). As shown in Table 1, all KD methods outperform the baseline model with AP of 82.28%, suggesting that the privileged multi-modal knowledge is beneficial for the pathological grading. More importantly, our proposed method with AP of 86.12% outperforms existing KD methods, indicating that our method enables more effective knowledge transfer from the multi-modal teacher to the pathology student.

To further measure the capacity of different KD methods in distilling the teacher's knowledge, we compared the distance between the correlation matrices of the teacher's and student's logits. Specifically, for the multi-modal teacher and pathology student, we computed the pairwise correlation matrices between the predicted logits of a batch of samples (batch size = 256). Then, the distance between the teacher's and student's correlation matrices is computed and shown in Fig. 2. A smaller distance (i.e., lighter color in Fig. 2) indicates the student is better aligned with the teacher due to effective knowledge transfer. It is clear that our method (Fig. 2 (e)) leads to smaller teacher-student difference than the baseline (Fig. 2 (a)) and other KD methods (Fig. 2 (b–d)), suggesting that our method distills the most correlation knowledge from the multi-modal teacher.

Ablation Study: We performed an ablation study to evaluate the independent contribution of our proposed DC-Distill and GK-Refine modules. As shown in Table 1, both DC-Distill and GK-Refine improve the results with respect to the baseline model and existing KD methods. Especially, the superiority of our DC-Distill over the existing contrastive KD [22] indicates the discrepancy-induced contrastive samples in our method lead to more effective knowledge transfer. Furthermore, we degrade the GK-Refine to GK-Refine$^-$ by keeping the knowledge bank while removing the knowledge refinement. The inferior performance of GK-Refine$^-$ (with AP of 84.65%) indicates that the knowledge refinement in our GK-Refine is crucial for the student to absorb beneficial knowledge.

Comparison with Multi-modal Methods: Owing to the effective knowledge transfer, our method with sole pathology slides achieves similar performance (with AP of 86.12%) as the multi-modal teacher (upper bound in Table 1, with AP of 86.54%). What's more, we surprisingly find that our unimodal method outperforms two multi-modal methods [6,7], which require paired pathology slides and genomic data for glioma grading. These results suggest that distilling the pathology-genomic knowledge is a promising direction to improve the pathological grading performance, which is of great significance in clinical applications.

4   Conclusion

In this paper, we present the first work that studies the distillation of the pathology-genomic knowledge towards more accurate glioma grading with pathology slides. For effective knowledge transfer, we propose the DC-Distill module that allows the teacher to provide knowledge via reliable contrastive samples with teacher-student discrepancy, and GK-Refine scheme which allows the student to selectively absorb the beneficial knowledge according to the gradient-based agreement. Experiments on the TCGA GBM-LGG dataset indicate that our proposed distillation framework improves the pathological glioma grading significantly and outperforms state-of-the-art KD paradigms. More importantly, our method with the sole pathology slides achieves comparable performance with existing multi-modal approaches, which is of great clinical significance.

Acknowledgements. This work is supported by Innovation and Technology Commission-Innovation and Technology Fund ITS/100/20 (CityU 9440276).

References

1. Louis, D.N., et al.: The 2016 world health organization classification of tumors of the central nervous system: a summary. Acta Neuropathol. **131**(6), 803–820 (2016). https://doi.org/10.1007/s00401-016-1545-1
2. Rathore, S., Niazi, T., Iftikhar, M.A., Chaddad, A.: Glioma grading via analysis of digital pathology images using machine learning. Cancers **12**(3), 578 (2020)
3. Aum, D.J., Kim, D.H., Beaumont, T.L., Leuthardt, E.C., Dunn, G.P., Kim, A.H.: Molecular and cellular heterogeneity: the hallmark of glioblastoma. Neurosurg. Focus **37**(6), E11 (2014)
4. Ertosun, M.G., Rubin, D.L.: Automated grading of gliomas using deep learning in digital pathology images: a modular approach with ensemble of convolutional neural networks. In: AMIA Annual Symposium Proceedings, vol. 2015, p. 1899. American Medical Informatics Association (2015)
5. Xing, X., et al.: An interpretable multi-level enhanced graph attention network for disease diagnosis with gene expression data. In Proceedings of BIBM, pp. 556–561. IEEE (2021)
6. Mobadersany, P., et al.: Predicting cancer outcomes from histology and genomics using convolutional networks. Proc. Natl. Acad. Sci. **115**(13), E2970–E2979 (2018)

7. Chen, R.J., et al.: Pathomic fusion: an integrated framework for fusing histopathology and genomic features for cancer diagnosis and prognosis. IEEE Trans. Med. Imaging **41**, 757–770 (2020)
8. Braman, N., Gordon, J.W.H., Goossens, E.T., Willis, C., Stumpe, M.C., Venkataraman, J.: Deep orthogonal fusion: multimodal prognostic biomarker discovery integrating radiology, pathology, genomic, and clinical data. In: de Bruijne, M., et al. (eds.) MICCAI 2021. LNCS, vol. 12905, pp. 667–677. Springer, Cham (2021). https://doi.org/10.1007/978-3-030-87240-3_64
9. Chen, R.J., et al.: Multimodal co-attention transformer for survival prediction in gigapixel whole slide images. In: Proceedings of ICCV, pp. 4015–4025 (2021)
10. Lathe, W., Williams, J., Mangan, M., Karolchik, D.: Genomic data resources: challenges and promises. Nat. Educ. **1**(3), 2 (2008)
11. Hinton, G., Vinyals, O., Dean, J., et al.: Distilling the knowledge in a neural network, vol. 2, no. 7. arXiv preprint arXiv:1503.02531 (2015)
12. Garcia, N.C., Morerio, P., Murino, V.: Learning with privileged information via adversarial discriminative modality distillation. IEEE Trans. Pattern Anal. Mach. Intell. **42**(10), 2581–2593 (2019)
13. Hu, M., et al.: Knowledge distillation from multi-modal to mono-modal segmentation networks. In: Martel, A.L., et al. (eds.) MICCAI 2020. LNCS, vol. 12261, pp. 772–781. Springer, Cham (2020). https://doi.org/10.1007/978-3-030-59710-8_75
14. Chen, C., Dou, Q., Jin, Y., Liu, Q., Heng, P.A.: Learning with privileged multimodal knowledge for unimodal segmentation. IEEE Trans. Med. Imaging **41**, 621–632 (2021)
15. Li, G., Li, X., Wang, Y., Zhang, S., Wu, Y., Liang, D.: Knowledge distillation for object detection via rank mimicking and prediction-guided feature imitation. arXiv preprint arXiv:2112.04840 (2021)
16. Peng, F., Wang, C., Liu, J., Yang, Z.: Active learning for lane detection: a knowledge distillation approach. In: Proceedings of ICCV, pp. 15152–15161 (2021)
17. Li, Z., et al.: Not all knowledge is created equal. arXiv preprint arXiv:2106.01489 (2021)
18. Shangchen, D., et al.: Agree to disagree: adaptive ensemble knowledge distillation in gradient space. In: Proceedings of NeurIPS, vol. 33, pp. 12345–12355 (2020)
19. Van Loan, C.F.: The ubiquitous Kronecker product. J. Comput. Appl. Math. **123**(1–2), 85–100 (2000)
20. Tarvainen, A., Valpola, H.: Mean teachers are better role models: weight-averaged consistency targets improve semi-supervised deep learning results. In: Proceedings of NeurIPS, vol. 30 (2017)
21. Xiang, L., Ding, G., Han, J.: Learning from multiple experts: self-paced knowledge distillation for long-tailed classification. In: Vedaldi, A., Bischof, H., Brox, T., Frahm, J.-M. (eds.) ECCV 2020. LNCS, vol. 12350, pp. 247–263. Springer, Cham (2020). https://doi.org/10.1007/978-3-030-58558-7_15
22. Tian, Y., Krishnan, D., Isola, P.: Contrastive representation distillation. In: Proceedings of ICLR (2019)
23. He, K., Fan, H., Wu, Y., Xie, S., Girshick, R.: Momentum contrast for unsupervised visual representation learning. In: Proceedings of CVPR, pp. 9729–9738 (2020)
24. Xing, X., Hou, Y., Li, H., Yuan, Y., Li, H., Meng, M.Q.-H.: Categorical relation-preserving contrastive knowledge distillation for medical image classification. In: de Bruijne, M., et al. (eds.) MICCAI 2021. LNCS, vol. 12905, pp. 163–173. Springer, Cham (2021). https://doi.org/10.1007/978-3-030-87240-3_16
25. You, S., Xu, C., Xu, C., Tao, D.: Learning from multiple teacher networks. In: Proceedings of KDD, pp. 1285–1294 (2017)

26. Tomczak, K., Czerwińska, P., Wiznerowicz, M.: The Cancer Genome Atlas (TCGA): an immeasurable source of knowledge. Contemp. Oncol. **19**(1A), A68 (2015)
27. Cerami, E., et al.: The cBio cancer genomics portal: an open platform for exploring multidimensional cancer genomics data (2012)
28. Paszke, A., et al.: PyTorch: an imperative style, high-performance deep learning library. In: Proceedings of NeurIPS, vol. 32 (2019)
29. He, K., Zhang, X., Ren, S., Sun, J.: Deep residual learning for image recognition. In: Proceedings of CVPR, pp. 770–778 (2016)
30. Passalis, N., Tefas, A.: Learning deep representations with probabilistic knowledge transfer. In: Proceedings of ECCV, pp. 268–284 (2018)
31. Tung, F., Mori, G.: Similarity-preserving knowledge distillation. In: Proceedings of ICCV, pp. 1365–1374 (2019)
32. Park, W., Kim, D., Lu, Y., Cho, M.: Relational knowledge distillation. In: Proceedings of CVPR, pp. 3967–3976 (2019)
33. Zhu, Y., Wang, Y.: Student customized knowledge distillation: bridging the gap between student and teacher. In: Proceedings of ICCV, pp. 5057–5066 (2021)

Radiological Reports Improve Pre-training for Localized Imaging Tasks on Chest X-Rays

Philip Müller[1(✉)], Georgios Kaissis[1,2,3], Congyu Zou[4], and Daniel Rueckert[1,3]

[1] Institute of Artificial Intelligence in Medicine, Technical University of Munich, 81675 Munich, Germany
`philip.j.mueller@tum.de`
[2] Institute of Radiology, Technical University of Munich, 81675 Munich, Germany
[3] Department of Computing, Imperial College London, London SW7 2BX, UK
[4] Department for Internal Medicine I, Klinikum Rechts der Isar, Technical University of Munich, 81675 Munich, Germany

Abstract. Self-supervised pre-training on unlabeled images has shown promising results in the medical domain. Recently, methods using text-supervision from companion text like radiological reports improved upon these results even further. However, most works in the medical domain focus on image classification downstream tasks and do not study more localized tasks like semantic segmentation or object detection. We therefore propose a novel evaluation framework consisting of 18 localized tasks, including semantic segmentation and object detection, on five public chest radiography datasets. Using our proposed evaluation framework, we study the effectiveness of existing text-supervised methods and compare them with image-only self-supervised methods and transfer from classification in more than 1200 evaluation runs. Our experiments show that text-supervised methods outperform all other methods on 13 out of 18 tasks making them the preferred method. In conclusion, image-only contrastive methods provide a strong baseline if no reports are available while transfer from classification, even in-domain, does not perform well in pre-training for localized tasks.

Keywords: Self-supervised representation learning · Radiology reports

1 Introduction and Motivation

In recent years, contrastive learning [1–4, 7, 10, 11, 13, 14, 21, 24, 25, 37, 39] has become state-of-the-art for self-supervised representation learning on images. It can effectively leverage unlabeled images to create pre-trained models for subsequent fine-tuning on labeled data. Recently, text-supervised methods

Supplementary Information The online version contains supplementary material available at https://doi.org/10.1007/978-3-031-16443-9_62.

Table 1. Comparison of studied pre-training methods.

	Training data	Supervision	Negatives	Local loss	Pooling
Random Init.	–	–	–	–	–
ImageNet [29]	Natural images	Class labels	–	✗	Avg
CheXpert [15]	Chest X-rays	Class labels	–	✗	Avg
BYOL [10]	Chest X-rays	Images	✗	✗	Avg
SimCLR [3]	Chest X-rays	Images	✓	✗	Avg
PixelPro [38]	Chest X-rays	Images	✗	✓	Avg
ConVIRT [40]	Chest X-rays	Reports	✓	✗	Avg
CLIP [26]*	Chest X-rays	Reports	✓	✗	Attention

* Modified to use the same image and text encoders as ConVIRT.

[5,17,22,23,26,30,40] have been proposed, which utilize companion text like radiological reports to pre-train image models, improving the results even further. In the medical domain such self- or text-supervised methods are very promising as high-quality annotated data is often expensive to acquire, requiring trained physicians to manually label samples. Self- or text-supervised pre-training can reduce the number of required labeled samples and it is therefore no surprise that they have been successfully applied for medical imaging tasks like image classification on chest X-rays [8,22,33,34,40]. However, most works focus on image-level downstream tasks like image classification and such pre-training methods are rarely applied on more localized medical tasks like semantic segmentation or object detection. In this work, we therefore study the effectiveness of text-supervised pre-training methods on localized tasks on chest X-rays.

Our contributions are as follows:

- We propose a downstream evaluation framework with 18 localized tasks on chest X-rays, including object detection and semantic segmentation on five public datasets. To our best knowledge this is the first localized evaluation framework for pre-training methods in medical imaging and allows to compare the performance of pre-training methods on localized tasks with medical data, specifically chest X-rays.[1]
- We conduct a comparative study of pre-training methods, including text-supervised and image-only contrastive methods as well as in- and cross-domain transfer from classification. We pre-train on MIMIC-CXR [9,18–20] and evaluate the studied methods on our localized chest X-ray evaluation framework in more than 1200 evaluation runs.
- We found that text-supervised methods outperform all other methods on 13 out of 18 tasks and are less sensitive to the downstream dataset size on some tasks. Image-only contrastive methods provide a strong baseline if no reports are available while transfer from classification, even in-domain, does not perform well in pre-training for localized tasks.

---

[1] Note that we do only contribute the selection of these datasets and the definition of tasks on them while we do not contribute any new datasets or ground truth labels.

2   Material and Methods

2.1   Studied Pre-training Methods

Table 1 gives and overview over the studied pre-training and initialization methods. Each of these methods provides an initialization for the image encoder (ResNet50 [12]) which is then evaluated on our evaluation framework. Details of the evaluation framework are given in Sect. 2.2. We use **Random Init.** as a baseline that is expected to be outperformed by all other methods. Here the image encoder is initialized using its default random initialization. For the study of cross-domain transfer from non-medical image classification, we use **ImageNet** [29] **Init.**, where we initialize the image encoder with weights pre-trained on the ImageNet ILSVRC-2012 task [29]. We pre-train all other studied methods on in-domain data, i.e. on frontal chest X-rays from version 2 of MIMIC-CXR [9,18–20]. For the study of in-domain transfer from classification, we pre-train the image encoder using supervised multi-label binary classification with **CheX-pert** [15] labels, i.e. we use image-level labels as supervision during pre-training. Additionally, we study two typically used (global, image-only) self-supervised methods, namely **BYOL** [10] and **SimCLR** [3]. Both methods use only the images, i.e. the frontal chest X-rays, but no additional supervision during training. Instead, they use contrastive losses on views generated from the training images using data augmentation. While SimCLR uses explicit negatives, i.e. views generated from different samples, to contrast the positive pairs of views generated from the same samples, BYOL solely relies on a momentum encoder and normalization for contrast but does not use explicit negatives. **PixelPro** [38] is an extension to BYOL that additionally uses a contrastive loss on local representations, i.e. on the last feature map of the image encoder before global pooling. For studying the effect of text-supervision, we use **ConVIRT** [40] and **CLIP** [26]. Both methods use the SimCLR framework with image views encoded by the image encoder and text views encoded by an additional text encoder. We use randomly sampled sentences from paired radiological reports as text views. Note that for comparability we adapted CLIP to use the same image and text encoders as ConVIRT such that the main difference between CLIP and Con-VIRT is that ConVIRT, like all other methods, uses the global average pooling layer of the ResNet50 model to compute the image representations for the loss function, while in CLIP the pooling layers is replaced by attention pooling.

2.2   Evaluation Framework

Pre-training Dataset. We pre-train all in-domain methods, i.e. all methods except ImageNet Init. and Random Init., on MIMIC-CXR 2 [9,18–20] as, to our best knowledge, it is the largest and most commonly used chest X-ray dataset that also includes reports that are relevant for text-supervised methods. As all downstream tasks only contain frontal views, we remove all lateral views, such that we have roughly 21000 training samples each with a report and one or more frontal images.

Table 2. Overview of downstream datasets.

	Type of targets	Target classes	Train/Val/Test
RSNA Pneumonia [31,36]	Object detection	Pneumonia	16010/5337/5337
COVID Rural [6,35]	Semantic Segmentation	COVID-19	133/44/44
SIIM-ACR Pneumothorax [32]	Semantic Segmentation	Pneumothorax	7229/2409/2409
Object CXR [16]	Object detection	Foreign objects	6400/1600/1000
NIH CXR [36]	Object detection	Atelectasis, Effusion, Cardiomegaly, Nodule, Pneumonia, Infiltrate, Pneumothorax, Mass	588/196/196

Evaluation Protocols. We use the ResNet50 [12] model as the (image) backbone in all pre-training methods. After pre-training, we evaluate the pre-trained ResNet50 on several semantic segmentation and object detection tasks. For semantic segmentation tasks we evaluate in the following settings: (i) **U-Net Finetune**: Here the ResNet50 is used as backbone of a U-Net [28] and is finetuned jointly with all other layers, (ii) **U-Net Frozen**: Here the ResNet50 is used as frozen backbone of a U-Net [28] and only the non-backbone layers are finetuned, and (iii) **Linear**: Here an element-wise linear layer is trained that is applied after the last feature map (before pooling) of the frozen ResNet50, before the results are upsampled to the segmentation resolution. For object detection tasks we evaluate in the following settings: (i) **YOLOv3 Finetune**: Here the ResNet50 is used as backbone of a YOLOv3 [27] model and is finetuned jointly with the non-backbone layers, (ii) **YOLOv3 Frozen**: Here the ResNet50 is used as frozen backbone of a YOLOv3 [27] model and only the non-backbone layers are finetuned, and (iii) **Linear**: Here the object detection ground truth is converted to segmentation masks and then the *Linear* segmentation protocol is used for evaluation.

The different evaluation protocols complement each other: While the *U-Net Finetune* and *YOLOv3 Finetune* protocols evaluate how well the pre-trained image models could be fine-tuned for practical applications, the *Linear* protocols directly evaluate the learned local representations (i.e. feature maps) while adding as few parameters as possible and therefore mostly omitting the variance introduced by random initialization during downstream evaluation. The *U-Net Frozen* and *YOLOv3 Frozen* protocols can be seen as middle ground between the two extremes, where representations are frozen but evaluated in a more practical setting (but with many randomly initialized layers). Overall this allows the analysis of many aspects of the pre-trained representations.

Downstream Datasets. We evaluate the pre-trained ResNet50 backbone on five public datasets of frontal-view chest X-rays, namely RSNA Pneumonia Detection [31,36], COVID Rural [6,35], SIIM-ACR Pneumothorax Segmentation [32], Object CXR [16], and NIH CXR [36].[2] These datasets include object detection or

---

[2] Note that NIH CXR is a small subset of the Chestx-ray8 [36] dataset that contains detection targets.

semantic segmentation targets for a variety of pathologies and for foreign objects, and they differ in the number of samples ranging from about 200 samples to more than 25000. Table 2 gives an overview over these datasets, the associated targets and the numbers of samples. We use the *U-Net Finetune, U-Net Frozen,* and *Linear* protocols on all segmentation datasets except on SIIM-ACR Pneumothorax, where we do not use the *Linear* protocol due to the fine-grained nature of the segmentation masks. We use the *YOLOv3 Finetune, YOLOv3 Frozen,* and *Linear* protocols on all object detection datasets except on NIH CXR, where we only use the *Linear* protocol due to the limited number of samples per class. Additionally, we evaluate on different subsets of the RSNA Pneumonia dataset, namely on 1%, 10%, and 100% of the downstream samples. This allows us to analyze the sensitivity of pre-training methods to the downstream dataset size.

Tuning and Evaluation Procedure. We tune all models on a single downstream task, *RSNA YOLOv3 Frozen 10%.* Other downstream tasks have not been evaluated during tuning to make sure that models are not biased towards the downstream tasks. After tuning, each model was evaluated on all downstream tasks. For each task the downstream learning rates were tuned individually per model (using single evaluation runs) before running five evaluations (all using the tuned learning rate). We report the average results of these five runs and their 95%-confidence interval.

3   Results

In Table 3 we present the downstream results of the studied methods on the RSNA Pneumonia and the SIIM-ACR Pneumothorax tasks and compare pre-training on 100% and 30% of MIMIC-CXR. Additional results on the other datasets and on subsets of RSNA Pneumonia are provided in the supplementary material.

Comparison of Methods. Our experiments show that there is no single best pre-training method for all evaluated downstream tasks. On 13 out of 18 tasks, text-supervised methods (i.e. CLIP or ConVIRT) outperform all other methods, on 10 of these tasks even by more than the confidence interval. CLIP outperforms non text-supervised methods and ConVIRT on 10 tasks (on nine of them significantly), while ConVIRT outperforms non text-supervised methods on 10 tasks (on five of them significantly) but outperforms CLIP on only four tasks (on two of them significantly). This shows that text-supervised methods are in general superior to other methods, performing best on the majority of tasks and making them the method of choice for localized tasks. Most notable, CLIP is better than ConVIRT on most tasks and should therefore be preferred as pre-training method.

Our experiments show that image-only self-supervised methods (i.e. BYOL, SimCLR, and PixelPro) provide very strong baselines, being the best methods on four tasks (BYOL on three and PixelPro on one task). They are therefore

Table 3. Results on the RSNA Pneumonia (object detection) and SIIM-ACR Pneumothorax (semantic segmentation) tasks. Detailed results on the remaining tasks can be found in the supplementary material: results on RSNA subset tasks (1% and 10% of the data) in Table 4 and results on COVID Rural, Object CXR and NIH CXR tasks in Table 5. All results are averaged over five evaluation runs and the 95%-confidence interval is shown. The best results per task are underlined, the second best results are dash-underlined and the best results per pre-training category (general initialization, pre-training on 30% and 100%) are highlighted in bold.

	RSNA Pneumonia			SIIM-ACR Pneumoth.	
	YOLOv3 finetune mAP (%)	YOLOv3 frozen mAP (%)	Linear Dice (%)	UNet finetune Dice (%)	UNet frozen Dice (%)
General initialization methods					
Random	14.9 ± 1.7	8.9 ± 0.9	5.3 ± 0.0	23.2 ± 1.0	23.9 ± 1.6
ImageNet [29]	**19.0 ± 0.2**	**15.7 ± 0.3**	**43.3 ± 0.0**	**38.5 ± 0.9**	**36.9 ± 0.7**
Pre-training on 30% of frontal MIMIC-CXR					
CheXpert [15]	**21.3 ± 0.3**	18.8 ± 0.4	48.1 ± 0.0	38.9 ± 0.9	40.7 ± 0.7
BYOL [10]	18.8 ± 0.2	**21.0 ± 0.2**	50.0 ± 0.0	**43.1 ± 0.6**	**42.9 ± 0.3**
SimCLR [3]	20.4 ± 1.8	19.9 ± 0.2	48.2 ± 0.0	42.6 ± 0.4	39.2 ± 0.7
PixelPro [38]	19.8 ± 0.4	13.4 ± 0.3	39.8 ± 0.1	39.3 ± 0.8	39.1 ± 0.3
ConVIRT [40]	18.3 ± 0.4	18.4 ± 1.1	50.2 ± 0.0	42.5 ± 1.0	42.5 ± 0.2
CLIP [26]*	19.7 ± 0.5	19.6 ± 1.4	**50.7 ± 0.0**	42.8 ± 1.5	42.5 ± 0.6
Pre-training on 100% of frontal MIMIC-CXR					
CheXpert [15]	**22.2 ± 0.4**	**20.0 ± 0.2**	46.9 ± 0.0	34.2 ± 0.8	37.7 ± 0.3
BYOL [10]	17.3 ± 1.1	15.9 ± 0.6	46.8 ± 0.0	42.6 ± 0.7	40.7 ± 0.7
SimCLR [3]	18.8 ± 1.0	17.3 ± 1.6	47.0 ± 0.0	41.2 ± 0.8	38.7 ± 0.5
PixelPro [38]	17.4 ± 1.7	12.6 ± 1.3	40.2 ± 0.1	39.4 ± 1.2	38.7 ± 0.6
ConVIRT [40]	18.5 ± 0.4	17.9 ± 0.3	50.4 ± 0.3	39.3 ± 0.3	43.1 ± 0.3
CLIP [26]*	19.9 ± 0.8	18.7 ± 0.0	**51.1 ± 0.0**	**44.0 ± 0.7**	**45.0 ± 0.5**

* Modified to use the same image and text encoders as ConVIRT.

very useful if no reports but only unlabeled images are available. However, they are outperformed by text-supervised methods on 13 tasks (by CLIP on 11 and by ConVIRT on 11 tasks) such that text-supervised methods should be preferred if text is available.

CheXpert, i.e. in-domain transfer from classification, is often outperformed by contrastive methods (image-only or text-supervised), only on one task it is the best method. On 16 tasks CheXpert is outperformed by text-supervised methods and on 13 tasks even by image-only methods. Therefore, contrastive pre-training should be preferred over transfer from supervised classification. This result is unexpected, as labels should provide additional semantics for pre-training. We suspect that common classification training methods, i.e. global average pooling and cross entropy loss, are unable to utilize the global labels effectively for pre-training localized representations. Utilizing global labels in pre-training for localized tasks is therefore a promising direction for future work.

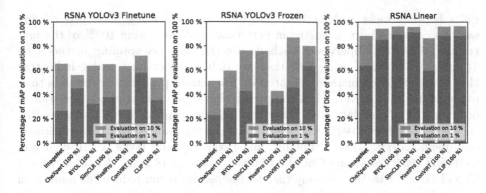

Fig. 1. Sensitivity of pre-training methods to the size of the downstream dataset, showcased on the RSNA Pneumonia Detection dataset. We plot the downstream results when evaluating on 1% and 10% of the data relative to the results when evaluating on 100%. In the *YOLOv3 Frozen* setting, we observe that for text-supervised methods (i.e. CLIP and ConVIRT), the downstream training set size is less relevant compared to other methods.

ImageNet initialization is always outperformed by some in-domain pre-training method (contrastive or CheXpert classification). This means that in-domain pre-training, even with smaller datasets, generally outperforms transfer learning from natural images. Random initialization, as expected, always performs worst and only provides a lower bound for evaluation results.

Comparison of Downstream Tasks. We found that text-supervised methods perform best on all *Linear* tasks and perform quite well on the *Frozen* evaluation protocol and the RSNA pneumonia detection dataset. Image-only self-supervised methods perform well when finetuned and on the COVID Rural dataset but do not perform well with the *Linear* evaluation protocol. We observe that Pixel-Pro (i.e. localized image-only pre-training) performs very well on the Object CXR finetuning task but is often outperformed, even by instance-level image-only methods (i.e. BYOL and SimCLR), on most other tasks. We assume that the localized image pre-training used in PixelPro is not effective for detecting high-level medical patterns like pathologies but works well for detecting simple patterns like foreign objects. CheXpert pre-training performs well on RSNA pneumonia detection with finetuning. On the Object CXR dataset ImageNet initialization provides a very strong baseline and outperforms CheXpert pre-training. Therefore, if the in-domain pre-training task is inherently different to the downstream task, the model performance may degrade. In the case of CheX-pert pre-training for the Object CXR tasks, CheXpert classification focuses on pathology detection while in Object CXR foreign objects are detected.

Relevance of Pre-training Dataset Size. In our experiments we do not observe a consistent benefit of using roughly 210000 pre-training samples (i.e. 100% of the data) over just using roughly 63000 samples (i.e. 30%). Instead benefits differ between tasks and methods and in some cases using more pre-training samples can even hurt downstream performance. When using 100% of the pre-training data text-supervised methods are the best pre-training methods on 14 tasks, while when using 30% they are only superior on 8 tasks, indicating that text-supervised methods profit more from larger pre-training datasets than other methods. However, when comparing text-supervised methods pre-trained on only 30% of the data with other methods pre-trained in both settings (i.e. 30% and 100%), we observe that text-supervised methods still outperform image-only methods (i.e. BYOL, SimCLR, and PixelPro) on 8 tasks (CLIP on 7 and ConVIRT on 6 tasks) and CheXpert classification on 14 tasks (CLIP on 14 and ConVIRT on 12 tasks), showing that text-supervision can reduce the number of required pre-training samples while achieving similar or better results.

Relevance of Downstream Dataset Size. As in practice the size of (localized) medical datasets is often very limited, we studied how much different pre-training methods affect the sensitivity to the size of the downstream dataset (i.e. how much reducing the datasets size decreases the performance). We show the sensitivities on the RSNA Pneumonia Detection dataset in Fig. 1. In the *YOLOv3 Finetune* setting, we do not observe a clear pattern but find that ConVIRT is less sensitive to the downstream dataset size than other methods, while CLIP is rather sensitive. In the *YOLOv3 Frozen* setting, we observe that for text-supervised methods (i.e. CLIP and ConVIRT), the downstream training set size is less relevant compared to other methods. Here, CLIP (100%) outperforms ImageNet initialization by 19% when using 100% of the downstream samples, while it outperforms ImageNet initialization by even 231% when only using 1% of the samples. In the *Linear* setting we observe little differences in most methods, except that ImageNet and PixelPro are much more sensitive to the downstream dataset size.

4   Discussion

Limitations of our Evaluation Procedure. We did not tune image encoder, downstream architectures, or preprocessing for downstream tasks, resized all inputs to only 224×224, and applied no data augmentation. Therefore, the presented downstream results are below results typically reported for these datasets. The evaluation procedure is kept simple to i) limit computational resources, ii) avoid bias induced by downstream tuning, and iii) allow for a fair comparison of pre-training methods, being the main purpose of this work. We assume that benefits observed in our evaluation procedure also indicate benefits for tuned real-world tasks, although they cannot be precisely quantified by our evaluation method.

5   Conclusion

We evaluate pre-training and initialization methods on a novel evaluation framework consisting of 18 localized tasks on chest X-rays. We found that text-supervised methods outperform all other methods on 13 out of 18 tasks and are less sensitive to the downstream dataset size on some tasks. This shows that radiology reports contain valuable information that can serve as supervision for localized medical image models. We assume that the nuanced descriptions in the reports provide information about localization of pathologies in the images, explaining the success. Existing text-supervised methods, even those proposed for natural images, utilize this information quite effectively and seem to work well on medical images like chest X-rays. Our experiments also show that transfer from pathology classification, even in-domain, does not perform well and common supervised classification methods seem to be unable to utilize image labels effectively for localized downstream tasks. Image-only supervised methods provide a strong baseline outperforming in-domain transfer from classification and should therefore be preferred if only images (with or without labels) are available for pre-training. We hope that our work provides valuable insights that encourage using text-supervised pre-training methods for localized medical imaging and that our evaluation framework simplifies the comparison of novel pre-training methods.

References

1. Bardes, A., Ponce, J., LeCun, Y.: VICReg: variance-invariance-covariance regularization for self-supervised learning. arXiv:2105.04906 (2021)
2. Caron, M., Touvron, H., Misra, I., et al.: Emerging properties in self-supervised vision transformers. In: ICCV, pp. 9630–9640 (2021). https://doi.org/10.1109/ICCV48922.2021.00951
3. Chen, T., Kornblith, S., Norouzi, M., Hinton, G.: A simple framework for contrastive learning of visual representations. In: ICML, pp. 1597–1607 (2020)
4. Chen, X., He, K.: Exploring simple Siamese representation learning. arXiv:2011.10566 (2020)
5. Desai, K., Johnson, J.: VirTex: learning visual representations from textual annotations. arXiv:2006.06666 (2020)
6. Desai, S., Baghal, A., Wongsurawat, T., et al.: Data from chest imaging with clinical and genomic correlates representing a rural COVID-19 positive population. Cancer Imaging Arch. (2020). https://doi.org/10.7937/tcia.2020.py71-5978
7. Ermolov, A., Siarohin, A., Sangineto, E., Sebe, N.: Whitening for self-supervised representation learning. arXiv:2007.06346 (2020)
8. Gazda, M., Gazda, J., Plavka, J., Drotar, P.: Self-supervised deep convolutional neural network for chest X-ray classification. arXiv:2103.03055 (2021)
9. Goldberger, A., Amaral, L., Glass, L., Hausdorff, J., et al.: PhysioBank, PhysioToolkit, and PhysioNet: components of a new research resource for complex physiologic signals. Circulation **101**(23), 215–220 (2000)
10. Grill, J.B., Strub, F., Altché, F., et al.: Bootstrap your own latent - a new approach to self-supervised learning. In: NeurIPS, pp. 21271–21284 (2020)

11. He, K., Fan, H., Wu, Y., et al.: Momentum contrast for unsupervised visual representation learning. In: CVPR, pp. 9726–9735 (2020). https://doi.org/10.1109/CVPR42600.2020.00975

12. He, K., Zhang, X., Ren, S., Sun, J.: Deep residual learning for image recognition. In: CVPR, pp. 770–778 (2016). https://doi.org/10.1109/CVPR.2016.90

13. Hjelm, R.D., Fedorov, A., Lavoie-Marchildon, S., et al.: Learning deep representations by mutual information estimation and maximization. arXiv:1808.06670 (2019)

14. Hénaff, O.J., Srinivas, A., et al.: Data-efficient image recognition with contrastive predictive coding. In: ICML, pp. 4182–4192 (2020)

15. Irvin, J., Rajpurkar, P., Ko, M., et al.: CheXpert: a large chest radiograph dataset with uncertainty labels and expert comparison. In: AAAI, pp. 590–597 (2019)

16. JF-Healthcare: object-CXR - automatic detection of foreign objects on chest X-rays. MIDL (2020). https://jfhealthcare.github.io/object-CXR/

17. Jia, C., Yang, Y., Xia, Y., et al.: Scaling up visual and vision-language representation learning with noisy text supervision. In: ICML, pp. 4904–4916 (2021)

18. Johnson, A., Lungren, M., Peng, Y., et al.: MIMIC-CXR-JPG - chest radiographs with structured labels (version 2.0.0). PhysioNet (2019). https://doi.org/10.13026/8360-t248

19. Johnson, A., Pollard, T., Berkowitz, S., et al.: MIMIC-CXR, a de-identified publicly available database of chest radiographs with free-text reports. Sci. Data 6(317), 1–8 (2019). https://doi.org/10.1038/s41597-019-0322-0

20. Johnson, A., Pollard, T., Mark, R., Berkowitz, S., Horng, S.: MIMIC-CXR database (version 2.0.0). PhysioNet (2019). https://doi.org/10.13026/C2JT1Q

21. Li, J., Zhou, P., Xiong, C., Hoi, S.C.H.: Prototypical contrastive learning of unsupervised representations. arXiv:2005.04966 (2021)

22. Liao, R., et al.: Multimodal representation learning via maximization of local mutual information. In: de Bruijne, M., et al. (eds.) MICCAI 2021. LNCS, vol. 12902, pp. 273–283. Springer, Cham (2021). https://doi.org/10.1007/978-3-030-87196-3_26

23. Liu, Z., Stent, S., Li, J., et al.: LocTex: learning data-efficient visual representations from localized textual supervision. arXiv:2108.11950 (2021)

24. Misra, I., van der Maaten, L.: Self-supervised learning of pretext-invariant representations. arXiv:1912.01991 (2019)

25. van den Oord, A., Li, Y., Vinyals, O.: Representation learning with contrastive predictive coding. arXiv:1807.03748 (2019)

26. Radford, A., Kim, J.W., Hallacy, C., et al.: Learning transferable visual models from natural language supervision. arXiv:2103.00020 (2021)

27. Redmon, J., Farhadi, A.: YOLOv3: an incremental improvement. arXiv:1804.02767 (2018)

28. Ronneberger, O., Fischer, P., Brox, T.: U-Net: convolutional networks for biomedical image segmentation. In: Navab, N., Hornegger, J., Wells, W.M., Frangi, A.F. (eds.) MICCAI 2015. LNCS, vol. 9351, pp. 234–241. Springer, Cham (2015). https://doi.org/10.1007/978-3-319-24574-4_28

29. Russakovsky, O., et al.: ImageNet large scale visual recognition challenge. Int. J. Comput. Vis. 115(3), 211–252 (2015). https://doi.org/10.1007/s11263-015-0816-y

30. Sariyildiz, M.B., Perez, J., Larlus, D.: Learning visual representations with caption annotations. In: Vedaldi, A., Bischof, H., Brox, T., Frahm, J.-M. (eds.) ECCV 2020. LNCS, vol. 12353, pp. 153–170. Springer, Cham (2020). https://doi.org/10.1007/978-3-030-58598-3_10

31. Shih, G., Wu, C.C., Halabi, S.S., et al.: Augmenting the national institutes of health chest radiograph dataset with expert annotations of possible pneumonia. Radiol. Artif. Intell. **1** (2019). https://doi.org/10.1148/ryai.2019180041

32. Society for Imaging Informatics in Medicine: SIIM-ACR pneumothorax segmentation (2019). https://www.kaggle.com/c/siim-acr-pneumothorax-segmentation

33. Sowrirajan, H., Yang, J., Ng, A.Y., Rajpurkar, P.: MoCo-CXR: MoCo pre-training improves representation and transferability of chest X-ray models. arXiv:2010.05352 (2021)

34. Sriram, A., Muckley, M., Sinha, K., et al.: COVID-19 prognosis via self-supervised representation learning and multi-image prediction. arXiv:2101.04909 (2021)

35. Tang, H., Sun, N., Li, Y.: Segmentation model of the opacity regions in the chest X-rays of the COVID-19 patients in the us rural areas and the application to the disease severity. medRxiv (2020). https://doi.org/10.1101/2020.10.19.20215483

36. Wang, X., Peng, Y., Lu, L., et al.: ChestX-ray8: hospital-scale chest X-ray database and benchmarks on weakly-supervised classification and localization of common thorax diseases. In: CVPR, pp. 3462–3471 (2017). https://doi.org/10.1109/CVPR.2017.369

37. Wu, Z., Xiong, Y., Yu, S., Lin, D.: Unsupervised feature learning via non-parametric instance discrimination. In: CVPR, pp. 3733–3742 (2018). https://doi.org/10.1109/CVPR.2018.00393

38. Xie, Z., Lin, Y., Zhang, Z., et al.: Propagate yourself: exploring pixel-level consistency for unsupervised visual representation learning. arXiv:2011.10043 (2020)

39. Zbontar, J., Jing, L., Misra, I., et al.: Barlow twins: self-supervised learning via redundancy reduction. arXiv:2103.03230 (2021)

40. Zhang, Y., Jiang, H., Miura, Y., et al.: Contrastive learning of medical visual representations from paired images and text. arXiv:2010.00747 (2020)

Anatomy-Guided Weakly-Supervised Abnormality Localization in Chest X-rays

Ke Yu[1]([⊠]), Shantanu Ghosh[1], Zhexiong Liu[1], Christopher Deible[2],
and Kayhan Batmanghelich[1]

[1] University of Pittsburgh, Pittsburgh, PA, USA
yu.ke@pitt.edu
[2] University of Pittsburgh Medical Center, Pittsburgh, PA, USA

Abstract. Creating a large-scale dataset of abnormality annotation on medical images is a labor-intensive and costly task. Leveraging *weak supervision* from readily available data such as radiology reports can compensate lack of large-scale data for anomaly detection methods. However, most of the current methods only use image-level pathological observations, failing to utilize the relevant *anatomy mentions* in reports. Furthermore, Natural Language Processing (NLP)-mined weak labels are noisy due to label sparsity and linguistic ambiguity. We propose an Anatomy-Guided chest X-ray Network (AGXNet) to address these issues of weak annotation. Our framework consists of a cascade of two networks, one responsible for identifying anatomical abnormalities and the second responsible for pathological observations. The critical component in our framework is an anatomy-guided attention module that aids the downstream observation network in focusing on the relevant anatomical regions generated by the anatomy network. We use Positive Unlabeled (PU) learning to account for the fact that lack of mention does not necessarily mean a negative label. Our quantitative and qualitative results on the MIMIC-CXR dataset demonstrate the effectiveness of AGXNet in disease and anatomical abnormality localization. Experiments on the NIH Chest X-ray dataset show that the learned feature representations are transferable and can achieve the state-of-the-art performances in disease classification and competitive disease localization results. Our code is available at https://github.com/batmanlab/AGXNet.

Keywords: Weakly-supervised learning · PU learning · Disease detection · Class activation map · Residual attention

1 Introduction

There is considerable interest in developing automated abnormality detection systems for chest X-rays (CXR) in order to improve radiologists' workflow effi-

Supplementary Information The online version contains supplementary material available at https://doi.org/10.1007/978-3-031-16443-9_63.

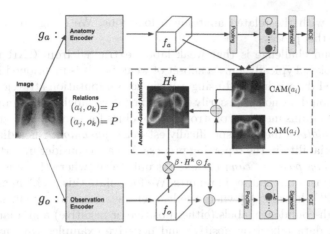

Fig. 1. Schematic diagram of the proposed AGXNet. Our architecture comprises of two classification networks, g_a for anatomical abnormalities and g_o for pathological observations. Relations $(a_i, o_k) = P$, $(a_j, o_k) = P$ are parsed from the report and represent that the observation o_k is annotated as *Present* in two anatomical landmarks a_i, a_j, respectively. We obtain the anatomy-guided attention map H^k by aggregating CAMs of a_i, a_j and incorporate the H^k as a residual attention into g_o. The symbols f_a, f_o denote the intermediate anatomy and observation features, respectively.

ciency and reduce observational oversights [3,6,13,19]. Typically, training a high-precision detection model using deep learning requires high-quality annotations. However, collecting large-scale annotations by clinical experts is time-consuming and prohibitively expensive. This motivates weakly-supervised learning (WSL) methods [2,14,16,24] that leverage weak supervision from paired CXR reports that are readily available on a large scale. There are, however, two unaddressed challenges: (1) The entire report is often summarized to a small set of disease labels, which misses the opportunity of incorporating anatomical context, and (2) Weak labels derived from radiology reports using Natural Language Processing (NLP) are noisy due to the sparsity of the labels and the ambiguity of the language. Improper handling of labeling noise can result in underperforming models. We propose a novel WSL framework to bridge these two gaps.

A variety of automated text labelers [9,11,18,24] have been developed to extract image-level observations from CXR reports. However, they do not consider *anatomy mentions* that provide important contexts for associated observations. Aiming to fill this gap, the recently developed pipelines, Chest ImaGenome [26] and RadGraph [10] extract fine-grained *"observation-located at-anatomy"* relations, (e.g., *"opacity in the right lower lung"*), from CXR reports. Yet none of them has been incorporated into a weakly-supervised disease detection method. We build our WSL framework based on the RadGraph dataset. Radiologists typically employ a systematic approach [21] when reading CXR images to ensure that no significant abnormality is missed. This approach is essentially documented in CXR reports which typically have imaging observations paired with the related anatomical locations. We design an architecture that reflects this process.

The second challenge is that weak labels extracted from CXR reports are inherently noisy. Typically, only a subset of weak labels is mentioned in a report, and unlabeled data is handled using a basic zero-imputation strategy. However, lack of mention does not necessarily mean a negative label. For example, a CXR report that contains mentions of strong visual clues for pneumonia, such as *opacity* or *consolidation*, may not specifically establish *pneumonia* as a diagnosis due to human variability in the reporting process. We also consider uncertainty mentions (e.g., *may, possible, can't exclude*) as unlabeled, where the noise originates from the intrinsic ambiguities in reports. We formulate this problem as a Positive Unlabeled (PU) learning [1] where the learner has access to a small set of examples with high-confidence labels (either positive or negative) and a large amount of unlabeled data mixed by positive and negative examples with an unknown mixture ratio.

In this paper, we present Anatomy-Guided chest X-ray Networks (AGXNet), a novel WSL framework that leverages information from pathological observations and their associated anatomical landmarks mentioned in CXR reports. Our architecture consists of a cascade of two networks, with the upstream anatomy network tasked with identifying anatomical abnormalities in CXR images and the downstream observation network tasked with identifying pathological observations. The inclusion of an anatomy-guided attention (AGA) module aims to aid the observation network paying attention to abnormalities in the context of associated anatomical landmarks. During training, the AGA module also provides a feedback loop to the upstream anatomy network, thus mutually improving the quality of both types of features. In addition, we adopt a PU learning technique to estimate the fraction of positive samples in the unlabeled data and use a self-training approach to iteratively reduce noise in labels. Our model is trained end-to-end. We evaluate the proposed framework on the MIMIC-CXR [12] dataset. Results show that the proposed AGXNet model outperforms both supervised and weakly-supervised baselines in disease localization. In-depth ablation studies demonstrate that both AGA module and PU learning can help to improve localization accuracy. We further evaluate the pre-trained encoders on the NIH Chest X-ray dataset [24] and show that the transferred models achieve the state-of-the-art (SOTA) results in disease classification and competitive performance in disease localization.

2   Methods

We parse the report into an adjacency matrix that encodes the relations between the observations and anatomical landmarks. The AGA module connects two networks that are used to predict the presence of observations and anatomical abnormalities in CXR images. We use PU learning to explicitly address noise in unlabeled observations. The proposed framework is illustrated in Fig. 1.

Report Representation. We use the RadGraph [10] to parse reports to anatomical landmark tokens $\{a_j\}$, observation tokens $\{o_k\}$, and encode their relations using an adjacency matrix A. An entry $A(j, k)$ can take one of the three variants P, N, or U, representing whether the observation o_k is *Present*, *Absent* or *Unlabeled* in the anatomical landmark a_j. Note, for o_k that is mentioned without a specified location (e.g., "*No evidence of pneumothorax*"), we link it to a special anatomical landmark token named *unspecified*. Figure A.1 shows an example adjacent matrix. We summarize the image-level labels for observation and anatomical landmark using the adjacency matrix. An observation o_k is assigned a positive label ($y(o_k) = 1$) if column $A(\cdot, k)$ has at least one P entry, a negative label ($y(o_k) = 0$) if column $A(\cdot, k)$ has no P entry and at least one N entry, and assigned as unlabeled ($y(o_k) = u$) if column $A(\cdot, k)$ only has U entries. For anatomy, an anatomical landmark is labeled as abnormal $y(a_j) = 1$ if row $A(j, \cdot)$ has at least one P entry, otherwise it is labeled as normal $y(a_j) = 0$.

Anatomy Network. The anatomy network g_a is responsible for identifying abnormalities from anatomical landmarks. The weighted binary cross-entropy (BCE) loss for a_j is given by:

$$\mathcal{L}_{a_j}(\theta_a, w_a^j) = -b_j^+ \sum_{y(a_j)=1} \ln(g_a(x)) - b_j^- \sum_{y(a_j)=0} \ln(1 - g_a(x)), \qquad (1)$$

where θ_a is the parameter of anatomy encoder network, w_a^j is the classification weight corresponding to a_j, x is an image, $g_a(x)$ is the predicted probability of being abnormal in a_j, b_j^+ and b_j^- are balancing factors introduced in [20].

AGA. We introduce the AGA module to guide the downstream observation network g_o focusing on the relevant anatomical regions mentioned in the radiology report. In detail, we construct the AGA map H^k for observation o_k by aggregating the class activation maps (CAMs) [27] for locations $\{a_j\}$ where o_k was positively observed, i.e., $A(a_j, o_k) = P$. Formally,

$$H^k(\theta_a, w_a^j) = \sum_j \mathbb{1}\{A(a_j, o_k) = P\} \underbrace{\sum_c w_a^j f_a(:, :, c; \theta_a)}_{\text{CAM}(a_j)}, \qquad (2)$$

where $f_a(:, :, c)$ represents the activation of channel c in anatomy feature map f_a, and $\mathbb{1}\{A(a_j, o_k) = P\}$ is an indicator function. We then transform the values of H^k to $[0, 1]$ using the min-max normalization.

Observation Network. The observation network g_o is responsible for identifying the presence of pathological observations in CXR images. We incorporate the H^k into g_o as a residual attention map and modify the observation feature map f_o as follows, $f'_{o_k} = (1 + \beta \cdot H^k) \odot f_o$, where $\odot$ indicates element-wise multiplication for spatial positions, and β is a scaling hyperparameter. The weighted

BCE loss for o_k is given by:

$$\mathcal{L}_{o_k}(\Theta) = -b_k^+ \sum_{y(o_k)=1} \ln(g_o(x, H^k)) - b_k^- \sum_{y(o_k)\in\{0,u\}} \ln(1 - g_o(x, H^k)), \qquad (3)$$

where $\Theta = (\theta_o, w_o^k, \theta_a, w_a^j)$, θ_o is the parameter of observation encoder network, w_o^k is the classification weight corresponding to o_k, $g_o(x, H^k)$ is the predicted probability of o_k being present in the image, b_k^+ and b_k^+ are balancing factors. It is important to note that mapping all unlabeled examples to negative ($|y(o_k) \in \{0, u\}| = |N| + |U|$) would overlook the noise in the unlabeled data, which can degrade model performance. We explicitly handle the randomness present in unlabeled data and formulate this problem using a PU learning approach.

PU Learning. The distribution of unlabeled data P_u can be decomposed as, $P_u = \alpha P_p + (1 - \alpha)P_n$, where α denotes the mixture proportion of positive examples in the unlabeled data, and P_p, P_n denotes the class-conditional distribution for positive and negative class, respectively. We adopt the method *Best Bin Estimation* (BBE) proposed by Garg *et al.* [5] to estimate α. In short, let X_p, X_n, X_u denote the positive, negative and unlabeled samples for o_k in the validation data set, $\hat{F}_p(z), \hat{F}_u(z)$ denote the empirical cumulative distributions of the predicted probabilities of the observation network, named Z_p, Z_u. The mixture proportion α is estimated by minimizing the upper confidence bound of the ratio, namely $(1 - \hat{F}_u(z))/(1 - \hat{F}_p(z))$. We integrate the BBE with an iterative self-training approach summarized as follows: (1) warm-start training with treating all unlabeled samples as negative, (2) estimating α using BBE, (3) removing α fraction of unlabeled training samples scored as most positive and relabeling the rest $1 - \alpha$ unlabeled samples as negative, (4) updating the model using positive samples ($|P|$) and provisional negative samples ($|N| + (1-\alpha)|U|$). We repeat steps 2 to 4 until the classifiers reach the best validation performance.

Optimization. The final loss is given by: $\mathcal{L} = \frac{1}{N_a}\sum_j^{N_a} \mathcal{L}_{a_j} + \frac{1}{N_o}\sum_k^{N_o} \mathcal{L}_{o_k}$, where N_a, N_o are the number of anatomy and observation labels used in training, respectively. The network is trained end-to-end. Importantly, during training, the gradients $\partial\mathcal{L}_{o_k}/\partial\theta$, $\partial\mathcal{L}_{o_k}/\partial w_a^j$ provide a feedback loop to the anatomy network through the AGA module, mutually reinforcing the anatomy and observation features. During inference time, we do not require any text input and set $H^k = 0$.

3   Experiments

Experiments are carried out to evaluate the performance of AGXNet in abnormality classification and localization. We conduct ablation studies to validate the efficiency of the AGA module and PU learning. We also test the robustness and transferability of the learned anatomy and observation features.

Experimental Details. We first evaluate the proposed AGXNet on the MIMIC-CXR dataset [12]. The RadGraph's inference dataset [10] provides annotations automatically generated by DYGIE++ [23] model for 220,763 MIMIC-CXR reports. We select the corresponding 220,763 frontal images from the

MIMIC-CXR dataset and obtain their adjacency matrix representations from the RadGraph annotations. The adjacency matrix of each sample has 48 rows (46 anatomical landmarks, 1 *unspecified*, 1 *other anatomies* representing anatomy tokens in the tail distribution) and 64 columns (63 mostly mentioned observations and 1 *other observations* representing observation tokens in the tail distribution). Appx. B provides additional details of anatomical labels. For evaluation of disease localization, we use a held out set [22] of MIMIC-CXR images with 390 bounding boxes (BBox) for pneumonia (196/390) and pneumothorax (194/390) annotated by board-certified radiologists. For evaluation of anatomical abnormality localization, we utilize the anatomy BBox from the Chest ImaGenome dataset [26], which were extracted by an atlas-based detection pipeline. We use a 80%-10%-10% train-validation-test split with no patient shared across splits. For both anatomy and observation encoders in AGXNet, we use DenseNet-121 [8] with pre-trained weights from ImageNet [4] as the backbone. We train our framework to predict presence of abnormality in the 46 anatomical landmarks and the presence of two diseases (i.e., pneumonia and pneumothorax) that have ground truth BBox in [22]. The β is set to be 0.1 based on validation results. We optimize the networks using SGD with momentum = 0.9, weight decay = $10e^{-4}$ and stop training once the validation error reaches minimum. The learning rate is set to be 0.01 and divided by 10 every 6 epochs. We resize all images to 512×512 without any data augmentation and set the batch size as 16. The model is implemented in PyTorch and trained on a single NVIDIA GPU with 32G of memory.

Evaluation Metric. We produce disease-specific CAMs for pneumonia and pneumothorax, and anatomy-specific CAMs for the anatomical landmarks. We apply a thresholding-based bounding box generation method and extract isolated regions in which pixel values are greater than the 95% quantile of the CAM's pixel value distribution. We evaluate the generated boxes against the ground truth BBox using intersection over union ratio (IoU). A generated box is considered as a true positive when IoU > T(IoU), where $T(*)$ is a threshold.

Table 1. Disease localization on MIMIC-CXR. Results are reported as the average over 5 independent runs with standard deviations. The highest values are highlighted in bold, and the best results without BBox annotation are underlined.

Disease	Model	Supervised	IoU @ 0.1		IoU @ 0.25		IoU @ 0.5	
			Recall	Precision	Recall	Precision	Recall	Precision
Pneumothorax	RetinaNet [15]	✓	0.52 ± 0.04	0.19 ± 0.00	0.41 ± 0.04	0.15 ± 0.00	$\mathbf{0.26 \pm 0.00}$	0.09 ± 0.00
	CheXpert [9]	✗	0.68 ± 0.01	0.40 ± 0.01	0.47 ± 0.01	0.28 ± 0.00	0.14 ± 0.01	0.08 ± 0.01
	AGXNet w/o AGA	✗	0.67 ± 0.04	0.41 ± 0.03	0.54 ± 0.04	0.33 ± 0.03	$\underline{0.16 \pm 0.02}$	$\underline{\mathbf{0.10 \pm 0.01}}$
	AGXNet w/ AGA	✗	$\underline{\mathbf{0.75 \pm 0.01}}$	$\underline{\mathbf{0.44 \pm 0.01}}$	$\underline{\mathbf{0.60 \pm 0.02}}$	$\underline{\mathbf{0.35 \pm 0.01}}$	$\underline{0.16 \pm 0.00}$	0.09 ± 0.00
	AGXNet w/ AGA + PU	✗	0.74 ± 0.03	0.43 ± 0.02	0.58 ± 0.03	0.34 ± 0.02	0.13 ± 0.01	0.07 ± 0.01
Pneumonia	RetinaNet [15]	✓	0.66 ± 0.04	0.27 ± 0.04	0.62 ± 0.04	0.26 ± 0.09	$\mathbf{0.42 \pm 0.04}$	$\mathbf{0.17 \pm 0.03}$
	CheXpert [9]	✗	0.73 ± 0.01	0.40 ± 0.01	0.51 ± 0.01	0.28 ± 0.01	0.19 ± 0.01	0.10 ± 0.01
	AGXNet w/o AGA	✗	0.69 ± 0.01	0.44 ± 0.01	0.59 ± 0.02	0.37 ± 0.02	0.18 ± 0.02	0.11 ± 0.01
	AGXNet w/ AGA	✗	0.67 ± 0.02	0.46 ± 0.01	0.58 ± 0.02	0.39 ± 0.02	0.18 ± 0.03	0.12 ± 0.02
	AGXNet w/ AGA + PU	✗	$\underline{\mathbf{0.73 \pm 0.01}}$	$\underline{\mathbf{0.47 \pm 0.01}}$	$\underline{\mathbf{0.62 \pm 0.01}}$	$\underline{\mathbf{0.40 \pm 0.01}}$	$\underline{0.20 \pm 0.01}$	$\underline{0.13 \pm 0.01}$

Evaluating Disease Localization. We trained a RetinaNet [15] using the annotated BBox from [22] as the supervised baseline and a DenseNet-121 using CheXpert [9] disease labels for CAM-based localization as the WSL baseline. We investigated three variants of AGXNet to understand the effects of AGA module and PU learning. Table 1 shows that AGXNet trained with PU learning achieved the best localization results for pneumonia at $T(\text{IoU}) = 0.1$ and 0.25, while AGXNet trained without PU learning performed best in pneumothorax localization at the same IoU thresholds. Note that the label noise in pneumothorax is intrinsically low, thus adding PU learning may not significantly improve its localization accuracy. RetinaNet achieved the best localization results at $T(\text{IoU}) = 0.5$, probably due to its direct predictions for the coordinates of BBox.

Table 2. Ablation studies on AGA and PU Learning. Results are averaged over 5 runs. α: mixture proportion of positive samples. AA: anatomic abnormality. LAL: left apical lung. LL: left lung. RAL: right apical lung. RLL: right lower lung. RL: right lung.

Model	Classification				AA localization				
	Pneumonia		Pneumothorax		Accuracy @ IoU = 0.25				
	α	AUPRC	α	AUPRC	LAL	LL	RAL	RLL	RL
AGXNet w/o AGA	–	0.57	–	0.55	0.02	0.10	0.08	0.43	0.12
AGXNet w/ AGA	–	0.57	–	**0.57**	0.28	0.22	0.49	0.63	**0.22**
AGXNet w/ AGA + PU	0.14	**0.62**	0.03	**0.57**	**0.38**	**0.23**	**0.50**	**0.69**	0.22

Ablation Studies. Table 2 shows additional results of ablation studies for AGXNet, including (1) disease classification performance on the positive and negative samples in the test set using the area under the precision-recall curve (AUPRC), and (2) anatomical abnormality localization accuracy at $T(\text{IoU}) = 0.25$ for $\{a_j\}$, where $A(a_j, o_k) = P$ and $o_k \in \{\text{pneumonia, pneumothorax}\}$. **Effect of AGA:** Results in Table 1 and Table 2 show that the AGA module significantly improved the results of pneumothorax detection and anatomical abnormality detection, suggesting that the attention module mutually enhanced both types of features. Figure 2 shows the qualitative comparison of two models with (M2) and without (M1) the AGA module. M2 correctly detects pneumothorax and the relevant abnormal anatomical landmarks in both examples, while M1 fails to do so. Furthermore, in the second example, a pigtail catheter is applied to treat pneumothorax and acts as the only discriminative feature used by M1, while M2 is more robust against this shortcut and detects disease in the correct anatomical location. **Effect of PU Learning:** Table 2 shows that the estimated mixture proportion of positive examples (α) is significantly higher in pneumonia (14%) than in pneumothorax (3%), reflecting the fact that there is considerable variability in pneumonia diagnoses [17]. Accordingly, results in Table 1 and Table 2 show that using the PU learning technique improves the performance of pneumonia classification and localization, while it is less effective for pneumothorax whose label noise is intrinsically low.

Table 3. The test AUROCs for the NIH Chest X-ray disease classification task.

Method	Atelectasis	Cardiomegaly	Effusion	Infiltration	Mass	Nodule	Pneumonia	Pneumothorax	Mean
Scratch	0.79	0.91	0.88	0.69	0.81	0.70	0.70	0.82	0.79
Wang et al. [24]	0.72	0.81	0.78	0.61	0.71	0.67	0.63	0.81	0.72
Wang et al. [25]	0.73	0.84	0.79	0.67	0.73	0.69	0.72	0.85	0.75
Liu et al. [16]	0.79	0.91	0.88	0.69	0.81	0.73	0.75	0.89	0.80
Rajpurkar et al. [20]	0.82	0.91	0.88	**0.72**	0.86	0.78	0.76	0.89	0.83
Han et al. [7]	0.84	**0.93**	0.88	**0.72**	**0.87**	0.79	**0.77**	**0.90**	**0.84**
AGXNet - Ana.	0.84	0.91	**0.90**	0.71	0.86	**0.80**	0.74	0.86	0.83
AGXNet - Obs.	**0.85**	0.92	**0.90**	**0.72**	0.86	**0.80**	0.76	0.87	**0.84**
AGXNet - Both	**0.85**	0.92	**0.90**	**0.72**	**0.87**	**0.80**	0.76	0.88	**0.84**

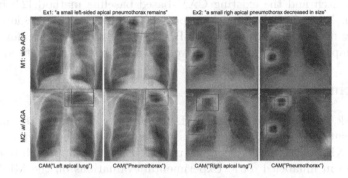

Fig. 2. AGXNet with (M2) or without (M1) the AGA module for pneumothorax detection. Heatmaps indicate the CAMs. The blue and green boxes stand for ground truth anatomy and disease annotations. The red boxes are generated from the CAMs. (Color figure online)

Table 4. The localization accuracy for the NIH Chest X-ray disease localization task. The highest values are highlighted in bold, and the best results without BBox annotation are underlined.

T(IoU)	Method	Supervised	Atelectasis	Cardiomegaly	Effusion	Infiltration	Mass	Nodule	Pneumonia	Pneumothorax	Mean
0.1	Scratch	✗	0.46	0.85	0.61	0.40	0.48	0.11	0.56	0.25	0.47
	Han et al. [7]	✓	**0.72**	0.96	**0.88**	**0.93**	**0.74**	**0.45**	0.65	**0.64**	**0.75**
	Wang et al. [24]	✗	0.69	0.94	0.66	<u>0.71</u>	0.40	0.14	0.63	0.38	0.57
	AGXNet - Ana.	✗	0.64	0.99	0.66	0.69	<u>0.69</u>	<u>0.29</u>	**0.73**	0.29	0.62
	AGXNet - Obs.	✗	0.61	0.99	<u>0.70</u>	0.70	<u>0.69</u>	0.23	0.72	0.55	0.65
	AGXNet - Both	✗	<u>0.71</u>	**1.00**	0.69	0.68	0.68	0.22	0.71	<u>0.62</u>	<u>0.66</u>
0.3	Scratch	✗	0.10	0.45	0.34	0.15	0.14	0.00	0.31	0.06	0.19
	Han et al. [7]	✓	**0.39**	**0.85**	**0.60**	**0.67**	**0.43**	**0.21**	0.40	**0.45**	**0.50**
	Wang et al. [24]	✗	0.24	0.46	0.30	0.28	0.15	<u>0.04</u>	0.17	0.13	0.22
	AGXNet - Ana.	✗	0.20	0.48	0.40	0.43	<u>0.28</u>	0.00	0.46	0.13	0.30
	AGXNet - Obs.	✗	0.22	<u>0.62</u>	0.45	<u>0.46</u>	0.27	0.01	<u>0.54</u>	0.31	<u>0.36</u>
	AGXNet - Both	✗	<u>0.27</u>	0.56	<u>0.47</u>	0.45	0.27	0.01	0.50	<u>0.36</u>	<u>0.36</u>

Transfer Learning on NIH Chest X-ray. We pre-trained an AGXNet + PU model on the MIMIC-CXR dataset using the 46 anatomy labels and 8 observation labels listed in C.1. We then fine-tuned the encoder(s) and re-trained a classifier using the NIH Chest X-ray dataset [24]. We investigated three variants of fine-tuning regimes: (1) only using anatomy encoder, (2) only using observation encoder, and (3) using both encoders and their concatenated embeddings. We compare our transferred models with a baseline model trained from scratch using the NIH Chest X-ray dataset and a series of relevant baselines. We evaluate the classification performance using the area under the ROC curve (AUROC) score and the localization accuracy at $T(\text{IoU}) = 0.1, 0.3$ across the 8 diseases. Table 3 shows that all variants of fine-tuned AGXNet models achieve disease classification performances comparable to the SOTA method [7], demonstrating that both learned anatomy and observation features are robust and transferable. Table 4 shows that all variants of fine-tuned AGXNet models outperform learning from scratch and the existing CAM-based baseline method [24] in disease localization. Note that the model proposed by Han et al. [7] achieved higher accuracy by utilizing the ground truth BBox during training, therefore it is not directly comparable and should be viewed as an upper bound method.

4   Conclusion

In this work, we propose a novel WSL framework to incorporate anatomical contexts mentioned in radiology reports to facilitate disease detection on corresponding CXR images. In addition, we use a PU learning approach to explicitly handle noise in unlabeled data. Experimental evaluations on the MIMIC-CXR dataset show that the addition of anatomic knowledge and the use of PU learning improve abnormality localization. Experiments on the NIH Chest X-ray datasets demonstrate that the learned anatomical and pathological features are transferable and encode robust classification and localization information.

Acknowledgements. This work was partially supported by NIH Award Number 1R01HL141813-01, NSF 1839332 Tripod+X, and SAP SE. We are grateful for the computational resources provided by Pittsburgh SuperComputing grant number TG-ASC170024.

References

1. Bekker, J., Davis, J.: Learning from positive and unlabeled data: a survey. Mach. Learn. **109**(4), 719–760 (2020). https://doi.org/10.1007/s10994-020-05877-5
2. Bhalodia, R., et al.: Improving pneumonia localization via cross-attention on medical images and reports. In: de Bruijne, M., et al. (eds.) MICCAI 2021. LNCS, vol. 12902, pp. 571–581. Springer, Cham (2021). https://doi.org/10.1007/978-3-030-87196-3_53
3. Castellino, R.A.: Computer aided detection (CAD): an overview. Cancer Imaging **5**(1), 17 (2005)
4. Deng, J., Dong, W., Socher, R., Li, L.J., Li, K., Fei-Fei, L.: ImageNet: a large-scale hierarchical image database. In: 2009 IEEE Conference on Computer Vision and Pattern Recognition, pp. 248–255. IEEE (2009)
5. Garg, S., Wu, Y., Smola, A.J., Balakrishnan, S., Lipton, Z.: Mixture proportion estimation and PU learning: a modern approach. In: Advances in Neural Information Processing Systems 34 (2021)

6. Gromet, M.: Comparison of computer-aided detection to double reading of screening mammograms: review of 231,221 mammograms. Am. J. Roentgenol. **190**(4), 854–859 (2008)

7. Han, Y., et al.: Knowledge-augmented contrastive learning for abnormality classification and localization in chest X-rays with radiomics using a feedback loop. In: Proceedings of the IEEE/CVF Winter Conference on Applications of Computer Vision, pp. 2465–2474 (2022)

8. Huang, G., Liu, Z., Van Der Maaten, L., Weinberger, K.Q.: Densely connected convolutional networks. In: Proceedings of the IEEE Conference on Computer Vision and Pattern Recognition, pp. 4700–4708 (2017)

9. Irvin, J., et al.: CheXpert: a large chest radiograph dataset with uncertainty labels and expert comparison. In: Proceedings of the AAAI Conference on Artificial Intelligence, vol. 33, pp. 590–597 (2019)

10. Jain, S., et al.: RadGraph: extracting clinical entities and relations from radiology reports (2021)

11. Jain, S., et al.: VisualCheXbert: addressing the discrepancy between radiology report labels and image labels. In: Proceedings of the Conference on Health, Inference, and Learning, pp. 105–115 (2021)

12. Johnson, A.E., et al.: MIMIC-CXR-JPG - chest radiographs with structured labels (version 2.0.0). PhysioNet (2019). https://doi.org/10.13026/8360-t248

13. Lakhani, P., Sundaram, B.: Deep learning at chest radiography: automated classification of pulmonary tuberculosis by using convolutional neural networks. Radiology **284**(2), 574–582 (2017)

14. Li, Z., et al.: Thoracic disease identification and localization with limited supervision. In: Proceedings of the IEEE Conference on Computer Vision and Pattern Recognition, pp. 8290–8299 (2018)

15. Lin, T.Y., Goyal, P., Girshick, R., He, K., Dollár, P.: Focal loss for dense object detection. In: Proceedings of the IEEE International Conference on Computer Vision, pp. 2980–2988 (2017)

16. Liu, J., Zhao, G., Fei, Y., Zhang, M., Wang, Y., Yu, Y.: Align, attend and locate: chest X-ray diagnosis via contrast induced attention network with limited supervision. In: Proceedings of the IEEE/CVF International Conference on Computer Vision, pp. 10632–10641 (2019)

17. Neuman, M.I., et al.: Variability in the interpretation of chest radiographs for the diagnosis of pneumonia in children. J. Hosp. Med. **7**(4), 294–298 (2012)

18. Peng, Y., Wang, X., Lu, L., Bagheri, M., Summers, R., Lu, Z.: NegBio: a high-performance tool for negation and uncertainty detection in radiology reports. AMIA Summits Transl. Sci. Proc. **2018**, 188 (2018)

19. Qin, C., Yao, D., Shi, Y., Song, Z.: Computer-aided detection in chest radiography based on artificial intelligence: a survey. Biomed. Eng. Online **17**(1), 1–23 (2018)

20. Rajpurkar, P., et al.: CheXNet: radiologist-level pneumonia detection on chest X-rays with deep learning. arXiv preprint arXiv:1711.05225 (2017)

21. Sait, S., Tombs, M.: Teaching medical students how to interpret chest X-rays: the design and development of an e-learning resource. Adv. Med. Educ. Pract. **12**, 123 (2021)

22. Tam, L.K., Wang, X., Turkbey, E., Lu, K., Wen, Y., Xu, D.: Weakly supervised one-stage vision and language disease detection using large scale pneumonia and pneumothorax studies. In: Martel, A.L., et al. (eds.) MICCAI 2020. LNCS, vol. 12264, pp. 45–55. Springer, Cham (2020). https://doi.org/10.1007/978-3-030-59719-1_5

23. Wadden, D., Wennberg, U., Luan, Y., Hajishirzi, H.: Entity, relation, and event extraction with contextualized span representations. In: Proceedings of the 2019 Conference on Empirical Methods in Natural Language Processing and the 9th International Joint Conference on Natural Language Processing (EMNLP-IJCNLP), pp. 5784–5789, November 2019. https://doi.org/10.18653/v1/D19-1585. https://aclanthology.org/D19-1585

24. Wang, X., Peng, Y., Lu, L., Lu, Z., Bagheri, M., Summers, R.M.: ChestX-ray8: hospital-scale chest X-ray database and benchmarks on weakly-supervised classification and localization of common thorax diseases. In: Proceedings of the IEEE Conference on Computer Vision and Pattern Recognition, pp. 2097–2106 (2017)

25. Wang, X., Peng, Y., Lu, L., Lu, Z., Summers, R.M.: TieNet: text-image embedding network for common thorax disease classification and reporting in chest X-rays. In: Proceedings of the IEEE Conference on Computer Vision and Pattern Recognition, pp. 9049–9058 (2018)

26. Wu, J.T., et al.: Chest ImaGenome dataset for clinical reasoning. In: Thirty-Fifth Conference on Neural Information Processing Systems Datasets and Benchmarks Track (Round 2) (2021)

27. Zhou, B., Khosla, A., Lapedriza, A., Oliva, A., Torralba, A.: Learning deep features for discriminative localization. In: Proceedings of the IEEE Conference on Computer Vision and Pattern Recognition, pp. 2921–2929 (2016)

Identification of Vascular Cognitive Impairment in Adult Moyamoya Disease via Integrated Graph Convolutional Network

Xi Chen[1], Wenwen Zeng[1], Guoqing Wu[1], Yu Lei[2], Wei Ni[2], Yuanyuan Wang[1], Yuxiang Gu[2], and Jinhua Yu[1]($\boxtimes$)

[1] School of Information Science and Technology, Fudan University, Shanghai, China
jhyu@fudan.edu.cn
[2] Department of Neurosurgery, Huashan Hospital, Fudan university, Shanghai, China

Abstract. As one of the common complications, vascular cognitive impairment (VCI) comprises a range of cognitive disorders related to cerebral vessel diseases like moyamoya disease (MMD), and it is reversible by surgical revascularization in its early stage. However, diagnosis of VCI is time-consuming and less accurate if it solely relies on neuropsychological examination. Even if some existing research connected VCI with medical image, most of them were solely statistical methods with single modality. Therefore, we propose a graph-based framework to integrate both dual-modal imaging information (rs-fMRI and DTI) and non-imaging information to identify VCI in adult MMDs. Unlike some previous studies based on node-level classification, the proposed graph-level model can fully utilize imaging information and improve interpretability of results. Specifically, we firstly design two different graphs for each subject based on characteristics of different modalities and feed them to a dual-modal graph convolution network to extract complementary imaging features and select important brain biomarkers for each subject. Node-based normalization and constraint item are further devised to weakening influence of over-smoothing and natural difference caused by non-imaging information. Experiments on a real dataset not only achieve accuracy of 80.0%, but also highlight some salient brain regions related to VCI in adult MMDs, demonstrating the effectiveness and clinical interpretability of our proposed method.

Keywords: Vascular cognitive impairment · Moyamoya disease · Functional magnetic resonance imaging · Graph convolution network · Brain biomarkers

---

X. Chen and W. Zeng—These authors contributed equally to this work.

Supplementary Information The online version contains supplementary material available at https://doi.org/10.1007/978-3-031-16443-9_64.

1 Introduction

Vascular cognitive impairment (VCI) is a heterogeneous group of cognitive disorders that share a presumed vascular cause [1], which is one of the common complications of cerebral vessel diseases like moyamoya disease (MMD). Patients with VCI usually have different impairment in cognitive domains including execution, memory, language, and visuospatial functions, bringing the heavy burden to the society and family. The morbidity of VCI is only inferior to Alzheimer's disease (AD) and it shows that there are nearly 2/3 MMD patients suffering from different degrees of VCI in a previous research, especially in adults [2]. Fortunately, VCI is reversible in its early stage of mild impairment through surgical revascularization [3,4], indicating that effective method of early VCI identification is important for disease prevention and treatment. However, the diagnosis of VCI at present is usually depended on neuropsychological examination, which is not only subjective and time-consuming, but also limited to the disability or uncooperativeness of patients. Therefore, identifying VCI in more objective and accurate way is of great clinical significance for both patients and clinicians.

In recent years, functional magnetic resonance imaging(fMRI) has been widely applied in neuroscience, providing people with more insight in neurological diseases. Among multiple imaging modalities, resting-state functional MRI (rs-fMRI) and diffusion tensor imaging (DTI) are the most popular. Specifically, rs-fMRI can describe spontaneous neuronal activity and intrinsic connections between brain regions while DTI is able to assess the integrity of white matter structures and detect abnormalities of neural fibers [5]. Lei et al. [6] studied the spatial pattern of functional brain activity in adult MMDs with VCI by examining the amplitude of low-frequency fluctuations (ALFF) of rs-fMRI, demonstrating that MMD exhibits a specific intrinsic pattern of ALFF. And Kazumata et al. [7] investigated the relationship between brain microstructural integrity and cognitive performance in MMD by analyzing DTI metrics, showing that disruption of white matter is important in the development of cognitive dysfunction. However, most of previous researches mainly focus on statistical analysis [8] rather than accurate diagnosis. Although a machine learning framework based on high-order functional connectivity (FC) was first proposed to recognize VCI in in adult MMDs [9], the radiomic process was less efficient to a certain extent.

Therefore, in order to achieve accurate and interpretable identification of VCI in adult MMDs, we propose a novel framework based on integrated GCN as shown in Fig. 1. The whole pipeline is on the basis of graph-level classification, which means a brain network can be characterized as a graph with a series of nodes and edges. **Our contribution**, when compared to existing methods includes: 1) we design different ways to extract complementary information from rs-fMRI and DTI when constructing graphs, which maximizes the utilization of characteristics of different modalities. 2) node-based normalization and similarity constraint item are proposed to improve performance by solving the problem of over-smoothing and integrating non-imaging information, respectively. 3) some salient biomarkers for VCI identification are selected by introducing self-attention graph pooling mechanism which combines node features and graph topology, showing the clinical value of the proposed model.

2   Method

2.1   Materials and Preprocessing

The datasets used in this paper were all obtained from the same hospital, composing of 218 adult MMDs with both rs-fMRI and DTI modalities. Detail about the dataset is shown in supplementary materials. According to the mini mental state examination (MMSE), 120 MMDs were diagnosed as VCI and the rest were not. The raw rs-fMRI and DTI data were mainly preprocessed by Data Processing & Analysis for Brain Imaging (DPABI [10]) and the pipeline for analyzing brain diffusion images (PANDA [11]), respectively. After that, both modalities were registered to standard space, and the whole brain was divided into 90 brain regions using the Automated Anatomical Labeling (AAL) brain mapping template as regions of interest (ROI).

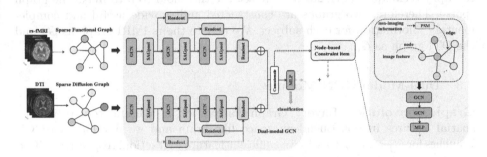

Fig. 1. Overview of the proposed method. The edge weight in node-based graph is calculated by pair similarity measurement (PSM).

2.2   Graph Construction with Different Modalities

Construction of graph plays an important role in a graph-level classification task since the features of each subject should be extracted and represented with a graph effectively and reasonably. We define an undirected weighted graph $G = \{\mathcal{V}, \mathcal{E}\}$ to describe brain regions and the associations between them, where $\mathcal{V} = \{v_i\}_{i=1}^{N}$ is the node set of brain ROIs, and $\mathcal{E} = \{e_{ij}\}$ is the edge set.

For rs-fMRI, temporal features of blood oxygen saturation and FC between ROIs are the most important characteristics, but many previous researches [12, 13] neglected the temporal information which has proven to be useful [14]. Therefore, we use the mean time series of each ROI as node feature and FC between corresponding nodes as the edge weights (Fig. 2a), fusing the functional connections and dynamic time series into a graph. It should be noticed that FC matrix used here is calculated through Pearson's correlation and 0.2 is the threshold to make this graph sparse instead of fully-connected.

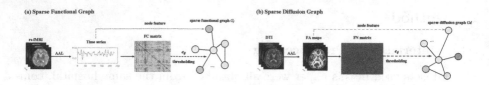

Fig. 2. Method of constructing functional and diffusion graph.

For DTI, some parametric maps such as fractional anisotropy (FA) generated from raw DTI data are widely used since they can reflect the growth of white matters explicitly. Thus we propose to extract radiomic features from FA maps for every ROI as the node feature, which includes the texture and grayscale features in this ROI. As for the edge, the FN matrix which describes the fiber number between two ROIs is used in our work according to deterministic trac- tography techniques (Fig. 2b) and we set the threshold to 5 to make the graph sparse. Therefore, two graphs are constructed and provide useful and comple- mentary information for each subject. We denote the rs-fMRI graph as G_f and the DTI graph as G_d.

2.3   Dual-Modal GCN Module

Graph Convolution Layer: Graph convolution is a useful tool to extract spatial features in non-Euclidean space like brain and we develop two GCN branches for G_f and G_d to achieve sufficient feature extraction, respectively. Tak- ing G_f as example, let $\tilde{A}_f$ and $\tilde{D}_f$ be the adjacency matrix with self-connections $\left(\tilde{A}_f = A + I_N\right)$ and the degree matrix of $\tilde{A}_f$, then the lth GCN layer can be calculated as:

$$H_f^{(l+1)} = \sigma\left(\tilde{D}_f^{-\frac{1}{2}}\tilde{A}_f\tilde{D}_f^{-\frac{1}{2}}H_f^{(l)}\Theta^{(l)}\right) \tag{1}$$

where $\Theta^{(l)}$ is the weight matrix in the lth layer and σ is the activation function.

Self-attention Graph Pooling Layer: Our objective is graph-level classifi- cation, but each graph still has high-dimension nodes or features after convo- lution. Therefore, a self-attention mechanism is introduced in pooling layer [15] to reduce the dimension of whole graph by keeping the most important nodes while dropping disturbing ones in each graph. The attention scores of nodes can also be calculated with Eq. (1). Both node features and graph topology are con- sidered with self-attention pooling, enhancing the reliability of selected nodes. More details are listed in supplementary materials.

Readout Layer: Every pooling layer is followed by a readout layer which is used to aggregate node features and transform them into a fixed-size representation by concatenating the max and mean value of all the node features in graph.

Therefore, a GCN module for each modality comprises three layers mentioned above. And the final representation of each subject with dual-modal GCN is the concatenation of output from the two GCN branches, which is $Z_{patient} = Z_{out,f} \parallel Z_{out,d} \in \mathbb{R}^{4F}$ and fed to the multilayer perceptron (MLP) for final classification (F is the dimension of node features in convolutional hidden layer).

2.4   Node-Based Normalization and Constrain

Even though the above backbone architecture can achieve automatic identification and biomarker selection, there are still some problems hindering the model from better performance. On the one hand, since GCN needs to aggregate information from neighbor nodes, features from different nodes tend to be more similar [16] after multi-layer convolution, making $Z_{patient}$ from different subjects less distinguishable. On the other hand, fMRI, especially rs-fMRI, is very sensitive to machine manufacturer, which means the extracted features from GCN may exist natural differences because images are generated from different machines. Many studies show that consideration of non-imaging information can effectively improve classification accuracy [12,13]. Therefore, in order to improve the final classification, we propose two improvements to reduce the impact of above problems. It is worth noting that we regard each subject as a node (not a graph anymore) in the following sections since each subject already has a row vector of imaging feature $Z_{patient}$ extracted previously.

Node-Based Normalization: Centralizing and scaling measures are adopted here to solve the indistinguishability of node features. According to Zhao et al. [17], we assume that normalization can expand the pairwise feature distance between different nodes, avoiding them from becoming more and more similar due to the aggregation of graph convolution. Let the features of all M subjects after dual-modal GCN denoted by $Z_{all} \in \mathbb{R}^{M \times 4F}$, the following equations are used for normalization:

$$centralizing : Z^c_{all,i\cdot} = Z_{all,i\cdot} - \frac{1}{M} \sum_{i=1}^{M} Z_{all,i\cdot} \tag{2}$$

$$scaling : \dot{Z}_{all,i\cdot} = s * \frac{Z^c_{all,i\cdot}}{\sum_{j=1}^{4F} \|Z_{all,ij}\|_2^2} \tag{3}$$

where $Z_{all,i\cdot}$ is the ith row of Z_{all}, i.e., the imaging feature extracted from the ith subject, $Z_{all,ij}$ is the element of the ith row and j^{th} column, and s is the scaling factor.

Similarity Constraint Mechanism: Normalization to some extent avoids different node features being too similar, but the influence of non-imaging information still needs to be considered. In order to fuse the imaging and non-imaging information (e.g., manufacturer, hemorrhage), we propose to use a node-based constraint item to restrict the result of above graph-level classification. As shown in Fig. 1, a node-based graph is constructed firstly: the node features are the normalized imaging features of each subject while the edge weights are influenced by the similarity of non-imaging information. Specifically, if we denote q non-imaging information of the uth node as $C_u = [C_u(1), C_u(2), \ldots, C_u(q)]$, then the edge weight between two nodes (i.e., subjects) can be calculated with following equations, where g_u is output of MLP:

$$g_u = \Phi^{(2)} \sigma \left(\Phi^{(1)} C_u + b \right) \tag{4}$$

$$w(u, v) = \cos(g_u, g_v) + 0.5 \tag{5}$$

Thus, both dual-modal imaging and non-imaging information are taken into account for all the subjects. After sending this whole graph into a simple GCN, a node-based loss from it can be regarded as a constraint item to force the main graph-level GCN to achieve final classification within those subjects with similar non-imaging information. The final loss function is $Loss = loss_1 + \lambda loss_2$. It should be noted that the final classification result is still based on the output of graph-level GCN from dual modalities.

3   Experiments and Results

3.1   Experimental Settings

In our experiment, we used two independent GCNs to implement graph-level classification and each one of them had three GCN modules. Design details are in supplementary materials. The SGD optimizing algorithm was used with initial learning rate $= 0.002$ and weight decay $= 0.001$. The number of subjects divided into training, validation and testing dataset were 156, 30 and 35, respectively. For comparison experiments, 10-fold cross-validation was also performed to verify the generalization of the proposed method. Classification accuracy (ACC), sensitivity (SEN), specificity (SPE), positive predictive value (PPV) and negative predictive value (NPV) were reported as final evaluation metrics.

3.2   Identification Results

As shown in the left part of Table 1, we first performed ablation experiments to demonstrate the effectiveness of the proposed model. It is obvious that combination of rs-fMRI and DTI performed better than a single modality, which is consistent with some previous studies since different modalities can provide complementary information. Based on the dual-modal GCN, we further implemented experiments in two aspects. On the one hand, we compared the performance by introducing our two improvements including node-based normalization and constraint item. It can be observed that the performance is better when adding the normalization and constraint item. In addition, we calculated the difference of node features before and after normalization mechanism, and the 11.3%

increment suggests that normalization can weaken the similarity between node features and make them more distinguishable. On the other hand, in order to investigate the significance of graph construction, we replaced our proposed G_f by taking each row of FC matrix as node features in the penultimate line of the left part of Table 1, which was a common method in other works [12]. The final performance indicates that graph construction is fundamental but important in a GCN model and other improvements can take effect only after reasonable graphs are constructed. The way we designed graphs is able to utilize imaging principles of two modalities effectively such as the dynamic information and functional correlation from rs-fMRI and white matter integrity from DTI.

Table 1. Identification results of experiments with independent testing set.

Ablation experiments				Comparison experiments			
	ACC	SEN	SPE		ACC	SEN	SPE
rs-fMRI	0.667	0.824	0.462	ResNet [18]	0.686	0.850	0.467
DTI	0.700	0.706	0.692	Population GCN [13]	0.629	0.600	0.667
Dual-modal	0.733	0.765	0.692	High-order FC [9]	0.714	0.800	0.600
Dual-modal+N	0.767	0.706	0.846	EV-GCN [19]	0.743	0.850	0.600
Dual-modal+N+C$^{(1)}$	0.667	0.706	0.615	BrainGNN [20]	0.743	0.762	0.714
Ours*	0.800	0.800	0.800	Ours*	0.800	0.800	0.800

We also compared the proposed model with some state-of-the-art methods as shown in the right part of Table 1 and the whole Table 2. The compared methods cover from the traditional radiomics [9] to different variants of GCN, and our model achieved the best performance for identifying VCI in adult MMDs. We attribute the outstanding performance to the complementary information extracted from dual-modal graphs and our node-based improvements.

Table 2. Identification results of comparison experiments with cross-validation.

Cross-validation (Mean ± SD)					
Methods	ACC	SEN	SPE	PPV	NPV
ResNet [18]	0.685 ± 0.052	0.800 ± 0.058	0.538 ± 0.104	0.689 ± 0.062	0.682 ± 0.083
Population GCN [13]	0.666 ± 0.045	0.717 ± 0.132	0.598 ± 0.158	0.699 ± 0.082	0.643 ± 0.082
High-order FC [9]	0.714 ± 0.067	0.775 ± 0.125	0.647 ± 0.161	0.741 ± 0.102	0.715 ± 0.124
EV-GCN [19]	0.737 ± 0.043	0.808 ± 0.079	0.659 ± 0.128	0.752 ± 0.090	0.740 ± 0.081
BrainGNN [20]	0.732 ± 0.053	0.714 ± 0.123	0.748 ± 0.097	0.779 ± 0.092	0.689 ± 0.096
Ours*	0.794 ± 0.039	0.825 ± 0.062	0.755 ± 0.070	0.808 ± 0.053	0.778 ± 0.070

3.3 Biomarker Interpretation

It is essential for a framework to be able to discover personal biomarkers for further clinical interpretation. Therefore, unlike using node-level classification GCN used in previous research, our proposed model established a graph-level classification GCN according to the imaging characteristics of rs-fMRI and DTI, projecting every node in a graph to a brain region of the subject. By adopting a self-attention graph pooling mechanism, the proposed model can retain some salient brain biomarkers for each subject. Since the final selected biomarkers may be varied among subjects, we analyzed them from two statistical perspectives.

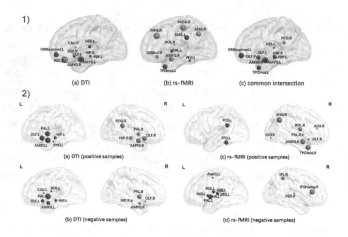

Fig. 3. Top 10 biomarkers selected with 1) frequency and 2) averaged scores. (Color and size indicate the located hemisphere and weight of corresponding biomarkers). (Color figure online)

The first one is based on the frequency distribution histogram of the final retained ROIs for all subjects, which means we calculated the frequency of all ROIs appeared in the final selected results across all the subjects and sorted them in descending order. Since each GCN branch corresponds to a single modality and generates several final selected ROIs from it, we visualized the top 10 ROIs with the highest frequency in Fig. 3 including ROIs retained by rs-fMRI branches, DTI branches, and their common intersection, respectively. It is apparent that ROIs from common intersection are very similar to those of DTI, which to some extent illustrates the dominant importance of DTI in this model. The second analyzing perspective is implemented by averaging attention scores of all ROIs in three pooling layers for positive and negative samples, respectively. Top 10 ROIs with the highest scores were also visualized in Fig. 3. We presented them from both left and right views to avoid overlapping of the same brain region from left and right hemispheres.

No matter viewed from which perspective, we find that biomarkers from basal ganglia, olfactory cortex and hippocampus-related regions play a vital role in ROI selection with DTI, which have been shown associated with execution, memory function and so on. As for rs-fMRI, brain regions mainly located at inferior parietal lobule (e.g., supramarginal gyrus, angular gyrus) and prefrontal lobe (e.g., middle frontal gyrus, inferior frontal gyrus). Moreover, the different weights of these biomarkers indicates that execution-related brain regions seem to be more sensitive than those related to memory, which is a difference between VCI in adult MMDs and other cognitive impairment diseases like AD.

4 Conclusion

In this paper, we proposed a novel framework to integrate imaging and non-imaging information for VCI identification in adult MMDs. We constructed functional and diffusion graphs based on two modalities and fed them into a dual-modal GCN to achieve graph-level classification. In order to solve problems including over-smoothing and natural difference of image caused by non-imaging disturbance, we adopted node-based normalization and similarity constraint item to enhance performance. Experiment results indicated that our method performed better than other existing methods. Moreover, some salient biomarkers for VCI identification were also investigated, increasing the clinical value of final results. In the future, we will explore information about specific cognitive domains and try to achieve more detailed identification of VCI in adult MMDs.

References

1. Moorhouse, P., Rockwood, K.: Vascular cognitive impairment: current concepts and clinical developments. Lancet Neurol. **7**(3), 246–255 (2008)
2. Araki, Y., Takagi, Y., Ueda, K., et al.: Cognitive function of patients with adult moyamoya disease. J. Stroke Cerebrovasc. Dis. **23**(7), 1789–1794 (2014)
3. Steffens, D.C., Otey, E., Alexopoulos, G.S., et al.: Perspectives on depression, mild cognitive impairment, and cognitive decline. Arch. Gen. Psychiatry **63**(2), 130–138 (2006)
4. Lei, Y., Li, Y.J., Guo, Q.H., et al.: Postoperative executive function in adult moyamoya disease: a preliminary study of its functional anatomy and behavioral correlates. J. Neurosurg. **126**(2), 527–536 (2017)
5. Kantarci, K., Murray, M.E., Schwarz, C.G., et al.: White-matter integrity on DTI and the pathologic staging of Alzheimer's disease. Neurobiol. Aging **56**, 172–179 (2017)
6. Lei, Y., Li, Y., Ni, W., et al.: Spontaneous brain activity in adult patients with moyamoya disease: a resting-state fMRI study. Brain Res. **1546**, 27–33 (2014)
7. Kazumata, K., Tha, K.K., Narita, H., et al.: Chronic ischemia alters brain microstructural integrity and cognitive performance in adult moyamoya disease. Stroke **46**(2), 354–360 (2015)
8. Liu, Z., He, S., Xu, Z., et al.: Association between white matter impairment and cognitive dysfunction in patients with ischemic Moyamoya disease. BMC Neurol. **20**(1), 302 (2020). https://doi.org/10.1186/s12883-020-01876-0

9. Lei, Y., Chen X., Su, J.B., et al.: Recognition of cognitive impairment in adult moyamoya disease: a classifier based on high-order resting-state functional connectivity network. Front. Neural Circuits **14**, 603208 (2020)

10. Yan, C.-G., Wang, X.-D., Zuo, X.-N., Zang, Y.-F.: DPABI: data processing & analysis for (resting-state) brain imaging. Neuroinformatics **14**(3), 339–351 (2016). https://doi.org/10.1007/s12021-016-9299-4

11. Cui, Z.X., Zhong, S.Y., Xu, P.F., et al.: PANDA: a pipeline toolbox for analyzing brain diffusion images. Front. Hum. Neurosci. **7**, 42 (2013)

12. Ktena, S.I., Parisot, S., Ferrante, E., et al.: Metric learning with spectral graph convolutions on brain connectivity networks. NeuroImage **169**, 431–442 (2018)

13. Parisot, S., Ktena, S.I., Ferrante, E., et al.: Disease prediction using graph convolutional networks: application to autism spectrum disorder and Alzheimer's disease. arXiv preprint arXiv:1806.01738 (2018)

14. Allen, E.A., Damaraju, E., Plis, S.M., et al.: Tracking whole-brain connectivity dynamics in the resting state. Cereb. Cortex **24**(3), 663–676 (2014)

15. Lee, J., Lee, I., Kang, J.: Self-attention graph pooling. In: International Conference on Machine Learning, pp. 3734–3743 (2019)

16. Li, Q., Han, Z., Wu, X.M.: Deeper insights into graph convolutional networks for semi-supervised learning. arXiv preprint arXiv:1801.07606 (2018)

17. Zhao, L., Akoglu, L.: PairNorm: tackling oversmoothing in GNNs. In: ICLR (2019)

18. He, K.M., Zhang, X.Y., Ren, S.Q., et al.: Deep residual learning for image recognition. In: CVPR (2016)

19. Huang, Y., Chung, A.C.S.: Edge-variational graph convolutional networks for uncertainty-aware disease prediction. In: Martel, A.L., et al. (eds.) MICCAI 2020. LNCS, vol. 12267, pp. 562–572. Springer, Cham (2020). https://doi.org/10.1007/978-3-030-59728-3_55

20. Li, X.X., Zhou, Y., Dvornek, N., et al.: BrainGNN: interpretable brain graph neural network for fMRI analysis. Med. Image Anal. **74**, 102233 (2021)

Multi-modal Masked Autoencoders for Medical Vision-and-Language Pre-training

Zhihong Chen[1], Yuhao Du[1], Jinpeng Hu[1], Yang Liu[1], Guanbin Li[2(✉)],
Xiang Wan[1,3(✉)], and Tsung-Hui Chang[1]

[1] Shenzhen Research Institute of Big Data, The Chinese University of Hong Kong,
Shenzhen, China
wanxiang@sribd.cn
[2] Sun Yat-sen University, Guangzhou, China
liguanbin@mail.sysu.edu.cn
[3] Pazhou Lab, Guangzhou, China

Abstract. Medical vision-and-language pre-training provides a feasible solution to extract effective vision-and-language representations from medical images and texts. However, few studies have been dedicated to this field to facilitate medical vision-and-language understanding. In this paper, we propose a self-supervised learning paradigm with multi-modal masked autoencoders (M^3AE), which learn cross-modal domain knowledge by reconstructing missing pixels and tokens from randomly masked images and texts. There are three key designs to make this simple approach work. First, considering the different information densities of vision and language, we adopt different masking ratios for the input image and text, where a considerably larger masking ratio is used for images. Second, we use visual and textual features from different layers to perform the reconstruction to deal with different levels of abstraction in visual and language. Third, we develop different designs for vision and language decoders (i.e., a Transformer for vision and a multi-layer perceptron for language). To perform a comprehensive evaluation and facilitate further research, we construct a medical vision-and-language benchmark including three tasks. Experimental results demonstrate the effectiveness of our approach, where state-of-the-art results are achieved on all downstream tasks. Besides, we conduct further analysis to better verify the effectiveness of different components of our approach and various settings of pre-training. The source code is available at https://github.com/zhjohnchan/M3AE.

Keywords: Multi-modal pre-training · Masked autoencoders · Medical vision-and-language analysis

1 Introduction

Medical data is inherently multi-modal, where vision (e.g., radiography, magnetic resonance imaging, and computed tomography) and language (e.g., radiology reports and medical texts) are two primary modalities. Medical vision-and-language pre-training (Med-VLP) aims to learn generic representations from large-scale medical image-text data, which can be transferred to various medical vision-and-language tasks (e.g., medical visual question answering (Med-VQA), medical image-text classification, and medical image-text retrieval). It is an essential technique for jointly understanding medical images and texts, which can be challenging due to the lack of large-scale labeled data and domain knowledge.

Although vision-and-language pre-training (VLP) has drawn sustaining attention [2,7,10,22,24], there are only several studies on VLP in the medical domain. [13] directly applied four VLP models (i.e., LXMERT [24], VisualBERT [12], UNITER [2], and PixelBERT [7]) to a medical image-text classification task yet found that these models did not perform as well as they did in the general domain when there is no domain-specific information integrated. Therefore, [8] proposed to perform the pre-training on medical image-text pairs to capture medical knowledge, but their evaluation was conducted only on Med-VQA despite the promising improvement is observed. The most related work to ours is [18], which pre-trained a Med-VLP model and verified its effectiveness on various downstream tasks. Yet it is limited to the chest X-ray, and more importantly, the pre-training was not performed in a self-supervised manner (i.e., using the diagnosis labels). Moreover, previous studies all used convolutional neural networks (CNNs) as their visual backbones, which limited their simplicity and effectiveness, and purely Transformer-based models [25] are not exploited. Therefore, it is essential to design an appropriate Med-VLP approach from four perspectives, including data (e.g., pre-training corpus), models (e.g., purely Transformer-based models), objectives (e.g., more suitable pre-training objectives), and evaluation (e.g., designs of the downstream benchmark) to promote Med-VLP.

In this paper, we propose an effective yet simple approach to Med-VLP by a multi-modal masked autoencoder (M^3AE) based on purely Transformer-based models. Our M^3AE masks random patches of the input image and random tokens of the input text and reconstructs the missing pixels and tokens. We develop the designs of M^3AE from three perspectives to make this simple approach work: (i) It uses different masking ratios for the input images and texts owing to different information densities of vision and language; (ii) It selects visual and textual features from distinct layers to perform the construction considering the different levels of abstraction in vision and language; (iii) It has two different decoder designs for vision and language, where a Transformer model and a multi-layer perceptron (MLP) are used for vision and language decoding, respectively. As a result, the proposed method is able to learn cross-modal domain-specific knowledge from large-scale medical image-text datasets in a self-supervised manner and does not require fine-grained annotations on either images or texts, resulting in better applicability. We perform the pre-training on two large-scale medical image-text datasets, i.e., ROCO [20] and MedICaT [23]. To verify the effectiveness of our approach and facilitate further research, we construct a medical vision-and-language understanding benchmark including three tasks (i.e., Med-VQA, medical image-text classification, and medical image-text retrieval).

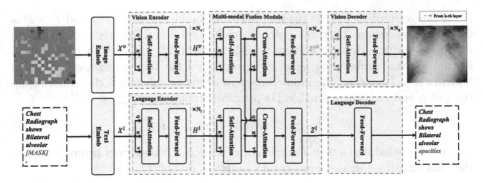

Fig. 1. The overall architecture of our proposed approach, where the vision encoder, language encoder, multi-modal fusion module, and decoders are shown in dash boxes.

Experimental results show that our approach outperforms previous studies on all downstream tasks. In addition, several analyses are also performed to analyze the effectiveness of different components and various settings of pre-training.

2 The Proposed Approach

We adopt the pre-train-and-fine-tune paradigm for medical vision-and-language understanding. In the pre-training stage, the framework develops a variety of pretext tasks to train the model using medical image-text pairs. Formally, given a medical image I and its corresponding description text T, the model is trained to minimize the objective through

$$\theta^*, \theta_1^*, ..., \theta_S^* = \underset{\theta, \theta_1, ..., \theta_S}{\arg\min} \sum_{s=1}^{S} L_s(Y_s, \mathcal{D}_{\theta_s}(\mathcal{M}_\theta(I, T))) \tag{1}$$

where S is the number of pretext tasks, L_s are the loss functions of pretext tasks, $\mathcal{D}_{\theta_s}$ are the decoders with their parameters $\theta_1, ..., \theta_S$, and $\mathcal{M}_\theta$ is the backbone model with its parameters θ. An overview of the proposed approach is demonstrated in Fig. 1, where the details of the backbone model architecture and multi-modal masked autoencoders are illustrated in the following subsections.

2.1 The Backbone Model Architecture

Our backbone model can be partitioned into three major components, i.e., the vision encoder, language encoder, and multi-modal fusion module.

Vision Encoder. In this paper, we focus on purely Transformer-based models and study the use of vision Transformer (ViT) for the vision encoder. In ViT, an image $I \in \mathbb{R}^{H \times W \times C}$ is first segmented into patches $\{p_1, p_2, ..., p_N\}$, where $H \times W$

is the image resolution, C is the number of channels, $p_n \in \mathbb{R}^{P^2 \times C}$ and $P \times P$ is the patch resolution. Then the patches are flattened and linearly projected into patch embeddings through a linear transformation $E^v \in \mathbb{R}^{P^2 C \times D}$ with a special learnable token embedding $p_I \in \mathbb{R}^D$ prepended for the aggregation of visual information. Therefore, the input representations are obtained via summing up the patch embeddings, learnable 1D position embeddings $E^v_{pos} \in \mathbb{R}^{(N+1) \times D}$:

$$X^v = [p_I; p_1 E^v; p_2 E^v; ...; p_N E^v] + E^v_{pos} \tag{2}$$

Finally, X^v is fed into a transformer model with N_v Transformer layers to obtain the contextualized image representations $H^v = [h^v_I; h^v_1; h^v_2; ...; h^v_N]$.

Language Encoder. In the language encoder, we follow BERT [3] to tokenize the input text to subword tokens $\{w_1, w_2, ..., w_M\}$ by WordPiece [27], where the tokens $w_m \in \mathbb{R}^V$ are represented in one-hot form and V is the vocabulary size. Then the tokens are linearly projected into embeddings through a linear transformation $E^l \in \mathbb{R}^{V \times D}$. Afterwards, a start-of-sequence token embedding $w_T \in \mathbb{R}^D$ and a special boundary token embedding $w_{SEP} \in \mathbb{R}^D$ are added to the text sequence. Therefore, the text input representations are computed via summing up the token embeddings and text position embeddings $E^l_{pos} \in \mathbb{R}^{(M+2) \times D}$:

$$X^l = [w_T; w_1 E^l; ...; w_M E^l; w_{SEP}] + E^l_{pos} \tag{3}$$

Finally, X^l is fed into a transformer model with N_l Transformer layers to obtain the contextualized text representations $H^l = [h^l_T; h^l_1; h^l_2; ...; h^l_M; h^l_{SEP}]$.

Multi-modal Fusion Module. We adopt the co-attention mechanism in the multi-modal fusion module to fuse the contextualized representations from images and texts. In detail, the multi-modal fusion module consists of two Transformer models, each of which is a stack of N_m Transformer layers. In each Transformer layer, there are three sub-layers, i.e., a self-attention sub-layer, a cross-attention sub-layer, and a feedforward sub-layer. The attention mechanism is applied in the self-attention and cross-attention sub-layers and it is defined as

$$\text{ATTN}(Q, K, V) = \text{softmax}\left(QK^\top\right) \cdot V \tag{4}$$

In the self-attention sub-layer, the representations interact within modalities:

$$H^{vs} = \text{ATTN}(H^v, H^v, H^v), \quad H^{ls} = \text{ATTN}(H^l, H^l, H^l) \tag{5}$$

In the cross-attention sub-layer, the representations interact across modalities to integrate cross-modal information into their representations:

$$H^{vc} = \text{ATTN}(H^{vs}, H^{ls}, H^{ls}), \quad H^{lc} = \text{ATTN}(H^{ls}, H^{vs}, H^{vs}) \tag{6}$$

Finally, H^{vc} and H^{lc} are input to the feedforward sub-layer (i.e., an MLP) to obtain the multi-modal representations $Z^v = [z^v_I; z^v_1; z^v_2; ...; z^v_N]$ for vision and $Z^l = [z^l_T; z^l_1; z^l_2; ...; z^l_M; z^l_{SEP}]$ for language.

2.2 Multi-modal Masked Autoencoders

The idea of masked autoencoders has achieved great success in natural language processing (i.e., BERT) and recently in computer vision (i.e., MAE [5]) as well. However, in the general VLP area, existing studies [4,10] mainly recovered the original tokens of masked texts (denoted as masked language modeling (MLM)) and demonstrated that reconstructing the original signals of masked images (denoted as masked image modeling (MIM)) hurts the pre-training performance. The reason is that vision and language have different characteristics, and appropriate designs are desired to make masked autoencoders work in such a multi-modal setting. In detail, we develop three essential yet straightforward designs.

Masking Strategy. Information density is different between vision and language. Languages are information-dense messages created by humans, and thus predicting only a few held-out tokens can induce a sophisticated language understanding task. On the contrary, images are spatial redundant.[1] As a result, we use random sampling with a much greater masking ratio for images (i.e., 75%) than for texts (i.e., 15%) to remove redundancy in images and enable the model to acquire valuable features from both images and texts.

Representation Selection for Reconstruction. Images and texts are abstracted at different levels, with pixels of images having a lower semantic level than tokens of texts. In our model, their representations are aggregated layer-by-layer in a hierarchical way. Therefore, to make the final learned representations of images at a high semantic level, we instead adopt the intermediate outputs of the multi-modal fusion module (i.e., the visual outputs from the k-th Transformer layer denoted as Z^{vk}) to perform the low-level construction task (i.e., MIM). For MLM, we still use the final output Z^l for the prediction of tokens since predicting missing words requires richer semantic information.

Decoder Designs. The vision and language decoders aim to map the representations Z^{vk} and Z^l back to their original input image and text, respectively. After being encoded by the backbone model (i.e., the encoder), Z^{vk} and Z^l are represented at a high semantic level. For the vision decoder, its output is in the pixel space, which is of a lower semantic level. Therefore, a Transformer model is introduced as the decoder to map Z^{vk} to lower semantic representations to perform the low-level reconstruction. For the language decoder, its targets (i.e., words) are abstracted at a high level, and thus its design is trivial (i.e., an MLP).

Finally, the MIM loss is computed using the mean squared error (MSE) between the reconstructed and original images in the pixel space, and the MLM loss is computed as the negative log-likelihood loss for the masked tokens.[2]

---

[1] A missing patch can be reconstructed easily from visible neighboring patches.

[2] Note that MLM and MIM are performed in different forward procedures.

3 Experiments

3.1 Pre-training Setup

We conduct our experiments on two datasets, i.e., ROCO [20] and MedICaT [23], where the former contains over 81,000 medical image-text pairs and the latter consists of over 217,000 medical images with their captions and inline textual references. For ROCO, we adopt their official splits, and for MedICaT, we randomly sample 1,000 images for validation, 1,000 images for testing, and the remaining images are used for training. For pre-training, we use the training set of ROCO and MedICaT to train models with the pre-training tasks presented in Sect. 2 together with the common image-text matching task [2] by default.

Table 1. Comparisons of our proposed approach with previous studies on the test sets of three Med-VQA datasets with respect to the accuracy metric.

Methods	VQA-RAD			SLACK			VQA-2019
	Open	Closed	Overall	Open	Closed	Overall	Overall
MFB [29]	14.50	74.30	50.60	72.20	75.00	73.30	–
SAN [28]	31.30	69.50	54.30	74.00	79.10	76.00	–
BAN [9]	37.40	72.10	58.30	74.60	79.10	76.30	–
MEVF-SAN [19]	49.20	73.90	64.10	75.30	78.40	76.50	68.90
MEVF-BAN [19]	49.20	77.20	66.10	77.80	79.80	78.60	77.86
CPRD-BAN [14]	52.50	77.90	67.80	79.50	83.40	81.10	–
Ours	**67.23**	**83.46**	**77.01**	**80.31**	**87.82**	**83.25**	**79.87**

For the implementation, we adopt the architecture of CLIP-ViT-B [21] for the vision encoder and the architecture of RoBERTa-base [16] for the language encoder. For the multi-modal fusion module, we set the number of Transformer layers $N_m = 6$ with the dimension of hidden states set to 768 and the number of heads set to 12. For all pre-training experiments, the models are trained with AdamW optimizer [17] for 100,000 steps. The learning rates for uni-modal encoders (i.e., the vision encoder and the language encoder) and the multi-modal fusion module are set to 1e−5 and 5e−5, respectively. We set the warm-up ratio to 10%, and use the linear learning rate scheduler after warm-up. We use center-crop to resize each image into the size of 288 × 288.

3.2 Vision-and-Language Transfer Tasks

We evaluate all models on three medical image-text understanding tasks (i.e., Med-VQA, medical image-text classification, and medical image-text retrieval).

Medical Visual Question Answering. This task requires answering natural language questions about medical images. We use the official dataset split of three publicly available datasets to train and evaluate the models: VQA-RAD [11], SLAKE [15], and VQA-2019 [1], where questions in VQA-RAD and SLAKE are both categorized into two types (i.e., closed-ended and opened-ended).

Medical Image-Text Classification. This task aims to produce the label given an image-text pair. We train and evaluate the models on MELINDA [26], a biomedical experiment method classification dataset, with its official split.

Medical Image-Caption Retrieval. There are two subtasks for this task, where image-to-text (I2T) retrieval requires retrieving the most relevant texts from a large pool of texts given an image and vice versa for text-to-image (T2I) retrieval. We train and evaluate the models on the official split of ROCO.

Table 2. Classification results on the MELINDA dataset.

Methods	Test
ResNet-101 [6]	63.84
RoBERTa [16]	74.60
NLF [26]	76.60
SAN [28]	72.30
Ours	**78.50**

Table 3. Image-to-text and text-to-image retrieval results on the ROCO test set.

Methods	T2I			I2T		
	R@1	R@5	R@10	R@1	R@5	R@10
ViLT [10]	9.75	28.95	41.40	11.90	31.90	43.20
METER [4]	11.30	27.25	39.60	14.45	33.30	45.10
Ours (ZS)	19.05	47.75	61.35	19.10	45.60	61.20
Ours (FT)	**22.20**	**52.50**	**66.65**	**22.90**	**51.05**	**65.80**

Table 4. Ablation study on the VQA-RAD test set.

MIM	MLM	Open	Closed	Overall
✗	✗	24.67	80.78	58.48
✓	✗	22.41	79.17	56.56
✗	✓	67.04	81.99	76.05
✓	✓	**67.23**	**83.46**	**77.01**

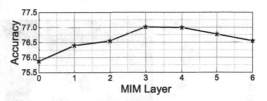

Fig. 2. Results of selecting the representations from different layers to perform MIM.

For the evaluation metrics, the models in Med-VQA and medical text-image classification tasks are evaluated w.r.t. accuracy, while those in the retrieval task are assessed using Recall@K (K = 1, 5, 10). We run each experiment three times with different random seeds and report the mean of its corresponding metric(s).

3.3 Comparisons with the State-of-the-Art

The main experimental results on all downstream tasks are shown in Table 1, 2, 3. Our proposed approach achieves state-of-the-art results on all datasets. For Med-VQA, compared with the advanced approach CPRD-BAN, the proposed method outperforms it by 14.7% and 5.5% w.r.t accuracy on open-ended and closed-ended questions on VQA-RAD. It also achieves 2.1% and 2.0% overall improvements on the SLACK and VQA-2019 datasets, respectively. For medical image-text classification, our method outperforms previous uni-modal and multi-modal methods under the non-continued pre-training setting, where it outperforms NLF by approximately 1.9%. For medical image-text retrieval, the proposed approach outperforms previous studies to a large extent in both the zero-shot (ZS) and fine-tuning (FT) settings.

3.4 Quantitative Analysis

Ablation Study. To demonstrate the effectiveness of different components of our approach, we perform an ablation study, with the results on VQA-RAD shown in Table 4. There are several observations drawn from the results. First, it is shown that only performing MIM in the pre-training can not bring an improvement when comparing between the 1st and 2nd rows. Second, adopting MLM as one of the pre-training objectives (i.e., the 3rd and 4th rows) can achieve considerably better results than the case without MLM (i.e., the 1st and 2nd rows). Third, the proposed M^3AE, with both MIM and MLM as its objectives, achieves the best result. This owes to the fact that with the help of both MIM and MLM, the critical mappings between medical images and texts can be implicitly modeled, which promotes the learning of multi-modal representations.

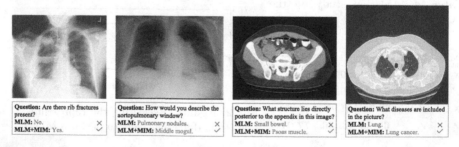

Fig. 3. Illustrations of four Med-VQA results from the models pre-trained with MLM and MLM+MIM on the VQA-RAD test set.

Effectiveness of Different Layers to Perform MIM. To analyze the impacts of representations from different layers to perform MIM, we select the representations from layer 0 to 6 to pre-train our model, with the results shown in Fig. 2. There are two observations drawn from the results. First, when using the representations from layer 0 (i.e., the visual features without any textual information), the method achieves the worst performance, verifying the importance of texts to perform MIM. Second, it is observed that using the representations from the intermediate layer (i.e., layer 3) can achieve the best results. This demonstrates that using the lower-level representations to perform MIM can help the model capture more hierarchical information in images and texts to keep the final learned image representations at a higher semantic level to promote representation learning.

3.5   Qualitative Analysis

To further investigate the effectiveness of our method, we perform qualitative analysis on four Med-VQA cases on the VQA-RAD test set as shown in Fig. 3. In the 1st case, the MLM+MIM model answered an "yes/no" question correctly, and in the 2nd and 3rd cases, it answered the organ-related questions correctly while the MLM model did not. Furthermore, in the last case, the MLM+MIM model answered the disease-related question correctly. This demonstrates that pre-training with MLM+MIM might aid the model in learning more fine-grained mappings between images and texts.

4   Conclusion

In this paper, we believe that massive medical image-text data contains rich context and structural information, which is significant for medical vision-and-language understanding. To this end, we propose an effective yet simple approach, i.e., multi-modal masked autoencoders, to perform pre-training on medical image-text pairs. We develop three simple designs from the perspectives of masking ratios, representation selection for reconstruction, and decoder designs to make this simple approach work. To evaluate our approach comprehensively, we construct a medical vision-and-language understanding benchmark, including three tasks. Experimental results on various datasets demonstrate the superior performance of our approach, where state-of-the-art results are achieved.

Acknowledgement. This work is supported in part by the Chinese Key-Area Research and Development Program of Guangdong Province (2020B0101350001), in part by the Guangdong Basic and Applied Basic Research Foundation (2020B1515020048), in part by the National Natural Science Foundation of China (61976250), in part by the Guangzhou Science and technology project (No. 202102020633), and is also supported by the Guangdong Provincial Key Laboratory of Big Data Computing, The Chinese University of Hong Kong, Shenzhen.

References

1. Abacha, A.B., Hasan, S.A., Datla, V.V., Liu, J., Demner-Fushman, D., Müller, H.: VQA-Med: overview of the medical visual question answering task at ImageCLEF 2019. CLEF (Working Notes), vol. 2 (2019)
2. Chen, Y.-C., et al.: UNITER: UNiversal Image-TExt Representation learning. In: Vedaldi, A., Bischof, H., Brox, T., Frahm, J.-M. (eds.) ECCV 2020. LNCS, vol. 12375, pp. 104–120. Springer, Cham (2020). https://doi.org/10.1007/978-3-030-58577-8_7
3. Devlin, J., Chang, M.W., Lee, K., Toutanova, K.: BERT: pre-training of deep bidirectional transformers for language understanding. In: Proceedings of the 2019 Conference of the North American Chapter of the Association for Computational Linguistics: Human Language Technologies, Volume 1 (Long and Short Papers), pp. 4171–4186 (2019)
4. Dou, Z.Y., et al.: An empirical study of training end-to-end vision-and-language transformers. arXiv preprint arXiv:2111.02387 (2021)
5. He, K., Chen, X., Xie, S., Li, Y., Dollár, P., Girshick, R.: Masked autoencoders are scalable vision learners. arXiv preprint arXiv:2111.06377 (2021)
6. He, K., Zhang, X., Ren, S., Sun, J.: Deep residual learning for image recognition. In: Proceedings of the IEEE Conference on Computer Vision and Pattern Recognition, pp. 770–778 (2016)
7. Huang, Z., Zeng, Z., Liu, B., Fu, D., Fu, J.: Pixel-BERT: aligning image pixels with text by deep multi-modal transformers. arXiv preprint arXiv:2004.00849 (2020)
8. Khare, Y., Bagal, V., Mathew, M., Devi, A., Priyakumar, U.D., Jawahar, C.: MMBERT: multimodal BERT pretraining for improved medical VQA. In: 2021 IEEE 18th International Symposium on Biomedical Imaging (ISBI), pp. 1033–1036. IEEE (2021)
9. Kim, J.H., Jun, J., Zhang, B.T.: Bilinear attention networks. In: Advances in Neural Information Processing Systems, vol. 31 (2018)
10. Kim, W., Son, B., Kim, I.: ViLT: vision-and-language transformer without convolution or region supervision. In: International Conference on Machine Learning, pp. 5583–5594. PMLR (2021)
11. Lau, J.J., Gayen, S., Abacha, A.B., Demner-Fushman, D.: A dataset of clinically generated visual questions and answers about radiology images. Sci. Data 5, 1–10 (2018)
12. Li, L.H., Yatskar, M., Yin, D., Hsieh, C.J., Chang, K.W.: VisualBERT: a simple and performant baseline for vision and language. arXiv preprint arXiv:1908.03557 (2019)
13. Li, Y., Wang, H., Luo, Y.: A comparison of pre-trained vision-and-language models for multimodal representation learning across medical images and reports. In: 2020 IEEE International Conference on Bioinformatics and Biomedicine (BIBM), pp. 1999–2004. IEEE (2020)
14. Liu, B., Zhan, L.-M., Wu, X.-M.: Contrastive pre-training and representation distillation for medical visual question answering based on radiology images. In: de Bruijne, M., et al. (eds.) MICCAI 2021. LNCS, vol. 12902, pp. 210–220. Springer, Cham (2021). https://doi.org/10.1007/978-3-030-87196-3_20
15. Liu, B., Zhan, L.M., Xu, L., Ma, L., Yang, Y., Wu, X.M.: SLAKE: a semantically-labeled knowledge-enhanced dataset for medical visual question answering. In: 2021 IEEE 18th International Symposium on Biomedical Imaging (ISBI), pp. 1650–1654. IEEE (2021)

16. Liu, Y., et al.: RoBERTa: a robustly optimized BERT pretraining approach. arXiv preprint arXiv:1907.11692 (2019)
17. Loshchilov, I., Hutter, F.: Decoupled weight decay regularization. In: International Conference on Learning Representations (2018)
18. Moon, J.H., Lee, H., Shin, W., Choi, E.: Multi-modal understanding and generation for medical images and text via vision-language pre-training. arXiv preprint arXiv:2105.11333 (2021)
19. Nguyen, B.D., Do, T.-T., Nguyen, B.X., Do, T., Tjiputra, E., Tran, Q.D.: Overcoming data limitation in medical visual question answering. In: Shen, D., et al. (eds.) MICCAI 2019. LNCS, vol. 11767, pp. 522–530. Springer, Cham (2019). https://doi.org/10.1007/978-3-030-32251-9_57
20. Pelka, O., Koitka, S., Rückert, J., Nensa, F., Friedrich, C.M.: Radiology Objects in COntext (ROCO): a multimodal image dataset. In: Stoyanov, D., et al. (eds.) LABELS/CVII/STENT -2018. LNCS, vol. 11043, pp. 180–189. Springer, Cham (2018). https://doi.org/10.1007/978-3-030-01364-6_20
21. Radford, A., et al.: Learning transferable visual models from natural language supervision. In: International Conference on Machine Learning, pp. 8748–8763. PMLR (2021)
22. Su, W., et al.: VL-BERT: pre-training of generic visual-linguistic representations. In: International Conference on Learning Representations (2019)
23. Subramanian, S., et al.: MedICaT: a dataset of medical images, captions, and textual references. In: Findings of the Association for Computational Linguistics: EMNLP 2020, pp. 2112–2120 (2020)
24. Tan, H., Bansal, M.: LXMERT: learning cross-modality encoder representations from transformers. In: Proceedings of the 2019 Conference on Empirical Methods in Natural Language Processing and the 9th International Joint Conference on Natural Language Processing (EMNLP-IJCNLP), pp. 5100–5111 (2019)
25. Vaswani, A., et al.: Attention is all you need. In: Advances in Neural Information Processing Systems, vol. 30 (2017)
26. Wu, T.L., Singh, S., Paul, S., Burns, G., Peng, N.: MELINDA: a multimodal dataset for biomedical experiment method classification. In: Proceedings of the AAAI Conference on Artificial Intelligence, vol. 35, pp. 14076–14084 (2021)
27. Wu, Y., et al.: Google's neural machine translation system: bridging the gap between human and machine translation. arXiv preprint arXiv:1609.08144 (2016)
28. Yang, Z., He, X., Gao, J., Deng, L., Smola, A.: Stacked attention networks for image question answering. In: Proceedings of the IEEE Conference on Computer Vision and Pattern Recognition, pp. 21–29 (2016)
29. Yu, Z., Yu, J., Fan, J., Tao, D.: Multi-modal factorized bilinear pooling with co-attention learning for visual question answering. In: Proceedings of the IEEE International Conference on Computer Vision, pp. 1821–1830 (2017)

Breaking with Fixed Set Pathology Recognition Through Report-Guided Contrastive Training

Constantin Seibold[1(✉)], Simon Reiß[1], M. Saquib Sarfraz[1],
Rainer Stiefelhagen[1], and Jens Kleesiek[2]

[1] Karlsruhe Institute of Technology, Karlsruhe, Germany
{constantin.seibold,simon.reiss,m.saquib.sarfraz,
rainer.stiefelhagen}@kit.edu
[2] University Medicine Essen, Essen, Germany
jens.kleesiek@uk-essen.de

Abstract. When reading images, radiologists generate text reports describing the findings therein. Current state-of-the-art computer-aided diagnosis tools utilize a fixed set of predefined categories automatically extracted from these medical reports for training. This form of supervision limits the potential usage of models as they are unable to pick up on anomalies outside of their predefined set, thus, making it a necessity to retrain the classifier with additional data when faced with novel classes. In contrast, we investigate direct text supervision to break away from this closed set assumption. By doing so, we avoid noisy label extraction via text classifiers and incorporate more contextual information. We employ a contrastive global-local dual-encoder architecture to learn concepts directly from unstructured medical reports while maintaining its ability to perform free form classification. We investigate relevant properties of open set recognition for radiological data and propose a method to employ currently weakly annotated data into training. We evaluate our approach on the large-scale chest X-Ray datasets MIMIC-CXR, CheXpert, and ChestX-Ray14 for disease classification. We show that despite using unstructured medical report supervision, we perform on par with direct label supervision through a sophisticated inference setting.

1 Introduction

Radiologists interpret a vast amount of imaging data and summarize their insights as medical reports. This documentation accumulates large databases of radiological imaging and accompanying findings, i.e., millions of collected chest radiographs annually [1]. Computer-aided-diagnosis (CAD) systems utilize these databases to streamline the clinical workflow and save time [8,11,17]. Modern CAD tools often rely on deep learning models [17] using large-scale data sets such

Supplementary Information The online version contains supplementary material available at https://doi.org/10.1007/978-3-031-16443-9_66.

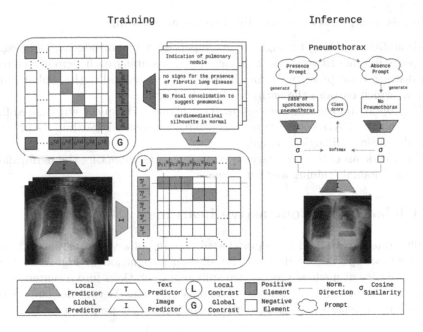

Fig. 1. Illustration of our proposed method. Training on the left, inference on the right.

as MIMIC-CXR [2,7,10,23] for training. Training for such tasks requires hand-designed supervision, typically by extracting a fixed set of predefined labels from the reports using rule- or deep learning-based models [7,20,23]. Such training typically requires hand-designed supervision in the form of extracting a set of labels from the reports using rule- or deep learning-based models [7,20,23]. While these tools can deliver acceptable performance on a subset of diseases [25], they lack generalization capabilities for diseases that were not part of the fixed label-set used for training. To add disease classes requires substantial effort annotating the data with extra labels and retraining the system. To circumvent this, one can approach training in a class-agnostic manner, however, it becomes unclear how models can still be applied to classify diseases.

Recent methods based on contrastive language-image pre-training [9,16,18, 24,26] indicate that by large-scale multi-modal representation learning, object recognition can be detached from prior fixed-set, hand-designed class definitions. These models learn joint feature spaces between images and textual descriptions and utilize text prompts to transform recognition from learned fix-set classification to a matching task between text and image embeddings. Radiological reports, in contrast to natural-image captions, have an inherently different structure, as they encompass multiple distinct sentences such that their entirety describes all relevant information. This shift makes a direct application of existing methods non-trivial.

In this work, we see our contributions as the following:

1. We address training through report supervision by considering radiological reports in one of two ways: The local level, assuming each sentence conveys a distinct concept relevant for the patient, and secondly, the global report view, which encodes the entirety of the findings.
2. We propose a novel inference setting that allows us to query any desired finding, and the CAD system generates a binary decision regarding its presence in the given radiological imagery.
3. We provide an extensive study on various factors impacting the performance of multi-modal training and inference.

2   Global-Local Contrastive Learning

We illustrate our method for report-based training of a vision model and inference protocol specifically designed for disease recognition in Fig. 1. To tackle the complexity of medical reports, we split representations into a sentence- and report-level from a shared visual and language encoder. We consider embeddings for both the presence and absence of a pathology for its prediction.

2.1   Model Overview

Contrastive language and image pretraining (CLIP) has shown immense potential for object recognition in natural images [18] through learning from image-caption pairs. In contrast to textual descriptions in CLIP-based models, medical reports typically consists of multiple sentences focusing on different parts in the image. As each sentence contains specific subset of information, we aim to capture sentence- (local) and report-level (global) context in our representations. Thus, our model builds on separate image- and text encoders ϕ and θ, which embed an image I via $z_I = \phi(I)$ and a sentence by $z_s = \theta(s)$, respectively.

In training, for a given report R, we capture the local context by splitting the full report into its sentences $R = \{s_1, \ldots, s_n\}$ and subsequently extract sentence-level embeddings $z_{s_i} = \theta(s_i), s_i \in R$. To generate global embeddings that contain the full information of the whole report, the sentence-level embeddings are aggregated through attention pooling: $z_R = \text{Attn}([z_{s_1}, \ldots, z_{s_{|R|}}])$ [21]. To embed z_I, z_s and z_R into shared multi-modal representations, we project sentences and reports via linear transformations p^S and p^R into two feature spaces. As the image encoder has access to global image information for report-level prediction as well as to local image patterns for selective sentence-dependent prediction, we project z_I twice: into a global representation $p^G(z_I)$ which shall align with $p^R(z_R)$ and a representation for local patterns $p^L(z_I)$ for alignment with $p^S(z_s)$.

During training, we are provided with a dataset of image-report pairs $(I_i, R_i) \in (I_1, R_1), \ldots, (I_N, R_N)$. For brevity and clarity in subsequent formulas, we will write projections, e.g. the global projection of an image I_i as p_i^G instead of $p_i^G(\theta(I_i))$, for the projection of the kth sentence from report R_i, we write p_{ik}^S.

2.2   Training Objectives

Local Contrast: While radiological reports describe the assessment of a patient's health, not every sentence is directly linked to specific findings, some sentences mention clinical procedures or required follow-up examinations. However, we can assume that all clinically relevant information is present in a subset of sentences in the report due to the doctors' obligation to document the findings. This property is the core of the multiple-instance learning (MIL) assumption. Therefore, it might seem natural to choose MILNCE [14] as the MIL-based objective for integrate sentences in training, yet, this assumption only holds when normalizing over sentences, as not every sentence has to match the image. However, if a sentence fits an image, it should match strictly that image, thus, we hold the regular formulation when normalizing over images. As such, MILNCE does not quite fit this use-case and we redesign it by splitting its symmetry:

$$\mathcal{L}_L(I_i, R_i) = -\log \frac{\sum_{k=1}^{n} \exp(\sigma(p_i^L, p_{ik}^S)/\tau_L)}{\sum_{j=1}^{N} \sum_{m=1}^{n} \exp(\sigma(p_i^L, p_{jm}^S)/\tau_L)}$$
$$-\sum_{k=1}^{n} \log \frac{\exp(\sigma(p_i^L, p_{ik}^S)/\tau_L)}{\sum_{j=1}^{N} \exp(\sigma(p_j^L, p_{ik}^S)/\tau_L)}, \quad (1)$$

with τ_L being a learned parameter and $\sigma(\cdot, \cdot)$ denoting the cosine similarity.

Global Contrast: For our batch we assume that an image-report pair is unique and formulate the following objective leveraging the attention-fused reports via:

$$\mathcal{L}_G(I_i, R_i) = -\log \frac{\exp(\sigma(p_i^G, p_i^R)/\tau_G)}{\sum_{j=1}^{N} \exp(\sigma(p_i^G, p_j^R)/\tau_G)} - \log \frac{\exp(\sigma(p_i^G, p_i^R)/\tau_G)}{\sum_{j=1}^{N} \exp(\sigma(p_j^G, p_i^R)/\tau_G)} \quad (2)$$

Self-supervision: CLIP has been established as a data-hungry algorithm [16, 18]. Several recent methods combine intrinsic supervision signals with the CLIP objective to make full use of the available data [12,15]. As we have access to severely smaller datasets in the medical domain as compared to the natural image domain, we follow Li et al. [12] and integrate SimSiam [4]. For this, we generate two augmented versions of the input image $A_1(I)$ and $A_2(I)$ and add a three-layer encoder-head p^E and a two-layer prediction-head p^P on top of the visual backbone ϕ to enforce similarity between the two views:

$$\mathcal{L}_S(A_1(I), A_2(I)) = -\sigma(p_{A_1(I)}^P, \text{detach}(p_{A_2(I)}^E)) - \sigma(p_{A_2(I)}^P, \text{detach}(p_{A_1(I)}^E)) \quad (3)$$

Furthermore, we utilize the augmented images used for the self-supervised objective to mirror our text-image objectives to the augmented samples.

$$\mathcal{L}_M(I_i, R_i) = \mathcal{L}_G(A_1(I_i), R_i) + \mathcal{L}_L(A_1(I_i), R_i) + \mathcal{L}_G(A_2(I_i), R_i) + \mathcal{L}_L(A_2(I_i), R_i) \quad (4)$$

Our final objective for report-based contrastive learning amounts to:

$$\mathcal{L}(I_i, R_i) = \lambda_1 * \mathcal{L}_L(I_i, R_i) + \lambda_2 * \mathcal{L}_G(I_i, R_i) + \lambda_3 * \mathcal{L}_S(I_i, R_i) + \lambda_4 * \mathcal{L}_M(I_i, R_i) \quad (5)$$

with $\lambda_1 = \lambda_2 = \lambda_3 = 0.5$ and $\lambda_4 = 0.25$.

2.3   Model Inference

For fixed set classification models, the inference process is straightforward: A given image I passed through a network with an activation in the final layer, returning class-wise pseudo probabilities. When model architecture and training procedure do not permit a classification layer, methods often resort to zero-shot-like inference [6, 18, 24] where a nearest neighbor search in semantic space is conducted [5]. In our considered design a text-based query is used to infer the presence or absence of a given disease. In similar CLIP-like models, the text embeddings (e.g., of the disease names) can be matched to an image embedding. The query with the maximum similarity can then be retained as matched.

Such a matching based disease discovery is feasible for detecting single disease class. The underlying assumption of having exclusively one dominant class to predict does not hold for chest radiographs as pathologies are not mutually exclusive. Similarly, modeling co-occurrence as individual classes is also infeasible due to the exponentially rising number of possible class combinations. As such, inference for a multi-label classification needs to be formulated for such contrastively trained methods.

We perform this, by querying an image with class-related textual prompts and interpreting their similarity scores as prediction probabilities for the respective disease class. In practice, we notice that a single query for class presence is ambiguous since the text embeddings of words and their negations may fall close to one another in the feature space. Due to this proximity of opposing semantics a query could be mistaken with the negation of its class.

To overcome this issue, we propose to perform inference over two sets of queries (q_c^p, q_c^n) for each class. While the query embedding q_c^p indicates the occurrence of a class c, q_c^n indicates its absence, *e.g.* OPACITIES CONSISTENT WITH PNEUMONIA for presence as opposed to THE LUNGS ARE CLEAR for its absence. Then, the cosine similarity between the image and both queries $\sigma(p^I, q_c^p)$ and $\sigma(p^I, q_c^n)$ is computed with the final prediction wrt. class c being defined as:

$$P(c, I) = \frac{\exp(\sigma(p^I, q_c^p)/\tau)}{\exp(\sigma(p^I, q_c^p)/\tau) + \exp(\sigma(p^I, q_c^n)/\tau)}, \tag{6}$$

where τ is the respective learned scaling factor depending on the used projection.

3   Prompt Engineering

Several works on zero-shot classification perform their inference by extracting the features of the class name through a word embedding model [22]. While sufficient for most zero-shot applications a lot of context regarding the class is lost. In order to effectively utilize language-vision models it is necessary to align the downstream task to the training [18]. As such we model a set of positive and negative prompts applicable for pathologies to enrich our matching process between visual and textual projections. While for our basic approach we consider ('{class}', 'No {class}') prompts, we found that a more detailed prompt design can overall

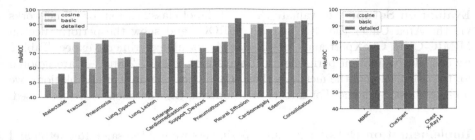

Fig. 2. Performance changes based on differences in prompt generation. Class wise performance on the left. Mean performance to the right. Models trained on MIMIC.

deliver improved performance. As such we consider a set of prompts following the templates ('{adverb}{indication_verb} {effect}* {location}* {class_synonym}', '{adverb} {indication_verb} {absence} {class_synonym}'). Hereby, we utilize all combinations over a small set of categories to gather a variety of different settings. During inference, features of all queries of the same set are averaged.

Prompt-Based Dataset Extension: Despite medical reports being the more common resource in the practical field, currently the majority of large-scale datasets are only publically available with fixed sets of labels. In order to investigate the effect of additional data in training of our method, we reverse our proposed prompt engineering to generate synthetic reports for the datasets Pad-Chest and ChestX-Ray14 based on their class-labels. Through this procedure, we are able to sample sentences indicating presence or absence of a class and generate more than 200k added image-report pairs.

4 Experiments

4.1 Experimental Setup

Datasets:

- **MIMIC-CXR:** It contains 377,110 chest X-rays taken from 65179 patients with 14 disease labels and 227,835 reports. We use the splits provided by [10]. Unless further specified all models were trained on this dataset.
- **CheXpert:** It contains 224,316 chest X-rays taken from 65,240 patients with 14 disease labels. The labels are shared with MIMIC-CXR. We only consider the validation split provided by [7].
- **ChestX-ray14:** It contains 112,120 frontal-view chest X-rays taken from 30,805 patients with 14 disease labels. We use the splits provided by [19].
- **PadChest:** It consists 160k chest X-rays of 67k patients with 174 findings.

Evaluation Setup: We evaluate the multi-label classification for all networks via the Area Under the ROC-curve (AUROC) and show the performance on MIMIC-CXR, CheXpert and ChestX-Ray14. For all experiments expect Table 2 we consider validation performance. Labels with value −2 and −1 are ignored for the calculation of the metric as their state is not certain. For all ablations, we use the "basic"-scheme. For further experiments, the "detailed"-scheme is used.

Implementation Details: For all experiments we use the same ResNet50 and Transformer as Redford et al. [18] as backbones. We optimize with AdamW [13], a learning rate of 0.0001 and a cosine schedule. We trained classification models with a learning rate of 0.0005. We trained all models for 40 epochs, while validating the models in every epoch on all datasets. After training, we evaluated the best validation model to their respective test sets. During training, we resize the images to the inference size of 320×320 and randomly crop by 288×288. For specific further augmentations, we follow SimSiam [4].

4.2   Results

Ablation - Effect of Heads: We investigate the impact of both prediction heads during inference. We start by showing the individual head performance and then go over to different fusion approaches on the left of Table 1. For feature fusion we consider the concatenation of local and global features of the same

Table 1. Left: Impact of chosen scores for inference. Right: Ablation of model parts.

Inf.	MIMIC	CheXpert	CXR14	Avg.	Parts	MIMIC	CXpert	CXR14	Avg.
Local	77.81	78.09	71.72	75.87	$\mathcal{L}_G$	75.47	77.24	69.22	73.97
Global	76.24	80.42	71.00	75.88	$\mathcal{L}_G+\mathcal{L}_L$	76.20	82.24	69.26	75.90
Max	76.85	71.29	78.22	75.45	$\mathcal{L}_G+\mathcal{L}_L+\mathcal{L}_S$	76.10	76.08	74.24	75.47
Cat	77.29	80.30	71.72	76.43	$\mathcal{L}_G+\mathcal{L}_L+\mathcal{L}_M$	77.03	77.36	71.72	75.37
Mean	77.06	81.08	71.50	76.54	Ours	77.06	81.08	71.50	76.54

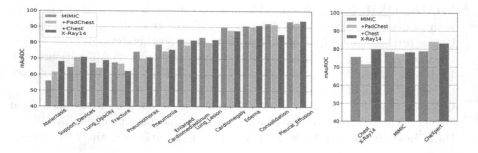

Fig. 3. Contributions of data scaling for chest radiograph dataset. Performance change of adding additional chest X-ray datasets with prompt-based captions.

Table 2. Classification performance on MIMIC, CheXpert and Chest-XRay14. * indicates that the model was trained with additional PadChest and ChestX-Ray14 data.

Method	MIMIC-CXR (in-domain)		CheXpert (out-of-domain)	ChestX-Ray14 (out-of-domain)	
	Val	Test	Val	Val	Test
Label-supervised	77.26	77.42	78.90	79.70	76.47
CLIP	73.23	70.25	75.85	68.03	63.34
SLIP	72.45	72.44	78.49	71.45	67.55
$MILNCE_{local}$	69.30	69.18	74.98	67.56	63.06
LoCo	77.03	78.15	81.71	71.92	68.14
GloCo	75.47	76.58	77.24	69.22	65.86
Ours	78.46	79.40	78.86	75.77	71.23
Ours*	78.30	80.40	83.24	79.90	78.33

modality. For score-fusions, we calculate scores as described above and aggregate the class predictions based on their maximum or average.

We see that for our method both the global and local head show nearly the same performance. While performing max-score fusion the across-dataset-performance drops by 0.4%, where mean-score fusion improves by 0.6%.

Ablation - Effect of Losses: On the right of Table 1, we show the impact of the objective functions. We see that adding local contrast improves the model by 2%. Adding the self-supervised and mirrored objective worsen performance by 0.45%. It can be noted that the self-supervised loss achieves the best performance on ChestX-Ray14 by more than 2%. Whereas adding both simultaneously improves across dataset performance by 0.6%.

Multi-label Inference and Prompt Engineering: We show the impact of our proposed inference scheme and prompts in Fig. 2. We see that performance overall improves with significant improvements for some classes such as fractures which were unable to be categorized just using cosine-similarity. When the detailed prompt the mean performance further improves.

Data Size Impact: We show the impact of using additional prompt-based reports during training in Fig. 3. We see that including artificial training data for ChestX-Ray14 significantly improves its validation performance. In general it seems that while for some classes performance seems to worsen, the overall performance improves when adding additional data.

Comparison with Other Approaches: We compare against the same vision network trained with label supervision on its respective dataset. All other methods were trained using the MIMIC-CXR dataset. SLIP [15] refers to a version of CLIP, which incorporates self-supervision in form of a SIMCLR-like objective [3]. $MILNCE_{local}$ refers to our local branch trained with the MILNCE [14] objective alone. LoCo and GloCo refer to our method trained with either just the local or global objective respectively. We evaluate using the "detailed"-prompt scheme. We show the results in Table 2.

We see that our formulation of the local contrastive loss outperforms the MILNCE version across all datasets. Our proposed method outperforms the considered contrastive language-image pretraining baselines in the form of CLIP and SLIP and manages to achieve similar performance as the supervised ResNet for domains similar to MIMIC, however, underperforms for the ChestX-Ray14 dataset. When adding the additional report datasets of PadChest and ChestX-Ray14 we manage to beat label-supervised performance across all datasets.

5 Conclusion

In this paper, we proposed an approach to make networks less reliant to label supervision through contrastive language-image pre-training on report level. In order to still maintain competitive levels of performance we introduced a novel way of constructing inference. Doing so we are able to offset issues stemming from explicit class similarities. We show that despite using unstructured medical report supervision, we perform on par with explicit label supervision through a sophisticated inference setting across different datasets.

Acknowledgements. The present contribution is supported by the Helmholtz Association under the joint research school "HIDSS4Health - Helmholtz Information and Data Science School for Health" and by the Helmholtz Association Initiative and Networking Fund on the HAICORE@KIT partition.

References

1. National health service. https://www.england.nhs.uk. Accessed 1 Mar 2022
2. Bustos, A., Pertusa, A., Salinas, J.M., de la Iglesia-Vayá, M.: PadChest: a large chest x-ray image dataset with multi-label annotated reports. Med. Image Anal. **66**, 101797 (2020)
3. Chen, T., Kornblith, S., Norouzi, M., Hinton, G.: A simple framework for contrastive learning of visual representations. In: International Conference on Machine Learning, pp. 1597–1607. PMLR (2020)
4. Chen, X., He, K.: Exploring simple Siamese representation learning. In: Proceedings of the IEEE/CVF Conference on Computer Vision and Pattern Recognition, pp. 15750–15758 (2021)
5. Frome, A., et al.: DeViSE: a deep visual-semantic embedding model. In: Advances in Neural Information Processing Systems, vol. 26 (2013)
6. Huang, S.C., Shen, L., Lungren, M.P., Yeung, S.: GLoRIA: a multimodal global-local representation learning framework for label-efficient medical image recognition. In: Proceedings of the IEEE/CVF International Conference on Computer Vision, pp. 3942–3951 (2021)

7. Irvin, J., et al.: CheXpert: a large chest radiograph dataset with uncertainty labels and expert comparison. In: Proceedings of the AAAI Conference on Artificial Intelligence, vol. 33, pp. 590–597 (2019)

8. Jaiswal, A.K., Tiwari, P., Kumar, S., Gupta, D., Khanna, A., Rodrigues, J.J.: Identifying pneumonia in chest X-rays: a deep learning approach. Measurement **145**, 511–518 (2019)

9. Jia, C., et al.: Scaling up visual and vision-language representation learning with noisy text supervision. In: International Conference on Machine Learning, pp. 4904–4916. PMLR (2021)

10. Johnson, A.E., et al.: MIMIC-CXR, a de-identified publicly available database of chest radiographs with free-text reports. Sci. Data **6**(1), 1–8 (2019)

11. Kim, M., et al.: Deep learning in medical imaging. Neurospine **16**(4), 657 (2019)

12. Li, Y., et al.: Supervision exists everywhere: a data efficient contrastive language-image pre-training paradigm. arXiv preprint arXiv:2110.05208 (2021)

13. Loshchilov, I., Hutter, F.: Decoupled weight decay regularization. arXiv preprint arXiv:1711.05101 (2017)

14. Miech, A., Alayrac, J.B., Smaira, L., Laptev, I., Sivic, J., Zisserman, A.: End-to-end learning of visual representations from uncurated instructional videos. In: Proceedings of the IEEE/CVF Conference on Computer Vision and Pattern Recognition, pp. 9879–9889 (2020)

15. Mu, N., Kirillov, A., Wagner, D., Xie, S.: SLIP: self-supervision meets language-image pre-training. arXiv preprint arXiv:2112.12750 (2021)

16. Pham, H., et al.: Combined scaling for zero-shot transfer learning. arXiv preprint arXiv:2111.10050 (2021)

17. Qin, C., Yao, D., Shi, Y., Song, Z.: Computer-aided detection in chest radiography based on artificial intelligence: a survey. Biomed. Eng. Online **17**(1), 1–23 (2018)

18. Radford, A., et al.: Learning transferable visual models from natural language supervision. In: International Conference on Machine Learning, pp. 8748–8763. PMLR (2021)

19. Seibold, C., Kleesiek, J., Schlemmer, H.-P., Stiefelhagen, R.: Self-guided multiple instance learning for weakly supervised disease classification and localization in chest radiographs. In: Ishikawa, H., Liu, C.-L., Pajdla, T., Shi, J. (eds.) ACCV 2020. LNCS, vol. 12626, pp. 617–634. Springer, Cham (2021). https://doi.org/10.1007/978-3-030-69541-5_37

20. Smit, A., Jain, S., Rajpurkar, P., Pareek, A., Ng, A.Y., Lungren, M.P.: CheXbert: combining automatic labelers and expert annotations for accurate radiology report labeling using BERT. arXiv preprint arXiv:2004.09167 (2020)

21. Vaswani, A., et al.: Attention is all you need. In: Guyon, I., et al. (eds.) Advances in Neural Information Processing Systems, vol. 30. Curran Associates, Inc. (2017). https://proceedings.neurips.cc/paper/2017/file/3f5ee243547dee91fbd053c1c4a845aa-Paper.pdf

22. Wang, W., Zheng, V.W., Yu, H., Miao, C.: A survey of zero-shot learning: settings, methods, and applications. ACM Trans. Intell. Syst. Technol. (TIST) **10**(2), 1–37 (2019)

23. Wang, X., Peng, Y., Lu, L., Lu, Z., Bagheri, M., Summers, R.M.: ChestX-ray8: hospital-scale chest X-ray database and benchmarks on weakly-supervised classification and localization of common thorax diseases. In: Proceedings of the IEEE Conference on Computer Vision and Pattern Recognition, pp. 2097–2106 (2017)

24. Wang, X., Xu, Z., Tam, L., Yang, D., Xu, D.: Self-supervised image-text pre-training with mixed data in chest X-rays. arXiv preprint arXiv:2103.16022 (2021)

25. Wu, J.T., et al.: Comparison of chest radiograph interpretations by artificial intelligence algorithm vs radiology residents. JAMA Netw. Open **3**(10), e2022779–e2022779 (2020)

26. Zhang, Y., Jiang, H., Miura, Y., Manning, C.D., Langlotz, C.P.: Contrastive learning of medical visual representations from paired images and text. arXiv preprint arXiv:2010.00747 (2020)

Explaining Chest X-Ray Pathologies
in Natural Language

Maxime Kayser[1]([⊠]), Cornelius Emde[1], Oana-Maria Camburu[2],
Guy Parsons[1,3], Bartlomiej Papiez[1], and Thomas Lukasiewicz[1,4]

[1] University of Oxford, Oxford, UK
maxime.kayser@cs.ox.ac.uk
[2] University College London, London, UK
[3] Thames Valley Deanery, Oxford, UK
[4] TU Wien, Vienna, Austria

Abstract. Most deep learning algorithms lack explanations for their predictions, which limits their deployment in clinical practice. Approaches to improve explainability, especially in medical imaging, have often been shown to convey limited information, be overly reassuring, or lack robustness. In this work, we introduce the task of generating natural language explanations (NLEs) to justify predictions made on medical images. NLEs are human-friendly and comprehensive, and enable the training of intrinsically explainable models. To this goal, we introduce MIMIC-NLE, the first, large-scale, medical imaging dataset with NLEs. It contains over 38,000 NLEs, which explain the presence of various thoracic pathologies and chest X-ray findings. We propose a general approach to solve the task and evaluate several architectures on this dataset, including via clinician assessment.

Keywords: Chest X-rays · Natural language explanations · XAI

1 Introduction

Deep learning (DL) has become the bedrock of modern computer vision algorithms. However, a major hurdle to adoption and regulatory approval of DL models in medical imaging is the lack of explanations for these models' predictions [21]. The combination of lack of model robustness [28], bias (algorithms are prone to amplifying inequalities that exist in the world) [9,27], and the high stakes in clinical applications [26,36] prevent black-box DL algorithms from being used in practice. In this work, we propose natural language explanations (NLEs) as a means to justify the predictions of medical imaging algorithms.

So far, the most commonly used form of explainability in medical imaging is saliency maps, which attribute importance weights to regions in an image. Saliency maps have many shortcomings, including being susceptible to adversarial attacks [8], conveying limited information and being prone to confirmation bias [1,3], as well as only telling us *how much* highlighted regions affect the

---

O.-M. Camburu—This work was done while Oana was at the University of Oxford.

L. Wang et al. (Eds.): MICCAI 2022, LNCS 13435, pp. 701–713, 2022.
https://doi.org/10.1007/978-3-031-16443-9_67

model's output, and not *why* [7]. NLEs, on the other hand, would be able to fully capture how the evidence in a scan relates to the diagnosis. Furthermore, saliency maps are post-hoc explainers, i.e., they do not constrain the model to learn in an explainable manner. In contrast, self-explaining models have many benefits, including being more robust and having a better prediction performance [35]. Alternative approaches for explainability in medical imaging include latent space disentanglement [30], counterfactual explanations [33], case-based explanations [16], and concept-based explanations [18]. NLEs are a valuable addition to the suite of self-explaining models for medical imaging, as they provide easy-to-understand explanations that are able to communicate complex decision-making processes and mimic the way in which radiologists explain diagnoses [6,25]. Previous attempts to augment medical image classification with textual information rely on template-generated sentences [6,22] or sentences from other images based on image similarity [20]. Furthermore, their focus lies mostly on adding descriptive information about a pathology (e.g., its location), instead of explaining the diagnoses. In this work, we leverage the free-text nature of chest X-ray radiology reports, which do not only provide additional details about pathologies, but also the degree of certainty of a diagnosis, as well as justifications of how other observations explain it. Our work builds on the growing work on NLEs approaches in natural language processing [4,19,32], natural image understanding [15,23,24,29], as well as task-oriented tasks such as self-driving cars [17] and fact-checking [19].

We propose MIMIC-NLE, the first dataset of NLEs in the medical domain. MIMIC-NLE extends the existing MIMIC-CXR dataset of chest X-rays [14] with diagnoses, evidence labels and NLEs for the diagnoses. We create MIMIC-NLE by using a BERT-based labeler, a set of clinical explanation keywords, and an empirically and clinically validated set of extraction rules. We extracted over 38,000 high-quality NLEs from the over 200,000 radiology reports present in MIMIC-CXR. Our extraction process introduces little noise, on-par or better than for NLE datasets in natural images [15]. Second, we establish an evaluation framework and compare three strong baselines. The evaluation by a clinician validates the feasibility of the task, but also shows that it is a challenging task requiring future research. The code and dataset are publicly available at https://github.com/maximek3/MIMIC-NLE.

2   MIMIC-NLE

Gathering NLEs is expensive, especially when radiologic expertise is required. To our knowledge, there is currently no NLE dataset for medical imaging. To address this, we show that it is possible to automatically distill NLEs from radiology reports, as radiologists typically explain their findings in these reports. We leverage the radiology reports from MIMIC-CXR [14], a publicly available chest X-ray dataset with 227,827 radiology reports. By applying various filters, labelers, and label hierarchies, we extract 38,003 *image-NLE* pairs, or 44,935 *image-diagnosis-NLE* triplets (as some NLEs explain multiple diagnoses). The extraction process is summarized in Fig. 1. Our filters consist mainly of removing sentences that contain anonymized data, or provide explanations based on patient history or technical details of the scan. More details are provided in the appendix. Furthermore, we only consider frontal, i.e., anteroposterior (AP) and posteroanterior (PA) scans, as these are most commonly used for diagnosis in routine clinical pathways and are most likely to contain the visual information required to generate NLEs.

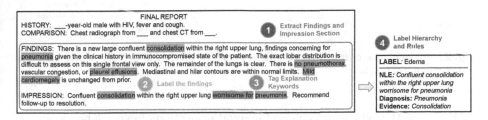

Fig. 1. The steps required to extract NLEs from raw radiology reports. We first extract the Findings and Impression sections, which contain the descriptive part of the report. Next, we identify the labels referred to in each sentence and the sentences that contain explanation keywords. Based on this information, we leverage the rules described in Table 1 to extract valid NLEs, as well as their diagnosis and evidence labels.

First, based on exploring the reports and discussions with clinicians, we observe that a small selection of phrases, such as "compatible with" or "worrisome for" (see full list in Appendix A), are very accurate identifiers of sentences where a potential pathology is explained by observations made on the scan (i.e., an *explanatory* sentence). Next, we make use of the CheXbert labeler [34], which can extract 14 different chest X-ray labels from clinical text, to identify the findings mentioned in each sentence. Thus, for each sentence in the MIMIC-CXR radiology reports, we know whether it is an explanatory sentence, which labels it refers to, and if the labels have a negative, positive, or uncertain (i.e., they are *maybe* present) mention. As the goal of NLEs is to explain predictions, we need to establish which of the labels mentioned in an NLE are being explained and which are part of the evidence. To be able to determine the evidence relationships present in NLEs, we need to restrict ourselves to a limited set of label combinations. For example, if an explanatory sentence in a radiology report indicates the presence of both *Atelectasis* and *Consolidation*, it is not obvious whether *Consolidation* is evidence for *Atelectasis* (e.g., "Right upper lobe new consolidation is compatible with atelectasis with possibly superimposed aspiration.") or whether they are both explained by a different finding (e.g., "A persistent left retrocardiac density is again seen reflecting left lower lobe atelectasis or consolidation."). However, for other label combinations, such as *Consolidation* and *Pneumonia*, the evidence relationship is usually clear, i.e., *Consolidation* is the evidence for *Pneumonia*. We therefore propose the evidence graph in Fig. 2, which depicts the label relationships that have a known and high-confidence evidence relationship (i.e., their co-occurrence in a sentence lets us deduct an evidence relationship with high probability). The graph was constructed with an external radiologist and by empirically validating the co-occurrences of these labels in MIMIC-CXR.

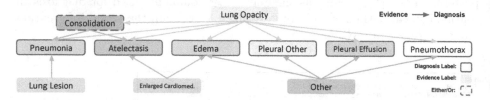

Fig. 2. Our evidence graph that visualizes which of the labels can act as evidence for which diagnosis labels.

Based on the evidence graph and manual inspection of every label combination that appears at least 25 times in an explanatory sentence (121 combinations in total), we established 12 mutually exclusive rules that mark a sentence as a valid NLE. The rules take into account the label combination, their uncertainty label (i.e., *Uncertain* or *Positive*), and the presence of explanation keywords. The rules are defined in Table 1. For each sentence in MIMIC-CXR, we use these rules to determine whether the label combination referred to in the sentence makes it a valid NLE and to determine which of the labels are part of the evidence and the diagnosis. Some label combinations, such as *Consolidation* and *Pneumonia*, are considered a valid NLE even if they do not contain an explanation keyword (as the keywords are not exhaustive). However, for label combinations with less frequent evidence relationships, such as *Enlarged Cardiomediastinum* and *Edema*, we also require the presence of an explanation keyword. It is worth noting that we focus on explanations for positive and uncertain cases only, as strictly negative findings will generally not require case-specific explanations.

Based on these rules, we obtain 38,003 NLEs. One of the authors evaluated a subset of 100 NLEs and found an accuracy of 92% (compared to between 49% to 91% for natural images NLE datasets [15]). We deemed an NLE as correct if it correctly uses the extracted evidence labels to explain the extracted diagnosis labels. The main reasons for incorrect NLEs were failures of the CheXbert labeler. Thus, with improved labelers, the accuracy could be further improved. Using the train, dev, and test splits in MIMIC-CXR [14], we get split sizes of 37,016, 273, and 714 for MIMIC-NLE, respectively.

3 Models

We propose self-explaining models that learn to detect lung conditions and explain their reasoning in natural language. The learned image representations are constrained by mapping them to language that explains the evidence backing a diagnosis. Our approach is illustrated in Fig. 1.

We denote the vision model, i.e., the image classification model, as task model M_T. In our case, $M_T(x) = Y$, where x is a radiographic scan, and Y is the prediction vector $Y \in \mathbb{R}^{n_{unc} \times n_{path}}$, with $n_{unc} = 3$ being the number of certainty levels, and $n_{path} = 10$ being the number of pathologies. This follows the *U-MultiClass* approach from Irvin et al. [13], i.e., for each pathology we classify the image as *negative*, *uncertain*, or *positive*.

Table 1. This table denotes all the included label combinations for NLEs, including which of the labels are being explained and which are the evidence. The column "*kw req.*" specifies which label combinations additionally require the presence of an explanation keyword to be considered an NLE. "*Other/misc.*" refers to evidence that has not been picked up by the CheXbert labeler. If not denoted by ^{U} or ^{P}, all labels can be either positive or uncertain. A^U and B^U are the sets A and B, where all labels are given as uncertain. $\mathcal{P}_{\geq 2}(A^U)$ is the power set of A^U, where each set has at least two labels (i.e., any combination of at least two labels from A^U).

MIMIC-NLE Label Combinations		
Evidence	Diagnosis label(s)	kw req.
Other/misc.	$d \in A = \{$Pleural Eff., Edema, Pleural Other, Pneumoth.$\}$	Yes
Other/misc.	$s \in \mathcal{P}_{\geq 2}(A^U)$	Yes
Lung Opacity	$d \in B = A \cup \{$Pneumonia, Atelectasis$\}$	No
Lung Opacity	$s \in \mathcal{P}_{\geq 2}(B^U)$	No
Lung Opacity	Consolidation	No
Consolidation	Pneumonia	No
{Lung Op., Cons.}	Pneumonia	No
Lung Lesion	Pneumonia	Yes
Lung Opacity	$\{$Atelectasis^P, Pneumonia$^U\}$	No
Consolidation	$\{$Atelectasis^U, Pneumonia$^U\}$	No
Enlarged Card.	Edema	Yes
Enlarged Card.	Atelectasis	Yes

To our knowledge, this is the first application of NLEs to multi-label classification. We address this by generating an NLE for every label that was predicted as *uncertain* or *positive* and is considered a diagnosis label in our evidence graph. Given a set of pathologies P (see Fig. 2), we generate an NLE for every pathology $p_j \in P$. We denote the explanation generator as M_E. For every pathology p_j, we condition the NLE generation on the M_T prediction, i.e., we have $M_E(x_{REP}, Y, p_j) = e_j$, where x_{REP} is the learned representation of the image x, and e_j is the NLE that explains the classification of p_j. By backpropagating the loss of M_E through M_T, M_T is constrained to learn representations that embed the correct reasoning for each diagnosis. During training, we condition M_E on the ground-truth (GT) pathology p_j and prediction vector Y.

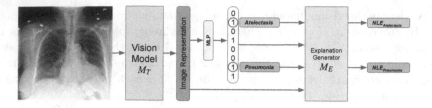

Fig. 3. The model pipeline to provide an NLE for a prediction.

4   Experimental Setup

We evaluate three baselines on our MIMIC-NLE dataset. We propose both automatic and expert-based evaluations.

Baselines. First, we adapt two state-of-the-art chest X-ray captioning architectures to follow the NLE generation approach outlined in Fig. 3. We re-implement TieNet [37], following their paper and a publicly available re-implementation[1] and RATCHET [11], a recent transformer-based captioning method. For each approach, we use DenseNet-121 [12] as the vision model M_T. DenseNet-121 is a convolutional neural networks (CNN) widely used for chest X-ray classification [13]. For TieNet, contrary to the original model, we do not predict the labels from the learned text representations, but condition the NLE generation on the labels. TieNet's decoder, an LSTM [10], is conditioned on the pooled features of the image x_{REP} via an attention layer, on the prediction vector Y by adding its embedding to x_{REP}, and on the diagnosis by initializing the hidden state of the decoder with an encoding of the predicted diagnosis p_j in textual form. RATCHET considers 49×1024 feature maps as x_{REP}, and the embedding of Y is modeled as an additional, 50-th, feature map. We condition on p_j in the same way as in TieNet, by initializing the hidden state of the decoder.

We also propose an additional baseline referred to as DPT (**D**enseNet-121 + **GPT**-2), which is inspired by state-of-the-art NLE approaches for natural image understanding [15,24]. DPT leverages a DenseNet-121 as M_T and a vision-enabled GPT-2 language model [31] as M_E. An attention layer enforces M_E to focus on the same regions in the scan as M_T. We get attention-weighted $7 \times 7 \times 1024$ feature maps from the last layer of DenseNet-121. We consider the NLE generation as a sequence-to-sequence translation from the 49 feature maps (to which we add the relative position of each feature map as a position encoding), a token for Y, and the diagnosis p_j in text form. For fair comparison, we use the same GPT-2 vocabulary for all models and initialize all word embeddings with the pre-trained GPT-2 weights. All models use DenseNet-121 from TorchXRayVision [5], which was pre-trained on CheXpert [13]. All text generation is done via greedy decoding.

---

[1] https://github.com/farrell236/RATCHET/tree/tienet.

Model Training. We use a maximum sequence length of 38 during training, which corresponds to the 98th percentile of the training set. We use the Adam optimizer with weight decay for the transformer models and without weight decay for TieNet (as it failed otherwise). We also use a linear scheduler with warmup for the learning rate. For each model, we experiment with different learning rates and batch sizes as hyperparameters. Using the dev set, we obtain the best hyperparameters as follows: for DPT, a learning rate of 5×10^{-4} and batch size of 16; for RATCHET, a batch size 16 and learning rate 5×10^{-5}; for TieNet, a batch size of 32 and learning rate 1×10^{-3}. We selected these based on the product of the task score S_T, CLEV score, and the average of the BERTScore [38] and METEOR [2] score, all outlined below.

Evaluation Metrics. We evaluate models both on their ability to solve the task, i.e., image multi-label classification, and their ability to explain how they solved the task, i.e., providing NLEs. The task score S_T is given by the weighted AUC score, where we consider *uncertain* and *positive* as one class (following Irvin et al. [13]). For the NLEs scores, we only consider the NLEs that explain correctly predicted labels, as in [4,15]. Previous works have shown that automated natural language generation (NLG) metrics are underperforming for NLEs, as the same answers can be explained in different syntactic forms and even different semantic meanings [15]. We therefore propose the CLEV (CLinical EVidence) score, which verifies whether an NLE refers to the right clinical evidence. For this, we leverage the CheXbert labeler, which extracts the evidence labels referred to in the GT and generated NLEs. For example, if the GT NLE mentions *Lung Opacity* and *Consolidation* as evidence for *Pneumonia*, we expect the generated NLE to contain the same findings. The CLEV score is the accuracy over all the generated NLEs, i.e., what share of them contains exactly the same evidence labels as the corresponding GT. We also provide the NLG metrics of BLEU, Rouge, CIDEr, SPICE, BERTScore, and METEOR as in [15]. For BERTScore, we initialize the weights with a clinical text pretrained BERT model.[2] It is worth noting that, out of the given suite of NLG metrics, BERTScore and METEOR have previously been shown to have the highest (although still low in absolute value) empirical correlation to human judgment in natural image tasks [15].

Clinical Evaluation. Given the difficulties of evaluating NLEs with automatic NLG metrics, we also provide an assessment of the NLEs by a clinician. They were presented with 50 X-ray scans, a diagnosis to be explained, and four different NLEs: the GT and one for each of our three models. The NLEs are shuffled for every image, and the clinician does not know which is which. They are then asked to judge how well each NLE explains the diagnosis given the image on a Likert scale of 1 (*very bad*) to 5 (*very good*).

5 Results and Discussion

Table 2 contains the AUC task score, the evaluation conducted by a clinician, and the automatic NLG metrics. We also provide AUC results for DenseNet-121 only,

---

[2] https://huggingface.co/emilyalsentzer/Bio_ClinicalBERT.

Table 2. The S_T score, clinical evaluation, and NLG scores for our baselines on the MIMIC-NLE test set. $\geq$GT reflects the share of generated NLEs that received a rating on-par or better than the GT. Clin.Sc. reflects the average rating of 1 (lowest) to 5 (highest) that was given to the NLEs by a clinician. R-L refers to Rouge-L, and Bn to the n-gram BLEU scores. Best results are in bold. As we only evaluate NLEs for correctly predicted diagnoses, our NLG metrics cover 534, 560, and 490 explanations for RATCHET, TieNet, and DPT, respectively.

	AUC	$\geq$GT	Clin.Sc.	CLEV	BERTS.	MET.	B1	B4	R–L	CIDEr	SPICE
GT	–	–	3.20	–	–	–	–	–	–	–	–
DenseNet-121	65.2	–	–	–	–	–	–	–	–	–	–
RATCHET	**66.4**	48%	**2.90**	74.7	77.6	**14.1**	**22.5**	**4.7**	**22.2**	**37.9**	**20.0**
TieNet	64.6	40%	2.60	**78.0**	**78.0**	12.4	17.3	3.5	19.4	33.9	17.2
DPT	62.5	48%	2.66	74.9	77.3	11.3	17.5	2.4	15.4	17.4	13.7

i.e., without an M_E module, which shows that providing NLEs can improve task performance, e.g., for RATCHET. We observed that the main reason why GT NLEs obtain an absolute rating score of 3.2/5 is inter-annotator disagreement between our clinician and the author of the reports (sometimes due to lack of patient context given in this scenario). Another reason is that some of the GT NLEs refer to a change in pathology with respect to a previous study, which is not something that can be assessed from the image. We also observe that the CLEV score neither correlates to expert evaluation nor NLG metrics. One explanation could be that the evidence labels in MIMIC-NLE are highly imbalanced, i.e., predominantly *Lung Opacity*. Therefore, as the CLEV score does not take into account much of the diversity that is inherent in our NLEs, such as the location of findings, and their size and appearance, a model that generates generic NLEs that make reference to *Lung Opacity* will yield a good CLEV score, as was the case for TieNet. Overall, the NLG metrics are on-par with the NLG metrics for report generation [11]. Hence, the difficulty of generating longer texts (reports) seems to be offset by the degree of difficulty of specific, but shorter NLEs. The results also indicate that the NLG metrics are generally poor at reflecting expert judgment. More precisely, BERTScore, which showed the highest correlation with human judgment for natural images [15], has poor indicative qualities for medical NLEs. One reason could be that automatic NLG metrics generally put equal emphasis on most words, while certain keywords, such as the location of a finding, can make an NLE clinically wrong, but only contribute little to the NLG score. While the GPT-2 based architecture of DPT proved very efficient for natural images [15], its performance is less convincing on medical images. A reason could be that GPT-2 is too large and relies on embedded commonsense knowledge to generate NLEs, which is less helpful on highly specific medical text. Example NLE generations are provided in the appendix.

6 Summary and Outlook

In this work, we introduced MIMIC-NLE, the first dataset of NLEs in the medical domain. We proposed and validated three baselines on MIMIC-NLE. Providing NLEs for medical imaging is a challenging and worthwhile task that is far from being solved, and we hope our contribution paves the way for future work. Open tasks include providing robust automatic metrics for NLEs and introducing more medical NLE datasets and better performing NLE models.

Acknowledgments. We thank Sarim Ather for useful discussions and feedback. M.K. is supported by the EPSRC Center for Doctoral Training in Health Data Science (EP/S02428X/1), and by Elsevier BV. This work has been partially funded by the ERC (853489—DEXIM) and by the DFG (2064/1—Project number 390727645). This work has also been supported by the Alan Turing Institute under the EPSRC grant EP/N510129/1, by the AXA Research Fund, the ESRC grant "Unlocking the Potential of AI for English Law", the EPSRC grant EP/R013667/1, and by the EU TAILOR grant. We also acknowledge the use of GPU computing support by Scan Computers International Ltd. BWP acknowledges a Nuffield Department of Population Health Research Fellowship.

A Supplementary

(See Fig. 4, Tables 3, 4, 5 and 6).

Table 3. Our different filters that were applied to extract image-diagnosis-NLE triplets from MIMIC-CXR. # of sentences corresponds to unique image-sentence pairs. "Non-descriptive aspects" are aspects that cannot be derived from the image itself. Duplicates are sentences that are from the same report and mention the same labels.

Filter	# of sentences
Extract Findings and Impression sections	1,383,533
Remove sentences with anonymized data	1,304,465
Remove sentences referring to non-descriptive aspects: **Patient history**: "prior", "compare", "change", "deteriorat", "increase", "decrease", "previous", "patient" **Recommendations**: "recommend", "perform", "follow" **Technical**: "CT", "technique", "position", "exam", "assess", "view", "imag"	1,007,002
Filter by rules from Table 1	43,612
Remove duplicates	39,094
Remove studies without AP or PA images	38,003

Table 4. The distribution of diagnosis labels in all our NLEs. Diagnosis label combinations are ordered by occurrence. # of sentences corresponds to unique image-sentence pairs. The table is displayed in two halves.

Diagnosis labels	# of sentences	Diagnosis labels	# of sentences
Atelectasis	10,616	Atel., Edema, Pneumonia	104
Pneumonia	9,032	Atelectasis, Edema	103
Edema	5,098	Pl. Eff., Pneumothorax	65
Atelectasis, Pneumonia	4,773	Pleural Effusion, Pneumonia	55
Pleural Effusion	4,585	Edema, Pleural Effusion	31
Consolidation	916	Atelectasis, Pleural Other	17
Pleural Other	846	Atel., Edema, Pl. Eff	9
Edema, Pneumonia	623	Edema, Pl. Eff., Pneumonia	8
Atelectasis, Pleural Effusion	397	Atel., Edema, Pl. Eff., Pneum	6
Pneumothorax	311	Edema, Pleural Other	6
Pleural Effusion, Pleural Other	195	Pleural Other, Pneumonia	4
Atel., Pl. Eff., Pneumonia	190	Other	13

Table 5. The distribution of evidence labels in all our NLEs. Evidence label combinations are ordered by occurrence. # of sentences corresponds to unique image-sentence pairs. The table is displayed in two halves.

Evidence labels	# of sentences	Evidence labels	# of sentences
Lung Opacity	29,115	Consolidation, Lung Opacity	692
Other/misc	5,842	Enlarged Cardiomediastinum	141
Consolidation	2,102	Lung Lesion	111

Table 6. The occurrence of different explanation keywords in the unfiltered sentences from MIMIC-CXR. The table is displayed in two halves.

Keyword	Count	Keyword	Count
suggest	19,787	relate	5,241
reflect	17,878	may represent	5,141
due	17,771	potentially	3,160
consistent with	14,654	worrisome for	1,685
concerning for	10,243	indicate	1,568
compatible with	6,025	account	1,193
likely represent	5,239	suspicious for	684

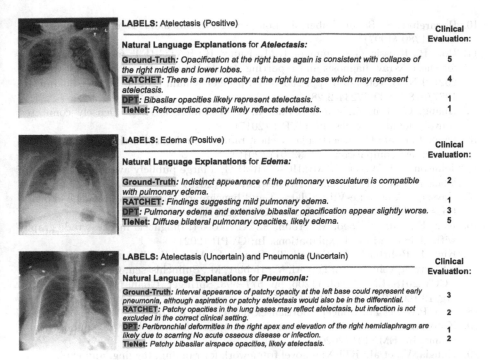

Fig. 4. The GT NLEs and three model-generated NLEs explaining three different diagnosis labels on three different scans. The clinical evaluation is given on a Likert scale, where 5 is the highest, and 1 is the lowest score.

References

1. Adebayo, J., Gilmer, J., Muelly, M., Goodfellow, I., Hardt, M., Kim, B.: Sanity checks for saliency maps. In: NeurIPS (2018)
2. Banerjee, S., Lavie, A.: METEOR: an automatic metric for MT evaluation with improved correlation with human judgments. In: Proceedings of the ACL Workshop on Intrinsic and Extrinsic Evaluation Measures for Machine Translation and/or Summarization (2005)
3. Bornstein, A.M.: Is artificial intelligence permanently inscrutable? (2016)
4. Camburu, O.M., Rocktäschel, T., Lukasiewicz, T., Blunsom, P.: e-SNLI: natural language inference with natural language explanations. In: NeurIPS (2018)
5. Cohen, J.P., et al.: TorchXRayVision: a library of chest X-ray datasets and models. arXiv preprint arXiv:2111.00595 (2020)
6. Gale, W., et al.: Producing radiologist-quality reports for interpretable artificial intelligence. arXiv preprint arXiv:1806.00340 (2018)
7. Ghassemi, M., Oakden-Rayner, L., Beam, A.L.: The false hope of current approaches to explainable artificial intelligence in health care. Lancet Digital Health **3**(11), e745–e750 (2021)
8. Gu, J., Tresp, V.: Saliency methods for explaining adversarial attacks. arXiv preprint arXiv: 1908.08413 (2019)
9. Hajian, S., Bonchi, F., Castillo, C.: Algorithmic bias: from discrimination discovery to fairness-aware data mining. In: SIGKDD (2016)

10. Hochreiter, S., Schmidhuber, J.: Long short-term memory. Neural Comput. **9**(8), 1735–1780 (1997)
11. Hou, B., Kaissis, G., Summers, R.M., Kainz, B.: RATCHET: medical transformer for chest X-ray diagnosis and reporting. In: de Bruijne, M., et al. (eds.) MICCAI 2021. LNCS, vol. 12907, pp. 293–303. Springer, Cham (2021). https://doi.org/10.1007/978-3-030-87234-2_28
12. Huang, G., Liu, Z., van der Maaten, L., Weinberger, K.Q.: Densely connected convolutional networks. In: CVPR (2017)
13. Irvin, J., et al.: CheXpert: a large chest radiograph dataset with uncertainty labels and expert comparison, vol. 33(01) (2019)
14. Johnson, A.E.W., et al.: MIMIC-CXR-JPG, a large publicly available database of labeled chest radiographs. arXiv:1901.07042 (2019)
15. Kayser, M., et al.: e-ViL: a dataset and benchmark for natural language explanations in vision-language tasks. In: ICCV (2021)
16. Kim, E., Kim, S., Seo, M., Yoon, S.: XProtoNet: diagnosis in chest radiography with global and local explanations. In: CVPR (2021)
17. Kim, J., Rohrbach, A., Darrell, T., Canny, J., Akata, Z.: Textual explanations for self-driving vehicles. In: Ferrari, V., Hebert, M., Sminchisescu, C., Weiss, Y. (eds.) ECCV 2018. LNCS, vol. 11206, pp. 577–593. Springer, Cham (2018). https://doi.org/10.1007/978-3-030-01216-8_35
18. Koh, P.W., et al.: Concept bottleneck models. In: ICML (2020)
19. Kotonya, N., Toni, F.: Explainable automated fact-checking for public health claims. In: EMNLP (2020)
20. Kougia, V., et al.: RTEX: a novel framework for ranking, tagging, and explanatory diagnostic captioning of radiography exams. J. Am. Med. Inform. Assoc. **28**(8), 1651–1659 (2021)
21. Langlotz, C.P., et al.: A roadmap for foundational research on artificial intelligence in medical imaging: from the 2018 NIH/RSNA/ACR/The Academy Workshop. Radiology **291**(3) (2019)
22. Lee, H., Kim, S.T., Ro, Y.M.: Generation of multimodal justification using visual word constraint model for explainable computer-aided diagnosis. In: Interpretability of Machine Intelligence in Medical Image Computing and Multimodal Learning for Clinical Decision Support (2019)
23. Majumder, B.P., Camburu, O.M., Lukasiewicz, T., McAuley, J.: Knowledge-grounded self-rationalization via extractive and natural language explanations. In: ICML (2022)
24. Marasović, A., et al.: Natural language rationales with full-stack visual reasoning: from pixels to semantic frames to commonsense graphs. In: EMNLP Findings (2020)
25. Miller, T.: Explanation in artificial intelligence: insights from the social sciences. Artif. Intell. **267**, 1–38 (2019)
26. Mozaffari-Kermani, M., et al.: Systematic poisoning attacks on and defenses for machine learning in healthcare. IEEE J. Biomed. Health Inform. **19**(6), 1893–1905 (2015)
27. Obermeyer, Z., et al.: Dissecting racial bias in an algorithm used to manage the health of populations. Science **366**(6464), 447–453 (2019)
28. Papernot, N., McDaniel, P., Goodfellow, I.: Transferability in machine learning: from phenomena to black-box attacks using adversarial samples. arXiv preprint arXiv:1605.07277 (2016)
29. Park, D.H., et al.: Multimodal explanations: justifying decisions and pointing to the evidence. In: ICCV (2018)

30. Puyol-Antón, E., et al.: Interpretable deep models for cardiac resynchronisation therapy response prediction. In: Martel, A.L., et al. (eds.) MICCAI 2020. LNCS, vol. 12261, pp. 284–293. Springer, Cham (2020). https://doi.org/10.1007/978-3-030-59710-8_28

31. Radford, A., Wu, J., Child, R., Luan, D., Amodei, D., Sutskever, I.: Language Models are Unsupervised Multitask Learners. openai.com (2019)

32. Rajani, N.F., McCann, B., Xiong, C., Socher, R.: Explain yourself! Leveraging language models for commonsense reasoning. In: ACL (2019)

33. Schutte, K., Moindrot, O., Hérent, P., Schiratti, J.B., Jégou, S.: Using StyleGAN for visual interpretability of deep learning models on medical images. arXiv preprint arXiv:2101.07563 (2021)

34. Smit, A., et al.: Combining automatic labelers and expert annotations for accurate radiology report labeling using BERT. In: EMNLP (2020)

35. Stacey, J., Belinkov, Y., Rei, M.: Natural language inference with a human touch: using human explanations to guide model attention. In: AAAI (2022)

36. Vayena, E., Blasimme, A., Cohen, I.G.: Machine learning in medicine: addressing ethical challenges. PLoS Med. 15(11), e1002689 (2018)

37. Wang, X., et al.: TieNet: text-image embedding network for common thorax disease classification and reporting in chest X-rays. In: CVPR (2018)

38. Zhang, T., Kishore, V., Wu, F., Weinberger, K.Q., Artzi, Y.: BERTScore: evaluating text generation with BERT. In: ICLR (2019)

RepsNet: Combining Vision with Language for Automated Medical Reports

Ajay K. Tanwani[(✉)], Joelle Barral, and Daniel Freedman

Verily & Google Research, San Francisco, USA
{ajaytanwani,jbarral,danielfreedman}@google.com

Abstract. Writing reports by analyzing medical images is error-prone for inexperienced practitioners and time consuming for experienced ones. In this work, we present RepsNet that adapts pre-trained vision and language models to interpret medical images and generate automated reports in natural language. RepsNet consists of an encoder-decoder model: the encoder aligns the images with natural language descriptions via contrastive learning, while the decoder predicts answers by conditioning on encoded images and prior context of descriptions retrieved by nearest neighbour search. We formulate the problem in a visual question answering setting to handle both categorical and descriptive natural language answers. We perform experiments on two challenging tasks of medical visual question answering (VQA-Rad) and report generation (IU-Xray) on radiology image datasets. Results show that RepsNet outperforms state-of-the-art methods with 81.08% classification accuracy on VQA-Rad 2018 and 0.58 BLEU-1 score on IU-Xray. Supplementary details are available at: https://sites.google.com/view/repsnet.

Keywords: Vision and language · Visual question answering · Report generation

1 Introduction

A long standing goal in artificial intelligence is to seamlessly interpret and describe medical images/videos with natural language. In this paper, we combine both vision and language modalities to interpret medical images in a visual question answering (VQA) setting, whereby we predict the answer to a given image and question using a novel encoder-decoder model (see Fig. 1). We present RepsNet that fuses the encoded image and question features by contrastive alignment, while the decoder learns the conditional probability distribution to generate descriptions with: 1) encoded image and question features, and 2) prior context retrieved from nearest neighbouring reports of the image. We leverage publicly available ResNeXt [37] and BERT [10] for warm-starting the encoder, and GPT-2 [28] as the base model for the natural language decoder.

Supplementary Information The online version contains supplementary material available at https://doi.org/10.1007/978-3-031-16443-9_68.

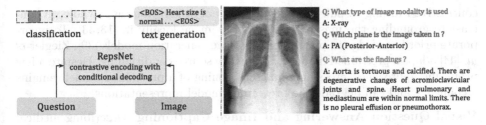

Fig. 1. *(left)* RepsNet analyzes medical images and automates report writing by providing answers to questions via classifying among known answer categories or generating natural language descriptions, *(right)* radiology report generation example with top two categorical answers and bottom one natural language descriptive answer.

We present its application to assist practitioners with automatic report generation from medical images [7,16,20,22,25]. Existing methods using hand-written notes, dictation services or electronic medical record templates are widely perceived to be time-consuming and cumbersome. To this end, we parse the medical report into a set of questions and handle both categorical (yes/no, multiple choices) and natural language descriptive answers (open-ended) in a visual question answering setting. We evaluate the proposed approach on two publicly available benchmark datasets: 1) visual question answering radiology (VQA-Rad) datasets in the span of 2018−2021 [19], 2) Indiana University x-ray (IU-Xray) dataset containing chest x-ray images paired with reports describing findings and impressions [8]. RepsNet outperforms state-of-the-art models across both VQA-Rad and IU-Xray datasets.

Contributions: This paper makes three contributions:

- We present RepsNet, an encoder-decoder model for writing reports that adapts pretrained models by contrastive alignment of images with answers in the encoding phase, and generates natural language descriptions by conditional decoding on images and prior context of retrieved reports.
- A visual question answering formulation to handle both categorical and natural language descriptive answers in generating automated reports.
- Experiments on publicly available VQA-Rad and IU-Xray datasets with 81.08% classification accuracy and 0.58 BLEU-1 score respectively, showing significant performance improvement over state-of-the-art methods.

2   Related Work

Vision and Language Pretraining: Self-supervised pre-training of language models such as BERT [10], GPT/GPT-2 [28], XLNet have shown promising results in transferring knowledge across related tasks [12]. This has led to combining both visual and language modalities by cross-alignment of domains in a joint embedding space [9,30]. Examples include LXMERT [35], ViLBERT [24], PixelBERT [14], VideoBERT [34] and VisualGPT [36]. Authors in [27,43] use contrastive learning to pair images with textual descriptions as a whole, in contrast to grounding the masked works in the image locally in [13,33]. To incorporate prior knowledge in pretrained language generation models [39], Ziegler *et al.* [44] adapt a pretrained model for arbitrary source conditioning. Despite a few promising approaches, cross-domain conditioning of a pretrained model remains a challenge and can degrade the pretrained model representations.

Visual Question Answering and Image Captioning: Describing medical images with visual question answering [4] or natural language [3,38] is difficult due to rare and diverse nature of abnormalities, weak association of image features with text in reports, lack of prior domain knowledge, case-based reasoning, and long descriptions of findings. Medical VQA has recently received attention with small scale datasets such as VQA-Rad to categorize answers by classification [5,11,21]. Several works have followed the image captioning line of work with an emphasis on generating long descriptions [2,16], incorporating domain-specific medical knowledge [22,42], retrieving description from a template [20,23], question answering [29,32], among others [25,40].

In this paper, we investigate the automated report writing under a novel visual question answering framework to handle both categorical and descriptive answers for a given image and a question. We use contrastive learning to align the paired images and report answers in an embedding space, and retrieve nearest neighbour report answers to incorporate prior knowledge in generating medical descriptions.

3   RepsNet: Proposed Approach

Problem Formulation: Given an image or a set of images $\mathbf{x} \in \mathbf{X}$, we are interested in generating a report comprising of s answers $\mathbf{y} = \{\mathbf{y}_1 \cdots \mathbf{y}_s\} \in \mathbf{Y}$, corresponding to the natural language questions $\mathbf{q} = \{\mathbf{q}_1 \cdots \mathbf{q}_s\} \in \mathbf{Q}$. Each answer $\mathbf{y}_i$ may be **close-ended** belonging to a fixed possible set of categories or **open-ended** comprising of multiple natural language sentences. Each word $\mathbf{w} \in \mathbf{V}$ in the open-ended answer belongs to a known natural language vocabulary. We seek to learn the model parameters $\boldsymbol{\Theta}$ to maximize the conditional likelihood $\mathcal{P}_{\boldsymbol{\Theta}}(\mathbf{y}_i \mid \mathbf{x}, \mathbf{q}_i)$ of predicting the answers for a given image and a set of questions,

$$\boldsymbol{\Theta} = \arg\max_{\boldsymbol{\Theta}} \sum_{i=1}^{s} \log \mathcal{P}_{\boldsymbol{\Theta}}(\mathbf{y}_i \mid \mathbf{x}, \mathbf{q}_i). \tag{1}$$

We formulate the problem with an encoder decoder model. The encoder $f_{\boldsymbol{\theta}_{\mathrm{enc}}} : \{\mathbf{X}, \mathbf{Q}\} \rightarrow \{\bar{\mathbf{X}}, \bar{\mathbf{Q}}\} \in \mathbb{R}^{\{n_{\mathbf{x}}, n_{\mathbf{q}}\} \times \{d_{\mathbf{x}}, d_{\mathbf{q}}\}}$ transforms the image and the input text sequence to a joint cross-aligned visual and language representation space with $n_{\mathbf{x}}$ image pixels/regions, $n_{\mathbf{q}}$ text tokens, and $\{d_{\mathbf{x}}, d_{\mathbf{q}}\}$ hidden space dimensions of image and text embeddings respectively. The decoder $h_{\boldsymbol{\theta}_{\mathrm{dec}}} : \{\bar{\mathbf{X}}, \bar{\mathbf{Q}}, \bar{\mathbf{C}}\} \rightarrow \mathcal{P}(\mathbf{Y})$ models the conditional probability distribution of predicting the target answer $\mathbf{Y}$ given the encoded hidden states $\{\bar{\mathbf{X}}, \bar{\mathbf{Q}}\}$, and the **prior context** $\bar{\mathbf{C}} \in \mathbb{R}^{n_c \times d_c}$ of n_c tokens with dimension d_c that represents the domain specific knowledge for controlled text generation (we discuss prior context further in the next section). Note that we only use the prior context for generating open-ended answers.

In this paper, we leverage large-scale pretrained models for warm-starting the encoder and the decoder model parameters. For close-ended answers, we map the combined image and question features to the output layer of all possible close-ended answers for classification. For open-ended answers, the decoder retrieves the prior context $\bar{C}$ as the nearest neighbouring answers of the encoded image features, and greedily maximizes the learned conditional distribution $\mathcal{P}_{\theta_{dec}}\left(Y_t|Y_{0:t-1}, \bar{X}, \bar{Q}, \bar{C}\right)$ to generate the answer sequence $Y_{1:t}$ in an auto-regressive manner (see Fig. 2).

3.1   Contrastive Image-Text Encoder

The encoder has four constituent parts: 1) **image encoder** to extract visual features, 2) **text encoder** to tokenize and contextualize natural language questions and answers features, 3) **bilinear attention network** to fuse the image and the question, and 4) **contrastive alignment** of visual features and textual answers.

Image Encoder: We use the `ResNeXt-101` [37] architecture as the base image encoder. We remove the last linear and pooling layer and add a 2D adaptive average pooling layer to resize the input image to a fixed feature space of $14 \times 14 \times 2048$ that preserves the correspondence between the visual features and the input image ($n_x = 196, d_x = 2048$). Moreover, we add image transformations, namely *color jittering, normalization, random erasing*, to augment the training data distribution within each batch before extracting the visual features.

Text Encoder: We adapt the BERT [10] model for the text encoder that is pre-trained to predict masked words locally based on the context provided by other non-masked words in the sequence. We filter out the punctuation marks and tokenize the text using WordPiece algorithm [10], before extracting the textual features.

Bilinear Attention Network (BAN): We use a BAN to fuse the cross-modal encoded question and image features [18]. The outer product or the bilinear product exhaustively combines the multi-modal features at the cost of higher computational complexity; in comparison to naive concatenation or inner product between the features. Compared to other co-attention mechanisms, BAN exploits bilinear interaction maps where each feature is pooled by low-rank bilinear approximations. Residual learning on top combines multiple bilinear attention maps for effective joint representation of question and image features.

For the sake of brevity and a slight abuse of notation, we denote $\bar{X}$ for both the image and the combined (image and question) features in describing the rest of the encoder and the decoder sections.

Contrastive Vision and Text Learning: We align images with natural language descriptions via bidirectional contrastive learning [6], that pulls together

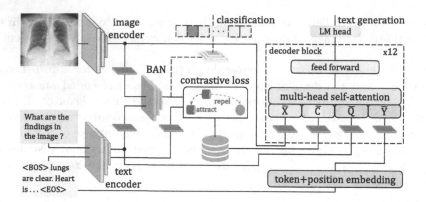

Fig. 2. RepsNet encoded image and question features are fused via bilinear attention network (BAN), before self-supervised contrastive alignment with natural language descriptions. The answer is categorized via classification among fixed answer categories or generated by conditional language decoding on image, question and prior context of answers retrieved by nearest neighbour search. Note that we omit the question features $\bar{\mathbf{Q}}$ in describing the conditional language decoder below for brevity.

a given image-answer pair, while pushing away observations that correspond to different image-answer pairs.

Given the encoded image (and question) $\bar{\mathbf{X}}$ and the natural language answer features $\bar{\mathbf{Y}} \in \mathbb{R}^{n_y \times d_y}$ with n_y tokens of dimension d_y, we first project them to a d-dimensional space with a linear transformation to $\hat{\mathbf{X}} \in \mathbb{R}^d$ and $\hat{\mathbf{Y}} \in \mathbb{R}^d$. During training, the loss operates on a mini-batch of N_T image-text pairs $\{\hat{\mathbf{x}}_i, \hat{\mathbf{y}}_i\}_{i=1}^{N_T}$, where each pair is in turn taken as a positive sample to maximize agreement against all other negative samples, i.e.,

$$\mathcal{L}_{\hat{\mathbf{x}} \to \hat{\mathbf{y}}} = -\frac{1}{N_T} \sum_{i=1}^{N_T} \log \frac{\exp\left(\langle \hat{\mathbf{x}}_i, \hat{\mathbf{y}}_i \rangle / \tau\right)}{\sum_{j=1}^{N_T} \exp(\langle \hat{\mathbf{x}}_i, \hat{\mathbf{y}}_j \rangle / \tau)}, \tag{2}$$

where $\langle \hat{\mathbf{x}}, \hat{\mathbf{y}} \rangle = \frac{\hat{\mathbf{x}}^\top \hat{\mathbf{y}}}{\|\hat{\mathbf{x}}\| \|\hat{\mathbf{y}}\|}$ represents the cosine similarity distance and $\tau \in \mathbb{R}^+$ represents the temperature parameter to scale the similarity metric. Similar to the image-to-text loss in Eq. (2), we also define the text-to-image loss $\mathcal{L}_{\hat{\mathbf{y}} \to \hat{\mathbf{x}}}$ to account for the asymmetry with respect to each input modality as in [27, 43]. Overall bidirectional encoder loss $\mathcal{L}_{\text{enc}}$ is the sum of the two constituent contrastive losses weighted by constant $\alpha_l \in \mathbb{R}^+$,

$$\mathcal{L}_{\text{enc}} = \alpha_l (\mathcal{L}_{\hat{\mathbf{x}} \to \hat{\mathbf{y}}} + \mathcal{L}_{\hat{\mathbf{y}} \to \hat{\mathbf{x}}}). \tag{3}$$

Prior Context Knowledge: We store the normalized natural language answers of the train set $\hat{\mathbf{Y}}_{\text{train}}$ during model training. We then compute topk nearest neighbours $\bar{\mathbf{C}}$ that maximize the cosine similarity between a given encoded image $\hat{\mathbf{X}}$ and the stored natural language answers $\hat{\mathbf{Y}}_{\text{train}}$. We use the FAISS library for scalable nearest neighbour search [17]. The prior context aids the decoder to attend to longer horizon dependencies and get additional case-based details for controlled text generation. This is particularly relevant in describing medical images with specific terminologies, writing style and class imbalanced abnormalities, i.e.,

$$\bar{\mathbf{C}} = \text{topk} \left[\max_{i \in \hat{\mathbf{Y}}_{\text{train}}} \langle \hat{\mathbf{X}}, \hat{\mathbf{Y}}^{(i)}_{\text{train}} \rangle \right]. \tag{4}$$

3.2 Conditional Language Decoder

The probability distribution of generating the output text sequence $\mathbf{Y}_{1:t}$ conditioned on the contextualized encoding sequence $\mathcal{P}_{\theta_{\text{dec}}}(\mathbf{Y}_{1:t}|\bar{\mathbf{X}}, \bar{\mathbf{C}})$ can be decomposed into a product of conditional distributions using Bayes' rule,

$$\mathcal{P}_{\theta_{\text{dec}}}(\mathbf{Y}_{1:t}|\bar{\mathbf{X}}, \bar{\mathbf{C}}) = \prod_{i=1}^{t} \mathcal{P}_{\theta_{\text{dec}}}(\mathbf{y}_i|\mathbf{y}_{0:i-1}, \bar{\mathbf{X}}, \bar{\mathbf{C}}), \tag{5}$$

where $\mathbf{y}_0 = \langle \mathbf{BOS} \rangle$ is a special token reserved for the beginning of a sentence. We model the conditional language generation with a stack of transformer-based blocks, using the GPT-2 model as the base pretrained language decoder [28]. We introduce modifications to the GPT-2 model for conditioning on image and prior context features by directly adding their attention outputs to the pretrained self-attention layers of the model, similar to [2,44], thereby adding the attention outputs for different conditional inputs with a parsimonious increase in the number of parameters only (see supplementary materials for details). The conditional probability distribution in Eq. (5) is maximized by optimizing the cross-entropy loss on the ground-truth and the predicted sequences.

Overall Approach: During training, we adapt the pretrained language and vision models in an end-to-end manner for contrastive encoding and conditional decoding with a small amount of image-text pairs. The overall training loss function comprises of the contrastive loss and the cross-entropy loss. During natural language generation, we predict the output sequence in an auto-regressive manner with greedy or beam search decoding, and stop generating the sequence once we predict a special end of text $\langle \mathbf{EOS} \rangle$ token.

4 Experiments, Results and Discussion

We evaluate the performance of RepsNet in interpreting visual concepts on publicly available VQA-Rad [19] for classification and IU-Xray [8] for natural language generation. We are interested in evaluating: 1) how feasible it is to adapt the pretrained language and vision models in describing small set of medical images, 2) what is the role of contrastive encoding in learning joint visual linguistic representations, 3) does conditional decoding on image features and prior context help with generating medical language descriptions, and 4) how does RepsNet fare in performance among the state-of-the-art approaches.

Table 1. Classification accuracy on the VQA-Rad datasets. Bottom three rows increase minimum occurrence threshold from 0 to 5 to 10 instances. RepsNet outperforms all other competing methods.

	2018	2019	All
MEVF [5]	66.10	–	–
MMQ [11]	67.00	–	–
QCR [41]	69.65	–	–
CLEF [1]	–	62.40	–
CRPD [21]	72.70	–	–
RepsNet-0	81.08	67.57	63.69
RepsNet-5	83.55	79.83	71.93
RepsNet-10	87.05	81.17	80.37

Table 2. BLEU scores (B1–B4) for medical report generation on IU-Xray dataset. RepsNet yields better scores than other methods.

	B1	B2	B3	B4
Co-Att [16]	0.45	0.29	0.20	0.15
HRGR [20]	0.44	0.30	0.21	0.15
CMAS [15]	0.46	0.30	0.21	0.15
Mem-T [7]	0.47	0.30	0.22	0.16
VTI [25]	0.49	0.36	0.29	0.15
PPKED [22]	0.48	0.31	0.22	0.17
RepsNet	**0.58**	**0.44**	**0.32**	**0.27**

4.1 Visual Question Answering

VQA-Rad Dataset: We use the VQA-Rad datasets [19] from $2018 - 2019$ and also introduce an aggregrated dataset, VQA-Rad All, that combines all the VQA-Rad datasets from $2018 - 2021$. Radiology images in the datasets are taken from open-access MedPix database, and the questions are predominantly posed from categories, such as image plane, imaging modality, organ system involved and image abnormalities. The VQA problem is posed as a multi-class classification over all possible set of answers, and classification accuracy on evaluation set is used as the performance metric. We use the standard training and evaluation splits provided with the datasets (see summary in supplementary materials).

Results: Table 1 shows that RepsNet outperforms all other competing methods across all the datasets. Similar to other methods, RepsNet uses bilinear attention mechanism to fuse the image and the question features. Contrary to other methods, RepsNet does not use fixed Glove word embeddings [26] or RNNs for sentence level representations; instead it learns the entire contextual embeddings using BERT-style transformer with WordPiece tokenization. Ablation study in Table 3 also shows that the performance increases the most with the use of pretrained models. We observe from Table 1 that simply filtering out instances and class categories with less than 5 and 10 instances per class category $M_o = \{5, 10\}$ proportionally increases the classification accuracy across all datasets, at the cost of reducing the overall number of instances and class categories to mitigate class imbalance in the datasets. Note that we do not take into account unseen class category instances of the evaluation set in computing the classification accuracy.

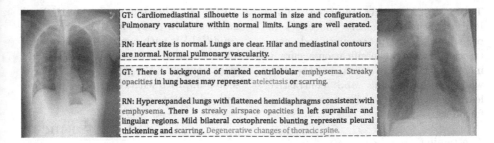

Fig. 3. Heatmap visualization and comparison between ground-truth (GT) and RepsNet generated (RN) report of: *(left)* normal case, *(right)* abnormal case. RepsNet shows strong alignment with ground-truth in describing medical findings. Text in blue shows abnormalities, text in red represents misalignment. (Color figure online)

4.2 Medical Report Generation

IU-XRay: The Indiana University x-ray dataset [8] comprises of frontal and lateral views of chest x-ray images that are associated with radiology report sections, namely *impressions, findings* and *manual tags*. For brevity, we only report results for populating the findings question in this work, i.e., we associate the same question for all the answers. After omitting the reports without findings section, we randomly split the remaining 3607 reports into 80% training and 20% evaluation sets. On average, each report instance has 5.7 sentences, while each sentence has 6.5 words that describe the image findings. Note that no classification labels are available to detect the anomalies. Max number of tokens for a report section is set to 200 and the report findings are zero-padded in case its length is less than max number of tokens. We use the sentence level BLEU scores as the performance metric computed using the nltk library that compares n-gram similarity between the ground-truth and the generated report where n varies from 1 to 4 (whereas for classification accuracy evaluation in VQA-Rad datasets, we compare the predicted and the ground-truth indices of the class categories).

Results: Results are summarized in Table 2. It can be seen that RepsNet performs significantly better than the state-of-the-art report generation methods across all the BLEU scores, suggesting the feasibility of adapting large-scale pretrained language and vision models on a small set of domain-specific medical data. Ablation study in Table 4 reveals that adding visual features, contrastive learning and prior context subsequently boosts the performance of the GPT2-decoder. Figure 3 provides a qualitative comparison between the ground-truth and the generated report findings, along with the heatmap visualizations using grad-cam [31] for an intuitive understanding of the approach. We observe a strong alignment in generating normal report findings, whereas part of the findings sometimes get omitted and/or added in describing the abnormalities, especially for rare cases (see supplementary materials for video demonstration and other examples). Systematic dealing of rare cases with external domain knowledge and past medical history of patients is a promising direction of our

Table 3. Ablation study on VQA-Rad dataset to quantify the effect of pre-training, pre-processing and contrastive learning. Classification accuracy increases the most with pre-training, while pre-processing and contrastive learning stage further improve the performance.

Pretraining		✓	✓	✓
Preprocess			✓	✓
Contrastive				✓
	74.47	79.12	80.09	**81.08**

Table 4. Ablation study on IU-Xray dataset with: *(top)* visual features - Vis, *(middle)* Vis with contrastive encoding - Vis + CE, and *(bottom)* Vis with CE and prior context - Vis + CE + PC. BLEU scores improve with contrastive learning and prior context.

	B1	B2	B3	B4
Vis	0.48	0.38	0.30	0.26
Vis + CE	0.55	0.42	0.32	0.27
Vis + CE + PC	0.58	0.44	0.32	0.27

future work. We are also interested in incorporating attention mechanisms for conditional visualization of generated text on image patches as a measure of uncertainty in the prediction. Making these reports self-explainable is critical for its wider adoption. Other areas of interest include reducing the liability of generated report errors, as well as working with medical experts to evaluate the generated reports.

5   Conclusion

In this paper, we have presented RepsNet that adapts pre-trained vision and language models for describing a small set of domain-specific medical images. We take a unified visual question answering approach to predict class categories or generate descriptive answers for writing automated medical reports. RepsNet is specifically tailored for contrastive alignment of images and text in the encoding phase, and combining visual and prior context of nearest neighboring reports with natural language generator in the decoding phase. This has enabled RepsNet to provide state-of-the-art results on challenging tasks of visual question answering and medical report generation on radiology images. In our future work, we plan to extend our approach to summarizing reports from videos, and transfer the developed methodology to clinical sites for automated reporting in gastroenterology.

References

1. Abacha, A.B., Hasan, S.A., Datla, V., Liu, J., Demner-Fushman, D., Müller, H.: VQA-MED: overview of the medical visual question answering task at ImageCLEF 2019. In: CLEF (2019)
2. Alfarghaly, O., Khaled, R., Elkorany, A., Helal, M., Fahmy, A.: Automated radiology report generation using conditioned transformers. Inform. Med. Unlocked **24**, 100557 (2021)

3. Anderson, P., et al.: Bottom-up and top-down attention for image captioning and VQA. CoRR abs/1707.07998 (2017)

4. Antol, S., et al.: VQA: visual question answering. CoRR abs/1505.00468 (2015)

5. Nguyen, B.D., Do, T.-T., Nguyen, B.X., Do, T., Tjiputra, E., Tran, Q.D.: Overcoming data limitation in medical visual question answering. In: Shen, D., et al. (eds.) MICCAI 2019. LNCS, vol. 11767, pp. 522–530. Springer, Cham (2019). https://doi.org/10.1007/978-3-030-32251-9_57

6. Chen, T., Kornblith, S., Norouzi, M., Hinton, G.E.: A simple framework for contrastive learning of visual representations. CoRR abs/2002.05709 (2020)

7. Chen, Z., Song, Y., Chang, T., Wan, X.: Generating radiology reports via memory-driven transformer. CoRR abs/2010.16056 (2020)

8. Demner-Fushman, D., et al.: Preparing a collection of radiology examinations for distribution and retrieval. J. Am. Med. Inform. Assoc. **23**(2), 304–310 (2016)

9. Desai, K., Johnson, J.: VirTex: learning visual representations from textual annotations. CoRR abs/2006.06666 (2020)

10. Devlin, J., Chang, M., Lee, K., Toutanova, K.: BERT: pre-training of deep bidirectional transformers for language understanding. CoRR abs/1810.04805 (2018)

11. Do, T., Nguyen, B.X., Tjiputra, E., Tran, M., Tran, Q.D., Nguyen, A.: Multiple meta-model quantifying for medical visual question answering. In: de Bruijne, M., et al. (eds.) MICCAI 2021. LNCS, vol. 12905, pp. 64–74. Springer, Cham (2021). https://doi.org/10.1007/978-3-030-87240-3_7

12. Dong, L., et al.: Unified language model pre-training for natural language understanding and generation. In: NeurIPS. vol. 32 (2019)

13. Gupta, T., Vahdat, A., Chechik, G., Yang, X., Kautz, J., Hoiem, D.: Contrastive learning for weakly supervised phrase grounding. CoRR abs/2006.09920 (2020)

14. Huang, Z., Zeng, Z., Liu, B., Fu, D., Fu, J.: Pixel-BERT: aligning image pixels with text by deep multi-modal transformers. CoRR abs/2004.00849 (2020)

15. Jing, B., Wang, Z., Xing, E.P.: Show, describe and conclude: on exploiting the structure information of chest X-ray reports. CoRR abs/2004.12274 (2020)

16. Jing, B., Xie, P., Xing, E.P.: On the automatic generation of medical imaging reports. CoRR abs/1711.08195 (2017), http://arxiv.org/abs/1711.08195

17. Johnson, J., Douze, M., Jégou, H.: Billion-scale similarity search with GPUS. arXiv preprint arXiv:1702.08734 (2017)

18. Kim, J.H., Jun, J., Zhang, B.T.: Bilinear attention networks. In: Advances in Neural Information Processing Systems, vol. 31 (2018)

19. Lau, J.J., Gayen, S., Ben Abacha, A., Demner-Fushman, D.: A dataset of clinically generated visual questions and answers about radiology images. Nat. Sci. Data **5** (2018)

20. Li, C.Y., Liang, X., Hu, Z., Xing, E.P.: Hybrid retrieval-generation reinforced agent for medical image report generation. CoRR abs/1805.08298 (2018)

21. Liu, B., Zhan, L.-M., Wu, X.-M.: Contrastive pre-training and representation distillation for medical visual question answering based on radiology images. In: de Bruijne, M., et al. (eds.) MICCAI 2021. LNCS, vol. 12902, pp. 210–220. Springer, Cham (2021). https://doi.org/10.1007/978-3-030-87196-3_20

22. Liu, F., Wu, X., Ge, S., Fan, W., Zou, Y.: Exploring and distilling posterior and prior knowledge for radiology report generation. In: CVPR, pp. 13753–13762 (2021)

23. Liu, G., et al.: Clinically accurate chest X-ray report generation. CoRR abs/1904.02633 (2019)

24. Lu, J., Batra, D., Parikh, D., Lee, S.: VILBERT: pretraining task-agnostic visiolinguistic representations for vision-and-language tasks. CoRR abs/1908.02265 (2019)

25. Najdenkoska, I., Zhen, X., Worring, M., Shao, L.: Variational topic inference for chest X-ray report generation. CoRR abs/2107.07314 (2021)
26. Pennington, J., Socher, R., Manning, C.: GloVe: global vectors for word representation. In: EMNLP, pp. 1532–1543 (2014)
27. Radford, A., et al.: Learning transferable visual models from natural language supervision. CoRR abs/2103.00020 (2021)
28. Radford, A., Wu, J., Child, R., Luan, D., Amodei, D., Sutskever, I.: Language models are unsupervised multitask learners (2019)
29. Ren, F., Zhou, Y.: CGMVQA: a new classification and generative model for medical visual question answering. IEEE Access **8**, 50626–50636 (2020)
30. Sariyildiz, M.B., Perez, J., Larlus, D.: Learning visual representations with caption annotations. CoRR abs/2008.01392 (2020)
31. Selvaraju, R.R., et al.: Grad-CAM: why did you say that? Visual explanations from deep networks via gradient-based localization. CoRR abs/1610.02391 (2016)
32. Sharma, D., Purushotham, S., Reddy, C.K.: MedFuseNet: an attention-based multimodal deep learning model for visual question answering in the medical domain. Sci. Rep. **11**(1) (2021). https://doi.org/10.1038/s41598-021-98390-1
33. Sun, C., Baradel, F., Murphy, K., Schmid, C.: Contrastive bidirectional transformer for temporal representation learning. CoRR abs/1906.05743 (2019)
34. Sun, C., Myers, A., Vondrick, C., Murphy, K., Schmid, C.: VideoBERT: a joint model for video and language representation learning. CoRR abs/1904.01766 (2019)
35. Tan, H., Bansal, M.: LXMERT: learning cross-modality encoder representations from transformers. CoRR abs/1908.07490 (2019)
36. Xia, Q., et al.: XGPT: cross-modal generative pre-training for image captioning. CoRR abs/2003.01473 (2020)
37. Xie, S., Girshick, R.B., Dollár, P., Tu, Z., He, K.: Aggregated residual transformations for deep neural networks. CoRR abs/1611.05431 (2016)
38. Xu, K., et al.: Show, attend and tell: neural image caption generation with visual attention. In: ICML, vol. 37, pp. 2048–2057 (2015)
39. Yu, W., et al.: A survey of knowledge-enhanced text generation. CoRR abs/2010.04389 (2020)
40. Yuan, J., Liao, H., Luo, R., Luo, J.: Automatic radiology report generation based on multi-view image fusion and medical concept enrichment. In: Shen, D., et al. (eds.) MICCAI 2019. LNCS, vol. 11769, pp. 721–729. Springer, Cham (2019). https://doi.org/10.1007/978-3-030-32226-7_80
41. Zhan, L.M., Liu, B., Fan, L., Chen, J., Wu, X.M.: Medical visual question answering via conditional reasoning. In: Proceedings of the 28th ACM International Conference on Multimedia, pp. 2345–2354 (2020)
42. Zhang, Y., Wang, X., Xu, Z., Yu, Q., Yuille, A., Xu, D.: When radiology report generation meets knowledge graph. In: AAAI, vol. 34, pp. 12910–12917 (2020)
43. Zhang, Y., Jiang, H., Miura, Y., Manning, C.D., Langlotz, C.P.: Contrastive learning of medical visual representations from paired images and text. CoRR abs/2010.00747 (2020)
44. Ziegler, Z.M., Melas-Kyriazi, L., Gehrmann, S., Rush, A.M.: Encoder-agnostic adaptation for conditional language generation. CoRR abs/1908.06938 (2019)

BERTHop: An Effective Vision-and-Language Model for Chest X-ray Disease Diagnosis

Masoud Monajatipoor[1]([✉]), Mozhdeh Rouhsedaghat[2], Liunian Harold Li[1],
C.-C. Jay Kuo[2], Aichi Chien[1], and Kai-Wei Chang[1]

[1] University of California, Los Angeles, CA 90095, USA
{monajati,aichi}@ucla.edu, {liunian.harold.li,kwchang}@cs.ucla.edu
[2] University of Southern California, Los Angeles, CA 90007, USA
{rouhseda,jckou}@usc.edu

Abstract. Vision-and-language (V&L) models take image and text as input and learn to capture the associations between them. These models can potentially deal with the tasks that involve understanding medical images along with their associated text. However, applying V&L models in the medical domain is challenging due to the expensiveness of data annotations and the requirements of domain knowledge. In this paper, we identify that the visual representation in general V&L models is not suitable for processing medical data. To overcome this limitation, we propose BERTHop, a transformer-based model based on PixelHop++ and VisualBERT for better capturing the associations between clinical notes and medical images.

Experiments on the OpenI dataset, a commonly used thoracic disease diagnosis benchmark, show that BERTHop achieves an average Area Under the Curve (AUC) of 98.12% which is 1.62% higher than state-of-the-art while it is trained on a 9× smaller dataset (https://github.com/monajati/BERTHop).

Keywords: Computer-Aided Diagnosis · Transfer learning · Vision & language model

1 Introduction

Computer-Aided Diagnosis (CADx) [10] systems provide valuable benefits for disease diagnosis including but not limited to improving the quality and consistency of the diagnosis and providing a second option to reduce medical mistakes. Although most existing studies focus on diagnosis based on medical images such as chest X-ray (CXR) images [1,2], the radiology reports often contain substantial information in the text(e.g. patient history and previous studies) that are difficult to be detected from the image alone. Besides, the diagnosis from both image and text is more closely aligned with disease diagnosis by human experts. Therefore, V&L models that take both images and text as input can be potentially more accurate for CADx.

L. Wang et al. (Eds.): MICCAI 2022, LNCS 13435, pp. 725–734, 2022.
https://doi.org/10.1007/978-3-031-16443-9_69

However, the shortage of annotated data in the medical domain makes utilizing V&L models challenging. Annotating medical data is an expensive process as it requires human experts. Although a couple of recent large-scale auto-labeled datasets have been provided for some medical tasks, e.g., chest X-ray [12,26], they are often noisy (low-quality) and degrade the performance of models. Besides, such datasets are not available for most medical tasks. Therefore, training V&L models with limited annotated data remains a key challenge.

Recently, pre-trained V&L models have been proposed to reduce the amount of labeled data required for training an accurate downstream model [6,14,25] in the general domain (transfer learning). These models are first trained on large-scale image caption data with self-supervision signals (e.g., using masked language model loss[1]) to learn the association between objects and text tokens. Then, the parameters of the pre-trained V&L models are used to initialize the downstream models and fine-tuned on the target tasks. In most V&L tasks, it has been reported that V&L pre-training is a major source of performance improvement However, we identify a key problem in applying common pre-trained V&L models for the medical domain: the large domain gap between the medical (target) and the general domain (source) makes such pre-train and fine-tune paradigm considerably less effective in the medical domain. Therefore, domain-specific designs need to be applied. Notably, V&L models mainly leverage object-centric feature extraction methods such as Faster R-CNN [20] which is pre-trained on general domain to detect everyday objects, e.g., cats, and dogs. However, the abnormalities in the X-ray images do not resemble everyday objects and will likely be ignored by a general-domain object detector.

To overcome this challenge, we propose BERTHop, a transformer-based V&L model designed for medical applications. In BERTHop, the visual encoder of the V&L architecture is redesigned leveraging PixelHop++ [7] and is fully unsupervised which significantly reduces the need for labeled data [21]. PixelHop++ can extract image representations at different frequency levels. This is significantly beneficial for highlighting abnormalities in different levels to be captured by the transformer in relation to the input text. Furthermore, BERTHop resolves the domain gap issue by leveraging a pre-trained language encoder, BlueBERT [18], a BERT [9] variant that has been trained on biomedical and clinical datasets.

2   Related Work

Transformer-Based V&L Models. Inspired by the success of BERT for NLP tasks, various transformer-based V&L models have been proposed [6,14,25]. They generally use an object detector pre-trained on Visual Genome [13] to extract visual features from an input image and then use a transformer to model visual features and input sentences. They are pre-trained on a massive amount of paired image-text data with a mask-and-predict objective similar to BERT.

---

[1] Part of the input is masked and the objective is to predict the masked words or image regions based on the remaining contexts.

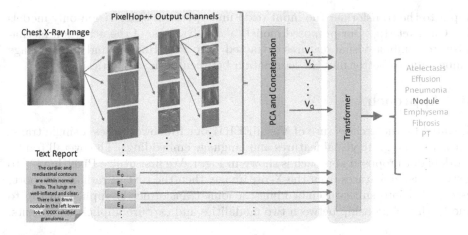

Fig. 1. The proposed BERTHop framework for CXR disease diagnosis. A PixelHop++ model followed by a "PCA and concatenation" block is used to generate Q feature vectors. These features along with language embedding are fed to the transformer that is initialized with BlueBERT.

Such models have been applied to many V&L applications[8,17,28]. However, for transferring the knowledge from these pre-trained models, the data distribution of source and target should be close enough or otherwise we need enough data for the target domain to properly transfer the knowledge.

V&L Models in the Medical Domain. Various V&L architectures have been proposed for disease diagnosis on CXR. TieNet is a CNN-RNN-based model for V&L embedding integrating multi-level attention layers into an end-to-end CNN-RNN framework for disease diagnosis TieNet uses a ResNet-50 pre-trained for general-domain visual feature extraction and an RNN for V&L fusion. As a result, it requires a large amount of in-domain training data (ChestX-ray14) for adapting to the medical domain, limiting its practical usage. Recently, Li *et al.* [15] evaluated the transferability of well-known pre-trained V&L models by fine-tuning them on MIMIC-CXR [12] and OpenI. However, the pre-trained models are designed and pre-trained for general-domain, and directly fine-tuning them with limited in-domain data leads to suboptimal performance. We refer to this method as VB w/ BUTD (Sect. 4.2).

PixelHop++ for Visual Feature Learning. PixelHop++ is originally proposed as an alternative to deep convolutional neural networks for feature extraction from images and video frames in resource-constrained environments. It is a multi-level model which generates output channels representing an image at different frequencies. PixelHop++ is used in various applications and shown to be highly effective on small datasets. These applications include face gender classification [22], face recognition [23], deep fake detection [5], and medical application [16]. To the best of our knowledge, this is the first study which integrates PixelHop++ and DNN models. Although using The PixelHop++ features alone as input to the transformer (no input text) underperform other vision-only models i.e. ChexNet [19], Our proposed model takes advantage of the attention mechanism to combine visual features extracted from PixelHop++ and the language embedding to better find the association between both modalities.

3   Approach

Inspired by the architecture of VisualBERT, our framework uses a single transformer to integrate visual features and language embeddings. The overall framework of our proposed approach is shown in Fig. 1. We first utilize PixelHop++ to extract visual features from the X-ray image; then the text (a radiology report) is encoded into subword embeddings; a joint transformer is applied on top to model the relationship between two modalities and capture implicit alignments.

3.1   Visual Encoder

We argue that extracting visual features from a general-domain object detector, i.e. the BUTD [3] approach that is dominant in most V&L tasks, is not suitable for the medical domain. Such an approach fails to detect medical abnormalities when applied to X-ray images. The reason is that the abnormalities in the image, which are of high importance for facilitating diagnosis, usually do not resemble the normal notion of an "object" and will likely be ignored by a general-domain object detector. Further, there exists no large-scale annotated dataset for disease abnormality detection from which to train a reliable detector [24].

We propose to adopt PixelHop++ [7] for unsupervised visual feature learning in the medical domain, which has been shown to be highly effective when trained on small-scale datasets. The key idea of PixelHop++ is computing the parameters of its model by a closed-form expression without using back-propagation [21]. As PixelHop++ leverages PCA for computing parameters, the model is able to extract image representations at various frequencies in an unsupervised manner. Inspired by the architecture of DNN models, PixelHop++ is a multi-level model in which each level consists of one or several PixelHop++ units followed by a max-pooling layer.

Suppose that we have N training images of size $s_1 \times s_2 \times d$, where d is 1 for gray-scale and 3 for color images. They are all fed into a single PixelHop++ unit in the first level of the model. The goal of training a PixelHop++ unit is to compute linearly independent projection vectors (kernels) which can extract strong features from its input data. There are one or more PixelHop++ units in each level of a PixelHop++ model.

In the first step of processing data in a PixelHop++ unit, using a sliding window of size $w \times w \times d$ and a stride of s, patches from each training image are extracted and flattened, i.e., $x_{i1}, x_{i2}, ..., x_{iM}$ where x_{ij} is the jth flattened patch for image i, and M is the number of extracted patches per image.

In the second step, the set of all patches extracted from training images are used to compute the kernels of the PixelHop++ unit. Kernels are computed as follows:

- The first kernel called DC is the mean filter, i.e., $\frac{1}{\sqrt{n}} \times (1, 1, ..., 1)$ where n is the size of the input vector, and extracts the mean of each input vector.
- After computing the mean of each vector, PCA kernels of the residuals are computed and stored as AC kernels. First k kernels are the top k orthogonal projection vectors which can capture the variation of residuals best.

Each image patch is then projected on computed kernels and a scalar bias is added to the projection result. By transforming $x_{i1}, x_{i2}, ..., x_{iM}$ by a kernel in a PixelHop++ unit, one output channel is generated. For example, in the first level of the model, the PixelHop++ unit generates 1 DC channel and $w \times w \times d - 1$ AC channels.

In the last step, model pruning is executed to remove the channels which include deficient data. The ratio of the variance explained by each kernel to the variance of training data is called the "energy ratio" of the kernel or its corresponding channel and is used as a criterion for pruning the model. An energy ratio threshold value, E, is selected and model pruning is performed using the following rule:

- If the energy ratio of a channel is less than E, it will be discarded as the variation of data along the corresponding kernel is very small.
- If the energy ratio of a channel is more than E, it is forwarded to the next level for further energy compaction.

Each output intermediate channel generated by a PixelHop++ unit will be fed into one separate PixelHop++ unit in the next level. So, except for the first level of the model, other levels contain more than one PixelHop++ unit.

3.2   In-Domain Text Pre-training

In BERTHop, the text report plays an important role in guiding the transformer to pay more attention to the right visual features in the attention mechanism. The report is written by an expert radiologist, who lists the normal and abnormal observations in the "finding" section and other important patient information including patient history, body parts, and previous studies in the "impression" section of the report. The text style of the report is drastically different from that of the pre-training corpora of BERT (Wikipedia and BookCorpus) or V&L models (MSCOCO and Conceptual Captions). Therefore, we propose to use BlueBERT [18] as the backbone in BERTHop to better capture the text report information. Pre-training with text-only corpora has been reported to how only marginal or no benefit [25]. In the medical domain, however, we find that using a transformer pre-trained on in-domain text corpora as our initialized backbone serves as a simpler yet stronger approach. Previous methods [15] do not take such a significant domain gap into consideration. Rather, they initialize the transformer with a model trained on general-domain image-text corpora, as in most V&L tasks.

4   Experiments

We evaluate BERTHop on the OpenI dataset and compare it with other existing models. To understand the effectiveness of the model designs, we also conduct detailed studies to verify the value of the visual encoder and the transformer initialization.

4.1   Experiment Setup

We focus on the OpenI dataset comprising 3,996 reports and 8,121 associated images from 3,996 unique patients. Its labels include 14 commonly occurring thoracic chest diseases, We first resize all images of OpenI to 206×206 and apply a three-level PixelHop++ for unsupervised feature learning from them. Then, we apply PCA to PixelHop++ output channels and concatenate the generated vectors to form a set of Q visual features of dimension D, i.e., $V = [v_1, v_2, ..., v_Q], v_i \in \mathbb{R}^D$. In BERTHop, D is set to be 2048. In our experiments setup, Q is equal to 15 but may vary depending on the size of the output channels of the PixelHop++ model and also the number of PCA components. As for the transformer backbone, we use BlueBERT-Base (Uncased, PubMed + MIMIC-III) from Huggingface [27], a transformer library. Having the visual features from the visual encoder and text embedding, we train the transformer on the training set of OpenI with 2,912 image-text pairs. We use batch size $= 18$, learning rate $= 1e-5$, max-seq-length $= 128$, and Stochastic Gradient Descent (SGD) as the optimizer with momentum $= 0.9$ and train it for 240 epochs.

4.2   Main Results

We train BERTHop on the OpenI training dataset and evaluate it on the corresponding test set. We evaluate BERTHop and baselines based on Receiver Operating Characteristic (ROC) and Area Under the ROC Curve (AUC) score using the same AUC implementation in scikit-learn [4] showsn in Fig. 2 b). Figure 2 a) summarizes the performance of BERTHop compared with existing methods. The results demonstrate that BERTHop outperforms TieNet, which is the current SOTA, in 11 out of 14 thoracic disease diagnoses and achieves an average AUC of 98.23% which is 14.37%, 12.83%, and 1.73% higher than VB w/ BUTD, TNNT, and TieNet, respectively. Note that TieNet has been trained on a much larger annotated dataset, i.e., the ChestX-ray14 dataset containing 108,948 training data while BERTHop is trained on 9× smaller case examples.

Regarding the VB w/ BUTD results, we reevaluate the results based on the released code[2] from the original authors. However, we cannot reproduce the results reported in the paper even after contacting the authors.

---

[2] https://github.com/YIKUAN8/Transformers-VQA.

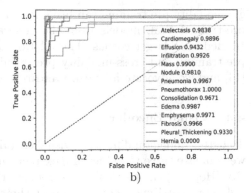

	TNNT	TieNet	VB w/ BUTD	BERTHop
Atelectasis	-	0.976	0.9247	**0.9838**
Cardiomegaly	-	0.962	0.9665	**0.9896**
Effusion	-	**0.977**	0.9049	0.9432
Infiltration	-	0.984	0.8867	**0.9926**
Mass	-	0.903	0.6428	**0.9900**
Nodule	-	0.960	0.8480	**0.9810**
Pneumonia	-	0.994	0.8537	**0.9967**
Pneumothorax	-	0.960	0.8931	**1.0000**
Consolidation	-	**0.989**	0.7870	0.9671
Edema	-	0.995	0.9500	**0.9987**
Emphysema	-	0.868	0.8565	**0.9971**
Fibrosis	-	0.960	0.6274	**0.9966**
PT	-	**0.953**	0.7612	0.9330
Hernia	-	-	-	-
AVG	0.854	0.965	0.8386	**0.9823**

a)   b)

Fig. 2. *a)* The AUC thoracic diseases diagnosis comparison of our model with other three methods on OpenI. BERTHop significantly outperforms models trained with a similar amount of data (e.g. VB w/ BUTD). *b)* The ROC curve of BERTHop for all 14 thoracic diseases.

Table 1. Effect of the transformer backbones when paired with different visual encoders. When using BUTD features, the model becomes insensitive to the transformer initialization and the expensive V&L pre-training brings little benefit compared to BERT. When using PixelHop++, the model benefits significantly from BlueBERT, which is pre-trained on in-domain text corpora.

Visual encoder	BUTD			PixelHop++	
Transformer backbone	VB	BERT	BlueBERT	BERT	BlueBERT
Atelectasis	0.9247	0.8677	0.8866	**0.9890**	0.9838
Cardiomegaly	0.9665	0.8877	0.8875	0.9772	**0.9896**
Effusion	0.9049	0.8940	0.9120	0.9013	**0.9432**
Mass	0.6428	0.7365	0.7373	0.8886	**0.9900**
Consolidation	0.7870	0.8766	0.8906	0.8949	**0.9671**
Emphysema	0.8565	0.7313	0.8261	0.9641	**0.9971**
AVG	0.8386	0.8309	0.8564	0.9177	**0.9823**

4.3   In-Domain Text Pre-training

We further investigate the influence of different transformer backbone initialization on model performance by pairing it with different visual encoders. The results are listed in Table 1. First, we find that the proposed initialization with a model pre-trained on in-domain text corpora (BlueBERT) brings significant performance boosts when paired with PixelHop++. Initializing with BlueBERT gives a 6.46% performance increase compared to initializing with BERT. Second, when using BUTD, the model is less sensitive to the transformer initialization and the performance is generally low (from 83.09% to 85.64%). In contrast to other V&L tasks [14], general-domain V&L pre-training is not instrumental. The above findings suggest that for medical V&L applications, in-domain single modality pre-training can bring larger performance improvement than using pre-trained V&L models from the general domain, even though the latter is trained on a larger corpus. The relation and trade-off between single-modality pre-training and cross-modality pre-training are overlooked by previous works [14] and we advocate for future research on this.

4.4   Visual Encoder

To better understanding what visual encoder is suitable for medical applications, we compare three visual feature extraction methods (BUTD, ChexNet [19], and PixelHop++). In particular, we replace the visual encoder of BERTHop with different visual encoders and report their performance. BUTD extracts visual features from a Faster R-CNN pre-trained on Visual Genome, which is prevailing in recent V&L models.

ChexNet is a CNN-based method that is proposed for pneumonia disease detection. It is a 121-layer DenseNet [11] trained on the ChestX-ray14 dataset for pneumonia detection having all pneumonia cases labeled as positive examples and all other cases as negative examples. By modifying the loss function, it is also trained to classify all 14 thoracic diseases and achieved state-of-the-art among existing vision-only models, e.g., [26]. By feeding each image into the network, it outputs a feature vector of dimension 1024. The output feature size of the ChexNet is 1024 and in order to make it compatible with our transformer framework, we apply a linear transformation that maps feature vectors of size 1,024 to 2,048. We fine-tune ChexNet along with the parameters of the linear transformation on the OpenI dataset. The results in Table 2 show that the visual encoder of BERTHop, PixelHop++, can provide more raw features of the CXR images as it uses a data-efficient method in an unsupervised manner and is capable of extracting task-agnostic image representations at different frequencies.

Table 2. Comparison between different visual encoders (BUTD, ChexNet, and PixelHop++) with BlueBERT as the transformer backbone. PixelHop++ outperforms BUTD and ChexNet, which is pre-trained on a large in-domain dataset.

	BUTD	ChexNet	PixelHop++
Atelectasis	0.8866	0.9787	**0.9838**
Cardiomegaly	0.8875	0.9797	**0.9896**
Effusion	0.9120	0.8894	**0.9432**
Mass	0.7373	0.7529	**0.9900**
Consolidation	0.8906	0.9000	**0.9671**
Emphysema	0.8261	0.9067	**0.9971**
AVG	0.8564	0.8798	**0.9823**

5   Discussion and Conclusion

We proposed a high-performance data-efficient multimodal model that jointly models X-ray images and clinical notes. In contrast with general V&L models which use an object detector to extract visual representations, our approach uses a parameter-effective encoder, PixelHop++, in an unsupervised setting. Our studies verify the effectiveness of PixelHop++ and illustrate that properly pre-training the transformer is of significance, which would provide valuable insight for designing future models. We urge our community to explore leveraging this method for other medical tasks suffering lack of annotated data. We believe that BERTHop is highly beneficial for reducing medical mistakes in disease diagnosis.

References

1. Abiyev, R.H., Ma'aitah, M.K.S.: Deep convolutional neural networks for chest diseases detection. J. Healthc. Eng. **2018** (2018)
2. Allaouzi, I., Ben Ahmed, M.: A novel approach for multi-label chest X-ray classification of common thorax diseases. IEEE Access **7**, 64279–64288 (2019)
3. Anderson, P., et al.: Bottom-up and top-down attention for image captioning and visual question answering. In: Proceedings of the IEEE Conference on Computer Vision and Pattern Recognition, pp. 6077–6086 (2018)
4. Buitinck, L., et al.: API design for machine learning software: experiences from the scikit-learn project. In: ECML PKDD Workshop: Languages for Data Mining and Machine Learning, pp. 108–122 (2013)
5. Chen, H.-S., Rouhsedaghat, M., Ghani, H., Shuowen, H., You, S., Jay Kuo, C.C.: DefakeHop: a light-weight high-performance Deepfake detector (2021)
6. Chen, Y.-C., et al.: UNITER: UNiversal Image-TExt Representation learning. In: Vedaldi, A., Bischof, H., Brox, T., Frahm, J.-M. (eds.) ECCV 2020. LNCS, vol. 12375, pp. 104–120. Springer, Cham (2020). https://doi.org/10.1007/978-3-030-58577-8_7
7. Chen, Y., Rouhsedaghat, M., You, S., Rao, R., Jay Kuo, C.-C.: PixelHop++: a small successive-subspace-learning-based (SSL-based) model for image classification. In: 2020 IEEE International Conference on Image Processing (ICIP), pp. 3294–3298. IEEE (2020)
8. Chou, S.-H., Chao, W.-L., Lai, W.-S., Sun, M., Yang, M.-H.: Visual question answering on 360deg images. In: Proceedings of the IEEE/CVF Winter Conference on Applications of Computer Vision, pp. 1607–1616 (2020)
9. Devlin, J., Chang, M.-W., Lee, K., Toutanova, K.: BERT: pre-training of deep bidirectional transformers for language understanding. arXiv preprint arXiv:1810.04805 (2018)
10. Giger, M.L., Suzuki, K.: Computer-aided diagnosis. In: Biomedical Information Technology, pp. 359-XXII. Elsevier (2008)
11. Huang, G., Liu, Z., Van Der Maaten, L., Weinberger, K.Q.: Densely connected convolutional networks. In: Proceedings of the IEEE Conference on Computer Vision and Pattern Recognition, pp. 4700–4708 (2017)
12. Johnson, A.E.W., et al.: MIMIC-CXR-JPG, a large publicly available database of labeled chest radiographs. arXiv preprint arXiv:1901.07042 (2019)
13. Krishna, R., et al.: Visual genome: Connecting language and vision using crowd-sourced dense image annotations. Int. J. Comput. Vis. **123**(1), 32–73 (2017)

14. Li, L.H., Yatskar, M., Yin, D., Hsieh, C.-J., Chang, K.-W.: VisualBERT: a simple and performant baseline for vision and language. arXiv preprint arXiv:1908.03557 (2019)

15. Li, Y., Wang, H., Luo, Y.: A comparison of pre-trained vision-and-language models for multimodal representation learning across medical images and reports. In: 2020 IEEE International Conference on Bioinformatics and Biomedicine (BIBM), pp. 1999–2004. IEEE (2020)

16. Liu, F., et al.: VoxelHop: successive subspace learning for ALS disease classification using structural MRI. arXiv preprint arXiv:2101.05131 (2021)

17. Lu, J., Goswami, V., Rohrbach, M., Parikh, D., Lee, S.: 12-in-1: multi-task vision and language representation learning. In: Proceedings of the IEEE/CVF Conference on Computer Vision and Pattern Recognition, pp. 10437–10446 (2020)

18. Peng, Y., Yan, S., Lu, Z.: Transfer learning in biomedical natural language processing: an evaluation of BERT and ELMO on ten benchmarking datasets. arXiv preprint arXiv:1906.05474 (2019)

19. Rajpurkar, P., et al.: CheXNet: radiologist-level pneumonia detection on chest X-rays with deep learning. arXiv preprint arXiv:1711.05225 (2017)

20. Ren, S., He, K., Girshick, R., Sun, J.: Faster R-CNN: towards real-time object detection with region proposal networks. IEEE Trans. Pattern Anal. Mach. Intell. **39**(6), 1137–1149 (2016)

21. Rouhsedaghat, M., Monajatipoor, M., Azizi, Z., Jay Kuo, C.-C.: Successive subspace learning: an overview. arXiv preprint arXiv:2103.00121 (2021)

22. Rouhsedaghat, M., Wang, Y., Ge, X., Hu, S., You, S., Jay Kuo, C.-C.: FaceHop: a light-weight low-resolution face gender classification method. arXiv preprint arXiv:2007.09510 (2020)

23. Rouhsedaghat, M., Wang, Y., Hu, S., You, S., Jay Kuo, C.-C.: Low-resolution face recognition in resource-constrained environments. arXiv preprint arXiv:2011.11674 (2020)

24. Shin, H.-C., et al.: Deep convolutional neural networks for computer-aided detection: CNN architectures, dataset characteristics and transfer learning. IEEE Trans. Med. Imaging **35**(5), 1285–1298 (2016)

25. Tan, H., Bansal, M.: LXMERT: learning cross-modality encoder representations from transformers. arXiv preprint arXiv:1908.07490, 2019

26. Wang, X., Peng, Y., Lu, L., Lu, Z., Bagheri, M., Summers, R.M.: ChestX-ray8: hospital-scale chest x-ray database and benchmarks on weakly-supervised classification and localization of common thorax diseases. In: Proceedings of the IEEE Conference on Computer Vision and Pattern Recognition, pp. 2097–2106 (2017)

27. Wolf, T., et al.: Huggingface's transformers: state-of-the-art natural language processing. arXiv preprint arXiv:1910.03771 (2019)

28. Zhou, L., Palangi, H., Zhang, L., Houdong, H., Corso, J., Gao, J.: Unified vision-language pre-training for image captioning and VQA. In: Proceedings of the AAAI Conference on Artificial Intelligence, vol. 34, pp. 13041–13049 (2020)

Author Index

Printed in the United States
by Baker & Taylor Publisher Services

Printed in the United States
by Baker & Taylor Publisher Services